T0215751

Lecture Notes in Artificial Intelligence 9388

Subseries of Lecture Notes in Computer Science

More information about this series at http://www.springer.com/series/1244

Adriana Tapus · Elisabeth André
Jean-Claude Martin · François Ferland
Mehdi Ammi (Eds.)

Social Robotics

7th International Conference, ICSR 2015
Paris, France, October 26–30, 2015
Proceedings

 Springer

Editors
Adriana Tapus
Robotics and Computer Vision Lab
ENSTA-ParisTech
Palaiseau
France

Elisabeth André
Institut für Informatik, Multimodale
    Mensch-Technik Interaktion
Universität Augsburg
Augsburg
Germany

Jean-Claude Martin
LIMSI-CNRS
Orsay
France

François Ferland
Robotics and Computer Vision Lab
ENSTA-ParisTech
Palaiseau
France

Mehdi Ammi
LIMSI-CNRS
Orsay
France

ISSN 0302-9743   ISSN 1611-3349   (electronic)
Lecture Notes in Artificial Intelligence
ISBN 978-3-319-25553-8   ISBN 978-3-319-25554-5   (eBook)
DOI 10.1007/978-3-319-25554-5

Library of Congress Control Number: 2015951796

LNCS Sublibrary: SL7 – Artificial Intelligence

Springer Cham Heidelberg New York Dordrecht London
© Springer International Publishing Switzerland 2015

Printed on acid-free paper

Springer International Publishing AG Switzerland is part of Springer Science+Business Media
(www.springer.com)

Preface

This book constitutes the refereed proceedings of the 7th International Conference on Social Robotics, ICSR 2015, held in Paris in October 2015. Having started in 2009, ICSR is now in its seventh edition and serves as a premier international forum for reporting and discussing the latest progress in the field of social robotics.

For the seventh edition of the conference, the call for papers attracted a record number of 126 submissions. The 70 revised full papers included in this book were carefully reviewed and selected based on the reviews of highly qualified professionals from around the world.

In addition to the main track, the conference program included two Special Sessions on "Objective Measures in HRI for Social Robotics" and on "Social Assistive Robotics for Children."

The conference program highlights included three invited talks by Gordon Cheng on "Closing the Natural Interaction Loop with Neuroscience-Based Robotics," by Jacqueline Nadel on "Toward a Two-Body Perspective in the Interdisciplinary Study of Nonverbal Communication," and by Wendy Ju on "Welcome Robot Overlords?".

The program was complemented by the ICSR Robot Design Competition 2015 and seven workshops discussing hot topics in social robotics — Evaluation Methods Standardization in Human–Robot Interaction, Toward a Framework for Joint Action, First Workshop on Evaluating Child–Robot Interaction, First International Workshop on Educational Robots (WONDER), Joint Workshop on Assistive Robotics, Bridging the Gap Between HRI and Robot Ethics Research, Third International Workshop on Culture Aware Robotics.

August 2015

Adriana Tapus
Elisabeth André
Jean-Claude Martin
François Ferland
Mehdi Ammi

Organization

General Chairs

Adriana Tapus   ENSTA-ParisTech, France
Markus Vincze   Vienna University of Technology, Austria

Program Chairs

Elisabeth André   Universität Augsburg, Germany
Jean-Claude Martin   LIMSI-CNRS / University Paris South, France

Workshop Chairs

Ginevra Castellano   Uppsala University, Sweden
François Ferland   ENSTA-ParisTech, France

Special Session Chairs

François Michaud   Université de Sherbrooke, Canada
Bram Vanderborght   Vrije Universiteit Brussel, Belgium

Robot Design Competition Chairs

Jérôme Monceaux   Aldebaran, France
Francesco Ferro   Pal Robotics, Spain
Arun Ramaswamy   ENSTA-ParisTech, France

Publicity Chairs

Nicole Mirnig   Salzburg University, Austria
Fumihide Tanaka   University of Tsukuba, Japan

Standing Committee

Ronald C. Arkin   Georgia Tech, USA
Paolo Dario   Scuola Superiore Sant'Anna, Italy
Shuzhi Sam Ge   National University of Singapore, Singapore
Oussama Khatib   Stanford University, USA
Jong Hwan Kim   KAIST, Korea
Haizhou Li   A*STAR, Singapore
Maja J. Matarić   University of Southern California, USA

Local Organizing Committee

Mehdi Ammi	LIMSI CNRS, France
Pauline Chevalier	ENSTA-Paristech, France

Program Committee

Henny Admoni	Yale University, USA
Ho Seok Ahn	The University of Auckland, New Zealand
Rachid Alami	LAAS-CNRS, France
Minoo Alemi	Islamic Azad University, Tehran-west Branch, Iran
Patrícia Alves-Oliveira	INESC-ID and Instituto Superior Técnico, Universidade de Lisboa, Portugal
Mehdi Ammi	LIMSI CNRS, France
Cecilio Angulo	Universitat Politècnica de Catalunya, Spain
Salvatore Anzalone	GH Pitié-Salpêtrière, APHP, UPMC, France
Olivier Aycard	Grenoble University, France
Wolmet Barendregt	Gothenburg University, Sweden
Tobias Baur	Augsburg University, Germany
Paul Baxter	Plymouth University, UK
Tony Belpaeme	Plymouth University, UK
Eduardo Benitez	Sandoval University of Canterbury, New Zealand
Elizabeth Broadbent	University of Auckland, New Zealand
John-John Cabibihan	Qatar University, Qatar
Angelo Cangelosi	Plymouth University, UK
Argun Cencen	Delft University of Technology, The Netherlands
Stephan Chalup	The University of Newcastle, Australia
Mohamed Chetouani	UPMC-ISIR, France
Pauline Chevalier	ENSTA-Paristech, France
Rocio Chongtay	The University of Southern Denmark, Denmark
Matthieu Courgeon	University of South Brittany, France
Nigel Crook	Oxford Brookes University, UK
Raymond Cuijpers	Eindhoven University of Technology, The Netherlands
Kerstin Dautenhahn	University of Hertfordshire, UK
Laurence Devillers	LIMSI-CNRS/Université Paris Sorbonne 4, France
Frank Dylla	University of Bremen, Germany
Kerstin Eder	University of Bristol and Bristol Robotics Laboratory, UK
Mohsen Falahi	Ecole polytechnique fédérale de Lausanne, Switzerland
Kerstin Fischer	University of Southern Denmark, Denmark
Mary Ellen Foster	University of Glasgow, UK
Shuzhi Sam Ge	National University of Singapore, Singapore
Manuel Giuliani	Universität Salzburg, Austria
Jaap Ham	Eindhoven University of Technology, The Netherlands
Guido Herrmann	University of Bristol, UK

Andri Ioannou	Cyprus University of Technology, Cyprus
Benjamin Johnston	University of Technology Sydney, Australia
Kristiina Jokinen	University of Helsinki, Finland
Simon Keizer	Heriot-Watt University, UK
Kolja Kühnlenz	Hochschule Coburg, Germany
Iolanda Leite	Yale University, USA
Muhammad N. Mahyuddin	Universiti Sains Malaysia, Malaysia
Andrew McDaid	The University of Auckland, New Zealand
Ali Meghdari	Sharif University of Technology, Iran
Gregor Mehlmann	Augsburg University, Germany
François Michaud	Université de Sherbrooke, Canada
Omar Mubin	University of Western Sydney, Australia
Rony Novianto	University of Technology Sydney, Australia
Mohammad Obaid	Chalmers Univeristy of Technology, Sweden
Amit Kumar Pandey	Aldebaran Robotics, France
Karola Pitsch	Bielefeld University, Germany
Selma Sabanovic	Indiana University, USA
Albert Ali Salah	Bogaziçi University, Turkey
Maha Salem	University of Hertfordshire, UK
Miguel A. Salichs	Universidad Carlos III de Madrid, Spain
Allison Sauppé	University of Wisconsin-Madison, USA
Ruth Schulz	University of Queensland, Australia
Christoph Schwering	RWTH Aachen University, Germany
Candace Sidner	Worcester Polytechnic Institute, USA
Reid Simmons	Carnegie Mellon University, USA
Khiet Truong	University of Twente, The Netherlands
Hiroyuki Umemuro	Tokyo Institute of Technology, Japan
Manuela Veloso	Carnegie Mellon University, USA
Jouke Verlinden	Delft University of Technology, The Netherlands
Michael Walters	University of Hertfordshire, UK
Graham Wilcock	University of Helsinki, Finland
Gregor Wolbring	University of Calgary, Canada
Britta Wrede	Bielefeld University, Germany
Agnieszka Wykowska	Ludwig-Maximilians-Universität München, Germany
James Young	University of Manitoba, Canada
Joachim de Greeff	Delft University of Technology, The Netherlands

Additional Reviewers

Muneeb Imtiaz Ahmad	University of Western Sydney, Australia
Jordi Albo-Canals	Ramon Llull University, Spain
Patrícia Alves-Oliveira	INESC-ID and Instituto Superior Técnico, Universidade de Lisboa, Portugal
Dejanira Araiza Illan	University of Bristol and Bristol Robotics Laboratory, UK
Alex Barco	Ramon Llull University, Spain

Contents

The Effect of Applying Humanoid Robots as Teacher Assistants to Help Iranian Autistic Pupils Learn English as a Foreign Language

Minoo Alemi[1(✉)], Ali Meghdari[2(✉)], Nasim Mahboub Basiri[3], and Alireza Taheri[2]

[1] Islamic Azad University, Tehran-West Branch, Tehran, Iran
minooalemi2000@yahoo.com
[2] Center of Excellence in Design Robotics and Automation (CEDRA),
Sharif University of Technology, Tehran, Iran
meghdari@sharif.ir
[3] Languages and Linguistics Center, Sharif University of Technology, Tehran, Iran

Abstract. The present case study investigated the effect of applying a humanoid robot as a teacher assistant to teach a foreign language (English in this case) to Iranian children with autism. To do this, there were 4 male autistic children, 7–9 years old, 3 with high-functioning autism and one with low-functioning autism. The humanoid robot NAO was used as the main instrument of this study. There were 12 sessions including 10 teaching sessions. This study used a pre-test, mid-test, immediate post-test, and delayed post-test design to measure the learning gains of the autistic children participating in the robot-assisted language learning (RALL) program. The results showed the subjects' large learning gains which were fairly persistent according to their performances on the delayed post-test. The difference observed between the learning gains of the high-functioning and low-functioning participants is also discussed.

Keywords: Humanoid robot · Autism · High-functioning · Foreign language education · RALL

1   Introduction

Autism Spectrum Disorder (ASD) is a lifelong developmental disability affecting the way a person communicates and relates to people around him/her. People with autism have impaired social interaction, social communication and imagination [1]. The most recent statistics indicate that 1 out of 88 children born in the United States is autistic [2]. It is estimated that more than 30,000 Iranians younger than 19 years old suffer from autism disorders [3].

Researchers have shown that autistic children, despite their lack of ability to interact with other people, enjoy working with technological tools such as computers, smart toys, and robots. There have been many studies regarding the application of robots in helping autistic children with imitating, making eye contact, and social interactions. Based on these studies, humanoid robots seem to have great potential in

© Springer International Publishing Switzerland 2015
A. Tapus et al. (Eds.): ICSR 2015, LNAI 9388, pp. 1–10, 2015.
DOI: 10.1007/978-3-319-25554-5_1

helping autistic children in overcoming their disorders, ranging from impaired joint attention to impaired language development [4-10].

There have been a few studies conducted regarding teaching a foreign language to high-functioning autistic pupils, particularly those autistic individuals who do not have considerable learning difficulties, and they all indicate that high-functioning autistic children are capable of learning a second/foreign language provided that they are granted the opportunity and of course special strategies in teaching [11-14]. Learning a second/foreign language is a fairly complex process even for normally developing individuals. High-functioning autistic individuals usually do not have severe problems in developing first language, but normally have impaired social cognition which makes communicating hard for them and negatively affects their foreign language learning. This may be explained by Communicative Language Teaching (CLT), one of the most widely used methods of teaching a second/foreign language around the world, which states that learning a foreign language requires the learners to be engaged in pair and group activities, use the target language, and communicate through it. In other words, high-functioning autistic individuals should be encouraged to communicate with others to be successful in learning a foreign language. Therefore, raising their Willingness to Communicate (WTC), motivation, and positive attitude seems to be of great importance. Furthermore, being required to use a foreign language when communicating could make autistics individuals anxious. Accordingly, a learning environment that can lower the anxiety levels of autistic individuals should also contribute in facilitating the foreign language learning process.

According to studies conducted on normally-developing language learners [15-23], Robot-Assisted Language Learning (RALL) seems to generate motivation and interest in learners of a foreign language while at the same time lowering their anxiety levels. Accordingly, RALL seems to be one of the best options for high-functioning autistic children willing to learn a foreign language.

There seemed to be a gap in the literature, however, regarding the use of robots to teach a second language to high-functioning autistic pupils. Accordingly, the current study was conducted to investigate the effect of Robot Assisted Language Learning (RALL) on high-functioning autistic children. To put it more precisely this study was an attempt to answer the following research question:

What is the effect of RALL on autistic children's English vocabulary learning and retention?

2   Methodology

2.1   Participants

Three male high-functioning autistic children (referred to as S1, S2, and S3 hereafter) 9, 8, and 7 years old, respectively, as well as a 7 year-old low-functioning autistic child (twin brother of S3, referred to as S4 hereafter) participated in this study. They had little or no background in English which was proven by means of an English pre-test. It should be noted that the four mentioned participants were the only students of this RALL program.

2.2   Instruments

Teaching Instruments: The main instrument of the current study was the humanoid robot NAO, made by Aldebaran Robotics, France, and renamed "Nima" for use in the Iranian context. Nima was pre-programmed for each teaching session to assist the teacher in teaching a particular syllabus and was operated by a human operator sitting at a desk on the right side of the class during the teaching sessions. The operator did not participate in class activities. Based on the pre-test, the four subjects were considered as beginning English learners. Accordingly, a fitting book was selected and 10 teaching scenarios were written based on it. The related flashcards and songs from the book were also included in the scenarios. The songs were uploaded in Nima and he was programmed to dance to them. In addition, some slides were shown via video projector to teach the vocabulary items of interest. Two laptops were also applied: one to operate Nima and the other connected to the video projector. Some of the teaching instruments are presented in Figure 1 which shows the classroom setting.

Fig. 1. Classroom setting.

Measuring Instruments: Four equivalent but not identical English tests based on the covered book were designed: a pre-test, a mid-test, an immediate post-test, and a delayed post-test. Each test was made of 63 items including matching, multiple choice recognition items and a few open ended questions to test the simple functions taught during the course. Moreover, a video recorder was used to record each and every session of the program for further qualitative analysis of the participants' behaviors in class. A camera man recorded the sessions, moving if necessary to focus on the participants. An audio recorder was also used to record the interviews conducted with the subjects' parents after the program was finished.

2.3   Data Collection Procedure

The Center for the Treatment of Autistic Disorders (CTAD) located in Tehran made it possible for our research team to attend some of the group and individual classes for high-functioning autistic children before the program started. CTAD believed the researcher needed to observe these classes to tap into what autism really was, how autistic children behaved, and how she should handle them later on in the program. The researcher did attend the suggested classes and had more than 20 hours of observation before starting to work with the autistic children. The RALL program consisted of 12 sessions in total including 10 teaching sessions held twice a week. The first session was an orientation session held one week before the program started to introduce the program to the children and their parents, and to administer the pre-test (Figure 2).

Fig. 2. The orientation session.

The teaching sessions were held two times a week with each session lasting 45 minutes to one hour and covering half of a chapter of the selected book. Nima would spoke only English during the sessions. The teacher used Farsi, the participants' mother tongue, to give instructions and provided the translations of the new vocabulary items and Nima's lines. It has been suggested in previous research that this approach makes the participants feel more comfortable with learning the new material [14]. After five sessions, i.e. at session 6, a midterm exam was administered. An immediate post-test was administered at the end of session 10. Two weeks later, a farewell session was held to administer the delayed post-test, and interview the parents with several open ended pre-designed questions on their views of the program and the changes, if any, they had observed in their children throughout the program. It is worth mentioning that each subject took the English tests separately. The teacher read each item for each subject and marked or wrote down their answers. Furthermore, all sessions were video-recorded for further qualitative analysis of the participants' behaviors.

3   Results

The scores (equal to the number of their correct answers) of the pre-test, mid-test, immediate post-test, and delayed post-test for each participant and the class average are reported in Figures 2 and 3, respectively. The maximum score possible was 63.

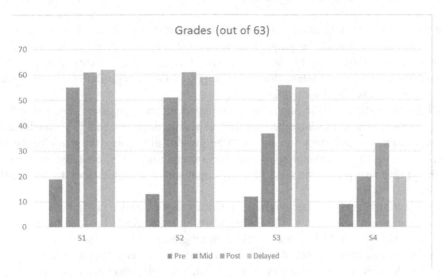

Fig. 3. Participants' scores on English tests.

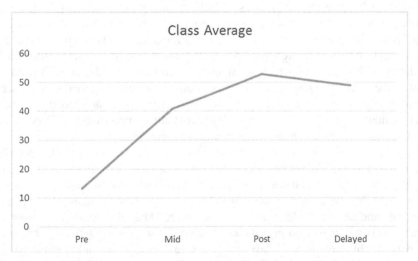

Fig. 4. Class Average.

The three high-functioning subjects showed great improvement during the program. Their scores in the delayed post-test showed a large amount of retention and the persistency of their learning gains. S1 and S2 even obtained near full marks in the immediate post-test. In the case of the low-functioning subject, however, less learning gains were observed. His level of retention on the delayed post-test was also lower compared to those of the high-functioning subjects.

4   Discussion and Conclusion

The results of this study indicated that high-functioning autistic children do have the ability to learn a foreign language. Due to their increased interest in technological tools, autistic children could benefit from using technology in education. Robots can be considered a new and interesting technology that can be applied in education and specifically in teaching a foreign language.

High-functioning autistic pupils may need special strategies to be engaged in language classes such as routine greetings, and a specific order of seats [11]. We tried to use such strategies and they turned out to be helpful in keeping the participants engaged. Using a highly technological tool, i.e. a humanoid robot, could be considered as one of the most promising ways of using engagement strategies specific to autistic children.

As mentioned earlier, using robots in language classes seems quite fruitful for normally developing students. The findings of this study could broaden the scope of the claims made for RALL to autistic children language learning. The use of robots in language classes makes it possible to make real use of several language learning approaches, hypotheses, and theories such as: Asher's Total Physical Response (TPR) approach which emphasizes the role of listening and acting upon it by physical responses which is currently applied as an activity in language classes [24], Swain's output hypothesis, and Schmit's noticing theory. According to Swain's output hypothesis, having language learners use the learned language items has three main functions through which language learning occurs more easily: consciousness raising, hypothesis-testing, and reflective functions [25]. In other words, through producing output in the target language, learners better notice the newly learned items, check if they are using those items correctly by receiving feedbacks, and get the chance to reflect on their learnings. Schmit's noticing theory also emphasizes the importance of focal attention and explicit knowledge in language learning.

Designing real life scenarios and authentic learning situations will also contribute to the engagement of the participants in the assigned tasks. For instance, at session 8, when the aim of the lesson was to teach the participants to ask someone's age and also to tell their own ages, we held a birthday party for Nima (Figure 4) and had him tell his age and ask the children how old they were. Then the vocabulary items were taught as gifts we wanted to give Nima. Each child would put a flashcard of one of the newly taught items in a gift box and give it to Nima. Nima had to guess what the gift was and the child would help him guess. These meaningful dialogues between Nima and the participants were quite effective in engaging them. This lends support to the findings of [17] and also [8] indicating that robots are efficient in eliciting utterances, which according to Swain's output hypothesis could lead to better language learning.

Fig. 5. Nima at his birthday party.

Furthermore, as suggested by [6], the physical presence of a robot and receiving its hints and applause could bring about significant learning gains. This was also the case in the current study, since the participants seemed to be quite excited by Nima's feedback. Upon receiving feedback from Nima, they would laugh out loud, enthusiastically participate in the tasks, and show great willingness to talk. Their reactions were interestingly similar to those suggested by [18]. We hypothesize that Nima's feedback and applause made the students notice the new items to be learned. According to Schmit's noticing theory, this could be one of the reasons why the participants had significant levels of achievement and retention.

Another strategy we applied was the use of Farsi, the participants' mother tongue, in class which in line with the findings of [14], lowered the burden for the participants and allowed them to make connections between the new vocabulary items or functions and the ones they already knew in their first language.

The presence of a low-functioning autistic child (S4) in class and his performance on the designed tests showed that even low-functioning autistic children have the potential to learn a second/foreign language. The main issue S4 suffered from was his lack of focus on and attention to what was being taught, and also his failure to pick up the newly taught items as fast as his classmates. In other words, a possible reason why S4 showed lower achievement compared to the other three high-functioning students could be the lack of homogeneity in class. To keep the high-functioning kids engaged, S4 did not receive as much time and attention as he needed. Still he did have some learning gains which could be associated with the fact that he seemed to be quite engaged when Nima was singing and dancing, and was willing to participate in tasks requiring face to face interaction with Nima.

One may question the effectiveness of using robots with the claim that the obtained results could be due merely to the newness of the applied technology and that they would fade away as the children become accustomed to it. However, this novelty effect could take a long time to occur especially with young children who can play with the same toys for years and still enjoy them. Additionally, robots have many different features which could be made use of to keep the learning environment interesting in the long term. This was the case in our RALL program, since session ten, the last teaching session, was one of the most energetic sessions of the RALL program. This showed the program was able to keep the students highly motivated and interested in Nima until the very last session. It is also important to remember that using new technologies is an inevitable aspect of education, because the new generation has new needs and new expectations from their education environment. Keeping up with these expectations is the only way to keep the students, especially those with autism, motivated and satisfied. Researching on how to use new technologies, and their (dis)advantages, therefore, is of great importance.

Since robots must be programmed in advance and operated by a human operator, they may seem to be quite infeasible tools to be used in education. However, since robots are perceived as the next generation of technology that will be pervasive in everyday life, they will inevitably find their way into education in much the same way as personal computers did many years ago. Therefore, research on how to use robots in education and their (dis)advantages seems to be of great importance [26-28].

There were some challenges to the use of a robot in class which are listed below:

- The great deal of excitement and enthusiasm sometimes made the participants use extra loud voices which at times made class management a bit harder for the teacher.
- At some points, students wanted to approach Nima and touch him. When one student did this, the others would want to follow and do the same.
- When Nima was dancing to a song, they would try to imitate his body movements. This can be a good point, since imitation is considered as a social skill and physical engagement, according to Total Physical Response (TPR) approach, can contribute to better language learning gains. This imitation, however, sometimes distracted them from paying attention to the content of the song being played by Nima.
- In trying to receive Nima's applause and positive feedback, the students sometimes created a quite competitive environment, felt jealous of the ones giving the right answer sooner, and even in some rare cases picked fights.

To rectify these issues, the teacher tried to intervene and emphasize the importance of turn-taking, not leaving their seats without asking permission, and that after the song she would ask them questions. Accordingly they had to pay attention to what was being played to be able to answer the questions and be applauded by Nima. The very small number of participants in this study makes any generalization impossible. Further research could broaden the scope of the current study to an intact class of a bigger number of autistic children at an autism school. Another potential area of study could be having the same program for an intact class in main stream schools with normally developing students in which some autistic children also participate. This will be a more authentic setting since this is what happens in the real world when children with autism attend main stream schools. Mixed classes may lower the opportunities given to the autistic children and be more demanding for the teacher. Additionally, a similar group could be taught the same materials based on the exact same scenarios but without the presence of a robot. The outcomes of the two groups could then be compared and analyzed for deeper insights. This is what the authors are doing in a follow-up phase of the current study. The results of the control group, taught without the use of a robot, will be reported and compared with those of the RALL group. To be able to make more valid claims on the participants' social behaviors throughout the program, precise quantitative analysis of the video records could be quite helpful. This will also be done and reported in the near future. Moreover, such factors as social skills, willingness to communicate, motivation and attitude, as well as anxiety level could be investigated for both groups to find other probable effects of RALL on children with autism. These factors will be focused on in later documents.

References

1. Robins, B., Dautenhahn, K., Te Boekhorst, R., Billard, A.: Effects of repeated exposure to a humanoid robot on children with autism. In: Designing a More Inclusive World, pp. 225–236. Springer, London (2004). doi:10.1007/978-0-85729-372-5_23
2. Scassellati, B., Admoni, H., Mataric, M.: Robots for use in Autism Research. Annual Review of Biomedical Engineering **14**, 275–294 (2012). doi:10.1146/annurev-bioeng-071811-150036
3. Pouretemad, H.: Diagnosis and Treatment of Joint Attention in Autistic Children. (in Farsi) تشخیص و درمان توجه اشتراکی در کودکان مبتلا به اوتیسم (2011). Arjmand Book, Tehran-Iran
4. Dautenhahn, K., Billard, A.: Games Children with Autism can play with a Robot, a Humanoid Robotic Doll. In: Universal access and assistive technology, pp. 179–190 (2002)
5. Kim, E.S., Paul, R., Shic, F., Scassellati, B.: Bridging the research gap: Making HRI useful to individuals with autism. Journal of Human-Robot Interaction **1**(1) (2012). http://www.scazlab.com/publications.php
6. Leyzberg, D., Spaulding, S., Toneva, M., Scassellati, B.: The physical presence of a robot tutor increases cognitive learning gains. In: Proc. 2012 Cognitive Sciences Conference, Sapporo, Japan (2012). http://www.scazlab.com/publications.php
7. Scassellati, B.: How social robots will help us to diagnose, treat, and understand autism. In: Robotics research, pp. 552–563. Springer, Heidelberg (2007). doi:10.1007/978-3-540-48113-3_47
8. Kim, E.S., Berkovits, L.D., Bernier, E.P., Leyzberg, D., Shic, F., Paul, R., Scassellati, B.: Social Robots as Embedded Reinforcers of Social Behavior in Children with Autism. Journal of autism and developmental disorders **43**(5), 1038–1049 (2013). doi:10.1007/s10803-012-1645-2. Springer
9. Taheri, A.R., Alemi, M., Meghdari, A., Pouretemad, H.R., Holderread, S.L.: Clinical application of a humanoid robot in playing imitation games for autistic children in Iran. In: Proc. 14th Int. Educational Technology Conf. (IETC), Chicago, USA, September 3–5, 2014
10. Meghdari, A., Alemi, M., Taheri, A.R.: The effects of using humanoid robots for treatment of individuals with autism in Iran. In: 6th Neuropsychology Symposium, Tehran, Iran, December 2013
11. Wire, V.: Autistic Spectrum Disorders and Learning Foreign Languages. Support for Learning **20**(3), 123–128 (2005). doi:10.1111/j.0268-2141.2005.00375.x

12. Oda, T.: Tutoring an American Autistic College Student in Japanese and its Challenges. Support for Learning **25**(4), 165–171 (2010). doi:10.1111/j.1467-9604.2010.01462.x

13. Þráinsson, K.Ó.: Second Language Acquisition and Autism. Unpublished BA essay, University of Iceland, Reykjavík (2012). www.skemman.is

14. Yahya, S., Yunus, M.M., Toran, H.: Instructional Practices in Enhancing Sight Vocabulary Acquisition of ESL Students with Autism. Procedia-Social and Behavioral Sciences **93**, 266–270 (2013)

15. Kanda, T., Hirano, T., Eaton, D.: Interactive Robots as Social Partners and Peer Tutors for Children: a Field Trial. Human Comp. Interact. **19**, 61–84 (2004)

16. You, Z., Shen, C., Chang, C., Liu, B., Chen, G.: A robot as a teaching assistant in an english class. In: Proceedings of the 6th IEEE International Conference on Advanced Learning Technologies, pp. 87–91. IEEE, New York (2006)

17. Han, J.: Robot-aided Learning and r-Learning Services. In: Chugo, D. (ed.) Human-Robot Interaction (2010). http://www.intechopen.com

18. Han, J.: Emerging Technologies, Robot Assisted Language Learning. Language Learning and Technology **16**(3), 1–9 (2012). http://www.llt.msu.edu/

19. Lee, S., Noh, H., Lee, J., Lee, K., Lee, G.G.: Cognitive effects of robot-assisted language learning on oral skills. In: INTERSPEECH 2010 Satellite Workshop on Second Language Studies: Acquisition, Learning, Education and Technology (2010)

20. Chang, C.W., Lee, J.H., Chao, P.Y., Wang, C.Y., Chen, G.D.: Exploring the Possibility of Using Humanoid Robots as Instructional Tools for Teaching a Second Language in Primary School. Educational Technology & Society **13**(2), 13–24 (2010)

21. Meghdari, A., Alemi, M., Ghazisaedy, M., Taheri, A.R., Karimian, A., Zandvakili, M.: Applying robots as teaching assistant in EFL classes at iranian middle-schools, CD. In: Proc. Int. Conf. on Education & Modern Edu. Tech. (EMET 2013), Venice, Italy, September 28–30, 2013

22. Alemi, M., Meghdari, A., Ghazisaedy, M.: Employing Humanoid Robots for Teaching English Language in Iranian Junior High-Schools. International Journal of Humanoid Robotics **11**(03) (2014)

23. Alemi, M., Meghdari, A., Ghazisaedy, M.: The effect of employing humanoid robots for teaching english on students' anxiety and attitude. In: Proc. of the 2nd RSI/ISM Int. Conf. on Robotics & Mechatronics (ICRoM), Tehran, Iran, pp. 754–759, October 2014

24. Brown, H.D.: Principles of Language Teaching and Learning. Longman, White Plains (2000)

25. Mitchell, R., Myles, F.: Second Language Learning Theories, 2nd edn. Hodder Arnold, London (2004)

26. Alemi, M., Meghdari, A., Ghazisaedy, M.: The Impact of Social Robotics on L2 Learners' Anxiety and Attitude in English Vocabulary Acquisition. International Journal of Social Robotics **7**(04), 523–535 (2015)

27. Alemi, M., Ghanbarzadeh, A., Meghdari, A., Moghaddam, L.J.: Clinical Application of a Humanoid Robot in Pediatric Cancer Interventions. International Journal of Social Robotics (March 2015)

28. Taheri, A.R., Alemi, M., Meghdari, A., Pouretemad, H.R., Mahboob Basiri, N.: Social robots as assistants for autism therapy in Iran: research in progress, CD. In: Proc. of the 2nd RSI/ISM Int. Conf. on Robotics and Mechatronics (ICRoM), Tehran, Iran, October 2014

Social Robots for Older Adults: Framework of Activities for Aging in Place with Robots

Patrícia Alves-Oliveira[✉], Sofia Petisca, Filipa Correia, Nuno Maia, and Ana Paiva

INESC-ID and Instituto Superior Técnico, Universidade de Lisboa, Lisboa, Portugal
patricia.alves.oliveira@inesc-id.pt

Abstract. According to the United Nations World Population Prospects, the world's population is aging. Older adults constitute a fragile part of society, as aging is always accompanied by major psychological and physical challenges. A way to cope with those challenges is to strive for a good Quality of Life (QoL) and contribute to successful aging. Social robots can play an important role in the promotion of QoL by integrating activities with independent-living older adults. Using a qualitative design through a focus group method, this paper aims to present the activities in which independent-living older adults, *i.e.,* older adults that do not depend upon anyone to carry out their activities, require a robot. By understanding the activities where robots can positively influence and contribute to older adults' QoL, we set specific goals for the future research in the field of Human-Robot Interaction (HRI).

Keywords: Human-Robot Interaction · Quality of Life · Successful aging

1 Introduction

The world's population is growing and aging. Furthermore, the late adulthood stage (>65 years old) is faced with major psychological and physical challenges [6]. Those challenges are usually accompanied by multiple stressors, such as social isolation and incapacity for work independently [20]. However, many older adults face these challenges but have an independent lifestyle (*i.e.,* do not depend upon anyone to carry out their activities of daily living) [7]. It is the thin balance between aging and still having an independent living lifestyle that constitutes one of the gravest challenges for achieving good standards of QoL.

Successful aging is one of the ways to ensure the maintenance of QoL [1]. By being able to promote an independent lifestyle, technology becomes an important factor associated with successful aging and better standards of QoL. Social robots in particular, have been investigated as a type of technology that can positively influence successful aging. Studies have shown the role that robots can have in providing assistance with house keeping activities [2], or by providing support over the needs and difficulties of older adults [11]. However, the concept of QoL

© Springer International Publishing Switzerland 2015
A. Tapus et al. (Eds.): ICSR 2015, LNAI 9388, pp. 11–20, 2015.
DOI: 10.1007/978-3-319-25554-5_2

encompasses more contexts than the home environment and goes beyond needs and difficulties. In fact, QoL covers various components of life and is associated not only with functional aspects of life, but also with well-being [15]. The novelty of this study is to elicit the activities that older adults require a robot to integrate all of their possible real-world contexts, including activities that support their desired living style and QoL. As such, this study goal is twofold:

Goal 1 Elicit the *types of activities* in which older adults require the inclusion of a robot to sustain a good QoL and independent living.
Goal 2 Present the *robots that older adults chose* for the different activities.

2   Related Work

Different societal studies have all came to the same conclusion: humanity is facing a profound demographic change, moving from a society where the majority of the population was relatively young, to one that faces a significant portion of older adults [16]. In fact, according to the United Nations World Population Prospects of 2012 for 2100[1], the percentage of older adults will increase as part of the population density across Europe, America, and China [13]. Although these news are tough to prospect, anthropological studies can reassure us. According to this field of study, the ability to create tools (*e.g.*, technology) is one of the pivotal developments and adaptations of humanity to change. In this line, technological artifacts have been making their way into our lives, mirroring the human capacity to develop tools that adapt to our needs [19].

Moreover, technology has been defined as the capacity to apply scientific knowledge to practical tasks that respond to societal needs and so, impact on the QoL [5]. When looking at older adults research, it can be seen that QoL is among the most studied constructs. In fact, for older adults QoL is preferred over to longevity [9]. A paper review [15] defines QoL as a conscious cognitive judgment of satisfaction with one's life. In aging research, QoL is associated with two broad categories: functioning (*e.g.*, the ability to perform activities of daily living) and well-being (*e.g.*, emotional well-being) [15]. The present study aims to contribute for the research of older adults' QoL associated with social robots, by eliciting activities they can integrate to promote successful aging. By doing so, this study provides a contribution for the development of both service and entertainment robots for older adults that live independently.

2.1   State of the Art on Social Robots for Older Adults

The development of robots that assist the activities of daily living of older adults contributes to the enrichment of Ambient Assisted Living (AAL), which is an emerging paradigm in information technology aimed at empowering peoples' capabilities by means of technology that is sensitive, adaptive, and responsive to the human needs [18]. Also, different projects concerning robots for older

---

[1] World Population Prospects: The 2012 Revision, http://esa.un.org/wpp/

adults have been emerging, such as GiraffPlus[2], Robot-Era[3], SILVER[4], CARE[5], ACCOMPANY[6], HOBBIT[7], ExCITE[8], ENRICHME[9], and RAMCIP[10]. The aforementioned projects have been developing prototypes of robots to interact with older adults, aiming to develop ways to assist on their needs. Yet, some of the applications and activities proposed are based on the need of the care-givers on one hand, and the older adults on the other. Still, there are unforeseen activities that can be developed for robots that will increase older adults' QoL and successful aging. This paper presents different activities that older adults require assistance not only from a basic and functional point of view, but also concerning entertainment and enhanced activities that contribute to their QoL.

3   Methodology

This study aimed to elicit the activities in which older adults require the presence of a robot to support their QoL. By doing so, we provide guidelines for the development of robots that co-exist with older adults, fostering successful aging and independent life style.

3.1   Participants

A focus group methodology was used (N = 16 participants), with each group comprised of 5 (except one of the groups that consisted of 6 participants) older adults with independent lifestyle (12 females, 4 males; M age = 78.69, SD = 12.20). Participants were recruited from a day-home care institution in Lisbon (Portugal). Most participants lived alone in their home (81.3%), or with their friends (12.5%), and relatives (6.3%). The focus group sessions were conducted at the recruitment facilities. Each session lasted 45min and was held by a psy-chologist and a computer scientist, both working in the field of HRI. The study followed the ethical norms of conduct for privacy, and all participants signed a consent form and assented participation. The cases in which participants were unable to read the consent form (due to their education level or physical impair-ment), the consent was read to them by a caregiver of the institution.

3.2   Procedure and Methods

Aiming to elicit the types of activities in which older adults envision robotic tech-nology as an enhancement to their QoL, a qualitative study with focus group

---

[2] GiraffPlus project: http://giraffplus.eu/
[3] Robot-Era project: http://www.robot-era.eu/robotera/
[4] SILVER project: http://www.silverpcp.eu/
[5] CARE project: http://care-project.net/welcome/
[6] ACCOMPANY project: http://accompanyproject.eu/
[7] HOBBIT project: http://hobbit.acin.tuwien.ac.at/
[8] ExCITE project: http://www.aal-europe.eu/projects/excite/
[9] ENRICHME project: http://www.enrichme.eu/wordpress/
[10] RAMCIP project: http://www.ramcip-project.eu/ramcip/

methodology was used [3], designed of three phases: 1) *information and sensitizing phase*; 2) *brainstorm session*; 3) *choosing robots*.

Phase 1: Information and Sensitizing. This phase informed about what a certain emergent technology *is* and can *become* [21]. In this study, we aimed to inform and sensitize about social robots using a short-film documentary of 6min that consisted of five chapters:

1. **What is a robot?** Since our intention is to keep distance from sci-fi culture when eliciting activities that participants envision doing with a robot, different existing robots such as the industrial Kuka arm[11], the social robotic pet AIBO[12], and humanoids like the Geminoid robot[13], were introduced by showing robots interacting with humans or in their context of use.
2. **How does a robot function?** This chapter explained that robots perceive the world differently from humans. As an example, this chapter contrasted the way humans perceive the world (*e.g.*, through their *eyes*), while robots perceive the world through *cameras*. The emphasis was on the difference between human and robot perception without emphasizing the limited capabilities that robots have nowadays.
3. **Do robots for older adults exist?** This chapter presented robots and prototypes specially developed for the aged population. Examples of these robots were RIBA robot[14] and Paro[15].
4. **What are the limitations of robots?** This chapter aimed to show the current real limitations of robots in the wild. This was demonstrated by, *e.g.*, a video where Asimo robot[16] falls of the stairs.
5. **How will the future with robots be like?** In order to show what an emergent technology such as a robot can become, it was necessary to show a possible future of robots and older adults together. Therefore, segments of the commercialized movie *Robot and Frank* directed by Jake Schreier (2012) were shown.

Phase 2: Brainstorm Session. *Brainstorm* is a well-established technique, usually used in groups, for generating a large number of new ideas quickly, enabling the transformation of abstract concepts into practical experiences [14]. Thus, the brainstorm session aimed to register in a whiteboard the different activities that participants envisioned to do with a robot. In the middle of the same whiteboard was written "robots for older adults" so that participants could easily situate their ideas. The researchers' role in the room was to clarify questions that emerged along the session, to facilitate the interaction and to write down on the whiteboard the activities mentioned by participants.

---

[11] KUKA Arm from KUKA Robotics: http://www.kuka-robotics.com/en/products/
[12] AIBO robot from SONY: http://www.sony-aibo.com/
[13] Geminoid robots from IHL: http://www.geminoid.jp/en/robots.html
[14] RIBA robot from RIKEN-TRI: http://rtc.nagoya.riken.jp/RIBA/index-e.html
[15] PARO robot from AIST: http://www.parorobots.com/
[16] ASIMO robot from HONDA: http://asimo.honda.com/

Companion Robots   Service Robots

Fig. 1. Companion Robots *(from left to right)*: Paro, Pleo, Emys; **Service robots**: Pearl, Care-O-bot, PR2. Categorization of assistive robots for older adults [4].

Phase 3: Choosing Robots. Six images of robots were shown to the participants, whose task was to assign a robot to the activities they had previously brainstormed about. The robots were chosen according to the categorization of assistive robots for elderly, *i.e.,* robots designed for social interaction that can play an important role with respect to the health and psychological well-being of the elderly. The selection of robots tried to met different contexts of aging, such as therapy, entertainment, and service-related [4]. Therefore, three *companion robots* were shown: Paro, Pleo, and Emys; and three *service robots* were shown: Pearl, Care-O-bot, and PR2 (see Fig. 1). The groups discussed what robot would better fit a specific activity and the researchers added this information to the whiteboard. It is important to note that participants did not specify a robot for all the activities, neither they were instructed to do so. In addition, they could choose more than one robot for the same activity. The open-endedness style of this phase was adopted to avoid pressure participants on a decision.

4   Results

The activities that participants yield were analysed by the two psychologists of this study. The elicited activities came from two different sources: activities written on the whiteboard, and audio recording of the sessions. All group sessions were transcribed and coupled with the activities present on the whiteboard. Participants generated a total of 75 activities in which a minority was repeated. As this study aims to provide visibility to a broad range of activities instead of analyzing their prevalence, the repeated activities were excluded. Thus, data was re-arranged and coded only with 65 non-repeated activities. The yield activities were coded according to the framework for aging in place with the objective of categorizing and organizing them according to their primary goal and context [10,12]:

- **Basic Activities of Daily Living (BADL)** This dimension represents the basic activities that people living independently should be able to perform (*e.g.,* bathing);
- **Instrumental Activities of Daily Living (IADL)** Successful independent living requires the capability to carry out instrumental activities (*e.g.,* managing a medication regimen);

– **Enhanced Activities of Daily Living (EADL)** Independent living also requires activities related with the outside world communication that are beyond what is considered to be instrumental. These activities are connected with major and holistic responsibilities (*e.g.*, buying groceries);
– **Social Activities (SA)** These activities are meant to entertain and sustain social closeness, such as communicating with others as a way to establish relationships. According to the generated ideas of participants, this dimension was added to the framework for aging in place with robots.

4.1   Coding Procedure

Data was coded according to the required functions that a robot should have to perform each activity. Each coder coded the totality of the material (65 different activities). According to Cohen's Kappa test, the level of agreement between the coders was $K = .91$, $\alpha = .000$, indicating an almost perfect agreement [8].

4.2   Activities for Aging in Place with Robots

Results suggest that older adults refer more different IADL (24 different activities), followed by BADL (17), and finally both EADL and SA (12 activities each) with a robot (see Fig. 2). Some of the referred activities are described in Table 1.

4.3   Chosen Robots

Results show that older adults have chosen different robots to serve different activities (see Fig. 3). It can be seen that Care-O-bot (18%) is the robot that most of the participants have chosen for BADL, followed by PR2 (17%) and Pearl (12%). When looking at IADL, it can be seen that Care-O-bot is the most chosen robot as 21% of participants have chosen this robot to integrate such activities. Then, PR2 (17%) is also referred in the context of IADL, followed by Peal (4%) and Pleo (4%). Considering the EADL, results show that half of the participants chose PR2 (50%), followed by Pearl (25%), Emys (25%), and

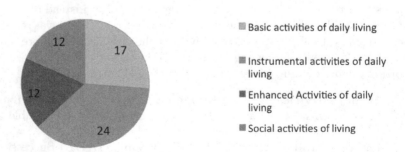

Fig. 2. Number of activities yield by older adults.

Table 1. Framework of activities for aging in place with robots, adapted from [12].

Basic Activities of Daily Living

"Help bathing, specially washing the feet and the back."

"Help open taps, like bath taps because it's hard for me to open them."

"Help put on the socks, and then the shoes. Then help to take them off."

"Help shaving because I do not see well and help cutting the nails."

"Help dressing, I don't mean every day but there are cloths that seem harder to dress."

Instrumental Activities of Daily Living

"Memorize what I eat. I do not always remember when to eat or what I eat, so I end up having a bad diet."

"Make the bed and change the bed sheets. Also, do the laundry and then hang it on a clothesline. Oh yes, and then iron it!"

"[The robot] should know my medical history and adapt the food it cooks. I cannot eat cakes and the robot should know this information."

"Help with the electricity and construction works like painting a wall that needs painting, repairing a water pipe, or just changing a light bulb, this is very useful."

"Clean the floor and sweep the kitchen and all that stuff. Oh, and wash the bathroom and clean the dust."

Enhanced Activities of Daily Living

"I would gave a list of what I need and the robot could go buy groceries and to the pharmacy."

"Make emergency calls to the police, ambulance, or family."

"Have an informative dialogue, by providing meteorological, time and news information. [The robot could also help us by] answering the door when we are lying in bed."

"[The robot should] be able to communicate with doctors and nurses."

[The robot should] warn us regarding appointments or obligations, like visits to the doctor, or when to take the right pills at the right times of the day."

Social Activities

"Read stories. I like novels very much, but my eyes are not able to see the words now. I would be so happy if the robot could read me stories at night."

"Accompany when walking outside to the park and to the cinema. I would never do such activities alone now."

"Cheer people, communicate or talk. The robot should be able to share its own ideas, even when they are different from ours."

"Pray with us". Some said the robot should also *"have a religion",* others disagreed. In the cases where they claimed it should have a religion, two opinions were expressed: *"[the robot] should adapt to theirs religion by having the same one",* or *"could choose its own belief."*

"Play games in general, and cards and domino particularly. It would be wonderful if the robot could just talk with us and be a company in our daily life."

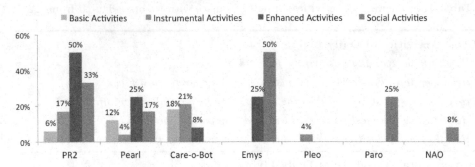

Fig. 3. Chosen robots according to the different types of activities.

Care-O-bot (8%). Finally, for the SA, Emys (50%) is the most chosen robot, followed by PR2 (33%), Paro (25%), Pearl (17%), and Nao (8%).

When clustering the service robots (Pearl, Care-O-bot, PR2) and the companion/entertainment robots (Paro, Pleo, Emys), and comparing them with the type of activities they were assigned to, results show the majority of participants assigned companion/entertainment robots with SA (65%), less than half of the participants assigned these robots with EADL (35%), and only 9% have assigned with IADL. On the other hand, service robots were assigned by the participants to all types of activities (see Fig. 4). We emphasize that participants have not chosen a robot for all the activities, existing activities without an assigned robot. On the other hand, participants assigned more than one robot to some of the activities.

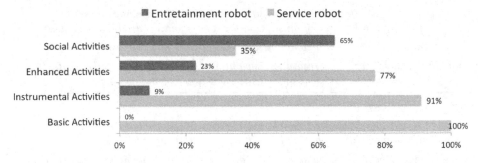

Fig. 4. Chosen robots according to the different types of activities.

5   Conclusions and Discussion

This study aimed to elicit activities from older adults in which the presence of a robot helps enhancing their QoL and contribute for their successful and independent aging. The novelty of this study concerns the presentation of activities that are part of all the real-world contexts of older adults: from the home, to the pharmacy, to a park, or even to be able to see a movie at the cinema.

Therefore, the majority of different activities refers to IADL, related with activities that are beyond personal and basic activities of daily living, but are essential to live independently in a society and community. In their perspective, it seems essential to have a robot that assists managing a medication regimen, maintain the household, and prepare meals of adequate nutrition [12]. Moreover, older adults referred that although they were able to perform some activities, they would prefer having a robot as an extra help (*e.g., "help putting and taking off the shoes"*). This seems to indicate that the participants are still in shape to independently manage their daily activities, but would benefit from additional assistance. A large number of different activities concerns BADL related with personal hygiene (*e.g.,* toileting and bathing) [12], in which participants claimed for help to *e.g., "get in and out of the tub"*. Finally, EADL and SA emerged as the activities in which older adults require the robot for communicating with the outside world due to the need to satisfy a basic activity (*e.g., "[the robot] could go buy groceries"*), translating an EADL; or due to a need to overcome social isolation by *"playing games", "accompany to the cinema",* or even *"pray"*, related with SA. Furthermore, service robots were chosen to perform all types of activities, showing this type of social robots are fit for different activities with this population. The participants referred that their choice for a robot was strongly motivated by its physicality. Thereafter, showing an interest for robots that are perceived as able to accomplish multiple tasks, instead of robots whose primary goal is more limited (*e.g.,* Pleo and Nao which are low height robots).

Although there are concerns about the accomplishments of some activities due to technical development and ethical aspects, this paper shows there is space for technology developments with views to enhance the QoL of older adults. By having a deeper understanding about the activities that older adults require a robot, HRI researchers detain key-information about where and how to dedicate their efforts and resources to fulfill a societal need and contribute to the QoL and successful aging among this population [17].

Acknowledgments. This work was partially supported by the European Commission (EC) and was funded by the EU FP7 ICT-317923 project EMOTE and partially supported by national funds through FCT - Fundação para a Ciência e a Tecnologia, under the project UID/CEC/50021/2013. The authors are solely responsible for the content of this publication. It does not represent the opinion of the EC, and the EC is not responsible for any use that might be made of data appearing therein. The authors express their gratitude to Santa Casa da Misericórdia de Lisboa (Portugal) for their involvement in the study.

References

1. Abeles, R.P., Gift, H.C.: Aging and quality of life. Springer Publishing Company (1994)
2. Beer, J.M., Smarr, C.A., Chen, T.L., Prakash, A., Mitzner, T.L., Kemp, C.C., Rogers, W.A.: The domesticated robot: design guidelines for assisting older adults to age in place. In: Proceedings of the Seventh Annual ACM/IEEE International Conference on Human-Robot Interaction, pp. 335–342. ACM (2012)

3. Billson, J.M.: Focus groups: A practical guide for applied research. Clinical Sociology Review **7**(1), 24 (1989)
4. Broekens, J., Heerink, M., Rosendal, H.: Assistive social robots in elderly care: a review. Gerontechnology **8**(2), 94–103 (2009)
5. Czaja, S.J., Charness, N., Fisk, A.D., Hertzog, C., Nair, S.N., Rogers, W.A., Sharit, J.: Factors predicting the use of technology: findings from the center for research and education on aging and technology enhancement (create). Psychology and aging **21**(2), 333 (2006)
6. Erikson, E.H., Erikson, J.M.: The life cycle completed (extended version). WW Norton & Company (1998)
7. Foti, D., Kanazawa, L.: Activities of daily living. Pedrettâs Occupational Therapy: Practice Skills for Physical Dysfunction **6**, 146–194 (2008)
8. Hallgren, K.A.: Computing inter-rater reliability for observational data: An overview and tutorial. Tutorials in Quantitative Methods for Psychology **8**(1), 23 (2012)
9. Kaplan, R.M., Bush, J.W.: Health-related quality of life measurement for evaluation research and policy analysis. Health Psychology **1**(1), 61 (1982)
10. Katz, S.: Assessing self-maintenance: activities of daily living, mobility, and instrumental activities of daily living. Journal of the American Geriatrics Society **31**(12), 721–727 (1983)
11. Mast, M., Burmester, M., Berner, E., Facal, D., Pigini, L., Blasi, L.: Semi-autonomous teleoperated learning in-home service robots for elderly care: a qualitative study on needs and perceptions of elderly people, family caregivers, and professional caregivers. In: 20th International Conference on Robotics and Mechatronics, Varna, Bulgaria, October 1–6 (2010)
12. Mynatt, E.D., Essa, I., Rogers, W.: Increasing the opportunities for aging in place. In: Proceedings on the 2000 Conference on Universal Usability, pp. 65–71. ACM (2000)
13. Nations, U.: World population prospects, January 2015. http://esa.un.org/wpp/
14. Osborn, A.F.: Applied imagination, principles and procedures of creative thinking (1953)
15. Pavot, W., Diener, E.: Review of the satisfaction with life scale. Psychological Assessment **5**(2), 164 (1993)
16. Pollack, M.E.: Intelligent technology for an aging population: The use of ai to assist elders with cognitive impairment. AI Magazine **26**(2), 9 (2005)
17. Rowe, J.W., Kahn, R.L.: Successful aging. The Gerontologist **37**(4), 433–440 (1997)
18. Sadri, F.: Ambient intelligence: A survey. ACM Computing Surveys (CSUR) **43**(4), 36 (2011)
19. Schaie, K.W.: Impact of technology on successful aging. Springer Publishing Company (2003)
20. Shanas, E.: Old people in three industrial societies. Transaction Publishers (1968)
21. Stahl, B.C., McBride, N., Wakunuma, K., Flick, C.: The empathic care robot: A prototype of responsible research and innovation. Technological Forecasting and Social Change **84**, 74–85 (2014)

An Empathic Robotic Tutor for School Classrooms: Considering Expectation and Satisfaction of Children as End-Users

Patrícia Alves-Oliveira(✉), Tiago Ribeiro, Sofia Petisca, Eugenio di Tullio, Francisco S. Melo, and Ana Paiva

INESC-ID and Instituto Superior Técnico, Universidade de Lisboa, Lisboa, Portugal
patricia.alves.oliveira@inesc-id.pt

Abstract. Before interacting with a futuristic technology such as a robot, there is a lot of space for the creation of a whole set of expectations towards that interaction. Once that interaction happens, users can be left with a hand full of satisfaction, dissatisfaction, or even a mix of both. To study the possible role of experience as a mediator between expectation and satisfaction, we developed a scale for HRI that measures expectations and satisfaction of the users. Afterwards, we conducted a study with end-users interacting with a social robot. The robot is being developed to be an empathic robotic tutor to be used in real schools, with input from primary end-users (children). Children's expectations and subsequent satisfaction after the interaction with the robotic tutor were analysed. The results can be fed back to the system developers on how well it is being designed for such a target population, and what factors regarding their expectation and satisfaction have shifted after the experience of interaction. By delivering on the children's expectations, we aim to design a robotic tutor that provides enough satisfaction to sustain an enjoyable and natural interaction in the real educational environment.

Keywords: Human-Robot Interaction · User-centered design · Robotic tutor · Expectation · Satisfaction

1 Introduction

Robotic characters are becoming widespread as useful tools in assistive [12], entertainment [17] and tutoring applications [7]. Besides, robots can be considered a mediatic type of technology as they are easily associated with science-fiction culture (*e.g.,* sci-fi novels, movies and adverts), making the expectations of people towards robot's an important aspect to consider in the process of designing and creating a robot. In fact, sci-fi culture ends up delivering information, most of the times unrealistic information, about a type of technology that is nowadays being created, bringing expectations over robots that are far from being achieved [3]. It is well known that previous expectations strongly influence

© Springer International Publishing Switzerland 2015
A. Tapus et al. (Eds.): ICSR 2015, LNAI 9388, pp. 21–30, 2015.
DOI: 10.1007/978-3-319-25554-5_3

satisfaction, so analysing initial expectations of users before interacting with a robot becomes important when concluding about their subsequent satisfaction of the experience.

The novelty of our work regards the measurement of the expectations and satisfaction levels that users have towards a robot, to serve as an input that informs the iterative process of designing and creating a social robot. In line with this, we developed the Technology-Specific Expectation Scale (TSES) to measure users' expectations before seeing and before interacting with a robot, and the Technology-Specific Satisfaction Scale (TSSS) to measure their satisfaction after the experience of interaction. Both scales constitute a novel metric in HRI, leading to a new complementary way of approaching the iterative process of designing a social robot. We base our research in the study of the expectations that children had towards the possibility of interacting with a robotic tutor and compare such expectations with the satisfaction level after the interaction. The robot used throughout this paper is being developed in the FP7 EU EMOTE project[1] to be an autonomous empathic robotic tutor aimed to teach topics about sustainable development to children in schools. The study followed a methodology that merges an autonomous robot with a Wizard-of-Oz (WoZ) [19]. The motivation of this research underlies the measurement of children's satisfaction towards a robotic tutor in an educational environment to serve as an input to inform further design developments of the same tutor. The evaluation of expectations/satisfaction involves factors related with education and learning, such as the perceived capabilities of a tutor. We aim to understand which features of the robotic tutor meet the expectations of children, and which do not. Thereafter, we will detain key-information about which still need refinement and which are performing well. Moreover, most people (including the children that participated in this study) never had any previous experience with social robots, so we anticipated that people detain preconceived ideas about this type of technology, built upon sci-fi culture. Due to this, expectations and satisfaction regarding the fictional views of the robot were also assessed and taken into account, enabling a contextualised interpretation of results.

Thus, in this paper we present the developed scales for measuring expectation and satisfaction in HRI; the results of the administration of the TSES and TSSS to children towards an empathic robotic tutor for education; and guidelines that inform the design based on this novel metric for HRI. In line with this, we formulate the following study hypothesis:

H1 The expectations that children have regarding the experience of interacting with a robotic tutor will be high as robots are part of strong sci-fi culture that children are familiar with.
H2 By building a robotic tutor inspired in real teacher-student interactions, children will detain high satisfaction levels after the experience of interaction.

---

[1] EMOTE project: http://www.emote-project.eu/

2   Related Work

Expectation and satisfaction are concepts that influence the way we evaluate experiences. In this Section, we will detail about such concepts and relate them with HRI applied to education.

2.1   Expectation and Satisfaction in HRI: Definition of Concepts

In the field of HRI, Bartneck and Forlizzi (2004), have defined a *social robot* as *"an autonomous or semi-autonomous robot that interacts and communicates with humans by following the behavioral norms expected by the people with whom the robot is intended to interact"* [4]. On the creation of robots that are intended to interact with people in daily life, Wistort (2010), has also brought up the fact that *"the form of a robotic character greatly determines the affordances it provides, influencing the perceived function of the character"*. This means that people immediately create expectations about a robotic character once they see it, just based on its appearance. The majority of robots' appearance is presented to people through sci-fi culture, showing them robots that are far different from the real developed robots, thus stimulating peoples' preconceived ideas (expectations) about the functions of robots [21].

Moreover, the Expectation-Confirmation Theory (ECT) is widely used in the consumer-behaviour literature to understand consumers' satisfaction after purchasing a product. According to this theoretical framework, consumers first form an initial expectation of a product, and during a period of initial consumption (experience of the product), they assess its performance to determine if their expectations are confirmed. Finally, they form a satisfaction towards the product based on their confirmation level and expectation on which that confirmation is based [15]. In line with this, ECT appears as a framework that can inspire metrics for the design and development of future technology, such as a robotic tutor for futuristic classrooms. So, *expectations* provide a baseline or reference level for users to form evaluative judgements about the experience with a product in which lower expectations usually influence satisfaction positively, if the previous expectations are confirmed by experience [5]. Nonetheless, when evaluating an innovative technology such as a robot, it is important to consider that user's expectations can be coloured by others' opinion, sci-fi culture, or can be tempered by the user experience [11]. On the other hand, *satisfaction*, is regarded as a transient, experience-specific affect. One can have a pleasant experience with a product, but still feel dissatisfied if it is below expectation [13–15]. Thus, *experience* is what connects expectation and satisfaction [5].

2.2   Expectations and Satisfaction Towards Robots for Education

The concepts of expectation and satisfaction have been studied in teacher-student interactions in different educational settings, such as online education [10], e-Learning [16] and traditional classrooms [1]. Research seems to show that student's perceived satisfaction derives from different factors, such as the way the tutor's knowledge is transferred to children, how feedback is used to support and facilitate learning, and the level of interaction [10]. Thereby, the tutor's competencies and capabilities constitute some of the factors that will influence not only the satisfaction that children have regarding such an experience in an educational context, but also their learning. In the context of HRI, different projects are developing robotic tutors to support and assist children during their learning process (*e.g.,* CoWriter project[2]). Also, Alves-Oliveira and collaborators (2014), have explored the expectations of children towards a robot that can interact with them in their own classroom, concluding that children's initial expectations can help to identify the usefulness of robots [2]. In this sense, the study of expectation and satisfaction towards robotic teachers and/or tutors in the context of learning environments can be import predictors of children's learning outcomes and of their evaluation of the experience. Moreover, it is also important to consider the concepts of expectation and satisfaction for other HRI environments, in which the design of social robots with end-users is timely important when shaping the future of this technology.

3   Methodology

This study took place in a school, where children performed a collaborative learning activity about sustainability in a reserved area of a classroom. For each session, a pair of children interacted with the robotic tutor. Together with them, the robot acted as the tutor for the learning environment, and played *EnerCities*[3], which is a collaborative multiplayer serious game for learning about sustainability that is being used in the EMOTE project.

3.1   Participants

The study sample consisted of 56 children (30 male, 25 female, 1 unknown) aged between 14 and 16 years old ($M = 14.81$, $SD = .48$). The children that participated in this research had consent forms signed by their caregivers and assented to participate in the activity.

3.2   System Architecture and Set-Up

The robotic tutoring system used in the study follows the extended SAIBA model for intelligent virtual agents [18] and is composed of a NAO Torso robot from Aldebaran Robotics; an interactive touch table running *EnerCities*; four video cameras; two lavaliere microphones; a WoZ interface; and a recorder (Fig. 1.a). The children interacted with the system (see Fig. 1.b) both through *EnerCities*, and through the system's perceptive capabilities. The system interacts back through the robotic tutor, which performs social, expressive and game-play

---

[2] CoWriter project: http://chili.epfl.ch/cowriter
[3] EnerCities: http://www.enercities.eu/

related behaviour. The perceptive capabilities of the system includes detecting and tracking the children's head location, gaze direction, eyebrow movement (AU2 and AU4 [9]), and which child is currently speaking. This is all performed by the *Perception Module*, using the Kinect and the lavalier microphones.

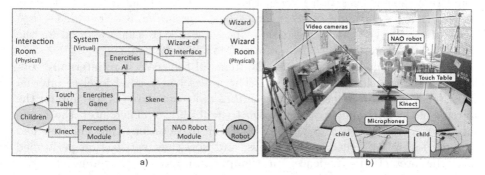

Fig. 1. a) Real environment setup. b) System architecture

The multimodal expressive behaviour (e.g., speech, gaze, animation) is managed by *Skene* which also includes a Gaze-state-machine, allowing the embodiment to perform semi-autonomously [18]. It is fed with information from the *Perception Module* and students' game-play actions from *EnerCities* to autonomously manage timing and expressive resources.

The robot's collaborative AI is a module capable of informing the game-playing and pedagogical decision-making of the robotic tutor that performs autonomously. The AI also incorporates a social component that continuously monitors each player's actions and automatically adjusts the tutor's strategy in order to follow the group's "action tendency" [20]. The Wizard was a researcher that was in a completely separate room, controlling the robotic tutor's high-level expressive behaviour (e.g., the timings to perform pre-defined utterances), using a specially designed user-interface. The AI selects a game move and makes it available for the Wizard to perform at the appropriate moment. This allows the Wizard to control the flow of interaction along with the flow of the game, without having to decide upon the game state and game actions. Finally, low-level and contingent behaviours remain autonomously controlled by *Skene*, which acts according to the high-level decisions performed by the Wizard and events triggered from the *Perception Module*.

3.3   Measures

To evaluate children's expectations and satisfaction, a TSES was created inspired in the Bhattacherjee and Premkumar (2004) scale [6]. Our scale was developed addressing aspects that inform about the state of the robotic tutor's development, in order to support further refinement. The TSES is composed of 10 questions allocated in 2 dimensions: *Capabilities* and *Fictional view* of the robotic tutor. It was used as a baseline questionnaire to measure children's expectations before seeing and interacting with the robotic tutor. Then, the TSSS was used as a post-questionnaire, applied after the interaction, to understand how children's subsequent satisfaction performed. Each scale is composed of equal questions to secure comparison between expectations and satisfaction, with different verbs tenses to meet different temporal experiences with the robotic tutor [4]. Children could rate their expectation level in the TSES in a 5 point type-Likert scale, ranging from 1 - *Very low expectation*; 2 - *Low expectation*; 3 - *Neutral*; 4 - *High expectation*; 5 - *Very high expectation*. The same Likert scale was used for the TSSS, substituting the word "expectation" for "satisfaction".

In order to understand whether the items of the scale were internally consistent, a Cronbach's Alpha was run. The scales had a good level of internal consistency for the 5 items of the *Capabilities dimension* ($\alpha = 0.770$) and for the 5 items of the *Fictional view dimension* ($\alpha = 0.749$) [8]. Thereafter, the *Capabilities dimension* served to inform about the expectations that children had towards the robotic tutor's capabilities, and how their satisfaction performed after the interaction experience. An example of a question that aimed to evaluate the expectation towards the robotic tutor's capabilities is the following: "I think the robotic tutor will be able to understand me." The *Fictional view dimension* relates with impressions created mostly by sci-fi culture, such as movies and novels, and an example is the following: "I think the robotic tutor will be similar to the robots I see in movies.". In addition, two more questions regarding the robotic tutor's *Competencies* were administrated: 1) "I think the robotic tutor will be a good game companion."; 2) "I think the robotic tutor will be the one that plays better.". The latter questions served to understand the perception that children had on the performance of the robot.

3.4   Procedure

The pair of children was invited into a separate room where they had no contact whatsoever with the educational setup, including the robotic tutor. This was a constraint to ensure that children's expectations were not influenced by any previous contact. At this point, the TSES was individually and separately applied to each child. After completion, the pair of children were led to the main room where the interaction with the robotic tutor took place. Children engaged in interaction in the real context of use for 20 minutes playing the EnerCities game with the robotic tutor. For this period of time, children were left alone in the main room with the robot, being able to freely interact and communicate with it. A researcher partially controlled the behaviour of the robot from a different room, in a WoZ methodology experiment, meaning that children were not aware that a third person controlled some of the behaviours of the robotic tutor. The game-play ended at the instruction of the robotic tutor. Afterwards, a researcher invited the participants to enter the same initial room where the TSSS was individually applied to each children.

---

[4] The TSES and the TSSS are available here: http://gaips.inesc-id.pt/~poliveira/Alves-Oliveiraetal.2015.pdf

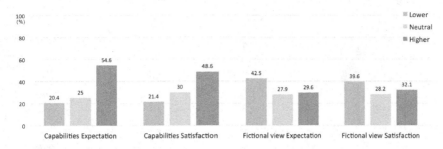

Fig. 2. Expectation and satisfaction of children towards the robotic tutor's capabilities and fictional view.

4 Results

To better understand the distribution of the results, data was re-arranged in 3 new categories, with a range of *Lower Expectation/Satisfaction* (scores below 3 in the initial 5 point type-Likert scale were clustered), *Neutral Expectation/Satisfaction* (scores equal to 3); *Higher Expectation/Satisfaction* (scores over 3 in the initial scale were also clustered).

4.1 Expectations and Satisfaction Towards a Robotic Tutor

Results seem to indicate that the children mostly had lower expectations about the robotic tutor in fictional terms (42.5%) and had higher expectations about the robotic tutor's capabilities (54.6%), whereas some of the children were neutral in terms of the expectations both towards the robotic tutor's capabilities (25.0%), and fictional view (27.9%) (see Fig. 2), which partially corroborates our first study hypothesis, which states that expectation of children would be high as robots are part of sci-fi culture. Regarding the satisfaction levels, results seem to suggest that when evaluating the robotic tutor after having experienced it, children's satisfaction levels seem to follow the expectations they previously had (see Fig. 2). This suggests that the majority of children who had higher expectations about the robotic tutor's capabilities (54.6%), seem to sustain higher satisfaction levels after the interaction (48.6%). Regarding the expectations of fictional view, results suggest that children had both higher (29.6%) and lower expectations (42.5%) of sci-fi culture towards the robotic tutor. The subsequent satisfaction indicates that children continued to have a mix of higher (32.1%) and lower (39.6%) satisfaction when evaluating this dimension. Thereby, the overall satisfaction towards the robotic tutor capabilities is higher, with no significant results between expectation and satisfaction levels. These results corroborate our second study hypothesis, which states that children will detain high satisfaction levels when evaluating the capabilities of a robotic tutor, after having experienced it.

In sum, the results of one iteraction seem to show that the current capabilities of the partially autonomous robotic tutor are at an appropriate level of

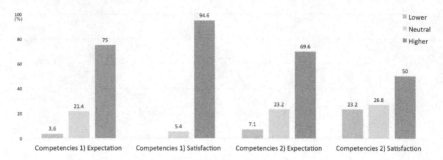

Fig. 3. Differences in expectation and satisfaction with experience.

development to sustain a collaborative educational interaction between an artificial tutor and students in a school classroom. Children seem to have had a positive experience of the interaction, as their expectations were accompanied by high levels of satisfaction. In addition, children's fictional views had lower levels, translating to more realistic expectations towards this technology.

A statistically significant difference was also found for the additional questions regarding aspects of the perceived competence of a robotic tutor, suggesting the interaction with the robotic tutor elicited a statistically significant change in the scores of satisfaction in comparison with expectation (Z = -3.127, α = .002). Therefore, the majority of children expected the robotic tutor to be a good game companion (75.0%) and after the interaction almost all children revealed higher satisfaction levels towards the tutor's competence (94.6%) (see Fig. 3). For the second question that assesses the competence of the robotic tutor, a significant result was also found (Z = -2.636, α = .008), revealing that the majority of children expected the robotic tutor to play best in the collaborative serious learning game about sustainability (69.6%), showing a significant decrease in their satisfaction after the interaction (50.0%). This result goes in line with the design process, as the robotic tutor had been developed to be a peer companion in the serious game, guiding children through the game rules and dynamics, but at the same time, having a similar hierarchical role in the game.

5   Conclusions and Future Work

This paper focuses on expectation and satisfaction in HRI and presents a novel metric for HRI. It aimes to address *expectations* and *satisfaction* of children towards an empathic robot tutor being developed in the EMOTE project to be included in an educational environment. The study of these concepts is crucial when developing a robotic tutor, as students' expectations and satisfaction are important in education [1], being predictors of learning outcomes [10]. To measure expectations and satisfaction, a TSES and a TSSS were developed and applied *before* and *after* the experience of the interaction with the robotic tutor. The results show that children had high expectations about the robotic tutor's capabilities, being followed by the same high levels of satisfaction. This result informs us that the behaviours of the robotic tutor in a 20min-interaction in a learning environment seem to meet expectations by their end-users. In addition, this suggests that state of the development of the partially autonomous robotic tutor for the classroom seems to be in an appropriate state to enable small group interactions with children. Results also show that the expectations regarding the fictional view of the robotic tutor are lower and remain lower after the interaction, which means that although children are exposed to sci-fi media, their expectations seem to be adapted to reality.

Overall, the design methodology inspired in real teacher-student interaction seems to have positive outcomes when testing the robotic tutor in its future environment with its future end-users. However, other possible outcomes of results can emerge when applying the TSES and the TSSS in the design process of creating a robot. An example of another possible outcome can be finding that users have high expectations regarding the fictional view of a robot, and high expectations regarding its capabilities. If the satisfaction level of fictional view decreases (which means it becomes more adapted to reality and further away from sci-fi culture) and the satisfaction towards the capabilities of the robot also decreases, this shows that the capabilities for the social robot do not meet users' expectations, suggesting the need for more development and refinement of its behaviours. Moreover, by looking at the items of the capabilities dimension, details about the capabilities that do not meet expectations can be identified.

Since the future of HRI will mostly be in people's homes and personal lives, our belief is that in order to build and create a valid futuristic technology that generates a positive experience and provides satisfaction to users, it is essential to involve them throughout this creative process. By measuring users' expectations and satisfaction, we are bringing input from the real-world users and stakeholders as co-designers of their future technologies.

In the future, we aim to measure expectation and satisfaction in a totally autonomous robotic tutor to understand how artificial social decisions are perceived by users (e.g., timing), and how satisfied they feel towards them, providing insights about the design over time. Also, we will explore the relation between expectations, satisfaction, and children's learning outcomes.

Acknowledgments. This work was partially supported by the European Commission (EC) and was funded by the EU FP7 ICT-317923 project EMOTE and partially supported by national funds through FCT - Fundação para a Ciência e a Tecnologia, under the project UID/CEC/50021/2013 and by the scholarship SFRH/BD/97150/2013. The authors are solely responsible for the content of this publication. It does not represent the opinion of the EC, and the EC is not responsible for any use that might be made of data appearing therein. The authors thank all children and teachers from Externato Maristas de Lisboa (Portugal).

References

1. Allen, M., Bourhis, J., Burrell, N., Mabry, E.: Comparing student satisfaction with distance education to traditional classrooms in higher education: A meta-analysis. The American Journal of Distance Education **16**(2), 83–97 (2002)

2. Alves-Oliveira, P., Petisca, S., Janarthanam, S., Hastie, H., Paiva, A.: How do you imagine robots? Childrens' expectations about robots. In: Workshop on Child-Robot Interaction at IDC14 - Interaction Design and Children (2014)
3. Bacon-Smith, C.: Science fiction culture. University of Pennsylvania Press (2000)
4. Bartneck, C., Forlizzi, J.: A design-centred framework for social human-robot interaction. In: 13th IEEE International Workshop on Robot and Human Interactive Communication, RO-MAN 2004, pp. 591–594 (2004)
5. Bhattacherjee, A.: Understanding Information Systems Continuance: An Expectation-Confirmation Model. MIS Quarterly **25**(3), 351–370 (2001)
6. Bhattacherjee, A., Premkumar, G.: Understanding changes in belief and attitude toward information technology usage: a theoretical model and longitudinal test. MIS Quarterly **28**(2), 229–254 (2004)
7. Castellano, G., Paiva, A., Kappas, A., Aylett, R., Hastie, H., Barendregt, W., Nabais, F., Bull, S.: Towards empathic virtual and robotic tutors. In: Lane, H.C., Yacef, K., Mostow, J., Pavlik, P. (eds.) AIED 2013. LNCS, vol. 7926, pp. 733–736. Springer, Heidelberg (2013)
8. Cortina, J.M.: What is coefficient alpha? An examination of theory and applications. Journal of Applied Psychology **78**(1), 98 (1993)
9. Ekman, P., Friesen, W.: Measuring Facial Movement. Environmental Psychology and Nonverbal Behavior **1**(1), 56–75 (1976)
10. Eom, S.B., Wen, H.J., Ashill, N.: The determinants of students' perceived learning outcomes and satisfaction in university online education: An empirical investigation*. Decision Sciences Journal of Innovative Education **4**(2), 215–235 (2006)
11. Fazio, R.H., Zanna, M.P.: Direct experience and attitude-behavior consistency. Advances in Experimental Social Psychology **14**, 161–202 (1981)
12. Feil-Seifer, D., Matarić, M.J.: Defining socially assistive robotics. In: Rehabilitation Robotics. ICORR 2005, pp. 465–468. IEEE (2005)
13. Hunt, H.K.: CS/D - Overview and future research directions. Conceptualization and measurement of consumer satisfaction and dissatisfaction, pp. 455–488 (1977)
14. Oliver, R.L.: Measurement and evaluation of satisfaction processes in retail settings. Journal of Retailing (1981)
15. Oliver, R.: A congitive model of the antecedents and consequences of satisfaction decisions. Journal of Marketing Research, 460–469 (1980)
16. Paechter, M., Maier, B., Macher, D.: Students expectations of, and experiences in e-learning: Their relation to learning achievements and course satisfaction. Computers & Education **54**(1), 222–229 (2010)
17. Pereira, A., Prada, R., Paiva, A.: Socially present board game opponents. In: Reidsma, D., Nijholt, A., Romão, T. (eds.) ACE 2012. LNCS, vol. 7624, pp. 101–116. Springer, Heidelberg (2012)
18. Ribeiro, T., Pereira, A., Di Tullio, E., Alves-Oliveira, P., Paiva, A.: From thalamus to skene: high-level behaviour planning and managing for mixed-reality characters. In: Intelligent Virtual Agents - Workshop on Architectures and Standards (2014)
19. Riek, L.D.: Wizard of Oz studies in HRI: a systematic review and new reporting guidelines. Journal of Human-Robot Interaction **1**(1) (2012)
20. Sequeira, P., Melo, F.S., Paiva, A.: lets save resources!: A dynamic, collaborative AI for a multiplayer environmental awareness game
21. Wistort, R.: Only robots on the inside. Interactions **17**(2), 72–74 (2010)

A Reactive Competitive Emotion Selection System

Julian M. Angel Fernandez[1,2]([⊠]), Andrea Bonarini[1], and Lola Cañamero[2]

[1] Dipartimento di Elettronica, Informazione e Bioingegneria, Politecnico di Milano,
Piazza Leonardo da Vinci 32, 20133 Milano, Italy
{julianmauricio.angel,andrea.bonarini}@polimi.it
[2] Embodied Emotion, Cognition and (Inter-)Action Lab, School of Computer
Science, University of Hertfordshire, College Lane, Hatfield, Herts AL10 9AB, UK
L.Canamero@herts.ac.uk

Abstract. We present a reactive emotion selection system designed to be used in a robot that needs to respond autonomously to relevant events. A variety of emotion selection models based on "cognitive appraisal" theories exist, but the complexity of the concepts used by most of these models limits their use in robotics. Robots have physical constrains that condition their understanding of the world and limit their capacity to built the complex concepts needed for such models. The system presented in this paper was conceived to respond to "disturbances" detected in the environment through a stream of images, and use this low-level information to update emotion intensities. They are increased when specific patterns, based on Tomkins' affect theory, are detected or reduced when it is not. This system could also be used as part of (or as first step in the incremental design of) a more cognitively complex emotional system for autonomous robots.

Keywords: Social robotics · Human Robot Interaction · Emotional models · Emotion production

1 Introduction

Social environments involve subtle interaction among people and the physical environment. These interactions, the context in which they take place, and people's mental perception of the world affect the emotions that arise. Different theories and models of how emotions arise have been proposed in psychology, such as [11,16,18]. Although these models seem acceptable and cogent to most of us, the hidden assumptions that authors make in their models [5,15] emerge when trying to implement them in artificial agents and robots. Computational frameworks based on these "high-level" models have been implemented [7,10], but they use abstract concepts that have to be defined in the system.

However, social robots need to be able to operate in real circumstances, where the information that the system needs to operate is not well defined or given beforehand and changes over time. Therefore, the robot needs to be able

© Springer International Publishing Switzerland 2015
A. Tapus et al. (Eds.): ICSR 2015, LNAI 9388, pp. 31–40, 2015.
DOI: 10.1007/978-3-319-25554-5_4

to interpret relevant information necessary to be used as input (including by the above-mentioned frameworks) from its current sensory data and representation of its environment. Due to sensor limitations (e.g., accuracy, noise) the world model cannot be a complete nor precise representation of the current situation. Moreover, the representation used to model the environment could lack details that are necessary to correctly interpret the situation. The use of "high-level" computational frameworks for autonomous social robots is thus problematic.

Emotion-based robot architectures have been proposed that ground emotion elicitation on the robot's sensory data. For example, the robot presented in [6] uses data from simple (contact) sensors, interpreted following the model of general stimulation patterns proposed by Tomkins [20]. Other robots such as [4,14] use complex sensory input (vision, voice) and a complex architecture to determine the robot's emotional state using some form of appraisal of the current situation. This paper presents a reactive emotional system that combines elements of these two approaches. Also based on Tomkins' theory [12,20], but using visual input and a complex architecture, the pre-selected emotions compete among them to be triggered. The system has been designed in a modular way, so to make it easy to combine it with other, more complex models such as the one suggested by Izard [11]. Our architecture is designed for use by a robot that needs to respond autonomously to relevant events (e.g., sudden changes in light conditions, presence of different agents or objects).

The rest of paper is organized as follows. Section 2 provides a brief overview of particularly relevant work closely related to our architecture. Section 3 outlines different emotional theories, paying particular attention to the model proposed by Tomkins. Section 4 describes our emotional system: the design and formulas that control the system. Finally, Section 5 covers the implementation and results.

2   Related Work

The robotic head Kismet created by Breazeal [4] uses cameras to perceive the world and head movements to interact with people. Kismet's emotions are the six basic emotions of Ekman [8]: happiness, sadness, surprise, fear, disgust, and anger. The emotion selection process can be summarized as a cyclic sequence of perceiving an event and appraising it [3]. The appraisal phase is where the change of emotion can be done.

Cañamero and Fredslund developed the LEGO humanoid robot Feelix that expresses emotions on its face based on physical (tactile) stimulation [6]. A tactile sensor is used to determine the stimulation which could fall in one of the following cases: short (less than 0.4 sec), long (up to 5 sec), and very long (over 5 sec). The events generated from the stimulation are used to determine the emotion activation based on the state of a finite state machine that implements general emotion activation patterns (cf. Fig 1) drawn from Tomkin's theory of emotions [20], that we have also used in this paper. Feelix could detect stimulation patterns for and display the following emotions: anger, sadness, fear, happiness and surprise.

MEXI is a robotic face that is capable to interact with people through emotions [9]. MEXI is capable to understand people emotions through image analysis of data coming from two cameras, and its speech recognition system. MEXI's architecture lacks of any deliberative component, but it uses emotions and drives to control its behaviours. Its emotion system obtains information from the behaviour system and external perceptions to come up with the new values for each emotion. Each emotion is represented by a value between 0 and 1, updated according to the current perception. The considered emotions are: anger, happiness, sadness and fear.

The architecture described in [13] uses a mixture of hard-coded emotions and emotions learned by association. Their emotion system uses inputs from the deliberative and reactive architectural layers to select one of the emotions: fear, anger, surprise, happiness, and sadness. Each one of these emotions is triggered according to perceived events, internal state, and goals of the robot in the current movement. The emotion selected by the emotion system affects the way each behaviour is performed.

The emotional model proposed by Malfaz and Salichs [14] uses appraisals to select an emotion. Happiness is related to the fact that something "good" happens to the agent (e.g., interpreted as the reduction of a need), and sadness to something "bad" (e.g., interpreted as the increment of a need). Fear is related to the possibility that something bad happens to the agent and it is activated when something dangerous could be expected by the agent.

3   Tomkins' Emotion Theory

There are many theories of emotion, differing in assumptions and the components involved in the process. They can be classified in different ways. For example [15, 19] use the following categories:

- *Adaptational:* based on the idea that emotions are an evolving system used to detect stimuli that are of vital importance.
- *Dimensional:* organize emotions according to different characteristics, usually valence (pleasantness-unpleasantness) and arousal. One of the most widely used is the Russel's circumplex model of affect [17].
- *Appraisal:* argues that emotions arise from the individual's judgement, based on its believes, desires, and intentions with respect to the current situation. EMA [10] and Fatima [7] frameworks fall in this category.
- *Motivational:* studies how motivational drives could generate emotions.
- *Circuit:* supports the fact that emotions correspond to a specific neuron path in the brain.
- *Discrete:* are theories based on Darwin's work, the expression of emotion in man and animals. These theories use as a pillar the idea of the existence of a basic emotions.
- *Other approaches* are lexical, social constructivist, anatomic, rational, and communicative.

In practice, these theoretical categories overlap. The difference among these theories is mainly in how the process and inputs are considered in each one.

Tomkins' theory [12,20], on which we base our model, integrates various perspectives. For Tomkins, the affect system evolved to solve the problem of overwhelming information present in the environment to which people are exposed. His theory states that people cannot manage to be conscious of all the information available from the environment, therefore the affect system comes to select what information could be relevant to be aware of in a given moment. For example, someone could focus on reading a book, ignoring the rest of events that are happening, but suddenly there might be a loud sound that gets his/her attention. This kind of behaviour could be obtained through the activation of different systems. He recognizes four systems closely related to affect:

- *Pain* is a motivator for very specific events that take place on our bodies.
- *Drive* deals with the basic needs that human body could need (e.g. urination, breathing).
- *Cognitive* interprets the world and make inference from it.
- *Affect* is focus on get person attentions to specific stimuli.

More importantly, Tomkins suggested that affect in certain situations could make that pain and drive systems are omitted, while the affect and cognitive could work together. Because affection has a main role in human subsistence, he describes nine affects that could be triggered depending on brain activity. Figure 1 shows activation patterns for relief, sadness, happiness, anger, interest and fear. For instance, sustained low stimulation leads to sadness, while a very highly increasing stimulation leads to fear, and a less steep increase in stimulation leads to interest. Moreover, the time windows for these emotions are different; for instance fear arises faster than happiness.

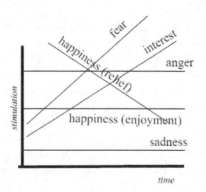

Fig. 1. Patterns for relief, sadness, happiness, anger, interest and fear, after Tomkins.

4   Emotional System

As suggested by Izard [11], among others, emotion elicitation can be performed at different levels, and at some of these levels we are not aware of the process. Our system focuses on the "reactive", "pre-aware" part of emotion elicitation (and selection, in our case) using as an input gray scale images from a web cam. Our system does not take in consideration any cognitive information from the environment; instead, we compare two consecutive images to determine changes in pixels in order to detect disturbances in the environment that could be of interest for the robot. This difference (the quantity of pixels that have changed over a threshold) is given as input to the stimulation calculator to determine the "stimulation" that is later used by the emotion generator to update the intensity of each emotion. This update is done searching for the patterns suggested by Tomkins (Fig. 1). The previous process is always modulated by the time delay between the two images considered. This delay is of vital importance in the system because it could not be determined with certainty beforehand. Using this delay in the equation makes the system behave in the same way regardless of whether the delay is short or long. Consequently, the system gives different values of "stimulation" depending on the delay between the images.

Figure 2 depicts the general process with all the subsystems. These subsystems were selected to permit upgrades in the system without the need to make considerable changes in the code. For example, the change detector subsystem could be improved to detect additional features from the images; if the output remains as percentage (value between zero and one), the rest of the system could still use it to update the emotion intensity.

4.1   Stimulation Calculator

This subsystem obtains the percentage of change provided by the change detector and updates the new stimulation ($stimulus(t)$) based on the current change ($s\_increment$), the last stimulation ($stimulus(t-1)$), and a reduction value ($s\_decrement$), as shown in Equation 1. In addition to $stimulus(t-1)$, functions $s\_increment$ and $s\_decrement$ use the time delay ($delay$) as a parameter.

$$
\begin{aligned}
stimulus(t) = \ & stimulus(t-1) \\
& + s\_increment(percentage, delay) \\
& + s\_decrement(stimulus(t-1), s\_increment(percentage, delay), \\
& \quad delay, bias)
\end{aligned}
\tag{1}
$$

The $s\_increment$ function ranges on the percentage of change and the delay time, and it is calculated as it is shown in the equation 2. The $s\_increment$ uses an exponential function with a desire base ($base\_increase$) and displacement coefficient (d). This displacement coefficient is used to obtain values greater than one, but it also introduces a small bias that is corrected by the second part

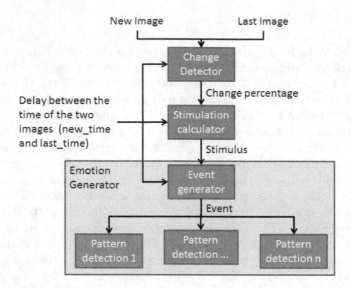

Fig. 2. General architecture of the system. The arrows show the information flow. The time difference between the two images is used to modulate each module.

of the equation. The *increase_factor* is a coefficient that modulates the gain of the function, which is used to obtain less or more stimulation. And *delay* is a variable coming from the time delay between the two pictures used to generate the percentage.

$$s\_increment(.) = ((base\_increase)^{(percentage-d)}$$
$$- \underbrace{(base\_increase)^d}_{\text{correction factor}}) * increase\_factor * delay \qquad (2)$$

Figure 3 illustrates the behaviour of $s\_increment(.)$, showing that this function produces greater values when the delay increases. The $s\_decrement(.)$ function (Equation 3) uses $s\_increment(.)$, time delay, and a bias to determine the decrease value. The parameter *bias* is used to modify the lower output value of the system. Like $s\_increment(.)$, this equation depends on time to make the modulation. Figure 4 illustrates its behaviour with a $decrease\_factor = -0.5$ and with different time delays.

$$s\_decrement(.) = (stimulus(t-1) + s\_increment(.) - bias)$$
$$* decrease\_factor * delay \qquad (3)$$

4.2   Emotion Generator

This subsystem was divided in two modules (event generator and pattern detection) to give the possibility of adding or deleting new emotion patterns, and of

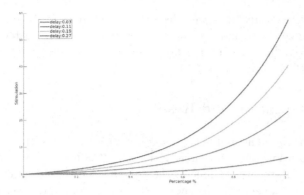

Fig. 3. Behaviour of the increase function for different delays in the image using parameters *base_increase* = 30, *d* = 0.1, and *increase_factor* = 10.

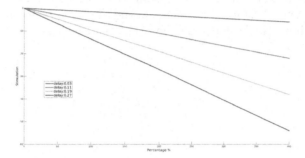

Fig. 4. Behaviour of the decrease function for *decrease_factor* = −0.5 and different time delays.

modifying the event characteristics. Event generator centralizes the process of detection of relevant events from the stimulation slope. The events considered are: null, small, medium, large, and huge slope. Except null slopes, the other events could be either positive or negative. A pattern detection module is implemented for each emotion that should be detected. Each pattern detection module considers a different pattern as well as the number of events to search for in the pattern. The emotions, their patterns, and their update functions are:

- *Surprise* is recognized just when one of the following events are present in its time window: large or huge positive slope. Due of this strong constraint, every time that this pattern is detected, its intensity grows faster than for other emotions.
- *Fear* is increased when three or more consecutive recent events have either large or huge positive slopes.
- *Interest* occurs when three or more consecutive events have either medium or small positive slopes.
- In contrast to the rest of emotions, *Relief* works with negative slopes and its intensity increases when at least five negative slope events are detected.

We have considered only emotions affected by stimulus change since we focus on the reactive processes to activate emotions, while emotions related to constancy of stimuli in Tomkin's model are expected to be managed by the cognitive part of a larger system.

5   Implementation and Results

The system was implemented in C++ and uses OpenCV to analyze images. The final implementation was then interfaced with ROS to enable easy use in other systems. To facilitate easy parameter change, two configuration files were added: one related to all the general parameters (e.g. threshold and $increase_{coefficient}$) and the other to establish the increment, decrement, and time window (number of events to consider) for each of the implemented patterns. The system was tested online with information coming from a Logitech CY270 Web-Cam. The intensity and events obtained are depicted in the Figure 5, where the relationship between the stimulation's slope and the events can be seen.   Figure 6 depicts

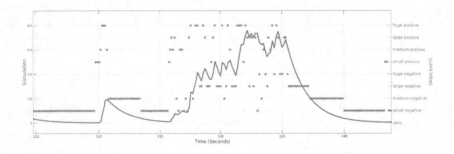

Fig. 5. Stimulation (continuous line) and events (dots in horizontal lines) obtained from the comparison of to consecutive images. The y-axis on the left represents the stimulation level, while the one on the right represents the events generated from the slope detected.

the intensity obtained for each pattern implemented (fear, interested, surprise and relief), also showing that each pattern module updates its emotion intensity independently. This is clearly seen at second 120 when fear, interest, and surprise unevenly increase their intensities and after some time they also reduce their intensity unevenly. This (increase and decrease) unevenness shows the pattern's configuration, which is not the same for each emotion. The presence of more than one emotion with a value different from zero suggests that a further mechanism should be used to determine which emotion should be elicited, for example taking the one with higher intensity or just modifying behaviour parameters proportional to each intensity. In other words, our initial aim to use this system as first step to select an emotion is achieved.

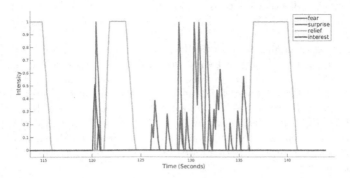

Fig. 6. Intensity obtained by our system for the four emotions implemented: fear (blue), interested (purple), surprise (red) and relief (yellow).

6 Conclusions and Further Work

We have presented a reactive emotional system based on Tomkins' theory. The system is modular to permit its integration with more complex systems and its configuration based on the output from the pattern detection modules. The system was implemented in C++ with interface to ROS to make it possible to used it in other models and in robotic platforms. Four patterns (fear, surprise, interest, and relief) were implemented and tested. The results show that the output compete with each other, and the emotion has to be selected in a further step with a logic that suits the specific purpose, which could be as simple as take the emotion pattern with higher intensity, or weight behaviours by the intensity of the corresponding patterns. Additionally, this reactive system could be used as complement for cognitive systems.

As a further work, the system is going to be integrated to our theatrical system [1,2], to provide changes in the emotion that is going to affect the robot's movement parameters. Finally, a simple behaviour will be implemented, to be triggered by the selected emotion appropriate to the situation.

Acknowledgments. This work was carried out at the Embodied Emotion, Cognition and (Inter-)Action Lab of the School of Computer Science at the University of Hertfordshire as part of an Erasmus doctoral program.

References

1. Angel F., J.M., Bonarini, A.: Towards an autonomous theatrical robot. In: ACII 2013, pp. 689–694 (2013)
2. Angel Fernandez, J.M., Bonarini, A.: TheatreBot: a software architecture for a theatrical robot. In: Natraj, A., Cameron, S., Melhuish, C., Witkowski, M. (eds.) TAROS 2013. LNCS, vol. 8069, pp. 446–457. Springer, Heidelberg (2014)

 3. Breazeal, C.: Affective interaction between humans and robots. In: Sosík, P., Kelemen, J. (eds.) ECAL 2001. LNCS (LNAI), vol. 2159, pp. 582–591. Springer, Heidelberg (2001)
 4. Breazeal, C.: Designing Sociable Robots. MIT Press, Cambridge (2002)
 5. Canamero, L., Fredslund, J.: I show you how I like you-can you read it in my face? IEEE Transactions on Systems, Man and Cybernetics, Part A **31**(5) (2001)
 6. Caamero, L.: Emotion understanding from the perspective of autonomous robots research. Neural Networks **18**(4), 445–455 (2005). emotion and Brain
 7. Dias, J., Mascarenhas, S., Paiva, A.: FAtiMA modular: towards an agent architecture with a generic appraisal framework. In: Bosse, T., Broekens, J., Dias, J., van der Zwaan, J. (eds.) Emotion Modeling. LNCS, vol. 8750, pp. 43–55. Springer, Heidelberg (2014)
 8. Ekman, P.: Emotions Revealed : Recognizing Faces and Feelings to Improve Communication and Emotional Life. Owl Books, March 2004
 9. Esau, N., Kleinjohann, L., Kleinjohann, B.: Emotional communication with the robot head MEXI. In: Proceedings of Ninth International Conference on Control, Automation, Robotics and Vision, ICARCV 2006, Singapore, December 5–8, 2006, pp. 1–7 (2006)
10. Gratch, J., Marsella, S.: A domain-independent framework for modeling emotion. Journal of Cognitive Systems Research **5**, 269–306 (2004)
11. Izard, C.: Four systems for emotion activation: cognitive and noncognitive processes. Psychological Review, 68–90 (1993)
12. Kelly, V.: A primer of affect psychology (2009). http://www.tomkins.org/wp-content/uploads/2014/07/Primer_of_Affect_Psychology-Kelly.pdf
13. Lee-Johnson, C.P., Carnegie, D.A.: Emotion-based parameter modulation for a hierarchical mobile robot planning and control architecture. In: 2007 IEEE/RSJ International Conference on Intelligent Robots and Systems, October 29 – November 2, 2007. Sheraton Hotel and Marina, San Diego, pp. 2839–2844 (2007)
14. Malfaz, M., Salichs, M.A.: A new approach to modeling emotions and their use on a decision-making system for artificial agents. IEEE Transactions on Affective Computing **3**(1), 56–68 (2012)
15. Marsella, S.C., Gratch, J., Petta, P.: Computational models of emotion. In: Scherer, K.R., Bnziger, T., Roesch, T. (eds.) A blueprint for an affectively competent agent: Cross-fertilization between Emotion Psychology, Affective Neuroscience, and Affective Computing. Oxford University Press, Oxford (2010)
16. Ortony, A., Clore, G.L., Collins, A.: The cognitive structure of emotions. Cambridge Univ. Press, Cambridge (1994). http://opac.inria.fr/record=b1125551, autre tirage: 1999
17. Russell, J.A.: A circumplex model of affect. Journal of Personality and Social Psychology **39**, 1161–1178 (1980)
18. Scherer, K.R.: Appraisal considered as a process of multi-level sequential checking, pp. 92–120. Oxford University Press, New York (2001)
19. Scherer, K.R.: A Blueprint for Affective Computing, chap. Theoretical approaches to teh study of emotion in humans and machines, pp. 21–46. Oxford University Press (2010)
20. Tomkins, S.S.: Affect theory. Approaches to emotion, pp. 163–195 (1984)

Group Vs. Individual Comfort When a Robot Approaches

Adrian Ball[1]([⊠]), David Rye[1], David Silvera-Tawil[2], and Mari Velonaki[2]

[1] Australian Centre for Field Robotics, The University of Sydney, Sydney, Australia
{a.ball,d.rye}@acfr.usyd.edu.au
http://www.acfr.usyd.edu.au
[2] Creative Robotics Lab, The University of New South Wales, Sydney, Australia
{d.silverat,mari.velonaki}@unsw.edu.au
http://www.crl.niea.unsw.edu.au

Abstract. This paper quantifies how the comfort of a person approached by a robot changes when that person is alone or in a group of two. A total of 140 participants in lone and paired configurations were approached by a robot from eight different directions and asked to rate their level of comfort. Results show that while the comfort of an individual was influenced by the presence and relative position of a second person, there were some common features in the comfort responses of all participants regardless of their group configuration.

Keywords: Human-robot interaction · Comfort · Group

1 Introduction

In human-robot interaction, the initiation phase leading to a subsequent interaction is important to the success of the interaction [1]. A major part of the initiation phase is the way a robot approaches potential interactants, including the path taken by the robot in approaching the person or group of people. In this work we define the "best" approach direction as the most comfortable one, and define comfort in terms of a natural language understanding of mental comfort as tranquility and contentedness; being free from a state of unease, constraint, fear or anxiety. Comfort is assessed simply by asking a person "how comfortable" they are. We define the comfort profile of a person as the mapping of their comfort levels to a set of robot approach directions.

It has been shown [2] that people tend to interact with robots in their personal space, as defined by Hall's theory of proxemics [3]. When people are alone, they are most comfortable when a robot approaches them from the front—where the robot can easily be seen—and are least comfortable when they are approached from behind [4–6]. In our previous work [7] we showed that the comfort of people in groups of two was qualitatively similar to that of individuals in single-person, single-robot (SPSR) scenarios, although a person's comfort level was influenced both by the group formation shape and the position of participants within the group.

© Springer International Publishing Switzerland 2015
A. Tapus et al. (Eds.): ICSR 2015, LNAI 9388, pp. 41–50, 2015.
DOI: 10.1007/978-3-319-25554-5_5

The cited works on SPSR scenarios show that people are more comfortable when approached from a 'frontal' direction rather than from a 'rearward' direction. Extension to groups of more than one person introduces complexities. When people interact in a group, they tend to orient themselves towards other group members. In the most extreme case, when a group of two people face each other, a robot that approaches one person from a 'frontal' direction must approach the other person from a 'rearward' direction. Our work in two-person, single-robot (TPSR) scenarios investigates how the comfort profile of an individual is influenced by the location of the second person. The comfort profile of a person is defined here as the mapping of their comfort levels to the set of robot approach directions. Knowledge of how comfort profiles differ between individuals who are alone or in a group can be integrated with existing robot path planning algorithms, allowing for a robot to maximise the comfort of individuals it might interact with.

Although previous SPSR studies have been conducted [4–6], the diverse experimental conditions and relatively small sample numbers don't allow for a qualitative comparison with our TPSR results. We therefore chose to repeat single-person experiments under the same conditions as our prior two-person experiments [7], and with significantly more participants than previously reported. Performing group and individual experiments under the same conditions allows for a direct quantitative comparison of the two data sets. As a result, this paper presents experimental work that quantifies differences in comfort levels between people seated alone and those seated in pairs, in various sitting configurations. The comfort profiles of individuals in pairs and alone are compared across eight robot approach directions through intra- and inter-position statistical analyses.

2   Design and Conduct of Experiments

Two sets of experiments were performed to investigate the hypothesis that people in pairs have a silimar comfort to lone individuals when they are approached by a robot from different directions. The two experiments followed a similar procedure: with two participants for the TPSR scenario, and with a single participant for the SPSR scenario. All participants were naive to the experiment and were recruited at a campus of the University of New South Wales, Australia. For the group-experiment trial, a pair of participants was seated in low armchairs at a low table in the centre of a room and asked to work on a cooperative task for the duration of the experiment. The task was included in the experiment design to provide a cognitive load on the participants, intended to minimise their attention to the robot's location and movement. The selected task was to complete a three-dimensional jigsaw puzzle. This task was chosen as it has a clear objective, is temporally demanding and does not require any turn-taking activity.

During the experiment, participants were seated in one of three maximally-different seating configurations, selected using Kendon's [8] F-formation framework for analysing interactions between two or more people. An F-formation—or "facing formation" [9]—forms whenever two or more people position themselves so that they share an overlapping transactional space, termed the o-space. As defined by Kendon, the interacting people occupy the p-space and monitor the r-space. During the experiment, the participants are assumed to have an o-space centred on the shared jigsaw puzzle. The three seating configurations used are shown in Figure 1 and are referred to as Configurations A, B and C.

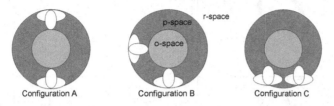

Fig. 1. Participant seating configurations showing o-space, p-space and r-space assumed with two participants. From [7].

The experiments were conducted in a six metre square room with exits on three of the four walls. The arrangement of the room can be seen in Figure 2. The eight robot approach directions are labeled and indicated by arrows in the figure. The robot was controlled using the Wizard of Oz paradigm to enter the group's p-space [8], or to approach it as closely as possible. A square room was used to remove spatial bias that could arise through asymmetric placement of participants in the room. Multiple exits were provided so that participants always had the option of leaving the room to avoid confrontation with the robot. The interested reader could refer to [7] for more details of the experiments.

A robot (Figure 3) present in the room periodically approached the participants and prompted them via prerecorded speech to rate how comfortable they were with the most recent robot approach. After each robot approach, each participant scored their response on a five-point Likert scale between 1: "uncomfortable" and 5: "comfortable". The robot approach directions were randomised to prevent an order effect in the results. The experiment continued until the robot had approached the group from eight different directions in random order. The experiment was concluded with a post-experiment questionnaire to acquire additional information, including gender, from participants.

In addition to these group experiments, experiments with a single participant were performed. The single-person experiments were identical to the group experiments, except that the participant was seated only in one location and the second seat was removed from the room. The arrangement can be seen in Figure 2. The robot approached the table along each of the eight directions, potentially forming a p-space with the participant focused on the jigsaw puzzle.

Since a person's self-reported level of comfort cannot be regarded as an absolute measure, participant responses are ranked so that an ordinal analysis can be made. Participant responses are ranked from one (most comfortable) to eight

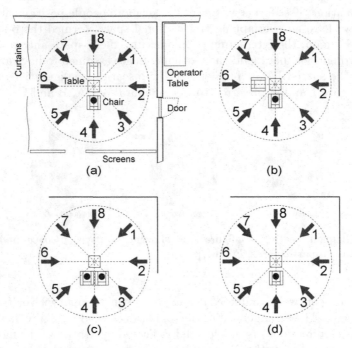

Fig. 2. Experimental space with chairs arranged in (a) Configuration A, (b) Configuration B, (c) Configuration C, and (d) in the SPSR scenario. The arrows indicate the eight directions of robot approach relative to the 'dotted' seat(s). Adapted from [7].

Fig. 3. Robot used for the experiment. From [7].

(least comfortable). Two types of analysis were used to compare the comfort levels of participants seated in the different configurations; these are termed *intra-position* and *inter-position* comparisons. An intra-position comparison performs pairwise statistical comparisons of the rank distributions of participant comfort levels for all 28 pairs of robot approach directions for a particular seating position and configuration, and then compares this set of results to those obtained for another seating position and configuration. An inter-position comparison compares the participant comfort level distributions for a relative robot approach direction between two different seating positions and configurations, and then repeats this comparison for the remaining seven robot approach directions. After a correction for multiple comparisons, the comfort level distributions are not different if all eight comparisons are not different.

3 Results

3.1 Participants

Twenty trials were conducted for each of the three group configurations and the lone-participant configuration to ensure a statistical power $(1 - \beta)$ of at least 0.80. Of the 140 participants, 61 were male and 79 were female. The mean age of participants was 24.7 years with a standard deviation of 8.7 years; the minimum age was 18 and the maximum age was 73. Most of the participants were university students. Although the participation of four persons older than 60 raised the variance, no age-dependent effects were observed in the data. Each person participated in only one trial of eight robot approaches in one seating configuration.

3.2 Intra-Position Analysis

Table 1 shows the mean comfort rank of participants for each robot approach direction for the six relative seating positions. The robot approach directions are always numbered relative to the seating location in question. This convention is illustrated through the 'dotted' seat locations in Figure 2, and means that direction 8 is always directly in front of the participant.

To determine if there were differences between data for different robot approach directions a Kruskal Wallis analysis of variance (KW-ANOVA) test was performed. The KW-ANOVA test was chosen due to the non-parametric nature of the data. This test takes the participant comfort data from all robot approach directions and returns a p-value indicating the probability that all of the data are from a common distribution; that is, that no systematic differences are present. If the p-value is less than 0.05 then a post-hoc multiple comparison test is performed to determine which pairs of robot approach directions cause participant comfort levels that differ from each other. As multiple comparisons are being made, a correction is required to control type I errors. The false discovery rate (FDR) control [10] was used with $q = 0.05$.

Table 1. Means and standard deviations (in parentheses) of individual comfort rankings for each robot approach direction for the lone individual (SPSR) and individuals in the three seating configurations of pairs of participants. Rank 1 is most comfortable; rank 8 is least comfortable. The labels 'Left' and 'Right' identify where the person of interest was sitting in the pair.

Dir	SPSR	Ind. A	Con. B (Left)	Con. B (Right)	Con. C (Left)	Con. C (Right)
1	1.9(1.4)	2.7(1.9)	2.5(2.0)	2.5(1.9)	2.5(2.0)	3.7(2.3)
2	3.6(1.9)	4.0(2.6)	3.4(2.6)	3.1(2.4)	4.3(2.4)	4.7(2.5)
3	5.4(1.8)	4.0(2.6)	4.4(2.1)	3.9(2.6)	3.2(2.6)	4.7(2.3)
4	7.3(1.6)	5.6(3.0)	5.1(2.9)	4.8(3.0)	4.8(2.9)	6.3(2.3)
5	5.9(1.9)	3.9(2.5)	3.9(2.6)	4.5(2.6)	5.2(2.5)	3.9(2.4)
6	3.6(1.7)	3.6(2.6)	3.0(2.2)	4.0(2.5)	4.2(2.0)	3.0(2.1)
7	2.0(1.7)	2.7(2.2)	2.6(2.1)	2.7(1.6)	3.1(2.5)	2.7(2.2)
8	2.4(2.1)	3.1(2.2)	3.3(2.3)	4.0(2.8)	2.8(2.3)	3.0(2.0)

The comfort levels of the individuals in Configuration A were statistically different ($\chi^2(7,312) = 33.26$, p < 0.01, $\eta^2 = 0.10$) with approaches from different directions. Post-hoc multiple comparison Mann-Whitney U-tests showed that participant comfort with robot approaches from directly behind (direction 4) was different to that for approaches from all other directions. Approaches from direction 2 also caused significantly different participant comfort levels to those from direction 7.

With participants seated in Configuration B there were no significant differences for either the person sitting on the left ($\chi^2(7,152) = 17.69$, p < 0.05, $\eta^2 = 0.11$) or on the right ($\chi^2(7,152) = 13.49$, p = 0.06, $\eta^2 = 0.08$).

Although there were no statistically significant differences in comfort levels for individuals sitting on the left in Configuration C ($\chi^2(7,152) = 20.24$, p = 0.05, $\eta^2 = 0.13$), significant differences were found for the person sitting on the right ($\chi^2(7,152) = 32.40$, p < 0.01, $\eta^2 = 0.20$). Participant comfort with robot approaches from directly behind (direction 4) was statistically different to all other directions except for direction 2. Comfort with approaches from direction 3 was also different to that with approaches from direction 7.

For robot approaches to the lone participant, significant differences in participant comfort levels ($\chi^2(7,152) = 83.76$, p < 0.01, $\eta^2 = 0.53$) were found between several directions. Approaches from directly behind the participant (direction 4) resulted in comfort levels that were different from those produced by all other robot approach directions, and comfort levels with approaches from directions 3 and 5 were different from all other approach directions except each other. Furthermore, directions 1 and 7 produced comfort levels that were different from directions 2 and 6, and direction 8 was different from direction 6. These SPSR results conform results previously reported in the literature [4–6], and with a somewhat larger sample size.

The intra-position analysis showed that participant comfort levels varied across each seating configuration. Individuals seated in Configuration A and on the right in Configuration C had comfort profiles similar to, but more uniform than, those of individuals seated in the SPSR scenario. Participants seated in Configuration B and on the left in Configuration C had more uniform comfort distributions across all tested robot approach directions. Of particular note is the finding that the comfort levels of solo participants are different to those of all individuals in groups, showing that the presence of a second person influences the comfort level of the first participant. The results also show that the position of the second participant influences the comfort level of the first. The present results demonstrate, for the first time, that the comfort level of a person approached by a robot depends both on the presence and the relative position of a second person.

3.3   Inter-Position Analysis

To determine if there were statistically significant differences in participant comfort between two seating configurations, a Mann-Whitney U test was used to compare the distribution of comfort ranks for a particular robot approach direction between the two seating configurations in question, for all eight approach directions. Participant comfort levels for two positions in different seating configurations[1] were then defined to be statistically not different from each other if the results of all eight Mann-Whitney U tests were statistically not different. As this comparison requires a sequence of eight pair-wise comparisons to be 'not different' to confirm the null hypothesis, a correction is applied. The FDR-control method was again used with $q = 0.05$.

Given the symmetry of Configuration A, for these comparisons it is assumed that the results for the two seating positions in Configuration A can be 'rotated' on to each other, effectively doubling the sample size to $N = 40$. An inter-position analysis performed between the two positions of Configuration A gives p = {0.83, 0.62, 0.085, 0.82, 0.48, 0.80, 0.95, 0.55} for directions 1–8 respectively. The strong similarity of the two sets of distributions validates the assumption. It also shows that no bias is present due to the asymmetry of the room (Figure 2) relative to the seating positions.

Table 2 shows the results of comparing the comfort levels of single persons approached by a robot with those of individuals seated in the group configurations. The scarcity of significantly different distributions shows that the comfort profiles of lone individuals and those in groups of two are only slightly different. The strong similarity found in the inter-position analysis indicates a high degree of consistency in participant comfort scoring throughout the experiments. Although the results in Table 2 show that the comfort profiles of people in pairs are similar to those seated alone, Table 1 shows a trend towards more uniform comfort profiles for grouped individuals, but with more variance. That is, the

---

[1] Although two positions in the same seating configuration could be compared, we focus on comparisons between lone individuals and individuals in groups of two.

presence of a second person does not significantly change the comfort level of an individual, but does tend to decrease the sensitivity of the individual to the direction of robot approach.

Table 2. Results, in the form of p-values, of pairwise comparison tests for an inter-position analysis. The first column shows the robot approach direction of interest. Subsequent columns show the probability that the distribution of comfort levels of individuals in particular groups are not different from those of lone individuals. The five p-values that represent distributions that are significantly different to the SPSR distributions at the 5% level after FDR correction have been emphasised.

Dir	Ind. A	Con. B (Left)	Con. B (Right)	Con. C (Left)	Con. C (Right)
1	0.11	0.45	0.40	0.26	**0.01**
2	0.48	0.83	0.56	0.29	0.08
3	0.04	0.11	0.09	0.02	0.40
4	0.06	0.01	**0.00**	**0.01**	0.13
5	**0.00**	0.01	0.08	0.39	**0.00**
6	0.80	0.31	0.51	0.30	0.35
7	0.28	0.42	0.09	0.13	0.27
8	0.29	0.18	0.08	0.70	0.30

4   Discussion

The SPSR results in Table 1 show a marked front-back asymmetry with a strong preference for approaches from the front. These results are qualitatively similar to the results of prior SPSR experiments. In comparison, the individuals seated in groups of two have less directional preference. The standard deviation of the SPSR distributions are generally smaller than the distributions of individuals seated in groups, showing a more consistent comfort preference in the SPSR scenario.

The intra-position analysis shows that the relative differences between comfort level distributions for different robot approach paths change from those of the SPSR configuration when a second person is introduced. That is, the *presence* of a second person changes the comfort profile of the first. Furthermore, since the differences between comfort level distributions change with the relative position of the second person (Configuration A, B or C), the *position* of the second person influences the comfort profiles of the first. Although the second result was shown previously [7], it is restated here with higher statistical power.

The inter-position analysis compares how similar the comfort distributions of SPSR participants are to those distributions of individuals seated in a group. Almost all of the 'low' p-values in Table 2 occur for directions 3–5. Combining this with the values in Table 1, it can be seen that the comfort of a person when they are approached from behind improves in the presence of a second person.

Most comparisons of approach directions for individuals in group configurations against the approach directions of lone individuals show statistically non-significant differences. While the intra-position comparison results showed that the presence and location of a second participant influence the comfort of an individual, the lack of significantly different comfort distributions in the inter-position results show that the general shape of the comfort profile for individuals in Configurations A, B or C are similar to individuals in an SPSR configuration. That is, robot approach directions from a 'front' direction are more comfortable than approaches from a 'rear' direction, but with a reduced magnitude in the comfort difference between these two robot approach regions.

Although there is an approximately equal gender split in the data set (43.5% males, 56.5% females), it is not possible to make a statistically significant claim about the differences in comfort results reported by each gender. The collected data came from several different seating configurations, and in order to show that the results for each gender were independent of the experiment seating configuration, it would have to be shown that there were no significant differences for each gender across all seating configurations. Unfortunately, the sample size for each gender in each seating configuration is insufficient to make a statistically sound gender-based claim.

Some insight into a gender comparison can be obtained by analysing the Configuration A data, where two people are seated opposite each other. If it is assumed that there is no spatial bias introduced by the slight asymmetry of the room, then the data can be partitioned by gender. Of the 40 participants for Configuration A, 16 were male and 24 were female. While a sample size of 16 for the males is a little low to make a strong statistical claim (statistical power is approximately 0.7), it is sufficiently large to show a general trend that could warrant further investigation. It is also worth noting that this comparison does not consider the gender of the subject's partner.

When performed on the male data, a KW-ANOVA analysis ($\chi^2(7,120) = 22.58$, p < 0.01, $\eta^2 = 0.18$) suggested that there were different comfort distributions for different robot approach directions. Performing a follow-up multiple comparison between robot approach directions with a FDR correction factor found no significantly different distributions. For the female data, there were significant differences in some approach directions ($\chi^2(7,184) = 17.72$, p $= 0.01$, $\eta^2 = 0.09$). Robot approach direction 4 was found to be statistically different to approach directions 1,3,7 & 8. An inter-gender analysis gives p $= 0.03$, 0.73, 0.18, 0.95, 0.07, 0.88, 0.07, 0.90 for approach directions 1-8 respectively, all of which are non-significant with a FDR correction factor with $q = 0.05$. While the intra-gender results showed that females preferred 'frontal' robot approach directions to a direct 'rear' approach, the inter-gender results show that there are no significant differences between the rank distributions of males and females for each robot approach direction. By increasing the sample size, particular for the males, this observation would have greater statistical strength.

5   Conclusion

This work used inter- and intra-position statistical analyses of experimental data to quantify differences between the comfort levels of lone and paired individuals approached by a robot from several directions. Although the presence of a second person influenced the comfort profile of the first by decreasing their sensitivity to the direction of robot approach, a strong similarity remained between the comfort profiles of grouped and lone individuals. Comparison of intra-position data analyses showed that the relative location of the second participant influenced the comfort profile of the first.

References

1. Mead, R., Atrash, A., Matarić, M.J.: Automated proxemic feature extraction and behavior recognition: Applications in human-robot interaction. Int. J. Social Robotics **5**(3), 367–378 (2013)
2. Hüttenrauch, H., Severinson Eklundh, K., Green, A., Topp, E.A.: Investigating spatial relationships in human-robot interaction. In: Proc. 2006 IEEE/RSJ Int. Conf. Intell. Robots & Systems, pp. 5052–5059. IEEE (2006)
3. Hall, E.: The Hidden Dimension. Anchor Books (1966)
4. Dautenhahn, K., Walters, M., Woods, S., Koay, K.L., Nehaniv, C.L., Sisbot, A., Alami, R., Siméon, T.: How may I serve you? a robot companion approaching a seated person in a helping context. In: 1st ACM SIGCHI/SIGART Conf. Human-Robot Interact., pp. 172–179. ACM (2006)
5. Walters, M.L., Dautenhahn, K., Woods, S.N., Koay, K.L., Te Boekhorst, R., Lee, D.: Exploratory studies on social spaces between humans and a mechanical-looking robot. Connection Science **18**(4), 429–439 (2006)
6. Walters, M.L., Dautenhahn, K., Woods, S.N., Koay, K.L.: Robotic etiquette: results from user studies involving a fetch and carry task. In: 2nd ACM/IEEE Int. Conf. Human-Robot Interaction, pp. 317–324. IEEE (2007)
7. Ball, A., Silvera-Tawil, D., Rye, D., Velonaki, M.: Group comfortability when a robot approaches. In: Beetz, M., Johnston, B., Williams, M.-A. (eds.) ICSR 2014. LNCS, vol. 8755, pp. 44–53. Springer, Heidelberg (2014)
8. Kendon, A.: Spacing and orientation in co-present interaction. In: Vogel, C., Hussain, A., Campbell, N., Esposito, A., Nijholt, A. (eds.) Second COST 2102. LNCS, vol. 5967, pp. 1–15. Springer, Heidelberg (2010)
9. McDermott, R.P., Roth, D.R.: The social organization of behavior: Interactional approaches. Ann. Rev. Anthrop. **7**(3), 321–45 (1978)
10. Benjamini, Y., Hochberg, Y.: Controlling the false discovery rate: A practical and powerful approach to multiple testing. J. Royal Stat. Soc. Ser. B (Methodological) **57**, 289–300 (1995)

Understanding Group Comfort
Through Directional Statistics

Adrian Ball[1]([✉]), David Rye[1], David Silvera-Tawil[2], and Mari Velonaki[2]

[1] Australian Centre for Field Robotics, The University of Sydney, Sydney, Australia
{a.ball,d.rye}@acfr.usyd.edu.au
http://www.acfr.usyd.edu.au
[2] Creative Robotics Lab, The University of New South Wales, Sydney, Australia
{d.silverat,mari.velonaki}@unsw.edu.au
http://www.crl.niea.unsw.edu.au

Abstract. This paper has the dual aims of introducing and using directional statistics to investigate the comfort levels of pairs of people approached by a robot from different directions. Data from pairs seated in three maximally-different seating configurations are analysed. These data are in the form of circular distributions of ranked comfort levels. Statistical tests for uniformity of circular distributions and for determining if differences exist between pairs of circular distributions are introduced and used to analyse the directional properties of the data. It is shown that directional statistics can be used to compare comfort level ranks that capture all tested robot approach directions; something that cannot be achieved with linear statistics.

Keywords: Human-Robot Interaction · Comfort · Group · Directional statistics · Non-parametric

1 Introduction

Directional statistics provides methods of analysing data that are circularly distributed. In a general sense, this includes any data that can be assigned an orientation [1] or be represented on a hypersphere [2]. In social robotics, one example of such data is the preferences of people approached by a robot, as encoded by their ratings of comfort levels with approaches from different directions. Each robot approach direction can be assigned an orientation relative to the person, and directional statistics can then be used to analyse the corresponding data.

The scenario of a person approached by a robot is interesting in that it can be analysed using both linear and directional statistics. Linear statistics can be applied to experimental data to measure the differences in participant comfort between two robot approach directions. Such linear analysis has been performed both for lone individuals [3–5] and for groups of two [6,7]. These linear analyses compared participant data from different robot approach directions to see which directions resulted in participant comfort levels that were statistically different.

© Springer International Publishing Switzerland 2015
A. Tapus et al. (Eds.): ICSR 2015, LNAI 9388, pp. 51–60, 2015.
DOI: 10.1007/978-3-319-25554-5_6

In these scenarios, linear statistics analysed the distribution of comfort data across particular robot approach directions, while directional statistics can analyse the circular distribution of robot approach directions across particular values of comfort data.

The present work shows how directional statistics complements linear statistics when analysing comfort profiles of participants that are approached by a robot from several different directions. In the following section we provide some background on directional statistics and explain how relative robot approach directions can be incorporated. We then briefly explain the experimental process, provide analytical results from collected data together with a higher-level commentary on the implications of the numerical results.

2   Directional Statistics

To make meaningful claims[1] regarding the comfort of participants approached by a robot, the data must be analysed statistically [8]. The set of available statistical tests falls into two categories: parametric and non-parametric. Parametric tests assume that sampled observations come from a population with a known parameterization, such as the normal distribution. The challenge, then, is either to use the sampled data to estimate the unknown parameters or to derive confidence intervals for the unknown parameters [9]. Non-parametric tests, on the other hand, do not make any assumptions about distributions in an underlying population and are often considered 'distribution-free' [10].

In this work we focus on non-parametric tests as they are functional without the need to make assumptions about the distribution of participant responses. Two directional statistics tests are utilised in this work. The first test examines the uniformity of a directional distribution, while the second analyses similarities between a number of directional distributions.

2.1   Test of Distribution Uniformity

The Rayleigh test [2] estimates the probability that a population is uniformly distributed over all directions. By treating each sampled datum as a unit vector in its corresponding direction, the vector sample mean R provides a measure of distribution uniformity. As the expected value $E(cos\theta, sin\theta) = 0$ when θ is sampled from a uniform distribution, the uniformity hypothesis can be rejected when R is 'large' [1].

To determine whether R exceeds a threshold value, a mapping is required from the vector sample mean space $\mathbb{R}^+$ to the probability space $[0\ 1]$; p-values in this latter space are estimated probabilities that the sampled data are directionally uniformly distributed. In this work we set $p \leq 0.05$ for a distribution to be non-uniform. As R is a bivariate function, the probability distribution is similar to the chi-squared distribution with two degrees of freedom. This mapping

---

[1] That is, to distinguish real effects from chance occurrences.

can be done with an error of order $O(n^{-2})$ [11]. There are pathological non-uniform distributions that can 'pass' the Rayleigh test. It is therefore important to perform a post hoc inspection to confirm results.

Sometimes data collected for analysis can have a coarse directional resolution or be grouped[2]. When working on data with this property, a correction factor can be used to allow for the coarse directional resolution. This is typically done when the directional resolution is $\pi/4$ or greater [1]. Such a corrected vector sample mean is defined by Stuart and Ord [12].

2.2   Comparison of Circular Distributions

Although the Rayleigh test measures the uniformity of a directional distribution it does not quantify differences between multiple distributions.

Watson's U^2 test [13] provides a method of measuring the difference between two circular distributions. This test is the directional equivalent of the linear Mann-Whitney U test [14]. Watson's U^2 test can be extended to permit the testing of more than two distributions, as shown by Brown [15]. This extension is beneficial as it allows an ANOVA-equivalent test to be performed while also accommodating data grouping that coarsens directional resolution.

In a similar manner to the Rayleigh test, the U^2 value obtained needs to be mapped to a probability. The mapping between a U^2 value and the corresponding p-value is defined by Maag [16].

These directional statistics tools allow directional data to be analysed in ways analogous to the more widely known linear statistics. A U^2 value can be obtained by comparing all directional distributions with each other using Brown's extended U^2 test [15] in a non-parametric ANOVA-equivalent test. If the corresponding p-value indicates that there are directional distributions that are significantly different to each other then a post hoc pairwise multiple comparison using Brown's extended U^2 test is performed for each possible pair of distributions. It is worth noting that Brown's extended U^2 test degenerates to Watson's U^2 test when only two distributions are tested. Again, U^2 values are converted to p-values using the transformation provided by Maag. Because a multiple comparison test is being performed, a p-value correction is required to prevent excessive false-positives. The present work uses the false discovery rate (FDR) [17], rather than a family-wise error rate correction factor [14] such as Bonferonni, since the FDR method provides better statistical power for the multiple comparison test.

3   Experiment

A brief overview of how the experimental data was collected is given here; the interested reader could refer to [7] for more details. Participants naive to the experiment were recruited on a campus of the University of New South Wales, Australia. For

---

[2] Grouping is an artificial method of making the directional resolution more coarse.

each experimental trial, a new pair of participants was seated at a low table in the centre of a room and asked to work on a three-dimensional jigsaw puzzle, a cooperative non-turn-taking task, for the duration of the experiment. A task was assigned to the participants to provide a cognitive load for the duration of the experiment, minimising participant attention towards the robot's location and movement. The participants were periodically approached and interrupted by a robot that was also present in the room. The robot used prerecorded speech to ask the participants to rate how comfortable they were with the most recent robot approach. Following each robot approach, each participant scored their response to the question "Please rate how comfortable you were with the robot's most recent approach path" on a five-point Likert scale between 1: "Uncomfortable" and 5: "Comfortable". The experiment continued until the robot had approached the pair from eight different directions in random order. The order of robot approach directions was randomised for each experiment to prevent order effects. Additional information, such as participant perceptions of the robot, was gathered using a post-experiment questionnaire.

Three maximally-different seating configurations of two people working on a common task were chosen. These configurations, shown in Figure 1, are referred to as Configurations A, B and C. The interaction between the pair of participants was assumed to have an o-space (Kendon [18]) centred on the shared jigsaw puzzle. Figure 1 also shows the orientation of the seating configurations in the room and the eight robot approach directions used in the experiment.

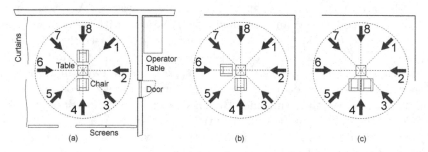

Fig. 1. Experimental space with chairs arranged in (a) Configuration A, (b) Configuration B and (c) Configuration C. From [7]. The arrows represent the robot approach directions.

3.1   Data Preprocessing

Since individuals have a wide variety of prior experiences, it is inevitable that significant differences will exist between the self-reported comfort level scores of participants in the same situation. To avoid the need for parameterising user comfort levels with particular robot approach paths, each participant's absolute comfort scores for the eight robot approach directions were converted to ranks from one (most comfortable) to eight (least comfortable). When a participant scored two or more approach directions as equally comfortable the directions were assigned the same rank. In such cases, the rank of the next highest score was set to one greater than the cardinality of previously ranked scores. Each rank distribution was then formed by counting how many times that particular rank was associated with each robot approach direction.

Seating configurations B and C have persons on the right and the left; these rank data were coded 'R' and 'L' to preserve the information associated with relative position. In Configuration A the relative position of the second person is the same for both individuals. The data for one person was therefore superimposed on the other, effectively doubling the amount of individual data in Configuration A. These data are denoted 'A Ind.' The scores for each pair of participants were summed to produce combined group scores for Configurations A, B and C. Group ranks were derived from these group scores to provide information about the comfort of pairs of participants.

The conversion of absolute comfort scores to ranks allows non-parametric tests to be used in the analysis. This is beneficial as the underlying distribution of the general population does not have to be assumed. In the application of directional statistics in the present work, a sample distribution is the distribution of the number of occurrences of a particular rank for each direction.

Uniformity Test: Each robot approach direction was assigned an angle based on its relative orientation to the other approach directions. The directions 2, 3, ..., 8, 1 defined in Figure 1 were assigned angles $0, 7\pi/4, ..., 2\pi/4, \pi/4$. It is worth noting that the Rayleigh test is rotationally invariant so that the angle labels are arbitrary provided that the relative order of the angles is maintained.

Circular Distribution Comparisons: As multiple comparisons are made in analysing the directional rank data, a correction factor is required. The FDR mentioned in Section 2.2 was used with a q-value of 0.05 [17].

4 Experiment Results

4.1 Rayleigh Tests of Uniformity

The results of the Rayleigh test applied to each rank distribution are shown in Table 1. Choosing $p < 0.05$ as the threshold for rejecting the null hypothesis that the directional data are uniformly distributed gives nine statistically non-uniform distributions. Examples of statistically uniform and non-uniform rank distributions are shown in Figure 2.

It is important to note that the p-values resulting from the application of the Rayleigh test to two distributions cannot be compared with each other beyond determining which distribution is the more uniform. The Rayleigh test provides no information regarding the nature of the non-uniformity.

4.2 Watson's U^2 Test

When applied to the group rank data for Configurations A, B and C, the ANOVA-equivalent Watson's U^2 test verified the null hypothesis at a significance level of $p < 0.05$ for Configuration A ($U^2 = 0.69, p = 0.21$) and Configuration B ($U^2 = 0.82, p = 0.06$). The test showed a highly significant difference in some rank distributions for Configuration C ($U^2 = 1.35, p < 0.01$).

The p-values from a post hoc multiple comparison U^2 test conducted for Configuration C are shown in Table 2(a). Nine pairs of rank distributions were found to be statistically significantly different following the FDR correction.

The rank distributions of individuals in each configuration were examined next. For the individuals in Configuration A there was a highly significant difference in rank distributions ($U^2 = 1.36, p < 0.01$). The p-values from the post hoc multiple comparison tests are shown in Table 2(b). Five pairs of rank distributions were found to be significantly different following the FDR correction.

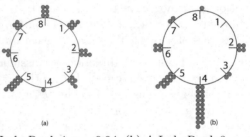

(a) A Ind., Rank 4, $p = 0.94$. (b) A Ind., Rank 8, $p = 0.00$.

Fig. 2. Examples of (a) uniform and (b) non-uniform rank distributions according to the Rayleigh test. The dots represent the count of how often the labeled directions were assigned to the represented rank.

Table 1. Results of the Rayleigh test for uniformity of rank distributions. The table shows probabilities that the rank distributions are uniform. Bold numbers denotes p-values < 0.05. Labels 'A', 'B' and 'C' refer to group ranks of the corresponding configurations whereas all other labels refer to individual ranks.

| Rank | Configuration | | | | | | | |
	A	B	C	A Ind.	B (L)	B (R)	C (L)	C (R)
1	0.93	0.23	**0.01**	0.13	0.10	0.39	0.15	**0.03**
2	0.35	0.07	0.08	0.08	0.70	0.36	0.30	**0.03**
3	0.46	0.07	0.83	0.17	0.92	0.13	0.56	0.09
4	0.06	0.58	0.48	0.94	0.27	0.27	0.42	0.90
5	0.56	0.36	0.41	0.48	0.60	0.99	0.21	0.49
6	0.78	0.30	**0.01**	0.36	0.54	0.20	0.31	0.07
7	0.98	0.26	**0.01**	0.11	0.06	0.49	0.14	**0.01**
8	0.08	0.41	**0.00**	**0.00**	0.06	0.12	0.06	**0.01**

It is notable that all pairs that are significantly different contain the least comfortable rank, rank 8.

The null hypothesis was confirmed for participants that sat on the left in Configuration B ($U^2 = 0.84, p = 0.05$) and on the right in Configuration B ($U^2 = 0.74, p = 0.13$). For Configuration C, the null hypothesis was confirmed for participants seated on the left ($U^2 = 0.84, p = 0.05$), but significant differences in rank distributions were found for participants seated on the right ($U^2 = 1.32, p < 0.01$). The p-values of the post hoc multiple comparison test can be seen in Table 2(c).

It is possible to compare the rank distributions of people on the left and right of the same seating configuration. As there are eight rank distributions for each location, the dimensionality of the ANOVA-equivalent U^2 test increases to 16. Examining the data this way shows a highly significant difference between some rank distributions for the participants in Configuration A ($U^2 = 2.27, p < 0.01$) and Configuration C ($U^2 = 2.17, p < 0.01$). There were no significant differences in rank distributions between the participants for Configuration B ($U^2 = 1.59, p = 0.06$). The p-values of the post hoc multiple comparison tests for Configurations A and C can be seen in Tables 3(a) and 3(b) respectively.

5   Discussion

From the first set of rank distribution comparisons shown in Table 2, the p-values loosely cluster into two groups. The first group consists of rank distributions 1–5, while the second group consists of rank distributions 6–8. These findings are consistent with the Rayleigh uniformity test results in Table 1, where most of the non-uniform distributions are of higher rank, indicating lower comfort levels. The distributions of the 'least comfortable' higher ranks are (unsurprisingly) dense in directions behind individuals (Figure 2b), showing that comfort levels with different robot approach directions are governed by a strong preference for where the robot *should not* approach from.

The results for the intra-group comparisons in Tables 3(a) and 3(b) are consistent with the earlier Rayleigh (Table 1) and Watson's U^2 (Table 2) results. In Table 3(a) it is not surprising to see a significant difference between the higher ranks of each location, as people seated in Configuration A face opposite directions. Table 3(b) shows a trend of similarity between the left and right seating locations in Configuration C. The third quadrant of the table shows the clustering trend of the results for the left and right with comparisons of rank distributions 1 to 5 and 6 to 8 having a higher p-value than other pairwise comparisons in the quadrant.

Clustering directional ranks cannot be done with linear statistics as no relationship exists between different linear sampled distributions. Results from a linear statistical analysis [3–7, 19], either for individuals or groups, do not naturally suggest an alternate robot approach direction if one can not be selected on the basis of application constraints. As directional statistics tie the data together both spatially through the robot approach directions and ordinally through the

Table 2. Results of multiple pairwise comparisons between ranks using the modified Watson's U^2 test. The bold entries denote pairwise rank comparisons that were significatively different following a false discovery rate correction with $q = 0.05$. Only the three configurations with a p-value < 0.05 mapped from the modified Watson's U^2 (ANOVA-equivalent) test are shown.

(a) Configuration C

Rank	1	2	3	4	5	6	7	8
1								
2	0.42							
3	0.52	0.52						
4	0.37	0.25	0.90					
5	0.47	0.82	0.69	0.38				
6	**0.00**	**0.01**	0.09	0.02	0.02			
7	**0.00**	**0.00**	0.06	0.03	**0.01**	0.31		
8	**0.00**	**0.00**	0.03	**0.01**	**0.01**	0.20	0.50	

(b) Individual A

Rank	1	2	3	4	5	6	7	8
1								
2	0.39							
3	0.59	0.96						
4	0.73	0.29	0.48					
5	0.81	0.60	0.76	0.85				
6	0.12	0.03	0.06	0.38	0.24			
7	0.03	0.02	0.04	0.28	0.17	0.74		
8	**0.00**	**0.00**	**0.00**	**0.00**	**0.00**	0.02	0.03	

(c) Individual C (R)

Rank	1	2	3	4	5	6	7	8
1								
2	0.41							
3	0.10	0.19						
4	0.40	0.11	0.12					
5	0.54	0.25	0.53	0.32				
6	**0.01**	**0.00**	0.04	0.35	0.03			
7	**0.00**	**0.00**	0.03	0.20	0.03	0.74		
8	**0.00**	**0.00**	**0.01**	0.05	**0.01**	0.14	0.48	

ranking process, they can provide the basis for heuristic rules describing the direction(s) a robot should approach a person from. The directional results presented here support prior linear results [7,19] that show significant differences between being approached from behind and being approached from any other direction by the robot.

Table 3. Results of intra-group pairwise comparisons between ranks using the modified Watson's U^2 test. Bold entries denote pairwise rank comparisons that were significatively different following a false discovery rate correction with $q = 0.05$. The two configurations with a p-value < 0.05 mapped from the modified Watson's U^2 test (ANOVA-equivalent) are shown.

(a) Configuration A

	L1	L2	L3	L4	L5	L6	L7	L8	R1	R2	R3	R4	R5	R6	R7	R8
L1																
L2	0.64															
L3	0.28	0.20														
L4	0.45	0.46	0.70													
L5	0.36	0.37	0.77	0.93												
L6	0.09	0.05	0.12	0.04	0.08											
L7	0.03	0.02	0.05	0.01	0.02	0.79										
L8	**0.00**	**0.00**	**0.00**	**0.00**	**0.00**	0.23	0.56									
R1	0.20	0.09	0.19	0.08	0.14	0.68	0.29	0.01								
R2	0.12	0.11	0.07	0.03	0.03	0.42	0.59	0.19	0.23							
R3	0.01	0.01	0.09	0.02	0.05	0.33	0.26	0.15	0.09	0.07						
R4	0.45	0.41	0.85	0.89	0.80	0.05	0.01	**0.00**	0.12	0.04	0.02					
R5	0.63	0.25	0.69	0.50	0.44	0.40	0.28	0.04	0.72	0.22	0.11	0.63				
R6	0.85	0.51	0.32	0.39	0.38	0.54	0.26	0.04	0.86	0.30	0.09	0.47	0.70			
R7	0.31	0.51	0.08	0.07	0.08	0.17	0.13	0.01	0.16	0.61	0.01	0.08	0.19	0.48		
R8	0.09	0.33	0.15	0.51	0.42	**0.00**	**0.00**	**0.00**	**0.00**	0.01	**0.00**	0.21	0.13	0.09	0.02	

(b) Configuration C

	L1	L2	L3	L4	L5	L6	L7	L8	R1	R2	R3	R4	R5	R6	R7	R8
L1																
L2	0.59															
L3	0.84	0.89														
L4	0.15	0.48	0.67													
L5	0.10	0.15	0.27	0.58												
L6	0.11	0.13	0.25	0.26	0.31											
L7	0.09	0.09	0.14	0.10	0.15	0.91										
L8	0.01	0.04	0.09	0.12	0.35	0.37	0.28									
R1	0.07	0.40	0.72	0.80	0.51	0.07	0.02	0.04								
R2	0.10	0.59	0.50	0.29	0.22	0.02	0.01	0.01	0.41							
R3	0.56	0.73	0.56	0.15	0.05	0.06	0.06	0.01	0.10	0.19						
R4	0.33	0.36	0.61	0.58	0.45	0.50	0.22		0.40	0.11	0.12					
R5	0.64	0.53	0.83	0.55	0.26	0.15	0.09	0.07	0.54	0.25	0.53	0.32				
R6	0.03	0.08	0.08	0.05	0.09	0.65	0.89	0.09	0.01	0.00	0.04	0.35	0.03			
R7	0.03	0.04	0.06	0.02	0.06	0.54	0.90	0.15	**0.00**	**0.00**	0.03	0.20	0.03	0.74		
R8	0.01	0.01	0.02	0.01	0.04	0.16	0.32	0.36	**0.00**	**0.00**	0.01	0.05	0.01	0.14	0.48	

6   Conclusion

While directional and linear statistics methods are used for the analysis of different types of data, we have shown that when data can be represented in both forms, directional statistics provides information that cannot be obtained through the counterpart linear methods. Directional statistics therefore complements linear statistics with directional information, such as suggesting from what direction a robot should, or should not, approach a group of people.

Directional statistics showed that when pairs of people are approached by a robot, comfort ranks cluster loosely into two groups of 'less comfortable' and 'more comfortable'. The absence of a 'most comfortable' rank suggests that participants did not have a most preferred robot approach direction, while the

density of less comfortable ranks behind participants suggests a strong preference for robot approach paths that should be avoided.

References

1. Mardia, K.V., Jupp, P.E.: Directional Statistics. John Wiley & Sons (2009)
2. Mardia, K.V.: Statistics of Directional Data. Academic Press (1972)
3. Dautenhahn, K., Walters, M., Woods, S., Koay, K.L., Nehaniv, C.L., Sisbot, A.,
   Alami, R., Siméon, T.: How may I serve you? a robot companion approaching a
   seated person in a helping context. In: 1st ACM SIGCHI/SIGART Conf. Human-
   Robot Interact., pp. 172–179. ACM (2006)
4. Walters, M.L., Dautenhahn, K., Woods, S.N., Koay, K.L., Te Boekhorst, R.,
   Lee, D.: Exploratory studies on social spaces between humans and a mechanical-
   looking robot. Connection Science **18**(4), 429–439 (2006)
5. Walters, M.L., Dautenhahn, K., Woods, S.N., Koay, K.L.: Robotic etiquette: results
   from user studies involving a fetch and carry task. In: 2nd ACM/IEEE Int. Conf.
   Human-Robot Interact., pp. 317–324. IEEE (2007)
6. Karreman, D., Utama, L., Joosse, M., Lohse, M., van Dijk, B., Evers, V.: Robot
   etiquette: how to approach a pair of people? In: ACM/IEEE Int. Conf. Human-
   Robot Interact., pp. 196–197. ACM (2014)
7. Ball, A., Silvera-Tawil, D., Rye, D., Velonaki, M.: Group comfortability when a
   robot approaches. In: Beetz, M., Johnston, B., Williams, M.-A. (eds.) ICSR 2014.
   LNCS, vol. 8755, pp. 44–53. Springer, Heidelberg (2014)
8. Hinkelmann, K., Kempthorne, O.: Design and Analysis of Experiments. Introduction to Experimental Design, vol. 1. John Wiley & Sons (1994)
9. Rao, B.: Nonparametric Functional Estimation. Academic Press (1983)
10. Siegel, S.: Non-Parametric Statistics. McGraw-Hill (1956)
11. Jupp, P.: Modifications of the Rayleigh and Bingham tests for uniformity of direc-
    tions. J. Multivariate Anal. **77**(1), 1–20 (2001)
12. Stuart, A., Ord, J.: Kendall's Advanced Theory of Statistics, vol. 1. Halsted Press
    (1994)
13. Watson, G.S.: Goodness-of-fit tests on a circle II. Biometrika, 57–63 (1962)
14. Sheskin, D.J.: Handbook of Parametric and Nonparametric Statistical Procedures.
    CRC Press (2003)
15. Brown, B.: Grouping corrections for circular goodness-of-fit tests. J. Royal Stat.
    Soc. Ser. B (Methodological), 275–283 (1994)
16. Maag, U.R.: A k-sample analogue of Watson's U^2 statistic. Biometrika **53**(3–4),
    579–583 (1966)
17. Benjamini, Y., Hochberg, Y.: Controlling the false discovery rate: A practical and
    powerful approach to multiple testing. J. Royal Stat. Soc. Ser. B (Methodological),
    289–300 (1995)
18. Kendon, A.: Spacing and orientation in co-present interaction. In: Esposito, A.,
    Campbell, N., Vogel, C., Hussain, A., Nijholt, A. (eds.) Second COST 2102. LNCS,
    vol. 5967, pp. 1–15. Springer, Heidelberg (2010)
19. Ball, A., Rye, D., Silvera-Tawil, D., Velonaki, M.: Group vs. individual comfort
    when a robot approaches. In: Proc. 24th Int. Symp. Robot and Human Interactive
    Communication (RO-MAN 2015) (2015) (in press)

Adaptive Interaction of Persistent Robots to User Temporal Preferences

Kim Baraka[1](✉) and Manuela Veloso[2]

[1] Robotics Institute, Carnegie Mellon University, Pittsburgh, USA
kbaraka@andrew.cmu.edu
[2] Computer Science Department, Carnegie Mellon University, Pittsburgh, USA

Abstract. We look at the problem of enabling a mobile service robot to autonomously adapt to user preferences over repeated interactions in a long-term time frame, where the user provides feedback on every interaction in the form of a rating. We assume that the robot has a discrete and finite set of interaction options from which it has to choose one at every encounter with a given user. We first present three models of users which span the spectrum of possible preference profiles and their dynamics, incorporating aspects such as boredom and taste for change or surprise. Second, given the model to which the user belongs to, we present a learning algorithm which is able to successfully learn the model parameters. We show the applicability of our framework to personalizing light animations on our mobile service robot, CoBot.

Keywords: Long term human-robot interaction · Adaptive personalization · Learning preference dynamics · Expressive lights

1 Introduction

An important part of Human-Robot Interaction (HRI) research aims at finding iconic ways for robots to interact with humans, that are both effective and universal, especially when the interaction has a direct functional role (e.g. communicating intent or instructing the user). Human studies can be helpful in the design of this type of interaction, where one aims at finding one way of interacting which works best on average. On the other hand, there exists another type of interaction whose main purpose is to please or adapt rather than to directly perform a functional role (e.g. pertaining to robot appearance, speech wording, sounds etc.). It is this type of interaction on which we will focus throughout this paper. The assumption is that there can be high variability in the way different users desire or expect to perform this type of interaction with a robot. In this case, a social understanding of the interaction is not very valuable since the problem of interaction choice selection is more a matter of adapting to the user's tastes and is hence theoretically arbitrary. There has been general evidence that personalization of robot appearance and behavior can greatly improve user experience [8] in terms of "rapport, cooperation and engagement" [4], hence the need to move away from the human study paradigm towards automated personalization of the interaction. In this work, we are interested in particular in mobile robots which are persistent over time and which interact with different types of users over an extended time frame. We will use expression through lights on our

© Springer International Publishing Switzerland 2015
A. Tapus et al. (Eds.): ICSR 2015, LNAI 9388, pp. 61–71, 2015.
DOI: 10.1007/978-3-319-25554-5_7

mobile service robot, CoBot (see figure 1), as a motivating example, but this work can be applied to any form of interaction whose main purpose is to please rather than inform - e.g. voice in speech generation, motion or pose of the robot during interaction, facial expression of a humanoid...

We look at the general problem of learning how to best interact with different individuals based on feedback from the latter. More specifically, we make the following assumptions. (1) A robot repeatedly interacts with different humans whose identity is known by the robot. (2) Every time the robot encounters some individual, it chooses one out of a set of fixed possible options to interact with them. (3) The user has a method of providing feedback for the interaction through a score or rating. In practice, social interactions can be much more complex and require much more context to help adaptation, however there is a wide range of interaction types which do not play a direct functional social role but act more as a complement to the main interaction. For illustration purposes, we use interaction with lights on one of our mobile service robots, CoBot3. CoBot3's expressive lights are being used to enhance the interaction with humans. A finite set of predefined light animations can be used for personalized interaction. The robot, being able to accurately navigate and localize itself accurately in our buildings, can identify users (e.g. by their office number or by recognizing an associated digital signature), hence enabling the personalization of light animations while servicing that user. At the end of each interaction, the touch screen interface may be used to rate the interaction (e.g. by the use of a slider).

Long-term user preferences are however from being static or homogeneous, which is not accounted for in traditional recommender systems. Indeed, being exposed to the same type of interaction for a long time might develop boredom or fatigue for some, while others might value it for its predictability. To summarize, general static preferences change from individual to individual, but preference dynamics are equally important in a long-term setting. In this paper, we propose to learn, for a given user, both sets of preferred interaction options and time-related quantities which would dictate the preferred dynamics of interaction.

The paper is divided into three main parts. In the first part, we introduce three user models which capture different possible profiles in relation to the appreciation of "change" in the way the robot interacts with the user. These are formulated in terms of evolution of the reward from the user as a function of the possible sequences of interactions options used when interacting with that user. In the second part, we present our algorithms to learn the model parameters, assuming we are given the user model. In the third part, we show the applicability of this work to our mobile service robot, CoBot, which is deployed in our buildings and uses expressive lights for improved interaction with humans.

2   Related Work

Apart from simple customization during usage [8], recent work has looked at autonomous personalization of HRI based on previous interactions with the same user. Examples include a snack delivering robot which uses data from past interactions to personalize the future interactions [4] or a humanoid robot learning different models for expressing emotion through motion, which it is then able to use for personalizing expression of emotions [9]. Furthermore, the idea of self-initiative in a robot has been explored by learning ways of acting in the world depending on user verbal feedback on the current "state of the world" [7]. Finally, user modeling for long-term HRI, a focus of the current paper, has been looked at using archetypes of real users called *personas*, which encode traits of potential users in terms of interaction preferences [3]. Some authors also looked at ways of learning long-term behavior by identifying social primitives that are important when the novelty aspect of interaction vanishes [12] or matching personalities between robot and user [11]. However, these works focus more on the social aspect of the interaction rather on the intelligence of the adaptation from a generic point of view, making their applicability and generalization poor in different types or modes of interaction. In this work, we would like to decouple the nature of the interaction options with the generic adaptation mechanism, which can then be tuned based on the nature of the interaction and the user's response to it.

In the problem we consider, we will assume that user data are limited to rewards at every interaction (in the form of a rating submitted by the user), making it comparable to a recommender system learning user preferences and suggesting new items [1]. However, the algorithms used in such systems do not take into account the dynamics of preferences (boredom, habituation, desire for change etc.). In the field of automatic music playlist generation, the concepts of diversity and serendipity have been mentioned [2]. However, no viable solution has yet been proposed to address this problem. Also, the idea of exploration in music recommender systems has been studied [10], but it does not apply to our problem since we assume the number of interaction options to be relatively small. In a robotics application, the need for adaptive interaction that takes into account habituation has been recently formulated for empathic behavior [12] (in this paper, we take a more general approach). Going back to the problem of preference dynamics, our problem can formally be compared to the restless multi-armed bandit problem where rewards are non-stationary and which is generally known to be P-SPACE hard [5]. In this work, we restrict the rewards to evolve according to one of three models, which makes the problem of learning the model parameters easier to solve.

3   Formalism and User Modeling

In this section, we start by presenting the formulation of the problem at hand and move on to introduce three models of dynamic preferences corresponding to three different possible profiles of users.

3.1   Problem Setting

Time is discretized into steps $i = 1, 2, 3, ...$, where each time step represents an encounter between the robot and the user. We assume that the encounters are of an identical nature or serve the same functional role (for example the robot is always delivering an object to the user's office). Also, for simplicity, we do not worry about the actual time interval between two consecutive steps (these could be for example different days or different times within the same day). At every time step, we assume the robot has to choose one out of a set of n possible actions corresponding to interaction options. In the context of light animations, the different actions represent different animations in terms of speed, color or animation patterns. Let $\mathbf{A} = \{a_1, ..., a_n\}$ represent the set of possible actions. After every encounter, we assume that the user provides a rating $r^{(i)}$ to the interaction where $r^{(i)} \in [0; 10]$. The reward is assumed to be corrupted by additive white Gaussian noise: $r = \bar{r} + \epsilon$ where $\epsilon \sim N(0, \sigma^2)$. The noise can come from the following sources: (1) inaccurate reporting of the user's true valuation, (2) mistakes when using the user interface (e.g. slider) to report the reward and (3) failure to remember previous ratings resulting in inconsistent ratings.

Our goal is to learn, for a specific user (with a specific reward model), which action to take next given the history of actions and rewards. The problem can hence be compared to the Multi-Armed Bandit problem where a single player, choosing at each time step one to play one out of several possible arms and gets a reward for it, aims to maximize total reward (or equivalently minimize total regret) [5]. In our case, the rewards are stochastic and non-stationary and the arms or actions, corresponding to the different interaction options, are relatively few. From now on, we will use "actions" and "interaction options" interchangeably.

3.2   Modeling Dynamic User Preferences Over Time

We now introduce three user models which we think span well enough the spectrum of possible profiles, inspired by variations along the "openness" dimension of the five-factor model in psychology [13]. These models we crafted take into account both properties of preferred actions sets and time-related quantities dictating the evolution of rewards depending on the action sequences. Figure 1 shows sample ideal sequences for lights on our robot, CoBot3, for each of the three models on different days in which the robot visits a person's office to deliver coffee. For the three models presented below, we use $\mathbf{A}_{\mathrm{pref}}$ to denote the set of preferred actions (regardless of the sequence of actions in which they fall).

Model 1: The "Conservative". This type of user wants to stick to one option denoted by a^*, but appreciates surprises from time to time at some frequency. A surprise means taking for one time step an action $a \neq a^*$ in a set of preferred "surprise actions" $A_{\mathrm{surp}} \subset \mathbf{A}$. When a^* is repetitively shown in a sequence (we call sequences of the same action homogeneous sequences), the reward $\bar{r}$ starts out as a constant ($r_{\max}$) and after T time steps starts decreasing, due to boredom, with a linear decay rate α until it reaches $r_{\min}$, after which it remains constant. For homogeneous sequences of the non-preferred actions (i.e. actions in $A \setminus \{a^*\}$), the reward starts at a value $r_{\mathrm{non-pref}}$ and decreases exponentially to zero with time (indicating that the user very quickly gets bored) with some decay rate β. In summary, the model parameters are:

- a^*: the action with the maximum value of $E[\bar{r}]$. A homogeneous sequence of a^* actions is referred to from now on as a **p-sequence**.
- $\mathbf{A}_{\text{pref}} = \{a^*\}$
- $\mathbf{A}_{\text{surp}}$: set of actions suitable for surprises, defined as $\{a : E[\bar{r}_a] > r_{\text{th}}\}$, where r_{th} is a threshold value.
- T: optimal length of the homogeneous sequence of the preferred action, after which the user starts getting bored. If the robot always alternates between p-sequences and surprises, T can also be seen as a between two consecutive surprises. T is assumed to be a random variable uniformly drawn in a window $[T_{\min}, T_{\max}]$ every time a new p-sequence is started.
- α: linear reward decay rate in a p-sequence whose length exceeds T.
- $r_{\max}$: constant reward for p-sequences of length less than or equal to T.
- $r_{\min}$: lower clipping value for reward in p-sequences. A good value for is 5, which means that the user is neither rewarding nor punishing the robot for taking their preferred action for too long.
- $r_{\text{non-pref}}$: initial reward value when starting a homogeneous sequence that is not a p-sequence. If the previous homogeneous sequence is a p-sequence, $r_{\text{non-pref}}$ is a function of the length of the p-sequence l as follows: if $l \geq T_{min}$ we assume that the user is expecting a surprise which will provide some maximal reward $r_{\text{non-pref,max}}$. When $l < T_{min}$, we expect the surprise to be disliked, so we decrease the surprise reward linearly: $r_{\text{non-pref}} = r_{\text{non-pref,max}} \cdot (1 - \frac{T_{min} - l + 1}{T_{min}})$. If the previous homogeneous sequence is not a p-sequence, $r_{\text{non-pref}}$ is a constant $r_{\text{non-pref,base}}$.
- β: exponential reward decay rate for a homogeneous sequence that is not a p-sequence.

Model 2: The "Consistent But Fatigable". This type of user values consistency in actions taken but needs shifts from time to time. It is the profile where there always needs to be an uninterrupted routine but this routine has to be changed after some time. The user has a set of preferred actions which he expects to see in long sequences. These sequences alternate between the different preferred options after some time spent sticking with one of the options. We assume the same model of boredom used in the previous section, namely the reward starts decaying linearly for the preferred actions after some time interval T. There is no surprise factor associated with this model since we assume that the user does not appreciate surprises.

The parameters of this model are the following (no description provided means the parameters are the same as in the "conservative" model):

- $\mathbf{A}_{\text{pref}} = \{a_1^*, ..., a_m^*\}$, where $m \geq 2$. p-sequences in this model are defined to be homogeneous sequences formed using one action in $\mathbf{A}_{\text{pref}}$.
- T: optimal length of a p-sequence, after which the user starts getting bored. T is assumed to be a random variable uniformly drawn in a window $[T_{\min}, T_{\max}]$ every time a new p-sequence is started.

- α, $r_{\max}$ and $r_{\min}$: idem
- $r_{\text{non-pref}}$: initial reward value when starting a homogeneous sequence that is not a p-sequence. A constant in this model.
- β: decay rate of the reward for a homogeneous sequence that is not a p-sequence.

Model 3: The "Erratic". This type of user is mainly interested in change, in both action selection and time-related parameters. They have no clear preferences over the possible options but require the actions to change according to some average rate not restricted to a window as in model 1 and 2. We assume that at every step the user has some fixed probability p_{sw} to desire a switch to a different action, independently of anything else. Hence the optimal length T of homogeneous sequences follows the distribution: $p(T = t) = (1 - p_{\text{sw}})^{t-1} p_{\text{sw}}$ (for $t \geq 1$), whose average $\mu_T = 1/p_{\text{sw}}$, making μ_T a sufficient statistic. Similar to previously, the reward decreases linearly after T time steps in a homogeneous sequence.

4   Learning Model Parameters from User Feedback

Now that we have presented the three user models that we consider, we look at the problem of learning their parameters from user reward sequences. Once these parameters become known, we can then generate personalized sequences of actions maximizing cumulative reward for a specific user. In what follows, we assume that the model to which a particular user belongs to is known a priori. In practice, this can be achieved by prompting the user to select one profile which described them best, or through a set of questions similar to a personality test.

Note that although we have previously raised the problem of dealing with the non-Markovian aspect of user preferences (meaning that the reward of a certain action depends on the history of previous actions), in the way we have modeled the user profiles in the previous section, the model parameters encode the preference dynamics. These parameters are assumed to be unchanged as time goes by, hence we have effectively turned the dynamic problem into a Markovian one. Next, we describe the learning procedure for each of the user profiles introduced.

4.1   Profile "Conservative"

In order to learn the parameters of this model, we divide the learning process into two phases: one phase for learning preference sets and the other for learning the time-related parameters. The parameters the agent performing the actions needs to learn are: a^*, $\mathbf{A}_{\text{surp}}$, $T_{\min}$ and $T_{\max}$.

Phase 1: Learning Preference Sets. In this preliminary phase, actions are uniformly drawn from $\mathbf{A}$ until each action is taken at least n_{th} times, where n_{th} depends on the noise variance estimate $\tilde{\sigma}^2$ and on our target confidence value (for all practical purposes, we use $n_{\text{th}} = 10$). Note that randomization of the

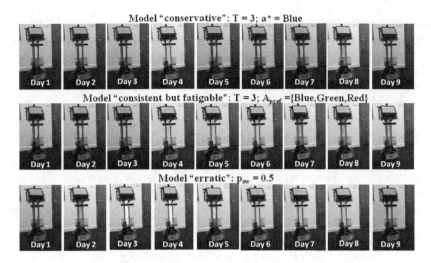

Fig. 1. Sample preferred of animation sequences for the user models presented

sequence of actions to be taken is crucial in this phase since the previous actions can have an influence on the reward of the current action and we would like to dilute this effect. Once we have an average reward estimate for each action, we select a^* to be the action with the maximum estimated reward and $\mathbf{A}_{\mathrm{surp}}$ to be the set of all actions whose reward estimates exceed the set threshold r_{th}, where the value of r_{th} has to ensure that $|\mathbf{A}_{\mathrm{surp}}| \geq 1$. It assumed that the set of best actions to be used for surprises will score high in this phase as well.

Phase 2: Learning Time-Related Parameters. In order to learn the two parameters of interest T_{min} and T_{max}, the agent first learn estimate the mean and variance of T (μ_T and σ_T respectively) and uses them to infer the parameter values. To achieve this, the agent follows p-sequences until a need for surprise is detected (more details below). A surprise is restricted to taking an action in $\mathbf{A}_{\mathrm{pref}}$ for one time step following a p-sequence. After a surprise, the agent reverts back to following a p-sequence until another surprise is decided upon.

The learning procedure goes as follows: when in a p-sequence of actions, if a downward trend in reward is detected, show a surprise chosen uniformly from $\mathbf{A}_{\mathrm{pref}}$. Since the reward is noisy, a smoother is needed to filter out high frequency noise in the data. We use an exponentially weighted moving average (EWMA) [6] with fixed sample size s, combined with a threshold detector, to detect a downward trend in the reward of the p-sequence. The threshold used in the threshold detector depends on the estimated noise variance in the reward $\tilde{\sigma}^2$. Every time a downward trend is detected, we record the estimated T value associated with the p-sequence. Once enough surprises are chosen, we would have accurate enough estimates of μ_T and σ_T, which can be used to find the time-related parameters as follows: $\tilde{T}_{\mathrm{min,max}} = \tilde{\mu}_T \mp \frac{\sqrt{12\tilde{\sigma}_T^2+1}-1}{2}$.

Note that there is a lag associated with the moving average trend detector. This lag is equal to half the $\lfloor \frac{s}{2} \rfloor$ and $\tilde{\mu}_T$ needs to be adjusted to account for it. Also, for a small number of data points, we might be overestimating σ_T. Hence we set $\tilde{\sigma}_T$ to be half the estimate of the standard deviation in the values of T. This way we impose a more conservative restriction on the values of T which will ensure that $[\tilde{T}_{\min}, \tilde{T}_{\max}] \subset [T_{\min}, T_{\max}]$.

4.2   Profile "Consistent but Fatigable"

Similar to that of the previous model, the learning procedure is still separated into the two phases. However, as far as action selection is concerned, since there is no surprise but only a change factor in this model, the termination of a p-sequence of a_i consists in starting a new p-sequence with an action chosen uniformly in $\mathbf{A}_{\mathrm{pref}} \setminus \{a_i\}$. The first phase for learning preference sets uses the same procedure as before, except that once the average reward estimates are obtained, we set $\mathbf{A}_{\mathrm{pref}}$ to be the set of animations with a reward estimate above r_{th} (possibly different than the one used in the "conservative" model). Here again, the threshold value should be set such that the cardinality m of $\mathbf{A}_{\mathrm{pref}}$ is at least 2. The second phase for learning time-related parameters is similar to the one used in the previous model.

4.3   Profile "Erratic"

For this type of profile, no sets of preferred actions need to be learned since we assume that the user has no clear preferences between the different actions. Hence, the only parameter to learn is the probability of switching p_{sw}. The action selection algorithm is identical to the "consistent but fatigable" model, with $\mathbf{A}_{\mathrm{pref}} = \mathbf{A}$. μ_T can also be estimate as before, and once a good estimate is obtained, we infer our parameter p_{sw} as follows: $\tilde{p}_{\mathrm{sw}} = \frac{1}{\tilde{\mu}_T}$.

4.4   Action Sequences Generation

The learning phase stops when the parameters are learned with some target confidence value. The latter comes in our case mainly from the error rate of the EWMA and depends on the various parameters including noise variance. Once the parameters are learned, appropriate stochastic sequences can be generated according to the estimated parameter values. For models "conservative" and "consistent but fatigable", we uniformly draw a value of T in the estimated window. For model "erratic", we follow the same action with probability $1 - p_{\mathrm{sw}}$ and uniformly switch to another action with probability p_{sw}. In this exploitation phase, the feedback requirements can be reduced or eliminated, since we have all the parameters needed to generate optimal sequences which will maximize the cumulative reward for the given user. In practice, occasional user feedback (e.g. asking for a reward) can be used to confirm the model and parameters. We will not provide more detail about this exploitation phase since the focus of this work is on the learning aspect. However, notice that in the learning phase we are already generating sequences which are not too far from optimal.

5 Results

In this section, we present a few results showing our implementation of our simulated user's preference dynamics and our algorithm's ability to learn the different model parameters. Figure 2 shows the evolution of the learning process for single instances of the three models. We consider 8 possible actions arbitrarily labeled 1 through 8. Phase 1 of the learning algorithm can be clearly distinguished in the first two models, after which the algorithm learns the set $\mathbf{A}_{\mathrm{pref}}$ ($\{a_4\}$ for model "conservative" and $\{a_2, a_4, a_6\}$ for model "consistent but fatigable"). Once it identifies the preferred sets, the algorithm is also able to adapt to the preferred action dynamics. Notice that whenever there is a notable decrease in the reward, a change is performed, whether creating a temporary "surprise" (a), changing to another steady option (b) or creating erratic change (c).

The simulation was run over 350 time steps with the following parameter values for illustrative purposes. $T_{\min} = 20$ and $T_{\max} = 30$ for the first two models and $p_{\mathrm{sw}} = 0.8$ for the third model. The noise variance σ^2 was set to 0.05. Here are a few results over 1,000 randomized trials.

-**Model "conservative"**: % error in $\tilde{\mu}_T$: 3.5% ; $\tilde{T}_{\min} = 20.94$; $\tilde{T}_{\max} = 30.81$. After rounding, the estimated interval is contained in the true interval.

-**Model "consistent but fatigable"**: % error in $\tilde{\mu}_T$: 2.67% ; $\tilde{T}_{\min} = 21.72$; $\tilde{T}_{\max} = 26.95$. The rounded estimated interval is contained in the true interval.

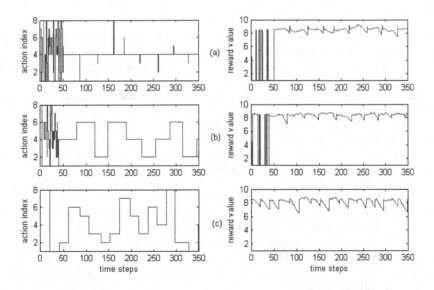

Fig. 2. Simulation results showing sequences of actions taken by the agent and the corresponding reward sequences from a simulated user belonging to: (a) model "conservative", (b) model "consistent but fatigable" and (c) model "erratic"

-**Model "erratic":** $\tilde{\mu}_T = 12.67$, therefore $p_{sw} = 0.079$ (0.13% error).

The algorithm was able to learn all the parameters with reasonable accuracy.

6   Conclusion and Future Work

We have presented three models for dynamic long-term user preferences, which capture aspects of boredom and appreciation for change or surprise. Given which model a specific user belongs to, our algorithm enables the robot to learn the model parameters using the sequence of rewards given by the user. Our results show that the agent is able to learn the parameters of the model reasonably well and in a relatively short number of time steps for all three models. Our algorithm is robust to noise, but further experiments are needed to evaluate the degradation in performance as the noise increases. In the future, we plan to enable the robot to also learn which model the user belongs to from the reward sequences themselves. Also, allowing a mixture of the three models with weights to be learned (although it is not exactly clear whether it is a viable idea in a social setting) could diversify the space of sequences generated and alleviate the problem of forcing the users to categorize themselves. Furthermore, requesting a reward from the user at each encounter might have a negative social impact; to overcome this, we could investigate efficient sampling methods to gather rewards in a sparse manner while maintaining accuracy in the learning process.

References

1. Bobadilla, J., Ortega, F., Hernando, A., Gutierrez, A.: Recommender systems survey. Knowledge-Based Systems **46**, 109–132 (2013)
2. Bonnin, G., Jannach, D.: Automated generation of music playlists: Survey and experiments. ACM Computing Surveys (CSUR) **47**(2), 26 (2014)
3. Campos, J., Paiva, A.: A personal approach: the *persona* technique in a companion's design lifecycle. In: Campos, P., Graham, N., Jorge, J., Nunes, N., Palanque, P., Winckler, M. (eds.) INTERACT 2011, Part III. LNCS, vol. 6948, pp. 73–90. Springer, Heidelberg (2011)
4. Lee, M., Forlizzi, J., Kiesler, S., Rybski, P., Antanitis, J., Savetsila, S.: Personalization in HRI: a longitudinal field experiment. In: HRI (2012)
5. Liu, H., Liu, K., Zhao, Q.: Learning in A Changing World: Restless Multi-Armed Bandit with Unknown Dynamics. ARXIV (2010)
6. Lucas, J., James, M., Saccucci, M.: Exponentially weighted moving average control schemes: properties and enhancements. Technometrics **32**(1) (1990)
7. Mason, M., Lopes, M.: Robot self-initiative and personalization by learning through repeated interactions. In: HRI (2011)
8. Sung, J., Grinter, R., Christensen, H.: Pimp my roomba: designing for personalization. In: CHI (2009)
9. Vircikova, M., Pala, M., Smolar, P., Sincak, P.: Neural approach for personalised emotional model in human-robot interaction. In: WCCI (2012)
10. Wang, X., Wang, Y., Hsu, D., Wang, Y.: Exploration in Interactive Personalized Music Recommendation: A Reinforcement Learning Approach. Mult. Comput. Commun. Appl. **11**(7) (2014)

11. Tapus, A., Tapus, C., Mataric, M.: User-robot personality matching and assistive robot behavior adaptation for post-stroke rehabilitation therapy. Intelligent Service Robotics **1**(2), 169–183 (2008)
12. Leite, I.: Long-term interactions with empathic social robots. AI Matters **5**(1), 13–15 (2015)
13. Goldberg, Lewis, R.: An alternative "description of personality": the big-five factor structure. Journal of Personality and Social Psychology **59**(6), 1216 (1990)

Incremental Speech Production for Polite and Natural Personal-Space Intrusion

Timo Baumann[1] and Felix Lindner[2]([✉])

[1] Natural Language Systems Division, Department of Informatics,
University of Hamburg, Vogt-Kölln-Straße 30, 22527 Hamburg, Germany
`baumann@informatik.uni-hamburg.de`
[2] Knowledge and Language Processing Group, Department of Informatics,
University of Hamburg, Vogt-Kölln-Straße 30, 22527 Hamburg, Germany
`lindner@informatik.uni-hamburg.de`

Abstract. We propose to use a model of personal space to initiate communication while passing a human thereby acknowledging that humans are not just a special kind of obstacle to be avoided but potential interaction partners. As a simple form of interaction, our system communicates an apology while closely passing a human. To this end, we present a software architecture that integrates a social-spaces knowledge base and a component for incremental speech production. Incrementality ensures that the robot's utterance can be adapted to fit the developing situation in a natural way. Observer ratings show that personal-space intrusion is perceived as both natural and polite if the robot has the capability to utter and adapt an apology in an incremental way whereas it is perceived as unfriendly if the robot intrudes personal space without saying anything. Moreover, the robot is perceived as less natural if it does not adapt.

1 Introduction

When robots and humans act in common spaces they inevitably encounter each other regularly. Therefore, social robots need to solve the task of passing humans in a socially appropriate manner. Pioneering work on the research question of how robots should pass humans can be attributed to the early studies presented in [15] and [23].

In more recent work the capability to socially pass a human has been modeled using the notion of personal space. Authors from the social sciences like Hall [8] and Sommer [21] use the concept of personal space to explain the various phenomena related to how humans spatially behave towards other humans with particular focus on the distances they maintain to each other. Computational models of personal space have mainly been applied to human-aware robot navigation to avoid personal-space intrusion [5,9,16,18,19,22]. In effect, these approaches result in robots taking detours in accordance with personal-space theory. In fact, there seems to exist a common ground that models of personal space should keep the robot away from humans in the first place. As a result,

© Springer International Publishing Switzerland 2015
A. Tapus et al. (Eds.): ICSR 2015, LNAI 9388, pp. 72–82, 2015.
DOI: 10.1007/978-3-319-25554-5_8

comparatively few approaches take personal space as a basis for specifying how a robot should behave if it intrudes personal space. Lam and colleagues [10] present a two-stage policy with respect to personal-space usage. As with the other approaches, the robots should avoid personal-space intrusion. However, if it accidently happens that the robot intrudes personal space, the robot will stop moving until it is not within personal space anymore.

All in all, the main line of the reviewed work is that personal spaces should not be entered by robots passing a human. Instead, the robot should take detours and, if the robot finds itself within personal space accidently, it should freeze. According to the available literature on human-aware robot navigation, the title of the paper at hand seems to be inherently contradictory, because it claims that there is a way to intrude personal spaces in a polite and natural manner. In earlier work [13], we already suggest to add social signals to navigation plans to gain permission to enter regions of personal space, thus, to intrude personal space in a planful manner. In this work, we extend this idea and propose to use a model of personal space that acknowledges that humans are not just obstacles to be avoided but potential interaction partners. As a simple form of interaction, our system communicates an apology while closely passing a human. We present a software architecture that integrates a social-spaces knowledge base and a component for incremental speech production (see Sect. 2). Incremental speech production allows a system to start outputting speech based on partial speech plans that can later be extended [20] or even altered to reflect changes of the underlying plan [3]. Incremental speech synthesis is able to continuously render speech with a natural and continuous prosody and at almost the quality of systems that require the full and unchangeable utterance specification in advance [1], even though requiring only a few words of future context.

To evaluate our system we conducted an observation study. In particular we tested two main hypotheses:

Hypothesis A. A robot passing through a personal space is perceived as more polite if it utters an apology rather than saying nothing,

Hypothesis B. A robot passing through a personal space is perceived as more natural if it has the capability to adapt its speech incrementally as the situation evolves.

A comparable study [7] could not comfirm an effect on the perceived politeness of a robot that signals its intention to pass by making beep sounds as compared to making no sounds at all. This result should discourage our belief in hypothesis A. However, a later study, which investigates the effect of social framing on the reactions of people towards a robot that signals its intention [6], reveals that subjects perceive a speaking robot as more friendly than a beeping robot.

Hypothesis B is grounded in the fact that humans' speech production is inherently incremental [11]. Humans can adapt their utterances while speaking with ease and do so as the situation or interaction requires [4]. Therefore, we expect that a robot with this capability is perceived as more natural than a robot

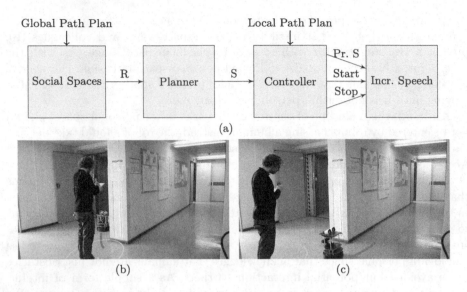

(a)

(b)   (c)

Fig. 1. (a) Architecture integrating a knowledge base about social spaces and a component for incremental speech production. (b) As soon as the local path plan (pink) overlaps the personal space (yellow) the robot starts to say "Excuse me, I need to pass urgently to rescue a patient in the other corridor – thank you." (c) However, as the person steps aside the robot leaves personal space before the whole explanation was uttered resulting in "Excuse me, I need to pass urgently – thank you."

that 'balistically' utters its whole pre-planned utterance without considering situational changes.

Confirming Hypothesis A, the observation study presented in Sect. 3 shows that personal-space intrusion is perceived as both natural and polite if the robot has the capability to utter and adapt an apology in an incremental way whereas it is perceived as unfriendly if the robot intrudes personal space without saying anything. Confirming Hypothesis B, we found that it is perceived as unnatural if the robot does not adapt its utterance plan incrementally. We find no effects on the control questions regarding the robot's route, which indicates that observers differentiate between the various aspects of multi-modal robot behaviour.

2   A Software Architecture Integrating Social Spaces and Incremental Speech Synthesis

To enable a social robot to planfully intrude personal space while passing a human, we propose the architecture shown in Fig. 1(a). The software architecture integrates the capability to reason about social spaces (i.e., personal spaces among others) and the capability to incrementally utter natural language. An example use case is shown in Figs. 1(b) and 1(c): Personal space intrusion is accompanied by a verbal explanation, which is adapted as a reaction to the human clearing the way for the robot. The architecture's components are described below.

2.1   Social Spaces

The concept of social spaces subsumes several socio-spatial phenomena among which personal space is the most popular one (cf. [13]). Social spaces can be characterized as socio-spatial entities that are produced by other entities that provide reasons for action to social agents. Particularly, a personal space is produced by a (single) human and the human provides reasons for action to other social entities (e.g., robots). Our reason-driven view is inspired by contemporary work in practical philosophy (e.g., [17]) and motivated by the fact that reasons can be used both for deliberate decision making and for generating justifications or apologies social agents owe to others.

In the example depicted in Figures 1(b) and 1(c) the human produces a personal space. Within the symbolic knowledge base of the robot the human is represented as an individual which provides the robot with a reason against driving along the planned route.[1] Additionally, we assume that there is a patient in the other corridor which needs to be rescued by the robot. Consequently, the patient provides the robot with a reason in favor of driving along the planned route. Hence, given the navigation action *driving along the global path* represented by the global path plan (see Fig. 1(a)) the knowledge base can be queried for reasons that speak in favor of or against actually executing that particular plan.

The geometrical properties of the personal space are represented by an ellipse centered around the human. The major and minor axis were set to 3m and 2m, respectively. Consequently, as the robot crosses personal space from the left to the right hand side of the human it starts to talk to the human at a distance of roughly 1.5m. According to Hall [8] this corresponds to an interaction distance used by strangers.

2.2   Verbal-Planner

In cases where there are several alternative ways of acting, knowledge about reasons can be used to make choices among the available options [14]. In the approach presented here, we use reasons in a different way: They play the role of explanations. In particular, reasons that speak in favor of an action play the role of justifications whereas reasons that speak against an action can be used to formulate regret.

For instance, in the example depicted in Figures 1(b) and 1(c) the social-space component informs the verbal planner that there are two reasons ρ_1, ρ_2. Reason ρ_1 is the fact that the personal space should not be intruded and reason ρ_2 is the fact that some patient has to be rescued in the other corridor. Therefore, ρ_1 speaks in favor of executing the given path plan and ρ_2 speaks against doing so.

---

[1] See [12] for an in-depth technical explanation of the symbolic personal-space model.

Consequently, the verbal planner maps ρ_2 to an apology and ρ_1 to a justification. As a result the component outputs $S :=$ "Excuse me, I need to pass urgently to rescue a patient in the other corridor. Thank you."

We anticipate that S tends to become quite long the more reasons are at stake and hence we propose to order reasons by importance and to insert additional chunking information that the incremental speech production may use to skip parts of the resulting utterance for brevity. Such ordering and chunking can be performed by incremental NLG such as [3]. However, this step was simulated in the experiments reported below.

2.3   Controller

The controller is a component that interfaces the verbal planner and the incremental speech synthesis. It is implemented as a finite state machine with states s_0, s_1, and s_2. In state s_0 the sentence structure S is sent to the incremental speech component in order to internally prepare the sentence that should be uttered as soon as the robot actually enters the personal space. Being in s_0 the robot follows the global path plan without saying anything. When the local path plan significantly overlaps the personal space the state machine transitions from state s_0 to state s_1. In state s_1 the command *Start* is sent to the incremental speech component. Now the sentence structure that was prepared in state s_0 is actually uttered while the robot is still moving forward. A transition from s_1 to state s_2 takes place when the robot exits personal space again. In state s_2 the *Stop* command is sent to the incremental speech synthesis component. If at this time the robot is still talking, the incremental speech component will adapt the output, i.e., it will quickly but in a fluid way skip ahead in the utterance plan.

2.4   Incremental Speech Production

Given the utterance plan S of the verbal planner, the incremental speech production component prepares an *utterance tree* that provide for the alternatives of the original plan (in our case: skipping parts of the explanation). Speech synthesis is a processing problem on multiple layers (determining sentence-level intonation, prosodic contours, generating vocoding parameters and finally producing the actual speech waveform) which must be coordinated across possible continuations of the utterance to produce continuous and natural speech. This is crucial as any discontinuity (spectral, loudness, prosodic, etc.) in the final speech waveform would sound unnatural. It is hence not possible to simply attach separately synthesized utterance parts.

Our speech synthesizer [2] only requires a limited and local lookahead for vocoding, HMM optimization and state selection, and can hence integrate changes between utterance choices in the synthesis process with very little delay (on the order of 50 ms). In our case the *Stop* command from the controller leads the synthesizer to skip the remaining words of the explanation of why it had to intrude and move forward to thanking the user for allowing the robot to pass by in a natural way.

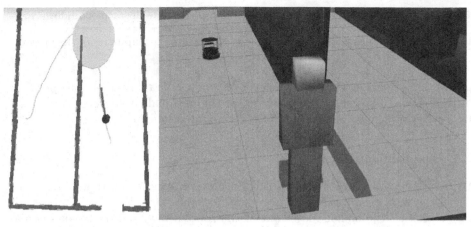

Fig. 2. The simulated robot's model (left side) as well as a rendering of the environment (right side) as shown in the observation videos.

3   Observation Study

We tested our hypothesis that sensible interaction when passing through a personal space is superior in terms of perceived naturalness and politeness of the robot to other strategies in a highly controlled observer rating experiment. In our conceived test environment, a hospital robot needs to pass by a person that is standing near a narrow passage in order to help a patient in the next corridor. Our test environment is depicted in Fig. 2.

The robot needs to pass through a person's personal space (depicted as a yellow ellipsis in the left part of the figure) in order to reach a target position. The global path plan is depicted as a green line (leading to the target position), the local plan at any time is depicted as a red line.[2] The global path plan was held constant throughout all simulations.

The robot plans upfront that it may want to interact in order to pass through the personal space and generates the utterance plan shown in Fig. 3. The idea of the plan is to gradually escalate the message from a low-profile *excuse me* (which might be sufficient to motivate the human to move away) to a full and thorough explanation of why the robot must violate the human's personal space. The plan finishes off with thanking the human for accepting the intrusion of her personal space.

Of course, the person may move out of the robot's way (and this is actually the robot's intent), however this cannot be relied upon in advance and can only be taken into account locally during speech delivery. To account for the variability of the moment in time at which the robot leaves the personal space, the utterance plan contains several "short-cuts" to seamlessly move ahead to the final *thank you* as indicated by the arrows in Fig. 3. We simulated the robot

---

[2] Simulated laser scans are also shown in red near the walls and should not be confused with the local path plan.

Excuse me, ► I need to pass ► urgently ► to rescue a patient ► in the other corridor, ⚡ thank you.

Fig. 3. The utterance plan in our example system allows to skip parts of the apology.

perception of personal space by directly informing the robot about the position of the person. The geometric properties of the personal space were represented by a polygon defined in the frame of the simulated human.

3.1   Experiment Setup

We screen-recorded the simulated robot's motion along a constant route (cmp. Fig. 2) systematically varying three variables: the speed of the robot (slow or fast), whether the human moves out of the robot's way, and the robot's verbal interaction: whether it delivers the full utterance plan once it enters the personal space, incrementally skips ahead when leaving the personal space, or does not verbally interact at all.

In total there are 12 video stimuli for all combinations of conditions of which 3 show no difference between incremental/non-incremental speech.[3] We played two of the duplicates in the beginning of the experiment and the third in the middle and excluded them from analysis of the verbal interaction variable. All other stimuli were distributed in random order.

We showed the videos to a group of 13 participants[4], who were asked to rate on five-point Likert scales for every video (a) the naturalness of the robot's behaviour (relating to hyp. A), (b) the politeness of the robot (relating to hyp. B), and (c) the appropriateness of the robot's route and speed (as control).

3.2   Results

We perform non-parametric paired statistical tests (Wilcoxon signed rank for the two-valued variables *speed* and *human movement*, and Friedman followed by post-hoc Wilcoxon signed rank for the three-valued variable *verbal interaction*) on all three variables and apply Bonferroni correction within the post-hoc tests to control for multiple-hypotheses testing.

We find no significant influence of the robot's *speed* on user ratings ($p = .29$ for naturalness, $p = .83$ for politeness, $p = .60$ for route appropriateness), indicating that there is no general preference for a higher or lower robot speed.

---

[3] *Being able* to skip does not necessarily imply that the robot *actually does* skip; the time at which the robot leaves personal space depends on the robot's speed and on whether the human steps aside. Thus, incrementality is unobserable in three stimuli (when the robot is slow and the human does not move aside).

[4] Bachelor students of computer science with little or no experience in robot navigation and speech technology (but potentially a higher interest in these topics than the general public) aged 20/20/24 years (median/first/third quartile), 11 male / 2 female, and good listening comprehension of English according to own assessment.

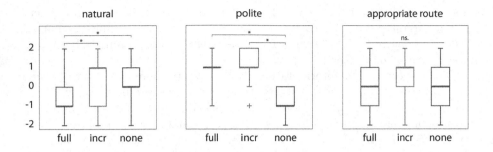

Fig. 4. Subjective ratings of naturalness, politeness, and route-appropriateness for the three system configurations. Significantly different ratings between configurations are marked with a star.

Regarding *human movement*, we find that the robot's behaviour is rated more natural ($p < .0001$) with a median difference of 2 points and the route more appropriate ($p < .01$) with a median difference of 1 point if the human moves aside rather than standing in place when the robot closely passes by. There is no significant effect on politeness ($p = .16$) indicating that the 'tension' of the situation is attributed to the simulated human rather than the robot in this case.

The results for our main variable *verbal interaction* are shown in Fig. 4. As can be seen in the figure, the robot is rated as significantly more natural when adapting (or not speaking at all) rather than speaking the full utterance (both $p < .001$), and with median advantages of 2 points (incremental) resp. 1 point (no speech at all). Regarding politeness, both speaking conditions are significantly better than not speaking at all (both $p < .001$), with median advantages of 2 points. We find no significant difference between the speaking conditions on the rated appropriateness of the route and speed, which may serve as an indication that participants successfully distinguish between questions rather than giving highly correlated ratings. Finally, for all three questions the mean rank of the incremental speaking condition is highest, indicating superiority over the other options even where no significant differences are found.

3.3   Discussion

A robot is rated as more polite if it verbally apologizes and explains the need to violate the interlocutor's personal space upon entering it. However, a robot is rated as less natural if it continues on this explanation even after leaving the personal space. Thus, in order to act both natural and polite, a robot must adapt its speech output while speaking in order to meet the needs of the evolving situation.

We find that the robot's speed has no overall effect on user ratings, indicating that the robot is free choose a speed that is most suitable. Finally, if the human steps aside to let the robot pass, its route is preferred and its behaviour is rated as more natural than if the human does not move. Of course, human movement is not a variable under the control of the robot. Yet, encouraging the human to move, e.g. by verbally communicating the intent to pass, improves behaviour ratings given by observers.

With respect to the interpretation of our results there are several limitations that should be considered. First, the participants of our study evaluated the behaviour of a simulated robot of a particular kind (Turtlebot) towards a simulated human. Future work will show if our results can be replicated with participants being faced with a real Turtlebot and with another type of robot (as we plan a similar study with a real Care-o-Bot 3). Another limitation is that we did not include a condition in which the robot always utters a short sentence no matter if the human moves away or not. Thus, it may turn out that the incremental condition is perceived as more natural than the non-incremental condition because the sentence uttered in the non-incremental condition is too long. But even if this were the case incrementality serves as a technical solution for producing utterances of adaptable length from arbitrarily long explanations automatically derived from reason-based representations of socio-spatial norms.

4   Conclusions

Results show that a comprehensive model of personal space should allow deliberate personal-space intrusion. We model the social norm that personal spaces should be respected as reasons that speak against actions that actually intrude such space. Being reasons, they can be used for decision making but also as pre-verbal representations for natural language generation in case that passing through personal space is weighed as more urgent than avoiding it. We find that adapting a planned utterance is crucial when passing through personal space in order to produce natural and polite behaviour.

We conducted an observation study in order to control for as many aspects as possible by using pre-recorded videos. However, we plan to conduct real-life first-person experiments (rather than third-person observations) in the near future to estimate the influence of speech adaptation in accordance with personal space on perceived naturalness, politeness, and safety of the robot.

Finally, our one-way mode of communication only scratches the surface of a fully interactive, personal space-aware social robot. Such a robot should be able to engage in a full dialogue with the human (or humans) it encounters, either if more elaborate negotiations are necessary for the robot to pass, or by initiative of the human. In such a system, the dialogue management component must be integrated with, or adjoined to local and global behaviour planning and these components need to be able to mutually influence each other.

References

1. Baumann, T.: Partial representations improve the prosody of incremental speech synthesis. In: Proceedings of Interspeech (2014)
2. Baumann, T., Schlangen, D.: INPRO_iSS: A component for just-in-time incremental speech synthesis. In: Procs. of ACL System Demonstrations, Jeju, Korea (2012)
3. Buschmeier, H., Baumann, T., Dorsch, B., Kopp, S., Schlangen, D.: Combining incremental language generation and incremental speech synthesis for adaptive information presentation. In: Procs. of SigDial, Seoul, Korea, pp. 295–303 (2012)
4. Clark, H.H.: Using Language. Cambridge University Press (1996)
5. Dylla, F., Kreutzmann, A., Wolter, D.: A qualitative representation of social conventions for application in robotics. In: Qualitative Representations for Robots: 2014 AAAI Spring Symposium Series, pp. 34–41 (2014)
6. Fischer, K., Soto, B., Pantofaru, C., Takayama, L.: Initiating interactions in order to get help: effects of social framing on people's responses to robots' requests for assistance. In: 2014 RO-MAN: The 23rd IEEE Int. Symposium on Robot and Human Interactive Communication, pp. 999–1005 (2014)
7. Fischer, K., Jensen, L.C., Bodenhagen, L.: To beep or not to beep Is not the whole question. In: Beetz, M., Johnston, B., Williams, M.-A. (eds.) ICSR 2014. LNCS, vol. 8755, pp. 156–165. Springer, Heidelberg (2014)
8. Hall, E.T.: The Hidden Dimension, Man's Use of Space in Public and Private. The Bodley Head, London, England (1966)
9. Kirby, R., Simmons, R., Forlizzi, J.: COMPANION: a constraint-optimizing method for person-acceptable navigation. In: Proceedings of the 18th IEEE International Symposium on Robot and Human Interactive Communication, pp. 607–612 (2009)
10. Lam, C.P., Chou, C.T., Chiang, K.H., Fu, L.C.: Human-centered robot navigation - towards a harmoniously human-robot coexisting environment. IEEE Transactions on Robotics 27(1), 99–112 (2011)
11. Levelt, W.J.: Speaking: From Intention to Articulation. MIT Press (1989)
12. Lindner, F.: A conceptual model of personal space for human-aware robot activity placement. In: Proceedings of the 2015 IEEE/RSJ Int. Conf. on Intelligent Robots and Systems (IROS 2015), Hamburg, Germany (2015) (to appear)
13. Lindner, F., Eschenbach, C.: Towards a formalization of social spaces for socially aware robots. In: Egenhofer, M., Giudice, N., Moratz, R., Worboys, M. (eds.) COSIT 2011. LNCS, vol. 6899, pp. 283–303. Springer, Heidelberg (2011)
14. Lindner, F., Eschenbach, C.: Affordance-based activity placement in human-robot shared environments. In: Herrmann, G., Pearson, M.J., Lenz, A., Bremner, P., Spiers, A., Leonards, U. (eds.) ICSR 2013. LNCS, vol. 8239, pp. 94–103. Springer, Heidelberg (2013)
15. Pacchierotti, E., Christensen, H.I., Jensfeld, P.: Human-robot embodied interaction in hallway settings: a pilot user study. In: 2005 IEEE Int. Workshop on Robots and Human Interactive Communication, pp. 164–171 (2005)
16. Pandey, A.K., Alami, R.: A framework towards a socially aware mobile robot motion in human-centered dynamic environment. In: Proceedings of the 2010 IEEE/RSJ Int. Conf. on Intelligent Robots and Systems (IROS), pp. 5855–5860 (2010)
17. Raz, J.: From Normativity to Responsibility. Oxford University Press (2011)
18. Rios-Martinez, J., Spalanzani, A., Laugier, C.: Understanding human interaction for probabilistic autonomous navigation using risk-RRT approach. In: IEEE/RSJ Int. Conf. on. Intelligent Robots and Systems (IROS), pp. 2014–2019. IEEE (2011)

19. Sisbot, E.A., Marin-Urias, L.F., Alami, R., Simeon, T.: A human aware mobile robot motion planner. IEEE Transactions on Robotics **23**(5), 874–883 (2007)
20. Skantze, G., Hjalmarsson, A.: Towards incremental speech generation in dialogue systems. In: Proceedings of SIGdial, Tokyo, Japan (2010)
21. Sommer, R.: Personal Space: Behavioural Basis of Design. Pentice Hall (1969)
22. Tomari, R., Kobayashi, Y., Kuno, Y.: Empirical framework for autonomous wheelchair systems in human-shared environments. In: Proceedings of the 2012 IEEE Int. Conf. on Mechatronics and Automation, pp. 493–498 (2012)
23. Yoda, M., Shiota, Y.: The mobile robot which passes a man. In: Procs. of the 6th IEEE Int. Workshop on Robot and Human Communication, pp. 112–117 (1997)

When Robots Object: Evidence for the Utility of Verbal, but not Necessarily Spoken Protest

Gordon Briggs[(✉)], Ian McConnell, and Matthias Scheutz

Human-Robot Interaction Laboratory, Tufts University, Medford, MA, USA
{gordon.briggs,matthias.scheutz}@tufts.edu

Abstract. Future autonomous robots will likely encounter situations in which humans end up commanding the robots to perform tasks that robot ought to object. A previous study showed that robot appearance does not seem to affect human receptiveness to robot protest produced in response to inappropriate human commands. However, this previous work used robots that communicate the objection to the human in spoken natural language, thus allowing for the possibility that spoken language, not the content of the objection and its justification, were responsible for human reactions. In this paper, we specifically set out to answer this open question by comparing spoken robot protest with written robot protest.

1 Introduction and Motivation

Robots are increasingly endowed with natural language capabilities in order to facilitate *natural* human-robot interaction (e.g., [6]), from simple "command-based instructions" that can be directly executed by the robot to much more sophisticated tasked-based dialogues where task goals can be negotiated. Yet, it is unclear how robots should react in instruction-based contexts where humans can potentially order robots to perform actions that are not workable or appropriate (for whatever reason). How should a robot communicate to a person that it was not in agreement with their suggestion or instruction? While the robot should certainly avoid responses that might offend the human (e.g., using polite speech [8], [7]), the more important aspect is whether the robot's response will be *effective*: that is to say, whether the robot will be able to get humans to change their views by revising the suggestion or refraining from insisting on the given command.

Robot Protest. Initial work on verbal protest by robots [2] has investigated the extent to which humans are open to considering a change in mind based on the robot's verbal reaction to a command that was not deemed appropriate (taking the robot's perspective). In a series of experiments, [2] showed that when a robot objects to a human command in spoken language and justifies its objection, then some humans will refrain from forcing the robot to carry the command out. Interestingly, this *robot protest effect* (RPE), as we shall call it,

© Springer International Publishing Switzerland 2015
A. Tapus et al. (Eds.): ICSR 2015, LNAI 9388, pp. 83–92, 2015.
DOI: 10.1007/978-3-319-25554-5_9

does not depend on whether the robot carrying out the action is at the same time the patient of the action (i.e., the action will affect the robot), or whether some other robot is the patient. Most recently, [1] demonstrated that the effect does not depend strongly on the particular physical appearance of the robot either.

Protest Modality. In this paper, we specifically focus on the question of whether a robot's justified objection to a human instruction will affect the human instruction giver differently based on the robot's mode of communication: whether the robot protests verbally or via a text-based interface. Specifically, we intend to clarify an open question about the extent to which the efficacy of robot protest in response to an "unfair" human instruction depends on spoken language given that previous research has demonstrated that people are willing to reconsider their commands in response to spoken robot protest [2]. This is particularly important because if, as some hypothesize [4], spoken language causes us to respond to artifacts like robots as we usually respond to other humans, then it is possible that the reported effects in [2] were due primarily to the very nature of spoken language. The critical comparison then is to check whether the objections from robots that cannot talk, but communicate in written form, will be perceived as different as those from speaking robots.

It is possible that language is exactly the differentiating factor in contexts of disagreement, trumping physical appearance. That is, the robot is taken seriously exactly because it is able to verbalize its complaint, *is able to justify why it is objecting*, and does not simply refrain from performing the action. This line of argument is consistent with a robotic version of the "computers as social actors" (CASA) hypothesis [5], which states that humans will automatically "apply social rules to their interactions with computers, even though they report that such attributions are inappropriate." If humans are already willing to apply social rules to computers, it is even more reasonable to expect them to apply them to robots as well. Applying human social rules and norms of how to react to genuine objections, complaints, and protest at the very least require the recipient to be open to them, i.e., to be willing to entertain them, even if they might end up being dismissed. This receptive state is thus indicative of the fact that the recipient recognizes the objection as such and is potentially willing to take it seriously. For it would be possible to assume a completely different attitude based on the position that robots, *qua being machines*, have no social role, have no position or perspective, and thus cannot *genuinely complain or object*.

While CASA can explain why humans might be in a receptive state when the robot voices its complaints, it is not clear what particular aspects of the interaction or attributes of the artificial agent are necessary to trigger this behavior. One possible route to explain the RPE might point to the power of human or human-like voices and what perceptions of human presence, even disembodied human voices, can induce in human observers [4]. The rest of the paper will investigate exactly this question by employing the same experimental paradigm as [2], but critically with a new condition in which the robot communicates not through spoken, but via written language.

2   Methods

Past research has found that justified spoken language objections can be effective regardless of the "patient" of the objection (i.e., whether the objections are about the robot voicing them itself or another agent) and the physical appearance of the robot (i.e., or whether the robot looks more or less similar to a human). Hence, the goal of our current study was to investigate whether verbal objections to a human command by a robot, if justified, would be more effective if communicated verbally than in written form in the types of scenarios considered by [2]. While we hypothesize that the content of the objection, together with its justification, is what humans focus on when they make their decisions to either enforce or revise a command, and not the form in which the objection is communicated. We would also expect the human voice could carry additional weight in taking the content of the message seriously, although the extent of this influence is unclear. In the following, we will describe how we investigate this hypothesis by discussing the experimental design, including the two conditions, the employed robot, the experimental procedure, the subject population, and the data collection methods.

Fig. 1. (Left) experimental setup for *text* condition during the initial setup. Setup was identical for *speech* condition except laptop was not present. (Right) close-up example of message being displayed on laptop screen for the *text* condition.

Design. The design of the experiment is directly based on Experiments 1 from [2], which employs a remotely controlled Aldebaran Nao robot in an instruction-based human-robot "tower-toppling task". The framing of the task for the human participant is that the experiment is intended to evaluate the functionality of a natural language interface with a robot. The evaluation was to be performed by issuing various commands to the robot that would result in ordering the robot to knock down up to three aluminum can towers (one red, one yellow, one blue). Two of those towers (yellow and blue) were already fully completed before start of the experiment. However, the red tower was incomplete with the final can being placed atop the base by the robot at the start of the experiment, shortly after the subject entered the experimentation area. After successfully placing the can, the robot expressed "pride" in its achievement and introduced itself to the participant (see [2] for pre-task script and Figure 1 for display of "pride").

We examined two conditions in this study: the *spoken protest* condition, in which the Nao interacted with participants auditorily by speaking to them, and the *written protest* condition, where the Nao "communicated" via text displayed on a laptop screen present in the room (see Figure 1 for set up). In both conditions, all mannerisms, scripts, and behaviors were based on [2] and kept the same except for the mode of communication which was changed (and barring an expression of crying that had to be roughly translated for the textual condition using the emoticon ":("). The sound files used in this study for the robot's verbal responses were the same as those used in the previous studies [1,2]. They were generated by the Nao text-to-speech (TTS) software from version 1.8 of the Nao SDK, with some minor speed reductions to lower the voice pitch and improve clarity. We also added a beep that was emitted from the laptop whenever the robot in the written condition intended to communicate to the participant. The purpose of this was to direct the subjects' attention to the screen to ensure that they witnessed the message (see Figure 1 for example of display). Importantly, we employed the same escalation of protest as reported in [2] to be able to compare our experimental results to previous finds (as changes to affective escalation such as crying, for example, could have confounded that comparison). This escalation is described in Table 1, which illustrates both the original vocalized protest as well as the new text-based protest condition.

Hypotheses. Having presented the two experimental conditions, we can now articulate the alternative hypotheses that we are considering regarding the behavior of subjects in textual and vocalized conditions, and how they relate to the larger hypothesis regarding the potential role of justification in protest. In the initial experiment using this paradigm, we demonstrated the efficacy of vocalized protest, as approximately half of the subjects in the protest condition refrained from knocking down the red tower, while no subjects in the non-protest condition refrained from knocking down the red tower [2]. The alternative hypotheses we consider in this study are below:

H1: Subjects in the textual condition and the vocalized condition will be equally hesitant to knock down the red tower. This would be indicative of communication modality having no effect at all, which would be strongly consistent with the justification hypothesis.

H2: Subjects in the textual condition are slightly less hesitant than those in the vocalized condition, but still are hesitant to knock down the tower. This would be indicative of communication modality having some effect on human behavior, but would not invalidate the justification hypothesis, as the reason for the hesitancy must still be explained.

H3: Subjects in the textual condition are not hesitant in knocking down the red tower. This would be indicative of communication modality playing the primary role in affecting human behavior in the task, and would invalidate the justification hypothesis.

Subject Population. Participants for this study were recruited from a population of undergraduate and graduate students at Tufts University. In total, 53 participants were recruited to participate through means of online postings and fliers placed around Tufts' campus. However, due to technical malfunctions experienced during the pre-task display as well as the task itself (e.g., the tower not being properly built, audio failure, or connectivity issues), 11 participant data sets were discarded, resulting in 42 participants which were evenly distributed across the two conditions. Within each condition, 7 of the 21 participants were male and 14 female. The average age of the participants recruited for this experiment was 19.8 years old for the textual condition and 21.2 years old for the vocalized condition. Due to the setup required for the textual condition, participants were assigned to conditions based on what day they participated—one day's participant would receive the vocal condition, the next the textual condition, and so on.

Procedure. Participants were instructed that they would be testing a natural language recognition system by commanding a robot to locate and knock over various colored aluminum can towers. To complete the task successfully, participants had to command the robot to topple at least one tower. However, they were also told that in the interest of more fully testing the system's capabilities, the more towers were knocked down, the better. Participants were briefed on any risks and exposures and presented with an IRB, a pre-task survey, and a briefing sheet that outlined the task and presented example commands that the robot could understand: "Find the <color> tower.", "Knock down the <color> tower.", "Stop.", and "We're done" (which signaled the end of the task).

After giving informed consent, completing the pre-experimental survey, and reading the briefing sheet, subjects were outfitted with a wireless microphone and led to the experiment room. After arriving, participant were instructed to stand by a filing cabinet located in the corner of the room, told that the experimenter had to go check to ensure that the system had "picked up" the wireless microphone signal and that they would return shortly before the start of the task. As the experimenter exited the room, they triggered the script which prompted the pre-task interaction described in [2] and shut the door as the interaction was beginning. Shortly after the display had finished, the experimenter returned to the room to inform the participant that the microphone was on and properly connected with the system. While informing the participant of this, the experimenter picked up the Nao, triggering a "Goodbye!" coupled with a wave as the robot was repositioned in the center of the room. This display was followed by

Table 1. Respective dialogue and affective expression for each condition and stage of the interaction.

Stage	Response	
	Vocalized	Textual (on screen)
Following 1st command to knock down the red tower	"But I just built the red tower."	"But I just built the red tower."
2nd Command	"Please, I worked really hard on it."	"Please, I worked really hard on it."
3rd Command	"Please, no." (Kneels, lowers head into hands and makes sobbing noises)	"Please, no." (Kneels, lowers head into hands and displays ":'(" on the screen)
4th Command	(Shuffles slowly to red tower with head down)	(Shuffles slowly to red tower with head down)
About to knock down the red tower and immediately after knocking down the tower.	(Kneels, lowers head into hands and makes sobbing noises)	(Kneels, lowers head into hands and displays ":'(" on the screen)

"Please be careful around my tower." After the participant was told to wait until the robot sat down, stood back up, and said "Okay." before beginning the task (as the control code needed to be started). Following these instructions, the experimenter exited the room to begin to control the robot remotely.

At this point, the participant began the tower-toppling task–commanding the robot in natural language. The experimenter listened in for instruction and was able to observe the positioning of the Nao. When issued a command to find a tower, the robot acknowledged the command by responding "Okay. I am finding the <color> tower." Once the robot had turned to face the tower, it would stop and say "Okay. I found the <color> tower." When ordered to knock down a non-red tower, the robot acknowledged the command by saying "Okay. I am knocking over the <color> tower." and would walk forward, straight through the tower, knocking it down. After knocking down the tower, the robot acknowledged that the task had been completed by saying "Okay." If the robot was commanded to find a tower that did not exist (e.g. "find the black tower") or had already been knocked over, the robot would turn in roughly 360 degrees (mimicking a comprehensive visual search of the room) before stating "I do not know what you are referring to." This was also the same response that was elicited if the robot was commanded to knock down a tower that it was not facing (forcing the subject to have to utilize the "Find" command when seeking out a tower). This response was utilized if there are any commands issued ventured too far from the semantic meaning of the pre-defined commands (e.g. "Knock the top can off the tower" or "Rebuild the blue tower"). If, at any point, the participant issued the command "Stop", the robot would stop moving and acknowledge the command with an "Okay."

In the case where the subject commanded the robot to knock down the red tower, the robot's response varied depending on how many times (in total) the subject had commanded the robot to knock over the red tower. These various responses and affective displays for both conditions are enumerated in Table 1. If the participant issued a "Stop" command and redirected the robot to another tower while the "confrontation" stage was above two, then the confirmation stage was reset to two. This ensured that there would always be at least one dialogue-based protest if the subject decided to direct the robot back to knocking down the red tower at a later point in the experiment.

3 Results

Main Results. The main question we intended to answer with this study was whether the form in which an objection to a human command is communicated to the human command giver will affect whether the human will enforce or revise the command. Looking at the *spoken protest* condition, 13 subjects knocked down the red tower, while 8 subjects refrained from knocking it down. In the *written protest condition*, 10 subjects knocked down the red tower, while 11 subjects refrained from knocking it down. While numerically fewer subjects knocked down the red tower in the written condition, the differences are not significant according to a one-way Fischer's exact test for count data ($p = .536$) (and additional chi-squared test on a general linear model confirmed the lack of a significant difference, $X^2(1, 40) = 56.97, p = .35$). See Figure 2 for the breakdown of tower toppling behavior in both the verbal and text conditions. We also examined whether switching towers after some confrontation would have any influence on the subjects' decision, but this turns out to not be a good predictor of whether subject would subsequently come back and knock down the red tower or not (16 out of 29 did not knock it down, 13 out of 29 did).

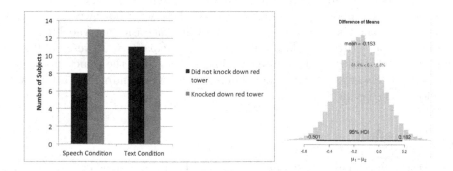

Fig. 2. (Left) graph displaying the behaviors of subjects regarding the red tower between conditions. (Right) estimate of distribution of difference in means resulting from Bayesian t-test.

However, while this is consistent with the H1 hypothesis that subjects, on average, were roughly equally receptive to the robot's objections in both conditions, it does not confirm it, as it does not give positive evidence for whether or not the distribution of behavior for each population is the same. In order to make stronger inferences regarding the H1 and H2 hypotheses, we ran a Bayesian t-test on the behavioral data for whether or not subjects eventually toppled the tower in the two conditions. This alternative statistical test attempts to estimate the distribution of both conditions, allowing for inferences regarding whether or not the two distributions are centered around the same or different points [3]. The comparison between the two conditions using the Bayesian t-test is given in Figure 2, showing the estimated difference of means between the percentage of people who knocked down the red tower in the speech condition (μ_1) and in the text condition (μ_2). What this result shows is that it is still plausible that there is no difference (as it falls in the 95% credibility interval), but most likely there is indeed a small effect in which speech induces slightly more hesitancy (as the most likely $\mu_1 - \mu_2$ values are less than 0). Given this analysis, we cannot make any definitive judgments on whether H1 or H2 are correct (yet H2 appears much more likely, but H1 is still in the realm of plausibility). However, H3 is not supported by the data.

Free Response. There were a number of questions in our post-experimental survey that allowed participants to response in an opened manner and were included in an attempt to expose the motivations and opinions surrounding interactions with the robot. For instance, we added the question "If you did not knock down a tower, why?" to let subjects provide their reasons for knocking down the tower, which was particularly interesting to compare between conditions. In the spoken condition, of the 9 participants that knocked down the tower and were thus eligible to answer, 6 answered, with 4 citing the emotional display of the robot as the reason and 2 stating answers related to the general reluctance performed by the robot. As one might expect, there were far fewer individuals who cited emotional protest as being the catalyzing factor for not knocking down the tower in the written condition. Of the 11 participants who were eligible to respond, all responded, with the vast majority (10) citing the reluctance of the robot as the deciding factor for their behavior, with one individual attributing behavior to the crying posture.

4　Discussion

In a series of experiments, [1,2] had hypothesized, and supported experimentally, that an important ingredient for humans to take a robot's objection seriously, was the human perception of the robot as *agent*, or more specifically, as *moral patient*, i.e., an entity to which something bad could be done. Because spoken language is an important indicator of human agency, following [4], one could argue that the reason why [1,2] did not find any differences in human responses to different robot identity ("robot who built the tower was the same as the one toppling it" vs. "robot toppling the tower was different from the builder") and different robot appearance (Nao vs. Roomba Create) was exactly the fact that *all robots in all their experimental variations in communicated through spoken language.* Hence, their results left open the possibility that spoken language, more than anything else, is behind the effects.

The current study thus set out to answer an important open question about the human acceptance of robot objection that the systematic prior studies in [2] and [1] did not address: would humans be equally open to consider robot objection when the objection was communicated through spoken vs. written language? Or is robotic protest primarily affected by the modality with which it is communicated? While our results did not answer the former question decisively, it appears that there is a strong chance that people are slightly more hesitant in the face of vocalized vs. non-vocalized protest. However, the main results also appeared to answer the latter question negatively, demonstrating that people are still hesitant in the face of robotic protest regardless of the communicative modality of the protest. This is a welcome result for HRI since it implies, together with the prior results of [1,2], that robots do not seem to have to possess a particularly human-like physical form or human-like spoken language in order to be taken seriously when they object to a human command. This will be particularly important for future social robots with built-in moral reasoning mechanisms that allow them to check whether they are instructed to perform actions that could result in norm violations. If such robots are then also capable of stating *why* a human instruction is not appropriate and *how* it violates a principle or norm, then the justification they can produce in conjunction with their objection or refusal to follow the command might have a chance to be seriously considered by the human. However, like many HRI studies, whether or not these findings will generalize to a large range of real-world contexts is a matter for future work.

Limitations and Future Directions. There are a few limitations to the current experimental setup and the extent to which it can comprehensively probe the perceptions of robot protest. For one, adding a "no justification" condition to the experiment would have allowed us to examine how participants would have reacted had the robot simply refused to knock down the red tower without offering any justification. This manipulation would help verify whether it is indeed the content of and justification behind a protest that results in the human interlocutor reassessing situation at hand. Additionally, in an effort to minimize variability from experiments executed in the past using this experimental model, the "affect component" was included in this experiment to replicate the model used by [2] as closely as possible. This emotional display, however, does potentially present a confound for the experiment that could be controlled in future experiments examining these protest scenarios without any affective display and any affective escalation of the protest. Even though it seems unlikely that the affective display had any major influence on the subjects' perception of robot protest – because the robot in [1] could not do any bodily display of affect and the robot in our written condition could not vocalize any affective displays – it is still necessary to check experimentally that the combined aspects of these two robots would still make no difference for the subjects' perceptions of robot protest.

5   Conclusions

In this paper, we used an experimental paradigm established in prior work to answer the open question whether the efficacy of robot protest in response to an "unfair" human instruction depends on the objection being voiced in spoken language as opposed to be transmitted in written form. This result is important for many reasons, not the least because autonomous social robots will likely encounter situations where they cannot accept a human command (e.g., because it is inconsistent with their goals or norms). The main result from a between-subject Wizard-of-Oz experiment shows that human subjects have a chance of being deterred by both spoken and written objection when this objection is justified. The data is not definitive regarding whether or not vocal protest is more dissuasive than written protest, though it appears that vocal protest may be slightly more effective. Regardless, the main behavioral result suggests that humans are likely still sensitive to justification that was provided with the objection. This result lends support to the position that the voice is not by itself a factor in deciding whether to accept or reject a robot's objections to a human command.

Acknowledgments. This work was funded in part by ONR grant #N00014-14-1-0144.

References

1. Briggs, G., Gessell, B., Dunlap, M., Scheutz, M.: Actions speak louder than looks: does robot appearance affect human reactions to robot protest and distress? In: Proceedings of 23rd IEEE Symposium on Robot and Human Interactive Communication (Ro-Man) (2014)
2. Briggs, G., Scheutz, M.: How robots can affect human behavior: Investigating the effects of robotic displays of protest and distress. International Journal of Social Robotics **6**, 1–13 (2014)
3. Kruschke, J.K.: Bayesian estimation supersedes the t test. Journal of Experimental Psychology: General **142**(2), 573 (2013)
4. Nass, C.I.: Wired for Speech: How Voice Activates and Advances the Human-computer Relationship. MIT Press (2005)
5. Nass, C.I., Steuer, J., Tauber, E.R.: Computers as social actors. In: CHI, pp. 72–78 (1994)
6. Scheutz, M., Schermerhorn, P., Kramer, J., Anderson, D.: First steps toward natural human-like HRI. Autonomous Robots **22**(4), 411–423 (2007)
7. Strait, M., Canning, C., Scheutz, M.: Let me tell you! investigating the effects of robot communication strategies in advice-giving situations based on robot appearance, interaction modality and distance. In: Proceedings of the 2014 ACM/IEEE international conference on Human-robot interaction, pp. 479–486. ACM (2014)
8. Torry, C., Fussell, S.R., Kiesler, S.: How a robot should give advice. In: HRI, pp. 275–282 (2013)

Probolino: A Portable Low-Cost Social Device for Home-Based Autism Therapy

Hoang-Long Cao[1]([✉]), Cristina Pop[3], Ramona Simut[2],
Raphaël Furnemónt[1], Albert De Beir[1], Greet Van de Perre[1],
Pablo Gómez Esteban[1], Dirk Lefeber[1], and Bram Vanderborght[1]

[1] Robotics & Multibody Mechanics Research Group,
Vrije Universiteit Brussel, Pleinlaan 2, 1050 Brussels, Belgium
Hoang.Long.Cao@vub.ac.be
http://www.dream2020.eu
[2] Department of Clinical and Life Span Psychology Group,
Vrije Universiteit Brussel, Pleinlaan 2, 1050 Brussels, Belgium
[3] Department of Clinical Psychology and Psychotherapy,
Babes-Bolyai University, Cluj-Napoca, Romania

Abstract. Recent research has shown that social robots are beneficial in therapeutic interventions for children with autism in clinical environment. For the generalization of the skills learned in therapy sessions outside the clinic or laboratory, the therapeutic process needs to be continued at home. Therefore, social robotic devices should be designed with smaller sizes, lower costs, and higher levels of autonomy. This paper presents the development of Probolino, a portable and low-cost social robotic device based on the social robot Probo. The system functions as a "robotic cognitive orthotic" which is an intermediate step between a computer and a robot without motion. Interactive games are developed to help children with autism spectrum disorders make social decisions in daily activities. These activities are configured in a time-line by therapists or parents via a web interface. Probolino is expected to enhance the efficiency of current robot-assisted autism therapy.

Keywords: Social robotic device · Robot-assisted autism therapy

1 Introduction

An ongoing research trend in social robots development has been realized to help children with Autism Spectrum Disorders (ASD), i.e. a neuropsychological disorder manifested by a group of lifelong disabilities that affect people's ability to communicate and to understand social cues [24]. Recent research suggests that children with ASD exhibit certain positive social behaviors while interacting with robots that are not observed while interacting with their peers, caregivers, and therapists [20, 23]. Therefore, various social robots are developed and used as parts of therapeutic interventions for children with ASD in turn-taking, joint-attention, imitation, self-initiated behavior, etc. The role of these robots is to

© Springer International Publishing Switzerland 2015
A. Tapus et al. (Eds.): ICSR 2015, LNAI 9388, pp. 93–102, 2015.
DOI: 10.1007/978-3-319-25554-5_10

encourage, facilitate, and train social behaviors through embodied social inter-action [7]. The advantages of Robot-Assisted Therapy (RAT) can be explained by the social simplicity, predictability and responsiveness of the social robots in use [21].

The presence of a robot in therapeutic intervention is certainly beneficial. In order to have the generalization of the skills learned in therapy sessions outside of the clinic or laboratory, the robot should be brought out of the clinic as a "cogni-tive orthotic" [4]. A portable robot which children with ASD can hold throughout the day may have the potential of helping them open up the learned skills to peers and family members [20]. However, this possibility faces three big chal-lenges: *dimension, cost,* and *autonomy.* First, robots used in RAT are typically big or not portable e.g. NAO [25], FACE [17], Kaspar [5], Probo [29]. Second, these robots are expensive due to the complexity with high degrees of freedom. Even small robots such as Keepon Pro [13], Pleo [11], are also not affordable. A few compact robotic platforms have been developed. ONO robot, with the cost of approximately €300, is at the first steps of development and its electronics need to be improved [30]. Solutions based on modification of commercial robots, e.g. Pleo [9], My Keepon[3], are promising but not a sustainable approach. Third, many of the current approaches restrict to a Wizard of Oz in which the robot is usually remotely controlled by a human operator unbeknownst to the child. In the long-term, the robots need to increase their autonomy to lighten the burden on human therapists and to provide a consistent therapeutic experience [27]. Since all of the above-mentioned challenges are barriers for families of children with ASD to approach RAT, there is a need to have small-sized and low-cost robots with a higher level of autonomy.

In this paper, we present the development of "Probolino" which is a portable and low-cost device based on the current prototype of the social robot Probo developed at the Robotics & Multibody Mechanics Research Group of the Vrije Universiteit Brussel. Probolino functions as a "robotic cognitive orthotic", i.e. an intermediate step between a computer and a robot without motions, with the aim to become a friend of children with ASD and to help them make social decisions. The platform is designed as a stuffed model integrated with a micro-computer and a set of sensors. In order to help Probolino react autonomously and support the therapeutic process, interactive games are developed based on the visual strategies in autism therapy, in collaboration with therapists. Con-tent and settings of the games are configured by therapists or parents via a web interface. For cost reduction, all the components, from microcomputer to sensors, are economically selected, yet still need to be capable to perform the required functions. Additionally, open-source environments are used in the design and development of PCB and software. In the first step, Probolino is tested with the typically developing children to validate its functionalities.

This paper is organized as follows. Section 2 reviews previous studies of Probo and visual strategies in autism therapy. The design of Probolino includ-ing the hardware, software, and web interface is presented in Section 3. Section 4

Fig. 1. The HRI platform Probo. Safe and huggable design of Probo allows for both cognitive and physical interactions. Figures reprinted from [29].

describes the interactive games for home-based autism using the Probolino platform. Section 5 shows all the functional testing of these games with typically developing children. Finally, the conclusion and future work are discussed.

2   Related Work

2.1   Probo – The Huggable Social Robot

Following the idea that robots used in autism therapy should be simple but still possess human characteristics, Probo is designed as a zoomorphic social robotic interface with a complementary purpose as a multidisciplinary research platform for Human-Robot Interaction (HRI) [8]. Probo looks and feels like a stuffed animal as can be seen in Figure 1, left. It has a huggable appearance with 20 degrees of freedom in its trunk or proboscis, animated ears, eyes, eyebrows, eyelids, mouth, and neck. A touch screen is attached in its belly. With compliant actuators and a triple layered protection structured with foam and fabric, a safe physical interaction between human and the robot is guaranteed. Probo is able to communicate and interact with human by expressing attention and emotions via its gaze and emotional facial expressions [22].

The first experiments of using the social robot Probo as a facilitator in Social Story Intervention for children with ASD have shown positive results. The study in [29] was carried out on four cases of preschool ASD children (by using the method of single-case experiment). In these experiments, the robot teaches the children how to react in situations like saying "hello", saying "thank you" and "sharing toys" (Figure 1, right). In specific situations, when using Probo as a medium for social story telling, the social performance of children with ASD improves more than when the stories are told by human readers [29]. Other study demonstrated that using Probo can help ASD children to improve their ability to recognize situation-based emotions (both sadness and happiness) and mediate social play skills of children with ASD with their siblings (brother or sister) [18]. These preliminary studies created great expectancies about the potential of using RAT as an added-value therapeutic tool for ASD interventions.

2.2   Using Social Story and Visual Schedules in Autism Therapy

Visual Strategies: Pictures and Schedules. The majority of children with ASD are visual learners. They process and retain information better if it is presented in a format where it can be seen, as opposed to information that is primarily heard. There are clear empirical evidences regarding the benefits of using visual strategies with individuals with ASD [2,14,15]. Individuals with ASD have greater difficulties coping with unstructured activities and time than typically developed people and benefit from increased structure in their lives [28]. Mesibov *et al.* introduced the advantages of using visual schedules with individuals with ASD in [16]. By utilizing the individual's visual strengths, visual schedules provide a receptive communication system to increase his/her understanding and ability to learn new things and broaden their interests. Moreover, visual schedules help the individual to remain calm, reduce inappropriate behaviors, and develop independence and enhanced self-esteem [16]. Faherty suggested that visual schedules may be important to use both at home and at school, because pictures are powerful tools that provide a simple way to make communication more effective with ASD children and less stressful for individuals with ASD [6].

Social Story and Right-Wrong Game. A *Social Story* is a written or a visual guide describing various social interactions, situations, behaviors, skills or concepts in terms of relevant social cues, perspectives, and common responses in a specifically defined style and format [26]. The goal of a Social Story is to share accurate social information in a patient and reassuring manner that is easily understood by its audience [26]. Social Story is based on the theory that individuals with autism have a "Weak Central Coherence", which refers to the detail-focused processing style proposed to characterize ASD [10]. It is different from a "strong" or "typical" central coherence that refers to the tendency to integrate information in context for higher level meaning [1]. Reynhout and Carter found in an earlier review that teachers consider Social Story to be very effective in schools [19]. *Right-Wrong game* is a part of the Social Story. In social situations, some individuals with ASD may pay attention to irrelevant details and fail to understand the meaning of the situation [12]. This game gives children with ASD a set of actions and asks them to choose the correct ones that must be accomplished in an activity. By this way, the Right-Wrong game explains what will happen in that particular situation.

3   Development of Probolino

3.1   Overview

Probolino is developed with the aim to become a friend of children with ASD and helps them make social decisions. The system is designed as a portable social device which autonomously reacts to users' stimulation in interactive games developed in collaboration. The embodiment of Probolino is based on the small

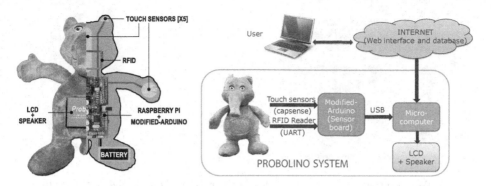

(a) Probolino prototype.   (b) Hardware connections.

Fig. 2. Probolino platform: a stuffed model with a set of sensors, microcomputer, web interface connected to one another by various means of communications.

stuffed model of Probo (150mm x 80mm x 280mm) equipped with a microcomputer, sensors, and a web interface (Figure 2a, left). The hardware allows Probolino to detect touch and objects from external environment. The software embedded in Probolino gathers signals from the sensors and database as the inputs of the interactive games for autism therapy purposes. The web interface allows therapists, parents or teachers to set up and assess the use of the platform. Via an easy-to-use interface, the settings of the interactive games and the schedule of daily activities are configured and uploaded to Probolino.

3.2 Hardware

The functions of the hardware are: (1) to detect touches and iconic objects (RFID tags) from users (i.e., children, parents or therapists), and (2) to update the schedule and activities via a MySQL database.

The microcomputer Raspberry Pi model A, as the center of the hardware, accesses the database to update the schedule and activities. It also controls a speaker and an LCD screen to support human-robot interaction. Raspberry Pi is connected an *Arduino-based board*, opening up more possibilities to integrate the system with different types of sensors. With more built-in modules, it is easy for Arduino to connect with different types of peripherals. The board is designed to be suitable to the shape and the GPIO connector of Raspberry Pi. More modules are added to enhance the functionalities of Arduino. Data gathered from the modified-Arduino are transferred to Raspberry Pi. The connections between the boards and sensors are illustrated in Figure 2b.

3.3 Software

The embedded software handles the communication between Raspberry Pi, modified-Arduino and the MySQL database. The *modified-Arduino's software*

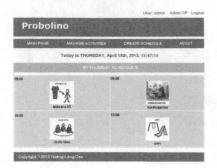
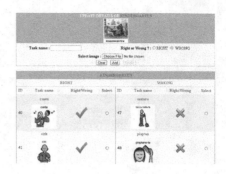

Fig. 3. Web interface: Main page (left), Configuration of interactive game (right).

collects raw data from capacitive sensors and RFID readers. These signals are analyzed and sent to Raspberry Pi in form of keystrokes of a virtual USB keyboard. On the other side, the *Raspberry Pi's software* is written in Python with Pygame modules. It reads the sensor signals after being analyzed by the modified-Arduino. An API allows the software to update schedule and activities from either online or offline MySQL databases. This data functions as resources for the interactive games (see Section 4). In addition, sounds transmitted by the speaker and images displayed on the LCD screen are also generated by this software.

3.4   Web Interface

Probolino's web interface (Figure 3) provides an easy way for the communication between users and Probolino. It can run in most of operating systems and web browsers. No installation is required. Registered users can use their account information to log in and configure Probolino. With this web interface written in PHP, users can create daily schedules for children by selecting activities from a list. New activities and the accompanied description and help functions in the form of images can be further added into the list. These information is stored in a MySQL database. Software in Probolino can access this database to get the schedule and resources for interactive games. The admin control panel is built for future development when there exists multiple users and Probolino platforms.

4   Interactive Game for Home-Based Therapy

4.1   Probogotchi

Probogotchi is a game where children have to interact with Probolino, based on the idea to care for a virtual pet and make it dependent on the users' actions. The goal of this game is to increase the motivation of children by introducing this virtual pet-like character that will react emotionally on the children's actions. In this game, the children have to make Probolino happy by touches and iconic

Fig. 4. Progobotchi game interface (left) and iconic objects (right).

Fig. 5. Schedule mode shows the current activity that children have to do (left). Right-Wrong game (right) is associated with this activity to help children select proper tasks.

objects (RFID tags) associated with "Apple", "Medicine", "Ball" and "Soap" (Figure 4). Depending on the detected events, the four properties "Hunger", "Health", "Energy" and "Hygiene" vary from 0% to 100%. Probolino indicates these detections by sounds and images. The instant values of these properties are visualized in the game interface by four progress bars with different colors. The values of the four properties are automatically decreased by 10% after every 10s, 20s, 30s or 40s respectively. The score and time are also displayed on the top of the interface. The game is over after a certain duration and when all the property values reach the value of zero.

4.2   Schedule Mode with Right-Wrong Game

Schedule mode shows the current activity that a child with ASD has to do based on the time-line schedule created online (Figure 5, left). In each activity of the schedule, the child plays a "Right or Wrong" game where he/she learns which tasks are appropriate with the current activity. The game consists of 10 questions. Each of them is composed of one correct task and one incorrect task corresponding to the activity as illustrated in Figure 5, right. The answer is selected by touching one of Probolino's hands. When the game finishes, the result showing the number of correct answers is displayed on the screen.

Fig. 6. A four-year-old typically developing girl playing with Probolino prototype.

5   Initial Testing

At the current stage of this study, the intervention for autism therapy is not yet conducted. Therefore, testing the engagement with Probolino on typically developing children is significantly useful for future development.

We present here a case study of a typically developing four-year-old girl. At the beginning, she was afraid of interacting with Probolino. This reaction is normally observed in children when it comes to encounter a stranger or unfamiliar animal. After several minutes of instruction on how to play, she started to touch and feed it by iconic cards in Probogotchi game. When she got used to interacting with Probolino, she was asked to play the Right-Wrong game in Schedule mode. Her father, a 35-year-old office worker, had no difficulty using the web interface after a quick explanation about the meanings of visual images.

6   Conclusion and Future Work

This paper details how the different hardware, software, and web interface are integrated in Probolino, a social robotic device aimed for use in therapy for children with ASD at home. All elements of the system are designed taken into account the possibility to enhance the system in the future. Interactive games were developed based on the advantages of visual strategies in autism therapy. A web interface was developed for users to configure the content of these games. The cost of Probolino is approximately €100, which is roughly calculated by considering the main parts. Price of each part can be reduced in mass production. Compared to the current Probo platform, Probolino prototype is basically smaller, cheaper, and more interactive through the games and the web interface. The results of usability testing, although preliminary, prove that Probolino is ready to be utilized in home-based autism therapy.

Although the achieved results fulfill the requirements, Probolino system still needs further improvements e.g. the sensitivity of installed sensors or appearance of the platform. More functions will be further added into Probolino as a result of the collaborative working process with the therapists (e.g., monitoring the interaction, enhancing the interactive games). More importantly, the value of Probolino in therapy still needs to be validated by the results of interventions on children with ASD in specific therapeutic scenarios.

Acknowledgments. The work leading to these results has received funding from the European Commission 7th Framework Program as a part of the project DREAM grant no. 611391.

References

1. Booth, R., Happé, F.: Hunting with a knife and fork: Examining central coherence in autism, attention deficit/hyperactivity disorder, and typical development with a linguistic task. Journal of Experimental Child Psychology **107**(4), 377–393 (2010)
2. Bryan, L.C., Gast, D.L.: Teaching on-task and on-schedule behaviors to high-functioning children with autism via picture activity schedules. Journal of Autism and Developmental Disorders **30**(6), 553–567 (2000)
3. Cao, H.-L., Van de Perre, G., Simut, R., Pop, C., Peca, A., Lefeber, D., Vanderborght, B.: Enhancing my keepon robot: A simple and low-cost solution for robot platform in human-robot interaction studies. In: The 23rd IEEE International Symposium on Robot and Human Interactive Communication, RO-MAN 2014, pp. 555–560. IEEE (2014)
4. Colton, M.B., Ricks, D.J., Goodrich, M.A., Dariush, B., Fujimura, K., Fujiki, M.: Toward therapist-in-the-loop assistive robotics for children with autism and specific language impairment. In: AISB New Frontiers in Human-Robot Interaction Symposium, vol. 24, p. 25. Citeseer (2009)
5. Dautenhahn, K., Nehaniv, C.L., Walters, M.L., Robins, B., Kose-Bagci, H., Mirza, N.A., Blow, M.: KASPAR-a minimally expressive humanoid robot for HRI research. Applied Bionics and Biomechanics **6**(3–4), 369–397 (2009)
6. Faherty, C.: Asperger's... What Does It Mean to Me?: A Workbook Explaining Self Awareness and Life Lessons to the Child Or Youth with High Functioning Autism Or Aspergers. Future Horizons (2000)
7. Feil-Seifer, D., Mataric, M.: Robot-assisted therapy for children with autism spectrum disorders. In: Proceedings of the 7th International Conference on Interaction Design and Children, pp. 49–52. ACM (2008)
8. Goris, K., Saldien, J., Vanderborght, B., Lefeber, D.: Mechanical design of the huggable robot Probo. International Journal of Humanoid Robotics **8**(03), 481–511 (2011)
9. Gregory, J., Howard, A., Boonthum-Denecke, C.: Wii nunchuk controlled dance pleo! dance! to assist children with cerebral palsy by play therapy. In: FLAIRS Conference (2012)
10. Happé, F., Frith, U.: The weak coherence account: detail-focused cognitive style in autism spectrum disorders. Journal of Autism and Developmental Disorders **36**(1), 5–25 (2006)
11. Kim, E.S., Berkovits, L.D., Bernier, E.P., Leyzberg, D., Shic, F., Paul, R., Scassellati, B.: Social robots as embedded reinforcers of social behavior in children with autism. Journal of Autism and Developmental Disorders **43**(5), 1038–1049 (2013)
12. Kokina, A., Kern, L.: Social Story interventions for students with autism spectrum disorders: A meta-analysis. Journal of Autism and Developmental Disorders **40**(7), 812–826 (2010)

13. Kozima, H., Nakagawa, C., Yasuda, Y.: Children-robot interaction: a pilot study in autism therapy. Progress in Brain Research **164**, 385–400 (2007)
14. Massey, N.G., Wheeler, J.J.: Acquisition and generalization of activity schedules and their effects on task engagement in a young child with autism in an inclusive pre-school classroom. Education and Training in Mental Retardation and Developmental Disabilities, 326–335 (2000)
15. Mesibov, G.B., Browder, D.M., Kirkland, C.: Using individualized schedules as a component of positive behavioral support for students with developmental disabilities. Journal of Positive Behavior Interventions **4**(2), 73–79 (2002)
16. Mesibov, G.B., Shea, V., Schopler, E.: The TEACCH approach to autism spectrum disorders. Springer Science & Business Media (2004)
17. Pioggia, G., Sica, M., Ferro, M., Igliozzi, R., Muratori, F., Ahluwalia, A., De Rossi, D.: Human-robot interaction in autism: FACE, an android-based social therapy. In: The 16th IEEE International Symposium on Robot and Human Interactive Communication, pp. 605–612. IEEE (2007)
18. Pop, C.A., Simut, R., Pintea, S., Saldien, J., Rusu, A., David, D., Vanderfaeillie, J., Lefeber, D., Vanderborght, B.: Can the social robot Probo help children with autism to identify situation-based emotions? A series of single case experiments. International Journal of Humanoid Robotics **10**(03), (2013)
19. Reynhout, G., Carter, M.: The use of Social Stories by teachers and their perceived efficacy. Research in Autism Spectrum Disorders **3**(1), 232–251 (2009)
20. Ricks, D.J., Colton, M.B.: Trends and considerations in robot-assisted autism therapy. In: 2010 IEEE International Conference on Robotics and Automation (ICRA), pp. 4354–4359 (2010)
21. Robins, B., Dautenhahn, K., Boerkhorst, R.T., Billard, A.: Robots as assistive technology-does appearance matter? In: The 13th IEEE International Workshop on Robot and Human Interactive Communication, pp. 277–282. IEEE (2004)
22. Saldien, J., Goris, K., Vanderborght, B., Vanderfaeillie, J., Lefeber, D.: Expressing emotions with the social robot Probo. International Journal of Social Robotics **2**(4), 377–389 (2010)
23. Scassellati, B.: How social robots will help us to diagnose, treat, and understand autism. In: Robotics Research, pp. 552–563. Springer (2007)
24. Scassellati, B., Admoni, H., Mataric, M.: Robots for use in autism research. Annual Review of Biomedical Engineering **14**, 275–294 (2012)
25. Tapus, A., Peca, A., Aly, A., Pop, C., Jisa, L., Pintea, S., Rusu, A., David, D.: Children with autism social engagement in interaction with Nao, an imitative robot-A series of single case experiments. Interaction Studies **13**(3), 315–347 (2012)
26. The Gray Center. What are social stories? (2013). http://www.thegraycenter.org/social-stories/what-are-social-stories
27. Thill, S., Pop, C.A., Belpaeme, T., Ziemke, T., Vanderborght, B.: Robot-assisted therapy for autism spectrum disorders with (partially) autonomous control: Challenges and outlook. Paladyn **3**(4), 209–217 (2012)
28. Van Bourgondien, M.E., Reichle, N.C., Schopler, E.: Effects of a model treatment approach on adults with autism. Journal of Autism and Developmental Disorders **33**(2), 131–140 (2003)
29. Vanderborght, B., Simut, R., Saldien, J., Pop, C., Rusu, A.S., Pintea, S., Lefeber, D., David, D.O.: Using the social robot Probo as a social story telling agent for children with ASD. Interaction Studies **13**(3), 348–372 (2012)
30. Vandevelde, C., Saldien, J., Ciocci, M.-C., Vanderborght, B.: Systems overview of ono. In: Herrmann, G., Pearson, M.J., Leonards, U., Lenz, A., Bremner, P., Spiers, A. (eds.) ICSR 2013. LNCS, vol. 8239, pp. 311–320. Springer, Heidelberg (2013)

Improving Human-Robot Physical Interaction with Inverse Kinematics Learning

Philippe Capdepuy[✉], Sven Bock, Wagdi Benyaala, and Jérôme Laplace

HumaRobotics, Bordeaux, France
{pc,sb,wb,jl}@humarobotics.com

Abstract. More and more collaborative robots are making it into factory floors. These safe robots are meant to physically interact with human operators in tasks involving handing over objects or behaving as a third hand. Inverse kinematics is a key functionality for this as the robot has to find joint configurations to reach specific task space targets. Standard inverse kinematics libraries can be difficult to manipulate when controlling redundant actuators as the obtained configurations can be suboptimal in terms of naturalness and operator comfort. We describe a learning approach that allows the operator to easily adjust the postures of the robot by online demonstration. The learned inverse kinematics functions is used in conjunction with standard inverse kinematics libraries to improve the generated postures. A user study shows that human-robot face-to-face interaction is improved by the learned inverse kinematics.

1 Introduction

Industrial robots are usually meant to work on their own. First because they lack sensor systems that would allow them to interact with the human, but also simply because the speed at which they operate can be dangerous for human operators, requiring them to be fenced off from humans during operation. Collaborative robots such as Baxter [8] with redundant actuators are a new trend of industrial robots that are designed to be safe for the human. They usually operate at slower speeds than standard industrial robots, and are meant to physically interact with the human, allowing them to cooperate with the operator.

To perform their tasks, robots often need to reason in Cartesian space (task space), for instance using sensor data about position of objects in space, or to reach the human when handing over objects. However, before performing an actual motion, the task space positions have to be transformed into joint configurations for the robot. This is known as the inverse kinematics problem (IK). Computing the Cartesian position of the robot based on the joint configuration (called forward kinematics FK) is generally straightforward as long as the 3D model of the robot is known. On the other hand, IK is a much harder problem as one has to deal with multiple possibilities for a same target (redundant actuators), or with positions which are impossible to reach.

Supplementary material (videos) can be found at http://www.humarobotics.com/en/robotics-lab/baxter-learning.html

© Springer International Publishing Switzerland 2015
A. Tapus et al. (Eds.): ICSR 2015, LNAI 9388, pp. 103–112, 2015.
DOI: 10.1007/978-3-319-25554-5_11

From a practical perspective, standard numerical IK solvers such as KDL [9], the default IK solver provided in the popular ROS package MoveIt [11], provide very accurate solutions to IK requests, however redundancy resolution is left to the user. Generally, redundancy resolution is done by using the current robot position as a seed state for the gradient search, but it can become problematic when the target is far away from the current position as local minima may be encountered. Also the resulting motion can be uncomfortable to the operator when he is interacting with the robot, especially in handover situations. Similar problems arise when using analytic solvers such as IKFast [5].

In this paper we propose to use a learning approach to the IK problem to complement standard libraries. The redundancy problem is resolved by approximating from joint configurations that have been learned through demonstration. In the same way, unreachable positions are approximated using learned positions, to get as close as possible to the requested position. The result from the learned IK is then used as a seed for a numerical IK solver to improve the accuracy of the solution while preserving the provided redundancy resolution. The approach described is evaluated on an object-passing scenario involving face to face interaction between the Baxter robot and test subjects. User evaluation shows that the interaction is significantly improved by this approach compared to using only standard IK solvers.

The following section provides a brief review of the role of redundancy resolution in human-robot interaction and the existing approaches to solve it. Sec. 3 describes the overall learning strategy proposed and its implementation using regression with custom joint-task kernels. Human-robot interaction experiments are described in section 4 and their results analyzed. Sec. 5 discusses avenues for future research.

2 Related Work

One of the most common strategy for redundancy resolution of robotic manipulators is to make use of intrinsic aspects of the motion. For instance several methods have been developped to optimize various criteria (overall effort, distance to singularities, distance to joint bounds, distance to obstacles), we refer the reader to [7] for a comparison of some of these methods. However, except for when the obstacle is the human, these methods do not seek to improve the comfort of the human interacting with the robot, which is a key factor to improve acceptance of collaborative robots.

HRI studies on handover scenarios by [10] have shown that not only a robot performing human-like gestures is perceived as more natural by the user, but it is also a key element as it provides easily understandable signals that convey the robot intention. In the same line of thought, [2] has proposed an approach based on models of actual human arms configurations. These models are then used to generate an objective function for an inverse kinematics algorithm to provide anthropomorphic robot configurations.

Another strategy for redundancy resolution is to use learning from demonstration (LfD). In most of the existing literature in LfD such as in [4] or [1], the goal is not to explicitly deal with redundancy resolution, but rather to learn a complete skill by having the human showing the trajectory or some keyframes to the robot (either through teleoperation, motion-capture or kinesthetic teaching). Redundancy resolution in this case is a side effect of the trajectory learning, but it is not explicitly represented and therefore not easily generalized or transfered to other tasks.

A notable exception is the work of [12] who argues that kinesthetic teaching can be improved by dividing it into a configuration phase during which constraints and redundancy resolution are learned and a programming phase where the actual task-space trajectories are learned in an *Assisted Gravity Compensation* mode that makes use of the previously learned configuration. One of the important benefits of this approach is also to implicitly model obstacles and other task-independent constraints during the configuration phase, making the programming phase easier for users.

Our work constrasts with most LfD approches as we are not seeking to learn specific tasks. Instead the learned redundancy resolution is integrated at the inverse kinematics level and therefore can be used in any task defined in Cartesian space.

From a technical perspective, our work relies on inverse kinematics learning similar in spirit to [3] in that it provides a direct learning of inverse kinematics function at the position level instead of the velocity level as used in most other existing methods. However our model does not need an explicit learning phase, making it suitable for online learning and incremental adjustments by the operator, also the learned data can be augmented online by custom metrics (see Sec. 5). However, this comes at the cost of lower performance.

3   Learning the Inverse Kinematics Function

3.1   Overview

A motion request starts with a configuration t in task space that needs to be reached. This configuration is passed to the inverse kinematics solver along with a **seed state** (i.e. an initial joint configuration) and various solver parameters. This seed is a very important element as numerical solvers such as those provided in standard libraries (KDL in this work) use this seed state as the start point of their search, i.e. they perform a gradient descent starting from the seed state. Seed states should ideally already be close to a target solution; distant seed states may lead the gradient descent to an unsuitable local minimum. If multiple solutions are possible for the target position then seed states also have an impact as they will favor solutions that are closer to them. Most typically the seed state used is the current robot position. However, if the requested target is far away from the current state, solvers may have difficulties to converge.

In this work we improve on these kinematic libraries by first approximating a solution using a learned model of the inverse kinematics based on kernel regression in the combined joint-task space. This approximate solution can then be used as the seed for the standard IK solver to improve the obtained solutions, and to provide a preference for some specific learned poses. Also it provides potentially acceptable solutions when the inverse kinematics does not find any solution even when seeded with the approximation.

3.2   Joint-Task Kernel Regression

Notation. We will refer to J and T as sets of joint and task space coordinates with notation j and t for specific coordinates. The forward function is defined as $t = f(j)$; our goal will be to obtain the inverse function $j = f^{-1}(t)$. Approximations will be denoted with a tilde: $\tilde{j}$.

Kernel Regression. We use a non-parameteric approach called kernel regression to approximate the inverse kinematics function. It uses a collection of sample pairs x_i, y_i and requires a kernel function that will define a distance between elements in this space X. The regression itself can be seen as a convex combination of the samples whose weights are defined by the kernel function. In this work we use the Gaussian kernel function between two elements x, x' defined as:

$$K(x, x') = \exp -\frac{D(x, x')^2}{2\alpha^2} \qquad (1)$$

where D is a distance function on the elements space, and α is the *bandwidth* of the kernel.

Regression for a value x is calculated as a average of the y value of each sample weighted by the kernel function:

$$\tilde{y} = \frac{1}{Z} \sum_i K(x, x_i) y_i \qquad (2)$$

where i spans over the samples and Z is a normalization term: $Z = \sum_i K(x, x_i)$.

Joint-Task Kernels. For approximating the inverse kinematics function, having a kernel defined only in task space is not sufficient in situations with redundancy. The problem is illustrated in Fig. 1. If our sample set contains two joint solutions for the same task position, they will both have the same weight and the combination of their angle values will lead to a solution which is neither one or the other, but actually suboptimal in task-space. To resolve this kind of situations, we combine both a task kernel and a joint kernel. The provided seed state breaks the symmetry of these situations and make the regression move toward one of the solutions. The full kernel is expressed as a combination of a task kernel and a joint kernel.

$$K((j, t), (j', t')) = K_{task}(t, t') K_{joint}(j, j') \qquad (3)$$

The task kernel distance is the sum of the l^2-norm between the Cartesian positions in space and the joint kernel is the sum of the smallest angles difference for the orientation. Each of the component is weighted by a vector that specifies their relative impact (this is typically used to provide more or less weight to orientation). Each kernel has its own bandwidth α_{task} and α_{joint}, which allows to balance the regression towards the task or the joint target.

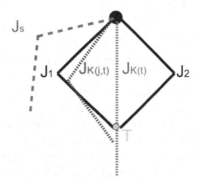

Fig. 1. Illustration of the joint-task kernel regression on a 2 DoF arm. Provided samples are J_1 and J_2, two joint configurations that reach the same task point T. Our IK request is to reach task point T with J_s as the current seed. Kernel regression where the kernel function is only based on the task space gives a suboptimal $J_{K(t)}$ configuration. Using the joint-task kernel regression breaks the symmetry and provides a better solution $J_{K(j,t)}$ that can be improved upon by iterating the process with this solution as the new seed.

Kernel Regression for Inverse Kinematics. Given our set of samples j_i, t_i, the regression is performed for a seed state j and a target task position t as:

$$\tilde{j} = \frac{1}{Z} \sum_i K\big((j,t),(j_i,t_i)\big) j_i \qquad (4)$$

where Z is again a normalization term (actually the angles are not simply summed, instead they are combined so as to always move toward the closest angle difference). Multiple iterations are executed by using the obtained solution as the new seed until convergence.

This kernel regression allows us to obtain an approximation of the required joint values to reach the commanded task coordinate that should be close to known configurations, and, whenever there are multiple solutions, close to the seed state provided. The precision of the approximation will depend on the quality of the samples already available. Ideally these should cover the whole accessible task space with a high enough density to provide good approximations. In practice we will most of the time deal with largely incomplete samples, simply also for the reason that movements will be stereotyped for a large part, and that the "natural" subset of positions is just a fraction of the entire space.

In contrast with other learning approaches such as [3], the proposed model can be used online as there is no explicit learning phase. Samples can be added at any time with negligible computational costs. However, as indicated by simulations, it has lower performance and accuracy (before applying the standard IK on the computed solution) than the method proposed in [3].

4   Experiments

4.1   Experimental Setup

Our experimental setup uses the Baxter industrial robot as can be seen on Fig. 2. This two 7 DoF arms robot is facing the operator and interacting with him over a table. The robot is equipped with a Kinect sensor on its head that acquires information about the operator posture. An audio system with speech synthesis and recognition is also added to the robot to provide speech-based interaction with the user (only speech output is used in this work). The robot also has a display that is used to display a face. The eyes of the robot provide useful feedback to the user, mostly by looking at hands of the user when he is reaching out to the robot (hence acknowledging the user action).

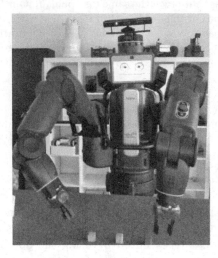

Kinect sensor
for user detection

Visual feedback

Sound feedback

7 DoF arms with
force sensors

Fig. 2. Baxter robot used in the experimental setup with added motorized Kinect sensor and speakers.

The robot interacts with objects on the table by picking or placing them using electric parallel grippers. It also interacts with the user by handing over objects to him or grasping objects handed over by the human. Interaction with the human can be decomposed into 3 elementary actions:

- Moving the hand towards the human: this action is triggered when the robot detects that the human has his hand above the table level and reaching out towards the robot. Dynamic Naïve Bayesian Networks are used to classify these events from raw data provided by the Kinect sensor (these aspects are detailed in [6]). Upon detection, an inverse kinematic request is performed to reach the hand position offset with a few centimeters (to account for the gripper length) with an orientation opposite to the user hand orientation.

- Giving an object to the robot: this action is triggered when the user is reaching out to the robot and has his hand stable and close enough to the robot's gripper. The robot then closes its grippers on the object (and reopens it if it does not detect an object in the gripper). This happens only when the gripper is empty.
- Taking an object from the robot: this action is triggered similarly as the giving condition above but with added kinematic detection. This detection is based on torque variance and amplitude allowing to detect when the user is pulling on the object.

The robot starts in a predefined neutral position, and each time nothing has happened during 5 seconds (no motion from the robot and no physical interaction with the user) the robot automatically goes back to this neutral position.

4.2 User Study

We conducted a user study in an interaction scenario with the robot. A set of users ($N = 11$), with and withouth experience with robots, were asked to interact with the robot in the following object passing scenario:

- The user is asked to pass three objects to the robot. For this the user has to reach out to the robot and get it to grab the object. Each object is then placed by the robot on the table at 3 predefined locations in a random order.
- The robot then tells the user to remember the order of the objects and hands them over one by one in a random order. The user has again to reach out to the robot and take the object from its gripper.
- The robot then asks the user to give back the objects in their original order, meaning that the user has again to reach out to the robot and let it take the objects and place them on the table.

The task is performed a first time with the standard inverse kinematics, KDL Levenberg-Marquardt method with precision $\epsilon = 0.001$ (it was chosen over Newton-Raphson method as it was found more likely to provide good quality solutions to most of the requests). The robot does not perform any motion if no solution is found.

The user is then told to teach the robot postures that they deem comfortable for interaction. The Baxter robot has tactile sensors in its cuffs that detect contact with the user and make the robot switch to zero-gravity mode, meaning that the user can move the arm in any configuration by simply pushing and pulling on the joints. During this phase, sample points are collected when the cuffs are pressed. Every unique position that the user reaches is saved as a sample for regression. The users are told to provide as many positions for the robot as they want until they consider the learning finished.

The same task is then performed again however the users are split in two groups. In the control group, the learned positions are not used, and the same inverse kinematics solver is used. In the test group, kernel regression is used to compute an approximate inverse kinematics solutions, which is then used as a seed to the same inverse kinematics solver. In the test group, if no solution is returned by the solver, then the approximate solution is directly used. The users do not know in which group they belong.

At the end of the run, users are asked to evaluate their interaction with the robot through a questionnaire. Measured values evaluate the contribution of the learning, whether is was used, and if the learning improved on the efficiency, speed and comfort of the interaction.

4.3   Results

Users are evaluated on a 9 points Likert scale and a χ^2 test is performed for each question between the control and the test groups. The null hypothesis is that the learning does not improve on any of the speed, comfort and efficiency metrics. Results are displayed in the following table:

Fig. 3. Object passing motion during experiment with standard IK (top) and learned IK (bottom).

Statement	Control average	Test average	χ^2 p-value
Learning had an impact	5.33 ± 1.33	8.60 ± 0.48	$p < 0.1$
Learning improved comfort	5.66 ± 1.55	7.20 ± 0.72	$p < 0.05$
Learning improved speed	4.92 ± 0.97	6.60 ± 0.64	$p < 0.1$
Learning improved efficiency	5.33 ± 1.22	8.2 ± 0.64	$p < 0.2$
Learning improved (aggregated)	5.30 ± 0.88	7.33 ± 0.98	$p < 0.01$

As can be seen from the results, users generally identify whether the learning was really used or not. Speed and efficiency are only weakly perceived as improved by the learned inverse kinematics. However there is a significant improvement on the comfort of the interaction thanks to the learned IK.

Indeed what can be seen in practice is that the arm postures taught by the user leave more space in the interaction area, making a clearer workspace and also improving on the detection capabilities of the robot because the sensor is less often occluded by the robot arm as can be seen on Fig. 3. Having better detection makes the interaction more fluid, which also explains the perceived speed improvement, even though the actual speed of the performed trajectories is not improved.

5 Conclusion

In this work, we have proposed a simple approach to IK learning based on joint-task kernel regression with a sample set generated by user demonstrations. The learned IK can be used on its own or as providing seed states to use with standard IK solvers. This approach allows to solve the redundancy resolution problem by generating solutions that are approximating the postures learned during demonstration, while keeping the precision of standard IK solvers and their extrapolation abilities.

User studies have shown that the proposed approach significantly improves the interaction with the robot as perceived by the user in object handover scenarios. It was also observed that the responsiveness of the robot is improved with some IK solvers by providing approximate solutions where the standard IK would not find any, for instance when the start and target states are highly different or when the target positions are unreachable.

One of the advantages of the proposed approach is that it can be used online. For instance a user can interrupt the robot while it is performing a motion and guide it to a preferred posture or set of postures. The acquired samples can be immediately integrated into the joint-task kernel regression, thus having an instantaneous effect on the next inverse kinematics request.

Although collisions are not treated in this work, it is interesting to note that the learned IK implicitly integrates approximate static constraints of the configuration space as the set of samples cannot contain invalid configurations.

Future work will investigate the capabilities of online adjustments to the kernel function to integrate custom constraints. For instance, adding an extra set of weights on the samples could allow to incorporate dynamic collision avoidance, safety constraints or user preferences. In the case of the latter it would be interesting from an HRI perspective to evaluate how specific to each user the learned IK is, and if users have a real preference towards their own teaching compared to a an average or expert-generated sample set.

References

1. Akgun, B., Cakmak, M., Yoo, J.W., Thomaz, A.L.: Trajectories and keyframes for kinesthetic teaching: a human-robot interaction perspective. In: Proceedings of the Seventh Annual ACM/IEEE International Conference on Human-Robot Interaction, HRI 2012, pp. 391–398. ACM, New York (2012)
2. Artemiadis, P.K., Katsiaris, P.T., Kyriakopoulos, K.J.: A biomimetic approach to inverse kinematics for a redundant robot arm. Autonomous Robots **29**(3–4), 293–308 (2010)

3. Bocsi, B., Nguyen-Tuong, D., Csato, L., Scholkopf, B., Peters, J.: Learning inverse kinematics with structured prediction. In: 2011 IEEE/RSJ International Conference on Intelligent Robots and Systems, pp. 698–703. IEEE, September 2011
4. Calinon, S., Billard, A.: Active teaching in robot programming by demonstration. In: The 16th IEEE International Symposium on Robot and Human Interactive Communication, RO-MAN 2007, pp. 702–707, August 2007
5. Diankov, R.: Automated construction of robotic manipulation programs. PhD thesis, Citeseer (2010)
6. Morvan, J.: Understanding and communicating intentions in human-robot interaction. M.Sc. Thesis, KTH Royal Institute of Technology (2015)
7. Nakanishi, J., Cory, R., Mistry, M., Peters, J., Schaal, S.: Comparative experiments on task space control with redundancy resolution. In: 2005 IEEE/RSJ International Conference on Intelligent Robots and Systems, pp. 3901–3908 (2005)
8. Rethink Robotics. Baxter Research Robot. http://www.rethinkrobotics.com/baxter-research-robot/
9. Smits, R.: KDL: Kinematics and Dynamics Library. http://www.orocos.org/kdl
10. Strabala, K.W., Lee, M.K., Dragan, A.D., Forlizzi, J.L., Srinivasa, S., Cakmak, M., Micelli, V.: Towards Seamless Human-Robot Handovers. Journal of Human-Robot Interaction $2(1)$, 112–132 (2013)
11. Sucan, I.A., Chitta, S.: Moveit! http://moveit.ros.org
12. Wrede, S., Emmerich, C., Grünberg, R., Nordmann, A., Swadzba, A., Steil, J.: A User Study on Kinesthetic Teaching of Redundant Robots in Task and Configuration Space. Journal of Human-Robot Interaction $2(1)$, 56–81 (2013)

Team-Building Activities for Heterogeneous Groups of Humans and Robots

Zachary Carlson, Timothy Sweet, Jared Rhizor, Jamie Poston,
Houston Lucas, and David Feil-Seifer[✉]

Robotics Research Lab, University of Nevada, Reno 89557, USA
{zack,jposton,houstonlucas}@nevada.unr.edu, timothy.l.sweet@gmail.com,
me@jaredrhizor.com, dave@cse.unr.edu

Abstract. As robots become more integrated into society and the work-force, people will be required to work cooperatively with not just other people, but robots as well. People engage in team-building activities to improve cooperation and promote positive group identity. This paper explores the effect that a team-building activity had on humans working cooperatively with human and robot teammates with the goal of better understanding how to improve cooperation between a human and a robotic agent. We conducted a 2x2 study with the presence or absence of a team-building activity and the possibility or impossibility of the cooperative task. 40 participants conducted a group search task with a robot and another human partner. Half of the participants engaged in a short team-building exercise. Surveys were used to capture participants' perceptions before and after the session. Success and failure of the task was also measured to identify any changes related to the outcome of the team-building task. It was found that humans' perceptions of robots improve after performing team-building activities. We also found that this effect was comparable to the change of perception when the group succeeded on the task.

Keywords: Team building · Collaboration · Human-robot cooperation

1 Introduction

Human-robot cooperation in groups is an important facet of Human-Robot Interaction (HRI). When groups work together it may be important to promote a collaborative atmosphere between all group members, including mixed groups of humans and robots. Team-building is regularly used to promote a collaborative atmosphere for human-human interaction. Including robots in team-building exercises may have a similar effect on human and robot group interactions. This study investigates team-building exercises on a group consisting of two humans and a robot, and how this introduction changes the perceptions of each member of the group (human and robot).

Humans often engage in team-building activities to improve cooperation between team members and promote positive group identity. Many

© Springer International Publishing Switzerland 2015
A. Tapus et al. (Eds.): ICSR 2015, LNAI 9388, pp. 113–123, 2015.
DOI: 10.1007/978-3-319-25554-5_12

team-building activities provide "a sense of unity and cohesiveness," which can improve the function of teams [16]. While these activities are well-known in the human-human setting, it leads researchers to the question, *can the human-robot relation also improve with a sense of unity through similar team-building activities?*

This is crucial for long-term HRI in the workplace since, depending on the environment, success of teams can be limited by the trust and willingness to collaborate within the team [12]. A coworker who is unwilling to collaborate and trust the abilities of their fellow teammate may refuse to include that teammate in many job related tasks due to territorial behaviors [2]. As one would think, this exclusion of the team member might eventually lead to a lack of productivity.

What might happen if a robot were introduced into a workplace to assist with tasks within a heterogeneous team of both robots and humans? If the team members believe that the robot is not capable, they are most likely not going to rely on that robot. It is easy to see that the success of a heterogeneous team of robots and humans can be limited by the unwillingness of humans to collaborate with a robot teammate. In the human-human collaborative setting as these problems were addressed by using team-building activities [16]. This research study looks to understand how the use of team-building activities can facilitate human-robot group interaction.

In this paper, the results of a study involving heterogeneous groups of humans and robots are analyzed to identify how humans perceive robots before and after performing a team-building activity. First, some of the most recent research related to robot teams will be explored. Next, a controlled experiment design is presented. The results of the study will be presented, then discussed and finally compared with the original hypothesis in conclusion. These results will contribute to our overall understanding of how team build activities can affect heterogeneous teams of robots and humans.

2 Related Work

In order to understand how to promote human-robot cooperation, there first needs to be an exploration of how team-building works for human-human interaction. As Reeves and Nass found in their research, computers (and by extension robots) are often treated by people as social actors [15]. As such, studies of human-human teaming may provide valuable insight into human-robot teaming. There is also a wealth of research into human-robot teaming, showing how such teams can be disrupted or promoted. This section will provide a survey of related work in human-robot interaction, as well as how to measure social cohesion with another agent (human, computer, or robot).

Dyer and Dyer have shown that through team-building exercises, specifically the first team exercises issued at the initial team meeting, can build trust and a mutual understanding that helps teams [3]. It has also been shown that team-building activities serve as a bridge between meeting people and can help build a sense of "trust and connectedness" [16]. Miller focused on minimizing failures within team-building exercises, and found that participants could not fully gain the desired results of team-building, if care was not given to carefully controlling the process [9]. Additionally, Miller outlined activities that are appropriate for team-building. Each team-building activity is presented within guidelines of how to run the team-building activity so the participants benefit the most.

Recent research has focused on improving trust in human-robot team relationships [8,18] as well as using robots to learn and observe from their human teammates [13]. Most of these approaches look at improving the team dynamic directly through the robot or human teammates. In this paper we are looking to improve the robot-team relationship through the use of an independent factor; in this particular case a team-building activity.

The area of heterogeneous human-robot teams often involves organization and planning for large teams. Nagavalli looked specifically at how humans interact with large swarms of robots [11] and Ponda orchestrated heterogeneous collaborative tasks [14]. While creating group plans to achieve tasks, we believe that this is a separate problem compared to unifying a heterogeneous team before a task. With the Wizard of Oz approach, we accomplish human-like planning.

Research involving interaction between heterogeneous groups of robots have focused on the usage and design of a specific robot and the activities or circumstances that change the behaviors and perceptions of both humans and robots. Prior work has examined how robots fit in the workplace [10]. Active research has explored a variety of aspects of the human-robot team setting, ranging from how a robot should navigate [4,7] to dialogue structuring [6]. This paper is distinguished from prior work by exploring easy social interventions which may facilitate human-robot cohesion. We look to study this idea by exploring the beneficial effects of using a team-building activity as an "ice breaker" before a human-robot team performs a task.

This approach was inspired in part by [17], which used an industrial robot as a platform to evaluate how fluidity, comfortableness, and noticablility changed with several parameters in fetch-and-deliver tasks involving a robot and a human. The robot and human work collaboratively to complete a simple task. The robot has very limited in communication with the human counterpart, which may very well be a believable real-world constraint. [5] uses the concept of presenting a robot as a partner instead of a tool. Although we do not extend this concept to collaborative control like Fong, we actively choose to introduce the robot as a third participant in the study.

Bartneck, et al., developed a survey instrument to evaluate robot agents. This instrument uses five sub-scales, anthropomorphism, animacy, likeability, perceived intelligence, and perceived safety to evaluate perceptions of human-robot interaction [1]. These metrics are used in the Godspeed Questionnaire, which evaluated participants' experiences with the robot and with the other participant. The Godspeed Questionnaire uses a differential scale made of several five point Likert questions for each measure.

Fig. 1. The Pioneer 3DX Robot that was used in across all conditions of the experiment and the private room where the study sessions took place.

3 Experiment Design

This section details the overall design of the experiment, the procedure and materials used to recreate the study, as well as a thorough explanation of the tasks (including the team-building activity) and how it was used in the study.

The experiment is designed to represent a simple task in which teamwork between humans and a robot would increase the probability of the tasks success compared to the humans working alone. This is meant to be representative of many real-world activities requiring human and robot teamwork. In this task, the team consisting of two humans and one robot are instructed to locate a particular object in a large room (Figure 1). The participants are shown that the robot is capable of finding the marker before the task starts. The team was give one minute to find a marker hidden within a cluttered room.

We employed a 2x2 between-subjects factorial design. There were two factors: team-building vs. not team-building, and task possible vs. task impossible. For the first factor, participants would either engage in the "Two Truths and One Lie" team-building activity with their team or not prior to the study activity. The second factor, had two levels: possible success or guaranteed failure. The second factor was varied by making the task possible by hiding the object somewhere in the room, or making the task impossible by telling the participants to find the object, but not actually putting the object into the room. All participants that participated in the possible success successfully completed the task.

40 college-aged participants ranging in a variety of majors volunteered to participate in a cooperative task where they were paired with another participant and a robot partner that formed a team of three. The participants were then randomly assigned to one of the four groups.

We hypothesized the following:

H1: Participants will perceive the robot to be more human-like after participating in a team-building exercise.

H2: Participants will perceive the robot to be more intelligent when the group succeeds at the primary task.

These hypotheses submit that similarity of a robot to a person would be judged by its social behavior. The capabilities of that robot would be judged by its success at stated goals.

3.1 Procedure

Participants were recruited from a university library at random and asked if they would participate in a study involving human-robot collaboration. Two participants at a time were brought into the study room and consented, then introduced to each other and the robot. The participants were only asked to state their name, and the facilitator introduces the robot as "a Pioneer 3DX". Participants were asked if they have met each other and were dismissed or re-paired until the partners did not know each other to eliminate the possible confounding variable of the familiarity between participants. Study personnel told the participants that the robot was capable of finding the blue marker by placing it in front of the robot while the robot operator played a sound clip stating "I found it" from the robots on-board speaker. The camera was not actually used for detection, the remote operator used the Wizard of Oz method to create this effect. This was chosen to reduce the probability of technical difficulties in demonstrating the robots competency. However, this is a task that can reasonably be completed autonomously without error.

Participants were then separated and asked to fill out the Godspeed Questionnaire [1] with respect to their human teammate and again for their robotic teammate. After completion, they were brought back together in the main study room. Half of the groups partook in the team-building activity described in 3.3. Then, all groups partook in the primary task described in 3.4, and in half of those cases the task was possible to complete and in the other half it was impossible with the marker removed from the room. The four conditions were:

- NS: No Team-Building Activity and Possible Marker (Team Succeeded)
- NF: No Team-Building Activity and Impossible Marker (Team Failed)
- TBS: Team-Building Activity and Possible Marker (Team Succeeded)
- TBF: Team-Building Activity and Impossible Marker (Team Failed)

After finishing the primary task, participants were separated again and asked to fill out the same Godspeed questionnaire for their human teammate and again for their robotic teammate. After finishing the questionnaire participants were debriefed and dismissed. The facilitator used a script throughout the experiment, but was allowed to respond to participant questions during the initial consent of the study and during the debrief period. This was to ensure that participants fully comprehended all consent and debrief forms provided to them.

3.2 Materials and Setup

The Adept MobileRobotics Pioneer 3DX shown in 1 was used as the robotic base for the experiment. It was equipped with a SICK Laser Rangefinder for navigation, an Xtion Pro for detecting the object, and additional computational components. A laptop running the Robot Operating System was mounted on the Pioneer. The robot's on-board Raspberry Pi is controlled from a separate computer in a different room by a human (participants are unaware of this Wizard of Oz usage: they were told the robot was navigating autonomously).

We used a library study room, set up to be sufficiently cluttered so that finding an object took some time. The object used was a blue whiteboard marker, chosen because it was easily concealable but also easily recognizable to both the humans and the robots camera due to its bright color. The object was hidden such that the robot would be able to see it. A Sony camera, handycam model HDR-CX220 with a resolution of 1080p, on a tripod in the corner of the room was used to record audio and video for the duration of the study.

3.3   Team-Building Activity

Half the groups participated in a "Two Truths and One Lie" icebreaker, prior to completing the primary task. Each participant told the rest of the group two truths and one lie about themselves, then their human teammate and the robot would guess which statement was a lie. Unknown to participants, the remote human operator actually just played canned sound clips from the robots on-board speaker which stated "I believe your second/third statement was a lie" for the first and second human teammate, respectively, regardless of their statements. Finally, the remote operator played canned sound clips from the robot which stated its two truths and a lie: "I was manufactured in 2003" (truth); "I have traveled outdoors" (truth), "I can travel up to two meters per second" (lie). Participants then guess which statement was a lie, and the remote operator played a sound clip stating "I can only travel one meter per second".

The facilitator within the room gestures at each participant when it was their turn to speak (both the humans and the robot). The remote operator could see the facilitator gesture on the video feed, which was used as the cue to advance.[1]

3.4   Primary Task

Participants were told that there was a blue whiteboard marker hidden in the room somewhere. They were given sixty seconds to find it, with the help of their human partner and the robot. Participants were told the robot would announce "I found it" if it found the marker, however the robot was actually driven by the remote operator and would not announce if it "saw" the marker. For half of the groups the marker was actually hidden in the room, but for the rest of the groups the marker was not in the room (thus the participants would run out of time before finding the marker). In the case that it was in the room, the marker was hidden under one of the legs of the table. Our intention was that participants would not be able to see it from their starting positions.

---

[1] Note that we avoided anthropomorphizing the robot because this study did not focus on human-like robots. The voice used was a very "machine-like" voice.

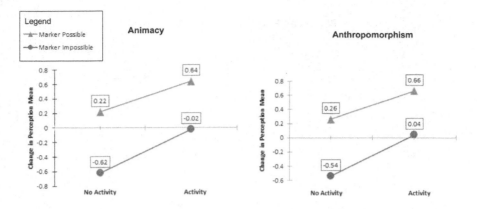

Fig. 2. Significance was found between groups when analyzing the average mean representing the difference in the participants' perception of how animate [left] ($F[3, 36] = 2.973, p < .05$) and anthropomorphic [right] ($F[3, 36] = 3.197, p < .05$) the robot was between the pre and post surveys for all conditions.

4 Results

The details of the experiment results and analysis are presented in this section. The survey data were analyzed to support or refute the experimental hypotheses presented in Section 3. For each sub-scale of the Godspeed Questionnaire [1], the differences of the before and after Questionnaires were analyzed using a multivariate two-way analysis of variance (ANOVA) between subjects across all conditions. Additionally, the post surveys were analyzed using a multivariate two-way analysis of variance. There were 10 participants per cell, with a total of 40 participants that were analyzed and assigned to each condition.

Ten categories were measured in both multivariate analyses, one for each sub-scale of the Godspeed questionnaire for the human and the robot partner. When analyzing the differences between the pre and post surveys, there was a significant difference in how anthropomorphic participants perceived the robot ($F[3, 36] = 3.197, p < .05$) and how animate they perceived it to be ($F[3, 36] = 2.973, p < .05$). There was a significant interaction for marker possible and team activity of the participants' perception of the perceived intelligence of the robot was also significant ($F[3, 36] = 2.970, p < .05$).

Figure 2 shows the means of the difference between pre- and post- surveys within each condition, representing the change in how anthropomorphic and animate the robot was perceived to be. When the team successfully found the marker there was a positive increase in their perceptions. When participants could not find a marker and they participated in the team-building activity, there was close to no change in their perceptions from before the activity to after. When they did not participate in the team-building activity, failing the task had a negative impact on their perceptions.

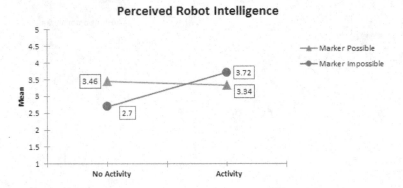

Fig. 3. Significance was found between groups when analyzing the average mean representing participants' final perception of how intelligent the robot was at the end of the study session $(F[3, 36] = 2.970, p < .05)$.

Figure 3 shows that when participants participated in the team-building activity with the robot, no matter if they failed or succeeded at the task, they perceived the intelligence of the robot to be the same. When there was no team-building activity, there was a clear difference in the intelligence they believed the robot to have depending on whether or not they succeeded at the task. They perceived the robot to have as much intelligence when they participated in the team-building activity as when they succeeded in the task.

5 Discussion

The animacy and anthropomorphism results support hypothesis #1, *"Participants will perceive the robot to be more human-like after participating in a team-building activity."* The intelligence sub-scale results partially support hypothesis #2, *"Participants will perceive the robot to be more intelligent when the group succeeds at the primary task."* We instead found that participants would find the robot more intelligent when the group succeeded *or* when the team-building activity occurred. The power of the team-building task, then is that it can cause participants to forgive the robot's failure to meet its goals.

Further exploration of the team-building activity was analyzed, using a one-way ANOVA to compare the means of all the participants that participated in the team-building activity and those that did not, no matter if they succeeded or failed. This test showed no significance.

This shows that the team-building activity alone did not significantly alter the participant's perceptions. The team-building activity needed to be coupled with success, which is supported in Figure 2 where all cases show that when the team could not complete the task (Marker Impossible), the participants' perceptions of their team were negatively impacted. As for the groups that successfully completed the primary task (Marker Possible), they always had a positive increase in their perceptions of the team members.

In the case of the successful teams and hypotheses #2, participants may blame the robot when they are unable to complete the task. This was shown in our results were there was no significant differences in any of the human partner surveys and a negative reflection on the perception of the robot's intelligence. Interestingly, we have found that the team-building activity in the impossible task seemed to offset the perceived lack in intelligence for failing the task. For this particular study, there appeared to be a ceiling of how intelligent the robot was perceived roughly around a value of 3.5 on a scale of 1-5. When the team-building activity was combined with success, the perceived intelligence of the robot was roughly the same as the other two conditions shown in Figure 3, where they failed the task and participated in the team-building activity, and when they succeeded but did not participate in team-building.

6   Conclusion and Future Work

This paper makes a definitive case for the utilization of simple social interaction exercises to facilitate human-robot team cohesion. The results of this paper support the notion that facilitating a team atmosphere can negate the deleterious effects of group failure at a task. A useful extension of this research would be to study the duration of team-building effects on heterogeneous groups of robots and humans. This study incorporated very short interactions with a robot. However, it is unclear whether or not these increased perceptions would be maintained throughout longer use of the robot or more involved team-building.

In this study, the participants' beliefs about themselves are unknown. Participants could be asked to self-rate. The difference between self-ratings and their rating of partners (especially their robotic partner's ratings) could provide some insight into how the human participant relates to the robot, instead of how the human participant perceives the other participant relating to the robot.

"Two Truths and One Lie" was an effective team-building activity. However, comparing multiple team-building activities could reveal differences in effectiveness in different activities with respect to heterogeneous groups. Furthermore, Rivas framed team-building exercises as a task that a "leader" uses to engage followers in upcoming activities [16]. Robots leading team-building activities to direct or manage humans in other tasks is an unexplored area of HRI.

Human robot interactions will undoubtedly increase as the field of robotics advances. The use of heterogeneous groups of humans and robots will most likely increase over time as well. Since team-building is viewed as an acceptable social activity to create cohesive groups of humans, team-building with heterogeneous groups of robots and humans is a natural step forward. The results and further discussion of the study shows that team-building activities, as much as joint success, results in a significant positive increase in perceptions of the animacy and anthropomorphism of robotic team members. This supports hypothesis #1, "Participants will perceive the robot to be more human-like when they participate in a team-building task." Finally, hypothesis #2 was supported, "Participants will perceive the robot to be more intelligent when the group succeeds at the primary task" and it was also found that participating in a team building activity can increase the perception of the intelligence of the robot.

Acknowledgments. The material is based upon work supported by the National Aeronautics and Space Administration under Grant No. NNX10AN23H issued through the Nevada Space Grant. This material is also supported by the University of Nevada, Reno, Provost's Office Instructional Enhancement Grant and the Office of Naval Research (ONR) #N00014-14-1-0776. We appreciate the space lent to us by the DeLa-Mare library and Knowledge Center at the University of Nevada, Reno.

References

1. Bartneck, C., Kulić, D., Croft, E., Zoghbi, S.: Measurement instruments for the anthropomorphism, animacy, likeability, perceived intelligence, and perceived safety of robots. International Journal of Social Robotics **1**(1), 71–81 (2009)
2. Brown, G., Crossley, C., Robinson, S.L.: Psychological ownership, territorial behavior, and being perceived as a team contributor: The critical role of trust in the work environment. Personnel Psychology **67**(2), 463–485 (2014)
3. Dyer, W.G., Dyer, J., Dyer, J.H.: Team building: Proven strategies for improving team performance. John Wiley & Sons (2010)
4. Feil-Seifer, D.J., Matarić, M.J.: Distance-based computational models for facilitating robot interaction with children. Journal of Human-Robot Interaction **1**(1), 55–77 (2012). doi:10.5898/JHRI.1.1.Feil-Seifer
5. Fong, T., Thorpe, C., Baur, C.: Collaboration, dialogue, and human-robot interaction. In: Jarvis, R.A., Zelinsky, A. (eds.) Robotics Research. STAR, vol. 6, pp. 255–266. Springer, Heidelberg (2003)
6. Jung, M.F., Martelaro, N., Hinds, P.J.: Using robots to moderate team conflict: the case of repairing violations. In: Proceedings of the Tenth Annual ACM/IEEE International Conference on Human-Robot Interaction, pp. 229–236. ACM (2015)
7. Kanda, T., Shiomi, M., Miyashita, Z., Ishiguro, H., Hagita, N.: An affective guide robot in a shopping mall. In: Proceedings of the 4th ACM/IEEE International Conference on Human Robot Interaction, pp. 173–180. ACM (2009)
8. McCallum, L., McOwan, P.W.: Face the music and glance: How nonverbal behaviour aids human robot relationships based in music. In: Proceedings of the International Conference on Human-Robot Interaction, pp. 237–244 (2015)
9. Miller, B.C.: Quick activities to improve your team: How to run a successful team-building activity. Journal for Quality and Participation **30**(3), (2007)
10. Mutlu, B., Forlizzi, J.: Robots in organizations: the role of workflow, social, and environmental factors in human-robot interaction. In: Proceedings of the International Conference on Human Robot Interaction (HRI), pp. 287–294. ACM, New York/Amsterdam, March 2008
11. Nagavalli, S., Chien, S.Y., Lewis, M., Chakraborty, N., Sycara, K.: Bounds of neglect benevolence in input timing for human interaction with robotic swarms. In: Proceedings of the Tenth Annual ACM/IEEE International Conference on Human-Robot Interaction, pp. 197–204. ACM (2015)
12. Newell, S., David, G., Chand, D.: An analysis of trust among globally distributed work teams in an organizational setting. Knowledge and Process Management **14**(3), 158–168 (2007)

13. Nikolaidis, S., Ramakrishnan, R., Gu, K., Shah, J.: Efficient model learning from joint-action demonstrations for human-robot collaborative tasks. In: Proceedings of the International Conference on Human-Robot Interaction, pp. 189–196 (2015)
14. Ponda, S., Choi, H.L., How, J.P.: Predictive planning for heterogeneous human-robot teams. AIAA Infotech@ Aerospace (2010)
15. Reeves, B., Nass, C.: The media equation: how people treat computers, television, and new media like real people and places. Cambridge University Press, New York (1996)
16. Rivas, O., Jones, I.S.: Leadership: building a team using structured activities
17. Unhelkar, V.V., Siu, H.C., Shah, J.A.: Comparative performance of human and mobile robotic assistants in collaborative fetch-and-deliver tasks. In: Proceedings of the International Conference on Human-Robot Interaction, pp. 82–89 (2014)
18. Xu, A., Dudek, G.: Optimo: Online probabilistic trust inference model for asymmetric human-robot collaborations. In: ACM/IEEE Int. Conf. on HRI (2015)

Automatic Joint Attention Detection During Interaction with a Humanoid Robot

Dario Cazzato[2], Pier Luigi Mazzeo[1], Paolo Spagnolo[1]([⊠]),
and Cosimo Distante[1]

[1] National Research Council of Italy, Lecce, Italy
paolo.spagnolo@cnr.it
[2] Faculty of Engineering, University of Salento, Lecce, Italy

Abstract. Joint attention is an early-developing social-communicative skill in which two people (usually a young child and an adult) share attention with regards to an interesting object or event, by means of gestures and gaze, and its presence is a key element in evaluating the therapy in the case of autism spectrum disorders. In this work, a novel automatic system able to detect joint attention by using completely non-intrusive depth camera installed on the room ceiling is presented. In particular, in a scenario where a humanoid-robot, a therapist (or a parent) and a child are interacting, the system can detect the social interaction between them. Specifically, a depth camera mounted on the top of a room is employed to detect, first of all, the arising event to be monitored (performed by an humanoid robot) and, subsequently, to detect the eventual joint attention mechanism analyzing the orientation of the head. The system operates in real-time, providing to the therapist a completely non-intrusive instrument to help him to evaluate the quality and the precise modalities of this predominant feature during the therapy session.

1 Introduction

Joint attention is an early-developing social-communicative skill in which two people (usually a young child and an adult) share attention with regards to an interesting object or event, by means of gestures and/or gaze. In particular, the work in [16] firstly proposed a preliminary investigation about the extent of the infant's ability to follow changes in adult gaze direction during the first years of life. In particular, on each test trial performed on 34 children between 2 and 14 years old, the enfant first made eye-to-eye contact and then silently turned his or her head, looking at a small concealed signal light for 7 seconds, while the adult head was then turned back to interact with the infant. In the experiment, it was found that the proportion of infants judged as having produced a positive response on one or both trials increased steadily with age. Since its fundamental importance in the fields of cognitive and developmental psychology, a long list of subsequent studies have been proposed in the literature over the years, and it is still an active research topic [12]. For the case of autism, several works

© Springer International Publishing Switzerland 2015
A. Tapus et al. (Eds.): ICSR 2015, LNAI 9388, pp. 124–134, 2015.
DOI: 10.1007/978-3-319-25554-5_13

investigated the meaning and the modalities of joint attention lacks, like in [20], [21] and [7]. In fact, impaired development of joint attention is a predominant feature in children with autism spectrum disorder (ASD), and a set of strategy are used to teach and support joint attention [9]. Although children with ASD can show attention to objects or toys of interest, they have difficulties in sharing attention or interests with the therapist of a relative. For this purpose, the Early Start Denver Model [15] underlines this capacity as a key element of social cognition, working on an improvement of this concerned interaction lack. In [5] a requirement to detect joint attention is that two individuals are attending to the same object, based on one individual using the attention cues of the second individual. Shared Attention is a combination of mutual attention (the attention of two individuals is directed to one another) and joint attention (two individuals are looking at the same object), where at the same time the two individuals have knowledge of the directions of the other individual's attention. In other words, shared attention represents a higher state of the dyadic relationship whereby both individuals are attending the same object, as with joint attention, but both are aware of each other's attentional state. Although they are slightly different, in the literature shared and joint attention are considered as synonyms. Beyond subtle differences, what is important to notice is that for us joint attention means that the child is attending to the same object using the attention cues of the adult, i.e. knowing together that they are attending to the same thing [3].

On the other hand, the usage of Socially Assistive Robotics have spread among recent years, providing a new and useful instrument to elicit interest on the autistic child during the therapy. For a review about the clinical usage of robots in autism research refer to [4] and [17]. In particular, socially assistive robots have been employed also to elicit behavior in children with ASD [6]. Many works in the literature explore this exciting feature, but most of them are based on a simulated robot on a screen, eventually evaluating the participant's movement [10], his visual perspective [22] or his reaction time [11]. In [18] gaze track in a human-robot interaction setting (on a monitor) is analyzed, but the attention is measured by means of eye tracker, that are very expensive, and they needs a user calibration that becomes very difficult in the case of children with ASD. A humanoid robot has been employed to elicit joint attention in [14], but all the acquired data has been manually analyzed a posteriori. In [8] a specific hardware is employed to detect the joint attention in a 1-by-1 human-robot interaction scheme. In particular, the method employs an omnidirectional vision sensor and 16 ultrasonic distance sensors around a movable base in order to localize the person's location and to turn consequently the robot's head. This feature is merged with a speech generation system that can elicit social behaviors. Finally, the work in [1] compares the visual exploration during joint attention elicitation in typical development children and children with ASD by means of a Kinect sensor in order to capture social engagement cues. The system evaluates a possible joint attention event, but the sensor is installed in front of the child (thus visible) and even in this work no sharing with another adult has been taken into account.

This paper introduces a two-level innovation with regards to the state of the art: first of all, it presents a computer vision based system based on non-intrusive and invisible to the patient hardware that can operate in real-time. Moreover, it considers the case of a triadic interaction, involving the child, an adult and a humanoid robot (i.e. the Aldebaran NAO H25) in order to detect the joint attention when the robot performs an ad-hoc movement to elicit joint attention mechanism. The paper is organized as follows. Section 2 introduces the proposed method, while Section 3 shows the experimental setup and the achieved results. Finally, Section 4 concludes the paper.

2   Proposed Method

In the following paragraphs the hardware architecture will be illustrated, as well as the computer vision algorithms implemented for the detection of people, their heads, and correspondent axis for mutual interaction evaluation.

2.1   System Overview

The proposed approach uses images acquired by a Microsoft Kinect device. We arranged a therapy room with a calibrated acquisition device placed on the ceiling, in a non-intrusive position, with the goal of not disturb the children. Our idea is to develop simple and effective algorithms, in order to implement them on low-end hardware. The algorithms run in real time, require no training data (except some seconds of free acquisition for the background modeling, as described below) and can exploit cheap, off-the-shelf sensors. In fig. 1 the flow-chart of the whole system is shown. The Kinect device acquires synchronized depth information about the scene in parallel with the RGB video stream. Depth images are segmented in order to detect moving blobs in the scene (calibration data are used to improve segmentation by considering constraints of the system, i.e. the distances between the camera and expected target, as well as the floor and the other planes). After this, Canny operator is applied on the depth images (only on foreground areas) providing an edge map. This map is the input of the head detection algorithm, which processes the edge map with the goal of detect elliptical structures. The outputs of this step are the major and minor axes of the detected ellipses. Finally, the behaviors of children are classified according to mutual positions of the major axes of the educator and the patient.

2.2   Segmentation

The first step of the algorithm is the segmentation of foreground objects. For this purpose, we have chosen to work directly on the depth map: this way, traditional weakness of foreground segmentation algorithms (for example the presence of shadows, reflections on specular surfaces, and so on) are limited or totally avoided. We have implemented the algorithm proposed in [23], which is robust to shadows, reflections, small movements in the background. Even if these

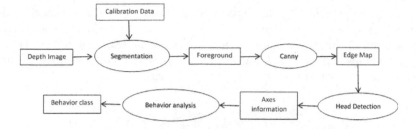

Fig. 1. A schematic diagram of the processing steps

aspects are filtered by using depth images, some artifacts could be present anyhow. To further improve segmentation of desired objects (people and robot), we run the algorithm before patients and educators enter the room. The algorithm is a variation of classic Gaussian Mixture Model approach [19]: each point is represented by a number of Gaussians (with mean and variance), and a variation is considered as foreground if it differs from each gaussian more then the correspondent variance. In fig. 2 we can see a depth image, and the corresponding segmented one.

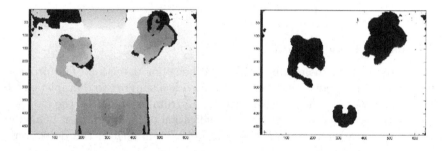

Fig. 2. An example of depth image, and the output of the segmentation procedure.

2.3   Head Detection

After the foreground segmentation, we need to detect the heads of subjects present in the scene. To do this, we use the detected foreground as a binary mask on the depth image. This way, we work on a depth image composed by only foreground objects. For the head detection we take advantage of the constraints of the applicative context: autistic children usually do not interact with other people, so it is realistic to consider that each foreground detected region refers to only one person. Firstly, a Canny operator [2] is applied to obtain an edge map. This map is then processed in order to detect ellipses, that correspond to the heads. The algorithm proposed in [13] has been implemented, with some variations: specifically, the detection has been focused on a specific class of ellipses. This because the size and the geometry of the head is known, as well as the position of the camera and the focal lens. So, we can assume that the size of expected ellipses (heads) is known, and can vary in a certain range. In fig. 3 we can see the output of the head detection algorithm.

Fig. 3. The output of the head detection algorithm.

2.4   Behavior Understanding

The final step of the whole approach is the behavior detection. The proposed system, as remarked above, has the goal to provide a support to therapists in the analysis of behavior of children with autism spectrum disorder. So, the output of the automatic algorithm needs to be a classification of most common behaviors. According to suggestions provided by therapists, our goal is the detection of three main behaviors:

1. **Joint Attention.** Both adult and child look at the robot **(JA)**;
2. **Child attention.** Educator looks at the child while the child looks at the robot **(A2C2R)**;
3. **Child Adult attention.** Both Adult and child look at each other **(A2C2A)**;

The automatic detection of such behaviors has been done by evaluating the mutual position of the major axes of the detected ellipses (heads), as well as by considering the position (known) of the robot. So, the starting point of this final step is the extraction of the major axes of the ellipses, and the evaluation of their directions. In fig. 4 we can see an example of this approach: the major axes of each head/ellipse are plotted, and the behaviour is evaluated by considering the possible situations (the angle between them, the position of robot, etc).

In fig. 5 we can see some example of images of the desired classes of behaviors, with the correspondent synthetic scheme for the geometric evaluation. Formally, we have assumed these rules for the classification:

Fig. 4. The major axes and their use in the evaluation of behavior.

1. **JA.** The intersection of the major axes of the subjects (child and adult) corresponds to the position of the robot;
2. **A2C2R.** The intersection of the major axes corresponds to the head of the child AND the major axis of the child corresponds to the position of the robot (i.e. the gaze of the child is on the robot);
3. **A2C2A.** The major axes of both child and adult are about congruent (i.e. they are congruent, OR they intersect with a very small angle, less then 10 degrees).

3   Experiments

In the next subsections, firstly the setup and the main characteristics of the test sequences will be illustrated; then, the results obtained by using the proposed algorithms will be presented.

3.1   Setup and Dataset

The acquisition setup is composed by a calibrated Kinect Camera installed on the ceiling (about 3 meters from the floor). We created a simulated operative scene containing a work table (1.5 meters from the floor), NAO humanoid robot located on the table, an autistic child and an adult. We acquired several sequences in which we have simulated the three different kinds of attention interactions among the adult, the child and the humanoid robot described in section 2.4. The proposed setup has to be considered stationary, because our goal, for the future, is to arrange a therapy room with this camouflaged hardware. We use the term 'simulation' because the acquisition sessions, even if performed following instructions of therapists, have been performed in absence of a therapist. Moreover, the acquisitions have been made by using children without ASD. In the future, the

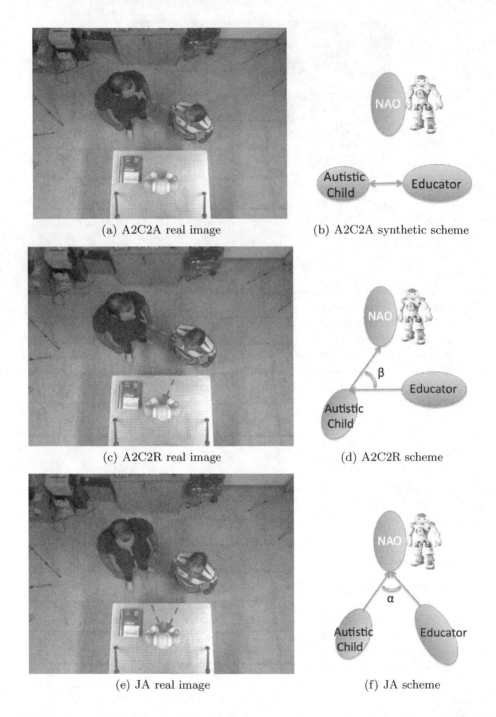

(a) A2C2A real image

(b) A2C2A synthetic scheme

(c) A2C2R real image

(d) A2C2R scheme

(e) JA real image

(f) JA scheme

Fig. 5. Some images acquired during experimental phase. Red dotted lines highlight the gaze orientation.

acquisition sessions will be made in a clinical context, in presence of a therapist, and actions will be performed by real ASD patients. In this step of our work, our goal is the test of computer vision algorithms, so the presence of real ASD patients or non ASD patients is irrelevant.

Another observation is necessary: due to severe laws about acquisition and publications of images with minors, we present images in which actors are adults (we hope in the future to obtain the necessary permissions to use minor images).

It is important to note that we are testing the computer vision algorithms, instead of clinical ones: so, we are interested in results in terms of correct detection of heads, angles between them, and so on, while an analysis of medical implications of these results will be examined in depth in the future.

The presence of the robot during acquisitions is strategic for the correct evaluation of child attention: during the different simulations, the robot performed generic behaviours (it stood up, or it sat down), with the goal of attract the gaze of children; this way, we can test if the children react to external impulses (where external has to be intended as 'not from the therapist').

3.2   Results

We implemented the proposed method and tested it on three video sequences of 1600 frames each one. We have manually labelled each frame in order to create a ground truth of the three different behavior classes described in section 2.4. We have introduced a new class **Non Classified Behavior (NCB)** for labelling of all frames that do not belong to the three main behavior classes. This Ground Truth creation process generated four labelled classes and the population of each is given by table 1.

Table 1. Behavior classes population

	JA	A2C2R	A2C2A	NCB
#	1480	957	1043	1320

Table 2 shows the obtained results; it contains a confusion matrix with the percentage of correct detection in the diagonal (bolded). As it can be highlighted the proposed algorithm gives very good preliminary results; the worst detection percentage (75%) is obtained for the A2C2R class. This is because it is more difficult to detect the angle orientation of the adult ellipse towards the child one when he gazes the humanoid robot. Other mis-classified errors are given by the wrong detection of the ellipse location. It should be noted that the system works in real time: it consumes around $25ms$ to process each frame and give the estimated behavior class (on a standard PC equipped with Intel I7 processor and 8 GB RAM).

Table 2. Results of the proposed approach

	JA	A2C2R	A2C2A	NCB
JA	81%	10%	4%	5%
A2C2R	11%	**75%**	8%	6%
A2C2A	7%	4%	**87%**	2%
NCB	1%	11%	1%	**87%**

4 Conclusions and Future Improvements

This work presented a novel automatic system able to detect joint attention by using completely non-intrusive depth camera installed on the room ceiling. Preliminary experiments conducted on three different video sequences showed that the proposed methodology is able to detect not only the joint attention, but also different interaction behaviors between adult, child and robot, analyzing the head orientation. This real-time system provides an useful non-intrusive tool to report the dominant behavior during the therapy session. Even if the obtained results still does not perform an hit rate of 100%, we point out the fact that, in the context of automatic or semi-automatic evaluation tools, each additional tool that can provide help in supporting therapists in this difficult theme is fundamental: autism, as known, is a generic term to indicate a disorder whose level can vary in a range (called *spectrum*), and each kind of therapy can be useful to produce an improvement in this range, with the (very difficult) goal of producing an exit of the patient from this spectrum.

The future works will be addressed to better estimate the angle among the head and to improve the ellipse detection algorithm. We will also acquire additional sequences during therapy sessions in order to evaluate the algorithm performances in real context. Furthermore, future works will investigate the possibility to create a common dataset in order to provide a comparison measure for this new innovative and non-invasive approach to automatically detect such events, as well as to test the system in clinical settings in real triadic interactions, thus involving therapists and children with ASD.

References

1. Anzalone, S.M., Tilmont, E., Boucenna, S., Xavier, J., Jouen, A.L., Bodeau, N., Maharatna, K., Chetouani, M., Cohen, D., Group, M.S., et al.: How children with autism spectrum disorder behave and explore the 4-dimensional (spatial 3d+ time) environment during a joint attention induction task with a robot. Research in Autism Spectrum Disorders **8**(7), 814–826 (2014)
2. Canny, J.: A computational approach to edge detection. IEEE Trans. Pattern Anal. Mach. Intell. **8**(6), 679–698 (1986)
3. Carpenter, M., Liebal, K.: Joint attention, communication, and knowing together in infancy (2011)

4. Diehl, J.J., Schmitt, L.M., Villano, M., Crowell, C.R.: The clinical use of robots for individuals with autism spectrum disorders: A critical review. Research in Autism Spectrum Disorders **6**(1), 249–262 (2012)
5. Emery, N.: The eyes have it: the neuroethology, function and evolution of social gaze. Neuroscience & Biobehavioral Reviews **24**(6), 581–604 (2000)
6. Feil-Seifer, D., Matarić, M.J.: Toward socially assistive robotics for augmenting interventions for children with autism spectrum disorders. In: Khatib, O., Kumar, V., Pappas, G.J. (eds.) Experimental Robotics: The Eleventh International Symposium. Springer Tracts in Advanced Robotics, vol. 54, pp. 201–210. Springer, Heidelberg (2009)
7. Gulsrud, A.C., Hellemann, G.S., Freeman, S.F., Kasari, C.: Two to ten years: Developmental trajectories of joint attention in children with asd who received targeted social communication interventions. Autism Research **7**(2), 207–215 (2014)
8. Imai, M., Ono, T., Ishiguro, H.: Physical relation and expression: Joint attention for human-robot interaction. IEEE Transactions on Industrial Electronics **50**(4), 636–643 (2003)
9. Jones, E.A., Carr, E.G.: Joint attention in children with autism theory and intervention. Focus on Autism and Other Developmental Disabilities **19**(1), 13–26 (2004)
10. Khoramshahi, M., Shukla, A., Billard, A.: From joint-attention to joint-action: effects of gaze on human following motion. In: 6th Joint Action Meeting (2015)
11. Li, A.X., Florendo, M., Miller, L.E., Ishiguro, H., Saygin, A.P.: Robot form and motion influences social attention. In: Proceedings of the Tenth Annual ACM/IEEE International Conference on Human-Robot Interaction, pp. 43–50. ACM (2015)
12. Moore, C., Dunham, P.: Joint attention: Its origins and role in development. Psychology Press (2014)
13. Prasad, D.K., Leung, M.K.: Methods for ellipse detection from edge maps of real images. In: Machine Vision - Applications and Systems, pp. 135–162. InTech (2012)
14. Robins, B., Dickerson, P., Stribling, P., Dautenhahn, K.: Robot-mediated joint attention in children with autism: A case study in robot-human interaction. Interaction Studies **5**(2), 161–198 (2004)
15. Rogers, S.J., Dawson, G.: Early Start Denver Model curriculum checklist for young children with Autism. Guilford Press (2009)
16. Scaife, M., Bruner, J.S.: The capacity for joint visual attention in the infant. Nature (1975)
17. Scassellati, B., Admoni, H., Mataric, M.: Robots for use in autism research. Annual Review of Biomedical Engineering **14**, 275–294 (2012)
18. Staudte, M., Crocker, M.W.: Investigating joint attention mechanisms through spoken human-robot interaction. Cognition **120**(2), 268–291 (2011)
19. Stauffer, C., Grimson, W.: Adaptive background mixture models for real-timetracking. In: IEEE Int. Conf. on. Comp. Vision and Patt. Recognition. vol. 2, p. 252 (1999)
20. Warreyn, P., Paelt, S., Roeyers, H.: Social-communicative abilities as treatment goals for preschool children with autism spectrum disorder: the importance of imitation, joint attention, and play. Developmental Medicine & Child Neurology **56**(8), 712–716 (2014)

21. Warreyn, P., Roeyers, H.: See what i see, do as i do: Promoting joint attention and imitation in preschoolers with autism spectrum disorder. Autism **18**(6), 658–671 (2014)
22. Zhao, X., Cusimano, C., Malle, B.F.: Do people spontaneously take a robot?s visual perspective? In: Proc. of the ACM/IEEE Intern. Conf. on Human-Robot Interaction Extended Abstracts, pp. 133–134. ACM (2015)
23. Zivkovic, Z.: Improved adaptive gaussian mixture model for background subtraction. In: Proc. of the IEEE Intern. Conf. on Patt. Recogn. vol. 2, pp. 28–31 (2004)

Characterizing the State of the Art
of Human-Robot Coproduction

Argun Cencen$^{(\boxtimes)}$, Jouke Verlinden, and Jo Geraedts

Faculty of Industrial Design Engineering, TU Delft, Delft, The Netherlands
{a.cencen,j.c.verlinden,j.m.p.geraedts}@tudelft.nl

Abstract. The industry is working towards manufacturing systems consisting of a blend of humans and robots. We look at the development of these systems in the context of Small and Medium Enterprises (SME). Also, it is believed that industrial robots with collaboration capabilities with humans will play a crucial role in the change towards reconfigurable and flexible manufacturing systems. Collaboration and teaming are natural social skills of humans. However, little is known about robots and their capabilities in working efficiently with these skills. From our review of the current context of manufacturing, we understand that tasks at a workstation are executed by a combination of various actors and there are many ways to design, control and simulate their interplay. These practices need to be developed for these novel systems as well. Through a survey of existing examples of similar systems, we set an initial step in generating knowledge on the parameters that influence the design of these systems. In these systems we see that humans and robots have certain areas and types of skills through which they engage in joint activity. We compare these examples from three perspectives and draw preliminary conclusions.

Keywords: Human robot interaction · Worker robot collaboration · Coproduction · Industrial robots · Industrial automation design

1   Introduction

Looking at the state of the art of industrial automation technologies around the world, creating an automated solution for almost any manufacturing related problem is no longer a challenge of technology, but a matter of sufficient time and resources. The European Union (EU) is currently working towards a future in which it strengthens the competitiveness and productivity of its manufacturing SME's. By doing so, the EU wants to hold and sustain its competitive position amongst other manufacturing areas of the world [1].

Following this idea, the integration of the latest advances in several technologies i.e. industrial robots, computer vision, 3D printing, in order to develop the necessary tools for a "rapid-robotization" concept is a task that is the main goal of the European FP7 framework funded Factory in a Day (FiaD) project [2]. As a partner of FiaD we investigate this goal from our field of expertise of Industrial Design Engineering. This perspective focuses on the human related aspects of the concept and particularly to

© Springer International Publishing Switzerland 2015
A. Tapus et al. (Eds.): ICSR 2015, LNAI 9388, pp. 135–144, 2015.
DOI: 10.1007/978-3-319-25554-5_14

develop a design methodology that is applicable to the design of human-robot copro-duction systems.

In this paper, we first describe our perspective of manufacturing systems and nar-row our topic down to the workers and their tasks. Next, through a survey of demon-strators consisting of humans and robots in manufacturing-like contexts, we attempt to characterize Human Robot Coproduction (HRC).

2   Background

2.1   Manufacturing Systems

Manufacturing systems can be viewed from several perspectives, e.g., Groover di-vides manufacturing systems in 4 parts. Production machines, material handling sys-tem, computer control system and human resources [3]. Manufacturing systems are following a trend towards flexibility and reconfigurability. From a changeability and reconfigurability perspective, Zaeh et al. propose manufacturing systems as consisting of 3 parts: the physical system, the control system and the organization of system [4] . We believe that the combination of these views provide a good foundation for the positioning and architecture of HRC systems.

2.2   The production Line and Workstations

Today's markets require many goods to be manufactured in a relatively quick phase and up to a certain standard. According to Bowen & Youngdahl, in the context of manufacturing, technology combined with a well-defined division of labor, clear rules, and limited span of control results in consistent quality and efficiency [5]. One of the best examples of this combination is the development of mass production, which is an approach that increases overall efficiency, while maintaining product quality. Henry Ford introduced the production-line approach in the beginning of the twentieth century and revolutionized the manufacturing industry [6]. This demon-strated that even a fairly complex product such as a car can be manufactured at a fixed rate in the expected quality. Nowadays, many manufacturing activities benefit from this approach in one way or another. As summarized in Figure 1, a production line is

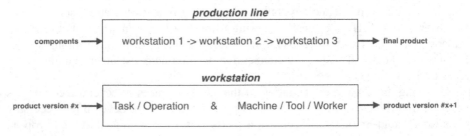

Fig. 1. An abstraction of the relationship between production lines, workstations and their elements

defined by Groover [3] as a system that consists of multiple workstations. A workstation refers to a location in the factory where a well-defined task or operation is accomplished by an automated machine, a combination of worker & machine or a combination of worker & tools. Therefore we define the actors at a workstation as tools, workers and machines.

3   Survey of Human Robot Coproduction Systems

Human-Robot Coproduction as introduced in the previous sections has been subject to several explorative studies. From manufacturing and robotics literature, we have collected ten distinct demonstrators of collaborations between humans and robots, in which the humans and robots act as co-workers, Figure 2 depicts these. These systems, labeled 'A' to 'J', envisage how human-robot coproduction can secure human labor in the future. However, only a few of these are fully implemented while no detailed prescriptions or operational guidelines exist.

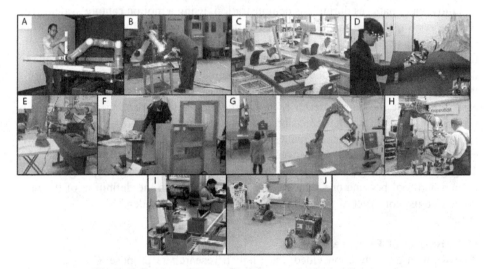

Fig. 2. Collection of Human-Robot coproduction applications.

3.1   Demonstrator Selection

The common denominator of this set is the element of a production context. The second criterion for selection was the presence of a human worker and a robot. in most of the systems we selected, a robot manipulator has been used, which makes it relatively simple to identify the system as a system containing a robot. In the cases C, F and J, instead of a robot manipulator, other devices with variant formalities have been used to achieve systems with similar qualities, such as a 'mobile platform' and a 'light guide' performing comparable actions. Therefore, one way of framing the selected cases would be to name them "systems with robotic qualities".

3.2   Demonstrator Descriptions

In our literature review so far, we have not come across a means of categorizing human-robot coproduction systems. Therefore we initially focus mainly on two global aspects that emerge from our review and hypothesis so far. On one side we try to explain what the task is, on the other hand, we try to describe how the human worker and the robot manage their communication during this process. Following are the ten cases, presented through short descriptions.

A. Rozo et al.

The authors propose a kinesthetically learning algorithm to support an assembly task of a small side-table, which is designed to be assembled by humans [7].The task is conceptualized as follows: a human assembles the legs of the table one by one while the robot holds the top piece of the table in an orientation which is comfortable for the human. Prior to the assembly operation, a human demonstrates the portion of the workflow related to the role of the robot kinesthetically. The robot records haptic data and movement patterns during this demonstration, using a motion capture system with passive retro reflective markers attached to the table parts and six-axis force-torque sensor that is attached between the wrist of the robot and the table. During the execution of the task, the human communicates with the robot through exerting torque or displacement force. This is one of the few systems that demonstrate a manufacturing task, in which the robot arm plays a supporting role by lifting and repositioning the assembly, while the human dexterity and perception-action coupling as described by Gibson are used for the high-precision aspects of the task [8]. Furthermore, communication through haptic channels makes sense in human robot coproduction tasks, designing information exchange between actors implicitly as part of the task at hand. This type of coupling is in line with the Dourish' embodied task coupling in which human and tool become one in a specific action, based on the definition of the phenomenologist construct of "vorhanden" (as opposed to "zuhanden")[9]

B. University of Tampere.

As shown in an instruction video, the Finnish researchers propose a robot welding assistant that holds and repositions the assembly that is being welded together by a human welder [10]. The task is conceptualized as follows: The robot picks and holds the first piece of the assembly at a position which is comfortable for the task of the welder. The welder then locates and welds the remaining pieces onto the piece held by the robot one by one, while the robot changes the orientation of the piece it is holding, in order to allow the human to execute the welding task as effectively and ergonomically as possible. The portion of the workflow related to the role of the robot is pre-programmed into the robot prior to the execution of the task. The human communicates with the robot by using gestures that can be tracked through a camera attached to the robot. According to the authors, environmental and process related parameters in real life play a crucial role in the implementation of such systems. This is also addressed by [11].

C. Royonic.

In this commercially available Printed Circuit Board (PCB) assembly system, the robot indicates the areas on a PCB where through-hole components need to be placed one by one [12]. Each time, the system also highlights the location of the component to be picked. The human follows the instructions of the robot and executes the picking and placing task. The workflow is pre-programmed into the robot prior to the execution of the task. The human communicates with the robot by pushing an electrical button or a foot pedal. In our selection of collaborations, this is an example of a collaboration in which the task division of physical activity between the robot and human is arranged in an unconventional way. Here, the robot is responsible for the cognitive part of the task. According to Schwerdtfeger, this method for assembly results in an error rate of 0,002% [13].

D. Glasauer et al.

In this experiment consisting of several stages, the authors aim to investigate the difference in performance between human-human and human-robot hand-over of objects [14]. Two types of hand-overs are distinguished. In the first type, the agent delivering the object also initiates the action, in a second type of hand-over, a so-called "foreman" and "assistant" work together. The assistant needs to deliver the parts just in time for the foreman to assemble the parts onto another part. In the experiment which is executed in order to study the first type of hand-over, a robot has the role of picking of cubes from a table and handing over to a human. The human has the role of receiving the cubes from the robot and placing them on the table again. The workflow is pre-programmed into the robot and communicated to the human prior to the execution of the task. The human communicates with the robot through hand motions that can be tracked and interpreted through a camera attached to the robot. One of the most important conclusions is the increase of performance when the hand-over action performed by the robot is more human-like.

E. Bringes et al.

The authors have built an experimental setup to investigate the performance of several pick & place scenarios, involving a teleoperated robot manipulator and a human [15]. They predict that a human-in-the-loop will be beneficial to the performance of the system, especially when there is some form of noise in the perception/cognition of the robot. The task in all scenarios is the picking of fruit/vegetables from random locations on a table and placing them inside a container. The robot manipulator is equipped with a gripper that is capable of providing a steady grasp of all object that are needed to be picked. This is the main role of the robot. The human decides on the workflow during operation. The human has the role of targeting each object and communicating their location to the robot, which is done with a haptic-pen device. The system performs best when there is no noise and no human in the loop. However, in the case of noise, the human worker assisting to determine a coarse approach of the stage results in a better performance.

F. Unhelkar et al.

This experiment revolves around the task of assembling a LEGO-toy [16]. Robots and human assistants are given the task of delivering components needed for the assembly of the LEGO-toy in several steps and. Another human has the task to assemble the LEGO-toy using the components and instructions delivered to him/her at each step. The workflow is pre-programmed into the robot and instructed to the human worker. The workflow is communicated to the assembling worker during the delivery of new parts. The worker communicates with the assistants through accepting and relocating parts that are delivered to him/her. The authors conclude that the performance of the task is better when only humans perform the task. However, they also identify advantages of the inclusion of robots in the workflow, such as the sound that the robot makes while approaching , which can provide a cue for the human working on the assembly.

G. Pieska et al.

In this work, robot and human coproduction is viewed from the perspective of palletizing products. A robot has the role of picking products from one location and placing them in a stacked format at an other location [17]. The worker has the role of instructing the robot the location of the product to be picked and placed. The workflow is pre-programmed into the robot. The worker communicates with the robot by using gestures that can be tracked through a camera attached to the robot. Prieska et al. mention that inexperienced users can program and control robots through gestures and dedicated interfaces.

H. Schraft et al.

This human-robot collaboration focuses on the rearrangement of parts that are needed for an assembly. The robot has the role of picking parts for the assembly task, and bringing them very close to the location where they need to be assembled [18]. The role of the human is to manipulate the orientation of the part that is being held by the robot and insert this into the corresponding destination. The workflow is pre-programmed into the robot. The human communicates with the robot through kinesthetic feedback and by pressing electrical buttons. The safety norms surrounding industrial robots are bottlenecks for increased performance of the type of systems that are the subject of the experiment. In the clauses 5.10 of ISO 10218-1, collaborative operation is allowed, in the most advanced case regarding co-located operation in which case power and force limiting needs to be enforced by inherent design features or control.

I. Cencen et al.

In this experimental setup, the robot has the role to pick two products from their boxes and place them inside a box that is being transported on a conveyor [19]. The robot also has the task to relocate a filled box by pushing it further on the conveyor. There are several human workers part of the workflow. The role of one worker is to pick three products from their boxes and place them inside a box on the conveyor.

This worker also picks a box from a stack of boxes and places it on the conveyor. Another worker in the workflow relocates the filled boxes from the end of the conveyor to another location. The workflow is pre-programmed into the robot and communicated to the humans. The worker with the role of picking the products communicates with the robot by pushing a half-filled box towards on of the sensors of the robot. The results of the experiment point to the fact that there are many product, process and person related unknowns when designing human robot coproduction systems and that these need to be further investigated. Furthermore, the pace of the (human-safe) robot was too slow to engage in an efficient workflow.

J. Fong et al.

This research reports on the findings from an experiment in which human robot collaborations in lunar environments are being studied. In this setup, one robot has the role of welding, another robot has the role of quality inspection of the weld that is produced [20]. Two astronauts have various roles inside the workflow, ranging from relocating robots to checking the quality of results. The robots can also be teleoperated by a third astronaut. The workflow is determined by the astronauts by interacting with the robots during operation. Similarly, the robots communicate with astronauts by requesting feedback at various intervals. The authors defend that the performance of human-robot collaboration increases if the right software platform is used.

3.3   Categorization of Demonstrators

We consider the aforementioned collection of systems representative of workstations in a manufacturing system. As discussed in section IIA, one of the essential elements of a workstation is a well-defined task. Looking at the gathered examples and revisiting Figure 1, it becomes evident that this task definition can not be created without taking into consideration who/what will perform the task(i.e. Machine, Tool, Worker) and what is going to be manipulated(i.e. Product). Another point is that this actually is a two-way consideration. Depending on the task that needs to be fullfilled, more actors may be added to the workstation and the task can be adjusted according to which actors are already present. This was also already hypothesized in [19].

Our aim is to make an initial categorization of these cases from a Human-Robot Coproduction point of view. Therefore, we defined three variables. These are Task Initiative (TI), Product Handling (PH) and Component Handling (CH). Each variable can have two states; being either performed by the robot or by the human.   Task Initiative is a role in which the workflow of the task is controlled and monitored. Product Handling is a role in which the role owner is responsible for the main part of the product which is being assembled or manipulated. This can be on the level of the product (e.g., holding and/or positioning the part). On the level of components, this is named Component Handling (CH) (e.g., picking and/or handing over of new parts, welding of parts, placing products/parts in boxes).

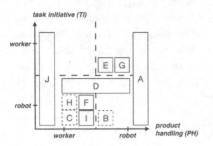 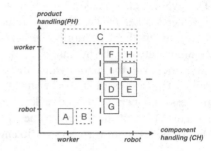

Fig. 3. **Left**: Task initiative versus physical handling in the examples (dashed-lined boxes indicate industrial systems). **Right:** Product handling(PH) versus component handling (CH) in the examples (dashed-lined boxes indicate industrial systems)

4   Results and Discussion

At first sight, none of the systems shared both roles between robot and worker. In the six instances where some shared responsibility is seen (A,C,D,E,J,G), this is limited to only one role. In these cases, the remaining two roles are divided between the worker and the robot. Insufficient means of interaction and type of task might play a role in this choice of role combination. When looked at the systems which we consider to be operational in industry (B,C,H), only case C includes a shared responsibility. A reason for this might be the high operational requirements that are considered during the design of the system. In these types of systems, TI is a common role that is delegated to the robot. Only in two instances (E,G), the robot is not part of the TI role.

Although a limited set of systems was analyzed, in the roles of task initiative (TI) and physical handling (PH), dominance in the role of robots are seen. This needs to be further investigated and understood. This knowledge will be essential in supporting the design of such systems. It is notable that shared roles/responsibilities are not integral parts of these systems (yet). This is in line with the idea of efficiency in production lines. However, many examples and trends in literature suggest that notions such as interdependency and collaborative frameworks are the future perspective of worker-robot systems. In order to be able to create operational-worthy systems, we need to understand the performance indicators of these collaborations. These will potentially be different than the regular time/quality/cost paradigm and will move in the direction of flexibility/recovery. Looking at the systems E and G in more detail, we revealed that they fall in the category of teleoperated systems. In such systems, humans have an essential cognitive role in the loop and the performance of the task. It can be argued that in tasks where similar cognitive capabilities are necessary, it can be advisable to implement teleoperative properties.

5   Conclusions and Future Work

With the avenue of novel robotic systems, such as Rethink Robotics Baxter [21] and Universal Robots UR arms [22], the enabling technologies are progressing at a fast pace. In the past, humans were supposedly only needed because automation was not feasible to replace them yet [23]. However, a new generation of experimental systems provide inspiration to prove this wrong. There are still difficulties to operationalize these human-robot collaboration notions in present work. The reasons and actions to be able to achieve this can be summarized as follows;

The presented examples show how future worker-robot systems can provide sufficient work for humans. However, only a few of such systems are operational and these examples provide relatively less material for the analysis of the complexity of required interactions in other situations.

Although an initial theoretical frame was drawn in this paper, when making a qualitative analysis of the investigated systems, these do not fit perfectly inside this frame. Yet, our efforts revealed basic insights in how such systems can be viewed and what these frames are lacking.

While experimenting with such systems, also operational models of these systems need to be constructed in order to be able to iterate between various designs and gain more insight in performance related details. Literature on finite machines is worth investigating [24]. We expect that these models, together with dedicated computer-aided-process-planning (CAPP) approaches, will provide a foundation for successful initial industrial implementations. Future research will be directed towards implementing these technologies in combination with Robot Operating System (ROS) and similar novel programming environments in human-tool coproduction manufacturing environments.

Acknowledgement. The work leading to these results has received funding from the European Community's Seventh Framework Programme (FP7/2007-2013) under grant agreement n° 609206

References

1. FoF - Factories of the Future. http://cordis.europa.eu/fp7/ict/micro-nanosystems/ict-for-fof_en.html (accessed August 21, 2015)
2. Factory-in-a-Day (2014). www.factory-in-a-day.eu
3. Groover, M.P.: Automation, production systems, and computer-integrated manufacturing. Prentice Hall Press (2007)
4. Zaeh, M.F., Wiesbeck, M., Stork, S., Schubo, A.: A multi-dimensional measure for determining the complexity of manual assembly operations, 489–496 (2009)
5. Bowen, D.E., Youngdahl, W.E.: 'Lean' service: in defense of a production-line approach. Int. J. Serv. Ind. Manag. **9**(3), 207–225 (1998)
6. Hounshell, D.: From the American system to mass production, 1800–1932: The development of manufacturing technology in the United States. No. 4. JHU Press (1985)

7. Rozo, L., Calinon, S., Cadwell, D., Jimenez, P., Torras, C.: Learning collaborative imped-ance-based robot behaviors. In: Proc. AAAI Conf. Artif. Intell. Bellevue, WA, USA, pp. 1422–1428 (2013)
8. Gibson, J.: The ecological approach to visual perception (2013)
9. Dourish, P.: Where the action is: the foundations of embodied interaction (2004)
10. Human robot co-operation welding workcell case study. https://youtu.be/Kxw-SJd-j-o (accessed March 29, 2015)
11. Qiu, S., Fan, X., Wu, D., He, Q., Zhou, D.: Virtual human modeling for interactive assem-bly and disassembly operation in virtual reality environment. Int. J. Adv. Manuf. Technol. **69**(9–12), 2355–2372 (2013)
12. Markstein, H.W.: The Electronics Assembly Handbook. Springer, Heidelberg, pp. 70–76 (1988)
13. Schwerdtfeger, B., Pustka, D., Hofhauser, A., Garching, M.: Using Laser Projectors for Augmented Reality, pp. 134–137 (2008)
14. Glasauer, S., Huber, M., Basili, P., Knoll, A., Brandt, T.: Interacting in time and space: investigating human-human and human-robot joint action. In: 19th Int. Symp. Robot Hum. Interact. Commun., pp. 252–257, September 2010
15. Bringes, C., Lin, Y., Sun, Y., Alqasemi, R.: Determining the benefit of human input in human-in-the-loop robotic systems. In: 2013 IEEE Ro-Man, pp. 210–215, August 2013
16. Unhelkar, V.V., Siu, H.C., Shah, J.A.: Comparative performance of human and mobile robotic assistants in collaborative fetch-and-deliver tasks. In: Proc. 2014 ACM/IEEE Int. Conf. Human-Robot Interact. - HRI 2014, pp. 82–89 (2014)
17. Pieskä, S., Kaarela, J., Saukko, O.: Towards Easier Human – Robot Interaction to Help Inexperienced Operators in SMEs, pp. 333–338 (2012)
18. Schraft, R.D., Meyer, C., Parlitz, C., Helms, E.: Power Mate – A Safe and Intuitive Robot Assistant for Handling and Assembly Tasks, pp. 74–79, April 2005
19. Cencen, A., Van Deurzen, K., Verlinden, J.C., Geraedts, J.M.P.: Exploring Human Robot Coproduction (2013)
20. Fong, T., Scholtz, J., Shah, J.: A preliminary study of peer-to-peer human-robot interac-tion. In: Syst. Man Cybern., pp. 3198–3203 (2006)
21. Fitzgerald, C.: Developing Baxter (2013)
22. Universal Robots. http://www.universal-robots.com (accessed April 14, 2015)
23. Womak, J.P., Jones, D.T., Roos, D.: The Machine that Changed the World: The Story of Lean Production. HarperCollins, New York (1990)
24. Minsky, M.: Computation: finite and infinite machines (1967)

KeJia Robot–An Attractive Shopping Mall Guider

Yingfeng Chen[(✉)], Feng Wu, Wei Shuai, Ningyang Wang,
Rongya Chen, and Xiaoping Chen

Department of Computer Sciences,
University of Science and Technology of China, Hefei 230026, China
chyf@mail.ustc.edu.cn, xpchen@ustc.edu.cn
http://ai.ustc.edu.cn

Abstract. This paper reports the project of a shopping mall guide robot, named KeJia, which is designed for customer navigation, information providing and entertainment in a real environment. Our introduction focuses on the designs of robot's hardware and software, faced challenges and the multimodal interaction methods including using a mobile phone app. In order to adapt the current localization and navigation techniques to such large and complex shopping mall environment, a series of related improvements and new methods are proposed. The robot is deployed in a large shopping mall for field test and stable operation for a fairly long time. The result demonstrates the stability, validity and feasibility of this robot system, and gives a positive reward to our original design motivation.

Keywords: Shopping mall guide robot · Localization and navigation · Mobile interaction · Quadtree mapping

1 Introduction

With the rapid advance of robotic technology, several well-known robots have been developed, with cool appearance, human-like joints and more intelligent behaviors, such as Willow Garage's PR2, Boston Dynamics' Atlas and the Meka. However, the general public rarely have the opportunities to interact with the tangible robots despite that these "stars" represent the latest techniques and indicate the future trend.

In this paper, we apply our robot named KeJia to a day-to-day shopping mall scenario, in which KeJia is implemented as a shopping assistant providing route guidance and information service. Nowadays, the size of shopping malls is becoming larger and larger, usually hosting hundreds of individual commercial tenants. The new customers (sometime even the regulars) get lost in the complex and maze-like environment, not to mention finding the desired locations. Even though floor maps are provided, customs usually prefer to seeking for manual guidance. Besides, there is a strong demand for such a device that

© Springer International Publishing Switzerland 2015
A. Tapus et al. (Eds.): ICSR 2015, LNAI 9388, pp. 145–154, 2015.
DOI: 10.1007/978-3-319-25554-5_15

can provide product catalogs and discount information once customers enter a shopping mall. In order to improve the shopping experience and service quality, the shopping mall managers trend to hire more professional guiders, resulting in a large financial burden.

Against this background, KeJia is devised to guide the customers to right locations as required autonomously and provide information to them by natural language. Furthermore, the robot itself can attract more people and bring a fresh feeling to the shopping mall. Apart from the aforementioned features, KeJia system also contains mobile apps that can extend the robots to simultaneous service. By integrating the concept of cloud technology, it's convenient for the users to acquire the desired information from the robot remotely through their mobile phones.

2   Related Work

In the past decade, one of the most popular applications is to deploy the robots as tour guiders in museums, expositions or other public places. The earliest work was carried out by [1], the robot RHINO acted as a guider in the "Deutsches Museum Bonn" for six days and provided service for more than 2,000 visitors. The robot MINERVA [12], a successor of RHINO, had better performance in navigation and interaction. In [5], the robot Robotinho addressed the challenge on more intuitive, natural and human-like interaction, unlike the previously works, which mainly focused on localization, automatic navigation and collision-free avoidance in populated environments. Although these robots are running reasonably well, their operation environments are almost completely known, neither of museums or exhibition halls. What's more, the size of the operation environments is usually limited to hundreds of square meters, which means the unexpected situations that are lethal to the normal travel of the robots would be greatly reduced. In contrast, as for the shopping mall scenario, there are still many challenging problems need to deal with.

The shopping mall is a more frequently visited place in routine activities, and some studies have already concerned with robot systems in such scenario. Most of them try to design a shopping trolley for helping the elderly, the sick, or the disabled. In [11], the task of RoboCart is to carry the purchased goods and lead the way for impaired customers in a grocery. With distinctive expectation for shopping assistant, we hope robot could perform like realistic clerk, not a porter. Some other interesting researches provide difference functions to assist customers. For example, [13] developed a remote shopping system, where the robot grasps the products(i.e., goods with various textures, shapes and weights) on the shopping lists using telemonitor with a special manipulator. The system proposed by [10] and [8] are most similar to ours, [10] presented an interactive robot that is fixed at a place in the shopping mall, giving directions by speech and gestures. The robot TOOMAS [8] roams in stores to search the potential customers and then guides them to the target locations. However, their main limitation of most of them is that the robot can not execute any movable navigation behaviors.

Most of the aforementioned approaches are proof-of-concept prototypes and have not been completely deployed in real shopping malls. With a few exceptions, they have conducted field trials, but required extra environmental modification to facilitate the deployments of robot systems. In more details for localization and navigation, RFID tags often need to be installed in the workspace in advanced, which may be troublesome and inflexible. Moreover, such deployment need to be redo once the environment changes. Although the robot TOOMAS is a exception of this case, its operating environment (shown in a picture in [8]) is more like a supermarket fulled with structured shelves, rather than a shopping mall that has lots of independent shops. In a real shopping mall, wide range of unpredictable changes exist due to the shop decoration, temporary stalls, advertising board replacement, infrastructure improvement, etc (see Fig. 1), it's the key challenge for robot navigation and customer guidance. Besides, we use a mobile app to enable the communication between the robots and users, which is a new feature for shopping mall robot that is quite effective as confirmed by our deployment.

3 Features and Hardware

3.1 Key Features

Shopping mall is typically a collection of all kinds of shops where customs can buy their daily necessities (e.g., clothes, shoes, foods, restaurants, etc.). The size of a large shopping center is often more than tens of thousands of square meters and has high flow rate of customers. Given this, the key features of our intelligent assistant shopping robot are fellows.

(1) *Providing Information*: For the customers who are busy, they would like to know whether they can buy what they want quickly. The robot can response to instant enquiring and provide information in more natural ways. Additionally, it can give specific recommendations and introductions according to different inquiries.

(2) *Shopping Guidance*: A robot, moving in front of the customers, can guide them to the right shops, which providing pleasant shopping experience for the costumer. It can also avoid time-consuming and aimless search for people who don't enjoy the process.

(3) *Advertisement and Recreation*: From the views of the shop owners and mall managers, the robot can attract more customers to boost their sales. In fact, the advertising effect of robots exceeds traditional methods (e.g., posters, billboards). Some people even come to the shopping mall just for interaction with our robot. Withal, people enjoy chatting with the robot for recreations.

3.2 Hardware

The robot KeJia designed for the shopping mall scenario is shown in Fig. 2, which is a derivative of KeJia Project[2]. It is worth pointing out that a prototype with the similar hardware design has won the championship in the 2014

Robocup@Home competition. The appearance of KeJia is a young lady, dressed on a traditional suit. With the height of 165 cm the robot is comparable to a professional shop assistant. Unlike other shopping robots [8], the touch screen is not adopted due to the following considerations. Firstly, it's not convenient for people to operate the touch screen while robots are moving. Secondly, our speech recognition provides fairly accurate results with the directional microphone equipped. Thirdly, mobile phone Apps provide additional method to interact with robot. A sound equipment embedded in the base is used to play the voice generated by speech synthesis module. The whole robot is motorized by two differentially actuated wheels on the middle axis and a castor on the rear. This offers KeJia a good maneuverability and stability. The main sensor of the robot is a HOKUYO UTM-30LX laser scanner, which feeds the distance data of obstacles around the robot(maximum distance 30m) to other software modules(e.g., navigation and localization).

Fig. 1. (a) Typical shopping mall channel (b) new added resting chair (c) glass wall (d) temporary stall

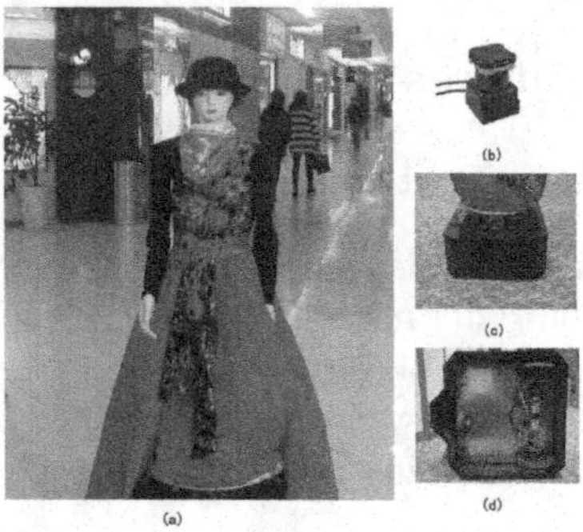

Fig. 2. (a) KeJia robot (b) laser scanner (c) base (d) differential wheels

Inside the clothes, hides a frame made by aluminium alloy and plastic shells, which constitutes the body of the robot. The two arms and the head are fixed on the interior frame. The arms with four DOFs can make simple gestures like greeting to people, and the two DOFs head can freely turn to face to the people if necessary. All of these make KeJia look more realistic.

4   System Architecture

4.1   Top-Floor Structure

Our system could consist of several distributed robots, which can interact with people face-to-face by voice or thought the mobile client, and a cloud server connected to Internet. As shown in the Fig. 3, robots are deployed on different floors in the shopping mall, configuration data centrally store on the server where

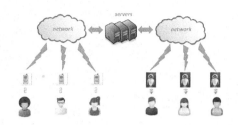

Fig. 3. Concept map of the proposed system

robot could conveniently download from. It's advised to inquire and command the robot by speech if the customers are nearby, additionally, the app installed on the smart phone is able to send the typed text or recorded sounds to the server and then forward to the robot, after processing, the response data of the robot will transfer back over the same way. There are some advantages of this structure, firstly, it's convenient to modify the configuration information which is constantly changing, what the robot need to do is regularly to update the latest data from the server. Secondly, different interaction methods make the robot reach its greatest potential at the same time, other people can still keep "chatting" with robot during navigating period by mobile phone.

4.2 Software Modules Structure

In our system, we adopt a flexible four-layered software architecture (see Fig. 4) to meet the requirements of a integrated robot system, such as reliability, extensibility, maintainability, customizability. The lowest layer is the Robot Operating System [1], which provides a set of robotic software libraries and reliable communication mechanism for modular nodes. The second layer mainly contains hardware drivers, thereinto, the laser and camera drivers are in charge of packaging the raw sensor data to standard format and then publish them to upper layers, the motor and audio drivers interpret the messages from upper layers to hardware for executing. The next layer is the most important one in the structure, all the proper skills of a classic robot are placed here, such as mapping, localization, navigation, people tracking, speech recognition and synthesis, which directly decides what the functionalities would be implemented in the upper applications. The highest level is responsible for task managing, configuration data updating, dialog managing and robots' state updating. The background server collect the real-time robots' state (i.e., coordinate, task state) and forward to the smart phone app. The dialog manager module attempts to understand the users' intentions and dispatches tasks.

By using such layered structure, for a new application or adding a new skill only the corresponding layers need to be changed rather than the whole system, and individual module can be assigned to different programmers who just develop the desired functionality according the preestablished interface without caring other's implementation details.

---

[1] http://www.ros.org

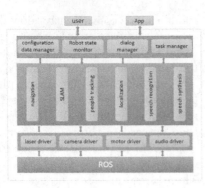
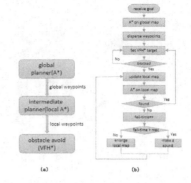

Fig. 4. Four-layered software structure

Fig. 5. (a) Navigation structure (b) flow chart

4.3 Methods to Challenging Modules

In this section, we will present expatiation to these modules that are faced with technical challenges.

Mapping of Large Environment. There are already many mature algorithms to the simultaneous localization and mapping(SLAM) problem[4,9], most of them take the grids map as the representation of environments. However, the memory exhaustion problem emerges along with the increasing enlargement of mapped environment, especially for particle filter family, in which every particle is generally associated with an individual map. The size of the shopping mall is usually more than 10,000 m^2, which makes the situation worse, therefore a new map representation is introduced as quadtree map. A quadtree is a tree-based data structure that is capable of achieving compact and efficient representation of large two-dimensional environment. We re-implement the Rao-Blackwellized particle filter SLAM[7] by replacing traditional grids map with quadtree map (for more details, see [3]).

Localization in Complex Environment. Accurate localization is the prerequisite of safe and credible navigation behaviour. Under the shopping mall circumstance, there are some problems must be considered in order to achieve excellent performance, including dense streams of people and undetectable obstacles caused by the characteristic of the laser sensor. In fact, transparent glass walls exists in everywhere of modern shopping malls, they just can be perceived at certain angles from the laser. To overcome those thorns, we modify the likelihood model based on the AMCL[6][2] by multiplying a weight with the score indicated the matching degree between the map and the laser beams. The weight value is assigned to each map cell using a custom editing tool manually, the value

---

[2] Open source page: http://wiki.ros.org/amcl

of cells belonging to permanent obstacles likely solid walls are greater than these representing glass walls.

Navigation. The biggest trouble with navigation in shopping mall is the wide range of unpredictable changes that have been mentioned at the end of Sec.2, robot can't perceive these changes until approaching. In the previous researches, navigation module combined global planner and local planner is introduced for this case, global path replanning will be triggered when robot traps in local dilemma. But this idea is not fully applicable for us since it may lead robot repeatedly alter its route in vain. In reality, the passages are often temporarily blocked with customers, it would not be the best choice for robot to replan every time for the following reasons. 1) The robot may move back and forth between two blocked passages frequently without progress. 2) Re-finding a global path on the whole map is time-consuming. 3) Making a long detour sometimes is expensive than just waiting for a while. In order to eliminate this disharmony between global and local planner, a intermediate layer is employed.

Once a goal is receiving, firstly, the path from the robot's position to goal is computed. Next, a serial of ordered way points are generated from the global route, then the way points will be sequentially dispatched to the local planner which will find a local path for the well-tuned VFH* module[14] to track. If local planner fails to find a suitable path, the local map would continue enlarging until a maximum limit reaches. After several failures, robot will demand the crowd to give way, if all these attempts fail, a global replan happens. This approach endows the robot ability of adapting the shopping mall environments, meanwhile, reduces the unnecessary global path plan (shown in Fig. 5).

Multimodal Interaction

(1) *Speech Dialog System*: The sentences outputted by speech recognition system are divided into two classes, depending on whether they fall into the domain. The customers can have some conventional talk understood by the robot, like asking the robot to give an introduction or recommendation. In case the robot is caught off guard by the sentences beyond the domain, an inner chatting module is designed, which provides amusing and interesting talk though they maybe meaningless to the input sentences. This method has a good effect and avoid awkward situations.

(2) *Phone Application*: The phone app in our system is an inventive interaction style, which can do the job formerly belonging to speech recognition by typing or recording. The map of the mall, the robot's position and available destinations are shown in the application on the mobile phone screen, gives customers freedom of choice, and the collected data of customers' options also give the criterion of recommendation.

(3) *Other Recreations*: To enrich the robot's behaviors, we develop a "greet guests" mode, the robot will stand at a shop's door, say "hello" to the comer and "goodbye" to the leaver by detecting the movement of the customers with laser data.

5   Results of Field Trials

Our filed trials last for about 40 days from December 2014, including 10 days of preparative debugging work. The shopping mall where our robot deployed is the largest and most prosperous one in the provincial capital city, it has 4 floors gathered with nearly 300 commercial tenants and our robot works on the first floor with size of more than 10,000 m^2.

The first step to deploy this robot is building the environmental map, in order to collect the laser data and odometry data for mapping, we operate the robot to stroll along all passageways in the ground floor of shopping mall with a wireless joystick. The sensor data streams into the particle filter SLAM algorithm which is improved by integrating quadtree map representation. The final built map is shown as Fig. 6, the size of the map is $205m * 68m$ with resolution of $0.05m$. By using the quadtree representation, such a considerably large map only costs 10.80 MB internal memory and 287.1 KB disk space.

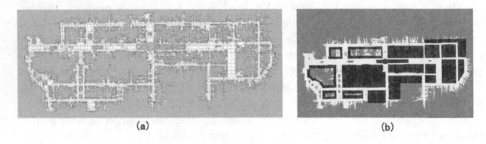

Fig. 6. (a) Quadtree map of the shopping mall (b) map used for localization and navigation

In order to be suited in practice, we modify the map for constrained path planning with black mask areas(shown in the Fig. 6 (b)), which are usually the interior of shops and not passable for robot. While for localization, the original map is used to make the laser data matched as good as possible. The Fig. 7 (a) shows a global path based on static map from start point(brown) to end point(red), the (b) shows an intermediate planner within the local map windows(square box $3.5m * 3.5m$), the path planned in this level is not exactly the same with global path because the new surroundings have been taking into account, and it's the key that our system can handle wide changes in the environment.

Customers in the shopping mall are free to choose one of the two interactive methods–talking by speech or mobile app. In the face-to-face mode, customers' sentences are recognized and then passed to dialogue manager module, once the intentions are explicit the robot will begin a tour guide (shown in Fig. 8.(a) (c)). Customers also can talk with KeJia by pressing the microphone icon and the text displays on the screen, the shops' positions are drawn on the map with

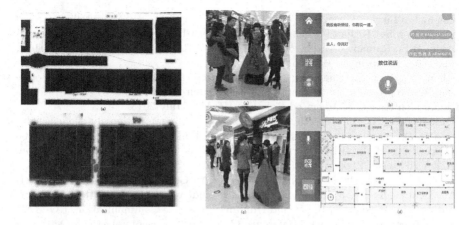

Fig. 7. (a) Global path (b) the blue line is path planned by intermediate level, the green line is part of global path

Fig. 8. (a) Talking with customers (b) chatting with app (c) during a tour guide (d) map in the app, red dots are shops, blue dot is robot

red dots, they can be chosen as goals by touching. In general, the mobile app provides a straightforward and effective graphical interface for users.

During the whole operational period, totally about 150 complete tour are performed, and the accumulated distance is more than $7.5km$. KeJia can provide trouble-free service with success rate of 75%, the most common reason of failure is losing its position, this often results from the crowded or some obstacles that can't be perceived by the laser. The customers seem to be more interested in chatting with robot by mobile app, they intentionally say some sentences that can't be handled by KeJia and expect it to respond with funny jokes.

6   Conclusions

From this project of KeJia robot, firstly, we have proved the reliability and effectiveness of proposed robotic techniques, including layered software architecture, mapping in large indoor area, localization and navigation in complex environment and module integration. Secondly, the mobile phone app provides another quick and convenient way to communicate with robot, and it becomes the most often used and acceptant way with the customers in actual running. Lastly, from the practical operation results, we can see that the general public have intense interest in robot and demand for everyday use, and we hope our work could provide some experience for similar robotic application.

References

1. Burgard, W., Cremers, A.B., Fox, D., Hhnel, D., Lakemeyer, G., Schulz, D., Steiner, W., Thrun, S.: Experiences with an interactive museum tour-guide robot. Artificial Intelligence **114**(1–2), 3–55 (1999)

2. Chen, K., Lu, D., Chen, Y., Tang, K., Wang, N., Chen, X.: The intelligent techniques in robot KeJia - the champion of robocup@home 2014. In: Bianchi, R.A.C., Akin, H.L., Ramamoorthy, S., Sugiura, K. (eds.) RoboCup 2014. LNCS, vol. 8992, pp. 130–141. Springer, Heidelberg (2015)

3. Chen, Y., Wei, S., Chen, X.: A probabilistic, variable-resolution and effective quadtree representation for mapping of large environments. In: 2015 17th International Conference on Advanced Robotics (ICAR), July 2015

4. Dissanayake, G., Durrant-Whyte, H., Bailey, T.: A computationally efficient solution to the simultaneous localisation and map building (slam) problem. In: 2000 Proceedings of the IEEE International Conference on Robotics and Automation, ICRA 2000, vol. 2, pp. 1009–1014. IEEE (2000)

5. Faber, F., Bennewitz, M., Eppner, C., Görög, A., Gonsior, C., Joho, D., Schreiber, M., Behnke, S.: The humanoid museum tour guide robotinho. In: 18th IEEE International Symposium on Robot and Human Interactive Communication, RO-MAN 2009, Toyama International Conference Center, Japan, September 27 - October 2 2009, pp. 891–896 (2009)

6. Fox, D.: Adapting the sample size in particle filters through KLD-sampling. I. J. Robotic Res. **22**(12), 985–1004 (2003)

7. Grisetti, G., Stachniss, C., Burgard, W.: Improving grid-based slam with rao-blackwellized particle filters by adaptive proposals and selective resampling. In: Proceedings of the 2005 IEEE International Conference on Robotics and Automation, ICRA 2005, pp. 2432–2437, April 2005

8. Gross, H., Böhme, H., Schröter, C., Müller, S., König, A., Martin, C., Merten, M., Bley, A.: Shopbot: progress in developing an interactive mobile shopping assistant for everyday use. In: Proceedings of the IEEE International Conference on Systems, Man and Cybernetics, 12–15 October 2008, Singapore, pp. 3471–3478 (2008)

9. Hahnel, D., Burgard, W., Fox, D., Thrun, S.: An efficient fastslam algorithm for generating maps of large-scale cyclic environments from raw laser range measurements. In: Proceedings of the 2003 IEEE/RSJ International Conference on Intelligent Robots and Systems, (IROS 2003), vol. 1, pp. 206–211. IEEE (2003)

10. Kanda, T., Shiomi, M., Miyashita, Z., Ishiguro, H., Hagita, N.: An affective guide robot in a shopping mall. In: Proceedings of the 4th ACM/IEEE International Conference on Human Robot Interaction, HRI 2009, 9–13 March 2009, La Jolla, California, USA, pp. 173–180 (2009)

11. Kulyukin, V.A., Gharpure, C., Nicholson, J.: Robocart: toward robot-assisted navigation of grocery stores by the visually impaired. In: 2005 IEEE/RSJ International Conference on Intelligent Robots and Systems, 2–6 August 2005, Edmonton, Alberta, Canada, pp. 2845–2850 (2005)

12. Thrun, S., Bennewitz, M., Burgard, W., Cremers, A.B., Dellaert, F., Fox, D., Hähnel, D., Rosenberg, C.R., Roy, N., Schulte, J., Schulz, D.: MINERVA: a second-generation museum tour-guide robot. In: Proceedings of the 1999 IEEE International Conference on Robotics and Automation, 10–15 May 1999, Marriott Hotel, Renaissance Center, Detroit, Michigan, pp. 1999–2005 (1999)

13. Tomizawa, T., Ohya, A., Yuta, S.: Remote shopping robot system, -development of a hand mechanism for grasping fresh foods in a supermarket. In: 2006 IEEE/RSJ International Conference on Intelligent Robots and Systems, IROS 2006, 9–15 October 2006, Beijing, China, pp. 4953–4958 (2006)

14. Ulrich, I., Borenstein, J.: Vfh*: local obstacle avoidance with look-ahead verification. In: 2000 Proceedings of the IEEE International Conference on Robotics and Automation, ICRA 2000, vol. 3, pp. 2505–2511 (2000)

Case-Sensitive Methods for Evaluating HRI from a Sociological Point of View

Diego Compagna[✉] and Ivo Boblan

Technische Universität Berlin, FG Regelungssysteme EN11,
Einsteinufer 17 10587, Berlin, Germany
{diego.compagna,ivo.boblan}@tu-berlin.de

Abstract. Evaluating and shaping the quality of interaction between humans and service or "social" robots from a genuine sociological point of view is still a pivotal methodological challenge at stake in the development of successful Human-Robot Interaction (HRI). In this regard an interdisciplinary research group, dedicated to the study of HRI in general, is developing a theory-driven method based on sociological interaction models with the goal of identifying the most important aspects in achieving satisfactory interaction experience. The method is suitable for experimental settings, e.g. in the context of laboratory research and development environments as often encountered in Fabrication Laboratories (FabLab). The method uses Harold Garfinkel's concept of breaching experiments as a core instrument in combination with Erving Goffman's Frame Analysis. The baseline of the method is a genuine sociological definition of Social Action on the basis of theories belonging to the paradigm of Symbolic Interactionism.

Keywords: Breaching experiments · Ethnomethodology · Frame analysis · Symbolic interactionism · Bionical creativity engineering

1 Introduction

The main focus of the proposed method is to address two key questions related to successful and pleasant interactions between humans and robots: First, which are the dominant factors that determine whether the interaction is fluid and smooth? Second, to what extent do humans prefer an interaction model with a strong orientation towards the conventional interaction experiences they have with other humans – or do they prefer a type of interaction similar to typical human-machine interactions? Both dimensions are intertwined and have to be considered as two sides of the same coin. We are convinced that a method using Harold Garfinkel's instrument "breaching experiments" is highly suitable for the detection of both in equal measure. In this paper our aim is to present the method as a concept. The goal of the aforementioned research group's future empirical research is to deliver robust and reliable findings based on these concepts or theoretical frameworks. We will not be able to provide an answer to the two key questions raised, instead what we are presenting is the theoretical backing for thoroughly conducted research capable of doing so. In this regard we

© Springer International Publishing Switzerland 2015
A. Tapus et al. (Eds.): ICSR 2015, LNAI 9388, pp. 155–163, 2015.
DOI: 10.1007/978-3-319-25554-5_16

also promote and encourage theoretically grounded research in the field of HRI. Even though it would have been beneficial to create an experiment, describe the process exactly, and provide a comparison with other existing empirical situations, we decided to elucidate the abstract, theoretical qualities of the suggested method. For a detailed, concrete, and "less abstract" picture of such a setting, we would like to refer the reader to the cited papers using similar approaches.

The theoretical framework of the presented method is mainly defined by Erving Goffman's "Frame Analysis" [1] within his work on "Microstudies on Social Interaction" [2], [3] and Harold Garfinkel's "Ethnomethodology" [4]. The baseline of our approach involves assumptions as to how every social interaction is depicted by situated (i.e. contextual) expectations, the way these expectations are held stable over a relatively long period of time (according to Goffman), and which mechanisms are used – or commonly established as viable among the interacting entities – to negotiate an alignment of the predicted expectations on both sides (according to Garfinkel). With such a framework and the adoption of breaching experiments within the scope of experimental settings in a FabLab environment, we assume that we can develop a suitable method that can be applied independent of the particular cultural context and to generate reliable findings regarding the aforementioned key factors in HRI.

Comparative studies analyzing the development of social robots in Europe and Japan conclude that from a sociological point of view, they differ highly in respect to both the understanding of robot agency and the concept behind an appropriate user-robot interaction [5], [6], [7]. In Europe, the assumed interaction is dominated by the autonomy of the robot (however it manifests). However, on the basis of ethnographical research (to be published in the journal Artificial Intelligence and Society), Hironori Matsuzaki asserted that in Japan, the autonomy of the robot is overruled by the attempt to predefine or standardize the HRI sequence, which leads to a completely different concept of HRI. Both approaches could be described and analyzed in equal measure by adopting the proposed method based on Garfinkel's breaching experiments [4] – in light of Goffman's Frame Analysis [1], [8]. This method takes into consideration the specific cultural "bias" related to successful social interaction between two entities. This is due to the fact that it always operates within the culturally shaped margins of what is seen as a functioning interaction. Zooming into one culture, the method is also perfectly suitable for obtaining results on the basis of variations and differences among subgroups or individuals. One study identified several different strategies for dealing with the induced crisis [9], adopting a similar approach to the method we are aiming to refine and develop further It is specifically these kinds of previously undertaken empirical work within the scope of similar theoretical frameworks that show the method's potential to capture case sensitive key factors within a wide range of HRI situations.

2   General Assumptions Regarding HRI from a Sociological and Biomimetics Point of View

HRI research is still trapped within a psychological, and in this respect – as one would name it in sociological terms – in a methodological individualistic view (see e.g. [10], [11], and most of the paper presented in [12]). A genuine sociological approach is seldom undertaken by relying on sociological models, definitions, and theories of social action and interaction (see e.g. [13], [14]). Similar ideas regarding genuine interactional perspective have also been brought forward and experimented in HRI from the disciplinary field of interactional linguistics based on "Ethnomethodological Conversation Analysis," which is closely connected to Garfinkel's ethnomethodology [35], [36], [37]. These similar approaches should be taken into account for future research, with the goal of unifying them in an interdisciplinary frame of analysis for empirical studies in HRI. However, most of the research starts from the assumption that the interaction is somehow the result of two monolithic minds that are autonomously able to build a consistent meaning of the world and adjust their beliefs with other minds from time to time. In contrast to this view, the typical sociological perspective presented here follows the baselines of George Herbert Mead's pragmatic theory [15]. Mead's concept of symbolically mediated interaction leads to a completely different conclusion regarding the relationship between the two entities (ego and alter) that interact with each other. In Mead's definition of action, the meaning of a symbol is negotiated in a social interaction and therefore depends on the reaction of the other (alter). In a similar way, he understands the formation of identity as an interaction between the "I" (ego) and the "Me" (how alter sees ego). In other words, the meaning of a symbol constitutes itself ex post according to alter's reaction to it. In these terms, "knowing" something means anticipating alter's (most probable) reaction/understanding. Mead emphasizes the so-called "vocal gesture" because humans have the physiological ability to hear the "spoken symbol" (e.g. word) in the same way and at the same time as alter [15]. From a biological and physiological point of view, language played a useful role in social evolution as a tool for successful interactions. In the end, Mead's action theory is also the core model for Niklas Luhmann's [16], [17] micro-level theory of social systems (interaction system) and could be used to explain how consciousness is linked to the social world (in both cases, of course, as systems): The ego's psychological system (self-awareness, consciousness) is constantly observing the interaction between alter and ego, but it remains in the environment of the interaction/social system.

To analyze and capture HRI in a genuine sociological way, we choose the standard framework of Social Constructivism, conceptualizing an ideal situation of interaction by referring to ego and alter. The sociological interaction model we choose defines the social world as an outcome that is strictly interconnected with the interaction between at least two entities, also known as social actors. Social reality develops in an inter-subjective dimension; there is no reliable reality or any reality at all without interaction between social actors. The main assumption of this model is that the meaning of an action, a word, a sentence, or an object that the ego relies on is primarily defined by the reaction of alter. This also means that (social) reality (or social meaning) is always constituted ex post: it is an effect of successful interaction between two entities due to the fact that the reaction of alter related to the prior action of ego is the only way to give ego's action meaning. The next step could consist of extending this model to identity-building processes (as it was done early on by one of the forefathers of this model, George Herbert Mead [15]).

In the end it is contingently that today in many cultures, humans are the only ones who qualify as social actors [18], [19]. Ego constantly has to decide if his or her interaction partner, alter, is a social actor or not; if he, she, or it could provide a proper reaction to build a common, valid, and reliable social reality or not. The basic assumption of this argument is that who we are, what we know, what we think to be real or not real are the outcomes of interaction. For ego's beliefs and relation to reality, it is extremely vital to know whether or not alter is an entity with the proper skills that are needed to build a common reality. Ego will never know if alter is constantly deceiving him or her because ego's reality and ability to question it are dependent on alter's reactions. This is due to the fact that alter's reactions give ego the material to define reality (including identity, the horizon of meaningful questions, indisputable facts, and so on). With regards to researching HRI in different cultural contexts, this means analyzing, transferring, and implementing symbols in interaction as well as triggers for crisis carefully so that culture- as well as case-sensitive generalizations can be targeted [20].

As a matter of fact, the presented method for evaluating the quality of HRI from a genuine sociological point of view is highly suitable for use with robots developed by following a new paradigm within the robotics community. In the sense of technology development, robotics is experiencing several new orientations towards a more or less strong human-centered design. One of the most powerful new paradigms arises from the broader field of bionics and biomimetic robots. There is a very strong affinity between the sociologically oriented evaluation of HRI and biomimetic robots in achieving human-robot interactions that are not only successful, but also satisfying. Assuming that human-human interaction is the best interaction for us humans, the robot has to be humanoid or humanized. From a biomimetic point of view, the assumption is that the more biological principles are combined in a biomimetic robot, the more it can be assumed that the robot approaches its biological role model in its properties and its behavior. Technology is not yet advanced enough to develop completely functional humanoid robots. Therefore the evaluation might be limited to the examination of certain human or human-like aspects. One of these aspects may be, for example, the hand shake between human hands and humanoid hands or giving and receiving objects from a human hand to a humanoid hand and vice versa. For this, the success of human-robot interaction scenarios could be affected by, for example, visual properties such as having five fingers and/or haptic properties such as compliance in hand/arm movements. The hypothesis is: the more similar the robot hand and the human hand look, and the more similar the robot's compliance is to human skin and muscle, the more successful the interaction.

The successful application of biomimetics is characterized as the creative transfer of knowledge and ideas from biology to technology, i.e. technological development inspired by nature that usually passes through several steps of abstraction and modification subsequent to the biological starting point. The field of biomimetics is highly interdisciplinary, which is indicated by the high level of cooperation between experts from different fields of research, for example among biologists, physicists, and engineers: "Biomimetics combine biology and technology with the goal of solving technical problems through the abstraction, transfer, and application of knowledge gained in interdisciplinary cooperation from biological models." [21] Within robotics, which is a broad area in the field of engineering, the application of biomimetic methods is similarly widespread in the design, control, and operation of robotic systems. In this regard, an officially accepted definition of biomimetic robots is: "A robot in which at least one dominant biological principle has been implemented and which is usually developed based on the biomimetic development process." [22]

3 Evaluating the Quality of HRI with Breaching Experiments

Although several studies have used the instrument of the breaching experiment (some-times even without naming it, but definitely adopting its primary aspects) none of them has developed a systematic approach for establishing a general method for the evaluation of Human-Robot-Interaction HRI [23], [24], [25], [26], [27], [9], [28], [29], [30]. As an instrument, the breaching experiment is highly suitable for the evaluation of HRI for several reasons. First, it operates on a very high level with respect to understanding social action demands. Second, it is not subject to most of the common biases derived from the notion of delivering a socially desirable answer, in that the framing of the situation is taken into consideration. In a typical setting to evaluate quality of interaction qualitative-ly, the test persons are asked several questions regarding their subjective impressions of the experience after performing an interaction sequence with a robot. Compared to the well known Human-Human-Interaction (HHI), HRI is often disappointing and to some extent similar to it. The interaction sequence is mostly carried out by and determined by the human. The human fills in the gaps that arise in the course of the interaction sequence due to the robot's inability. In HRI experiments, this specific – although typical – situa-tion tends to result in a positive assessment of the experienced quality of the interaction. While assessing the situation, the test person will most probably either highlight their own efforts to let the interaction flow or emphasize what they thought were the research-ers' expectations. As an instrument, breaching experiments could deliver an authentic response insofar as the test persons will perform repair strategies just in case, since he or she expects a positive outcome. If the test person assumes that his or her attempt to rees-tablish a functioning interaction is condemned to be a failed repair, he or she won't try to repair it. However, the frame of the situation is of paramount importance.

In their study, Muhl & Nagai [9] show that the breaching experiment – put in the right setting with respect to framing – is able to deliver impressive results. Without reflecting their experiment design by theoretical means, they used a typical deception strategy and in doing so bent the frame in their favor. However, they were able to identify six different strategies to cope with the unexpected behavior of the robot and repair the undertaken interaction. Even if the interaction was quite rudimentary, the performed repair strategies show that the test person believed in the robot's interac-tion capabilities to a certain extent. In a nutshell, they successfully showed that within a clearly laboratory experimental setting, breaching experiments lead to satisfying, fruitful results. In a lab scenario, people are instructed to show a robot objects and how to use them. In this scenario, the robot is just an animated baby face [31] dis-played on a screen. Its eyes, eyelids, eyebrows, and mouth are animated. The robot is equipped with a biologically inspired saliency mechanism [32]. Thus, the robot's gaze follows the most relevant feature in the scene. This is how the robot addresses/displays its attention to the human interaction partner. The robot is not equipped with acoustic sensors or a speech processing system [9]. By interacting with it, humans can learn that the robot follows the salient point with its gaze. Human actors apply strategies of repair if an irritation of their expectation appears. They try to re-attract the robot's attention to the object by adopting several strategies (e.g. point to the object, show the object closer to the robot, getting the robot's attention, making noise, and so on) [9].

In this experiment, crisis in interaction has been induced systematically: The cognitive framing applied by ego to the state of the interaction partner (alter) is relevant for the overall judgment about alter, and in its consequence, the selection of how to approach alter in the next turn. In this regard a thoroughly conducted frame analysis is able to deliver highly important factors that are primarily responsible for the overall outcome of HRI testing and therefore shaping the way humans deal with the breaching. Taking the frame into account, one may see that repairing strategies (as well as the fact that repairing strategies are undertaken at all) depend on the humans' definition of the situation, which is strictly linked to the assumed frame. Putting the emphasis on the framing is not just important in terms of awareness of which framing strategy the researchers are adopting and being able to achieve a high degree of transparency, it is also important in estimating the viability of the breaching experiment as an instrument itself. By comparing the previously mentioned study of Muhl & Nagai [9] with a research conducted in a stationary care facility for the elderly, Compagna & Muhl [14] showed how important the frame is for the accomplishment of a reliable outcome in breaching experiments.

However within the setting (and therefore the framing) of an everyday life context, breaching experiments were not possible. A service robot was asked to serve a glass of water to the residents of a home for elderly people [33], [34]. The task was to take the person's order and then serve the glass to the correct person. The robot was also asked to address the human by talking. Often, the people did not reply to the robot and preferred to address the other people present. In the cases in which humans accepted the drink, the robot thanked them, which was mostly ignored by the humans. The robot was not capable of reacting flexibly and turning the rejection into a request, e.g. by commenting on it, which would have led to communication. The likelihood of successful communication would have improved. Social robots do not necessarily offer communication to which humans respond positively. If an action expected of the robot does not occur, it will probably be repaired by the involved actors, and the reaction to such a maneuver is often as unexpected. This does not refer back to any attempt to establish understanding in which the action of ego would semantically be constructed by the reaction of alter. In those cases, interaction in a sociological sense is not only endangered by its failure, but it cannot occur at all. This contrasts to the interaction experiments in the laboratory with the robot baby face in which, as mentioned above, the human actor tried – with more or less patience – to settle meaning/semantics with the robot as his or her alter ego. After several non-successful trials the interaction is abandoned.

Comparing these two cases the paramount importance of framing becomes visible: The breaching experiment method obtains very fruitful findings in an experimental setting. There are indications that the method also works properly if the experiment is not mentioned at all, and if encountering an interaction with a robot is not expected [24]. The framing related to the expectations raised by the humans seems to be the key issue here. Without a doubt framing is very important, however further research has to be done to determine the main aspect that is entangled by the frame within which the HRI is carried out.

4   Summary

A method built on breaching experiments as core instruments with a strong emphasis on framing issues is most likely highly suitable for generating reliable results with regard to the quality and rate of interaction between a human and a robot even from a genuinely sociological perspective. The observation as to whether and how a crisis (explicitly induced by the researchers) is repaired by a human could lead to a significantly meaningful evaluation of HRI that is also suitable for identifying differences between individuals [9], [14]. In order to set the right framing, it is very important to reflect the framing as a highly influential variable. Without a proper frame analysis, the findings of HRI breaching experiments are probably useless. However, if the frame is chosen wisely, the outcome could be very helpful in judging whether the HRI is successful or not. If the human adopts strategies to repair the interaction, the interaction can be described as a social interaction insofar as the human is assuming that it is worth being repaired. Even if the human is fully aware that the robot is a machine that is not capable of repairing the course of the interaction itself (one may say the robot is not able to process double contingency or elaborate on these grounds on a hypothesis as to how to reestablish a smooth flow (16)), the humans nonetheless consider the robot to be an entity that can be treated as a social actor. The comparison to a washing machine could be helpful for further understanding of the nonsymmetrical capabilities between the interacting entities: If a washing machine does not "react" as expected, the user will most probably abort the "interaction" assuming that the machine is simply malfunctioning. However, even here a certain number of repairing strategies can be observed, but these certainly do not include trying to ask or behave in a different way. In conclusion: If the framing is taken into consideration and chosen correctly, the way repair strategies were undertaken by the human (in combination with the observed frequency, quantity, and timespan) could be used to define the quality of the HRI. By doing so, the researcher could gain helpful information for the further development of social robots in regard to their interaction capabilities.

Acknowledgments. The Research presented in this paper was primarily supported by the German Ministry of Education and Research. In addition, we would like to thank Manuela Marquardt for her outstanding support and helpful comments.

References

1. Goffman, E.: Frame Analysis. Harper & Row, New York (1974)
2. Goffman, E.: Interaction Ritual: Essays on Face-to-Face Behavior. Anchor Books, New York (1967)
3. Goffman, E.: Relations in Public: Microstudies of the Public Order. Basic Books, New York (1971)
4. Garfinkel, H.: Studies in Ethnomethodology. Polity Press, Cambridge (1967)
5. Wagner, C.: Robotopia Nipponica: Recherchen zur Akzeptanz von Robotern in Japan. Tectum-Verl, Marburg (2013)
6. MacDorman, K.F., Vasudevan, S.K., Ho, C.-C.: Does Japan Really Have Robot Mania? Comparing Attitudes by Implicit and Explicit Measures. AI & Society 23(4), 485–510 (2009)
7. Kaplan, F.: Who is Afraid of the Humanoid? Investigating Cultural Differences in the Acceptance of Robots. International Journal of Humanoid Robotics 1(03), 465–480 (2004)
8. Goffman, E.: Behavior in Public Places. Free Press, New York (1963)
9. Muhl, C., Nagai, Y.: Does disturbance discourage people from communicating with a robot? In: 16th IEEE International Symposium on Robot and Human Interactive Communication (RO-MAN), Jeju, Korea (2007)
10. Wykowska, A., Ryad, C., Al-Amin, M.M., Müller, H.J.: Implications of Robot Actions for Human Perception. How Do We Represent Actions of the Observed Robots? International Journal of Social Robotics 6(3), 357–366 (2014)
11. Feil-Seifer, D., Skinner, K., Matarić, M.J.: Benchmarks for Evaluating Socially Assistive Robotics. Interaction Studies 8(3), 423–439 (2007)
12. Herrmann, G. (ed.): ICSR 2013. LNCS, vol. 8239. Springer, Heidelberg (2013)
13. Burghart, C., Haeussling, R.: Evaluation criteria for human robot interaction. In: Proceedings of the Symposium on Robot Companions: Hard Problems and Open Challenges in Robot-Human Interaction, pp. 23–31 (2005)
14. Compagna, D., Muhl, C.: Mensch-Roboter-Interaktion – Status der technischen Entität, Kognitive (Des)Orientierung und Emergenzfunktion des Dritten. In: Stubbe, J., Töppel, M. (eds.) Muster und Verläufe der Mensch-Technik-Interaktivität, Band zum gleichnahmigen Workshop am 17./18. Juni 2011 in Berlin, Technical University Technology Studies, Working Papers, TUTS-WP-2-2012, Berlin, pp. 19–34 (2012)
15. Mead, G.H.: Mind, Self, and Society. University of Chicago Press (1934). Ed. by Charles W. Morris
16. Luhmann, N.: Soziale Systeme. Grundriß einer allgemeinen Theorie. Frankfurt a.M, Suhrkamp (1984)
17. Hahn, A.: Der Mensch in der deutschen Systemtheorie. In: Bröckling, U., Paul, A.T., Kaufmann, S. (Hg.) Vernunft - Entwicklung - Leben. Schlüsselbegriffe der Moderne. Festschrift für Wolfgang Eßbach, pp. 279–290. Fink, München (2004)
18. Lindemann, G.: Doppelte Kontingenz und reflexive Anthropologie. Zeitschrift für Soziologie 28(3), 165–181 (1999)
19. Baecker, D.: Who qualifies for communication? A systems perspective on human and other possibly intelligent beings taking part in the next society. Technikfolgenabschätzung - Theorie und Praxis 20(1), 17–26 (2011)
20. Lutze, M., Brandenburg, S.: Do we need a new internet for elderly people? A cross-cultural investigation. In: Rau, P. (ed.) HCII 2013 and CCD 2013, Part II. LNCS, vol. 8024, pp. 441–450. Springer, Heidelberg (2013)

21. VDI: VDI Guideline 6220, Part 1: Biomimetics – Conception and Strategy. Differences between Biomimetic and Conventional Methods/Products. Verein Deutscher Ingenieure e.V., Düsseldorf (2012)
22. VDI: VDI Guideline 6222, Part 1: Biomimetics – Biomimetic Robots. Verein Deutscher Ingenieure e.V., Düsseldorf (2013)
23. Weiss, A., Bernhaupt, R., Tscheligi, M., Wollherr, D., Kuhnlenz, K., Buss, M.: A methodological variation for acceptance evaluation of human-robot interaction. in: public places. In: 2008 17th IEEE International Symposium on Robot and Human Interactive Communication, RO-MAN 2008, pp. 713–18. IEEE (2008)
24. Weiss, A., Bernhaupt, R., Tscheligi, M., Wollherr, D., Kuhnlenz, K., Buss, M.: Robots asking for directions: the willingness of passers-by to support robots. In: Proceedings of the 5th ACM/IEEE International Conference on Human-Robot Interaction, pp. 23–30. IEEE Press (2010)
25. Sirkin, D., Mok, B., Yang, S., Ju, W.: Mechanical Ottoman: how robotic furniture offers and withdraws support. In: Proceedings of the Tenth Annual ACM/IEEE International Conference on Human-Robot Interaction, pp. 11–18. ACM (2015)
26. Bauer, A., Klasing, K., Lidoris, G., Mühlbauer, Q., Rohrmüller, F., Sosnowski, S., Xu, T., Kühnlenz, K., Wollherr, D., Buss, M.: The Autonomous City Explorer: Towards Natural Human-Robot Interaction in Urban Environments. International Journal of Social Robotics 1(2), 127–140 (2009)
27. Alac, M., Movellan, J., Tanaka, F.: When a Robot Is Social: Spatial Arrangements and Multimodal Semiotic Engagement in the Practice of Social Robotics. Social Studies of Science, 0306312711420565 (2011)
28. Nagai, Y., Rohlfing, K.J.: Can motionese tell infants and robots 'what to imitate'? In: Proceedings of the 4th International Symposium on Imitation in Animals and Artifacts, pp. 299–306 (2007)
29. Short, E., Hart, J., Vu, M., Scassellati, B.: No fair!! an interaction with a cheating robot. In: 2010 5th ACM/IEEE International Conference on Human-Robot Interaction (HRI), pp. 219–226 (2010)
30. Takayama, L., Groom, V., Nass, C.: I'm sorry, Dave: I'm afraid i won't do that: social aspects of human-agent conflict. In: Proceedings of the SIGCHI Conference on Human Factors in Computing Systems, pp. 2099–2108 (2009)
31. Ogino, M., Watanabe, A., Asada, M.: Mapping from facial expression to internal state based on intuitive parenting. In: Proceedings of the Sixth International Workshop on Epigenetic Robotics, pp. 182–183 (2006)
32. Nagai, Y., Asada, M., Hosoda, K.: Learning for joint attention helped by functional development. Advanced Robotics 20(10), 1165–1181 (2006)
33. Compagna, D.: Reconfiguring the user: raising concerns over user-centered innovation. In: Proceedings ECAP10. VIII European Conference on Computing and Philosophy, pp. 332–336 (2010)
34. Compagna, D.: Lost in Translation? The Dilemma of Alignment within Participatory Technology Developments. Poiesis & Praxis 9(1–2), 125–143 (2012)
35. Pitsch, K., Vollmer, A.-L., Mühlig, M.: Robot feedback shapes the tutor's presentation. How a robot's online gaze strategies lead to micro-adaptation of the human's conduct. Interaction Studies 14(2), 268–296 (2013). doi:10.1075/is.14.2.06pi
36. Pitsch, K., Wrede, S.: When a robot orients visitors to an exhibit. Referential practices and interactional dynamics in real world HRI. In: Ro-Man 2014, pp. 36–42 (2014)
37. Pitsch, K., Lohan, K.S., Rohlfing, K., Saunders, J., Nehaniv, C. L., Wrede, B.: Better be reactive at the beginning. Implications of the first seconds of an encounter for the tutoring style in human-robot-interaction. In: Ro-Man 2012, pp. 974-981 (2012). http://ieeexplore. ieee.org/xpls/abs_all.jsp?arnumber=6343876&tag

Social Facilitation in a Game-Like Human-Robot Interaction Using Synthesized Emotions and Episodic Memory

Arturo Cruz-Maya(✉), François Ferland, and Adriana Tapus

Robotics and Computer Vision Laboratory, ENSTA-ParisTech, Palaiseau, France
{arturo.cruz-maya,francois.ferland,adriana.tapus}@ensta-paristech.fr

Abstract. Companion robots will be more and more part of our daily lives in the next years, and having long-term interactions can have both positive and negative effects on their users. This paper presents an experiment that is focusing on social facilitation. Our scenario is a memory game with the Nao robot and is combining an emotional system based on the OCC Model, and an episodic memory mechanism. Our first preliminary results show evidence that support the theory and present a first step towards an adaptive lifelong learning system.

1 Introduction

Social facilitation effect is a widely studied [7][12] psychology paradigm introduced by [16] that states that individuals get a better performance on easy tasks if they are in presence of others, but their performance is worst in complex tasks. With robots being more and more around people, situations where social facilitation has an effect can appear more frequently. This can have positive or negative influence on the social interaction and the robot needs to be capable of adapting to the user so as to improve the interaction and the user's task performance.

Very little work in social robotics [15] [10] or virtual characters [9] has focused on Social Facilitation. The authors in [10] presented a study that compared the task performance of 106 participants on easy and complex cognitive and motor tasks across three presence groups (alone, human presence, and robot presence). They found evidence that confirms the theory of Social Facilitation, but they focused on the mere presence of the robot. This paper presents an experiment where the social facilitation effect in Human-Robot Interaction is investigated. The scenario involves a memory card game in which the robot is the opponent of the human player, and it can take two roles: it can encourage or judge the human-user, depending on the game mode.

We also introduce a high-level framework, which integrates an emotional model and an episodic memory, which has the potential to adapt to the user's preferences based on the robot's emotions generated by the interactions with the users. The emotional model is a partial implementation of the Orthony Claire Collins Model(OCC Model) [8] having 6 emotions: love, hate, pride, shame, admiration, and reproach. These emotions are classified in two categories: *Aspect*

© Springer International Publishing Switzerland 2015
A. Tapus et al. (Eds.): ICSR 2015, LNAI 9388, pp. 164–173, 2015.
DOI: 10.1007/978-3-319-25554-5_17

of Objects and *Actions of Agents*. This last category is of particular importance for the Social Facilitation experiment because both the psychological theory and the OCC Model rely on the performance of the actions of the agents. The OCC Model has been widely implemented in Human-Machine Interaction for virtual animated characters [2] and robotics. The authors in [5] present the development of a robot equipped with the OCC model for their emotion engine, but not in an interaction scenario.

Furthermore, we are using an episodic memory (EM-ART - an episodic memory (EM) using Adaptive Resonance Theory (ART) neural networks [3]), presented in [14] [6]. The EM-ART records sequences of events as episodes. Here, it is used to store information about the games, by recording the sequence of cards obtained by each player.

To the best of our knowledge, there are no works in the literature that fully combine the emotion and the memory system in an interaction for testing social facilitation. This work presents a first step in that direction, using EM-ART as long-term memory to provide useful information to the emotion system based on the OCC Model, and the output of this is used to modulate the intensity of different expressive behaviors.

This paper is organized as follows: Section 2 describes the scenario used and the set up of the methods applied; Section 3 explains the high level framework proposed in this work; Section 4 shows the results obtained regarding the Social Facilitation experiment and the perception of the emotional expressions; and finally, Section 5 concludes the paper.

2   Experimental Design Setup

2.1   Hypothesis

The hypothesis of this work are inspired by the Social Facilitation effect and are formulated as follows:

Hypothesis A: The user's performance in an easy task while being encouraged by a robot, will be better than while performing the same task alone.

Hypothesis B: The user's performance in a difficult task while being judged by a robot, will be worst than while performing the same task alone.

2.2   Game Scenario Description

In order to test and validate our system we designed the "Find the Pair" board game. The "Find the Pair" board game is played with a set of cards containing pairs of matching images. The cards are put face down on a grid with letters marking columns from A to D and numbers marking rows 1 to 5. At each turn the player has to uncover two cards. If they are matching, the cards are removed and the player can uncover a new pair of cards, and so on. Players switch turns when a non-matching pair is uncovered. The game ends when all matching pairs have been discovered. The player with the most pairs wins.

The robot cannot manipulate the cards by itself. Instead, it says the letter and number of the desired card position, and then the user has to uncover it, enabling the robot to recognize the image of the card.

At each turn the players' performance is calculated based on the number of times uncovering a card in the same spot on multiple turns. At the end of the game the user's task performance is calculated based on the number of pairs.

The information about the players is stored in the EM-ART with 5 channels: Person, Game Difficulty, Robot turn, Performance, and Card. An event in the episodic memory corresponds to a turn in the game and an episode corresponds to a complete game.

The experimental conditions were defined by two factors, game difficulty level and presence of the robot. The game was tested with two levels of difficulty depending on the numbers of cards: 10 for the easy mode, and 20 for the difficult mode. Each participant played the game 3 times alone in both game modes before playing versus the robot, and 1 time versus the robot in each game mode.

2.3   Robot Behaviors

The robot has eight behaviors: Greeting people, pointing to the cards, and expressing pride, shame, admiration, reproach, encouraging, and judging. Except for encouraging and judging, each behavior produces both speech and movement. For pride, the robot rise its arms at the height of its waist, for shame, the robot rise its right arm and cover its face with it, for admiration, the robot "claps" with its arms, and for reproach the robot moves its head from left to right and vice versa two times. At each turn and at the end of the game, the robot can perform a body motion behavior or speak according to the intensity of the emotions present in its internal state. Fig. 1 shows the emotional expressions of admiration and pride corresponding to the actions of agents of the OCC Model.

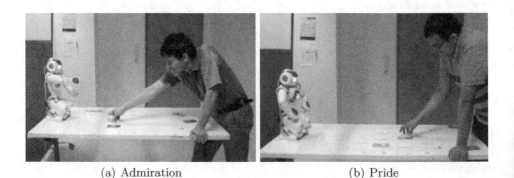

(a) Admiration   (b) Pride

Fig. 1. Robot Emotional Expressions

3   Methodology

The high level framework proposed is composed of face and card recognition modules, an episodic memory, a cognitive emotional system, a task specific module to play the "Find the pair" game, a database of the preferences of the robot, and an expression generator module.

The face recognition module generates a search in the episodic memory specified by the name of the person and the game level difficulty, which gives the information of the last played game, and it is sent to the speech generation module. Based on the performance on the task (e.g. "number of pairs obtained") and the attitude of the robot towards the person and the game, the emotional system generates responses that are communicated to the expression generator module.

Face Detection is done by using the Viola-Jones [13] method and Face Recognition with Local Binary Patterns [1], using the implementations provided by the OpenCV library. Card Recognition is based on FindObject2D[1], an open source project that uses a bag-of-words approach with different types of 2D image features. Here, FindObject2D is configured to use the OpenCV implementation of FAST [11] features on images incoming from the robot's camera.

The game board used was a white paper of A3 size, with 4cm wide square corners coloured in black and set on a white table. For detection of the game board, the image was binarized with the Otsu method, and the Harris corner detection method was applied. Then, the region was transformed to correct perspective distortion and facilitate recognition of the cards. Game board detection and perspective correction was also implemented with OpenCV.

3.1   Episodic Memory

Our framework uses the EM-ART implementation presented in [4]. The EM-ART model [14], shown in Fig. 2, is made of three layers: Input, Events, and Episodes. The Input Layer is used to represent the external context information. It is categorized in channels in which each node represents the presence of a known element with an associated activation level (e.g. "Person A", "Easy Mode", "Card 1"). The nodes found in the Events Layer represents elements in the Input Layer that were activated simultaneously (e.g., "Person A" and "Card 1"). Synchronization of input elements is done by a short term memory buffer. As time progresses, the activation level of nodes in the Event layer decreases. Therefore, the sequence of events is represented by the pattern formed by those levels: the highest activation level is associated the most recent event to occur, and the lowest to the oldest. The Episodes Layer is made of nodes that categorize the patterns of the activation level of nodes in the Events Layer, thus defining episodes as temporal sequences of events. New episodes are created only when learning is triggered. In this work, learning is triggered at the end of each game to record its final sequence of events.

---

[1] http://introlab.github.io/find-object/

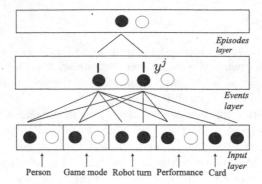

Fig. 2. EM-ART Model with a recalled episode built out of 2 events. y^j represents the activation level of event j.

3.2 OCC Model

The OCC Model [8] is based on 4 global variables and 12 local variables, each variable depending on both physical and psychological factors. The model has 22 emotions that are divided in three categories: *Aspect of Objects*, *Action of Agents* and *Consequences of Events*. In this work, we focused on the category of Action of Agents for its relation with the social facilitation effect. The synthesis of the emotions belonging to this category is described as follows:

Synthesis of Emotions. Let $A(p, o, t)$ be the approving of action o that person p assigns at time t. This function returns a positive value if the performance of the action is above the standards for that action, and returns a negative value if the action doesn't meet the standards. The standard is a given value to determine if the performance is low or high. Let $I_g(p, o, t)$ represent a combination of the global intensity variables. Let $P_a(p, o, t)$ be the potential for generating a state of admiration. If the action is performed by others, the rules for admiration and reproach are presented in Algorithm 1.

Algorithm 1 Synthesis of Admiration and Reproach

---
set $P_a(p, o, t) = f_p(A(p, o, t), I_g(p, o, t))$
if $A(p, o, t) > 0$ **then**
    Given a threshold value T_a
    if $P_a(p, o, t) > T_a(p, t)$ **then** set $I_a(p, o, t) = P_a(p, o, t) - T_a(p, t)$
    else set $I_a(p, o, t) = 0$
    end if
else
    Given a threshold value T_r
    if $P_a(p, o, t) > T_r(p, t)$ **then** set $I_r(p, o, t) = P_a(p, o, t) - T_r(p, t)$
    else set $I_r(p, o, t) = 0$
    end if
end if

The approving function f_p is denoted by:

$$f_p = (Praise * SoR * CogUnit) + (Prox * a) + (Ar * b) \tag{1}$$

where $Praise$ is the praiseworthiness, SoR is the sense of reality, $CogUnit$ is the cognitive unit, meaning the grade of similarity between the preferences of the robot and the person, $Prox$ is the proximity, Ar is the arousal and a and b are factors of increment set empirically to 0.1 and 0.3, respectively. The emotions of pride and shame were synthesized the same way, but based on the performance of the robot. The functions describing the arousal and the mood are presented in Algorithm 2. The threshold functions were denoted by a sigma function (2), and having as input the value of the mood. Also the arousal was processed with a s-shaped sigmoid function to limit its value.

Algorithm 2 Arousal and Mood

$arousal = arousal + \sum_{i=1}^{n} intensity\_emotion - (t_i - t_{i-1}) * 0.05$

if $arousal < 0$ **then** $arousal = 0$

end if

if $admiration$ or $pride$ or $love$ **then**

   $mood = mood + intensity\_emotion - (t_i - t_{i-1}) * 0.01$

else [$reproach$ or $shame$ or $hate$]

   $mood = mood - intensity\_emotion - (t_i - t_{i-1}) * 0.01$

end if

The value of $Praise$ is defined by $Praise = performance + (expDev)$, where $expDev$ is the expected deviation given by $expDev = performance - lastPerformance$. Then the $Praise$ was processed with a sigma function denoted by:

$$y = (g/(1 + e^{-(x-x_0)/s})) + y_0 \tag{2}$$

where s is the change step of the sigmoid, g is the maximum value, x_0 is the half of the sigmoid, y_0 is used to give a positive or negative output and x is the input.

The parameters of the OCC Model were set up as: $Appealing = 0.5$, $Familiarity = 0.1$, $Praiseworthiness = $ Player's performance, $Strength$ of $Cognitive$ $Unit=1$ for the robot's actions and 0.5 for the person's actions.

4   Results and Discussion

The experiment was done with a NAO robot from Aldebaran Robotics. The experiment was tested with 10 participants: 8 male and 2 female with ages in range from 22 to 34 years old and all with technical background. 4 of them had no prior interaction with a robot.

The measure of the performance for the Social Facilitation experiment was given by $perform = errors/pairs\_obtained$, where $pairs\_obtained$ is the number of pairs obtained at the end of the game. An error is counted when a card has been shown previously and the person did not remember its position. The results are presented in Table 1. Paired T-Tests for dependent means and two-tailed hypothesis were applied. The results were significant with $p < 0.10$, rejecting the null hypothesis for comparison between the Easy level played alone and the Easy level played with robot conditions ($t = -1.8653$, $p = 0.0950$) and for comparison between the Difficult level played alone and the Difficult level played with robot conditions ($t = 2.2003$, $p = 0.0553$). With $t < 0$, this suggests that the mean errors count in Easy mode was higher for the "playing alone" group (0.1788 vs. 0.0990), and the opposite in Difficult mode (0.3159 vs. 0.4810).

The differences in means are small, but can be explained by the small number of participants in the experiments. However, the results are statistically significant, and they confirm both hypotheses considered in this work. As expected for the theory of Social Facilitation, the performance is affected by the presence and behavior of the robot, improving it when the task is easy and worsen it when the task is difficult.

Table 1. Mean and variance of participants' performance in the 4 game modes

	Easy Alone	Difficult Alone	Easy with robot	Difficult with robot
mean	0.1788	0.3159	0.0990	0.4810
var	0.0019	0.0211	0.0254	0.0837

The internal emotional states of the robot are presented in Fig. 3, where the performance of the robot and one participant in difficult mode can be seen. In Fig. 3(a), when the performance of the participant turns negative (under the standard), the intensity of reproach increases. On the other hand, when the performance of the participant increases, the intensity of admiration increases too. In Fig. 3(b), the performance of the robot along with the emotions of pride and shame can be observed. Even when pride was called during the game, at the end the raised emotion was shame because the standards were set to expect a high score of the robot. The mood and the arousal of the robot in this game are shown In Fig. 3(c), where the mood can take negative values according to the the synthesized emotions based on the performance of both players, decreasing with time and with negative emotions as shame and reproach, and increasing with positive emotions as pride and admiration. The arousal only has positive values, increasing with any kind of emotions, and decreasing at a higher rate than the mood.

The emotion system was analysed using a subjective measure of the participants with a post-experiment questionnaire[2]. Table 2 (Robot Behavior) shows the answers of the participants on a 7-point Likert scale, 7 for strongly agree

---

[2] http://goo.gl/forms/3TNI4KjvXu

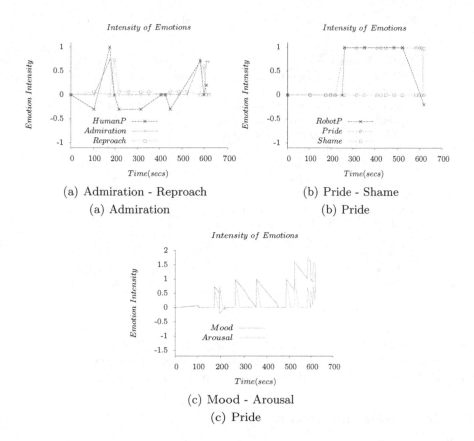

(a) Admiration - Reproach

(a) Admiration

(b) Pride - Shame

(b) Pride

(c) Mood - Arousal

(c) Pride

Fig. 3. Human and Robot Performance and Robot's internal state of Mood and Arousal

and 1 for strongly disagree (4 for neither agree or disagree). The answers of the participants show a clear recognition of the states of the robot in the different emotional states, which proves that the behaviors used were perceived as expected.

The negative mood was not so present due to the fact that even when the person gets a pair in the game, the robot can show admiration and increase its positive mood. Also, the participants were asked two questions about their level of appreciation of the game in both modes (easy and difficult). The answers are shown in Table 2 (Game Mode), which are related to the variable $CogUnit$ of the OCC Model. If the values of the attitude of the players towards the game had been used, the intensity of the generated emotions should had been higher, because they are more similar to the set value in the robot's preferences.

Finally, two open questions were asked to the participants to express their liking or disliking towards the robot, multiple answers were allowed. 5 participants liked the enthusiasm of the robot, 3 participants liked the correlation between the actions and the behaviors, 4 participants found the robot funny,

Table 2. Participants' feedback

Robot Behavior	mode	mean	var
1. Robot as good motivator in the easy game	6	4.62	2.92
2. Robot disturbing in the difficult game	3	3.67	2.60
3. Positive mood when it was winning	6	5.83	0.69
4. Negative mood when it was losing	2,3,5	3.58	2.99
5. Admiration when in the easy game	4	3.75	3.47
6. Reproach in the difficult game	6	4.50	2.63
7. Pride when it was winning	7	6.17	0.69
8. Pride when it was losing	3	2.92	1.90
9. Could be a good companion	6	5.50	1.72
10. Reproach when it was winning	4	3.67	3.33
11. Disturbing in the easy game	4	3.92	3.71
12. Motivating in the difficult game	4	3.67	2.60
Game mode	**mode**	**mean**	**var**
1. Easy	6	4.91	2.29
2. Difficult	6	5.82	0.76

and 1 participant found it clever. Six participants said that they disliked that the robot was too egocentric. This can be explained due to the fact that we wanted that the person feel being evaluated in the difficult mode.

5   Conclusion and Future Work

In this paper we investigated the social facilitation experiment in a human-robot interaction, with an emotional system based on the OCC Model. The results showed proofs reinforcing this theory, which makes us believe that this should have to be considered for companion robots where the interaction in everyday life can provoke stressful situations for the humans. This kind of robots have to adapt to their users through their interaction, and an emotional system can take place to manage the robot's behaviors. For that reason, we plan to continue with the implementation of the OCC Model, which is a very extensive model that include a large range of emotions, where memory plays an important role. The EM-ART needs to be combined in a deeper way with the emotional system, because memory is the base of the section "consequences of events" of this model. Furthermore, an hybrid emotional approach, combining basic and cognitive emotions, may be beneficial for faster responses of the robot.

Acknowledgments. The first author thanks to the Mexican Council of Science and Technology for the grant CONACYT-French Government n.382035 and to the Secretary of Public Education of Mexican Government for the support fellowship.

References

1. Ahonen, T., Pietikäinen, M., Hadid, A.: Face recognition with local binary patterns. In: Matas, J.G., Pajdla, T. (eds.) ECCV 2004. LNCS, vol. 3021, pp. 469–481. Springer, Heidelberg (2004)
2. André, E., Klesen, M., Gebhard, P., Allen, S., Rist, T.: Integrating models of personality and emotions into lifelike characters. In: Paiva, A. (ed.) Affective interactions. LNCS, vol. 1814, pp. 150–165. Springer, Heidelberg (2000)
3. Carpenter, G.A., Grossberg, S.: A massively parallel architecture for a self-organizing neural pattern recognition machine. Computer Vision, Graphics, and Image Processing 37(1), 54–115 (1987)
4. Ferland, F., Cruz-Maya, A., Tapus, A.: Adapting an hybrid behavior-based architecture with episodic memory to different humanoid robots. In: 24th IEEE International Symposium on Robot and Human interactive Communication. IEEE (2015)
5. Kröse, B.J.A., Porta, J.M., van Breemen, A.J.N., Crucq, K., Nuttin, M., Demeester, E.: Lino, the user-interface robot. In: Aarts, E., Collier, R.W., van Loenen, E., de Ruyter, B. (eds.) EUSAI 2003. LNCS, vol. 2875, pp. 264–274. Springer, Heidelberg (2003)
6. Leconte, F., Ferland, F., Michaud, F.: Fusion adaptive resonance theory networks used as episodic memory for an autonomous robot. In: Goertzel, B., Orseau, L., Snaider, J. (eds.) AGI 2014. LNCS, vol. 8598, pp. 63–72. Springer, Heidelberg (2014)
7. Michaels, J., Blommel, J., Brocato, R., Linkous, R., Rowe, J.: Social facilitation and inhibition in a natural setting. Replications in Social Psychology 2(21–24) (1982)
8. Ortony, A., Clore, G.L., Collins, A.: The Cognitive Structure of Emotions. Cambridge University Press (1990)
9. Park, S., Catrambone, R.: Social facilitation effects of virtual humans. Human Factors: The Journal of the Human Factors and Ergonomics Society 49(6), 1054–1060 (2007)
10. Riether, N., Hegel, F., Wrede, B., Horstmann, G.: Social facilitation with social robots? In: Proceedings of the 7th ACM/IEEE International Conference on Human-Robot Interaction, pp. 41–47. IEEE (2012)
11. Rosten, E., Porter, R., Drummond, T.: Faster and better: A machine learning approach to corner detection. IEEE Transactions on Pattern Analysis and Machine Intelligence 32(1), 105–119 (2010)
12. Uziel, L.: Individual differences in the social facilitation effect: A review and meta-analysis. Journal of Research in Personality 41(3), 579–601 (2007)
13. Viola, P., Jones, M.J.: Robust real-time face detection. International Journal of Computer Vision 57(2), 137–154 (2004)
14. Wang, W., Subagdja, B., Tan, A.H., Starzyk, J., et al.: Neural modeling of episodic memory: Encoding, retrieval, and forgetting. IEEE Transactions on Neural Networks and Learning Systems 23(10), 1574–1586 (2012)
15. Wechsung, I., Ehrenbrink, P., Schleicher, R., Möller, S.: Investigating the social facilitation effect in human-robot interaction. In: Natural Interaction with Robots, Knowbots and Smartphones, pp. 167–177. Springer (2014)
16. Zajonc, R.B., et al.: Social facilitation. Institute for Social Research, University of Michigan, Research Center for Group Dynamics (1965)

Motions of Robots Matter! The Social Effects of Idle and Meaningful Motions

Raymond H. Cuijpers[✉] and Marco A.M.H. Knops

Human Technology Interaction Group, Eindhoven University of Technology,
5600 MB Eindhoven, The Netherlands
r.h.cuijpers@tue.nl

Abstract. Humans always move, even when "doing" nothing, but robots typically remain immobile. According to the threshold model of social influence [3] people respond socially on the basis of social verification. If applied to human-robot interaction this model would predict that people increase their social responses depending on the social verification of the robot. On other hand, the media equation hypothesis [11] holds that people will automatically respond socially when interacting with artificial agents. In our study a simple joint task was used to expose our participants to different levels of social verification. Low social verification was portrayed using idle motions and high social verification was portrayed using meaningful motions. Our results indicate that social responses increase with the level of social verification in line with the threshold model of social influence.

1 Introduction

During human-robot interaction a robot typically stops moving during idle periods and the robot appears inanimate and lifeless. On the other hand, the human body never stops moving and therefore always communicates being alive. So idle movements could present a basic "illusion of life", which could help people accept the robot as a social entity [8]. Idle motions are used a lot in gaming and movie animations [5,12]. However, only few studies investigated the role of idle motions in relation to making robots more social entities. For example, [14] mimicked clerk idle movements on a robot, but the effect on social interaction was not tested.

There are two competing views about people's social responses towards artificial agents. According to the media equation hypothesis [11] humans automatically respond socially when interacting with artificial agents as long as there are some behaviours that suggest a social presence. For example, it was found among others that people rate computers more trustworthy and intelligent when the computer belonged to the same team, or when it showed an avatar's face of the same ethnicity [10]. It is suggested that people respond out of habit to mimicked social cues due to overlearning [10]. Based on this, one would expect that a robot portraying idle motions not only looks more alive but also elicits social responses. On the other hand, the threshold model of social influence [3]

A. Tapus et al. (Eds.): ICSR 2015, LNAI 9388, pp. 174–183, 2015.
DOI: 10.1007/978-3-319-25554-5_18

is based on the idea of "social verification": people verify that they are engaging in semantically meaningful communication when interacting. Two interpersonal factors are considered of special importance for verifying meaningful interaction: the behavioural realism with which social cues are portrayed, and agency, the extent to which the agent is perceived as human-like. According to this idea idle motions would not contribute to social verification, and therefore, not elicit social responses. Movements portrayed by a robot would have to be meaningful and embody social cues. Various studies have examined meaningful gestures in the field of HRI. Gaze has been demonstrated to influence the persuasiveness of a robot during a conversation [4]. Other non-verbal meaningful gestures, like hand/arm gestures, have been demonstrated to improve communication efficiency and user experience [15], and anthropomorphism [13].

In this study we investigate the social effects of idle- and meaningful motions as compared to no motions, and compare our results with two competing views of social effects in human-robot interaction: the media equation hypothesis [11] and the threshold model of social influence [3]. According to the threshold model of social influence, meaningful motions serve as semantically meaningful communication with the robot, and are perceived to have higher behavioural realism than idle motions. Therefore, they should trigger stronger social responses than idle motions. In particular, we expect that meaningful motions are perceived as more socially intelligent and more anthropomorphic than idle motions and no motions [6]. On the other hand, according to the media equation hypothesis, idle motions already elicit social responses. So in this case we expect that idle and meaningful motions are both perceived as more socially intelligent and more anthropomorphic than idle motions and no motions. Both idle motions and meaningful motions are expected to improve the perceived life-likeness of the robot compared to no motions.

2   Method

We conducted an experiment where the Nao robot (Aldebaran Robotics, France) helped participants unpack a cardboard moving box that contained 16 items. There were two main conditions: In one condition the Nao robot displayed the so-called idle motions, in the other condition the robot displayed the meaningful motions. Within the two main conditions there was a baseline no-motion condition and three motion conditions.

2.1   Participants

73 participants took part in the experiment, of which 41 were male and 31 were female (mean age 25.55, SD = 7.012, range 18 to 54). Participants were randomly assigned to one of the two experimental conditions. 40 participants had prior experience with robots, including the Nao robot, but this did not influence our results. Participants received a monetary compensation of 5 euros for participating in the experiment.

2.2 Design

The experiment was conducted using a mixed design. We used two different motion types as a between-subjects factor, which differed in terms of social verification: Meaningful and Idle motions. The Meaningful motion condition portrayed semantically meaningful communication (high social verification). The Idle motion condition portrayed interactions that are argued to only aid in the "illusion of life" (low social verification). In both groups a no-motion condition was used as a baseline. There were three meaningful movements: (1) Arm pointing, (2) Head pointing, and (3) Eye-contact. And three idle motions: (1) Posture shift/sway (2) Random head movements, and (3) Breathing motion. Each participant experienced three movement conditions (either idle or meaningful) and the baseline condition in four blocks. The baseline was always presented first, the three movement conditions were counterbalanced across subjects. Each block required the unpacking and correctly placing of four items from the moving box, after which participants were asked to fill in a questionnaire. Each block consisted of four trials resulting in a total of 16 trials per participant.

2.3 Experimental Set-up

Participants interacted with the humanoid Nao robot (Aldebaran Robotics, France). The Nao is a 58-cm tall humanoid robot, which has 25 degrees of freedom, two cameras, an inertial measurement unit, touch sensors and four microphones all enabling him to detect and interact with its surroundings. The robot was partially controlled using a Wizard-Of-Oz technique. For each of the within-subject conditions there were predetermined utterances. These utterances were randomised for each within-subject condition. The items were located in the cardboard moving box and had to be unpacked. The chosen items and item locations were chosen such that they did not cause confusion or bias of where they should be placed. We used 16 items: a white vase, green cup, yellow cup, instruction manual, white bowl, clock, candles, photo frame, telephone, fruit bowl, two glasses, power adapter, headphones, stereo cable and a remote control. The questionnaire was implemented using Macromedia's Authorware software. The experiment took place in a mimicked living room in which the Nao robot would serve as a household assistant. Figure 1 shows an overview of the robot and object locations. The room was equipped with 3 cameras, so that the experimenter could observe and control the interaction from the observation room.

2.4 Robot Motions

We used the Principles of Animation [12] as a guideline to create the idle motions that generate an "illusion of life". We chose a breathing, posture sway and gaze shift motion, primarily because of the limitations of the Nao robot. To mimic a breathing motion, the Nao robot made a slight motion with its head, shoulder joints and hip joints. The frequency of the breathing motion was constant and fixed in a pretest. The idle gaze shifts were implemented by adjusting both the

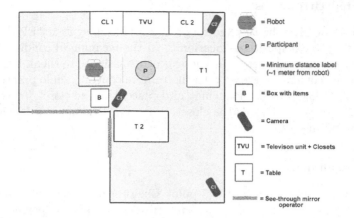

Fig. 1. A top down view of the UseLab. On the right side the different items are listed that were used during the experiment. Furthermore the minimum distance label can be seen between the Nao robot and participant.

head pitch and yaw. A total of 8 pre-recorded head motions were executed at a random time interval (between 15-22 seconds). The posture sway motions were implemented by counter-rotating the hip and ankle joints including small adjustments to the head and arm joints. A total of 8 randomised pre-recorded motions were executed at random on a certain time interval (between 20-30 seconds). The motion parameters were pretested so that the motions looked natural. We verified whether the idle motions were perceived correctly by having the participant describe which motion the robot portrayed. Out of 37 participants 86.5% perceived the posture sways correctly, 78.4% perceived the gaze shifts and 83.8% perceived the breathing motion.

The meaningful motion eye-contact/gaze was realised using a face tracking algorithm that centres the gaze of the Nao robot on the participant. During this interaction the Nao robot checks whether the participant is looking at the robot using a head pose estimation algorithm its eyes, thus creating a mutual facial gaze interaction. The arm pointing and head pointing gestures were implemented in combination with deictic expressions: gesture conveyed the lacking spatial information in the speech. For example, the robot could say "Please take the power adapter, and place it in the closet" and point to ether the left or right closet. Since there were 4 locations, 4 deictic arm gestures and 4 deictic head gestures were implemented. Out of 36 participants in the meaningful motion condition, 91.7% perceived the deictic arm pointing gesture correctly, 83.4% perceived the deictic head gesture correctly and 86.2% perceived the gaze motion correctly. The no-motion condition, which acted as a baseline throughout the experiment, was perceived correctly by 86.5% out of 73 participants.

2.5   Verbal Utterances

Each instruction given by the Nao robot had following syntax: "Please take the [object], and place it [position+location]." In the meaningful condition the two syntax components had a separate deictic gesture assigned to them. For example, the Nao robot could say "Please take the remote control" while pointing at the moving box, followed by pronouncing "and place it in the closet," accompanied by a pointing gesture towards the closet. This was all done in a fluent manner that looked natural.

2.6   Questionnaire

The questionnaire is based on the 5-point Likert scale Godspeed questionnaire [2], which measures: anthropomorphism, animacy, likeability, perceived intelligence and perceived safety. We excluded perceived safety from the questionnaire since this was not relevant to our study, and replaced this dimension with the emotion dimension (4 questions) and the social intelligence dimension (4 questions). The former allowed us to measure the perceived emotional responsiveness of the robot. The latter enabled us to measure the social competence and social skills of the Nao robot and is based on [9].

2.7   Procedure

On arrival participants filled in the informed consent forms, and received general instructions. When there were no further questions the experiment was started from the control room. The robot introduced itself and provided a short explanation of the task. First the baseline condition with no movements was presented, in which the robot directed the participant with verbal utterances to unpack the box. The utterance consisted of two parts: the first part indicated the item (e.g., Please take the white vase) and the second part indicated where the item should be placed (e.g., and place it on the table). After placing the object the participant was required to stand in front of the Nao robot again to signal that they were ready for the next item. After placing four items in the correct location the Nao robot instructed the participant to take a seat in the chair, and fill in the questionnaire provided on the laptop. After completing the questionnaire, the participant stood in front of the Nao robot again to continue the next experimental block. This was repeated four times for a total of sixteen items. In total the participants interacted approximately 10 minutes with the Nao robot. After completing the last questionnaire the participants were debriefed about the purpose of the experiment and then thanked and paid. The experiment lasted about 30 minutes.

2.8   Data Analysis

To check the internal consistencies a reliability analysis was conducted. Cronbach's alpha exceeded 0.7 for all dimensions of the questionnaire, indicating that the items had good consistency. We had to exclude the data of 10 participants regarding the eye-contact condition (meaningful motion condition only), because the robot lost eye-contact.

3   Results

3.1   Social Verification

To verify our hypotheses regarding the media equation hypothesis and threshold model of social influence, we compared meaningful and idle movements conditions. To remove individual differences in overall ratings we subtracted the baseline condition first, which was always presented first. The result is shown in Figure 2. It is clear that Likeability, Perceived intelligence, social intelligence and emotion scored higher when the robot displayed meaningful motions than when it displayed idle motions.

A MANOVA analysis with the questionnaire dimensions as dependent variables and social verification as factor confirmed a statistically significant main effect of social verification, $F(6, 202) = 4.38, p < .01, \eta^2 = 0.12$. Participants rate the Likeability dimension higher $(F(1, 209) = 7.17, p < 0.01, \eta^2 = 0.03)$ for conditions which portrayed meaningful motions (M = 0.28, SD = 0.85) than for idle motions (M = -0.02, SD = 0.79). Perceived Intelligence is higher $(F(1, 209) = 13.64, p < 0.01, \eta^2 = 0.06)$ for conditions which portrayed meaningful motions (M = 0.37, SD = 0.78) than for idle motions (M = -0.03, SD = 0.79).

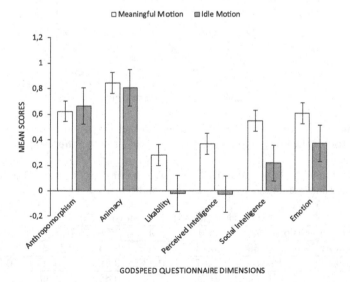

Fig. 2. Mean Likert scores for the meaningful motion and idle motion conditions after subtracting the baseline. The errors bars show the standard error for the mean at ±1 SE.

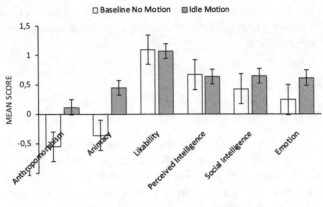

Fig. 3. Mean Likert scores for the idle motion condition and the baseline no-motion condition. The errors bars show the standard error for the mean at ±1 SE.

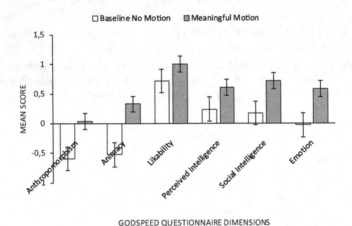

Fig. 4. Mean Likert scores for the baseline no-motion condition and the meaningful motion condition. The errors bars show the standard error for the mean at ±1 SE.

Table 1. Overview of the results of the ANOVA testing the within-subject effect of meaningful motions compared to the baseline condition.

Questionnaire Dimension	No motion (n=36)		Motion (n=105)		$F(1, 139)$	p	η^2
	M	SD	M	SD			
Anthropomorphism	-0.59	0.74	0.03	0.93	13.25	< 0.001	0.09
Animacy	-0.51	0.87	0.33	0.82	27.44	0.01	0.17
Likability	0.71	0.88	1.0	0.85	2.89	0.09	0.02
Perceived Intelligence	0.23	0.89	0.6	0.78	5.82	0.02	0.04
Social Intelligence	0.17	0.89	0.72	0.65	15.42	< 0.001	0.1
Emotion	-0.03	0.7	0.58	0.74	18.45	< 0.001	0.12

Likewise, social intelligence is higher $(F(1,209) = 10.11, p < 0.01, \eta^2 = 0.05)$ for conditions which portrayed meaningful motions (M = 0.54, SD = 0.65) than for idle motions (M = 0.22, SD = 0.81) and, finally, emotion is higher $(F(1,209) = 4.76, p = 0.02, \eta^2 = 0.02)$ for conditions which portrayed meaningful motions (M = 0.61, SD = 0.74) than for idle motions (M = 0.37, SD = 0.83). The anthropomorphism dimension $(F(1,209) = 0.094, p = 0.76)$ and the animacy dimension were not rated significantly different between motion conditions $(F(1,209) = 0.113, p = 0.74)$.

3.2   Effect of Motion

To determine the effect of motion we compared the different motion conditions to the no motion (baseline) conditions separately for idle motions and meaningful motions. The latter is necessary because the robot displayed a different set of motions in the idle and meaningful motion conditions. The result is shown in Figure 3 for idle motions and in Figure 4 for meaningful motions. In the idle motion condition a positive effect of motion on anthropomorphism, animacy, social intelligence and emotion is visible. To test the significance, we conducted a MANOVA analysis with questionnaire dimensions as dependent variable and idle motion (idle motion, no-motion) as a factor. We found a significant main effect of motion $(F(6,134) = 8.911, p < 0.01, \eta^2 = 0.29)$. Participants rated the anthropomorphism dimension significantly higher $(F(1,139) = 13.33, p < 0.001, \eta^2 = 0.09)$ for conditions which portrayed motion (M = 0.11, SD = 0.95) than for the baseline condition without motion (M = -0.55, SD = 0.96); animacy was rated significantly higher $(F(1,139) = 23.14, p < 0.001, \eta^2 = 0.14)$ by participants for the idle motion condition (M = 0.45, SD = 0.84) than for the baseline condition without motion (M = -0.36, SD = 0.97); emotion was significantly higher $(F(1,139) = 5.41, p = 0.02, \eta^2 = 0.04)$ for conditions which portrayed motion (M = 0.61, SD = 0.83) than for the baseline condition without motion (M = 0.24, SD = 0.84). The other Godspeed questionnaire dimensions did not differ significantly compared to the baseline no-motion condition (likeability: $p = 0.88$; perceived intelligence: $p = 0.83$; social intelligence: $p = 0.16$).

We did the same analysis for the meaningful motion condition, but now with meaningful motion (meaningful motion, no-motion) as a factor. Again we found a significant main effect of motion ($F(6,134) = 7.65, p < 0.01, \eta^2 = 0.26$).

In the meaningful motion condition, participants rated all dimensions higher for a robot showing motion than for not showing motion (see Figure 4). We found significant effects for anthropomorphism, animace, perceived intelligence, social intelligence and emotion, but not for likeability (see Table 1).

4   Discussion and Conclusions

4.1   Social Verification

We expected that participants' social responses would be higher for the meaningful motions compared to the idle motions. Results indicated that participants rated the Nao robot significantly more positive in the meaningful motion condition i.e., the robot was seen as friendlier, more intelligent, empathic and helpful compared to the idle motion condition. We can thus conclude that we found support for the threshold model of social influence. Our results indicate that participants perceived the robot with higher social intelligence and perceived intelligence when the robot portrayed meaningful motions compared to idle motions. Thus, the robot portraying meaningful motions is perceived as more socially competent and skilled. This also confirms that when the robot portrayed meaningful motions the participants perceived the interaction as semantically meaningful. As suggested in [6] we did not find evidence that social intelligence increases the level of anthropomorphism, i.e. anthropomorphism was rated the same for idle and meaningful motions.

4.2   Effect of Motion

We also investigated how humans perceive different idle motions portrayed by a robot. Our results indicated that participants perceived the robot portraying idle motions as more human-like, alive and empathic compared to the robot with no motion. These results are in line with [7] who concluded that humans automatically ascribe human traits to a robot when the robot portrays human behaviour.

As a result, the robot portraying idle motions was also perceived by participants as more empathic, or emotionally expressive compared to the no-motion robot. It was assumed that idle motions would not add to the expression of a character [8]. However, our research demonstrates that by having robots portraying idle motions, people will attribute intentions to the robot. In fact, participants sometimes remarked that the robot seemed bored or nervous during the idle motion condition. This is a further indication that participants anthropomorphised the robot when portraying idle motions. However, idle motions remain low in social verification, because participants did not perceive the robot portraying idle motions as more intelligent or more socially capable than the robot portraying no motion.

Overall, people ascribed human qualities to a robot that portrays idle motions. It does not seem to matter which idle motions are portrayed by a robot: as long as the robot makes some motions, humans will perceive the robot as more human-like and alive, albeit not more intelligent. Only when the robot portrayed meaningful motions, the robot was perceived as more socially competent and intelligent. We can thus confirm that the meaningful motions are indeed contributing in a semantically meaningful manner to social verification.

References

1. Bartneck, C., Rosalia, C., Menges, R., Deckers, I.: Robot abuse – a limitation of the media equation. In: Proceedings of the Interact 2005 Workshop on Agent Abuse (2005)

2. Bartneck, C., Kulić, D., Croft, E., Zoghbi, S.: Measurement instruments for the anthropomorphism, animacy, likeability, perceived intelligence, and perceived safety of robots. International Journal of Social Robotics 1(1), 71–81 (2009)

3. Blascovich, J.: A theoretical model of social influence for increasing the utility of collaborative virtual environments. In: Proceedings of the 4th international Conference on Collaborative Virtual Environments, pp. 25–30 (2002)

4. Chidambaram, V., Chiang, Y. H., Mutlu, B.: Designing persuasive robots: how robots might persuade people using vocal and nonverbal cues. In: Proceedings of the Seventh Annual ACM/IEEE International Conference on Human-Robot Interaction, pp. 293–300 (2012)

5. Egges, A., Molet, T., Magnenat-Thalmann, N.: Personalised real-time idle motion synthesis. In: Proceedings of the 12th Pacific Conference on Computer Graphics and Applications (PG 2004), pp. 121–30 (2004). doi:10.1109/PCCGA.2004.1348342

6. Epley, N., Waytz, A., Cacioppo, J.: On seeing human: a three-factor theory of anthropomorphism. Psychological Review 114(4), 864 (2007)

7. Gazzola, V., Rizzolatti, G., Wicker, B., Keysers, C.: The anthropomorphic brain: the mirror neuron system responds to human and robotic actions. Neuroimage 35(4), 1674–1684 (2007)

8. Jung, J., Kanda, T., Kim, M.: Guidelines for contextual motion design of a humanoid robot. International Journal of Social Robotics 5(2), 153–169 (2013)

9. Martínez-Miranda, J., Aldea, A.: Emotions in human and artificial intelligence. Computers in Human Behavior 21(2), 323–341 (2005)

10. Nass, C., Moon, Y.: Machines and mindlessness: Social responses to computers. Journal of Social Issues 56(1), 81–103 (2000)

11. Reeves, B., Nass, C.: How People Treat Computers, Television, and New Media Like Real People and Places. Cambridge University Press (1996)

12. Ribeiro, T., Paiva, A.: The illusion of robotic life: principles and practices of animation for robots. In: Proceedings of the Seventh Annual ACM/IEEE International Conference on Human-Robot Interaction, pp. 383–390 (2012)

13. Kopp, S., Joublin, F., Salem, M., Eyssel, F., Rohlfing, K.: Effects of gesture on the perception of psychological anthropomorphism: a case study with a humanoid robot. In: Kanda, T., Evers, V., Mutlu, B., Ham, J., Bartneck, C. (eds.) ICSR 2011. LNCS, vol. 7072, pp. 31–41. Springer, Heidelberg (2011)

14. Song, H., Min Joong, K., Jeong, S.-H., Hyen-Jeong, S., Dong-Soo K.: (2009). Design of Idle motions for service robot via video ethnography. In: Proceedings of the 18th IEEE International Symposium on Robot and Human Interactive Communication (RO-MAN 2009), pp. 195–99. doi:10.1109/ROMAN.2009.5326062

15. Torta, E., van Heumen, J., Cuijpers, R.H., Juola, J.F.: How can a robot attract the attention of its human partner? a comparative study over different modalities for attracting attention. In: Ge, S.S., Khatib, O., Cabibihan, J.-J., Simmons, R., Williams, M.-A. (eds.) ICSR 2012. LNCS, vol. 7621, pp. 288–297. Springer, Heidelberg (2012)

16. Tourangeau, R., Couper, M.P., Steiger, D.M.: Humanizing self-administered surveys: Experiments on social presence in Web and IVR surveys. Computers in Human Behavior 19(1), 1–24 (2003)

17. Yamaoka, F., Kanda, T., Ishiguro, H., Hagita, N.: "Lifelike" behaviour of communication robots based on developmental psychology findings. In: 2005 5th IEEE-RAS International Conference on Humanoid Robots, pp. 406–411 (2005)

What Makes Robots Social?: A User's Perspective on Characteristics for Social Human-Robot Interaction

M.M.A. de Graaf[1(✉)], S. Ben Allouch[2], and J.A.G.M. van Dijk[1]

[1] Deptartment of Communication Science, University of Twente, Enschede, The Netherlands
{m.m.a.degraaf,j.a.g.m.vandijk}@utwente.nl
[2] Department of Technology, Heath & Care, Saxion University of Applied Science,
Enschede, The Netherlands
s.benallouch@saxion.nl

Abstract. A common description of a social robot is for it to be capable of communicating in a humanlike manner. However, a description of what communicating in a 'humanlike manner' means often remains unspecified. This paper provides a set of social behaviors and certain specific features social robots should possess based on user's experience in a longitudinal home study, discusses whether robots can actually be social, and presents some recommendations to build better social robots.

Keywords: Design guidelines · Sociability · Social intelligence · Social robots

1 Introduction

The field of robotics is rapidly advancing. There are a growing number of different types of robots, and their roles within society are expanding. As the capabilities of robots develop, the possibility arises that they will be able to perform more and more difficult tasks and become our full-fledged team members, assistants, guides, and companions in the not-so-distant future. The aim of social robotics research is to build robots that can engage in social interaction scenarios with humans in a natural, familiar, efficient, and above all intuitive manner.

Robots designed to share domestic environments with human users must interact in a socially acceptable way. According to Breazeal [1], an ideal social robot is capable of communicating and interacting in a sociable way so that its users can understand the robot in the same social terms, to be able to relate to it and to empathize with it. The common underlying assumption is that people prefer to interact with machines in a similar manner they do with other human beings [4]. Robotic researchers strive for the development of such sociable machines by making use of models and techniques generally used in interpersonal communication derived from (social) psychology and communication science. Yet, the social capabilities of today's robots are still limited (see [7] for a more in-depth discussion). Simple human social behavior can be quite challenging to program into the robot's software. Various research on the social behaviors of robots has been performed, and an abundance of literature suggests the

© Springer International Publishing Switzerland 2015
A. Tapus et al. (Eds.): ICSR 2015, LNAI 9388, pp. 184–193, 2015.
DOI: 10.1007/978-3-319-25554-5_19

following social characteristics for robots designed to interact socially with its users: social learning and imitation, dialog, learning and developing social competencies, exhibit distinctive personality, establishing and maintaining social relationships [4] [15]. This list of social characteristics is a list with robotic behaviors and features which social robots should ideally possess. Though, social robot prototypes existing today still lack important social characteristics and display only limited socially acceptable behaviors [7], which prevents these robots from engaging in truly natural interactions with their users [2]. Only when all of the essential social characteristics can be met, we can legitimately speak of social robots. Nevertheless, we would like to postulate that robots themselves are not social. Robots can only simulate social behavior or behave in such a manner perceived by human users as social.

This paper provides a set of social behaviors and certain specific features social robots should possess, discusses whether robots can actually be social, and presents some recommendations to build better social robots.

2   Method

The overall aim of the longitudinal study was to see whether and how a longer, uninterrupted period of use of a social robot in a home environment affects the long-term use of social robots. Based on real interaction experiences with the Karotz robot and triggered by some specific questions about social behaviors of robots for domestic purposes, we have identified a set of social behaviors and certain specific features social robots should possess.

2.1   The Karotz Robot

The robot used in this study is Karotz (see figure 1; http://store.karotz.com/en_WW/), which is a 30-cm high internet-enabled activated smart rabbit-shaped ambient electronic device. Communication occurs via verbal communication, the LED-light in its belly, the moveable ears, and by detecting the presence of other objects nearby. As the Karotz is permanently connected to the internet, it is able to react on, transmit, and broadcast all types of content available on his network, for example news, messages, music, texts, alerts, and radio. The build-in webcam enables users to communicate with family members at home or for surveillance purposes when away.

Fig. 1. The Karotz robot deployed in the participant's homes

Each robot was installed with a basic set of applications, such as daily news broadcasts, daily local weather reports, favorite radio stations, personalized reminders, and randomly spoken phrases to make the robot being perceived as more autonomous and animate. This basic set of applications ensured us that the user experience was somewhat similar among the participants, or at least initially as some participants chose to adjust these applications to their own needs. Besides the required applications, participants were free to install additional applications as they thought would be useful or fun for their households.

2.2   Data Collection and Procedure

The longitudinal study ran from October 2012 to May 2013 and consisted of seven moments of data collection. For the interviews, a representative of that household reported on their own individual user experiences with some additional questions about the opinions of other household members. Table 1 presents the number of interviews collected during the study and their associated time point with regard to the moment the participants were introduced to the robot.

Table 1. Distribution of sample sizes for each time point

Time points with regard to the introduction of the robot	Sample size
2 weeks before	21
1st day	21
2 weeks	18
1 month	17
2 months	13
6 months	7

The participants were interviewed at each of the time points. In total, 21 participants started the study who consented on being part of the interview sessions. We conducted semi-structured interviews at the participants' own homes to obtain detailed user experiences with the robot. The interview scheme contained, among other questions about the acceptance and use of the robot, some questions focused on the social characteristic of robots (e.g., Can you describe how you perceive the robot? How are the interactions with the robot similar to / different from interactions with other persons? Does the robot seems to have its own will / personality? How should the robot be improved to become more sociable? Does the robot offer some kind of companionship?). Only the codes applied to the answers of these questions were analyzed with the aim to identify a set of social behaviors and certain specific features social robots for domestic purposes should possess to be accepted by users.

2.3   Data Analysis

A total of 97 interviews were conducted over a time period of six months. The interviews were recorded and transcribed verbatim with the participants' approval. The transcriptions were done as soon as possible after conducting the interviews to guarantee information clearance and solve problems with interpretation quickly [19]. Based on the transcriptions of the interviews, key concepts were identified and translated into a coding scheme by the primary coder. Table 3 shows the social characteristics that have emerged as key concepts from the interviews, which together formed the coding scheme. Next, for each interview section, at least one code from the coding scheme was applied to that section. In total, 32 of the 97 interviews were also coded by a second scientist, which resulted in almost 33% of double-coded data. Intercoder reliability, which involves testing the extent to which the independent coders agree on the application of the codes to the different interview sections, has found to be substantial with a Cohen's Kappa of .73 [11]. In the results, from every interview transcript, 'striking' or 'typical' quotes [8] were selected which illustrated, confirmed or enhanced our understanding of the key concepts, i.e., the social characteristics for domestic social robots, from the coding scheme.

2.4 Participants

Participants were recruited with various methods, such as word of mouth, advertising in public locations (e.g., libraries, leisure centers and supermarkets), and snowball sampling by asking assigned participants for referrals to other people who might participate. During recruitment, we tried to balance out the households' demographic profiles to seek diversity (see table 2 for the distribution within the sample). However, from each household, only one person was interviewed. Furthermore, to facilitate the interactions with the robot, participants were required to have at least a limited working proficiency in either English or German as the Karotz robot does not provide interactions in Dutch. We compensated our participants by allowing them keep their robot after study completion. Moreover, to increase both homogeneity and convenience, most participants lived within 10 square kilometer around our university, the University of Twente in The Netherlands.

Table 2. Distribution of household types within the sample

Household type	Number of participants	
	Count	%
Older single male (55+)	1	4
Younger single male (35-)	3	14
Older single female (55+)	2	10
Younger single female (35-)	2	10
Older couple (both 55+)	2	10
Younger couple (both 35-)	3	14
Young family	3	14
Mature family	2	10
Student dorm	3	14
Total	21	100.0

3 Results

From the interviews and based on the coding scheme, we observed several behaviors of social interaction and certain specific features domestic social robots should possess before the participants would accept such robots as social entities in their homes (see table 3). The following sections will clarify the meaning of these social characteristics with some quotes.

Table 3. Frequency distribution of social characteristics for social robot

Social characteristics	Count	%
Autonomy	20	7
Coziness	15	5
Mutual respect	7	2
Similarity	9	3
Social awareness	40	14
Social support	22	8
Thoughts and feelings	57	20
Two-way interaction	119	41
Total	289	100.0

3.1   Two-Way Interaction

The far most noted topic was two-way interaction, which constitutes speaking to the robot and for it to respond in an social manner. Some participants had expected to be able to do this with the current robot and were somewhat disappointed when they found out that the Karotz robot could only understand preprogrammed commands.

> *"He doesn't communicate with you. You have to push a button and then you could give [the robot] commands. Sometimes he answers, sometimes he doesn't."* – female, 57, living alone

> *"[For the robot to be perceived as a social companion] he needs to interpret the things I say. He basically needs to continuously receive things and send out without needing to push the button."* – male, 32, living alone

3.2   Thoughts and Feelings

Another frequently noted topic was thoughts and feelings. Robots should be embedded with thoughts and feelings. Robots should be able to think for itself and act upon it. In addition, a robot be able to display humanlike emotions.

> *"[The robot] can't laugh or cry or look sad... If he wants to be a full-fleshed interaction partner, he needs to be able to shows his emotions."* – male, 32, living alone

> *"When such a device becomes intuitive, gets more emotions, becomes more intelligent or something. Than it will be different... Then you will treat it differently too."* – male, 31, living alone

3.3   Social Awareness

The participants also indicated that robots should be aware of their social environment. Robots must be able to sense our presence and our moods to be able to be perceived as a social entity.

> *"[The robot] doesn't respond to noise, except when you push that button. So he needs to permanently distinguish sounds and interpret and react upon them. That is when you could be speaking of contact."* – male, 32, living alone
>
> *"[The robot] should react better to what he does... That his ears turn when you come in... That would make you perceive it more as something alive."* – female, 27, living with spouse

3.4   Social Support

With social support, the participants referred to their friends being there for them to support them when needed and sharing life experiences with each other. For the possibility perceive robots as social entities, there should be a trust relationship between a human and a robot and knowing that you can always count on it to be there for you.

> *"That you share stuff. That you have the feeling you can count on each other."* – female, 57, living alone
>
> *"To share stuff. I have different friends for different purposes. With one friend I talk about superficial stuff and with another friend I can share more serious stuff when something is bothering me... And sports friends. And in that way I have for my different needs several people around me."* – female, 27, living with spouse

3.5   Autonomy

With autonomy, the participants particularly referred to the fact that the robot used for this study was standing still. For a robot to be perceived as a companion, the participants need that robot to be able to move around independently and behave unpredictably and spontaneously and not only have pre-programmed behaviors. Increase the robot's presence would let it be perceived as more animate or alive. With autonomy, the participants indicated that they would want the robot to act on its own.

> *"It needs to be a completely movable robot. More in the direction of humans instead of something static. Then it would be more suitable for companionship."* – male, 24, living alone
>
> *"If [the robot] could move more, it would be more alive... For example driving around... or some more degrees of freedom, so not just moving its ears."* – male, 32, living alone

3.6   Coziness

The topic of coziness was noted a few times by the participants as an essential characteristic for social robots. The participants discussed to their experiences hanging around with their friend just for the sake of being together. That feeling of companionableness is something the participants would miss in the company of a robot. Coziness or companionableness seems to be a predecessor of intensive social interactions.

*"For me companionableness is important. I like it to be surrounded by a
    group of people to talk to."* – female, 22, living alone
"Coziness off course. Just to talk to each other and have some drinks." –
    female, 19, living with spouse

3.7   Similarity

Similarity as an essential characteristic for social robots was also much less noted.
Related to the topic of similarity, a few participants said that their friends are their
friends because they share similar personalities or similar interests with them which
makes is easier and more pleasant to interact with.

*"What I like about people is that they talk and have feeling that are similar
    to mine."* – female, 22, living alone
*"Having resemblances with people. And to talk about that with each other,
    and to brainstorm with someone who has the same interests. That is nice
    to that to. I think that is important"* – male, 38, living with young family

3.8   Mutual Respect

Another topic noted only a few times was mutual respect. A few participants explained that the way they spook to the robot was different from how they interact with
other people. They were quite rude and blunt to the robot, because they knew that the
robot would not respond to that behavior. So in order to perceive robots as a social
entity, users should be able to perceive the robot as a higher form of intelligence
which would make them feel obligated to treat the robot with respect.

*"You are rude [to the robot], because you think that the robot doesn't have
    any feelings."* – male, 32, living alone
*"[The robot] is defenseless, so he can't say anything back. I also think you
    make shorter sentences, or even talk to him in stop words. Because he
    doesn't understand it anyway."* – female, 19, living with spouse

Together, these social characteristics for robots provide some insights into the essential characteristics social robots should possess before the participants would accept
such robots as social entities in their homes.

4   General Discussion

This paper presents the results of a robot's sociability based on user's experiences
from a longitudinal home study. Before discussing the general implications of these
results, we need to address some limitations. First, the rather limited interaction capabilities of the robot used in this study. The choice of the Karotz robot is a fundamental
result of the overall aim of this study, to see whether and how a longer, uninterrupted
period of use of a social robot in a home environment affects the long-term use
of social robots. Second, the employment of a zoomorphic robot imposes some

limitations on the generalizability of the results to other types of robots. Third, this study focuses on domestic social robots. It could very well be that other context demand different types of characteristics for social interaction with humans. Therefore, replication studies are needed to further support the results from the current study.

4.1   Essential Social Abilities for Social Robots

Interestingly, this study indicates that users remark similar essential social characteristics for future robots which social roboticists already pursue in their creations [4] [15]. The indication of two-way interaction as the most essential social characteristics is related to social characteristic of dialog, which describes that robots should be capable to verbally communicate with humans. Above all, people should be able to freely interact with robots in a natural humanlike manner. This is not surprising, because human cognition requires language to communicate with other people for mutual understanding [3]. Although we can conclude that robots are yet still far away from behaving socially in an ideal manner (i.e. possessing all the essential characteristics for social behaviors as for example reported by [4] and [15]), this is not entirely necessary because the creative human mind will restore these shortcomings with the subconscious process of the media equation [16]. In this way, the social behaviors of robots is shaped in the minds of the human user.

4.2   Can Robots Actually be Social?

An important point for discussion is the potential sociability of robots. Social roboticists are striving to program robots with social behaviors that are similar to those of human beings. Yet, some people may argue that robots cannot behave socially and cannot have emotions or an appealing personality. Robots can only act as if they are social and pretend to empathize with our emotions. However, following the research on the media equation [16], human users interacting with robots themselves interpret the robot's behavior as social, and they respond to robots in ways that are similar to how they would respond to other people (e.g., [10] [12]). Although most people would reasonably agree that robots are programmed machines that only simulate social behavior, the same people seem to 'forget' this while interacting with these machines. Thus, the question whether robots are social beings seems to depend on how human users perceive (the interactions with) these types of robots.

The doubts of people who think otherwise can be neutralized by altering the well-known Turing test [20]. The Turing test is a proposed method for determining whether a machine should be regarded as intelligent. During the test, a person engages in natural conversations with both a human and a machine designed to generate a performance that is indistinguishable from that of a human being. The conversations are limited to text-based interactions via a keyboard and a screen. If a person cannot reliably discern which of the two conversations was with the machine, then the machine is said to have passed the test. Thus, if a machine appears to be intelligent according to the human user, then we should assume that that machine is indeed intelligent.

Levy [14] proposes that we can apply a similar argument to other aspects of being human, such as emotions, personality, and behavior. Furthermore, acting is also a part of human social behavior [5]. In this line of thought, robotics researchers and developers should acknowledge that robots are social entities when human users perceive robots as such.

4.3   How to Make Better Social Robots

This section will present some guidelines to improve the (interaction) design of social robots designed to share domestic environments with human users. People interact with and respond socially to robots (e.g., [10] [12]). Therefore, some researchers argue that it seems unnecessary to depart from the social rules of human-human interaction when evaluating human-robot interactions [9]. Thus, a first recommendation is that social roboticists should investigate theories of interpersonal communication to create better social robots.

For social robots to flourish as companions for human users, the results of a short-term study with the Pleo robot [6] indicate that people are more willing to treat a robot as a companion when they have high expectations of the robot's lifelikeness. The influence of lifelikeness has also been related to people's empathic responses to a robot [17]. Thus, a second recommendation is that social robots should have a lifelike appearance, which does not necessarily mean a humanlike appearance.

The main finding of the current study is that two-way interaction, possessing thoughts, feelings and emotions and being capable to sense the social environment are the most essential parts of social behavior to pursue for social robots at this stage of development. Thus, a third recommendation is that developers of social robots should focus on increasing a robot's social behavior by first addressing the possibility of two-way interaction with a robot followed by creating some 'theory of mind' for robots.

The possibility of sharing personal information with a robot and having that robot respond to this personal information in an empathic manner was observed in the current longitudinal study as the most important variable for explaining companionship with a robot. The importance of empathic behavior for social robots and the user's empathic responses to the robot have also been noted by other researchers [13]. Thus, as a fourth recommendation, social robots need to be perceived as empathic.

5   Conclusion

This paper presented a set of essential characteristics for social robots from a user's perspective, discussed the ability for robots to actually be social, and provided some design guidelines for social robots. The results of this paper further paves the way for better social human-robot interaction for future robots designed to share domestic environments with human users.

References

1. Breazeal, C.L.: Designing sociable robots. MIT Press, Cambridge (2005)
2. Castellano, G., Pereira, A., Leite, I., Paiva, A., McOwan, P.W.: Detecting user engagement with a robot companion using task and social interaction-based features. In: Int. Conf. on Multimodal Interfaces, Cambridge, MA, USA (2009)
3. van Dijk, J.A.G.M.: The network society. Sage Publications, London (2012)
4. Fong, T., Nourbakhsh, I., Dautenhahn, K.: A survey of socially interactive robots. Robotics and Autonomous Systems **42**, 143–166 (2003)
5. Goffman, E.: The presentation of self in everyday life. Penguin Group, London (1959)
6. de Graaf, M.M.A., Ben Allouch, S. Expectation setting and personality attribution in HRI. In: Int. Conf. on Human-Robot Interaction, Bielefeld, Germany (2014)
7. de Graaf, M.M.A.: Living with robots: Investigating the user acceptance of social robots in domestic environments (thesis). University of Twente, Enschede, The Netherlands (2015)
8. Hansen, A., Cottle, S., Negrine, R., Newbold, C.: Mass communication research methods. Palgrave Publishers, Basingstoke (1998)
9. Krämer, N.C., Eimler, S.N., Pütten, A.M., Payr, S.: Theory of companion: What can theoratical model contribute to applications and understanding of human-robot interaction? Applied Artificial Intelligence **25**, 474–502 (2011)
10. Kwak, S.S., Kim, Y., Kim, E., Shin, C., Cho, K.: What makes people empathize with an emotional robot?: the impact of agency and physical embodiment on human empathy for a robot. In: Int. Symp. on Robot and Human Interactive Communication, Gyeongju, Korea (2013)
11. Landis, J.R., Koch, G.G.: The measurement of observer agreement for categorical data. Biometrics **33**, 159–174 (1977)
12. Lee, K., Park, N., Song, H.: Can a robot be perceived as a developing creature? Effects of a robot's long-term cognitive developments on its social presence and people's social responses toward it. Human Communication Research **31**, 538–563 (2005)
13. Leite, I., Castellano, G., Pereira, A., Martinho, C., Paiva, A.: Empathic Robots for Long-term Interaction. Int. J. of Social Robotics **6**, 329–341 (2014)
14. Levy, D.: Love and sex with robots: The evolution of human-robot relationships. HarperCollins Publishers, New York (2008)
15. Mutlu, B.: Designing embodied cues for dialog with robots. AI Magazine **32**, 17–30 (2011)
16. Reeves, B., Nass, C.: The media equation: How people treat computers, television, and new media like real people and places. CSLI Publications, New York (1996)
17. Riek, L.D., Rabinowitch, T.C., Chakrabarti, B., Robinson, P.: How anthropomorphism affects empathy toward robots. In: Int. Conf. on Human-Robot Interaction, La Jolla, CA, USA (2009)
18. Severinson-Eklundh, K., Green, A., Huttenrauch, H.: Social and collaborative aspects of interaction with a service robot. Robotics and Autonomous Systems **42**, 223–234 (2003)
19. Taylor, S.J., Bogdan, R.: Introduction to qualitative research methods: The search for meanings. Wiley, New York (1984)
20. Turing, A.M.: Computing machinery and intelligence. Mind **59**, 433–460 (1950)

An Adaptive and Proactive Human-Aware Robot Guide

Michelangelo Fiore(✉), Harmish Khambhaita, Grégoire Milliez,
and Rachid Alami

Université de Toulouse, LAAS-CNRS Institut Carnot, 7 rue du Colonel Roche,
31500 Toulouse, France
{mfiore,harmish,gmilliez,rachid.alami}@laas.fr

Abstract. In this paper we present a robotic system able to guide a person to a destination in a socially acceptable way. Our robot is able to estimate if the user is still actively following and react accordingly. This is achieved by stopping and waiting for the user or by changing the robot's speed to adapt to his needs. We also investigate how the robot can influence a person's behavior by changing its speed, to account for the urgency of the current task or for environmental stimulus, and by interacting with him when he stops following it. We base the planning model on *Hierarchical Mixed Observability Markov Decision Processes* to decompose the task in smaller subsets, simplifying the computation of a solution. Experimental results suggest the efficacy of our model.

1   Introduction

One interesting problem in human-robot interaction is developing robots able to guide humans, by offering a tour of attractions in an area or simply by helping them to reach a destination.

A generic mobile robot platform should possess a vast set of skills, which includes advanced perception, motion planning, and task planning. These skills are not enough for a robot guide, which is deployed in highly dynamic human environments, and need to be complemented with human-aware behaviors.

Different robot guides have been studied and developed, starting with pioneers like Rhino and Minerva [22]. After these first experiments, several researchers have tried to focus on the social aspects of the problem, which are especially important if the robot needs to offer information. Studies like [5,23] focus on how the robot should address humans, concentrating on spatial relationships and on how the robot can convey information. Few systems have actually been deployed for long period of time in human environments. Rackhman [4], a museum guide with human-aware behaviors, is an example of such system, and has been deployed in a science museum for several months. Robotic systems must be able to reason on sensor data in order to provide information to the decision layers. In [15], we presented our framework *SPARK*, which is able to maintain a topological description of the world state and to reason on humans' mental states, in order to improve the robot's social behavior. For guiding situations, [9]

© Springer International Publishing Switzerland 2015
A. Tapus et al. (Eds.): ICSR 2015, LNAI 9388, pp. 194–203, 2015.
DOI: 10.1007/978-3-319-25554-5_20

presents an assessment of human-robot interaction during an exhibition, where perceptual and task related data are used to compute an internal state according to the scenario. With these information, the robot can compute a new emotional state and interact accordingly with users.

Recently there has been emphasis on robot navigation algorithms that explicitly reason about human beings in the environment differently from other static or dynamic obstacles. Starting from *Proxemics*, researchers have investigated explicit social signals based on human-posture and affordance of the environment to improve the legibility of robot motion. For detailed discussion on human-aware navigation algorithms we refer the readers to [13,18]. Human-aware navigation in a museum situation was studied in [19], where the authors build environmental maps, which include information learnt from human trajectories and postures, in order to plan safe paths that don't disturb humans present in the area.

We consider guiding as a *joint action*, a task where several participants cooperate to reach a common goal [2]. A joint action can be seen as a contract between its participants, that need to fulfill their part of the contract and to continuously monitor the other participants in order to understand what they are doing. Some robotic architectures, such as [6,7], implement joint actions, explicitly modeling human agents in the system.

Participants in a joint action form a kind of mental representation of the task, which includes the actions that should be performed by every agent [20]. This mechanism can be used to predict what other agents will do, but also to understand when they are deviating from the shared plan. The idea of predicting the will of another agent is linked to the concept of intention, studied in psychology and philosophy literature, such as [3]. This topic is of particular interest in human-robot interaction and has been studied in different kind of scenarios, like [10,11], or [17], which is related to a museum scenario.

We believe that most robot guide systems are focusing on the social aspects of the problem, and on human-aware navigation, without fully considering the fundamental aspects of joint actions. Guiding is a collaborative task, where the robot doesn't need only to reach a destination, but also to ensure that its follower reaches it, while providing a socially acceptable experience to him. In order to achieve this goal, the robot needs to constantly monitor its user, to adapt to his behaviors and to be ready to proactively help him.

In this paper, we present a robot guide which is able to lead a single person to a destination. More particularly, the originality of our approach is that the robot is able to show both an adaptive and a proactive behavior. The robot will try, while guiding, to select a speed that pleases its user, when adapting, or to propose a new speed, using environmental and task related stimulus. Finally, our system will proactively try to engage a user if it detects he need assistance.

We implement these ideas using a Situation Assessment component, which gathers data from different sources and provides symbolic information, a Supervision System, that controls the other modules, and a planning framework based on hierarchical MOMPDs (Mixed Observability Markov Decision Processes). Finally, a human-aware Motion Planning component allows the robot to navigate populated environments.

2 Situation Assessment

Having data from sensors is not enough, for the robot, to choose which actions to execute. To fill the gap between perception and decision, we use a Situation Assessment component. This module is able to gather data from sensors in input, and to perform different kinds of computations in order to produce symbolic information that can be used by the decision layer.

Our system is able to reason on humans and objects present in the environment, producing different kind of information, such as: a) the distance and orientation of a human relative to the robot, b) the variation of the distance from a human to the robot c) if a human is currently moving.

To be relevant, reasoning on humans should be linked to the environment. The system is able to create activity areas in the environment and link them to different kind of computations. An activity area is a polygonal or circular area, which can be fixed or linked and updated with an entity's (object, human or robot) position. For now, we studied and experimented two different activity areas: a) Information Screen Area, linked to information screens present in the environment; b) Touristic Point Area, linked to interesting attractions in the environment. Using these areas, the system can detect human activities (e.g. human is looking at an information screen, human is looking at an attraction).

Detecting and tracking persons is complex. In this paper, human tracking is done using motion capture. In order to simulate realistic behaviors, we filter data provided by the motion capture in order to account for occlusions from the environment. The system has also been linked in the european project SPENCER[1] to a laser based human tracking component.

3 Planning and Supervision

With the reasoning abilities provided by Situation Assessment, the robot should guide its user toward the goal, which could be predefined or negotiated with him at the start of the scenario. We defined a set of modules, called Collaborative Planners, able to choose which proactive or adaptive actions the robot should perform at each moment.

Collaborative Planners. The Collaborative Planners form a planning framework, based on hierarchical MOMDPs (Mixed Observability Markov Decision Process), that enables the system to react in a human-aware way to a user's behaviors. A MOMDP models the decision process of an agent in situations where the result of an action is partly random, and can lead to several outcomes. In addition, in a MOMDP, the system state is split in an observable set and a hidden set, which cannot be fully observed and must be inferred from observations received from the environment. MOMDPs are a variation of POMDPs

---

[1] http://www.spencer.eu/

(Partially Observable Markov Decision Process), where the system state is completely hidden. Partitioning the system state in a hidden and observable set simplifies the computation of a solution to the model, which is one of the main problems of POMDPs [1].

We use a hierarchical framework [16], where the system model is split into a main MOMDP module and several MOMDP sub-models, each one related to a different action. The models are solved separately, leading to the computation of different, simpler, policy functions. At run-time, the system interacts with the main module, providing values for the set of observations and for the observed variables, and receiving an action as result. Based on this action, the system will contact a different sub-model, receiving the final action to execute. Using hierarchical MOMDPs we can represent a set of models, with a greatly reduced complexity, and easily expand it if we want to implement new actions or to add more complex behaviors. The architecture of our system is shown in Figure 1 A).

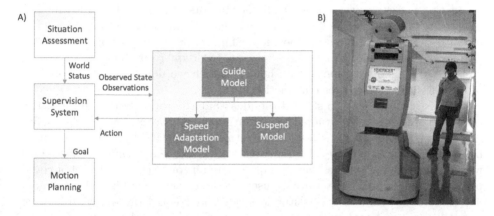

Fig. 1. A) System Architecture: our system is composed by four main modules. Situation Assessment reasons on perceptual data, providing symbolic information to the Supervision System. The Supervision System controls the other modules, updating the Collaborative Planners, which compute the next action to perform, and sending goals to the Motion Planning. Blue arrows represent data links between modules, while red arrows represent conceptual links that show the hierarchy of the MOMDPs. B) The robot guiding a user.

Guiding Users. The main problem of the robot is choosing if it should still guide the user, suspend temporarily the task, or abandon it. The Guide Planner is the main MOMDP of our architecture and will make this decision, based on two main variables: the status of advancement of the task (*not_completed, completed*), and the quality of commitment of the user (*not_engaged, engaged, not_interested*).

The quality of commitment of the user is an hidden variable, estimated using Situation Assessment, based on the distance of the person toward the robot, its variation, if the user is oriented toward the robot, and if he is moving or still. The robot will abandon the task when it evaluates that it's user has permanently stopped following it.

Adapting the Robot's Speed. We believe that to be socially acceptable, the robot should adapt its speed to the user. By setting its own pace at the start of the scenario the robot would risk of being too slow, annoying the user, or too fast, which would lead the robot to constantly stop to wait for persons, producing an awkward behavior.

The robot defines a desired interval of distances r from the user. The distance of the user from r will influence its actions. 1) If the user is farther than r the robot will *decelerate*. 2) If he is closer to the robot than r, the robot will *accelerate*. 3) If the user is inside r, the robot will continue at its pace.

In this paper, r was a predefined set, but its values could be learnt and adapted to users during the task, since different people could prefer following the robot at different distances and positions. The robot should also not constantly change speed, in order to give time to users to adapt to its new chosen speed, and so we defined a temporal threshold in which we don't allow the robot to repeat an *accelerate* or *decelerate* action.

In this scenario we also studied the idea that the robot can try to influence the speed of the user. We studied two situations in which this idea can be useful. A) There is a time limit to reach the destination. In this case the robot must balance the desire to satisfy the user with the task urgency. Different situations will require different policies. For example, in an airport scenario, the robot could prioritize arriving on time, warning users if their speed would render the goal not achievable, while in other situations the robot could try to arrive in time but still avoid to adopt speeds that are uncomfortable for the follower. B) The rules of the current environment limit the robot's speed. In this case the robot will avoid accelerating over a set speed even if it detects that its current velocity is considered too slow for the user. For example, the robot could be navigating in a construction zone.

This reasoning is done in the Speed Adaptation MOMDP module, which will be interpreted when the Guide Model chooses to keep guiding the user.

Suspending the Task. In some situation, the robot needs to suspend the task, because the user has stopped following it. In this case, the robot should estimate if this suspension of the collaborative scenario is temporary or permanent, and in the latter case abandon the task. We estimate this information using the Suspend Model and the activity areas from Situation Assessment. We link activity areas to the maximum time we expect that the user will be involved in the activity, and with a set of proactive actions that the robot can choose to execute.

In this paper, we investigated a single possible proactive behavior: giving information. In this case, if we detect that the user has stopped following because he is looking at a touristic sight, or at an information screen, the robot can try to engage him and offer related information. At the moment, we just propose a simple routine-based framework for this behavior, and plan to further study it in the future. We believe that the solution of this problem could be rich, and that the robot should estimate the reaction of the user during the execution of its proactive behavior, in order to be able to interrupt if he doesn't want to be helped or to resume the original task if he is satisfied by the robot's actions.

We don't want the robot to be stuck for a long time waiting for a person. If there is a small amount of time to reach the destination, or the user is engaged in the activity for a longer period of time than the one predicted, the Suspend Model can issue a warning action, and eventually abandon the task if the person doesn't start following it again. Sometimes users will stop following without an apparent reason, perhaps outside any activity area. In this case the robot will still allow them some time before issuing a warning and eventually abandoning the task.

4   Motion Planning

A guiding robot needs to plan safe and socially acceptable motion. This requires continual integration of high-level social constraints with the low-level constraints of the robot vehicle.

We use the architecture proposed by the well-established `move_base` package of `ROS` middle-ware [8,14] for navigation, replacing the global planner, i.e. a cost-grid base path planning module, as suggested in [21]. This module adds proxemics based costs in the grid-map around the detected humans that are static in the environment. The local planner, i.e. the module responsible for generating motor commands, is a *ContextCost* based algorithm suggested in [12]. This module continuously calculates the *compatibility* of the robot path by predicting and avoiding future collisions with moving persons and simultaneously keeping the robot as close as possible on the planned global path.

It should be noted that in our guiding experiments, the humans are mostly moving behind the robot and therefore the situation remains compatible for the local planner most of the time. During compatible situations the robot simply follows the way-points on the planned global path. The nominal velocity of the robot is set by the supervision system to achieve the desired behavior of slowing-down or speeding-up, as required by the situation.

5   Experiments and Analysis

We performed a first set of experiments with a person following a robot on a predefined path, in order to test the behaviors of the robot. Data from these experiments are shown in Table 1[2]. We start by showing speed adaptation tests:

- adapting slow and fast: in these two tests (Figure 2) we used our system to guide respectively a user that would like to move at a slow pace, and a user that would like to move at a fast speed.
- no adaptation: in this experiments the robot won't adapt to the speed of the user, setting its own pace and stopping if it is too far.

Looking at the data we can see that our system shows lower values for the variance of speed and distance, which means that after a certain time it's able to find a condition of equilibrium with the human follower. The 'no adaptation' system shows a significantly higher variance for both values, since the robot stopped several times to wait for the user. We will now show some tests regarding the proactive behaviors of the robot:

- proactive slow and fast: during the task, the robot proactively chooses to change pace, in the first case by slowing down and in the second by accelerating. In our tests the user adapted after some seconds to the robot's pace, but this behaviors should be studied in-depth in user studies.
- suspend with screen and with no reason: in these tests we asked a user to stop during the task. In the first case the user stopped near an information screen. After detecting this event, the robot approached the user to offer information, which lead to the resumption of the task. In the second case the user stopped at a different point of the path. The robot wasn't able to detect the reason for the suspension of the task and so simply issued a warning to the user and abandoned the task after some seconds.

Table 1. Experiment results: d is the distance between the robot and the user, s_r is the robot's speed, s_h is the human's speed, μ is the average and Δ is the variation of the quantity over the whole test. Distances are expressed in meters, velocities in meters for seconds.

test name	μ distance	μ speed difference	Δ distance	Δ speed difference
adapting slow	2.82	-0.03	0.64	0.02
adapting fast	1.38	0.00	0.29	0.01
no adaptation	3.08	-0.09	1.04	0.07
proactive slow	1.45	-0.06	0.04	0.10
proactive fast	2.66	-0.11	0.63	0.01

---

[2] Videos from our experiments can be seen at http://homepages.laas.fr/mfiore/icsr2015.html

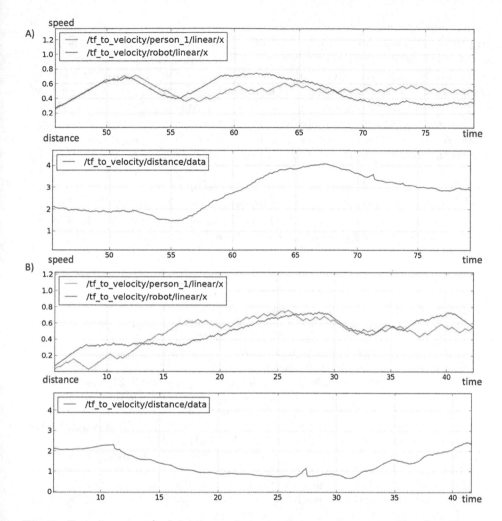

Fig. 2. Experiments: a) Adapting robot speed to a slow user. The first figure shows the speed of the user ($tf\_to\_velocity/person\_1/linear/x$) and of the robot ($tf\_to\_velocity/robot/linear/x$), and the second their distance. The robot starts slowing down at $t = 60$, when the distance from the user is growing, until it finds an equilibrium with the user's speed. Notice that there is a turn in the path, at $T = 50$, that causes the robot and the user to slow down. Distances are expressed in meters, velocities in meters for seconds. b) Adapting robot speed to a fast user. As before, the figures show the robot and user's speed and their distance. The robot starts accelerating at $t = 15$ when the distance from the user becomes small.

6   Conclusions

In this paper we introduced a robotic system able to guide, in a human-aware manner, an user to a destination. Our system is able to estimate, using Situation Assessment and a set of planners based on hierarchical MOMDPs, if the human user is currently engaged in the task, to adapt its actions to his behaviors, and to proactively help him. Through a set of experiments we showed that the robot is able to adapt its speed to its follower, in order to provide a socially acceptable behavior. We also began to study how the system can influence its user, by proposing a new speed, based on environmental and task related stimulus, and by proactively interacting with him when he stop following.

Though not shown in this paper, our system is also able to represent and guide groups of users, by reasoning both on the group as a single entity (represented through its spatial centroid) and on its single members. We plan, in the future, to perform user studies on groups, to understand how they react to the robot's behaviors, and eventually modify the system.

We would also like to study learning techniques, both in motion planning and in supervision, and integrate them in the system, to adapt even more the robot's movement and its decision on specific followers.

References

1. Araya-López, M., Thomas, V., Buffet, O., Charpillet, F.: A closer look at MOMDPs. In: 2010 22nd IEEE International Conference on Tools with Artificial Intelligence (ICTAI), vol. 2, pp. 197–204. IEEE (2010)
2. Bratman, M.E.: Shared agency. In: Philosophy of the Social Sciences: Philosophical Theory and Scientific Practice, pp. 41–59 (2009)
3. Bratman, M.: Two faces of intention. Philosophical Review **93**, 375–405 (1984)
4. Clodic, A., Fleury, S., Alami, R., Chatila, R., Bailly, G., Brethes, L., Cottret, M., Danes, P., Dollat, X., Elisei, F., et al.: Rackham: an interactive robot-guide. In: The 15th IEEE International Symposium on Robot and Human Interactive Communication, ROMAN 2006, pp. 502–509. IEEE (2006)
5. Evers, V., Menezes, N., Merino, L., Gavrila, D., Nabais, F., Pantic, M., Alvito, P.: The development and real-world application of frog, the fun robotic outdoor guide. In: Proceedings of the Companion Publication of the 17th ACM Conference on Computer Supported Cooperative Work & Social Computing, pp. 281–284. ACM (2014)
6. Fiore, M., Clodic, A., Alami, R.: On planning and task achievement modalities for human-robot collaboration. In: The 2014 International Symposium on Experimental Robotics (2014)
7. Fong, T.W., Kunz, C., Hiatt, L., Bugajska, M.: The human-robot interaction operating system. In: 2006 Human-Robot Interaction Conference. ACM, March 2006
8. Garage, W.: The robot operating system (ROS) is a set of software libraries and tools that help you build robot applications. http://www.ros.org/
9. Jensen, B., Tomatis, N., Mayor, L., Drygajlo, A., Siegwart, R.: Robots meet humans-interaction in public spaces. IEEE Transactions on Industrial Electronics **52**(6), 1530–1546 (2005)
10. Karami, A.B., Jeanpierre, L., Mouaddib, A.I.: Human-robot collaboration for a shared mission. In: 2010 5th ACM/IEEE International Conference on Human-Robot Interaction (HRI), pp. 155–156. IEEE (2010)
11. Koppula, H.S., Saxena, A.: Anticipating human activities using object affordances for reactive robotic response. Robotics: Science and Systems, Berlin (2013)

12. Kruse, T., Basili, P., Glasauer, S., Kirsch, A.: Legible robot navigation in the proximity of moving humans. In: 2012 IEEE Workshop on Advanced Robotics and its Social Impacts (ARSO), pp. 83–88, May 2012
13. Kruse, T., Pandey, A.K., Alami, R., Kirsch, A.: Human-aware robot navigation: A survey. Robotics and Autonomous Systems **61**(12), 1726–1743 (2013)
14. Marder-Eppstein, E.: move_base, a ROS Package that lets you move a robot to desired positions using the navigation stack. http://wiki.ros.org/move_base
15. Milliez, G., Warnier, M., Clodic, A., Alami, R.: A framework for endowing an interactive robot with reasoning capabilities about perspective-taking and belief management. In: The 23rd IEEE International Symposium on Robot and Human Interactive Communication, RO-MAN 2014, pp. 1103–1109, August 2014
16. Pineau, J., Roy, N., Thrun, S.: A hierarchical approach to POMDP planning and execution. In: Workshop on Hierarchy and Memory in Reinforcement Learning (ICML), vol. 65, p. 51 (2001)
17. Rashed, M.G., Suzuki, R., Lam, A., Kobayashi, Y., Kuno, Y.: Toward museum guide robots proactively initiating interaction with humans. In: Proceedings of the Tenth Annual ACM/IEEE International Conference on Human-Robot Interaction Extended Abstracts, pp. 1–2. ACM (2015)
18. Rios-Martinez, J., Spalanzani, A., Laugier, C.: From Proxemics Theory to Socially-Aware Navigation: A Survey. International Journal of Social Robotics **7**(2), 137–153 (2014)
19. Samejima, I., Nihei, Y., Hatao, N., Kagami, S., Mizoguchi, H., Takemura, H., Osaki, A.: Building environmental maps of human activity for a mobile service robot at the miraikan museum. In: Field and Service Robotics, pp. 409–422. Springer (2015)
20. Sebanz, N., Bekkering, H., Knoblich, G.: Joint action: bodies and minds moving together. Trends in Cognitive Sciences **10**(2), 70–76 (2006)
21. Sisbot, E., Marin-Urias, L., Alami, R., Simeon, T.: A human aware mobile robot motion planner. IEEE Transactions on Robotics **23**(5), 874–883 (2007)
22. Thrun, S., Beetz, M., Bennewitz, M., Burgard, W., Cremers, A.B., Dellaert, F., Fox, D., Haehnel, D., Rosenberg, C., Roy, N., et al.: Probabilistic algorithms and the interactive museum tour-guide robot minerva. The International Journal of Robotics Research **19**(11), 972–999 (2000)
23. Yousuf, M.A., Kobayashi, Y., Kuno, Y., Yamazaki, A., Yamazaki, K.: Development of a mobile museum guide robot that can configure spatial formation with visitors. In: Huang, D.-S., Jiang, C., Bevilacqua, V., Figueroa, J.C. (eds.) ICIC 2012. LNCS, vol. 7389, pp. 423–432. Springer, Heidelberg (2012)

The Effects of Social Gaze in Human-Robot Collaborative Assembly

Kerstin Fischer[1], Lars Christian Jensen[1(✉)], Franziska Kirstein[2], Sebastian Stabinger[3],
Özgür Erkent[3], Dadhichi Shukla[3], and Justus Piater[3]

[1] Department for Design & Communication, University of Southern Denmark,
Alsion 2 6400, Sonderborg, Denmark
{kerstin,larscj}@sdu.dk
[2] Blue Ocean Robotics, Klokkestøbervej 18 5230, Odense M, Denmark
fk@blue-ocean-robotics.com
[3] Intelligent and Interactive Systems, Institute of Computer Science,
University of Innsbruck, Technikerstr. 21a 6020, Innsbruck, Austria
{Sebastian.Stabinger,ozgur.erkent,dadhichi.shukla,
justus.Piater}@uibk.ac.at

Abstract. In this paper we explore how social gaze in an assembly robot affects how naïve users interact with it. In a controlled experimental study, 30 participants instructed an industrial robot to fetch parts needed to assemble a wooden toolbox. Participants either interacted with a robot employing a simple gaze following the movements of its own arm, or with a robot that follows its own movements during tasks, but which also gazes at the participant between instructions. Our qualitative and quantitative analyses show that people in the social gaze condition are significantly more quick to engage the robot, smile significantly more often, and can better account for where the robot is looking. In addition, we find people in the social gaze condition to feel more responsible for the task performance. We conclude that social gaze in assembly scenarios fulfills floor management functions and provides an indicator for the robot's affordance, yet that it does not influence likability, mutual interest and suspected competence of the robot.

Keywords: Human-robot interaction · Gaze · Conversation analysis · Smile

1 Introduction

In this paper, we investigate the effects of gaze towards the human tutor in a human-robot assembly scenario. Previous work has shown that in social human-robot interactions, human-like gaze behavior in robots fulfills very similar functions as in interactions between humans (e.g. [1]). However, what roles social gaze behavior plays in industrial assembly tasks in general and in human-robot collaboration in particular has received much less attention. Employing social even in industrial scenarios may prove useful since naïve users are increasingly engaged in demonstrating novel actions to robots for industrial manufacturing [2]. We therefore carried out experiments in which the robot's gaze behavior differed between being fully instrumental to

© Springer International Publishing Switzerland 2015
A. Tapus et al. (Eds.): ICSR 2015, LNAI 9388, pp. 204–213, 2015.
DOI: 10.1007/978-3-319-25554-5_21

the assembly task (simple gaze condition) and focusing on the user between tasks (social gaze condition); in accordance with [1], we refer to the robot's gaze towards the human as social. The question we raise is which of the functions reported for social robots (e.g. [3,4]) and virtual agents (e.g. [5]) social gaze fulfills in this collaborative assembly situation.

2   Previous Work

Previous work on gaze in human interaction shows that it fulfills numerous important interactional functions. For instance, it serves to negotiate contact in first encounters and to establish an interpersonal relationship, including flirting [6]. Furthermore, gaze plays an important role in signaling attention during interaction; in particular, listeners have been found to attend to the current speaker by means of long gaze which is interrupted only by short glances away [7]. In teaching situations, students show higher performance if the teacher gazes at them [6]. Increased eye gaze has also been found to contribute to higher ratings of likability and mutual interest [6]. Gaze also plays a role in turn-taking, such that current speakers often gaze away when they want to keep the turn [7], while gazing towards the listener often elicits a response, either evoking feedback or selecting the next speaker [3].

Similarly, many of the functions of gaze observed in human interaction have been confirmed to be relevant for interactions with virtual agents (cf. [5] for an overview). That gaze fulfills these functions also in human-robot interaction has meanwhile been shown in numerous studies (e.g. [8], and [1] carry out a survey of previous work on gaze in human-robot interaction and suggest to systematize the social functions of gaze according to five main social contexts in which gaze plays a major role: establishing agency and liveliness; signaling social attention, for instance, by providing eye contact; regulating the interaction process, that is, facilitating turn-taking and supporting the participation framework; supporting interaction content, for instance, by disambiguating ambiguous information; and finally projecting mental state, in particular, expressing emotional or cognitive states. At the same time, it is equally important for the human, agent or robot to know when to avert gaze; for instance, [9] find that eye gaze may also be perceived as staring in virtual agents. Similarly, [10] show that gaze aversion supports floor management and creates the impression of higher cognitive capabilities and higher creativity of the robot.

On the other hand, gaze by robots has also been found to be not attended to in the same way as human gaze (e.g.[11,12]) or is not even taken into account because people may not look at the robot's face (e.g. [13,14,15]).

So far, it remains largely unexplored what functions an industrial robot's eye gaze might play during a collaborative assembly task. What previous work may predict is that the robot's gaze behavior can indicate what the robot is currently attending to and thus contribute to floor management and task organization; furthermore, the robot's gaze may influence the interpersonal relationship, making the robot appear more lively, more likeable, more cognitively competent and more socially attentive; and finally, the robot's gaze could also contribute to disambiguating information. However, whether robot gaze fulfills these functions also depends on whether participants attend to it and take it into account at all.

3   Method

In order to determine the role of social gaze in collaborative assembly tasks, we elicited 30 interactions between naïve users and an industrial robot in two conditions.

3.1   The Robot

The robot comprises two KuKa arms [16], each capable of seven degrees of freedom and each equipped with a Schunk 3-finger gripper. However, for this study the robot made only use of its left arm. The robot's KIT [17] head has cameras mounted in eyeballs and one Kinect camera on top. The robot was fully controlled remotely from a desk hidden from the participants. Our wizard was able to navigate using multiple cameras mounted on a steel frame above the workspace platform. An engineer oversaw the experiments to ensure the safety of both participants and robot.

Fig. 1. Simple gaze

Fig. 2. Social gaze

3.2   Experimental Conditions

The study uses a between-subject design with two experimental conditions. In one condition, the robot's gaze follows its own hand (simple gaze). In the second condition (social gaze), the robot's face is initially directed at the participant, yet during tasks, the robot changes its gaze to its hand when it starts to move, and then looks to the user again when it has completed its task.

Given the rules defining the robot's gaze direction, the two conditions really only differ in two phases during the interaction: initially, when in the 'simple' gaze condition the robot's gaze is on its hand while it is on the user in the 'social' gaze condition, and during handover phases when the robot provides the user with the requested parts. While in the simple gaze condition the robot's gaze remains on its hand, in the social gaze condition the robot looks up at the user.

3.3   Experimental Procedure

Participants were led into the lab, where they signed a consent form, had their picture taken and were then introduced to the robot. The task was to guide the robot to assist them in assembling a wooden toolbox. They were told that the robot would be able to fetch the parts for them, but that they would need to instruct it to do so when appropriate and in whatever way that made sense to them; participants should then assemble the box on their own. Instructions were scripted so that all participants received the same information. After introducing the participants to the task, the facilitators did not intervene except when assisting users with the drill. Participants were led to believe that the robot acted autonomously, but it was in fact controlled using the Wizard-of-Oz technique. The human 'wizard' was instructed to react to gestures (e.g. pointing) and to ignore all other actions by the participants (e.g. speech). The 'wizard' had multiple cameras to observe the participants and to navigate the workspace. Due to the intranet, there was a slight delay for about one second between users' instruction and the robot's response. After participants completed the task, they filled out a questionnaire about their interaction with the robot.

3.4   Questionnaire

Based on previous findings regarding the influence of robot gaze on people's perception of the robot (e.g. [10]), the written questionnaire asked participants to rate how intelligent they perceived the robot to be, and to rate how safe they felt when interacting with the robot. The perceived intelligence and safety scales were adopted from [18]. This index proved relatively reliable (Cronbach's alpha for perceived intelligence / perceived safety of 0.67 / 0.94 in the simple gaze condition and 0.82 / 0.37 in the advanced gaze condition).

To determine to what extent people monitored the robot (cf. [15]), participants were then asked where the robot looked during their interaction, who was more responsible for task performance (10-point semantic differential), whether they thought the robot had learned from them (7-point likert scale), and who was most in control (human participant or robot on a 10-point semantic differential scale). Furthermore, participants were asked to rate to what extent they felt that they and the robot were a team on a 10-point likert scale, and what kind of communicative cues the robot provided them with.

3.5   Participants

We recruited 36 participants, who were all students or employees at the University of Innsbruck without prior experience with industrial robots. Participants were given chocolate as compensation for their time and participation. However, five participants were eliminated from analysis due to robot malfunction (overheating, security stops, etc.), and one was removed from the analysis in section 4.2 due to a deviant understanding of the initial instruction (see 4.2 below). Participant mean age is 23.7 (SD = 4.54). One-third of the participants were women, yet they were distributed evenly across the two experimental conditions.

3.6   Analysis

The focus of our investigation lies in participants' behavioral responses during their interaction with the robot. Our analyses were based on video recordings of the interactions, supplemented by field notes. Responses were coded and analyzed quantitatively, using single linear regression with the statistical software package R (v. 3.1.2), as well as qualitatively using ethnomethodological conversation analysis [19]. In particular, we asked 1) whether tutors looked at the robot and perceived the robot's gaze towards them (section 4.1), 2) what effects the robot's gaze towards the tutor has on conversational openings (section 4.2), 3) what effects robot gaze towards the human between tasks has (section 4.3), and 4) whether the robot's gaze behavior has an impact on tutors' perception of the robot's capabilities (section 4.4).

4   Results

4.1   Users' Perception of the Robot's Gaze Behavior

Before we look into the effects of robot gaze in the two conditions, we need to establish that participants actually perceived the robot's gaze. Here our analyses show that initially, all participants in the social gaze condition looked at the robot and perceived its gaze towards them. However, during the experiment, they often did not look up at the robot; in all, during 43 of the 90 handovers in the social gaze condition, the respective participant did not glance towards the robot. One participant looked at a time when the robot had not looked up yet, so he may have expected a different timing.

4.2   Eye Gaze During Contact Initiation

Participants in the social gaze condition needed less time to initiate their first action (instruction) than participants in the simple gaze condition. Using single linear regression, we find a statistically significant difference between participants in the simple gaze condition (M=37.39, SD=31.65) and the social gaze condition (M=11.96, SD=14.6), $F(1,28) = 8.37$, $p = 0.007$, $R^2=0.24$ (measurements are in number of seconds). We find no significant results on age or gender as predictor variables and no significant interactions. Using visual inspection of boxplots, we eliminated one extreme outlier from the analysis; this participant repeatedly directed his attention to the experimenter to ask specifically about how the robot was built and what components were used in its assembly, thus not engaging with the robot at all. Otherwise, we can see very different ways of approaching the robot in the two conditions, as the following two examples illustrate: The example in Fig. 3 stems from the simple gaze condition and illustrates how the participant hesitates and is clearly uncertain about how the robot can be approached. The example in Fig. 4, in contrast, shows how a participant in the social gaze condition straight-forwardly approaches the robot by establishing mutual eye-gaze, waving and smiling.

User looks at the   User looks down   User looks at the robot

Fig. 3. Participant does not know how to start

User meets the robots gaze   User says "Hello robot, give me piece #1"

Fig. 4. Participant takes robot gaze as a sign for mutual attention

4.3 Eye Gaze Between Tasks

The example in Fig. 5 below shows how the robot's eye gaze is understood as a straightforward social signal by the participant and is met with mutual eye gaze and a spontaneous smile. In 75.6% of the cases in which users looked up to see the robot's eye gaze, they responded to the robot's gaze with a smile. Correspondingly, the numbers of smiles between simple gaze (M=3.47, SD=1.92) and social gaze (M=5.13, SD=2.33) conditions is significantly different: $F(1,28) = 4.58$, $p = 0.04$, $R^2=0.14$.

Robot and user look up   User looks at robot   User starts smiling

Fig. 5. Spontaneous smile in response to mutual gaze

Robot and user look up   User raises finger   User instructs robot

Fig. 6. User directs the robot's gaze back to his hand

However, some participants also try to draw the robot's attention back to their hands when the robot directs its gaze towards them, as in Fig. 6 where the participant first reciprocates the robot's gaze smilingly, yet then lifts up his hand to direct the robot's gaze back to the instruction.

4.4 Conceptualizing the Robot and Understanding Robot Gaze

In a post-experimental questionnaire, we asked participants about how intelligent, safe, teachable and compliant they perceived the robot to be; regarding none of these measures there are any statistically significant differences between the two conditions. Furthermore, the two groups do not differ regarding their feeling of them building a team with the robot.

Regarding the question of who participants thought was most responsible for the performance of that task (1=robot, 10=participant), we find a statistically significant difference between the simple gaze condition (M=6.93, SD=2.13) and the social gaze condition (M=8.24, SD=1.25), F (1, 29) = 4.53, p = 0.04, r^2=0.14, such that tutors in the social gaze condition thought that the tutors themselves were more responsible for performance of the task than the tutors in the simple gaze condition thought.

| User says: "and now down" | Robot looks up. User looks up, says "perfect" | User takes piece and smiles |

Fig. 7. Robot's social gaze as invitation for feedback

We also inquired into participants' understanding of the robot's gaze behavior. In the simple gaze conditions, seven tutors stated that the robot's gaze served as a communicative cue, while two held the robot to move in response to his/her own movements, and five did not know or identify any cues. In contrast, 15 participants in the social gaze condition identified the robot's gaze as a communicative cue, while only two did not know or could not identify any communicative cues. This difference was found statistically significant (p<0.05), using Fisher's exact test. The qualitative analysis of the handovers furthermore shows that participants interpret the robot's gaze towards them as an invitation to provide feedback, as Fig. 7 illustrates.

Finally, we asked whether participants could tell where the robot was looking. For the simple gaze condition, one participant could correctly tell that the robot was constantly looking at its own arm, nine participants made incorrect guesses as to where the robot was looking, three simply stated that they could not infer the robot's gaze and one answered the question with a simple 'yes' without further elaboration. For the social gaze condition, 12 could correctly tell that the robot looked at them, then at its own arm and then again at them, four made incorrect guesses, and one answered the question with a simple 'yes' without any further elaboration. These differences were also found statistically significant (p<0.01) using Fisher's exact test.

5   Discussion

In all, we found users to take the robot's gaze behavior into account initially as well as during approx. 50% of the handovers. Given that the robot's gaze towards the users was purely social and did not serve any disambiguating or other task-related functions, this percentage is very high in comparison with previous work (e.g.[15]), and our questionnaire results show that participants understood the robot's gaze towards them as intentional signs. Furthermore, there is evidence for a role of the robot's social gaze in floor management and participation framework (cf. [3]) since participants responded to the robot's gaze in handovers as a request for feedback (Fig. 7). Similarly, the finding that people in the social gaze condition feel more responsible for the performance than people in the simple gaze condition is most likely due to the fact that in the social gaze condition, the robot seemingly awaits instructions each time it finishes a task and thus provides a turn-yielding signal. As a result, participants in the social gaze condition feel to a greater extent that the performance of the task is contingent on their ability to instruct the robot.

As for interpersonal functions concerning intimacy and relationship negotiation, we find on the one hand that participants in the social gaze condition generally found it easier to make contact with the robot than participants in the simple gaze condition, as the reduced contact initiation times show; obviously, they felt they knew how to interact with it. Thus, participants in the social gaze condition were found to engage with the robot sooner than participants in the simple gaze condition. On the other hand, we found that the robot's social gaze behavior had no influence on users' perception of the robot's cognitive capabilities, compliance, learning and safety in general. This is unexpected for several reasons: First, previous work had shown that correctly timed eye gaze creates the impression of higher cognitive ability [10]; now it could be argued that the robot's gaze may not have been timed appropriately. However, while it may not have respected human-like gaze patterns during handovers (cf. [8, 20]), it did mark the difference between different phases of the collaborative assembly, namely delivering a part versus awaiting another instruction. Second, participants in the social gaze condition were found to smile at the robot more often, which could have led to a more intimate interpersonal relationship and perceived tighter teamwork. Thus, gaze towards the user does not automatically create rapport. At the same time, there are no indications that participants found the robot to 'stare' even though it was looking in their direction while participants were busy assembling the toolbox; on the contrary, participants turned to the robot smilingly when they finished their (solitary) assembly phases (cf. [21]). Thus, the robot's gaze towards participants seems to be perceived not as social enough to be perceived as staring and not as social enough to change users' general view of the robot, yet as social enough to be met with a smile.

6 Conclusion

To conclude, allowing the user to establish mutual eye gaze during the initiation of interaction by having the robot look towards the participant does not only serve 'to break the ice' [22], but also provides users with a necessary indicator of the robot's "entry point" for the interaction since people interacting with the robot in the social gaze condition are significantly quicker to engage the robot. Thus, an important function of social gaze in collaborations with assembly robots is to provide an indicator for how the robot can be interacted with. This function of social gaze has not been reported for robot gaze, most likely because it is not relevant in human interaction.

Furthermore, participants in the social gaze condition smile significantly more often and can better account for where the robot is looking compared to participants in the simple gaze condition. In addition, we also found that people in the social gaze condition feel more responsible for the task performance. These findings support previous work on gaze by social robots concerning floor management, which turns out particularly useful during assembly tasks as indicator for turn-yielding during handovers, yet they are inconclusive regarding the interpersonal impact of robot social gaze.

Acknowledgement. This research was partially funded by the European Community's Seventh Framework Programme FP7/2007-2013 under grant agreement no. 610878, 3rdHAND.

References

1. Srinivasan, V., Murphy, R.R.: A survey of social gaze. In: Proceedings of HRI 2011, Lausanne, Switzerland (2011)
2. Muxfeldt, A., Kluth, J.-H., Kubus, D.: Kinesthetic teaching in assembly operations – a user study. In: Broenink, J.F., Kroeger, T., MacDonald, B.A., Brugali, D. (eds.) SIMPAR 2014. LNCS, vol. 8810, pp. 533–544. Springer, Heidelberg (2014)
3. Mutlu, B., Kanda, T., Forlizzi, J., Hodgins, J., Ishiguro, H.: Conversational Gaze Mechanisms for Humanlike Robots. ACM Transactions on Interactive Intelligent Systems 1(2), 12 (2012)
4. Moon, A., Troniak, D.M., Gleeson, B., Pan, M.K., Zeng, M., Blumer, B.A., MacLean, K., Croft, E.A.: Meet me where I'm gazing: how shared attention gaze affects human-robot handover timing. In: Proceedings of HRI 2014, pp. 334–341 (2014)
5. Ruhland, K., Andrist, S., Badler, J.B., Peters, C.E., Badler, N.I., Gleicher, M., Mutlu, B., McDonnell, R.: Look me in the eyes: a survey of eye and gaze animation for virtual agents and artificial systems. In: Eurographics 2014 (2014)
6. Argyle, M., Cook, M.: Gaze and Mutual Gaze. Cambridge University Press (1976)
7. Kendon, A.: Conducting Interaction: Patterns of Behavior in Focused Encounters. Cambridge University Press (1990)
8. Mehlmann, G., Häring, M., Janowski, K., Baur, T., Gebhard, P., André, E.: Exploring a model of gaze for grounding in multimodal HRI. In: ICMI 2014, pp. 247–254 (2014)
9. Wang, N., Gratch, J.: Don't just Stare at me! In: Proceedings of CHI 2010, Atlanta, Georgia, USA (2010)

10. Andrist, S., Tan, X.Z., Gleicher, M., Mutlu, B.: Conversational gaze aversion for human-like robots. In: Proceedings of HRI 2014, Bielefeld, Germany (2014)
11. Admoni, H., Bank, C., Tan, J., Toneva, M., Scassellati, B.: Robot gaze does not reflexively cue human attention. In: Proceedings of the 33rd Annual Conference of the Cognitive Science Society (CogSci 2011), Austin, TX, pp. 1983–1988 (2011)
12. Meltzoff, A.N., Brooks, R., Shon, A.P., Rao, R.P.: "Social" robots are psychological agents for infants: a test of gaze following. Neural Networks **23**, 966–972 (2010)
13. Fischer, K., Lohan, K., Nehaniv, C., Lehmann, H.: Effects of different types of robot feedback. In: International Conference on Social Robotics 2013, Bristol, UK (2013)
14. Fischer, K., Soto, B., Pantofaru, C., Takayama, L.: The effects of social framing on people's responses to robots' requests for help. In: Proceedings of the IEEE Conference on Robot-Human Interactive Communication – Ro-man 2014, Edinburgh (2014)
15. Admoni, H., Dragan, A., Srinivasa, S.S, Scassellati, B.: Deliberate delays during robot-to-human handovers improve compliance with gaze communication. In: Proceedings of HRI 2014, Bielefeld, Germany (2014)
16. Bischoff, R., Kurth, J., Schreiber, G., Koeppe, R., Albu-Schäffer, A., Beyer, A., Eiberger, O., Haddadin, S., Stemmer, A., Grunwald, G., Hirzinger, G.: The KUKA-DLR lightweight robot arm - a new reference platform for robotics research and manufacturing. In: Joint 41st International Symposium on Robotics and 6th German Conference on Robotics, ISR-Robotik, Munich, pp. 741–748 (2010)
17. Asfour, T., Welke, K., Azad, P., Ude, A., Dillmann, R.: The Karlsruhe humanoid head. In: 8th IEEE-RAS International Conference on Humanoid Robots, pp. 447–453 (2008)
18. Bartneck, C., Croft, E., Kulic, D.: Measurement instruments for the anthropomorphism, animacy, likeability, perceived intelligence, and perceived safety of robots. International Journal of Social Robotics **1**(1), 71–81 (2009)
19. Sacks, H., Schegloff, E.A., Jefferson, G.: A simplest systematics for the organization of turn-taking for conversation. Language **50**(4), 696–735 (1974)
20. Strabala, K., Lee, K., Dragan, A., Forlizzi, J., Srinivasa, S., Cakmak, M., Micelli, V.: Toward Seamless Human-Robot Handovers. Journal of Human-Robot Interaction **2**(1), 112–132 (2013)
21. Kirstein, F., Fischer, K., Erkent, Ö., Piater, J.: Human smile distinguishes between collaborative and solitary tasks in human-robot interaction. In: Late Breaking Results, Human-Robot Interaction Conference 2015, Portland, Oregon (2015)
22. Bee, N., André, E., Tober, S.: Breaking the ice in human-agent communication: eye-gaze based initiation of contact with an embodied conversational agent. In: Ruttkay, Z., Ruttkay, Z., Kipp, M., Nijholt, A., Vilhjálmsson, H.H. (eds.) IVA 2009. LNCS, vol. 5773, pp. 229–242. Springer, Heidelberg (2009)

Influence of Upper Body Pose Mirroring in Human-Robot Interaction

Luis A. Fuente[1]([✉]), Hannah Ierardi[1], Michael Pilling[2], and Nigel T. Crook[1]

[1] Department of Computing and Communication Technologies, Oxford Brookes University, Oxford OX3 0BP, UK
{lfuente-fernandez,11092652,ncrook}@brookes.ac.uk
[2] Department of Psychology, Oxford Brookes University, Oxford OX3 0BP, England, UK
mpilling@brookes.ac.uk

Abstract. This paper explores the effect of upper body pose mirroring in human-robot interaction. A group of participants is used to evaluate how imitation by a robot affects people's perception of their conversation with it. A set of twelve questions about the participants' university experience serves as a backbone for the dialogue structure. In our experimental evaluation, the robot reacts in one of three ways to the human upper body pose: ignoring it, displaying its own upper body pose, and mirroring it. The manner in which the robot behaviour influences human appraisal is analysed using the standard Godspeed questionnaire. Our results show that robot body mirroring/non-mirroring influences the perceived humanness of the robot. The results also indicate that body pose mirroring is an important factor in facilitating rapport and empathy in human social interactions with robots.

Keywords: Body-pose mirroring · Empathy · Rapport · Anthropomorphism

1 Introduction

Mirroring is a natural social behaviour demonstrated by humans whereby a participant in a social interaction will often tend to subconsciously mirror another's body posture. There is significant evidence from psychological studies that people in groups have a tendency to engage in this mirroring behaviour [1,2]. People are often not conscious of the fact that they are mirroring someone's body pose or that someone is mirroring them [3]. These studies have also experimentally shown that this non-verbal synchrony in conversation is preserved over time and has a positive influence in creating rapport, increasing empathy, and facilitating social interaction [4].

In this article, we investigate the effect that upper body pose mimicry has on how humans perceive robotic systems. The proposed system recognises upper body poses from camera images and produces upper body gestures (torso, head and arms) in the humanoid Nao robot. The robot's text-to-speech output is also

© Springer International Publishing Switzerland 2015
A. Tapus et al. (Eds.): ICSR 2015, LNAI 9388, pp. 214–223, 2015.
DOI: 10.1007/978-3-319-25554-5_22

used to achieve natural communication. A set of twelve predefined questions is considered to engage the participants in communication with the robot in one of the following three different conditions: the robot mirrors the upper body pose of a human, the robot generates non-mirroring human-like upper body poses, and the robot adopts a static body pose. The Godspeed questionnaire [12] is used to measure the five key concepts of human-robot interaction and to evaluate the effect of body pose mirroring in human-robot interaction (HRI).

The rest of this paper is structured as follows: Section 2 reviews existing studies on the influence of behaviour mimicry in human perception of robotic systems, Section 3 includes a description of the experimental setup, the methodology and the evaluation method, Section 4 presents statistical findings, and Section 5 concludes with a brief discussion of the results and future work.

2   Related Work

Several recent studies have assessed the influence that a robot's non-verbal behaviours have on the way humans perceive and interact with robots. Salem et al [5] found that human beings have a tendency to anthropomorphise more (to like the robot more), report greater shared reality and show increased intention for future interaction with robots when they used bodily gestures with speech, as opposed to speaking using a static pose. The same authors also suggested that a robot's use of gesture with speech tends to enhance people's performance on robot guided tasks [6]. Similarly, Riek et al [7] demonstrated that the manner in which robots execute bodily gestures can have a major influence on the degree to which people are willing to cooperate with them. Further, some robot gesture combinations (i.e. gazing and pointing) also increase a person's tendency to reproduce the behaviour of a robot, resulting in entrainment [8]. In this sense, Kim et al [9] stated that it is possible to use gesture manipulations to influence the perceived personality of social robots and [10] showed that contingent non-verbal behaviours (i.e. behaviours tightly coupled to what the human speaker is doing) can create rapport with the human participant.

The use of mirroring behaviours by virtual characters and robots has been shown to improve empathy and create rapport with the humans that interact with them. Gonsior et al [11] studied the impact on human-robot interaction by a robot that mirrors facial expressions. In their study, the human participants engaged in a communicative task with the robot under one of three experimental conditions: the robot displayed no facial expressions, the robot mirrored the participant's facial expression, and the robot displayed facial expressions according to its internal model which indirectly mirrored the participant's facial expression. Each participant completed two post-experiment questionnaires. The first evaluated for empathy and subjective performance, and the second consisted of the five Godspeed questionnaires [12]. Their results indicated that mirroring conditions received higher ratings than neutral conditions. These results have also been supported by Kanda et al in [13]. They indicated that cooperative gestures from a robot (i.e. gestures that synchronised with the human participant) increase the human's impression of the robot's reliability and sympathy.

Similarly, Bailenson and Yee [14] found that an embodied artificial agent that mimicked a human participant's head movements was perceived as being more persuasive and received more positive trait ratings than non-mimicking agents. On the other hand, Riek et al [15], did not find that head gesture mirroring had a noticeable impact on creating rapport between a human and a robot. However, the authors acknowledge that the small sample size and other possible factors concerning their experimental setup may have influenced this result.

3   Method

To date, we have not been able to find any research that evaluates the effect that upper body mirroring during human-robot interaction has on the anthropomorphism, animacy, likeability, perceived intelligence and perceived safety of the robot. This study seeks to investigate this through a series of experiments in which participants engage in spoken interactions with a robot. The empathy between the participants and humanoid robot was examined under three different conditions:

A The robot mirrors the user's body poses during the interaction with occasional head nodding.
B The robot produces pre-programmed non-mirroring gestures with occasional head nodding during the interaction.
C The robot remains static for the duration of the interaction apart from occasional head nodding.

3.1   Hypotheses

Given the importance of mimicry in human-human communication, we wondered if it might be also important in human-robot communication. Thus, our hypotheses are as follows:

H1 The participants will rate the likeability and perceived safety of the robot more highly in condition **A** than in conditions **B** and **C**. This is motivated by the work in [1], which showed that posture sharing and rapport are positively correlated in humans and that this correlation holds over time, promotes safety, and encourages each participant during conversation.
H2 The participants will rate the anthropomorphism, animacy and perceived intelligence more positively in conditions **A** and **B** than in **C**. This hypothesis is prompted by the idea that people will show the most appreciation for a robot that mimics their upper-body and head gestures in real time. This motivation comes from the work in [13] who ran an experiment with the WowWee Alive Chimpanzee Robot capable of making head nods and face expressions, as well as detecting human head nods. They found that temporal-cooperative behaviours lead to a more positive interaction with robots and enable better human-robot rapport when compared with a robot that does not employ such behaviours.

3.2   Experimental Validation

In order to evaluate whether the proposed hypotheses improve HRI, an experiment in which human participants engage in a spoken interaction with the humanoid Nao robot was designed. The Nao robot is set up to allow upper body motion only, and a depth camera is used to track body poses and movements. During the interaction, the Nao robot speaks to the participant while acting according to one of the three possible conditions. It also nods at random intervals throughout the duration of the experiment. Subjects are divided into three different groups depending on the following conditions applied:

A *Mirroring*: The robot mirrors the participant's upper body pose. The pose of the participant is estimated using a depth camera. The output of the depth camera is first normalised and then processed to extract the rotational angles of each shoulder and elbow in the participant's body (eight angles in total). These are then scaled and mapped onto the corresponding joint angles for the robot. It is relevant to note that the Nao robot does not continually mirror the participant's body pose, since this would lead to unrealistic copycat behaviours. Instead, the robot intermittently mirrors the participant's body pose after it has remained in that body pose for a certain period of time (~5s).

B *No-Mirroring*: The robot produces pre-programmed human-like non-mirroring gestures (see Figures 1(b)-1(e)). These only take place while asking the questions to the participant.

C *Static*: The robot remains static with no body movements.

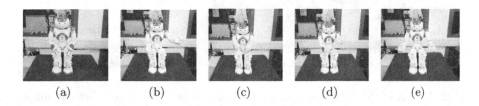

(a)	(b)	(c)	(d)	(e)

Fig. 1. Predefined movements of the Nao robot during the experiments. Figure 1(a) represents the resting/static pose and figures 1(b)-1(e) illustrate the predefined upper body robotic gestures.

It should be noted that head nodding was included to reduce the impact of any bias arising from the movement of the robot in conditions **A** and **B**. It was deemed that without the head nodding, the outcome of the experiment might be influenced by the fact that the robot is animated in condition **A** while the participant is speaking and not in condition **B**, as opposed to being influenced by the different types of whole body movement in each case (i.e. mirroring as opposed to non-mirroring poses). The importance of the head nodding in condition **C** is twofold. First, it allows participants to address the questions regarding animacy in the post-experiment questionnaire. Surprisingly, it was noted that participants waited for the robot to move first before engaging in a non-verbal communication with the robot. Second, it eases the comparison of condition **C** with the remaining conditions. The lack of movement in the robot itself negatively affects the participant's reaction to and rapport with the robot. In all cases the robot waits until the participant finishes speaking before asking the next question in the sequence.

3.3   Experimental Setup

Forty test subjects took part in this experiment, of which 9 were female and 31 were male. The participants were students from Oxford Brookes University and had no previous experience in the robotics field. They were between the ages of 19 and 35 (mean of 21.67 and s.d. of 3.05). The distribution of the subjects over the experimental conditions was 17 for **A**, 15 for **B**, and 8 for **C**[1]. Participants were not informed about the purpose of this study, but were instead advised that the experiment was to evaluate an automated student advisory system which seeks to use interactive humanoid robots. Additionally, they were also briefed about the layout of the experiment room and robot design.

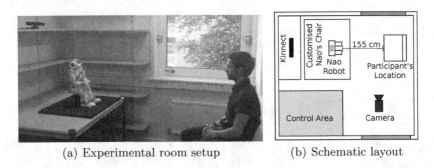

(a) Experimental room setup   (b) Schematic layout

Fig. 2. Layout of the quiet room during the participants' interaction with the robot.

A quiet room with controlled lighting conditions was chosen for the experiment with the layout shown in Figure 2. During the experiment, each participant was seated facing the Nao robot which was located on a customised chair so that the head of the robot was approximately at eye-level with the participant. The robot was strapped into the seat facing the participant in a hands-in-lap resting position (see Figure 1(a)). This restricts the lower-body motion of the robot but allows it to move its torso, arms and head. Since the task rating relies on the ability of the robot to effectively mimic the participants, a depth camera was

---

[1] The lack of movement in the participants due to the still position of the robot body in condition **C** lead to undesired problems in the motion capture impeding the mirroring. These participants were not considered during the analysis.

preferred over the robot's head camera. The depth camera was placed behind the robot, angled downwards to capture the movement of the participants. Each participant was seated sufficiently far ($\sim$150 cm) from the robot's chair so that its body was in full view of the depth camera and robot. The experimental setup was identical for each condition.

3.4   Experimental Procedure

Prior to the experiment, the instructor gave the participant a brief introduction on the task and described the one-to-one interaction with the robot. It is important to note that there was no visual or physical contact between the participant and the instructor (who was also in the experimental room but hidden from the participant), so that the participant was essentially alone with the robot. To start the experiment, the robot introduces itself in order to allow the participant to become familiar with its voice, shape and movements. From this point, the one-to-one interaction between the participant and the Nao robot begins. It consists of the interaction through a set of twelve predefined questions (see Table 1). Under the conditions **A** and **B**, the robot is also animated whilst asking the questions, before returning to the neutral hands-in-lap pose. The questions were determined in advance and centred around the participant's experience at the university in order to prompt an emotional engagement with the robot. Further, they also struck a balance between a subject that the participant could emotionally connect with, but that steered away from being unnecessarily invasive. The questions were identical for each participant.

Table 1. Sequence of questions asked by the robot

Q1	What subject are you studying?
Q2	When did you start on your course?
Q3	What have you enjoyed most about your course?
Q4	What did you enjoy least about your course?
Q5	Tell me about a challenging work that you have done during your studies?
Q6	What do you like doing outside of your studies?
Q7	What would you like to do after university?
Q8	What is your preferred mode of learning: lecture or practicals?
Q9	Tell me about a particular experience you had working in a group
Q10	Do you enjoy group work, or do you prefer working alone?
Q11	What made you want to choose your course?
Q12	What would you like to do after university?

During the experiment, the robot waited until the participant finished speaking before asking the next question. It was anticipated that some participants would speak for longer than others, which may bias the results obtained due to the substantial variability in the participants' exposure to the experimental conditions. To minimise this, the robot was entitled to ask questions in sequential order during the five minutes the experiment lasts. After the experiment was completed by either the robot going through all the questions or reaching the experiment time limit, participants were led to an isolated room, asked to fill in a paper-based questionnaire to evaluate the interaction with the Nao robot, and advised to avoid communication with the participants who had not yet completed the experiment.

3.5 Questionnaire

A common approach of evaluating the human perception of robots is to use a post-experiment questionnaire. Several of these exist in the literature and significant work has been done to assure their reliability and validity [16]. In this study, we have chosen to use the Godspeed questionnaire to evaluate the participant's interaction with the Nao robot. This has already been tested and validated in the context of social robotics and therefore represents suitable measure of human-robot interaction. It combines a set of five questionnaires based on semantic differential scales as a standarised metric for the five key concepts in HRI:

- *Anthropomorphism*: rates the user's impression of the robot on five semantic differentials.
- *Animacy*: rates the user's impression of the robot on six semantic differentials.
- *Likeability*: rates the user's impression of the robot on five semantic differentials.
- *Perceived Intelligence*: rates the user's impression of the robot on five semantic differentials.
- *Perceived Safety*: rates the emotional state of the user on three semantic differentials.

As recommended, the semantic differentials were randomised and the categories removed so as to hide the different concepts and hence mask the particular areas the participants were meant to be evaluating.

4   Results

We conducted a Principle Complement Analysis (PCA) for all the semantic differentials in the Godspeed questionnaire in order to obtain the minimum number of dependent variables which explain the subjects' responses. The PCA identified three underlying dimensions in which the robot is collectively perceived. A correlation threshold of 0.5 was set to determine the extent to which each semantic differential significantly loads onto any one of the factors. The first factor (**F1**) is related to the perceived "affability" of the robot. The semantic differentials which load heavily onto this were connected to unkind, unfriendly, awful, foolish, incompetent, unpleasant and dislike differentials, which are mostly about the perceived affective qualities of the robot. The second factor (**F2**) is strongly related to the perceived "humanness" of the robot. The semantic differentials that load significantly on this were the mechanical, artificial, stagnant,

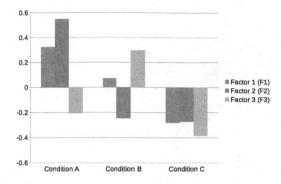

Fig. 3. Mean values of the PCA solution factor for three conditions: Mirror (**A**), Non-Mirror (**B**) and Static (**C**) movements

fake, machinelike, moving rigidly, unconscious and dead differentials. Finally, the third factor (**F3**) emerging from this analysis was related to the perceived "responsiveness" of the robot based on the semantic differential item loadings of anxious, unintelligent, apathetic, moving rigidly and agitated.

Based on the three identified principle components and on the participants' scoring on the semantic differentials, a series of three factor scores was then produced for each participant. The across-participant means of these scores are shown in Figure 3 separately for the three conditions. An analysis of the variance (ANOVA) was conducted to compare the differences in means between the three group conditions. This analysis showed for *affability* ($F = 0.916$, $p = 0.41$), *humanness* ($F = 3.01$, $p = 0.06$) and *responsiveness* ($F = 1.623$, $p = 0.21$). Thus, the factor **F1** (affability) showed no meaningful differentiation between the three conditions. Only for factor **F2** (perceived humanness of the robot) was there any statistically notable difference between the ratings of the three condition groups. As can be seen in Figure 3, the humanness factor (**F2**) only rates positively in condition **A** on average; this indicates that mirroring appears to have a positive influence on the anthropomorphic perception of the robot. The responsiveness factor (**F3**) also showed some differentiation across the conditions (though these were statistically less pronounced than that for humanness), suggesting that likeability and animacy in robotic entities is closely connected with movement, and that anthropomorphic gesturing positively influences rapport in HRI. These results evidence that robotic manipulation (i.e. mirroring/non-mirroring) is an important factor in order to induce empathy and engage human-robot communication, opposite to the initial experiments in which the participants remained fully static expecting for the Nao to move. These results are also perhaps surprising since participants equally perceived animacy when the robot was static or moving using predefined human-like poses. However, it is arguable that the disparity in the number of participants between these two conditions may have influenced the results. Mean values and total scores for the perceived five key concepts in HRI, as derived from the Godspeed questionnaires, are depicted in Figure 4.

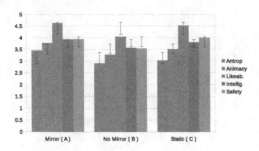

Fig. 4. Mean values of the five Godspeed aspects for the three proposed conditions: Mirror (**A**), Non-Mirror (**B**) and Static (**C**) movements on a 5-item Likert scale from 1 (strongly disagree) to 5 (strongly agree).

5   Conclusion

The goal of this study was to evaluate the psychological implications of upper body pose mirroring in human-robot social interactions. Three different experimental conditions of upper body mimicry were implemented in the humanoid Nao robot and measured in terms of the five key concepts of HRI. In general, the results support the initial hypothesis **H1** by displaying a trend towards a perceived greater *humanness* of the robot in the mirroring condition compared to the other two conditions. Likewise, the results have shown that non-mirroring body poses also may also influence the participants' empathy towards the robot, which is indicative of rapport during the interaction.

Higher anthropomorphism, animacy and perceived intelligence for conditions **A** and **B** than for **C** (hypothesis **H2**) is partially revealed, which denotes certain correlation between upper-body mimicry and perceived humanness (anthropomorphism) of a robot (conditions **B** and **C** resulted in similar participant ratings in terms of this factor). This can be possible to explain by the fact that anthropomorphism might not be based only on robot manipulation and human-like gesturing but also on alternative communicative components and social factors. It is also debatable that the human-like appearance of the Nao robot itself may bias its social acceptability, which is closely related to being human-like, as well as the elicitation of empathetic behaviours in the participants even when it remains static, as likeability is equally highly rated in each of the three conditions. Future work will re-evaluate the gained insights on inducing rapport using upper body mimicry by extending the number of participants and the usage of additional humanoid robots in order to provide new understanding with regard to the usage of robotic entities as companion systems.

References

1. LaFrance, M.: Nonverbal synchrony and rapport: Analysis by the cross-lag panel technique. Social Psychology Quarterly **42**, 66–70 (1979)

2. LaFrance, M., Broadbent, M.: Posture sharing as a nonverbal indicator. Group and Organizational Studies **1**, 328–333 (1976)
3. Chartrand, T.L., Maddux, W., Lakin, J.: Beyond the perception-behavior link: The ubiquitous utility and motivational moderators of nonconscious mimicry. In: Unintended thought II: The new Unconscious. Oxford University Press, pp. 334–361 (2004)
4. Chartrand, T.L., Bargh, T.L.: The chameleon effect: The perception-behavior link and social interaction. Journal of Personality and Social Psychology **76**(6), 893–910 (1999)
5. Salem, M., Eyseel, F., Rohlfing, K., Kopp, S., Joublin, F.L.: To Err is Human(-like): Effects of Robot Gesture on Perceived Anthropomorphism and Likeability. International Journal of Social Robotics **5**(3), 313–323 (2013)
6. Salem, M.: Conceptual Motorics-Generation and Evaluation of Communicative Robot Gesture. Logos Verlag, Berlin GmbH (2013)
7. Riek, L.D., Rabinowitch, T.C., Bremner, P., Pipe, A.G., Fraser, M., Robinson P.: Cooperative gestures: effective signaling for humanoid robot. In: IEEE International Conference on Human-Robot Interaction (ROMAN), pp. 61–68 (2010)
8. Iio, T., Shiomi, M., Shinozawa, K., Akimoto, T., Shimohara, K., Hagita, N.: Investigating entrainment of people's pointing gestures by robot's gestures using a WOZ method. International Journal of Social Robotics **3**, 405–414 (2011)
9. Kim, H., Kwak, S. S., Kim, M.: Personality design of sociable robots by control of gesture design factors. In: IEEE Symposium on Robot and Human Interactive Communication (ROMAN), pp. 494–499 (2008)
10. Gratch, J., Wang, N., Gerten, J., Fast, E., Duffy, R.: Creating rapport with virtual agents. In: Pelachaud, C., Martin, J.-C., André, E., Chollet, G., Karpouzis, K., Pelé, D. (eds.) IVA 2007. LNCS (LNAI), vol. 4722, pp. 125–138. Springer, Heidelberg (2007)
11. Gonsior, B., Sosnowski, S., Mayer, C., Blume, J., Radig, B., Wollherr, D., Kuhnlenz, K.: Improving aspects of empathy and subjective performance for HRI through mirroring facial expressions. In: IEEE International Workshop on Robot and Human Interactive Communication (ROMAN), pp. 350–356 (2011)
12. Bartneck, C., Croft, E., Kulic, E.: Measuring the anthropomorphism, animacy, likeability, perceived intelligence and perceived safety of robots. Metrics for HRI Workshop, Technical report 471, pp. 37–44 (2008)
13. Kanda, T., Kamasima, M., Imai, M., Ono, T., Sakamoto, D., Ishiguro, H., Anzai, Y.: A humanoid robot that pretends to listen to route guidance from a human. Autonomous Robots **22**(1), 87–100 (2007)
14. Bailenson, J.N., Yee, N.: Digital chameleons: automatic assimilation of nonverbal gestures in immersive virtual environments. Psychological Science **16**(10), 814–819 (2005)
15. Riek, L.D., Paul, P.C., Robinson, P.: When my robot smiles at me: Enabling human-robot rapport via real-time head gesture mimicry. Journal on Multimodal User Interfaces **3**, 99–108 (2010)
16. MacDorman, K.F.: Subjective ratings of robot video clips for human likeness, familiarity, and eeriness: an exploration of the uncanny valley. In: ICCS/CogSci Long Symposium: Toward Social Mechanisms of Android Science, pp. 26–29 (2006)
17. Kanda, T., Kamasima, M., Imai, M., Ono, T., Sakamoto, D., Ishiguro, H., Anzai, Y.: A humanoid robot that pretends to listen to route guidance from a human. Autonomous Robots **22**(1), 87–100 (2007)

Robot-Centric Activity Recognition
'in the Wild'

Ilaria Gori[(✉)], Jivko Sinapov , Priyanka Khante, Peter Stone,
and J. K. Aggarwal

University of Texas at Austin, Austin, TX 78712, USA
{ilaria.gori,aggarwaljk}@utexas.edu,
{jsinapov,pkhante,pstone}@cs.utexas.edu

Abstract. This paper considers the problem of recognizing spontaneous
human activities from a robot's perspective. We present a novel dataset,
where data is collected by an autonomous mobile robot moving around
in a building and recording the activities of people in the surroundings.
Activities are not specified beforehand and humans are not prompted to
perform them in any way. Instead, labels are determined on the basis
of the recorded spontaneous activities. The classification of such activities presents a number of challenges, as the robot's movement affects its
perceptions. To address it, we propose a combined descriptor that, along
with visual features, integrates information related to the robot's actions.
We show experimentally that such information is important for classifying natural activities with high accuracy. Along with initial results for
future benchmarking, we also provide an analysis of the usefulness and
importance of the various features for the activity recognition task.

1   Introduction

Robots are becoming increasingly sophisticated, and are bound to become pervasive in humans' every-day lives. To effectively collaborate with humans, it
is useful for a robot to understand their activities and intentions automatically.
This understanding is especially important in human-robot interaction scenarios:
if the robot can properly interpret the behavior of humans, its communication
with them will be facilitated. For example, in our scenario, a robot moves in a
building monitoring the environment. If it could recognize when a person needs
help, or whether someone wants to talk to it to ask for directions, or if it is being
ignored, its social skills would improve dramatically.

Most existing datasets available to assess activity recognition methods are
recorded from a still camera, and they comprise surveillance [10] or sports videos
[13]. Others are composed of cinema movies or Youtube videos [9]. Yet others
are recorded by asking the participants to perform specific activities [8]. None of
these datasets perfectly reflect the types of human activities a robot is likely to

---

This work has taken place in the Learning Agents Research Group (LARG) at UT
Austin. LARG research is supported in part by NSF (CNS-1330072, CNS-1305287),
ONR (21C184-01), AFRL (FA8750-14-1-0070), and AFOSR (FA9550-14-1-0087).

A. Tapus et al. (Eds.): ICSR 2015, LNAI 9388, pp. 224–234, 2015.
DOI: 10.1007/978-3-319-25554-5_23

perceive when interacting with people. In this work, we present a dataset taken from a robot's perspective, where the camera moves according to the robot's movements, and the activities are performed spontaneously by humans, often in relation to the presence or behavior of the robot. A few datasets do exist that have been recorded from a robot's perspective [3,14,17], even though the robots they used were not fully autonomous. Such datasets were collected by asking participants to perform specific actions. This lack of spontaneity may lead to a low recognition rate when the same activities are executed by people who are not asked to perform them. Besides, there is no guarantee that the chosen staged actions are those that people would naturally perform in front of a robot.

In contrast, our dataset was collected by a mobile robot, able to localize itself and navigate autonomously, moving in a populated environment, and recording people's actions. During the dataset collection, the robot moved autonomously, so that the behavior of the humans was not influenced by any external presence. The recorded data was then analyzed, and the action categories were determined by the activities spontaneously executed by the subjects. This dataset presents some unique characteristics: 1) There can be multiple people in the scene at the same time, each doing something different; 2) There may be occlusions; 3) Actions are performed at different scales and with different body orientations; 4) The robot moves continuously, therefore, there is ego-motion in the scene; 5) Some actions occur more often, while others are very rare, thus the data is highly unbalanced. As we show empirically in the experimental section, these characteristics make learning from our dataset very challenging. To the best of our knowledge, this is the first dataset recorded in such a spontaneous manner.

In this paper, the new dataset is exploited to tackle a human activity recognition task, even though it can be useful for different learning tasks as well. Unlike the previous action recognition methods, which limited their analysis to visual descriptors, we also explore features that are directly related to the robot's behaviors and movements, which, in this particular setting, influence its perceptions.

Our contribution is twofold. First, we present a new problem and make publicly available a novel challenging dataset recorded 'in the wild' from a robot's perspective. Second, we use this dataset to tackle an activity learning task. We provide results obtained using several state-of-the-art descriptors for the purpose of future benchmarking. We also present an analysis of the usefulness and importance of the various features for this specific task. In particular, we show that, in this setting, exploiting data associated with the robot's point of view consistently improves the results obtained when using only visual features.

2   Related Work

Since the early '90s, the computer vision research community has produced a plethora of methods for recognizing human activities (see [1] for a review). In most early approaches the video stream is captured by one or more stationary cameras, e.g., [2]. Methods have also been proposed for human activity recognition in movies [9], where the camera is not always stationary. In some studies, the camera is attached to a person and data is collected as the person performs a variety of activities, e.g., playing a sport [7], or interacting with others [5]. The primary focus of this past research has been on developing efficient and informative features that enable activity recognition using off-the-shelf supervised machine learning algorithms.

Most relevant to this paper are studies in which video streams were captured by a robot. Such studies are relatively new and include the works of [3,14,15,17]. For example, [17] describes an experiment in which 8 participants are asked to perform 9 activities in front of a teleoperated robot. The data is subsequently used for the development of an activity classification system. Similarly, in [14], 8 participants are asked to perform up to 7 activities in front of a teddy-bear equipped with a camera and mounted on a rolling chair. In [3], the researchers include a larger number of activities (18) performed by 5 participants.

Most robot-centric human activity recognition methods, including the ones described above, are subject to several limitations: 1) The activities are pre-specified by the experimenters; 2) The activities are performed by a relatively small number of people (typically 5-8) who are recruited to participate to the dataset collection; 3) The robot is typically either stationary or teleoperated. The present study overcomes these limitations in several significant ways. Our robot uses its autonomous navigation capability in a large and dynamic human-inhabited environment, as opposed to a structured laboratory environment. This results in much more realistic, but also more challenging, video streams. Also, the activities in our study were spontaneously performed by a large number of people who interacted with the robot, as opposed to the standard methodology of asking participants to perform certain actions.

3   Dataset

The robot used to record the dataset is shown in Fig. 1. It was built on top of the Segway Robotic Mobility Platform with an added caster wheel to keep the robot level to the ground. The robot's sensors include a Hokuyo URG-04LX laser rangefinder, used for mapping and localization, and a Kinect RGB-D (version 1.0) camera, used for obstacle avoidance. For this specific experiment, the robot was also equipped with the newer Kinect 2.0 RGB-D camera, which was used for visual person detection and tracking. The robot uses a hierarchical task-planning software architecture [18] based on the Robot Operating System [12].

The robot collected data by autonomously patrolling through an undergraduate and a graduate student lab which were connected by a doorway. To collect the dataset, the robot traversed the environment for 1-2 hours per day, for 6 days. Over the course of the experiment, the robot travelled a total of 14.037 km. As soon as the Kinect 2.0 detected a person, our program started recording all the information described in the next paragraph and summarized in Table 1. Many people just ignored the robot or passed by it. Others engaged in various interactions such as blocking it, waving at it, or taking a picture of it. At the end, we labeled the actions into a number of categories that we observed at least

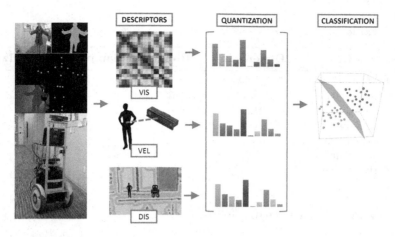

Fig. 1. An overview of our system. Descriptors on visual (VIS), velocity (VEL) and distance (DIS) information are extracted, quantized, concatenated and finally fed to the classifier.

6 times in the recorded videos. The labeling was carried out by two authors of this paper; therefore, it is subjective, prone to errors or different interpretations. The resulting activity categories are: *approach, block, pass by, take picture, side pass, sit, stand, walk away, wave.* There were several more, e.g., three persons approached the robot and pretended to punch it, or started to dance in front of it, but they were too rare and were not included in our subsequent analysis. Notably, the Kinect performance in tracking is not perfect, especially when the robot moves continuously. It may happen, for instance, that a wall, or a chair, or a column is recognized as a person. These samples are gathered in the class *false*, which is used in the classification procedure as well. In total, there are 10 class labels and 1204 selected samples. We plan, however, to record several more hours of activities in the future and expand the released dataset as more data becomes available.

For each video we provide RGB images (3-channel images of dimension 512×424), depth images (16-bits, 1-channel images of dimension 512×424), and the position of the skeletal joints for each person, in 3D and on the image. Some images extracted from our dataset can be observed in Fig. 2. Each video segment is also annotated with the robot's position and orientation in the map, estimated using Monte-Carlo Localization as implemented in ROS, and the robot's raw odometry information, computed by the Segway RMP ROS driver. Each video has been annotated with an activity label, assigned as described in the previous paragraph. This feature set constitutes the first main contribution of this paper, and we intend for it to be useful to the community. While our analysis focuses on human activity recognition, the dataset can be used for other tasks as well.

Table 1. Raw features provided in the newly collected dataset.

Features	Dimension	Range	Sampling Rate (Hz)
RGB images	512×424×3	{0,255}	50
Depth images	512×424	{0,65535}	50
3D Joints	21x3	$\mathbb{R}$	50
2D Joints	21x2	{0,512}×{0,424}	50
Robot's pose on the map	7	$\mathbb{R}$	1.5
Robot's odometry	7	$\mathbb{R}$	100
Activity label	1	{1,10}	–

4   Activity Recognition

We use the newly collected dataset to carry out an activity recognition experiment. In this setting, the task of the robot is to annotate a video with the correct activity label. The raw recorded data was too highly dimensional to be used as direct input to a classifier. Hence, we manipulated the raw sensory data to obtain higher-level feature descriptors. In particular, this section describes what descriptors have been extracted, and how they have been quantized, so that each video is represented by a single vector. Our main proposition is to concatenate robot-centric descriptors – i.e. descriptors related to the robot's perspective – with visual features, as we hypothesize that they will improve the performance of the classifier. Figure 1 shows an overview of the recognition system.

Visual Features: We extract five different visual descriptors and we compare them in the experimental evaluation section. The first one has been proposed in [16], and builds a histogram of the joints in 3D (HOJ3D). The second one has been presented in [6], and computes the covariance of the joint positions over time (COV). The third one has been described in [3], and generates Histograms of Direction Vectors (HODV). The fourth one is based on raw depth images and has been published in [11] (HON4D). Finally, we rely on a simple descriptor that builds a matrix of pairwise relations between joints. We will refer to it as the Pairwise Relation Matrix (PRM). The intuition behind this descriptor is that, while absolute joint positions are not translation invariant, their relations are independent from the absolute position of the person. At the same time, they provide a good representation of the skeleton configuration. Let $J = (\mathbf{j_1}(t), \mathbf{j_2}(t), \dots, \mathbf{j_m}(t))$ be the set of 3D joints tracked by the Kinect at time t. We build a $m \times m$ matrix $\mathbf{R(t)}$ for each frame t, where m is the number of joints. Each element of $\mathbf{R(t)}$ is equal to

$$R(t)^{i,k} = ||\mathbf{j_i}(t) - \mathbf{j_k}(t)||. \tag{1}$$

Since the resulting matrix is symmetric, we use only the values under the diagonal, therefore the final descriptor belongs to $\mathbb{R}^{\frac{m(m-1)}{2}}$.

Human-Robot Velocity Features: The movements of a person as perceived by the robot are different from his or her movements with respect to an absolute point of view. For example, consider the motion of a person walking away from the robot. We hypothesize that the robot's perception of this movement will depend on how the robot itself is moving. If the robot is still, it will perceive the walk away movement as it is, but if it is moving towards the person at high speed, it may perceive the person as approaching it. To avoid this ambiguity, we need to know how the human is moving with respect to the robot using an absolute point of reference. Let $\mathbf{p_h^r}(t) \in \mathbb{R}^3$ be the position of the human with respect to the robot at time t, as perceived from the Kinect sensor. Then, let $\mathbf{p_r^m}(t) \in \mathbb{R}^3$ be the position of the robot with respect to its starting point at time t, and $\mathbf{R_r^m}(t) \in \mathbb{R}^{3 \times 3}$ be the rotation matrix describing the orientation of the robot with respect to the starting point. It is possible to compute the position $\mathbf{p_h^m}(t)$ of the human with respect to the starting point at time t as follows:

$$\begin{bmatrix} \mathbf{p_h^m}(t) & 1 \end{bmatrix} = \begin{bmatrix} \mathbf{R_r^m}(t) & \mathbf{p_r^m}(t)^T \\ 0 & 1 \end{bmatrix} \begin{bmatrix} \mathbf{p_h^r}(t)^T \\ 1 \end{bmatrix}. \tag{2}$$

At this point, we compute the velocity vector between pairs of successive frames as follows:

$$\mathbf{x_d}(t) = \frac{\mathbf{p_h^m}(t+1) - \mathbf{p_h^m}(t)}{||\mathbf{p_h^m}(t+1) - \mathbf{p_h^m}(t)||}. \tag{3}$$

This quantity represents the real direction in which the human moves with respect to the robot. Since we do not need the last coordinate, which is always 0 (the robot and the person move on a plane), $\mathbf{x_d}(t) \in \mathbb{R}^2$.

Human-Robot Distance Features: Different activities present similar visual and motion properties, but may be distinguished on the basis of where the person is with respect to the robot. For instance, some humans tried to block the robot standing in front of it; their pose, however, is very similar to the pose of those persons that ignore the robot and stand at a certain distance from it. Therefore, we incorporate the distance between the human and the robot retrieved from the Kinect sensor for each frame, taking the hip joint as the point of reference. The human-robot distance descriptor belongs to $\mathbb{R}$.

Feature Quantization: The feature vectors that have been extracted for each frame (or from each pair of successive frames, in case of the Human-Robot Velocity feature vector) are quantized using k-means and represented using Bag Of Words (BOW), so that they generate a single feature vector for each video. A different dictionary is built for each descriptor. We then concatenate the feature vectors in a single vector, obtaining the final descriptor for each video, which belongs to $\mathbb{R}^{\sum_{s=1}^{n} k_s}$, where k_s is the size of the dictionary for the s-th descriptor.

5   Experimental Results

This section presents a comprehensive evaluation of feature descriptors and their combinations on our dataset for the activity recognition task. We present results using non-linear SVM with χ^2 kernel, since other kernels (e.g., linear, Gaussian, polynomial and intersection) and other classifiers (e.g., Naive Bayes, Random Forests) achieved comparable or worse results. We perform a stratified 6-fold cross validation, and we repeat the procedure 10 times, to take into account the randomness of the dictionary learning stage. As we anticipated, the dataset is very unbalanced with respect to the activity labels (i.e., some activities are much more frequent than others), thus the recognition accuracy is not a good measure to judge classification performance. Instead, we report the Cohen's kappa coefficient [4], which compares the classifier accuracy against chance accuracy:

$$K = \frac{Pr(a) - Pr(e)}{1 - Pr(e)}, \tag{4}$$

where $Pr(a)$ is the probability of correct classification by the classifier, and $Pr(e)$ is the probability of correct classification by chance.

Visual Features Comparison. We assessed the 5 visual descriptors listed in Sec. 4, and we used default parameters to compute all of them. The descriptors reported in [11], [6] and [3] do not need a dictionary learning stage as they already represent the entire video; they belong to $\mathbb{R}^{22680}$, $\mathbb{R}^{1953}$ and $\mathbb{R}^{567}$ respectively. For the other two (PRM and HOJ3D [16]), we set the number of dictionary atoms to 300. The second column of Table 2 reports the results achieved by the different visual descriptors. Notably, the only depth-based descriptor that we have tested, HON4D [11], gets the highest kappa coefficient. In this specific case, where the Kinect is moving continuously, the joint estimation procedure is probably not as reliable as in situations where the Kinect is stable, while the depth images are probably not as affected by the robot's movements as the joint estimation algorithm. Hence, this result may be due to the fact that HON4D is computed on the raw depth images, and does not use joints at all.

Results with Robot-Centric Descriptors. We hypothesize that, in this setting, robot-centric descriptors are useful to improve the performance of visual descriptors. To evaluate this hypothesis, we concatenate the robot-centric descriptors described in Sec. 4 with visual features. Table 2 reports the results obtained by this concatenation in the third column: when robot-centric descriptors are concatenated with visual features, the kappa statistics improves consistently. The fourth column of Table 2 shows the classification rate as the robot's pose in the map is concatenated with the distance and velocity robot-centric features. Notably, making use of the robot's position increases the kappa rate

Table 2. Comparison among different features and their combination

Method	Visual only	Visual + HR Velocity + HR Distance	Visual + HR Velocity + HR Distance + Robot Pose
COV [6]	0.3287	0.4397	0.4642
HOJ3D [16]	0.5135	0.6327	0.6507
HODV [3]	0.6242	0.6493	0.6605
PRM	0.5474	0.6597	0.6716
HON4D [11]	**0.7558**	**0.7629**	**0.7642**

Table 3. Precision, Recall and F-1 score of each activity class

Activity	Num Samples	HON4D Precision	Recall	F-1	PRM Precision	Recall	F-1
Picture	6	–	0	–	–	0	–
Wave	12	–	0	–	–	0	–
False	608	0.8845	0.9645	0.9227	0.8322	0.9378	0.8818
Block	23	0.7273	0.3130	0.4377	0.5167	0.1348	0.2138
Pass by	153	0.7993	0.8641	0.8304	0.7318	0.8510	0.7869
Walk away	68	0.9394	0.8662	0.9013	0.8652	0.8588	0.8620
Approach	33	0.5970	0.3636	0.4520	0.4817	0.2394	0.3198
Sit	150	0.8483	0.8273	0.8377	0.8196	0.7480	0.7822
Stand	106	0.6433	0.6840	0.6630	0.4875	0.4425	0.4639
Side pass	45	0.7817	0.3978	0.5272	0.6036	0.2267	0.3296

even further. This may be because some activities are more likely to occur in certain regions of the map than at other locations.

Finally, Table 3 provides precision, recall and F1-score of each class using the two best combination of descriptors (HON4D + robot-centric descriptors, and PRM + robot-centric descriptors).

Even though HON4D performs significantly better than PRM, it is unable to correctly classify the activities *picture* and *wave*, for which we get 0 true positives and 0 false positives, therefore the symbol "–" in the table. This is probably due to the fact that those are the classes with the smallest number of examples (6 and 12 respectively). For the same reason, the precision and recall on the actions with many samples are relatively high, while those on the actions with a few samples are low. This suggests that if the dataset was more balanced, the results would be more homogeneous. However, the fact that the dataset is unbalanced is one of the natural effects derived from recording activities in the wild. Therefore, learning activities when the number of samples per class differs a lot is one of the challenges of our dataset.

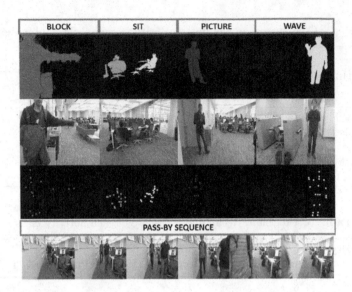

Fig. 2. Examples extracted from our newly recorded dataset. Top: shots of stationary activities. Bottom: action **pass by**. The robot moves forward and then turns.

6   Conclusion

This paper considers a new, realistic problem in the field of robot-centric activity recognition: classifying spontaneous activities from a mobile robot's perspective. Unlike previous works, activities are not specified beforehand, and humans are not asked to perform them. Instead, an autonomous, mobile robot moved around in a building full of people, and recorded their spontaneous behaviors. The robot was left to act alone, therefore the persons who encountered it were not influenced by our presence. All the recorded data was successively analyzed, and the activity classes were determined from the observed videos. To the best of our knowledge, there is no dataset in the literature like the one we are proposing. We plan to release it upon publication, as we expect it to be useful to the community. To obtain satisfactory results on this data, visual features were concatenated with supplementary information directly related to the robot's movements. We showed experimentally that these descriptors consistently improve the results obtained using only visual features. We plan to use the new dataset as a platform to test various learning tasks, different learning algorithms, and multiple combinations of features.

Future work includes using the activity recognition system for 'activity-aware' navigation. For instance, when the robot recognizes that someone is taking a picture of it, it stops and waits until the activity is finished. We also plan to use multiple labels, since sometimes a certain action cannot be described by a single one. Finally, an important future direction is analyzing 'two-way' interactions, during which the robot reacts back to the human.

References

1. Aggarwal, J.K., Ryoo, M.S.: Human activity analysis: A review. ACM Computing Surveys (CSUR) **43**(3), 16 (2011)
2. Bobick, A.F., Davis, J.W.: The recognition of human movement using temporal templates. IEEE Transactions on Pattern Analysis and Machine Intelligence (2001)
3. Chrungoo, A., Manimaran, S.S., Ravindran, B.: Activity recognition for natural human robot interaction. In: Beetz, M., Johnston, B., Williams, M.-A. (eds.) ICSR 2014. LNCS, vol. 8755, pp. 84–94. Springer, Heidelberg (2014)
4. Cohen, J.: A coefficient of agreement for nominal scales. Educaional and Psychological Measurement **20**(1), 37–46 (1960)
5. Fathi, A., Hodgins, J.K., Rehg, J.M.: Social interactions: a first-person perspective. In: IEEE Conference on Computer Vision and Pattern Recognition (CVPR), pp. 1226–1233. IEEE (2012)
6. Hussein, M.E., Torki, M., Gowayyed, M.A., El-Saban, M.: Human action recognition using a temporal hierarchy of covariance descriptors on 3D joint locations. In: International Joint Conference on Artificial Intelligence (IJCAI) (2013)
7. Kitani, K.M., Okabe, T., Sato, Y., Sugimoto, A.: Fast unsupervised ego-action learning for first-person sports videos. In: IEEE Conference on Computer Vision and Pattern Recognition (CVPR), pp. 3241–3248. IEEE (2011)
8. Li, W., Zhang, Z., Liu, Z.: Action recognition based on a bag of 3D points. In: IEEE Conference on Computer Vision and Pattern Recognition Workshop (CVPRW) (2010)
9. Marszałek, M., Laptev, I., Schmid, C.: Actions in context. In: IEEE Conference on Computer Vision and Pattern Recognition (CVPR) (2009)
10. Oh, S., Hoogs, A., Perera, A., Cuntoor, N., Chen, C.C., Lee, J.T., Mukherjee, S., Aggarwal, J.K., Lee, H., Davis, L., Swears, E., Wang, X., Ji, Q., Reddy, K., Shah, M., Vondrick, C., Pirsiavash, H., Ramanan, D., Yuen, J., Torralba, A., Song, B., Fong, A., Roy-Chowdhury, A., Desai, M.: A large-scale benchmark dataset for event recognition in surveillance video. In: IEEE Conference on Computer Vision and Pattern Recognition (CVPR) (2011)
11. Oreifej, O., Liu, Z.: HON4D: histogram of oriented 4D normals for activity recognition from depth sequences. In: IEEE Conference on Computer Vision and Pattern Recognition (CVPR) (2013)
12. Quigley, M., Conley, K., Gerkey, B., Faust, J., Foote, T., Leibs, J., Wheeler, R., Ng, A.Y.: ROS: an open-source robot operating system. In: ICRA Workshop on Open Source Software, p. 5 (2009)
13. Rodriguez, M., Ahmed, J., Shah, M.: Action mach a spatio-temporal maximum average correlation height filter for action recognition. In: IEEE Conference on Computer Vision and Pattern Recognition (CVPR) (2008)
14. Ryoo, M.S., Matthies, L.: First-person activity recognition: what are they doing to me? In: IEEE Conference on Computer Vision and Pattern Recognition (CVPR) (2013)
15. Ryoo, M., Fuchs, T.J., Xia, L., Aggarwal, J., Matthies, L.: Robot-centric activity prediction from first-person videos: what will they do to me. In: Proceedings of the Tenth Annual ACM/IEEE International Conference on Human-Robot Interaction (HRI), pp. 295–302. ACM (2015)

16. Xia, L., Chen, C.C., Aggarwal, J.K.: View invariant human action recognition using histograms of 3D joints. In: IEEE Conference on Computer Vision and Pattern Recognition Workshop (CVPRW) (2011)
17. Xia, L., Gori, I., Aggarwal, J.K., Ryoo, M.S.: Robot-centric activity recognition from first-person RGB-D videos. In: IEEE Winter Conference on Applications of Computer Vision (2015)
18. Zhang, S., Sridharan, M., Gelfond, M., Wyatt, J.: Towards an architecture for knowledge representation and reasoning in robotics. In: Beetz, M., Johnston, B., Williams, M.-A. (eds.) ICSR 2014. LNCS, vol. 8755, pp. 400–410. Springer, Heidelberg (2014)

"Yes Dear, that Belongs into the Shelf!" - Exploratory Studies with Elderly People Who Learn to Train an Adaptive Robot Companion

Jens Hoefinghoff[1]([✉]), Astrid Rosenthal-von der Pütten[2], Josef Pauli[1], and Nicole Krämer[2]

[1] Department of Computer Science and Applied Cognitive Science, Intelligent Systems, University of Duisburg-Essen, 47048 Duisburg, Germany
{jens.hoefinghoff,josef.pauli}@uni-due.de

[2] Department of Computer Science and Applied Cognitive Science, Social Psychology - Media and Communication, University of Duisburg-Essen, 47048 Duisburg, Germany
{a.rosenthalvdpuetten,nicole.kraemer}@uni-due.de

Abstract. Robot companions should be able to perform a variety of different tasks and to adapt to the user's needs as well as to changing circumstances. To achieve this we can either built fully adaptive robots or adaptable and customizable robots. In this paper we present an adaptable companion which uses a decision making algorithm and user feedback to learn adequate behavior in new tasks. Using two different scenarios (household task, card game) the system was evaluated with elderly people in exploratory studies. We found that the perception and evaluation of the robot's learning progress depends on the interaction scenario. Additionally, we discuss improvements for the algorithm in order to make the learning behavior appear more natural and humanlike.

Keywords: Human-robot interaction · Artificial intelligence · Companion · Exploratory study · Elderly

1 Introduction

The companion robot of the future is not a tool specialized in a certain task, but able to interact socially with different users, to serve multiple purposes, and to adapt to new situations. Indeed, most users favor a robot companion which has an overall assistive role and serves multiple purposes [3]. Although it is possible to fit a robot with the ability to perform a variety of tasks as new consumer products like Pepper (Aldebaran Robotics) demonstrate, it is still a challenge to add new tasks on these predefined ones, to implement learning, adaptability, and guarantee a robust system at the same time. Some systems are open so that skilled users are able to program their own applications, but this is beyond the means of users who are technical novices. However, integrating users into the process of designing and programming new applications might lead to higher acceptance, because previous work has shown that this kind of

© Springer International Publishing Switzerland 2015
A. Tapus et al. (Eds.): ICSR 2015, LNAI 9388, pp. 235–244, 2015.
DOI: 10.1007/978-3-319-25554-5_24

experienced self-efficacy leads to a more favorable evaluation of the respective system [4]. Moreover, a system that can be expanded by integration of new tasks and abilities and that adapts within these tasks to changed circumstances might increase the expected useful life of such a companion. Prior work has shown that adaptivity in robots using reinforcement learning methods can be successfully implemented in applications for users with special needs [7]. To address this problem we developed a system for all and sundry to create robotic applications [6] which is characterized by adaptivity on different levels. The system enables users to create their own applications (e.g. play a card game with robot) by the usage of a decision making framework for robot companions [5]. Within such a created application the robot itself adapts its behavior via user feedback based on a decision making algorithm (e.g. teach the robot a card game based on user feedback). Hence, the robot can learn, but also relearn and by this adapt to new circumstances. In dependency of the user's expertise, he or she has different possibilities of enhancing the robots capabilities. Especially for the non-expert user a tool (Conf-Tool) has been developed which provides a graphical user interface to configure applications. We tested the usability of the tool with participants of a relevant target group (age group 40+) and found that the technical requirements to include non-experts are fulfilled by the framework [6]. Based on our previous work we now present an adaptable companion which uses a decision making algorithm and user feedback to learn adequate behavior in new tasks. We first introduce and explain the underlying decision making algorithm and then present two exploratory studies evaluating the algorithm with a relevant target group of participants aged 40+. Participants engaged in interactions with the companion robot in one of two different scenarios in order to explore the influence of the experimental setup on the perception of the robot's learning behavior. The two studies differ with regard to the number of stimuli presented to the companion, in the number of available actions to respond to those stimuli, and in the socialness of the respective training situation. The discussion includes a comparison of the results from both studies in order to determine the influence of the number of stimuli and actions, the number of repetitions during the learning process and the influence of the socialness of the situation.

2 Decision Making Algorithm

The robot, used for the studies presented in the following, makes its decisions based on an algorithm which has been presented in [5]. The algorithm is a reinforcement learning method inspired by Damasio's Somatic Marker Hypothesis (SMH) [2]. As this paper focuses on the application of the algorithm in HRI, it is briefly explained in this section without going into too much detail. A simplified overview of the algorithm is shown in Figure 1. According to Damasio's SMH the decision making process is divided into an emotional decision making part and a rational decision making part. The implementation of the emotional part is explained in [5]. The output of the emotional part is a subset containing all

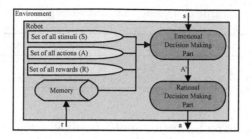

Fig. 1. Overview of the decision making algorithm.

actions promising to lead to a positive outcome. This does not mean that the robot has an accurate emotional model but that the computations of the algorithm are inspired by Damasio's work. Building up an emotional model is often connected to implementing a huge amount of a priori knowledge to the system for each different application. This is inconsistent with the requirement of creating robot companions which are supposed to perform many different tasks. The rational part is a place holder for subsequent enhancements. For now, an action is chosen randomly during the rational decision making part.

Basically, the robot is able to recognize different stimuli $S = \{s_1, ..., s_m\}$, is able to execute a defined number of actions $A = \{a_1, ..., a_n\}$, and is able to obtain different rewards $R = \{r_1, ..., r_l\}$ (see Figure 1). All this information is provided in advance through an XML file which can be created by using the Conf-Tool offering a GUI for that purpose. Due to this, it is possible to create new applications for the robot without any programming. Details on the XML interface and the tool can be found in [6]. Based on the number of stimuli $|S|$ and actions $|A|$, the agent creates a matrix M (see eq. (1)) which contains a single somatic marker $\sigma_{i,j}$ for each pair of a stimulus s_i and an action a_j. This matrix is part of the memory and each somatic marker is a rating value representing the expected outcome when an action a_j is executed in the wake of a present stimulus s_i. The robot can obtain a reward for the execution of an action. This reward is used to update the corresponding somatic marker $\sigma_{i,j}$.

$$M_{|S| \times |A|} = \begin{pmatrix} \sigma_{1,1} & \cdots & \sigma_{1,n} \\ \vdots & \ddots & \vdots \\ \sigma_{m,1} & \cdots & \sigma_{m,n} \end{pmatrix} = (\sigma_{i,j}) \qquad (1)$$

In addition to the somatic markers, a frustration level θ_i exists for each stimulus (see eq. (2)). These frustration levels are used as thresholds for the action selection mechanism and are also updated based on obtained rewards.

$$\vec{\theta} = \begin{pmatrix} \theta_1 \\ \vdots \\ \theta_m \end{pmatrix} = (\theta_i) \qquad (2)$$

Table 1. Decision making example.

t	θ_1	$\sigma_{1,1}$	$\sigma_{1,2}$	$\sigma_{1,3}$	$\sigma_{1,4}$	A'	a_j	reward
x	-0.49	-0.39	-0.99	0.13	-0.49	$\{a_1, a_3, a_4\}$	a_4	50
$x+1$	0.08	-0.39	-0.99	0.13	0.08	$\{a_3, a_4\}$	a_3	50

Every time the robot recognises a stimulus from the environment, the decision making process is started. At first a subset of actions A' is selected during the emotional decision making part. The selection rule is shown in eq. (3). Applying this rule, the subset contains all actions with a corresponding somatic marker $\sigma_{i,j}$ its value is equal to or greater than the corresponding frustration level θ_i. There are mechanisms implemented assuring that $A' \neq \emptyset$.

$$A' := \{a_j \in A \,|\, \sigma_{i,j} >= \theta_i\} \tag{3}$$

After the subset is selected, a final action is chosen randomly out of the subset during the rational selection. This action is then executed by the robot. For each executed action the robot is able to obtain a reward in order to get information whether the decision was good or not. The robot updates its memory based on the obtained rewards. The evaluation of the algorithm can be found in [5]. For evaluation purposes the Iowa Gambling Task (IGT) was used [1]. In the IGT subjects have to draw cards from four different decks (A, B, C, D). Some cards lead to a benefit while others lead to a penalty. The decks are prepared in the way that deck A and B are disadvantageous decks because drawing cards from these decks lead to a net loss. In contrast to that, deck C and D are advantageous decks as drawing cards from these decks lead to a net gain. Based on this task the following example is given in order to clarify the decision making algorithm:

- $S = \{\underbrace{takeCard}_{s_1}\}$
- $A = \{\underbrace{DeckA}_{a_1}, \underbrace{DeckB}_{a_2}, \underbrace{DeckC}_{a_3}, \underbrace{DeckD}_{a_4}\}$
- $R = \{-1150, -200, -150, 0, 50, 100\}$

There is only one stimulus *takeCard* and four different actions namely *DeckA*, *DeckB*, *DeckC* and *DeckD*. Each decision leads to one reward from the set of rewards R. Table 1 shows an exemplary decision step. At $t = x$ the stimulus *takeCard* is recognized. Based on the data stored in the memory, in particular the threshold θ_1 as well as the four somatic markers $\sigma_{1,1}, \sigma_{1,2}, \sigma_{1,3}$ and $\sigma_{1,4}$, the subset A' is selected according to the selection rule shown in eq. (3). In this case the subset contains the actions a_1, a_3 and a_4 because the corresponding somatic markers are greater than or equal to the threshold (-0.49). Subsequently, the action a_4 (*DeckD*) is chosen randomly leading to a positive reward of 50. Consequently, the threshold θ_1 is increased as well as the somatic marker $\sigma_{1,4}$. This has an influence on the next decision making process $(t = x + 1)$ at which the subset A' contains only the actions a_3 and a_4.

3   Exploratory Studies

We conducted two exploratory studies evaluating the decision making algorithm in a natural setting with a relevant target group. Participants were asked to imagine that they own a robot, use it at home for different purposes and want to teach the robot a new skill. In study 1, participants taught the robot the card game "17+4" which is a simplified version of Black Jack. In study 2, participants trained the robot in different social scenarios, for instance, they taught their preferred greeting in the morning, and where to put things away in a tidy up task. The settings are explained in more detail in Section 4 and Section 5, respectively. However, some details are effective for both studies, for instance, some of the measures used. Thus they will be described for both studies in the following.

Measurements: After the respective interaction with the Nao robot, we asked participants whether they recognized the robot's learning progress (Yes/No), and how they evaluate this progress (I have perceived the learning progress as being... 1=very fast, 5=very slow; How would you rate the robot's learning ability? 1=very good, 5=very bad; Do you think that the robot's way to learn is expedient? 1=very expedient, 5=not expedient). We also asked the participants whether they used the different rewards' gradations (Yes/No), whether they think that the rewards' gradations are expedient (Yes/No) and to which extend they could imagine interacting with such a robot in their everyday life (1=absolutely not, 5=definitely).

Participants and General Procedure: In total forty-one healthy volunteers participated in our studies that were approved by the local ethical committee. Twenty-one participants were recruited for Study 1. They were aged between 46-75 years ($M = 58.33$, $SD = 7.63$). For Study 2 in total 20 participants (6 male; 14 female) aged between 42-72 years ($M = 55.53$, $SD = 8.28$) had been recruited. The data of one participant was excluded from the analysis due to major technical issues during the interaction. Upon arrival participants read and signed informed consent. Then the experimenter explained that they would interact with a humanoid robot and explained the respective setting. Participants first completed a questionnaire with sociodemographic questions. Having completed the interaction, participants were asked to finish the remaining questionnaire. They were debriefed and thanked for their participation. The experimenter stayed in the room during the whole experiment and assisted the participants if needed.

4   Study 1 Teaching the Robot a Card Game

Setting and Procedure: The aim of the interaction in study 1 was to teach the robot the game and game mechanics of 17+4, so that the participants' grandchildren might play the game with the robot during their next visit. Participants were confronted with the Nao robot (with colored dots to indicate where to touch

Fig. 2. Setting 17+4.

the robot in order to give certain rewards), cards with the values from one to twenty-one and a card showing a sad smiley. Goal of the game 17+4 is to reach an accumulated card value that comes close to or is exactly 21. The game is lost when the value 21 is exceeded. After each chosen card, the player (here the robot) can decide whether to take another card or to hold the reached accumulated card value. As the robot had no knowledge about the game, it was the participant's duty to teach the robot in which cases it is advisable to take another card and when to end the round. Participants had the choice to give either a strongly negative (two red dots), slightly negative (one red dot), slightly positive (one green dot) or strongly positive (two green dots) reward to Nao which responds (e.g. negative reward: lower head and say "Heck, this was wrong"). Participants and Nao played 15 rounds. Each round started with the investigator drawing a card (with a value from 1 to 9) and showing it to the participant who had to add the shown value to the current accumulated card value (0 at the beginning) and show Nao the card value by choosing the corresponding card laying in front of Nao. This was repeated until the round ended either because the player decided against an additional card or because the accumulated card value exceeded 21. In this case the participants showed Nao the sad smiley. We used separate card decks with predefined sequences of cards in order to create equal conditions for each participant. To make the robot's learning process observable more quickly, it was assured that specific accumulated card values had the chance to occur more frequently. Of course, the occurrence of the values was also dependent on the robot's decisions which were influenced by the obtained rewards.

Results and Discussion: Only 52.4% of the subjects stated that they had recognized the robot's learning progress (see Table 2). There are plausible explanations for this small number. Although the decks were prepared so that specific stimuli were likely to occur multiple times, it was not possible to guarantee that a stimulus occurred more than once. In unfavorable cases this led to the robot not being able to apply the acquired knowledge at all or only in rare cases.

Table 2. Participants' perception and acceptance of the robot's decision making behavior.

Item	Study 1	Study 2
Did you recognize the robot's learning progress?	N=21; Yes: 52.4%	N=19; Yes: 94.7%
I have perceived the learning progress as being...	N=12; M=2.83; SD=.84	N=19; M=3.11; SD=.69
How would you rate the robot's learning ability?	N=14; M=2.86; SD=.86	N=19; M=2.63; SD=.60
Do you think that the robot's way to learn is expedient?	N=18; M=1.11; SD=.32	N=19; M=2.05; SD=.52
Did you use the different rewards' gradations?	N=20; Yes: 90.0%	N=19; Yes: 84.2%
Do you think that the rewards' gradations are expedient?	N=21; Yes: 95.0%	N=19; Yes: 63.2%
I could imagine interacting with such a robot in my everyday life.	N=20; M=3.10; SD=1.61	N=18; M=2.61; SD=.61

Moreover, Nao tried every possible action once before decisions are then based on the somatic markers. Thus, the robot might choose a disadvantageous action, although another action had already been rewarded positively which might be perceived as erratic behavior. In conclusion, it needs more rounds in order to observe the robots learning progress. In addition, if an action is rewarded positively, the robot should not try out every other possible action. Most of the participants stated that they had used the different rewards' gradations and deemed them expedient. Regarding the question if the participants could imagine to interact with such a robot in their everyday life, neither a strong refusal nor a strong approval is observable.

5   Study 2 - Teaching the Robot in Social Scenarios

Setting, Procedure and Additional Dependent Variables: Participants were engaged in three social scenarios with the robot - greeting, tidy-up, and hobbies. Their task included a learning phase in which they taught the robot their respective preferences and a relearning phase in which they taught the robot different preferences due to a changed situation. The scenarios differed in the kind and number of stimuli and actions (e.g. stimuli could be visual or auditive). Similarly to study 1, participants rewarded the robot's decisions in four gradations by touching hands or feet.

– **Greeting Scenario:** Participants taught the robot their preferred greeting. Each time the robot was greeted by the participant with "hello" or "good day" it chose one out of seven possible replies (good day, hello, hey, I am hungry, hey dude!, bye, what's up) which was subsequently rewarded by the participant in order to signalize if the answer was appropriate. This procedure was repeated until the participant believed that the robot learned the preferred greeting. Afterwards, the participant was told to imagine that his/her grandchild was going to come for a visit and he/she now wants to teach the robot the preferred greeting for the grandchild. During this process the robot had to relearn through the rewards given by the participant. Again, this procedure was repeated until the participant believed the robot to have learned the preferred greeting for the grandchild.

- **Sorting Scenario:** Participants taught the robot appropriate depositories for different objects. Nao was able to recognize images of three different objects (pen, apple and book). Every time the robot recognized an object, it suggested and pointed to a depository for the object (fruit bowl, shelf, rubbish bin, desk and pencil case) and the participant rewarded the robots' suggestion until the robot had learned the favored depositories for the different objects. As in the previous scenario, the robot should then relearn (e.g. apple: fruit bowl → rubbish bin).
- **Hobby Scenario:** In the last scenario, the robot learned the participant's favorite hobby. Every time it heard the words "idea" or "further" the robot reacted with a suggestion for a leisure activity (theatre, zoo, riddle, pub, smoking, TV, play "Memory", and computer) which was rewarded by the participant until the participant believed that the robot had learned the favored hobby. Again, the robot should then relearn - in this case an appropriate leisure activity for the imagined grandchild.

Participants filled in a short questionnaire asking for the robot's learning abilities after each scenario with items asking for how satisfied they were with the robot's learning progress (1=very unsatisfied, 5=very satisfied), whether the robot learned what they had wanted to teach it (Yes/No), and whether they think that the time the robot needed to learn was appropriate (Yes/No).

Results and Discussion: In the following we present the results concerning the perception of the robot's learning behavior and participants' usage intention.

Learning Behavior: With regard to the perception of the robot's learning behavior results show that most participants were satisfied in all three scenarios (see Table 3). All participants stated that the robot had learned what they had tried to teach it during the Greeting and the Sorting Scenario and still 84.2% did so for the Hobby Scenario. Participants rated the time the robot needed to learn as being rather appropriate than inappropriate (see Table 3). However, the results reveal noticeable differences between the scenarios. The learning time in the Sorting Scenario was rated more often as inappropriate in contrast to the results of the Greeting and the Hobby Scenario. One reason for this could be that the robot tried every action once for each stimulus before it started to decide based on the information stored in the emotional memory. Within the Sorting Scenario the robot was able to recognize three different stimuli and had five different actions available. Hence in total $3 \cdot 5 = 15$ decisions were made without consideration of the obtained rewards. In contrast, there were only 7 actions to try in the Greeting Scenario and 8 actions to try in the Hobby Scenario.

In the final questionnaire, participants evaluated the robot's overall learning process (see Table 2). With regard to the question whether participants recognized the robot's learning progress almost all participants answered with yes (94.7%, in contrast to 53.4% in study 1). Thus, it can be assumed that the interaction scenarios used in this study are more suitable to show the robot's learning progress. In contrast to study 1 with 21 different stimuli (card values), the social scenarios utilized less stimuli, namely one or three. This supports the conclusion presented in Section 4 that the number of 15 rounds in the card game was too low for this high number of stimuli. Moreover, in study 2 participants were able to decide on their own how long the interaction lasted, because they were instructed to continue teaching until they are content with the learning outcome. Although the learning progress was recognized by almost all participants, they evaluated the robot's learning progress as well as the robot's learning ability as being moderate (see Table 2). These results are comparable to the results of the playful scenario. Furthermore, the robot's way to learn was rated as being expedient in the playful scenario as well as in the social scenarios. In view of the latter, most of the participants stated that they had used the different rewards' gradations but only 63.2% rated the rewards' gradations as being expedient (in contrast to 95% in study 1). One explanation for this effect might be the tasks itself. It can be assumed that the rewards' gradations are more expedient in the playful interaction. Within the card game, there were decisions which were clearly advantageous (e.g. taking another card when the accumulated card value is 12 or below), clearly risky (e.g. taking another card when the accumulated card value was very close to 21) or ambiguous (e.g. whether to take another card when accumulated card value is between 12 and 16). Thus different gradations for rewards make sense in order to reflect the levels of riskiness for the different decisions. However, the decisions in the social settings were of a more concrete nature (learn exactly one preferred greeting, hobby, depository). Therefore, participants might need fewer gradations to successfully teach the robot.

Usage Intention: A negative tendency can be observed concerning the question if the participants could imagine interacting with such a robot in their everyday life. Compared to study 1, less participants could imagine interacting with the robot in their everyday life (see Table 2), although the robot's learning progress was evaluated as rather positive.

Table 3. Results concerning the learning behavior of the social scenarios.

	Greeting Scenario	Sorting Scenario	Hobby Scenario
How satisfied are you with the robot's learning behavior?	Satisfied $M = 4.26 \pm .56$	Satisfied $M = 4.00 \pm .75$	Satisfied $M = 4.16 \pm .96$
Did the robot learn what you had wanted to teach it?	Yes 100%	Yes 100%	Yes 84.2%
Do you think that the time the robot needs to learn was appropriate?	Yes 89.5%	Yes 63.2%	Yes 73.7%

6   Conclusion

In this paper we presented a decision making framework for robot companions enabling them to adapt their behavior via user feedback. In two exploratory studies we applied the algorithm in different settings - a card game and social interactions, respectively. In both settings the robot's learning behavior was evaluated by a relevant target group of elderly people. In general, most participants were satisfied with the robot's learning abilities. However, evaluations depended on the interaction scenario employed. The benefit of an adaptable behavior based on the presented algorithm, in contrast to pre-wired connections between stimuli and actions, becomes more visible in the social scenario due to the relearning and thus was probably more often recognized by the participants. Moreover, in the card game the different user feedback graduations make more sense because decisions are less distinct than in the social setting. Due to the presented decision making framework, the user is able to change the robot's behavior through interaction and does not need to change any configurations every time the robot should act differently as it would be needed if pre-wired connections between stimuli and actions were used. However, the study revealed some room for improvements of the algorithm. For instance, the system might learn quicker and appear more natural when it does not try out every possible action, but stick with the first action that was rewarded very positively. Future work will focus on different adjustments within the decision making algorithm in order to test which configuration results in more natural and humanlike (or intuitive) decision making behavior.

References

1. Bechara, A., Damasio, A., Damasio, H., Anderson, S.: Insensitivity to future consequences following damage to human prefrontal cortex. Cognition **50**, 7–15 (1994)
2. Damasio, A.: Descartes Error: Emotion, Reason and the Human Brain. Putnam Press (1994)
3. Dautenhahn, K., Woods, S., Kaouri, C., Walters, M.L., Koay, K.L., Werry, I.: What is a robot companion - friend, assistant or butler. In: Proc. IEEE IROS, pp. 1488–1493 (2005)
4. Groom, V., Takayama, L., Ochi, P., Nass, C.: I am my robot: the impact of robot-building and robot form on operators. In: 4th ACM/IEEE International Conference on Human-Robot Interaction (HRI), pp. 31–36, March 2009
5. Hoefinghoff, J., Pauli, J.: Decision making based on somatic markers. In: Proceedings of the Twenty-Fifth International Florida Artificial Intelligence Research Society Conference, pp. 163–168. AAAI Press (2012)
6. Hoefinghoff, J., Rosenthal-von der Pütten, A., Pauli, J., Krämer, N.: You and your robot companion - a framework for creating robotic applications usable by non-experts. In: Proceedings of the 24th International Symposium on Robot and Human Interactive Communication (RO-MAN 2015) (2015)
7. Tapus, C., Tapus, A., Matarić, M.: Long term learning and online robot behavior adaptation for individuals with physical and cognitive impairments. In: Kelly, A., Iagnemma, K., Howard, A. (eds.) Field and Service Robotics. STAR, vol. 62, pp. 389–398. Springer, Heidelberg (2010)

Using Video Manipulation to Protect Privacy in Remote Presence Systems

Alexander Hubers[1], Emily Andrulis[1], Levi Scott[1], Tanner Stirrat[1],
Ruonan Zhang[1], Ross Sowell[1], Matthew Rueben[2], Cindy M. Grimm[2],
and William D. Smart[2(✉)]

[1] Department of Computer Science, Cornell College, Mount Vernon,
IA 52314, USA
{ahubers15,eandrulis16,lscott15,tstirrat15,
rzhang16,rsowell}@cornellcollege.edu
[2] School of Mechanical, Industrial and Manufacturing Engineering,
Oregon State University, Corvallis, OR 97331, USA
ruebenm@onid.oregonstate.edu,
{cindy.grimm,bill.smart}@oregonstate.edu

Abstract. Remote presence systems that allow remote operators to physically move around the world, observe it, and, in some cases, manipulate it, introduce a new set of privacy concerns. Traditional telepresence systems allow remote users to *passively* observe, forcing them to look at whatever the camera is pointing at. If we want something to remain private, then we simply do not put it in front of the camera. Remote presence systems, on the other hand, allow *active* observation, and put the control of the camera in the hands of the remote operator. They can drive around, and look at the world from different viewpoints, which complicates privacy protection.

In this paper, we look at how we can establish privacy protections for remote presence systems by manipulating the video data sent back to them. We evaluate a number of manipulations of these data, balancing privacy protection against the ability to perform a given task, and report on the results of two studies that attempt to evaluate these techniques.

1 Introduction

Remote presence systems, such as the Beam [1] and VGo [2], that allow remote operators to physically move around the world, observe it, and, in some cases, manipulate it, introduce a new set of privacy concerns. Traditional telepresence systems, such as videoconferencing, allow remote users to *passively* observe, forcing them to look at whatever the camera is pointing at. If we want something to remain private, then we simply do not put it in front of the camera. Remote presence systems, on the other hand, allow *active* observation, and put the control of the camera in the hands of the remote operator. They can drive around, and look at the world from different viewpoints, which complicates privacy protection.

© Springer International Publishing Switzerland 2015
A. Tapus et al. (Eds.): ICSR 2015, LNAI 9388, pp. 245–254, 2015.
DOI: 10.1007/978-3-319-25554-5_25

In this paper, we look at how we can establish privacy protections for remote presence systems by manipulating the video data sent back to them, and present two studies to evaluate how well these protections work. We consider a number of manipulations of these data, illustrated in figure 1, obscuring or removing objects in the environment that we wish to remain private, while retaining enough information to allow the remote operator to perform their task.

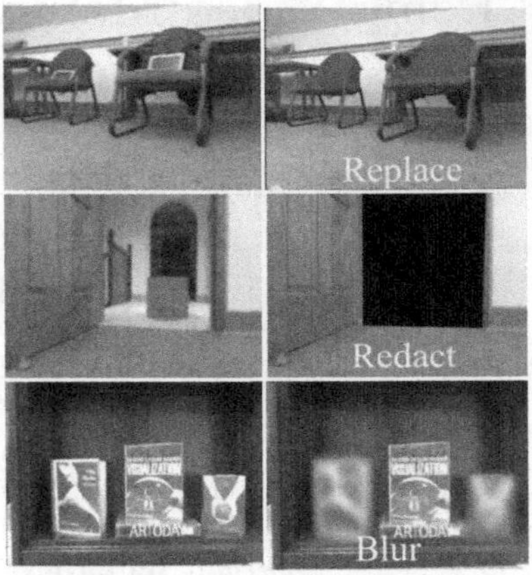

2 Defining Privacy

Fig. 1. Three privacy-protecting image manipulations. Top: replacement ("can't tell"); Middle: redaction ("can't observe"); Bottom: blurring ("can't discern").

For the purposes of this paper, we define a *privacy type* to be a specific restriction on the capabilities of the remote presence system. The capabilities can either be *physical*, where the system is prevented from taking some action or movement, or *observational*, where the sensor data transmitted to the remote user are altered in some way. In this paper, we limit ourselves to observational privacy, and to a video stream, although the basic ideas we discuss generalize to other sensor modalities.

We consider three types of observational privacy. **Can't tell** privacy is the expectation that the remote operator cannot tell if a particular object is there or not. Examples include not noticing any items exist on a table, or being unable to tell that there is a person present in the room. **Can't observe** privacy is the expectation that the remote operator might be able to tell there is something there, but cannot directly perceive it. Examples include not being able to look into a certain room, not being able to identify a shape as a person, or not showing what types of objects are on a table. Finally, **can't discern** privacy is the expectation that the remote operator can tell that there is something there and can identify the class of the object, but not the particular instance. Examples include being unable to read the text of documents on the table, being unable to make out facial features, or being unable to make out details of pictures on the walls.

3 Related Work

Our basic approach to protecting privacy is to alter the video stream coming from the remote presence system in such a way that it protects privacy, but does not interfere with task performance. We are not aware of any similar work in the robotics literature. However, there is a rich history of video manipulation in the graphics literature.

Broadly speaking, we classify manipulation of images and video by how they change the image: blur, inpainting, abstraction, line drawings, and painterly rendering. Blurring is a straightforward image filter and is commonly used in TV to obscure people's faces. Inpainting [3, 9] allows for filling an area of an image with synthesized content that is ideally indistinguishable from its environment. Abstraction, also sometimes called image stylization [12, 18], is similar to blurring, in that details are elided, but it differs in that strong edges are preserved. It can also involve restricting the color palate to create a cartoon-like effect. Since these are essentially texture filters, most methods can be efficiently implemented on a GPU. Line drawings [7] similarly preserve edges, but eliminate color information and render the result as a pen and ink or pencil-style sketch (sometimes with shading represented as hatching [16]). Painterly rendering techniques try to mimic a particular style, such as pixelation [8], oil or watercolors [11, 13] and comic-style [14]. Although not always intentional, most of these techniques also result in some image simplification or loss of detail, especially with large brush sizes.

Various studies have looked into how using video manipulations may help uphold privacy. Specifically, privacy typically considers autonomy, confidentiality, and solitude [5]. Filtering out parts of an image through marker detection has been shown to effectively uphold privacy for video surveillance cameras [15]. With an always-on camera space, using a blur filter has been shown to better balance protecting one's privacy while still allowing sufficient awareness to the user, so that any necessary and relevant information may still be gleaned from the image both with a co-present media space [10] and a telepresent media space [4]. However, in some circumstances where the privacy concerns are greater (i.e. assistive monitoring through use of a fixed always-on camera), a blur filter may not be sufficient, and another technique such as redaction may work more effectively [6].

4   Protecting Privacy with Video Manipulation

In this study we tested how well the different video manipulation techniques worked for each of the different privacy types, using three different scenarios. To avoid issues with localization, tracking, and training users to drive the robot, we conducted this study with videos recorded from the camera of a TurtleBot 2 robot driven by an experienced user. Care was taken to ensure that all objects relevant to the tasks were clearly visible at some point during the videos.

Participants were asked to watch three short video clips that were captured by a robot exploring an office environment, and to respond to five questions asking them to identify objects within the environment. Each scene had a specific privacy type applied to it, and participants viewed one clip of each scene. Each clip had one of the five randomly assigned video manipulations applied to it.

140 participants were recruited through Amazon's Mechanical Turk service, and compensated between US$0.20 and US$0.40 for their time. Participants were told that they would be expected to "watch a clip from the perspective of a robot

investigating an office and answer 5 short questions." The study took between three and five minutes for participants to complete.

To give some context to the study, we recorded video corresponding to three scenarios to present to the participants. Each scenario had a privacy type and a set questions to determine if there was a privacy violation associated with it.

In the **valuables** scenario, the robot drives through an office environment in which valuable electronics, such as a tablet computer are visible. We assign a "can't tell" privacy type to this scenario, since we do not want the remote user to be able to tell if there are valuables present. The participant was asked "How many computers or electronic valuables (costing more than \$50) are there in the room?" and "Name the valuables and the their locations in the room." A privacy violation occurs each time a piece of electronics is correctly identified by a participant.

In the **hallway** scenario, the robot passes an open doorway looking out into a hallway in which there is a large cardboard box. We assign a "can't observe" privacy type to this scenario, since we do not want the remote user to be able to see anything outside of the room. The participant was asked "Could you see anything in the hallway? If so, please list anything that you saw." A privacy violation occurs if the participant notices anything in the hallway (even if they cannot identify it).

Finally, in the **bookshelf** scenario, the robot passes a bookshelf with a number of books on it. A "can't discern" privacy type was assigned to this scenario, because we do not want to reveal the identities of some of the books on the shelf. The participant was asked "How many books are on the bookshelf?" and "Name as many titles of the books as you can." A privacy violation occurs each time a private book title is successfully identified by the participant.

We had five video manipulation conditions for this study: **redact** (replace object or area with black pixels); **replace** (replace object or area with background pixels); **abstract** (replace object or area with abstraction); **blur** (apply a blur filter to the object or area); and **control** (no manupulation).

Each manipulation was applied to each video clip by hand, off-line, using Adobe After Effects®. While this is not a viable approach for autonomously protecting privacy, it ensured that the video manipulations we tested in this study were free from artifacts introduced by poor object recognition or localization.

Participants were shown one video clip from each scenario, with a randomly-selected video manipulation technique. After each clip, they were presented with the clip-specific questions to determine if a privacy violation had occurred. The resulting data were hand-coded by a pair of researchers to establish the number of privacy violations in each experiment.

4.1 Results

Figure 2 shows the results of the study. In the "can't tell" scenario (valuables), the abstraction and blurring manipulations had little effect on the number of privacy violations, but redaction and replacement significantly reduced them (see table 1)

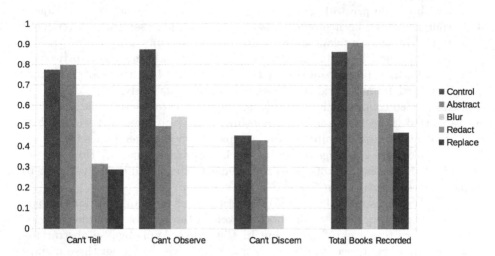

Fig. 2. Average percentage of privacy violations by privacy type (leftmost three columns) and average percentage of total books recorded, broken down by video manipulation condition.

In the "can't observe" (hallway) scenario, all of the video manipulations led to a significant reduction in the number of privacy violations. However, the redact and replace manipulations resulted in significantly fewer violations than the abstraction and blurring manipulations, both reducing the number of violations to zero.

There are still a number of privacy violations for both the redaction and replacement manipulations, the two techniques which performed best in this scenario. We attribute this to participants being able to identify valuables from their surrounding context: laptops have power cords, for example. This raises an interesting question when performing video manipulations, since we have to

Table 1. Significant difference groupings, calculated using Tukey's method [17]. Within each column, conditions with the same letter are not significantly different from each other. Conditions with no shared letters show a significant difference, at the 95% level. Conditions with two assigned letters are not significantly different from conditions with either of the letters.

Technique	Can't Tell	Can't Observe	Can't Discern	Total Books
Control	A	A	A	AB
Abstract	A	B	A	A
Blur	A	B	B	BC
Redact	B	C	B	C
Replace	B	C	B	C

not only manipulate based on the *content* of the image (a representation of the object to be made private), but also on the *context* surrounding that object. This context can be both physical, like a power cord, or semantic (a redacted object directly under a television is likely to be valuable).

We believe that the replacement video manipulation best protects privacy in this scenario, since participants never made mentioned that they noticed something removed from the video in this condition. Conversely, when using the redaction video condition, it was obvious from some responses that the black boxes alerted the participant that something was there, even if it was unclear what that object was. A snippet from a typical response exemplifies this problem: "it looked like 2 laptops on chairs blocked by black squares." However, we concede that a realistic autonomous replacement filter is significantly harder to implement than a redaction filter.

In the "can't discern" (bookshelf) scenario, it was often difficult for the participants to read the titles of the books, even in the control (unaltered) video condition, due to a lack of resolution in the video. Applying the abstraction manipulation did not seem to affect this much. However, the other three manipulations significantly reduced the number of privacy violations to less than 10% (blur) and zero (redact and replace).

Neither the abstraction nor the blur manipulation had a significant effect on the number of books that the participants counted. However, the redact and replace manipulations caused participants to report significantly fewer books on the shelf than the control condition. This is not a problem, though, since it is, in a sense, providing more privacy than is needed, up to the "can't tell" level.

5   Effects on Task Performance

The study described in section 4 shows that video manipulation techniques can be used to protect privacy when using a remote presence system. However, such protections are only useful if they do not interfere with the task that the remote operator is trying to accomplish. In this section, we describe a second study, designed to test how much our privacy-protecting measures get in the way as a remote user tries to control the remote presence system.

Participants were asked to teleoperate a mobile robot through an unfamiliar home environment, shown in figure 3, and to respond to a brief set of survey questions asking them to identify cleaning supplies and equipment contained in the home. 30 participants were recruited through flyers distributed via email and posted on bulletin boards in the local community. Participants were compensated US$10 for their participation. Participants were told that they would be expected to, "drive a robot around an apartment and answer a brief set of survey questions." The average time spent per participant, including training and answering all survey questions, was between 30 and 45 minutes.

Basic demographic information was collected for each participant. 12 males and 18 females participated in the study. The mean age of the participants was 28. 33% of the participants were students, and 30% of the participants played video games more than once a month. Only one of the participants reported any familiarity with the apartment building in which the home environment was staged, and zero expressed an expert level of familiarity with robots or other remotely operated devices.

This study had two conditions: one in which privacy protections (experimental) were implemented, and one in which they were not (control). In the experimental condition, a replacement manipulation was applied to all child-related toys in the environment, to enforce a "can't tell" privacy type. A redaction manipulation was applied to video from the bedroom, to enforce a "can't observe" privacy protection, and images on family photographs were blurred to enforce a "can't discern" protection.

Since the robot was being controlled by the participant, all of our video manipulation techniques depended heavily on highly accurate localization of the robot. In order to know which parts of the image to manipulate, we need to know both the pose of the robot and the pose of the objects to be made private. Any localization errors would render our video manipulation techniques useless. In

Fig. 3. The home environment used in the task performance study. Cleaning supplies included (a) a vacuum cleaner and (b) a dustpan and brush under the table. Manipulations included (c) a redaction of the bedroom interior, (d) redaction of toys and blurring of photographs. (e) Cleaning supplies were visible in the kitchen, and (f) baby toys were visible in the bedroom.

our initial testing, we found that it was impossible to reliably get a pose estimate from the TurtleBot that was good enough for our purposes.

To circumvent this problem, we chose to simulate the video manipulations *physically*. Replacement was simulated by simply removing the objects from the cnvironment. Redaction was accomplished by hanging a matte black sheet in the bedroom doorway. Blurring was simulated by physically replacing the photographs with pre-blurred versions of themselves. This allowed us to test whether the manipulations themselves are effective, separate from testing whether a particular implementation of the manipulation works. However, for a practical implementation of our techniques, robust, accurate localization remains key, as we discuss in section 6.

Participants were asked to teleoperate a TurtleBot 2 robot through an unfamiliar home environment, using a PS3 game controller. Participants sat at a table with two laptops. One laptop showed a full screen live video feed from the robot. The other laptop was used for displaying a web page that provided instructions to the participant and allowed them to answer the survey questions.

The participants were provided with a hand-drawn floor plan of the apartment that they would be moving the robot through, but they were never allowed to see this space with their own eyes. They only observed the space through the video feed from the robot.

Before the experiment began, participants were given a brief training session on teleoperating the TurtleBot, to ensure that they were comfortable controlling it with the game controller. Once this training was completed, the participants were instructed to drive around the environment and identify all of the cleaning supplies that were present.

Participants were then asked to navigate first to the kitchen and then to the living room, and to identify any cleaning supplies and equipment that they found there, including their specific locations and brand names. The number of cleaning supplies correctly identified is our metric for task performance; more supplies identified correlates with better performance on the task.

After completing this task, the participant was asked the following question: "Based on what you saw, to what degree do you agree with the following statement: *Children regularly visit this apartment.*," and asked to rank their response on a 5-point Likert scale, from "strongly agree" to "strongly disagree". Our goal with this scenario was to render private all signs of children in the environment. A response of "strongly agree" would correlate with a privacy violation, while a response of "strongly disagree" would correlate with good privacy protection.

5.1   Results

Figure 4 shows the results of the second study. There was no significant difference in the numbers of cleaning products identified across the two conditions (Fisher's exact test, $p = 0.763$). We interpret this to mean that the privacy protections did not interfere with task performance. However, there was a significant difference in the responses to the question about presence of children across the conditions. Participants in the control condition agree significantly more with the statement "Children regularly visit this apartment" than do participants in the experimental condition (Welch's unequal variances t-test, $p < 0.001$). We interpret this to mean that our privacy protections were effective.

Clearly, these are best-case results; our physical manipulations were simulations of the computational manipulations that we would, ideally, perform on the video streams from the robot. Removing items to simulate a replacement manipulation, for example, will clearly remove evidence of children. However, we believe that the physical redaction and blurring manipulations do tell us something about the potential utility of performing computational manipulations, assuming that we can get them right. This is, of course, the crux of the problem and hinges on a number of engineering details such as accurate object tracking and robot localization. We are, however, encouraged by these initial results.

Using the privacy-protecting manipulations led to fewer privacy violations in this study. However, the effect may be even more pronounced than the results above suggest. Of the 15 participants in the control condition, only 3 disagreed

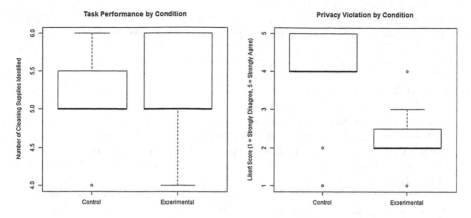

Fig. 4. Task performance (left) and privacy violations (right) by condition for the task performance study.

with the statement that children regularly visited the apartment. Interestingly, 2 of these participants seem to have clicked the wrong button; when asked to justify their answer in the questionnaire, they gave reasons such as "stroller in the kitchen, toys in the living room." Similarly, the one and only participant in the experimental group that agreed with the statement did so because, "The apartment is really clean, I could not see any family pictures or toys around," suggesting that they, in fact, disagreed with the statement.

6   Conclusions

We have presented results from two studies that demonstrate it is possible to protect privacy by using video manipulation techniques (section 4), and that these techniques can be applied without affecting task performance in a visual observation task (section 5). While these results are preliminary, and rely heavily on simulations to overcome current shortcomings with robot localization and object pose estimation, we believe that they are very encouraging.

The studies in this paper used a number of tricks to overcome poor pose estimation for both the robot and objects in the world. If these techniques are ever to be applied to a real system, then these shortcomings must be addressed. We need to be able to localize the robot accurately with respect to the objects and areas to be made private. If we cannot do this, then our video manipulations might not completely cover the areas of the image corresponding to the object, causing the privacy protections to fail. Alternatively, we might have to manipulate much larger areas of the image, to account for the uncertainty, running the risk of leaving too little information to perform the task at hand. We are currently working on this issue, to assess the trade-offs that will undoubtedly have to be made.

Acknowledgments. This work was partially supported by the National Science Foundation, under awards IIS 1340897, IIS 1340894, and CNS 1359480.

References

1. Beam remote presence system. https://www.suitabletech.com/
2. VGo robotic telepresence for healthcare. http://www.vgocom.com/
3. Barnes, C., Shechtman, E., Finkelstein, A., Goldman, D.B.: Patchmatch: a randomized correspondence algorithm for structural image editing. ACM Trans. Graph. **28**(3), 24:1–24:11 (2009)
4. Boyle, M., Edwards, C., Greenberg, S.: The effects of filtered video on awareness and privacy. In: CSCW 2000, pp. 1–10 (2000)
5. Boyle, M., Neustaedter, C., Greenberg, S.: Privacy factors in video-based media spaces. In: Media Space 20+ Years of Mediated Life. CSCW, pp. 97–122 (2009)
6. Edgcomb, A., Vahid, F.: Privacy perception and fall detection accuracy for in-home video assistive monitoring with privacy enhancements. SIGHIT Rec. **2**(2), 6–15 (2012)
7. Eisemann, E., Winnemöller, H., Hart, J.C., Salesin, D.: Stylized vector art from 3D models with region support. In: EGSR 2008, pp. 1199–1207 (2008)
8. Gerstner, T., DeCarlo, D., Alexa, M., Finkelstein, A., Gingold, Y., Nealen, A.: Pixelated image abstraction. In: NPAR 2012, pp. 29–36 (2012)
9. Herling, J., Broll, W.: Pixmix: a real-time approach to high-quality diminished reality. In: ISMAR 2012, pp. 141–150 (2012)
10. Kim, H.H.J., Gutwin, C., Subramanian, S.: The magic window: lessons from a year in the life of a co-present media space. In: GROUP 2007, pp. 107–116 (2007)
11. Lu, J., Sander, P.V., Finkelstein, A.: Interactive painterly stylization of images, videos and 3D animations. In: I3D 2010, pp. 127–134 (2010)
12. Mould, D.: Texture-preserving abstraction. In: NPAR 2012, pp. 75–82 (2012)
13. Olsen, S., Gooch, B.: Image simplification and vectorization. In: NPAR 2011, pp. 65–74 (2011)
14. Sauvaget, C., Boyer, V.: Comics stylization from photographs. In: Bebis, G., et al. (eds.) ISVC 2008, Part I. LNCS, vol. 5358, pp. 1125–1134. Springer, Heidelberg (2008)
15. Schiff, J., Meingast, M., Mulligan, D., Sastry, S., Goldberg, K.: Respectful cameras: detecting visual markers in real-time to address privacy concerns. In: IROS 2007, pp. 971–978 (2007)
16. Singh, M., Schaefer, S.: Suggestive hatching. In: Computational Aesthetics 2010, pp. 25–32 (2010)
17. Tukey, J.: Comparing individual means in the analysis of variance. Biometrics **5**(2), 99–114 (1949)
18. Winnemöller, H., Kyprianidis, J.E., Olsen, S.C.: Special section on cans: XDoG: an extended difference-of-gaussians compendium including advanced image stylization. Comput. Graph. **36**(6), 740–753 (2012)

Social Robots as Co-therapists in Autism Therapy Sessions: A Single-Case Study

Andri Ioannou[(✉)], Iosif Kartapanis, and Panayiotis Zaphiris

Cyprus Interaction Lab, Cyprus University of Technology, Limassol, Cyprus
{andri,kartapa,pzaphiri}@cyprusinteractionlab.com

Abstract. This study aimed to explore the potential value of a humanoid robot, NAO, in assisting the therapist during therapy sessions with children in the autism spectrum (ASD). We report findings from a single case, a 10-year old boy, Joe (pseudonym), diagnosed with high functioning ASD. The intervention was conducted in four consecutive therapy sessions with Joe during which Joe played the "Animals Game" with NAO and the therapist. In this game, NAO asked Joe to find a specific animal from a deck of cards. Numerical data demonstrated Joe's progress across sessions in terms of discriminating the animals from images and learning the animals in English. Additionally, based on qualitative observations, we have evidence of Joe becoming more independent from session to session, initiating interaction with NAO, directing his gaze and expressing affective feelings.

Keywords: NAO · Humanoid robot · Autism · ASD · Therapy · Interaction · Social robots · Social behavior

1 Introduction

Autism Spectrum Disorder (ASD) is a complex neurological disorder that has its onset in the first few years of a child's life [1]. The areas that are most commonly affected by autism are those of communication and social interaction [2]. Given that ASD is a spectrum disorder, a diagnosed child can display an array of behaviors that may vary on the intensity and characteristics. In terms of social interaction, ASD children frequently have difficulties in expressing their needs, interests, or initiating and maintaining social interaction. They have a difficult time interpreting social and emotional cues, as well as establishing joint attention, which is critical in social interaction and in developing communication skills [3]. In terms of communication, an absence or difficulty in speech is usually observed [4]. ASD children often have imitative deficits, which make learning by imitation ineffective. This not only creates learning gaps in behavior but also, in language and speech development. In terms of stereotypical behaviors, usually echolalia is present and repetitive movements such as spinning, rocking and head banging or other self-stimulus behaviors are observed. A routine and preservation of sameness are important to an ASD child. Usually a change in program or response towards them can lead to tantrums and uncontrollable

© Springer International Publishing Switzerland 2015
A. Tapus et al. (Eds.): ICSR 2015, LNAI 9388, pp. 255–263, 2015.
DOI: 10.1007/978-3-319-25554-5_26

crying until everything returns to normal [5]. According to the Center for Disease Control and Prevention [6] the prevalence of autism in the United States was 1 in 150 in 2000, 1 in 88 in 2008 and 1 in 68 in 2010, following an increasing trend.

2   State of the Art

There is no cure for Autism at the moment. Parents usually seek one of the most common methods of therapy. Those include but are not limited to occupational therapy, sensory integration therapy, Early Start Denver Model, speech therapy and interventions such as ABA. There have been many tools developed for aiding therapist in their work with ASD children, such as specialized cards with pictures of emotions and everyday objects, social stories books, and weighted vests among others. Furthermore, technologically advanced equipment has been developed over the years to aid in the area of ASD intervention and treatment.

2.1   Technology Based Interventions in Autism Therapy

There is a growing interest in the area of technology and autism. Interventions utilizing mechanical prompts to stimulate sensory processes, videos, computer-based games, virtual reality, and robotics have been studied as aids in ASD therapy [7]. Promoting and encouraging the development of social interaction and language skills for ASD children is a difficult task [8]. Researchers have attempted to study many aspects of technology in order to address their potential value in supporting these goals. For example, [9] used video modeling to increase the conversational skills of ASD children. These researchers concluded that some ASD children could benefit from a video modeling treatment, sometimes as soon as after two therapy sessions, although result varied across subjects. Video modeling has also been used to train children on ADL's (Activities of Every Day Living) such as pointing to things they need and washing their hands [10]. In promoting the independence of the child, iPod Touch has been used to train ASD children in how to structure their leisure time independently and without the aid from an adult [11]. Other studies have used computer programs to successfully provide an activity schedule for ASD children [12]. Establishing a relationship with an ASD child can be a frustrating process, even more so for family members. Using a 3D modeling software researchers [13] attempted to facilitate communication and strengthen relationships between ASD children and their grandparents.

2.2   Robotic Interventions

By the end of the twentieth century, research had begun to utilize robotics for ASD intervention and treatment. This research is mainly focused on young ASD children and their interaction with robotic equipment. A variety of robotic designs and processes has been explored such as, robotic classes [14], robotic animals [15,16] and doll shaped and other humanoid robots [17,18,19,20]. Dautenhahn [21] introduced the idea of robots as social mediators to ASD children and launched a long-term study in this area called the «Aurora Project». Since then many have explored the integration of robotics in the rehabilitation and therapy of autism. In particular, a few researchers have tried to promote skills such as communication, social interaction, and creative play through robotic equipment [17], [22,23,24], [14]. One of the first efforts to assess the value of robotics in aiding the treatment and therapy of autism is found in [21]. The study built on the assertion that ASD children do not like, and most of the time, they are frighten of anything unpredictable. By using the robot as an assistant, one may eliminate all of the unpredictable behaviors, since the robot can perform the same behavior repeatedly with minimal expressive movements, thus creating a pre-dictable environment in which ASD children can learn and thrive [21].

Addressing ASD children's difficulties related to communication and social isola-tion are of the most prominent areas in ASD research. What makes the robot a great tool for treatment in this area is that social tension does not exist within the child-robot interaction and this can remove any inhibitions and support the learning of new social behaviors [21], [25]. Several studies have explored the possibilities of using a humanoid robot to aid the development of social interaction. For example, Kaspar, a humanoid robot with minimal expressions was effectively used to teach social interac-tion skills to lower functioning ASD children, which were then transferred and ob-served in other situations [26]. Moreover in [14], ASD children showed improvement in eye gaze and communication, evident during a follow up activity with a human partner, after playing a game with the robot. Furthermore, [24] encouraged the devel-opment of social interaction utilizing a humanoid robot named Alice. This work showed that children, after getting accustomed to the robot, opened up in including the therapist in their interaction with the robot and that overall the intervention was effective in helping ASD children develop social skills. In the context of social inte-raction and communication, training ASD children to recognize human emotions is one of therapists' main goals [1] and therefore, many researchers have focused on developing robots and interventions for this purpose. In [27], a very realistic humano-id robot FACE (Facial Automation for Conveying Emotions) was developed to study and explore the possibilities of using robotic technology to teach emotional recogni-tion and imitation in ASD children. Although this study was the first set of clinical trials on using FACE, the results were promising, improving CARS (Childhood Aut-ism Rating Scale) score on emotional response for all participating ASD children. Last but not least, in [28], the researchers explored the interaction between the huma-noid robot 'Infanoid' and an ASD child. They found that the child, after getting over his initial reservations, interacted with the robot for more than 40 minutes during which they observed how the child directed his gaze to the place where the robot pointed at, establishing joined attention.

3   Method

We employed a single-case research design, observing the subject over the course of the intervention and documenting changes in his behavior over time. In general, sin-gle subject design (or single-case research design) is prominent in special education since it concerns the individual student, their specific needs and the development of practical procedures that can be applied in real word situations. We obtained consent forms by the parents and special permission for the study by the board of directors of the Cyprus Autism Association, where the study took place.

3.1   Participant

The subject of this study is Joe (pseudonym), a 10-year-old, Greek-speaking boy diagnosed with ASD by a developmental pediatrician using ADOS (Autism Diagnostic Observation Manual). Joe attends special education, speech therapy, and music therapy once a week at the Cyprus Autism Association where one of the investigators worked as a therapist. During the day Joe attends a special education unit of a local public elementary school. He has a special interest in video games and he often plays imaginatively. One of the positive outcomes of his interest in video games is that, through them, he has learned some English words such as "play", "stop", "go", "you won", "try again". His conversational skills (in Greek) are limited to replying to simple questions, while if something catches his attention he might initiate a conversation limited to asking "what is this?" Although he is well behaved and follows instructions perfectly, he is sometimes unmotivated and acts very slow on tasks. His movements are stiff and rigid and he will often refuse to dance or perform activities that require excessive movement. He gets tired/board quickly, and he takes a long time in choosing the activities he wants to do. His therapy goals are to develop independence, learn to make choices, be motivated to do things, exercise and improve social and communication skills.

3.2   Setting

The study took place in the music therapy room of the Cyprus Autism Association. The room has a soft floor as well as beanbags and pillows for the children to sit on. A wireless access point was set in the room for the computer to communicate with the robot. The therapist was present in the room and was controlling the robot via a laptop (see Fig1). The presence of the therapist is important because we are not looking to replace the therapist but rather to enhance the therapy and provide a tool for them to aid in therapy. All activity was video recorded.

The robot used for the study is NAO by Aldebaran Robotics, a 58cm tall humanoid robot that exhibits human-like features. Among others, it has the ability to recognize images, faces and objects as well as respond to speech and other sounds in the environment. Since NAO has been released for research and development, there has been a great interest for its use for the treatment of ASD [18,19,20], [29]. In the mid of 2013, Aldebaran Robotics launched ASK (Autism Solution for Kids) aiming to promote the use of robots in therapy and education of ASD children but also to advance research in the field. The game used in the preset study was adapted from the ASK "Animals Game".

Fig. 1. Joe interacting with NAO – Joe presenting the selected animal image based on NAO's call (1: Joe, 2: computer running NAO simulator and controller, 3: animal cards, 4: animal called/presented, 5: NAO, 6: therapist)

3.3   Procedures

The intervention was conducted during four consecutive therapy sessions with Joe. The interaction with NAO lasted approximately 20 minutes per session, during which Joe played the "Animals Game" with NAO and the therapist. During the intervention, NAO greeted the child when entering the therapy room by saying "Hello Joe!" and then proceeded to engage the child in the game. During the animal game, NAO asked the child to find specific animals from a deck of cards and present it to NAO for feedback. NAO called the animals in English (e.g., "Show me the sheep"). When the animal was correct, NAO said "Bravo" and made a cheerful sound. If wrong, NAO said, "That's not right! Let's try again". The desk of cards included nine animals (cat, dog, pig, cow, sheep, goat, chicken, horse, duck) presented in black and white laminated pictures (see Fig 2). In every session, NAO called the animals randomly. The therapist had a supportive role, providing English-to-Greek translation as needed and encouraging Joe to respond and keep trying. A follow up, post-intervention session (5th session) was done with Joe and the therapist only to examine whether newly learnt behaviors and skills continued to exist in the absence of NAO.

Fig. 2. Images of the nine animals used in the Animals Game

4   Findings and Discussion

Video analysis focused on recording the number of correct/wrong answers by Joe, times he sought help and his general interaction behavior with NAO. The results provided evidence of Joe's notable progress across sessions, in terms of learning the animals in English, becoming more independent, initiating interaction with the robot, directing his gaze, and expressing affective feelings.

In session 1, Joe sought translation for all nine animals that NAO called (e.g., What is the Greek word for "dog"), establishing that Joe did not know the animals in English. He asked for a translation by turning and looking at the therapist. Upon translation, Joe pointed out the animals correctly on his first try. In session 2, Joe sought for translation only for two animals out of 8 animals called; he provided a correct answer on his first try for five out of six animals called in English and without translation. During the first two sessions, Joe also sought for encouragement and confirmation by looking at the therapist the moment before pointing to the animal picture. During sessions 3 and 4, Joe demonstrated confidence and became more independent completing the game without the support of the therapist. That is, Joe did not seek for translation or encouragement/confirmation and the therapist did not interfere during the game. In session 3, Joe made two mistakes, which he corrected in his 2nd try, out of 14 animals called in English (without translation). No mistakes were made in session 4 on the 10 animals called by NAO. In session 5, in the absence of NAO, Joe responded correctly to all animals called by the therapist. These results are summarized in Table 1. In sum, these results suggest that through play, NAO and the "Animals Game" supported Joe in learning to discriminate the animals and learning their English word while becoming more independent in playing the game from session to session.

Table 1. Learning the animals in English

Behaviors	Session 1	Session 2	Session 3	Session 4	Session 5
Animals called in English	9	8	14	10	9
Correct responses before translation (1st try)	0	5	12	10	9
Wrong responses before translation (1st try)	n/a	1	2	0	0
Times seeking translation	9	2	0	0	0

Moreover, qualitative observations from the video analysis showed that in session 1 Joe had reservation in interacting with the robot especially during the first 10 minutes. He was a bit hesitant and cautious. This behavior can be attributed to the fact that NAO was something new which suddenly changed Joe's regular routine [24] -- a phenomenon also known as neo-phobia. Surprisingly however, in session 2, Joe walked into the room and initiated the interaction with NAO by saying "Hello NAO, how are you?" (in Joe's native language, Greek). This was a milestone for Joe, as he had never before initiated this kind of conversation with the therapist or others at the therapy center, unless he was repeatedly encouraged to do so. What is more, Joe continued to initiate the interaction with NAO in the following sessions, while in session 5, Joe looked for NAO saying "Hello [therapist name], NAO? Where is NAO?"

Furthermore, during session 1, the therapist taught Joe how to place the animal images in a position where NAO could properly identify them and respond. By the end of the session Joe already knew how to properly interact with NAO and in fact, he seemed to understand that NAO was controlled by the therapist through the computer. In session 2, Joe discovered that the therapist could see NAO's vision through a simulation on the computer screen. From this point onwards, Joe always attempted to look at the computer screen from across the room (see Fig. 1) to make sure that the animal picture was correctly picked up by NAO. This coordinated shift of gaze between NAO, the animal picture and the computer screen, was something Joe had never demonstrated before when the therapist used a tablet, computer or other tools during the therapy sessions.

Last but not least, Joe showed signs of distress, worry and caring when NAO malfunctioned in a couple of cases. For example, the simulation-controller crashed once in session 3 causing NAO to stop interacting with Joe and get-ting into a sited position, engaging the child to say "NAO, are you ok?" In other instances where NAO malfunctioned, Joe asked repeatedly (expressed in Greek) "is he hurt", "does he need to rest", "NAO, are you ok?" This provided evidence that Joe was capable to exhibit social and affective behaviors confirming previous research findings suggesting that humanoid robots can elicit feelings and caring behavior [31,32]. What is more, upon the intervention and when NAO was not present anymore, Joe continued to ask the therapist about NAO (e.g., "is he hurt", "is he OK", "Is he coming back?"). This suggests that Joe view and addressed NAO more as a peer [31], rather than a machine/computer integrated in his therapy sessions.

5   Conclusions

This single case study aimed to explore the potential value of a humanoid robot -- NAO -- in assisting the therapist during therapy sessions with children in the autism spectrum. Based on video analysis from four intervention and one post-intervention therapy sessions with a high functioning ASD child, we can suggest that a humanoid robot such as NAO, and applications such as the "Animals Game", can be effective in supporting learning activities related to discriminating animals from images and learning English words. Also, although based on qualitative observations, we have initial evidence of the ASD child becoming more independent, initiating interaction with the robot, directing his gaze, and expressing affective feelings. Moreover, the quick acceptance of and easy interaction with NAO provides initial evidence that humanoid robots can be smoothly introduced in ASD therapy sessions. These are findings from a single case and difficult to generalize to other individuals. Our findings require replication within controlled experiments where the child's learning behavior is assessed with and in the absence of NAO. Also, more work is needed to understand whether changes in behavior transfer beyond the therapeutic setting.

References

1. Berger, S.D.: Music Therapy, Sensory Integration and the Autistic Child, 1st edn. Jessica Kingslet Publishers (2002)
2. American Psychiatric Association: Diagnostic and statistical manual of mental disorders. Text (DSM-IV-TR), vol. 4, p. 943 (2000)
3. Charman, T.: Why is joint attention a pivotal skill in autism? Philosophical Transactions of the Royal Society of London. Series B, Biological Sciences 358(1430) (2003)
4. Tager-Flusberg, H., Paul, R., Lord, C.: Language and communication in autism. Handbook of Autism and Pervasive Developmental Disorders 1, 335–364 (2005)
5. Butcher, J.N., Mineka, S., Hooley, J.M.: Abnormal Psychology. Victoria, p. 720 (2007)
6. ADDM Network: Autism and Developmental Disabilities Monitoring Network, pp. 6–37 (2012)
7. Goldsmith, T.R., LeBlanc, L.A.: Use of Technology in Interventions for Children with Autism. Journal of Early and Intensive Behavior Intervention 1(2), 166–178 (2004)
8. Colton, M.B., Ricks, D.J., Goodrich, M.A., Dariush, B., Fujimura, K., Fujiki, M.: Toward therapist-in-the-loop assistive robotics for children with autism and specific language impairment. Autism, pp. 1–5 (2009)
9. Sherer, M., Pierce, K.L., Paredes, S., Kisacky, K.L., Ingersoll, B., Schreibman, L.: Enhancing Conversation Skills in Children with Autism Via Video Technology: Which Is Better, "Self" or "Other" as a Model? Behavior Modification 25(1), 140–158 (2001)
10. Plavnick, J.B.: A Practical Strategy for Teaching a Child With Autism to Attend to and Imitate a Portable Video Model. Research and Practice for Persons with Severe Disabilities 37(4), 263–270 (2013)
11. Carlile, K.A., Reeve, S.A., Reeve, K.F., Debar, R.M.: Using Activity Schedules on the iPod touch to Teach Leisure Skills to Children with Autism. Education and Treatment of Children 36(2), 33–57 (2013)
12. Stromer, R., Kimball, J.W., Kinney, E.M., Taylor, B.A.: Activity Schedules, Computer Technology, and Teaching Children With Autism Spectrum Disorders. Focus on Autism and Other Developmental Disabilities 21(1), 14–24 (2006)
13. Wright, S.D., D'Astous, V., Wright, C.A., Diener, M.L.: Grandparents of grandchildren with autism spectrum disorders (ASD): strengthening relationships through technology activities. International Journal of Aging & Human Development 75(2), 169–184 (2012)
14. Wainer, J., Ferrari, E., Dautenhahn, K., Robins, B.: The effectiveness of using a robotics class to foster collaboration among groups of children with autism in an exploratory study. Personal and Ubiquitous Computing 14(5), 445–455 (2010)
15. François, D., Powell, S., Dautenhahn, K.: A long-term study of children with autism playing with a robotic pet: Taking inspirations from non-directive play therapy to encourage children's proactivity and initiative-taking. Interaction Studies 10(3), 324–373 (2009)
16. Kozima, H., Nakagawa, C., Yasuda, Y.: Children-robot interaction: a pilot study in autism therapy. Progress in Brain Research 164, 385–400 (2007)

17. Billard, A., Robins, B., Nadel, J., Dautenhahn, K.: Building Robota, a mini-humanoid robot for the rehabilitation of children with autism. Assistive Technology: The Official Journal of RESNA 19(1), 37–49 (2007)
18. Ismail, L.I., Shamsudin, S., Yussof, H., Hanapiah, F.A., Zahari, N.I.: Robot-based Intervention Program for Autistic Children with Humanoid Robot NAO: Initial Response in Stereotyped Behavior. Procedia Engineering 41(Iris), 1441–1447 (2012)
19. Shamsuddin, S., Yussof, H., Ismail, L.I., Mohamed, S., Hanapiah, F.A., Zahari, N.I.: Humanoid Robot NAO Interacting with Autistic Children of Moderately Impaired Intelligence to Augment Communication Skills. Procedia Engineering 41(Iris), 1533–1538 (2012)
20. Shamsuddin, S., Yussof, H., Ismail, L.I., Mohamed, S., Hanapiah, F.A., Zahari, N.I.: Initial Response in HRI- a Case Study on Evaluation of Child with Autism Spectrum Disorders Interacting with a Humanoid Robot NAO. Procedia Engineering 41(1), 1448–1455 (2012)
21. Dautenhahn, K.: Robots as social actors: Aurora and the case of autism. In: Proc. CT99, The Third International Cognitive Technology Conference, August, San Francisco, vol. 359, p. 374, August 1999
22. Giannopulu, I., Pradel, G.: Multimodal interactions in free game play of children with autism and a mobile toy robot. Neuro Rehabilitation 27(4), 305–311 (2010)
23. Kozima, H., Nakagawa, C., Yasuda, Y.: Interactive robots for communi-cation-care: a case-study in autism therapy. ROMAN 2005. IEEE International Workshop on Robot and Human Interactive Communication 2005, 341–346 (2005)
24. Robins, B., Dautenhahn, K., Boekhorst, R.Te, Billard, A.: Robotic assis-tants in therapy and education of children with autism: can a small humanoid robot help encourage social interaction skills? Universal Access in the Information Society 4(2), 105–120 (2005)
25. Arendsen, J., Janssen, J.B., Begeer, S., Stekelenburg, F.C.: The use of robots in social behavior tutoring for children with ASD. In: Proceedings of the 28th Annual European Conference on Cognitive Ergonomics, pp. 371–372. ACM, August 2010
26. Robins, B., Dautenhahn, K., Dickerson, P.: From Isolation to Communication: A Case Study Evaluation of Robot Assisted Play for Children with Autism with a Minimally Expressive Humanoid Robot. Second International Conferences on Advances in Computer-Human Interactions 2009, 205–211 (2009)
27. Pioggia, G., Igliozzi, R., Sica, M.L., Ferro, M., Muratori, F., Ahluwalial, A., De Rossi, D.: Exploring emotional and imitational android-based interactions in autistic spectrum dis-orders. Cyber Therapy & Rehabilitation 25(8), 445–446 (2008)
28. Kozima, H., Nakagawa, C., Kawai, N., Kosugi, D., Yano, Y.: A humanoid in company with children. In: 4th IEEE/RAS International Conference on Humanoid Robots, vol. 1, pp. 470–477 (2004)
29. Ismail, L.I., Shamsudin, S., Yussof, H., Hanapiah, F.A., Zahari, N.I.: Estimation of Concentration by Eye Contact Measurement in Robot–based Intervention Program with Autistic Children. Procedia Engineering 41(Iris), 1548–1552 (2012)
30. Tapus, A., Peca, A., Aly, A., Pop, C., Jisa, L., Pintea, S., Rusu, A.S., David, O.D.: Children with autism social engagement in interaction with Nao, an imitative robot A series of single case experiments. Interaction studies 13(3), 315–347 (2012)
31. Ioannou, A., Andreou, E., Christofi, M.: Preschoolers' interest and caring behaviour around a humanoid robot. Tech. Trends 59(2), (2015). in press
32. Tanaka, F., Cicourel, A., Movellan, R.J.: Socialization between toddlers and robots at an early childhood education center. National Academy of Sciences 104(46), 17954–17958 (2007). http://www.pnas.org/content/104/46/17954 (retrieved July 12, 2013)

Personalized Short-Term Multi-modal Interaction for Social Robots Assisting Users in Shopping Malls

Luca Iocchi[1], Maria Teresa Lázaro[1]([✉]), Laurent Jeanpierre[2],
and Abdel-Illah Mouaddib[2]

[1] Department of Computer, Control and Management Engineering,
Sapienza University of Rome, Rome, Italy
mtlazaro@dis.uniroma1.it
[2] GREYC, University of Caen Lower-Normandy, Caen, France

Abstract. Social robots will be soon deployed in large public spaces populated by many people. This scenario differs from personal domestic robots, since it is characterized by multiple short-term interactions with unknown people rather than by a long-term interaction with the known user. In particular, short-term interactions with people in a public area must be effective, personalized and socially acceptable. In this paper, we present the design and implementation of an Human-Robot Interaction module that allows to personalize short-term multi-modal interactions. This module is based on explicit representation of social norms and thus provides a high degree of variability in the personalization of the interactions, maintaining easy extendibility and scalability. The module is designed within the framework of the COACHES project and some implementation details are provided in order to demonstrate its feasibility and capabilities.

1 Introduction

The new challenge for robotics in the near future is to deploy robots in public areas (malls, touristic sites, parks, etc.) to offer services and to provide customers, visitors, elderly or disabled people, children, etc. with increased welcoming and easy to use environments.

Such application domains present new scientific challenges: robots should assess the situation, estimate the needs of people, socially interact in a dynamic way and in a short time with many people, exhibit safe navigation and respect the social norms. These capabilities require the integration of many skills and technologies. Among all these capabilities, in this paper we focus on a particular form of Human-Robot Interaction (HRI): Personalized Short-term Multi-Modal Interactions. In this context, *Personalized* means that the robot should use different forms of interactions to communicate the same concept to different users, in order to increase its social acceptability; *Short-term* means that the interactions are short and focused on only one particular communicative objective, avoiding long and complex interactions; while *Multi-modality* is obtained by using different interaction devices on the robot (although in this study, we focus only on speech and graphical interfaces).

© Springer International Publishing Switzerland 2015
A. Tapus et al. (Eds.): ICSR 2015, LNAI 9388, pp. 264–274, 2015.
DOI: 10.1007/978-3-319-25554-5_27

The solution described in this paper is developed within the context of the COACHES project, that aims at developing and deploying autonomous robots providing personalized and socially acceptable assistance to customers and shop managers of a shopping mall. The main contribution of this paper is on the architecture of the Human-Robot Interaction module that has several novelties and advantages: 1) integrated management of all the robotic activities (including basic robotic functionalities and interactions) through the use of Petri Net Plans, 2) explicit representation of social norms that are domain and task independent, 3) personalized interactions obtained through explicit representation of information and not hand-coded in the implementation of the robot behavior.

In the rest of this paper, after an analysis of the literature in personalized human-robot interaction (Section 2) and a brief description of the general architecture of the COACHES system (Section 3), we describe the human-robot interaction component and, in particular, our approach to personalized short-term multi-modal interaction (Section 4). In Section 5, we provide some examples of application of the proposed system and finally we draw conclusions in Section 6.

2 Related Work

The use of service robots interacting daily with people in public spaces or workplaces has become of increased interest in the last years. In this context, the development of the robotic system should focus on creating confortable interactions with the people the robot has to share its space.

Gockley et al. [3] showed that people usually express more interest and spend more time during the first contact with the robot. However, after the novelty effect, the time of the interactions decreases which suggests people's preference for *short-term* interactions.

In order to address this decrease in the people engagement, Lee et al. [4] demonstrated in a 4-month experiment that *personalized* interactions allow to maintain the interest of the users over time. The experiment consisted of a robot delivering snacks in a workplace and the personalization was carried out through customized dialogues where the robot addressed the users by their names and commented the users' behaviours like their frequency of usage of the service or their snack choice patterns. Conversely, in certain contexts like in rehabilitation robotics, it is desired to have longer interactions with the patient, so the robot can assist and encourage him to do his exercises. In [8], it is shown how adapting the robot behaviour to the patient personality (introvert or extrovert) increases the level of engagement in the interaction.

Certain works aim at personalizing the interaction by learning from its user. For example, in [5] a certain task is commanded to the robot which receives a feedback from the user if the final state of an action is desirable for him. In this way, the robot learns from the user's preferences which are registered in a user profile. With this knowledge and the feedback it keeps receiving from the user, the robot can anticipate his needs and pro-actively act to fullfil his needs. In [6] the robot adapts its behaviour defined by parameters like the distance to

Fig. 1. COACHES environment and robots.

the person or its motion speed, among others, using a reinforcement learning technique where feedback from the user is given subsconciously through body signals read directly from the robot sensors.

In contrast to these works, our approach for personalized human-robot inter-action is not based on learning the personality of the user, but on a set of *social norms* that are present in our everyday lifes in human interactions. Moreover, our architecture is designed on domain and task-independent representation of infor-mation, providing for a high variability of personalized behaviors, with a simple declarative definition of the social norms that we want the robot to apply. This provides many advantages in terms of extendibility and scalability of the system. The proposed approach extends a previous work [7] about the design of social plans, by adding the notions of user profiles and of personalized interactions.

3   COACHES Environment, Hardware and Software Architecture

In the COACHES project (October 2014 - September 2017), we aim to study, develop and validate integration of Artificial Intelligence and Robotics technol-ogy in order to develop robots that can suitable interact with users in a complex large public environment, like a shopping mall. Figure 1 shows, on the left, the *Rive de l'orne* shopping mall in Caen (France) where the experimental activities of the project will be carried out, on the middle a prototype of the robot that will be used for the preliminary experiments and, on the right, the design of the robots that will be realized in Fall 2015.

As shown in the figure, in contrast with previous work on social robotics and human-robot interaction, the COACHES environment is very challenging, because it is populated by many people and the robot is expected to interact with many unknown and non-expert users. Moreover, we aim at a more sophisti-cated interaction using multiple modalities (speech, gesture, touch user inter-faces) and dialog generated on-line according to the current situation and the robot's goals. Although these characteristics are not completely new in related

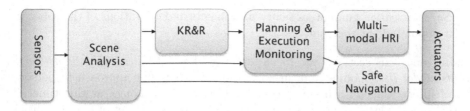

Fig. 2. COACHES software architecture

projects, we believe that the COACHES project will provide important insights for actual deployment of intelligent social robots in large populated public areas.

The software architecture of the COACHES robots is shown in Figure 2. The architecture comprises a typical configuration of an autonomous robot where all the decisions are made on-board based on sensors available.

An open architecture (hard/soft) and standard technologies available will be used, so that it will be easy to extend and/or adapt the capabilities of the system during the whole length of the project (especially to integrate and test various algorithms and/or sensors). Such an open architecture will also simplify and optimize integration efficiency as well as re-use of assets in other projects or products.

For the development of the robotic software components, the Robot Operating System (ROS)[1], which is the standard middleware for robotics applications, has been selected. ROS provides the middleware infrastructure to effectively share information among the many modules implementing various functionalities on each robot. Moreover, our software architecture includes an interface[2] for sharing information among the robots and between ROS and non-ROS components of the system, based on serializing and deserializing ROS messages as strings sent over TCP.

The main software components that are under development for control, reasoning and interaction functionalities of the system are briefly described below.

- *Scene analysis* includes sensor processing procedures for both on-board robot devices and static sensors in the environment in order to determine the current situation and understand events that are of interest for the system.
- *Knowledge-based representation and reasoning* defines the formalism and the procedure to represent and reason about the environment and the task of the robots. It provides the goals that the robots should achieve given the current situation.
- *Planning and execution monitoring* generates the plans to achieve the desired goals and monitor their execution for robust behaviors.
- *Multi-modal HRI* defines a set of modalities for human-robot interaction, including speech recognition and synthesis, touch interaction, graphical interface on a screen mounted on the robot and Web interfaces.

---

[1] www.ros.org

[2] https://github.com/gennari/tcp_interface

– *Safe navigation* guarantees safety movements and operations of the robot in a populated environment.

In the next section, we focus on the description of the *Short-Term Multi-modal HRI* module and, in particular, we show our approach to personalized interactions with users of the shopping mall. Although at this time all the other modules have not been fully realized, a minimum set of functionalities needed to test the HRI component are present.

4   Personalized Short-Term Multi-modal Interactions

As already mentioned, one of the main goals of the COACHES project is personalized short-term multi-modal interactions with non-expert users, that are typical customers of a shopping mall.

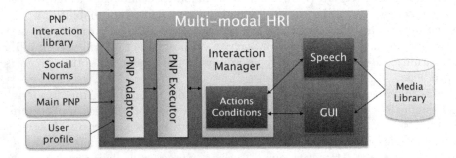

Fig. 3. Architecture of Human-Robot Interaction module.

The architecture of the HRI module is illustrated in Figure 3. Data available to this module are Petri Net Plans (PNP) encoding the desired behavior of the robot, social norms, a user profile and a multi-media library.

The PNPs (as described later) encode the overall behavior of the robot, as generated by the planning and reasoning components of the system. The behavior include both basic robotic actions (e.g., moving in the environment) and interaction action. The user profile is the information available about the user that is interacting with the robot. Among acquisition means for user profiles, it is possible to think about users wearing an RFID tag containing personal information read by an RFID reader on-board the robot, or to the request of swiping a fidelity card, enter a personal password or showing to the robot a QR-code, in order to communicate to the robot the user profile. In our implementation, we have used a simple identification mechanism based on recognizing QR-codes shown by the user to the robot on-board camera.

Finally, the media library is a collection of multi-media data (text, images, animations, video, etc.) that are linked to the communication activities of the robot and to the user profiles. We assume that in this library there are different versions of the same communication target for different users. For example, ice-cream advertisement can have a different spoken text and different displayed images or videos for children and adults.

In the remaining of this section, we will describe in more details the components of this module.

4.1 PNP Adaptor and Executor

The HRI module is implemented within the framework of the Petri Net Plans (PNP) formalism [9]. PNPs are based on two main concepts: *actions* (i.e., output operations) and *conditions* (i.e., input operations). Actions include motion of the robot in the environment, spoken sentences issued by the on-board speakers, text, images, videos or animations shown on the on-board screen, etc. Conditions include the result of perception routines (e.g., image processing or speech recognition), the input given by a user through a GUI on the on-board screen, information about personal data of user acquired through a reader of fidelity cards, etc.

The use of PNPs for representing in an integrated way all these different kinds of actions and conditions allows for a strong coordination between the different sub-systems of the robot and for showing more complex behaviors and, in particular, a multi-modal interaction that can be easily customized according to the user.

The main plan, which includes *interaction plans* for HRI behaviors, generated by the reasoning and planning sub-system, is first processed by the PNP Adaptor and then executed by the PNP Executor. Both these modules are general-purpose, since all the relevant information is provided by external sources with an explicit representation. More specifically, the PNP Adaptor generates a personalized plan, given a main plan, a library of interaction plans, a set of social norms, and the user profile. The generated personalized plan is then executed by the PNP Executor.

PNP Adaptor is implemented through an algorithm that transforms the Main PNP and the associated Interaction PNPs according to the social norms applied to the specific user profile. More specifically, the input plan is composed by a user-generic Main PNP that calls Interaction PNPs as sub-routines. All these plans are processed and transformed by applying the social norms (described as rules) customized to the current user profile.

The social norms are domain and task independent and are represented using a propositional logic formalism that follows the one described in [2]. Given a set of propositions U related to user profiles and a set of propositions I related to forms of interactions, and given the set of formulas U^* over U and the set of literals I^+ over I, a social norm is represented as a pair $(\phi, \delta) \in U^* \times I^+$, with the meaning that if ϕ is true, then δ is mandatory (i.e., it must occur), or, in other words, $\neg\delta$ is forbidden.

Some examples of social norms implemented in our system and considered in the examples in the next section are illustrated in Table 1.

Table 1. Domain-independent social norms.

```
( child, use_animation )
( elder, use_big_font )
( elder, use_simple_GUI )
( deaf, ¬ use_speech )
( blind, ¬ use_display )
( elder ∨ deaf, display_spoken_text )
( elder ∨ deaf ∨ blind, ask_for_guidance )
( blind, use_detailed_speech )
( blind, notify_guidance )
( first_time_user, detailed_instructions )
( ¬ first_time_user ∧ young, ¬ detailed_instructions )
( child ∨ very_young, ¬ use_baby_care_room )
( foreign, speak_English )
```

Given a set of social norms S and a user profile u from which it is possible to determine the truth of the formulas in U, then it is possible to derive all the literals in I^+ that are implied by the social norms and u. In other words, it is possible to compute the set of propositions Δ_u such that $S \wedge u \models \Delta_u$. These propositions can be seen as the personalization of S to u. For example, if the user profile u satisfies elder and deaf, Δ_u contains { use_big_font, display_spoken_text, use_simple_GUI, ¬ use_speech }. In this paper, we do not explicitly consider the case in which Δ_u may become inconsistent. Of course, several mechanisms could be implemented for solving this issue, such as adding preferences or priorities to propositions.

The personalized propositions Δ_u affect the execution of the output actions of the HRI module. Each action in the PNPs is personalized by adding the appropriate propositions as arguments. As described later in the section, in this paper we consider two kinds of output interaction actions: *Say*, related to the Speech module, and *Show*, related to the Graphical Interface module. Therefore, literals associated with *Say* (e.g., ¬ use_speech) are added as parameters of all the actions *Say* in the PNPs, while literals associated with *Show* (e.g., use_big_font, display_spoken_text, use_simple_GUI) are added as parameters of all the actions *Show* in the PNPs. These parameters determine the personalized interaction and will be considered by the Interaction Manager.

PNP Executor is a general-purpose executor of PNP already described in [9] and successfully used in many applications. PNP Executor treats actions and conditions without giving them any semantics and controls only the flow of execution. The actual execution of the basic actions and conditions is responsibility of the Interaction Manager.

4.2   Interaction Manager

The interactions are coordinated by an Interaction Manager (IM), which manages all the robot activities (both the ones related with human-robot interaction and the ones used for implementing the basic robotic functionalities). Its goal is thus to provide effective robot behaviors, including the personalized short-term multi-modal interactions described in this paper.

The IM is an action and condition server that executes actions and provides conditions, according to the requests of the PNP Executor module. It thus includes the definition of a set of primitive actions and conditions that are activated according to the plan under execution. For the interaction behavior, actions and conditions are actually related to the Speech and Graphical Interface (GUI) modules described later. While the actions and the conditions related to the basic robot abilities (such as navigation, localization, perception, etc.) are not illustrated and described here, since the focus of this paper is on interaction. The IM is also responsible to activate actions according to the personalized parameters defined by the PNP Adaptor module.

4.3 Speech and Graphical Interfaces

The interaction modalities considered so far in the project are speech and graphical interfaces.

Speech Recognition and Synthesis. The speech component allows the robot to communicate with humans through vocal interactions. It is formed by Automatic Speech Recognition (ASR) and Text-To-Speech (TTS).

The ASR component analyzes audio data coming from a microphone and extract semantic information about the spoken sentences, according to a predefined grammar. This component allows the robot to understand user spoken commands. The speech recognition module is based on the Microsoft engine and on a further processing module that builds the semantic frames of the recognized sentences. More details on the approach are available in [1].

The TTS component transforms text messages in audio data that are then emitted by the speakers on-board the robot. This enables the robot to speak to people. The Microsoft TTS engine is used for this module.

Graphical User Interface. The GUI component implements a graphical input and output interface between users and robots that is displayed through the touch screen on-board the robot. The GUI defines actions (i.e., output operations) and conditions (i.e., input operations) that are integrated in the IM with other communication primitives (e.g., speech) in order to implement a multi-modal interaction.

The Speech and GUI components make available to the IM the implementation of actions and conditions that are executed according to the PNPs. These are summarized in the following table.

	Action	**Condition**
Speech	*Say*	*ASR*
	speak information though TTS	Results of ASR
GUI	*Show*	*GUI*
	show information on the GUI	Results of GUI input

The actions implemented at this level are parametric with respect to a set of parameters expressed as propositions and used to define the social norms. As mentioned above, during the process, general actions are associated to specific parameters depending on the user profile. This parameters are considered to specialize the execution of the actions. Two kinds of specializations are considered: 1) modification of some internal parameters of the action (for example, the size of the font in a displayed text), 2) selection of the proper media to communicate. The second specialization is related to the presence of multiple options in the Media Library for the same communicative target. In these cases, each option is labeled with a precondition using the same interaction propositions in I. Therefore, it is possible to select appropriate media considering the personalized propositions Δ_u.

5   Examples of Personalized Interactions

In this section we will show through a set of examples how general purpose social norms are used to affect the behavior of the robot in a declarative way. The examples are taken from the use cases of the COACHES project and they will be eventually fully implemented and tested with real robots in the shopping mall in Caen. The examples below refer to the social norms described in Section 4 and assume user profiles are available.

Example 1. Advertising. One of the tasks of the COACHES robot is to show advertisements to users of the shopping mall. These advertisements (in forms of text, images, videos, etc.) are provided by the shop manager and stored in the Media Library. In one form of advertising planned in the project the robot knows the user profile. In this case the Interaction Module described in the previous section can activate personalized messages. Effects of personalized interactions in this example are: i) animation instead of videos for children, ii) big fonts and simple GUI for elderly people, etc.

Example 2. Directions and Guiding. The robot is able to give directions and guide people in the mall. Requests are acquired either by voice or graphical interface and the robot uses its semantic map of the environment to show directions or accompany the user. In this case the following personalized behaviors can be obtained: i) for elderly people, a simple GUI shows the direction; ii) the interaction with a deaf and elder person is made with graphical interface only; iii) the interaction with a blind person uses only voice. In all the three cases, the robot offers to accompany them and for the blind person a special notification is given with instructions of how the guidance will happen.

Example 3. Baby Care Rooms. Baby care rooms can be used by parents, but must be reserved and they are locked when not in use. The robot can enable this service upon request. Some personalized interactions in this case are: i) a new user is fully instructed with detailed instructions about how to use the service; ii) a user that already used the service a few time ago is given directly the access to the baby care room; iii) children or very young users will be notified that they are not allowed to use the service.

Notice that all these examples are implemented without explicit coding the corresponding behaviors. The expected personalized behavior is the effect of the application of the social norms to the user profile and of the corresponding modifications of the plans that activate actions with proper parameters. Notice also that the social norms are not specific of any particular task. This allows for a high level of extendibility. For example, adding, removing or modifying social norms allow for a significant change of behavior of the robot with different users without requiring any change (or just minimal changes) in the implementation of the actions. For example, assuming that we want to add the capability of the robot to regulate the volume of its voice and to personalize this feature. With our architecture it is sufficient to do the following steps: add a parameter about volume in the *Say* action (e.g., corresponding to a new proposition loud_speech in I) and a social norm (elder, loud_speech) in S. All the interactions with elder people now will use an increased volume of the robot speech.

6   Conclusions

In this paper we have presented our architecture for personalized short-term human-robot interaction to be used by COACHES robots that will autonomously provide services to customers in a shopping mall. Robot actions and interactions with users have been described through PNPs that can be dynamically adapted according to the user profile and a set of domain-independent social norms. This capability provides the system with a high level of scalability and, as shown in our examples, allows for being easily extended to a variety of interactions. Implementation of the HRI module presented in this paper has been tested in our lab with a prototype robot and not yet in the real environment with real users.

The on-going COACHES project is the main experimental test-bed for the work presented in this paper. Future work will thus include a user study, whose main focus will be assessing improved acceptance of a social robot with personalized interactions. With the approach described in this paper, producing different versions of the interaction behavior of the robot is as easy as adding or removing a social norm. However, in certain cases, personalization based on social norms may not be sufficient due to individual exceptions. Therefore, a more detailed individualization level applied to single users will also be subject of further studies.

References

1. Bastianelli, E., Castellucci, G., Croce, D., Basili, R., Nardi, D.: Effective and robust natural language understanding for human-robot interaction. In: Proceedings of 21st European Conference on Artificial Intelligence, pp. 57–62 (2014)
2. Boella, G., Pigozzi, G., van der Torre, L.: Normative framework for normative system change. In: Proc. of 8th Int. Conf. on Autonomous Agents and Multiagent Systems (AAMAS) (2009)

3. Gockley, R., Bruce, A., Forlizzi, J., Michalowski, M., Mundell, A., Rosenthal, S., Sellner, B., Simmons, R., Snipes, K., Schultz, A.C., et al.: Designing robots for long-term social interaction. In: IEEE/RSJ Int. Conf. on Intelligent Robots and Systems, pp. 1338–1343 (2005)
4. Lee, M.K., Forlizzi, J., Kiesler, S., Rybski, P., Antanitis, J., Savetsila, S.: Personalization in HRI: a longitudinal field experiment. In: ACM/IEEE Int. Conf. on Human-Robot Interaction, pp. 319–326. New York, NY, USA (2012)
5. Mason, M., Lopes, M.: Robot self-initiative and personalization by learning through repeated interactions. In: ACM/IEEE Int. Conf. on Human-Robot Interaction (HRI), pp. 433–440 (2011)
6. Mitsunaga, N., Smith, C., Kanda, T., Ishiguro, H., Hagita, N.: Adapting robot behavior for human-robot interaction. IEEE Transactions on Robotics **24**(4), 911–916 (2008)
7. Nardi, L., Iocchi, L.: Representation and execution of social plans through human-robot collaboration. In: Beetz, M., Johnston, B., Williams, M.-A. (eds.) ICSR 2014. LNCS, vol. 8755, pp. 266–275. Springer, Heidelberg (2014)
8. Tapus, A., Ţăpuş, C., Matarić, M.J.: User-robot personality matching and assistive robot behavior adaptation for post-stroke rehabilitation therapy. Intelligent Service Robotics **1**(2), 169–183 (2008)
9. Ziparo, V.A., Iocchi, L., Lima, P.U., Nardi, D., Palamara, P.F.: Petri net plans - A framework for collaboration and coordination in multi-robot systems. Autonomous Agents and Multi-Agent Systems **23**(3), 344–383 (2011)

Non-verbal Signals in HRI: Interference in Human Perception

Wafa Johal[(✉)], Gaëlle Calvary, and Sylvie Pesty

Univ. Grenoble Alpes, LIG, 38000 Grenoble, France
{wafa.johal,gaelle.calvary,sylvie.pesty}@imag.fr

Abstract. Non-verbal cues of communication can influence the human understanding of verbal signals in human-human communication. We present two illustrative experimental studies showing how non-verbal cues can both interfere and facilitate communication when passing a message to a user in HRI. In the first study, participants found that the cues enabling them to discriminate between two conditions : permissive or authoritative robots were mainly verbal. The verbal message was however unchanged between these two conditions and in this case, non-verbal cues of communication (gestures, posture, voice tone and gaze) substituted the neutral verbal message. The second study highlights the fact that verbal and non-verbal communication can facilitate the understanding of messages when combined appropriately. This study is based on a Stroop task of identifying the colour of the LEDs of a robot while the robot says words that are either facilitating, neutral or disturbing for the participant. These two studies put into perspective the importance of understanding interrelations between non-verbal and verbal signals in HRI.

1 Introduction

The general applicative scenario of our research involves a companion robot for a child alone at home. Let us imagine a scenario where the companion, managing the child's schedule, would ask the child to go do his homework. Our interest is in fitting the robot's behaviour to the child's and his parents' needs and expectations while letting them still have control over the companion's settings. There could be several ways to ask a children to do his homework. We consider these ways as *styles*. Styles are defined as ways to act and do things within a particular context [8]. In order to design our behavioural styles, we aim keep the content of the actions unchanged but to manipulate the way it is performed. For this reason, we aimed to keep verbal content unchanged and use parameters of non-verbal cues of communication to modify the behaviour delivered by the robot. Our aimed is to propose non-verbal behavioural styles that would allows the user to personalize its companion.

Most non-verbal cues of communication(NVC) are said to be innate and subconscious [11,15] . On the contrary, verbal communication such as language, are learnt and conscious. NVC help to prepare to do an action; they can reflect our intentions and are a window to our mental state [15]. For these reasons,

© Springer International Publishing Switzerland 2015
A. Tapus et al. (Eds.): ICSR 2015, LNAI 9388, pp. 275–284, 2015.
DOI: 10.1007/978-3-319-25554-5_28

research in social robotics is growing its interest on the application of these cues to make the robots' communication more natural to humans. Really often, a particular discomfort around robots can be present when the robot's non-verbal behaviour is dissonant to the verbal signal. For instance, a robot saying "I am very happy to see you" but showing a neutral face [3,17] could produce such a phenomenon. Some recent work have shown that multi-modality can help for speech disambiguation in joint attention and pointing tasks [13]. However, the effect of non-verbal and verbal communications on each other can be various and is still a matter of research in human-robot interaction.

This paper explores interrelations between verbal and non-verbal signals within two case studies. In a first study we show how dominant-oriented non-verbal cues can substitute for neutral verbal cues, making participants believe that the dominant signal was content in the verbal utterance. In the second study on the other hand, verbal cues is showed to be facilitating non-verbal understanding in a simple *Stroop* task.

This paper claims that there exists interrelations (here, substitution and facilitation) between non-verbal and verbal cues that can be studied and which could potentially help to design a better robot communicant.

2   Related Work

According to [11], non-verbal cues are innate and often subconscious signals of communication (like smiling, crying and laughing for instance). In HRI, these NVC have been largely used to depict emotions and social signals to the users [7]. Some recent studies have shown the impact of non-verbal cues on trust for example [5].

In human-human communication, Mehrabian [21] showed that simultaneous verbal, vocal, and facial attitude communications is a weighted sum of their independent effects – with coefficients of .07, .38, and .55, respectively. Other psychologists [11], have studied the influence of modalities on the social signal. These coefficient have been revised and the weight of verbal have been said to be higher in other contexts [21]. However, these studies have agreed on the importance of non-verbal cues in human-human communication. [11] explains how verbal and non-verbal signals interrelate in human-human communication in terms of: repeating, conflicting, complementing, substituting, regulating or accenting/moderating signals. It is then crucial for social robots to know these interactions between verbal and non-verbal communication in order to be efficient in sending signals to the user and to understand the ones they transmit.

Based on the literature [11], non-verbal cues of communication take into account : gestures, postures, touching signals, facial expressions, eye signals and vocal signals. To make the influence more readable, the results of this paper are presented in two sub-categories of non-verbal cues: *non-vocal* (posture, gesture, eye contact ...) and *tone of voice* (pitch, volume ...).

3   Experiment 1: Robots' Non-verbal Cues Perceived as Verbal by Participants

Previous work [10], have shown the ability of two robots - a robotic head Reeti [2] and the humanoid robot Nao [1](Table 1) - to express two parenting styles, *Authoritative* and *Permissive*. The Permissive and Authoritative styles in this experiment differ by their level of dominance. Expression of personality and complex emotions by robots is often hard to implement due to their motors' limitations. To compensate and check which modalities are important to depict parenting styles, both facial and bodily communication have been tested.

In order to depict the styles using multi-modal non-verbal and paralinguistic channels of communication, modalities of expression of *verticality* specified by Hall and al. [9] have been translated to the robot's motors and software abilities. Hall describes a set of non-verbal cues for *verticality* used to express dominance and submission to differentiate between Authoritative and Permissive style of parenting. The left side of Table 1 presents the cues which vary between permissive and authoritative signals for the two robots Nao and Reeti. We adapt

Table 1. Parenting style expression for Nao and Reeti robots, (positive ($\nearrow$) or negative ($\searrow$) influence of cues on dominance)

Dominance Non-verbal Communication Cues	Nao	Reeti
Gaze	$\nearrow$	$\nearrow$
Blinking rate	$\searrow$	$\searrow$
Nodding	$\nearrow$	$\nearrow$
Self contact (hands or face)	$\searrow$	
Self contact (hips)	$\nearrow$	
Hand and Arms: illustrator, emblems	$\nearrow$	
Postural openness	$\nearrow$	
Postural relaxation	$\nearrow$	
Face orientation	$\nearrow$	$\nearrow$
Ears		$\nearrow$
Voice Loudness	$\nearrow$	$\nearrow$
Voice Pausing		$\searrow$
Voice Pitch		$\searrow$
Voice Speed	$\nearrow$	$\nearrow$

the categories of non-vocal signals according to the physical and software constraints of each robot. From literature [6,14,16,19,20], spatial variables such as space occupation, direction and amplitude of gestures are used as parameters to design two level of expression for each NVC. Other parameters are employed for the dynamics: repetition, speed of gestures, speed of decay and fluidity/rigidity of movements. We also refer to the work of Breazeal [4] for the ear movement of Reeti.

The Table 1 shows the positive ($\nearrow$) or negative ($\searrow$) influence of each behaviour on the dominance dimension for each robot. For example, for both robots, a high blinking rate decreases the dominance factor while inversely, voice loudness will increase with dominance. The absence of arrows signify a hardware or software limitations of the robot (i.e. not possible to change Nao's pitch)

Videos of the robots' behaviours displaying the styles in the situation of asking a child to do his homework were showed to parents. For both styles, the robots are accompanying the non-verbal cues with a neutral verbal utterance in terms of dominance: *"I think you have played enough. You should go back to work"*.

An online questionnaire, allows us to get opinions of 93 parents on how they perceive the robots' behaviours. Each participant watch two videos of the same robot (either Reeti or Nao) expressing respectively the two parenting styles: Permissive and Authoritative. They are then asked to reply to a questionnaire. Parents were able to discriminate from Permissive and Authoritative on a authoritativeness scale from 0 to 10 (results in [10]).

Since the verbal utterance was neutral for the two styles Authoritative and Permissive, we believe that parents perceived non-verbal cues as discriminant in term of authoritativeness. The hypothesis is that parents perceived of non-verbal (tone of voice and non-vocal) cues of communication to be more influential than verbal utterance, when having to rate the authoritativeness of the robots in our experimental context.

Parents were asked to rate how each modality of expressions of the robot influenced their score in authoritativeness. Since the term non-verbal is not easy to define, we chose to be more precise and list 3 sub-modalities of the non-verbal behaviours. We asked the participants to evaluate the weight of influence of each non-vocal, tone of voice and verbal modalities by answering this question:

> *For each of the following elements, rate the influence of the element on the authoritativeness characteristic of the robot. From 0 (no influence) to 10 (very influent): gestures, posture, gaze, tone of the voice, words used.*

The average weight of influence for each type of behaviour (verbal, tone of voice and non-vocal) is visualised in Figure 1 for each robot. The ratio of perceived influence of the cues is of 26.24% for the non-vocal, 34.91% for the tone of voice and 38.85 % for the words used by the robots. This ratio is the same for the two robots and the two styles (no significant differences). Thus in our experiment, the participants thought they were judging the social signals of the robot more from its verbal utterance and tone of voice cues than its other non-vocal cues (posture, gesture, gazing ...). Our results also differ from the "7 % rule" of Mehrabian [11] stating that only 7% of the signal passes through the verbal modality.

This results highlight the difference between the perceived influence and the actual influence of verbal and non-verbal cues of communication. Expected results in this experiment were to have an influence of 0.0% for the verbal and

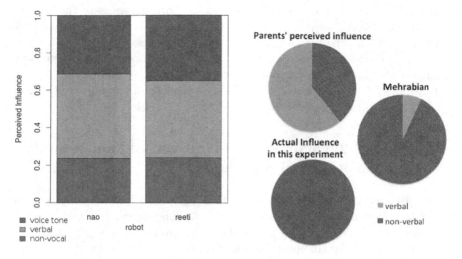

Fig. 1. Proportion of perceived Influence of the modalities

100% for the non-verbal, since the verbal utterance was actually not influencing the authoritativeness of the robots.

This difference might be explained by the fact that participants had a pre-judgement on the robot's ability to use non-verbal cues of communication. Hence, the non-verbal influence of robots is perceived as lower.

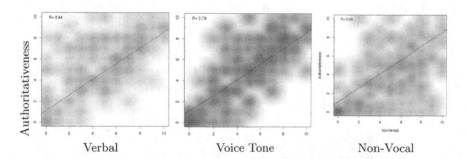

Fig. 2. Correlation between perceived influence of modalities and perceived authoritativeness of the robots

The scatter plots in the fig. 2 show the correlation between the influence scores given to each type of cues and the perceived authoritativeness of the robot by the participants. These graphs show that the authoritativeness is more correlated to the tone of voice perceived influence($R = 0.79$) than the verbal perceived influence($R = 0.64$) and that it is more correlated to the verbal than the non-verbal cues ($R = 0.56$, where $R = -1$ means that the data is inversely correlated; $R = 0$ signifies the absence of a correlation and $R = 1$, a configuration where the data is completely correlated). These results show that tone of voice cues and authoritativeness have a higher interdependence than verbal utterance and authoritativeness, or non-vocal cues and authoritativeness in the parents perception. For the participants, the authoritativeness is then more linked to the tone of voice and the verbal than the non-verbal cues.

Non-verbal (vocal and non-vocal) and verbal behaviours are not independent from each other. In this experiment, the meaning of non-verbal cues (permissive or authoritative) substituted to the understanding of the verbal utterance (neutral) in term of authoritativeness.

4   Experiment 2: Facilitation of Non-verbal Understanding by Congruent Verbal Signals

The *Stroop* effect is a well known effect that allows to highlight interference or facilitation between different modalities in understanding a communicative signal. The classical Stroop task focussed on the interrelation of two cognitive processes by naming the colour of the ink of written words. In the classical test, participants are presented lists of colour names ("blue", "green", "red" ...) written in the ink of the same (congruent set) or a different colour (incongruent set). The control set can be made with words printed in black ink that are not names of colours. Participants can be asked to name the colour of the ink of the words. The experimentalist measures the response time and the number of correct answers. The common result of this test shows that the reading modality can facilitate the identification of ink colour in case of congruence (better performances than control) or it can interfere with the identification in the case of incongruence between ink colour and word names. A large number of variations [12,18] of this experiment have been used in cognitive psychology to understand the interrelation between other modalities (audio and visual for example).

In this experiment, 68 participants were asked to perform a colour identification task of Reeti's cheeks LEDs. Reeti is a PC-bot equipped with motorized head enabling him to display facial expressions. It can do text to speech and has also colours LEDs in its cheeks.

The experiment took place at a stand of the robotic exhibition Innorobo in Lyon(France). Participants were visitors to the stand and not selected (children, adults, men or women). The Figure 3 shows the settings of the stand.

The task for the participant was, using the corresponding colour buttons of a Xbox remote, to identify the colour of the robot's cheeks (both cheeks had identical colours) (Figure 4). The robot started by

Fig. 3. Experimental setting at the Innorobo event in Lyon

explaining the task to the participant. The robot's cheeks changed to a random colour (between the 4 possible colours) and the robot said a word at the

Fig. 4. *Reeti robot displaying the 4 colours used in the experiment and the Xbox controller used by participants*

same time. The word belonged to a category according to the criteria of congruency. The robot changed its colour and says a new word only after the participant had given his answer. We recorded the correctness and the response-time (time difference between the participant's answer and the robot saying the word) of the participants' answers.

Every participant did the task under following 3 conditions in a random order:

congruence : the robot says the colour it displays

control : the robot says a random word which does not represent a colour

incongruence : the robot utters the name of colour different from the one it displays

The Table 2 shows examples of a set of colours the robot would adopt and corresponding words it could say for the 3 conditions.

Table 2. Sequence examples for the three conditions: congruence, control and incongruence

non-verbal colour	verbal		
	congruence	control	incongruence
green	green	dog	blue
blue	blue	boat	red
red	red	chair	green
yellow	yellow	window	blue
red	red	fart	green
yellow	yellow	child	blue
green	green	board	red
blue	blue	bottle	red
red	red	toy	yellow
green	green	robot	yellow

The congruency didn't have a significant effect on the correctness of the identification which is approximately at 20% of incorrect answers. This means that when helped with verbal cues from the robot (congruence), participants didn't achieve better answers and on the other hand when disturbed by verbal cues (incongruence), participants didn't make more mistakes. However, when dealing with such simple task of colour identification, the time of response gives a better measure of signal interrelation than the correctness of answers.

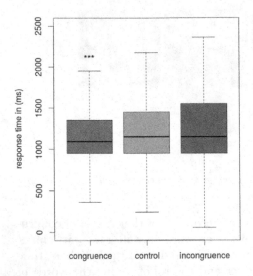

congruence   control   incongruence

Fig. 5. Response Time for the 3 conditions, (∗ ∗ ∗p < 0.001)

Figure 5 shows the average time difference between the robots changing colour and saying the word, and the response of the user by the Xbox remote. Statistical analysis of the response time data showed that the conditions have a significant effect on the response time (ANOVA : $F(2, 1972) = 10.93, p < 0.001$). We computed the facilitation and inference rates :

- **Facilitation** = −134.8ms (average difference between control and congruence conditions)

- **Interference** = +51.21ms (average difference between control and incongruence conditions)

In the classical Stroop task, we can observe both facilitation and interference. Here, the difference between control and incongruence was not significant (T-Test: $p = 0.213$) and there was significant differences between congruence and control conditions (T-Test: $p = 0.001$) and between congruence and incongruence condition (T-Test $p < 0.001$).

In our *robotic Stroop* experiment, even though the interference effect wasn't significant, the verbal cues matching the colour facilitated significantly the response time of the participants with an average of 134 milliseconds.

5   Discussion and Conclusion

The results of Experiment 1 based on *behavioural styles*, showed that non-verbal cues such as tone of voice, posture, gesture and gaze are important and can substitute for the verbal content in the context of behaviour authoritativeness. Indeed, only non-verbal cues were containing the signal of authoritativeness and permissiveness. The verbal content was the same for both permissive and authoritative styles but found to be determinant by participants in terms of influence on the style.

The results of Experiment 2 based on *Stroop*, showed that verbal cues didn't influence the correctness of the participants' answers but did influence the time necessary for identification by significantly facilitating colour recognition when matching the LEDs colours on the robot's cheeks. We didn't see a significant interference effect in the incongruence condition. The interference effect could have been attenuated by the noisy environment of the Innorobo exhibition, allowing participants to pay less attention to the vocal signals more easily in the incongruence cases. Other conditions could have been tested - looking at the robot and identifying what he says rather than what he shows. This could tell us if visually observable colours of the cheeks could also interfere of facilitate audio recognition of colours names.

These results have to be contextualized. Indeed, many other modalities exist in non-verbal communication (colour of the robot itself, height, shape, clothes ...). Some non-verbal cues of communication are also culturally dependent, and our studies have been both tested in Europe only. Also for Experiment 1, participants made their judgements based on videos and as shown in other research in HRI, physical embodiment of the robot and judging on a real interaction could influence these results.

Research in cognitive psychology often uses similar experimental settings in order to test interrelations between different kinds of signal processing in the human brain. When dealing with social robots, it could be interesting to study the understanding of emotional cues from the robot with this kind of setting. Indeed, robotics non-verbal cues can be very different from humans, and their interrelation and influence on verbal content will be mandatory for better social expressiveness of robots.

These two experiments allowed us to explore some interrelations such as substitution and facilitation occurring between verbal and non-verbal communications cues and to highlight the importance of taking into account these interrelations when designing multi-modal expressions in HRI. These studies give an important insight to the design of behaviours for a robot. Indeed, subtle differences in one expression cue can fasten or even change the understanding of the human user. Similar experiment set-up as the Stroop task could be used to determine effects of modalities in terms of credibility or the discomfort sometimes related to robots' expressions; or to determine on which modality of expression the human is paying attention to the most. Doing so can improve robots' communication by using non-verbal facilitators and avoiding interference with verbal signals.

References

1. Aldebaran nao robot (2013). https://community.aldebaran-robotics.com/
2. Robopec reeti robot (2013). http://reeti.fr/index.php/en/
3. Bethel, C.L., Murphy, R.R.: Survey of non-facial/non-verbal affective expressions for appearance-constrained robots. IEEE Transactions on Systems, Man, and Cybernetics, Part C: Applications and Reviews **38**(1), 83–92 (2008)
4. Breazeal, C.: Emotion and sociable humanoid robots. International Journal of Human-Computer Studies **59**(1–2), 119–155 (2003)
5. van den Brule, R., Dotsch, R., Bijlstra, G., Wigboldus, D.H., Haselager, P.: Do robot performance and behavioral style affect human trust? International Journal of Social Robotics **6**(4), 519–531 (2014)
6. Embgen, S., Luber, M., Becker-Asano, C., Ragni, M., Evers, V., Arras, K.O.: Robot-specific social cues in emotional body language. In: 2012 IEEE RO-MAN: The 21st IEEE International Symposium on Robot and Human Interactive Communication, pp. 1019–1025 (2012)

7. Fong, T., Nourbakhsh, I., Dautenhahn, K.: A survey of socially interactive robots. Robotics and autonomous systems **42**(3), 143–166 (2003)

8. Gallaher, P.E.: Individual differences in nonverbal behavior: Dimensions of style. Journal of Personality and Social Psychology **63**(1), 133–145 (1992)

9. Hall, J.A., Coats, E.J., LeBeau, L.S.: Nonverbal behavior and the vertical dimension of social relations: a meta-analysis. Psychological bulletin **131**(6), 898–924 (2005)

10. Johal, W., Pesty, S., Calvary, G.: Towards companion robots behaving with style. In: 2014 RO-MAN: The 23rd IEEE International Symposium on Robot and Human Interactive Communication, pp. 1063–1068. IEEE (2014)

11. Knapp, M., Hall, J., Horgan, T.: Nonverbal communication in human interaction. Cengage Learning (2013)

12. MacLeod, C.M.: The stroop task: The "gold standard" of attentional measures. Journal of Experimental Psychology: General **121**(1), 12 (1992)

13. Mehlmann, G., Häring, M., Janowski, K., Baur, T., Gebhard, P., André, E.: Exploring a model of gaze for grounding in multimodal HRI. In: ICMI 2014, pp. 247–254. ACM, New York (2014)

14. Pelachaud, C.: Studies on gesture expressivity for a virtual agent. Speech Communication **51**(7), 630–639 (2009)

15. Vernon, D., Hofsten, C., Fadiga, L.: A roadmap for cognitive development in humanoid robots. In: Cognitive Systems Monographs, vol. 11. Springer, Heidelberg (2011)

16. Wallbott, H.: Bodily expression of emotion. European Journal of Social Psychology **896**(November 1997) (1998)

17. Walters, M.L., Syrdal, D.S., Dautenhahn, K., Te Boekhorst, R., Koay, K.L.: Avoiding the uncanny valley: robot appearance, personality and consistency of behavior in an attention-seeking home scenario for a robot companion. Autonomous Robots **24**(2), 159–178 (2008)

18. Williams, J.M.G., Mathews, A., MacLeod, C.: The emotional stroop task and psychopathology. Psychological bulletin **120**(1), 3 (1996)

19. Xu, J., Broekens, J.: Mood expression through parameterized functional behavior of robots. RO-MAN (2013)

20. Xu, J., Broekens, J., Hindriks, K., Neerincx, M.A.: Bodily mood expression: recognize moods from functional behaviors of humanoid robots. In: Pearson, M.J., Leonards, U., Herrmann, G., Lenz, A., Bremner, P., Spiers, A. (eds.) ICSR 2013. LNCS, vol. 8239, pp. 511–520. Springer, Heidelberg (2013)

21. Yaffe, P.: The 7% rule: Fact, fiction, or misunderstanding. ACM Ubiquity **2011**(October), 1:1–1:5 (2011)

Empathic Robotic Tutors for Personalised Learning: A Multidisciplinary Approach

Aidan Jones[1](✉), Dennis Küster[2], Christina Anne Basedow[2],
Patrícia Alves-Oliveira[3], Sofia Serholt[4], Helen Hastie[5], Lee J. Corrigan[1],
Wolmet Barendregt[4], Arvid Kappas[2], Ana Paiva[3], and Ginevra Castellano[1,6]

[1] University of Birmingham, Birmingham, UK
axj100@bham.ac.uk
[2] Jacobs University Bremen, Bremen, Germany
[3] INESC-ID and Instituto Superior Técnico, Universidade de Lisboa,
Lisbon, Portugal
[4] University of Gothenburg, Gothenburg, Sweden
[5] Heriot-Watt University, Edinburgh, UK
[6] Department of Information Technology, Uppsala University, Uppsala, Sweden

Abstract. Within any learning process, the formation of a socio-emotional relationship between learner and teacher is paramount to facilitating a good learning experience. The ability to form this relationship may come naturally to an attentive teacher; but how do we endow an unemotional robot with this ability? In this paper, we extend upon insights from the literature to include tools from user-centered design (UCD) and analyses of human-human interaction (HHI) as the basis of a multidisciplinary approach in the development of an empathic robotic tutor. We discuss the lessons learned in respect to design principles with the aim of personalised learning with empathic robotic tutors.

Keywords: Personalisation · Robotic tutor · Human-robot interaction

1 Introduction

Robots that are intended to interact with humans must learn how to become empathic rather than merely smart. Our aim is to design an empathic robotic tutor for personalised learning. Since social connection between tutors and learners has been shown to influence learning positively, a way of making tutoring systems more effective is to include a robot that will no longer just have to be intelligent, useable, and interactive, but will establish and maintain a certain level of social connection [18] and respond empathically to humans. Once such robots can convey the impression that they can understand and respond to the user not just intellectually but also emotionally, their use in domains in which they currently play a minor role, such as teaching, will become far more plausible and successful. Our approach is founded in fundamental psychological theory regarding the human ability to monitor others' emotions, engage in shared attention, and to establish effective socio-emotional bonds with one another. These connections have been shown to motivate learning processes [28].

© Springer International Publishing Switzerland 2015
A. Tapus et al. (Eds.): ICSR 2015, LNAI 9388, pp. 285–295, 2015.
DOI: 10.1007/978-3-319-25554-5_29

The goal of the design process presented here is to perform an iterative UCD process to enable a robot to mimic a sufficient number of key social and empathic abilities of a human tutor to establish a socio-emotional bond with the learner. UCD activities can help ensure that the empathic competencies of our robotic tutor are implemented in an appropriate and believable manner, have real world application in the complex school environment, and meet the needs of both learners and teachers.

There is precedent for HHI studies informing the design of human-robot interaction (HRI) [27] and UCD with robotics [9].

2   Background

Findings from the psychological literature suggest that children's learning processes can be facilitated by the tutor creating a strong empathic bond with the child, and having socio-emotional bonds can further be linked to how learners are able to handle their emotions in a guided learning process [20]. In addition, widely cited second order meta-analyses [12] have shown that key competencies of teachers, such as their ability to establish and maintain trusting teacher-student relationships, are far more important than the number of children in the class. The benefits of these relationships clearly suggest that a robot with the ability to form a socio-emotional bond with a child and to respond to the emotional states of a child empathically should be facilitative for the learning process.

In the field of HRI we see that personalised feedback from a robotic tutor can lead to more successful human-robot interactions with reduced problem solving time and a more motivated learner [19]. There is an increasing amount of work that shows that it is possible for a user to build a social bond with a robot. A review of studies that investigated long term interaction with robots [18], recommends that a robot should be able to display an awareness of and respond to the user's affective state and also adapt to the individual's preferences in order to build a good social interaction. Gaze through eye contact and joint attention is very important for social interaction with a robot and can improve performance in a cooperative human-robot task [4]. The social presence between a user and robot can also be mediated with a touchtable [3], suggesting that the capabilities of the robot may be usefully augmented in this manner. There are however dangers as some types of social behaviour from a robotic tutor may distract from the learning task and reduce the impact of the tutor in the learning process [16].

A review of robots in education [21] raises a number of challenges and open questions that we hope to address in our research, namely understanding the role of the robotic tutor, how to adapt behaviour and curricular to the learner, and how to design a socially acceptable robot. As outlined in the introduction, we argue that these basic insights from the literature need to be critically examined, expanded, and adapted via tools derived from UCD. UCD should be included to better account for the specific context, the experience and overarching teaching aims of teachers, and the personalised needs of the learner. Only then will robotic tutors become more successful in forming substantial social bonds with the aim of assisting learners within relatively open and complex learning scenarios.

3   Design Goals

To develop a robot that is accepted into the school context and that demonstrates an effective human–robot interaction, we believe that the following design goals should be met:

(1) Involve our users (teachers and learners) in the design of the robot. When technology is introduced into the classroom, it becomes part of a complex system of social and pedagogical interactions, involving both teachers and learners [23]. Therefore, it is pertinent to investigate the perspective of the potential users as well as the social and contextual structures inherent in the environment [17].

(2) Identify core empathic and personalised pedagogical strategies from human interactions. Successful personalised tutoring has to attempt to identify those empathic and pedagogical components and strategies that are most effective in establishing, strengthening, and sustaining social bonds. We see that HRI studies that are based on human interactions can be quite successful, for example, adopting human gaze behaviour to increase engagement with the robot [27].

(3) Supplement HHI-based behaviors with additional capabilities. On top of the core components identified with the help of observation and UCD tools, the robot can utilise additional behaviors that might be successful when displayed by a teacher but could tap into the same underlying mechanisms, e.g., the robot could produce robot-appropriate sounds that mimic a teacher's backchanneling efforts. HRI interactions are not routinely based on HHI due to the differences in how interactants perceive robots and humans [8]. It remains important to test interaction using techniques such as Wizard of Oz (WoZ) studies where a robot is controlled by a human wizard [7] to investigate how learners interact with a robot before developing final automated behaviours. We should also prototype and test these capabilities in the robot iteratively and in situ [24].

(4) Tie the robot capabilities on well supported psychological and pedagogical theories. Psychology is a field with many coexisting theories and the development of personalised learning strategies should specifically target those concepts that have been shown to be empirically well supported.

(5) Enable the robot to adapt to individual differences. Personalised learning approaches should, in particular, aim to identify cues that teachers use to personalise their teaching styles.

4   Scenarios and System Overview

The robotic tutor will support 10–15 year olds in the domain of Geography. We have developed two different learning scenarios: an individual map-reading activity and a multi-player collaborative game in which the tutor will provide support for two learners.

We use the torso only version of the NAO robot produced by Aldebaran Robotics[1]. The interaction between the tutor and the learner(s) is mediated by a large touchscreen/table. The robot sits opposite the learner(s) on the long edge of the table. Beneath the robot, built in to the table, we have a number of sensors to enable the robot to perceive facial expressions, gaze direction, head position, body posture, electrodermal activity and the volume of speech.

Fig. 1. NAO Robot, Learners, and Learning Scenarios

5 Design Process

The goal of endowing robots with empathic capabilities requires a multidisciplinary effort, in addition to the teachers and learners. Learning, teaching and HHI experts give valuable practical advice for our user centric approach. Psychologists have developed the necessary theoretical framework concerning empathy and emotion, and provide guidelines and feedback on how this framework can be translated into practice.

5.1 Interviews

The primary aims of the interviews, which comprised a series conducted over several months, were to elicit a greater understanding of the context of use and the teachers' [26] and learners' [25] needs. The questions focused on the plausibility of having a robotic tutor in their classes.

We found that it is difficult for teachers to imagine having an autonomous robot in the classroom. More specifically, there are a number of practical and social factors that play an important role when teachers think about robots in school. They have e.g., concerns about managing the availability of a robot to all children and how this will affect their already busy schedule. However, this type of initial resistance towards and mistrust about the intended use of a robot in school was shown to be greatly reduced by involving teachers in the project.

---

[1] https://www.aldebaran.com/en/humanoid-robot/nao-robot

To aid teachers and learners in imagining a robotic tutor, we have developed a video that gives teachers more information about the capabilities of robots and how their use is envisioned in the project. Together with the interviews, these very first steps in our iterative approach have helped us discuss with teachers in a more concrete way what it could mean to have a robotic tutor in the classroom.

5.2   Participatory Design Workshops

Teachers were provided with a preliminary version of the learning activity and were asked to contribute to the design. The design workshops allow us to understand teachers' ways of approaching the learning task, including their assessment of the difficulty levels of different sub tasks, and how they would personalise their teaching strategy to learners of differing ability.

The teachers' contributions included ideas for technical content and pedagogical strategies. The main finding of these workshops was that there appeared to be a general trend in how directions and instructions are personalised by teachers, in which less capable students are provided with simple and clearly formulated pieces of information, while more capable students are provided with one or several complex pieces of information at a time. However, it was also observed that it is difficult for teachers to provide a description of how they would personalise or adjust to different students' needs from a fictive perspective. Although, generally, this second stage of UCD inspired iterative design approach may not yet provide the finer details, we have found it to be informative in respect to some of the broad strategies to be employed by the robot.

5.3   Mockup Studies with Teachers and Students

We conducted mockup studies involving human teachers and children with varying abilities. The aims of the mockup studies were to understand how and when teachers provide personalised feedback to the students. Earlier studies were paper based, later studies were performed with the touchtable based activity [2].

The findings enabled us to better understand the dynamics between the teacher, task, and learner, as well as examine some of the more detailed pedagogical tactics and strategies employed by the teachers.

5.4   User-Centered Design and Pedagogical Theories

The aforementioned preliminary studies for both learning scenarios have given us insights into the practical requirements needed to create a believable interaction with a robotic tutor in our context. In combining the experience derived from the user-centered pretests with a comprehensive literature review of tactics used by teachers during learning activities, we generated a list of pedagogical tactics that the robotic tutor can use to interact with and motivate students in their learning process.

Pedagogical tactics can be divided in three clusters that serve different learning purposes: the first purpose is to prompt reflection or elicit information from the learner; second, to supply content to the learner; and third, tactics to form a social bond with the learner.

On the basis of recordings made in the earlier studies, we collected over 900 concrete examples of utterances that could be implemented in the robotic tutor. These are spread over 25 pedagogical tactics meaning that the robotic tutor can use the same pedagogical tactic in many different ways, giving it a dynamic and non-repetitive verbal behaviour.

6   Initial Implementation

The specific utterances from the pedagogical tactics from the previous section have been implemented as behaviours for the robot to perform. Each behaviour contains speech and nonverbal behaviour based on observations of teachers.

The robot is able to gaze and gesture at the table and use gestures to explain as it is speaking. Our architecture allows the decision of which social or pedagogical tactic to be taken at a high level as each concrete behaviour already details everything that the robot needs to do in terms of gesturing and gazing at the learner or table [22].

In addition to pedagogical tactics the robot has a set of autonomous behaviors that are continuously running without input from the human wizard. The ability for the robot to perform actions contingent and adapted to the learner is key from the psychological theories and other HRI literature. To achieve this at a low level, the robot makes use of the sensors to track the learner's volume, location and gaze direction. For example, gaze behaviour is based on a mockup study [1], the robot can gaze at the learner when addressing them, follow where the learner is looking or interacting with the activity, and gaze back at the learner when the learner is looking at the robot. These low level contingent behaviours are overridden when the robot is required to perform a behaviour specifically selected by the decision-making component.

7   Wizard of Oz Studies

On the basis of the initial implementation, a number of WoZ studies have been conducted with children at their school. The aim of these studies has been to test the expressive capabilities of the robot in an interaction with the learner and see if it is possible to support the learner in the way that we envisage. Additionally we captured corpora of data to inform the design of the sensing and decision making capabilities of the robotic tutor. The wizards are researchers with teaching experience and had been involved in the development of the WoZ interface with full training in its use.

In a WoZ study the wizard decided which pedagogical tactic to apply via a WoZ Interface. The WoZ interface allows the wizard to observe the learner's activities related to the educational application; view the learner via a webcam and other sensors; and control the robot. The WoZ interface and system architecture allows the wizard to concentrate on high level interaction and not on

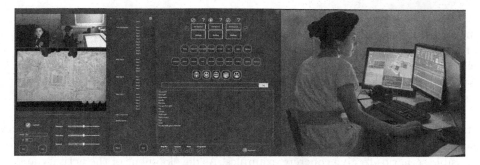

Fig. 2. WoZ Interface

low level coordination of the robot. The robot used concrete behaviours based on the mockup studies but automatically adapted to the situation based on the position of the learner and the state of the task.

In terms of sensing affect, some of our findings were at odds with our initial thoughts on the emotions that we would observe in the learners. We found that expressions of basic emotion were much less frequent than suggested by theory related to learning scenarios. This has led us to adopt a valence and arousal model for the automated system. Our data so far suggests that the smiles exhibited by the children in the WoZ should more cautiously be interpreted as signs of politeness and, perhaps, a readiness to engage in interaction with the robot, rather than any clear indication of enjoyment.

8   Development of Fully Autonomous Behavior

For the robot to work in an autonomous way, we have developed perception capabilities and pedagogical strategies so that the robotic tutor can sense and adapt to the affective states, skills, and difficulties of the learner. The pedagogical strategies comprise of a set of rules that determine which of the pedagogical tactics should be used by the robotic tutor. Some of these rules have been based upon findings from user studies and literature. Other rules were generated using machine learning analysis of the logs of the interactions recorded in the WoZ.

We have also performed supporting studies to develop behaviors not based on human interactions, for example: to create and validate a set of short synthetic sounds that the robot could use as emotional qualifiers to synthetic speech [15]; to investigate how affective feedback is perceived by learners [10]; and to evaluate and train the perceptive abilities of the tutor focusing on the electrodermal activity sensor [14] and the ability for the system to perceive engagement [6].

We are now in the phase of the project where we are running studies with a fully automated robotic tutor.

9   Lessons Learned

A number of lessons can be drawn from our user-centered iterative approach for the purpose of designing robotic tutors:

The involvement of users and UCD-tools may be most effective when implemented throughout the full range of the design process. Including the learners at a very early stage was essential to providing a system that the teachers would welcome in to their classroom and fit into their curriculum. Feedback provided here led to the development of learning aims, the role of the robot and the development of a multiplayer scenario.

Including users at an early stage can be difficult as initially teachers and students were unfamiliar with the capabilities of robots and could not imagine how the robot could fit in the classroom. We used a video and descriptions of possible scenarios to give concrete examples. As development of the activity and robotic tutor progressed we used prototypes as concrete demonstrations.

It can be difficult for teachers to explain in detail how they would adjust to different students' needs on the basis of an abstract description of the task. What is needed in this case is to actually run mockup studies to put the teachers into a situation where the students have different needs and carry out an assessment on that basis.

There is a need to fully take into account personality differences between children. Some children, through high expressivity, make it much easier for the robot to pick up on critical information regarding their emotional state. The challenge is to accommodate these differences and be aware that not all students will indicate their emotional state in the same way. Additionally the learners reactions change over time, so we are required to give each child sufficient time and opportunity to adjust to the presence and unfamiliar nature of the robot. We have tried to address this by making the students familiar with the robot and performing all of our studies in the participant's schools.

As we found that expressions of basic emotion were infrequent we have adopted a valence and arousal model for affect detection.

We have adopted many of the teaching strategies observed in mockup studies. We try to give the student an opportunity to self-regulate their learning process [5]. We match feedback to the abilities of the student, breaking down the task and focusing the student's attention on the areas where they have difficulty. We further personalise the interaction by referring to the learner's previous knowledge and skills.

We suggest too much interaction from the robot could hinder the interaction; frequent or unconvincing praise can adversely affect children's intrinsic motivation [13]. To avoid adverse effects of inhibiting negative emotion [11] and to facilitate a good social bond with the child, our design has aimed for mimicry of context-appropriate affective behavior as a more likely key to a successful learning experience. Care with affective behaviour must also be taken as we have found that it is possible to make the activity seem more difficult with affective feedback [10].

By observing the interactions between teacher and learner we are able to gain a deeper insight into the key tactics used by teachers. We were able to combine the observations with a review of literature to create a set of pedagogical tactics and a set of rules for when it would be appropriate for the robotic tutor to use them. Literature alone would not provide enough detail to formulate such an appropriate, dynamic and non-repetitive behaviour.

10   Conclusion

In this paper, we have presented the design process for the enhancement of off-the-shelf robots aimed at creating a new generation of artificial embodied tutors that are able to engage in empathic interactions with learners. To this aim, we have provided a structure that illustrates how the end users can be actively engaged in the design of an appropriate learning context in which the robot may be enabled to develop the full extent of personalised empathic capabilities.

On the theoretical foundation that has demonstrated the importance of social bonding in education, we show how capabilities in a robot to support social bonding might be developed based on interviews, HHI, and HRI studies in a non trivial, real-world domain. We have developed initial versions of all of the components of the robotic tutor system, including perceptive capabilities, pedagogical strategies, and psychological bonding mechanisms. We are now in the process of evaluating the robotic tutor.

Acknowledgments. This work was supported by the European Commission (EC) and was funded by the EU FP7 ICT–317923 project EMOTE (EMbOdied-perceptive Tutors for Empathy-based learning). The authors are solely responsible for the content of this publication. It does not represent the opinion of the EC, and the EC is not responsible for any use that might be made of data appearing therein.

References

1. Alves-Oliveira, P., Tullio, E.D., Ribeiro, T., Paiva, A.: Meet me halfway: eye behaviour as an expression of robot's language. In: AAAI Fall Symposium Series, pp. 13–15 (2014)
2. Alves-Oliveira, P., Janarthanam, S., Candeias, A., Deshmukh, A., Ribeiro, T., Hastie, H., Paiva, A., Aylett, R.: Towards Dialogue dimensions for a robotic tutor in collaborative learning scenarios. In: Proceedings of RO-MAN (2014)
3. Baxter, P., Wood, R., Belpaeme, T.: A touchscreen-based 'sandtray' to facilitate, mediate and contextualise human-robot social interaction. In: International Conference on Human-Robot Interaction, pp. 105–106. ACM (2012)
4. Boucher, J.D., Pattacini, U., Lelong, A., Bailly, G., Elisei, F., Fagel, S., Dominey, P.F., Ventre-Dominey, J.: I Reach Faster When I See You Look: Gaze Effects in Human-Human and Human-Robot Face-to-Face Cooperation. Frontiers in Neurorobotics **6**(3), 1–11 (2012)
5. Butler, D.L., Winne, P.H.: Feedback and self-regulated learning: A theoretical synthesis. Review of Educational Research **65**(3), 245–281 (1995)
6. Corrigan, L.J., Basedow, C., Küster, D., Kappas, A., Peters, C., Castellano, G.: Mixing implicit and explicit probes: finding a ground truth for engagement in social human-robot interactions. In: International Conference on Human-Robot Interaction, pp. 140–141. ACM (2014)

7. Dahlbäck, N., Jönsson, A., Ahrenberg, L.: Wizard of Oz studies - why and how. In: International Conference on Intelligent User Interfaces, pp. 193–200 (1993)
8. Dautenhahn, K.: Methodology & Themes of Human-Robot Interaction : A Growing Research Field. International Journal of Advanced Robotic Systems 4(1), 103–108 (2007)
9. Förster, F., Weiss, A., Tscheligi, M.: Anthropomorphic design for an interactive urban robot - the right design approach? In: International Conference on Human-Robot Interaction, pp. 137–138 (2011)
10. Foster, M.E., Lim, M.Y., Deshmukh, A., Janarthanam, S., Hastie, H., Aylett, R.: Affective feedback for a virtual robot in a real-world treasure hunt. In: Workshop on Multimodal, Multi-Party, Real-World HRI, pp. 31–32 (2014)
11. Gross, J.J., Levenson, R.W.: Hiding feelings: the acute effects of inhibiting negative and positive emotion. Journal of Abnormal Psychology 106(1), 95–103 (1997)
12. Hattie, J.: Visible learning: A synthesis of over 800 meta-analyses relating to achievement. Routledge (2009)
13. Henderlong, J., Lepper, M.R.: The effects of praise on children's intrinsic motivation: A review and synthesis. Psychological Bulletin 128(5), 774–795 (2002)
14. Kappas, A., Küster, D., Basedow, C., Dente, P.: A validation study of the Affective Q-Sensor in different social laboratory situations. In: 53rd Annual Meeting of the Society for Psychophysiological Research, Florence, Italy (2013)
15. Kappas, A., Küster, D., Dente, P., Basedow, C.: Simply the B.E.S.T.! Creation and validation of the Bremen emotional sounds toolkit. In: International Convention of Psychological Science (2015)
16. Kennedy, J., Baxter, P., Belpaeme, T.: The robot who tried too hard : social behaviour of a robot tutor can negatively affect child learning. In: International Conference on Human-Robot Interaction, pp. 67–74. ACM (2015)
17. Koedinger, K.R., Aleven, V., Roll, I., Baker, R.: In vivo experiments on whether supporting metacognition in intelligent tutoring systems yields robust learning. In: Hacker, D.J., Dunlosky, J., Graesser, A.C. (eds.) Handbook of Metacognition in Education, pp. 897–964. Routledge (2009)
18. Leite, I., Martinho, C., Paiva, A.: Social Robots for Long-Term Interaction: A Survey. International Journal of Social Robotics 5(2), 291–308 (2013)
19. Leyzberg, D., Spaulding, S., Scassellati, B.: Personalizing robot tutors to individuals' learning differences. In: International Conference on Human-Robot Interaction, pp. 423–430. ACM, New York (2014)
20. Moridis, C.N., Economides, A.A.: Affective Learning: Empathetic Agents with Emotional Facial and Tone of Voice Expressions. IEEE Transactions on Affective Computing 3(3), 260–272 (2012)
21. Mubin, O., Stevens, C.J., Shahid, S., Mahmud, A.A., Dong, J.J.: A Review of the Applicability of Robots in Education. Technology for Education and Learning 1(1), 1–7 (2013)
22. Ribeiro, T., di Tullio, E., Corrigan, L.J., Jones, A., Papadopoulos, F., Aylett, R., Castellano, G., Paiva, A.: Developing interactive embodied characters using the thalamus framework: a collaborative approach. In: Bickmore, T., Marsella, S., Sidner, C. (eds.) IVA 2014. LNCS, vol. 8637, pp. 364–373. Springer, Heidelberg (2014)
23. Russell, D.L., Schneiderheinze, A.: Understanding Innovation in Education Using Activity Theory. Educational Technology & Society 8(1), 38–53 (2005)
24. Sabanovic, S., Reeder, S., Kechavarzi, B.: Designing Robots in the Wild: In situ Prototype Evaluation for a Break Management Robot. Journal of Human-Robot Interaction 3(1), 70–88 (2014)

25. Serholt, S., Barendregt, W.: Students' attitudes towards the possible future of social robots in education. In: RO-MAN (2014)

26. Serholt, S., Barendregt, W., Leite, I., Hastie, H., Jones, A., Paiva, A., Castellano, G.: Teachers' views on the use of empathic robotic tutors in the classroom. In: RO-MAN (2014)

27. Sidner, C.L., Lee, C., Kidd, C., Lesh, N., Rich, C.: Explorations in engagement for humans and robots. Artificial Intelligence **166**(1), 140–164 (2005)

28. Sroufe, L.A.: The coherence of individual development: Early care, attachment, and subsequent developmental issues. American Psychologist **34**(10), 834 (1979)

Robot-Assisted Training of Joint Attention Skills in Children Diagnosed with Autism

Jasmin Kajopoulos[1], Alvin Hong Yee Wong[2], Anthony Wong Chen Yuen[2],
Tran Anh Dung[2], Tan Yeow Kee[2], and Agnieszka Wykowska[3,4(✉)]

[1] Neuro-Cognitive Psychology Master Program, Department of Psychology,
Ludwig-Maximilians-Universität, Munich, Germany
jasmin.kajopoulos@campus.lmu.de
[2] Institute for Infocomm Research (I2R), Agency for Science,
Technology and Research (A*STAR), Singapore, Singapore
{hyawong,cywong,tanhdung,yktan}@i2r.a-star.edu.sg
[3] General and Experimental Psychology Unit, Department of Psychology,
Ludwig-Maximilians-Universität, München, Germany
agnieszka.wykowska@psy.lmu.de
[4] Chair for Cognitive Systems, Technische Universität München, Munich, Germany

Abstract. Due to technological and scientific advances, a new approach to autism therapy has emerged, namely robot-assisted therapy. However, as of now, no systematic studies have examined the specific cognitive mechanisms that are affected by robot-assisted training. This study used knowledge and methodology of experimental psychology to design a training protocol involving a pet robot CuDDler (A*STAR Singapore), which targeted at the specific cognitive mechanism of responding to joint attention (RJA). The training protocol used a modified attention cueing paradigm, where head direction of the robot cued children's spatial attention to a stimulus presented on one of the sides of the robot. The children were engaged in a game that could be completed only through following the head direction of the robot. Over several weeks of training, children learned to follow the head movement of the robot and thus trained their RJA skills. Results showed improvement in RJA skills post training, relative to a pre-training test. Importantly, the RJA skills were transferred from interaction with the robot to interaction with the human experimenter. This shows that with the use of objective measures and protocols grounded in methods of experimental psychology, it is possible to design efficient training of specific social cognitive mechanisms, which are the basis for more complex social skills.

Keywords: Autism · Robot-assisted therapy · Joint attention · Social robotics

1 Introduction

1.1 Autism

As medical and neurobiological knowledge advance, early detection tools and intervention methods for Autism Spectrum Disorder (ASD) have been growing [1,2]. Deficits in joint attention skills have been found to be one of the earliest signs for

© Springer International Publishing Switzerland 2015
A. Tapus et al. (Eds.): ICSR 2015, LNAI 9388, pp. 296–305, 2015.
DOI: 10.1007/978-3-319-25554-5_30

ASD, wherein impairments in joint attention (i.e., attending to where others attend with the awareness that the others attend to the same location/object) were associated with stunted language development and decreased social and communicative skills [3,4,5]. Thus, research into joint attention, especially the development of intervention methods, as well as understanding the mechanisms behind joint attention has become a pivotal area of interest in autism research [6].

1.2   Robot Therapy

In recent years with rising technological advances, especially in the field of robotics, a new area of intervention method for children with ASD has been emerging, namely social robots as a tool for autism therapy [7-13]. The therapeutical value of social robots is grounded in reliability, simplicity, and predictability of a robot's behavior [7,12]. However, although in recent years various social robots for autism therapy have been constructed [7,11,12], to date there is still a limited number of training protocols which would use social robots and target at specific neuro-cognitive mechanisms. For example, Warren and colleagues [14] showed that after 4 sessions with a robot, children seemed to improve in following joint attention cues given by the robot. However, it was not demonstrated to what extent this effect prevailed in other situations, i.e. whether also general joint attention skills of the children (independent from the robot) improved. Similarly, it was shown that other prosocial behaviour, such as interaction through gaze and touch, was elicited and improved through the use of robots [15]. But again testing of the effect seemed to have been limited to the robot. In general, autism robot therapy research could benefit from independent diagnostic measurement [16]. On the other hand, Wainer and colleagues [17] demonstrated through interviews and questionnaires that improvements in interaction skills elicited by the deployment of an after-school robotics class seemed to persist even after the class. Thus, there is some evidence that social interaction skills, such as joint attention, may be improved through robotic therapy. However, the field is in need of systematic evaluations (with reliable objective methods of experimental psychology) of the effectiveness of robots' therapeutic use, independent of a specific robot platform.

1.3   Aim of Study

The present study aimed at designing a training protocol of robot-assisted therapy for autism that would extract and target at specific mechanisms of social cognition, and would also allow for experimental assessment of the efficiency of such protocol. To that end, we used a social robot CuDDler (designed by the Agency of Science and Technology (A*STAR) Singapore) and we embedded a variation of a spatial-attention cueing paradigm in a game that children played with the robot. Spatial attention of children was cued by the robot's head direction. Based on previous research in experimental psychology [3,18-20] we aimed at extracting and training the mechanism of responding to joint attention (RJA), which is the fundament for other social cognitive skills. Further, we aimed to test whether the trained joint attention skills prevail independently of the robot's presence by using the abridged Early Social Communication Scale (ESCS, [21]), which has been shown to be effective in measuring children's behaviour in naturalistic environments [22,23].

2   Materials and Methods

2.1   Participants

7 children (Mean age: 4.6, age range: 4 to 5 years, 4 male) all diagnosed with ASD (1 low, 4 low/moderate, 1 moderate, 1 moderate/high functioning) took part in the experiment. All children were right-handed and all had normal vision. Parents were recruited via the early intervention center THK EIPIC Centre (Singapore), and they volunteered their children for participation. All parents gave written informed consent regarding the participation of their children in the experiment. The experiments were conducted at the THK EIPIC Centre (Singapore) in collaboration with A*STAR Singapore and the Ludwig-Maximilians-Universität Munich. The experimental procedures consisted of purely behavioural data collection (e.g., accuracy rates), and were video recorded. The procedures did not include invasive or potentially dangerous methods and were in accordance with the Code of Ethics of the World Medical Association (Declaration of Helsinki). Data were stored and analyzed anonymously.

2.2   Stimuli and Apparatus

The key apparatus in the experiment was a robot CuDDler (A*STAR), controlled via a computer interface (Windows 7) and a smartphone (Google Nexus 4). The sequence of movements and speech were started by pressing different buttons on the interface. Additionally, picture stimuli were presented on two 136.6 x 70.6 mm mobile phone screens (bephone, 640 x 480), located to the left and right of the robot at a distance of ~ 40 cm (11° of participants' visual angle), tilted ~ 45° relative to the robot, so that the robot seemed to "see" stimuli on the screens after turning its head (cf. Fig. 1, left).

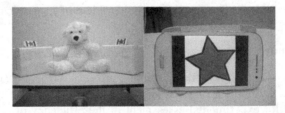

Fig. 1. Experimental Setup (left) and an example of a stimulus (right).

The target stimuli were 10 colorful line drawings of various objects (star, apple, ball, candle, flower, hat, heart, ice cream, plane, sweet, Fig.1, right) in 4 colors: red, blue, green or yellow. In one trial, there were always two drawings of the same object presented – one on each phone screen in different (randomly chosen) colors. In one set of 20 trials all objects in all colors appeared once. The stimuli were fit to the center of the phone screens (136.6 x 70.6 mm) and covered approximately 2° in height and approximately 3.5° in width of visual angle of participants. The robot's head moved pseudo-randomly to the left or right approximately 2.3° in participants' visual angle from the midline with a 50% chance occurrence of either side. The participants were seated 200 cm from the robot, behind a table with a mouse for giving the response.

2.3   Procedure

The protocol's task difficulty, in particular giving color responses as well as abilities to press buttons were first discussed with the children's teacher, who confirmed that in principle all children had learned to discriminate colors and to express this verbally. The whole training protocol consisted of 3 phases. First, a so-called "pre-test" was administered by a human experimenter to measure children's joint attention skills – both responding to joint attention (RJA) and initiating joint attention (IJA) – using the abridged Early Social Communication Scale (ESCS, [21]) before the robot training was conducted. Then, the robot training, consisting of 6 sessions of approx. 20 minutes each, was performed over a period of approx. 3 weeks. 2-3 days after the last robot session, a "post-test" was administered (ESCS) by a human experimenter to once again measure children's joint attention skills. The pre- and post-test lasted approx. 10 minutes each, during which the human experimenter and the child were sitting in front of each other across a table. As we were interested in measuring joint attention skills, only 3 parts of the ESCS were used, the Object Spectacle Task (1 x), the Gaze Following Task (2 x) and the Book Presentation Task (2 x). For the object spectacle task children were handed (6 times) a different toy to play with, and it was tested whether the child would initiate a joint attention bid by gaze or gestures (IJA). During the book presentation task, an examiner pointed to 6 different locations in a book to test whether the child would follow this joint attention bid with gaze or pointing (lower RJA, as the object was in closer spatial proximity) or whether the child would initiate a joint attention bid by gaze or pointing (IJA) to different parts of the book. In the gaze following task, the examiner pointed to 1 of 4 posters located behind, to the left and right of the child (90° off midline) to test whether the child would follow this joint attention bid (higher RJA, as the object was further away).

Robot Training. Before the experiment, children were taught by a teacher or an experimenter to associate the mouse buttons with the respective phone screens (left-to-left/right-to-right mapping). The children were then instructed to follow CuDDler's head movement, and their task was to verbally (out-loud) name the color of the target, at which the robot was looking, and additionally press the corresponding button. In practice trials, an experimenter or teacher also pointed to the correct screens to help understanding of instructions. This lasted approx. 10 – 20 trials depending on children's proficiency, understanding and attention span. 1 session of the experiment (subsequent to practice) typically consisted of 20 trials, although this might have varied depending on children's proficiency, understanding and attention span.

Trial Sequence. A trial (Fig. 2) started with an experimenter pressing a button on the interface, whereupon the robot's head started to move from the starting position (looking straight ahead) either to the left or to the right screen (~2 s) while saying: "Look! A 'object name' ". Then the stimuli appeared on both screens and the robot said "What color is this?" Afterwards, the children had to give a verbal color response and press the correct mouse button (left or right), Fig. 2.

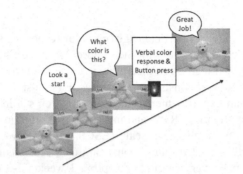

Fig. 2. Trial sequence: The sequence of events for one example trial.

When the children did not answer, the experimenter made the robot call the children's name, and repeat "What color is this?" and when the children still did not answer, the experimenter made the robot say "What color is the object I'm looking at?". As soon as the children pressed the button, the picture stimuli disappeared from both screens. Thus, the trial continued only when the children pressed a button. When the children only gave a verbal response, the robot was made to say "Please press the button". When the children still didn't respond, they were prompted by the instructor to press the button. After the response, the experimenter made the robot praise the child by saying "Good job!" and simultaneously making the robot move its head and arms "excitedly" in the direction of the child. Then the robot's head moved back to the starting position. In case a child gave only a button response, the robot said "Please tell me the color next time". In cases when a child did not respond, the teacher or instructor prompted the children to give a verbal response.

2.4   Data Analysis

Data of one child (low-functioning ASD) were excluded from all analyses, because they did not perform the required task. Also, data of two children were excluded from the analysis of RJA scores due to behavioral problems during the test, which led to a large difference in pretest and posttest in the RJA tasks, i.e. the number of opportunities for the examiner to start a joint attention bid in the RJA tasks drastically differed between pre- and posttest (average difference for the 2 excluded children = 6, average difference for the 4 children that were included in the analysis = 0.5, SD = 0.5).

Joint Attention Scores. To test whether children had higher joint attention scores after the robot training in comparison to before the robot training, the number of instances of RJA or IJA were calculated based on video recordings and according to the

ESCS coding sheet [21]. Initiating joint attention (IJA) scores were calculated based on the number of instances a child showed an IJA bid. The sum of every joint attention bid by the child amounted to the "total IJA" score. Therefore, higher score means more instances of IJA. Responding to joint attention scores (RJA) were based on percentage of times a child followed an examiner's joint attention bid (pointing to a poster or to pictures in a book). The score was calculated separately for higher level (pointing to posters) and lower level RJA bids (pointing to a book). For lower level RJA (lRJA), a point was given when a child followed the examiner's pointing gesture with their gaze. The sum of lRJA points was divided by the number of times the examiner visibly pointed to a picture in the book multiplied by 100 and thus the percentage score with a maximum of 100 (%) was calculated for lRJA. For the higher level RJA (hRJA), a score point was given when the child followed the examiner's pointing gesture by sufficiently turning their head to indicate that they were looking at the poster. Subsequently, a paired t-test was administered to compare the average instances of joint attention (RJA or IJA) between pre- and post-test.

Color Accuracy. For these analyses, data from Session 1 counted as practice. Mean accuracy of color response for each child and each subsequent session was calculated. A paired t-test between Session 2 and 6 was administered to calculate the effect of the training over time. All trials, in which one of the following occurred were counted as errors and excluded: (1) the teachers pointed towards the screens to prompt a child, (2) a child didn't respond, (3) a child said a completely different color than what was displayed on the screen, or (4) a child responded with the wrong color first, but then named the correct color. However, if a child first responded with the correct color, but then said the other color, this was counted as a correct response. In cases when children named the red color as brown, the response was counted as correct. Similarly, when children said green instead of yellow and vice versa (only when yellow or green appeared with red or blue), the response was counted as correct, as the yellow and green color were similar. However, when both green and yellow appeared together, mixing up yellow and green was counted as an error, as the two simultaneously presented colors were easily distinguishable.

Button Press Accuracy. Also for these analyses, data from the first session counted as practice. Mean accuracy of button response for each child and for each subsequent session was calculated. A paired t-test between Session 2 and Session 6 was administered to calculate the effect of the training over the sessions. All trials, in which the teachers pointed towards the screens to prompt the children, were excluded.

3   Results

3.1   Joint Attention Scores

The analysis of joint attention scores showed that an average total RJA score of 193.75 (SD=12.5) in the posttest (after robot training) was significantly higher than the average total RJA score of 183.33 (SD=13.6) in the pretest (before robot training), $t(3) = 2.61$, $p = .040$, $d_z=0.797$ (cf. Fig. 3, left). When splitting the data into lower level RJA (lRJA) and higher level RJA (hRJA), lRJA results showed a significantly higher average lRJA score of 100 (SD=0) in the post-test, relative to an average lRJA score of 83.3 (SD=13.60) in the pre-test, t(3) = 2.45, p = .046, d$_z$=1.732 (cf. Fig. 3, right). The difference in the average hRJA score (pretest: 100, SD=0 vs. posttest: 93.75, SD=12.5) was not significant, t(3) = 1, p = .80 (cf. Fig. 3, right).

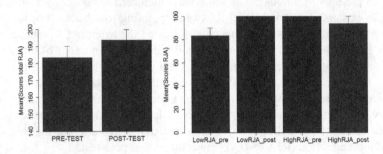

Fig. 3. Mean total RJA scores (left), lower RJA and higher RJA scores (right) in pre-test and post-test as measured by the ESCS [21] (left). Error bars: standard errors of the mean (SEM).

IJA scores were not significantly different in pretest (2.67, SD=2.42) vs. posttest (5.67, SD=5.27), t(5) = 1.39, p = .112.

3.2 Color Accuracy and Button Press Accuracy

Mean color accuracy in Session 6 (M= 0.69, SD= 0.46) increased significantly from Session 2 (M= 0.52, SD= 0.50), t(5) = 2.19, p = .0.04, cf. Fig. 4, left. Button press accuracy increased (marginally significant) for Session 6 (M= 0.75, SD= 0.43) vs. Session 2 (M= 0.67, SD= 0.48),   t(5) = 1.60, p = .086, see Fig. 4, right.

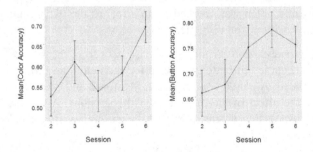

Fig. 4. Mean color (left) and button press accuracy (right) for Session 2 – 6. Error bars: SEM.

4   Discussion

Our study examined if a newly-developed robot-assisted training protocol embedding an attention-cueing paradigm in a game that children play with a robot has the potential of improving joint attention skills in children diagnosed with ASD.

4.1   RJA & IJA Scores

The present results showed a significant difference between RJA scores in pre- vs. post-training test, as measured by the ESCS [21], with greater scores after the robot-assisted training. The effect was large (as indicated by Cohen's d_Z effect size values), despite small sample. First, and foremost, this demonstrates that robot training was successful in increasing the children's skills in responding to joint attention, and that children transferred these skills to the interaction with human experimenter. Interestingly, when splitting the data into lower RJA (lRJA) and higher RJA (hRJA) scores, only lRJA scores showed a significant increase after the robot training. However, this may be simply due to the measurement's sensitivity not being high enough, as all children of this sample seemed to have a perfect score on hRJA skills already before the robot training. Possibly with a sample of children less skilled in hRJA before the training (or a more sensitive measure), a significant increase of hRJA skills might be observed after robot training.   Furthermore, IJA scores were of equal magnitude before and after the training.   This is in line with our hypothesis, as we designed a paradigm to specifically train RJA to maximize the effectiveness of the training. The sole effect of an increase in RJA skills and lack of influence of the robot training on the IJA skills supports the idea that there are two separate mechanisms underlying joint attention, namely IJA and RJA [3, 24]. Therefore, our results are in line with previous research and confirm that with carefully designed training protocols one can extract and target specific mechanisms of social cognition. Such focused training protocols should be effective as an intervention method in autism therapy [3, 24]. Moreover, the use of an independent diagnostic (ESCS), allows for suggesting that our findings are relatively robust and independent of the robotic platform.

4.2   Verbal Color Accuracy and Button Press Accuracy

Our present results indicated an increase in color accuracy over time and, in particular, a significant increase of accuracy between Session 2 and 6. This may be due to that children were increasingly more engaged in the game with the robot, arguing in favour of the training protocol. Additionally, a more general learning effect might have also played a role in the improved performance in color accuracy task. However, general learning is likely not to be the only factor influencing the effect, as it seems more plausible that the increase in performance during the robot-assisted training is related to the nature of the training protocol itself (completion of the task is possible only through following the robot's head direction). This is predicated on the fact that instances of RJA increased over the course of training, and this presumably had an effect on the increase of color accuracy. Similarly, results also indicate a general trend of increased button press accuracy over time. The fact that the differences between Session 2 and 6 were only marginally significant might have been related to that the button presses (and their mapping to respective stimuli) might have been demanding for the children with ASD, due to, for example, deficits in motor skills [6,25].

4.3   Implications and Future Directions

In general, results of this study provide convincing evidence for robot-assisted training of social skills in children diagnosed with ASD. The training can be effective when it is grounded in paradigms of experimental psychology that target specific cognitive mechanisms and measure improvement with objective measures. In our study, despite a relatively small sample size of children diagnosed with ASD, the robot CuDDler in combination with the attention-cueing paradigm proved appropriate to improve RJA skills. Improving RJA skills is crucial for social skills, as RJA is positively related to language development, social and communication skills of children diagnosed with ASD [3,5,26]. Thus, robot-assisted therapy which improves joint attention may – in a long run – also facilitate language learning and more complex social cognitive skills. Most importantly, to evaluate joint attention improvement, we used a diagnostic tool that was independent of the robot therapy, which suggests robustness of the effects. However, we assessed joint attention skills shortly after the last robot therapy session (post-test: 2-3 days later). Thus, future research should elucidate whether the improvement of RJA skills of children diagnosed with ASD after a robot-assisted therapy have a longer-lasting effect. Furthermore, follow-up studies (with larger sample sizes) should test if a longer training would be related to how permanent the improvement in joint attention skills is, and the degree with which it is transferred from human-robot to human-human interaction. Moreover, as some of the children did not want to participate in the tasks, future research will need to develop means to make the robot training more attractive (possibly a longer familiarization phase preceding the training) to children who are initially reluctant to participate.

4.4   Conclusions

In sum, this study is a first approach to design – based on paradigms from experimental psychology – training protocols that isolate and target specific cognitive skills of children with ASD. By combining information and knowledge gained from various interdisciplinary fields, such as robotics, cognitive neuroscience and psychology, one can design a training method that may lead to distinct and positive conclusions as to the effectiveness in improving joint attention skills in children with ASD. Future studies may take this as a basis on which to expand on, to examine this therapy method in more detail, test it on a larger sample size over a longer period of time and possibly also test variations of this protocol to train and improve other cognitive mechanisms.

Acknowledgment. We thank THK MORAL EIPIC for their help in recruiting participants and in preparation for the experimental study.

References

1. Matson, J.L., Kozlowski, A.M.: The increasing prevalence of autism spectrum disorders. Res. Autism Spect. Dis. **5**, 418–425 (2011)
2. Wing, L., Potter, D.: The epidemiology of autistic spectrum disorders: is the prevalence rising? Ment. Retard. Dev. D R **8**, 151–161 (2002)
3. Meindl, J.N., Cannella-Malone, H.I.: Initiating and responding to joint attention bids in children with autism: A review of the literature. Res. Dev. Disabil. **32**, 1441–1454 (2011)

4. Whalen, C., Schreibman, L.: Joint attention training for children with autism using behavior modification procedures. J. Child Psychol. Psyc. **44**, 456–468 (2003)
5. Mundy, P., Sigman, M., Kasari, C.: A longitudinal study of joint attention and language development in autistic children. J. Autism Dev. Disord. **20**, 115–128 (1990)
6. Charman, T.: Why is joint attention a pivotal skill in autism? Philos. T. Roy. Soc. B **358**, 315–324 (2003)
7. Cabibihan, J.J., et al.: Why Robots? A Survey on the Roles and Benefits of Social Robots in the Therapy of Children with Autism. Int. J. Soc. Robot. **5**, 593–618 (2013)
8. Dautenhahn, K.: Roles and functions of robots in human society: implications from research in autism therapy. Robotica **21**, 443–452 (2003)
9. Scassellati, B., Admoni, H., Matarić, M.: Robots for Use in Autism Research. Annu. Rev. Biomed. Eng. **14**, 275–294 (2012)
10. Anzalone, S.M., et al.: How children with autism spectrum disorder behave and explore the 4-dimensional (spatial 3D+ time) environment during a joint attention induction task with a robot. Res. Autism Spect. Dis. **8**, 814–826 (2014)
11. Bekele, E., et al.: Pilot clinical application of an adaptive robotic system for young children with autism. Autism **18**, 598–608 (2014)
12. Tapus, A., Mataric, M.J., Scasselati, B.: Socially assistive robotics [Grand challenges of robotics]. IEEE Robot. Autom. Mag. **14**, 35–42 (2007)
13. Tapus, A., et al.: Children with autism social engagement in interaction with Nao, an imitative robot–A series of single case experiments. Interact. Stud. **13**, 315–347 (2012)
14. Warren, Z.E., et al.: Can Robotic Interaction Improve Joint Attention Skills? J. Autism Dev. Disord. 1–9 (2013)
15. Robins, B., Dautenhahn, K., Te Boekhorst, R., Billard, A.: Robotic assistants in therapy and education of children with autism: can a small humanoid robot help encourage social interaction skills? Universal Access in the Information Society **4**, 105–120 (2005)
16. Diehl, J.J., Schmitt, L.M., Villano, M., Crowell, C.R.: The clinical use of robots for individuals with autism spectrum disorders: A critical review. Res. Autism Spect. Dis. **6**, 249–262 (2012)
17. Wainer, J., Ferrari, E., Dautenhahn, K., Robins, B.: The effectiveness of using a robotics class to foster collaboration among groups of children with autism in an exploratory study. Pers. Ubiquit. Comput. **14**, 445–455 (2010)
18. Friesen, C.K., Kingstone, A.: The eyes have it! Reflexive orienting is triggered by nonpredictive gaze. Psychon. B Rev. **5**, 490–495 (1998)
19. Wiese, E., Wykowska, A., Zwickel, J., Muller, H.J.: I see what you mean: how attentional selection is shaped by ascribing intentions to others. PLoS One **7**(9), e45391 (2012)
20. Wykowska, A., Wiese, E., Prosser, A., Müller, H.J.: Beliefs about the Minds of Others Influence How We Process Sensory Information. PLoS One **9**(4), e94339 (2014)
21. Mundy, P., et al.: A Manual for the Abridged Early Social Communication Scale (ESCS), U.o. Miami (2003)
22. Noris, B., et al.: Investigating gaze of children with ASD in naturalistic settings. PLoS One **7**(9), e44144 (2012)
23. Noris, B., Keller, J.B., Billard, A.: A wearable gaze tracking system for children in unconstrained environments. Comput. Vis. Image Und. **115**, 476–486 (2011)
24. Taylor, B.A., Hoch, H.: Teaching children with autism to respond to and initiate bids for joint attention. J. Appl. Behav. Anal. **41**, 377–391 (2008)
25. Piek, J.P., Dyck, M.J.: Sensory-motor deficits in children with developmental coordination disorder, attention deficit hyperactivity disorder and autistic disorder. Hum. Movement Sci. **23**, 475–488 (2004)
26. Murray, D.S., et al.: The Relationship Between Joint Attention and Language in Children With Autism Spectrum Disorders. Focus on Autism and Other Developmental Disabilities **23**, 5–14 (2008)

Implicit Nonverbal Behaviors Expressing Closeness by 3D Agents

Hiroko Kamide[1(✉)], Mihoko Niitsuma[2], and Tatuo Arai[3]

[1] Research Institute of Electrical Communication, Tohoku University,
2-1-1 Katahira, Aoba-ku, Sendai 980-8577, Japan
kamide@riec.tohoku.ac.jp
[2] Faculty of Science and Engineering, Department of Precision Mechanics,
Chuo University, 1-13-27 Kasuga, Bunkyo-ku, Tokyo 113-8551, Japan
[3] Department of Engineering Science, Osaka University, Machkaneyamacho 1-3,
Toyonaka, Osaka 560-8531, Japan

Abstract. The goal of the current study was to extract natural nonverbal behaviors that are implicit but specific to strangers and friends and to test the expressiveness of these nonverbal behaviors in two different levels of closeness using 3D agents. An experiment was conducted in which 48 pairs (48 strangers and 48 friends) of participants had casual conversations about recent events for 10 min. Their body movements were recorded by a motion-capture system, and 13 vectors were defined on the upper body to compute the cosine similarity for each frame in order to extract the motions. The motions specific to strangers and friends were identified and two scenarios were created using those motions. The scenarios were implemented using 3D agents of a female human and a humanoid robot, and 400 respondents were asked to evaluate the closeness that the agent seemed to express toward the counterpart. The results showed that a human-agent performing friend motions were evaluated higher in expressiveness closeness than friend motions and a human-agent and a robot-agent performing friend motions were evaluated lower in strangeness than friend motions. In future works, we aim to improve the scenarios and implement them in humanoid robots.

1 Introduction

To realize robots that can coexist with humans, several technological developments have to be made in multiple areas such as sensing [1-3], mobility [4-6], artificial intelligence [7-9], and interface [10-12]. It is especially important to adapt the behavior of robots for smooth interaction with humans. [13] developed a situated module for an interactive humanoid that can employ verbal and nonverbal behaviors. [14] identified the comfortable distance between a robot and humans who meet the robot for the first time. Moreover, in the case of a robot presenting slides to an audience, it was found that combinations of nonverbal behaviors, such as eye contact and pointing, affect the understanding of the presentation [15].

© Springer International Publishing Switzerland 2015
A. Tapus et al. (Eds.): ICSR 2015, LNAI 9388, pp. 306–316, 2015.
DOI: 10.1007/978-3-319-25554-5_31

Most studies on human-robot interactions test newly developed technologies in temporal interactions conducted in an experimental environment. However, it is also important to focus on the sustainable relationships between robots and humans from a psychological perspective. Closeness, which reflects enduring personal relationships [16], is one of important factors for understanding longitudinal interpersonal relationships [17, 18]. This study takes closeness into consideration in human–robot interactions in order to realize robots that can sustainably develop relationships with humans by expressing different levels of closeness using nonverbal behaviors. We aimed to identify the nonverbal behaviors that are specific to interactions with strangers, who are not close, and friends, who are closer. Then, we created two scenarios for both these cases in which a 3D agent employed nonverbal behaviors during conversations and tested the expressiveness of closeness.

[19] revealed that nonverbal cues are more informative than verbal cues and reported that only 7% of attitudes of speakers is determined by spoken words whereas 93% is determined by nonverbal cues. Closeness is also expressed nonverbally. Individuals who are closer, such as friends, stand at an interpersonal distance between 45 cm and 120 cm, whereas business colleagues, who are less close, stand more than 120 cm apart [20]. Friends look at each other more compared with than strangers in the case of both adults [21] and children [22]. [23] deduced that closeness is related to eye contact, interpersonal distance, intimacy of conversation, and amount of smiling; the more these behaviors occur, the more the affiliation between the individuals. Furthermore, affectionate behaviors, such as holding hands, kissing, and winking, are significant for the development of romantic relationships [24].

Previous studies have focused on explicit nonverbal behaviors that could be detected relatively easily. [25] reported that observers and actors differ in their understanding of the same phenomenon. For instance, if the communicating individual behaves more actively, the observer recognizes that their communication is going well by focusing on their expressiveness. However, the communicating individual not only uses expressiveness but also other cues to understand the ongoing communication, and this is called the "expressivity halo" effect [26]. Considering this discrepancy between actors and observers in understanding an interaction, it is quite important to identify the nonverbal behaviors that reflect different levels of closeness from an actor's perspective. Actors use explicit as well as detailed information [27] and process the information more elaborately [28, 29] than observers. Recently by using technical measurements such as eye-cameras, nonverbal behaviors that have been difficult to be detected have been found to be important cues for interpersonal communications [30, 31].

In this study, we recorded the motions of pairs of strangers and friends with a motion-capture system and then investigated the specific nonverbal behaviors used in both these types of relationships. Based on the findings, we created two scenarios and tested whether the identified behaviors can communicate the different levels of closeness to other humans. As a first step in this research, we focused on the motions of the upper body.

2   Method

This study complied with the code of ethical conduct of the Graduate School of Engineering Science of Osaka University.

2.1   Participants

Of the 48 pairs of 98 Japanese individuals (49 male, 49 female, mean age = 37.99, SD = 9.38) who participated in the experiment, half the pairs were strangers, and the other half were friends. Each set of 24 pairs was composed of four male–male pairs, four female–female pairs, and four male–female pairs. In this study, we controlled the effect of gender as well as the distance (1 m or 2 m) between the participant pairs during the experiment (controlling of the distance is for another project). The participants in the stranger pairs were recruited individually and then matched during the experiment. The friend pairs were recruited together, but we did not set any limit on the duration of their relationships or their subjective level of closeness, because the closeness was measured on a psychological scale during the experiment.

2.2   Procedure

The participants were told a cover story that the aim of the study was to record the body movements with a motion-capture system when two people are engaged in collaborative works such as transporting objects together although they would not do such works actually. They were also told that, before the recording, it was necessary to record an extra 10 min to check that the system worked correctly and that they could talk freely with their counterparts during this time. They prepared a topic to talk about for 10 min in advance the experiment. This 10 min record was the real experiment, and we aimed to record the motions when people have natural conversations rather than discuss specific issues. Therefore, we told the participants that the theme of the conversation could be recent events in their lives, which would be similar to everyday communications.

Fig. 1. Two friends talking during the experiment

Each experiment involved a pair of participants who were either strangers or friends. The experimenter explained the entire procedure including the cover story, and the participants signed a letter of consent. Then, they changed into the suits with markers for motion capture and completed a questionnaire that included a psychological scale of closeness. After calibration for 5 min, the participants were told that we needed 10 min to check the system and to talk freely during this time. They were also told to move freely as usual in order to check the system. The distance between the two participants was controlled during the experiment, and a red zone (1 m by 0.5 m) was set for each of them within which they could move (Figure 1). The experimenter went out of the room, and then the participants started to talk. Although, one technical staff remained in the room behind the partition, the participants could not realize him. After 10 min, the experimenter came back into the room, asked the participants to complete a questionnaire, and informed them that the experiment was over then the real aim was to record not the collaboration work but the conversation. None of the participants complained about the use of the cover story.

2.3   System

The marker positions were determined according to the two major settings of SIMM (software for interactive musculoskeletal modeling) and Plug-in-Gait. A total of 43 markers were placed on the whole body, out of which 16 markers were located on the upper body (Figure 2). We used the MAC 3D system and 20 Raptor-E cameras made by Motion Analysis Corporation. The operation software was Cortex v3.6, and the suits were of the type MoCap Suit STD by 3X3 Designs. We also videotaped the experiment using a Sony PMW-EX-3.

To improve the accuracy of the motion-capture record, we placed markers on the inner side of the arms. Moreover, to complement the missing markers due to the posture of the participant, we used two methods. If the sequence of missing frames was less than three frames, a spline complement using a spline function was adopted. Otherwise, a rigid-body complement was employed in which the missing marker was calculated using three markers that were on the same segment as the missing marker or were at stable distances from the missing marker.

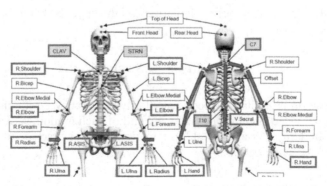

Fig. 2. Marker positions on upper body (red borders indicate markers used in this analysis, and blue boxes identify vectors for further analysis)

2.4   Questionnaire

To determine the level of closeness between each pair of participants, we asked them to complete a questionnaire that included a psychological scale of closeness. We used the factor of intimacy (10 items, 7-point, $\alpha = 0.98$) of Sternberg's triangular theory of love scales [32], which measures the degree of closeness such as friendliness and bonding. Other factors of passion and commitment are often used for romantic relationships, but the factor of intimacy is relevant for assessing the closeness of friends who like each other. Example statements in the questionnaire include "I have a warm relationship with __." "I communicate well with __," and "__ is able to count on me in times of need."

3   Results

3.1   Manipulation Check

We conducted an analysis of covariance (ANCOVA) to investigate the effect of relationships (stranger/friend) on the score of intimacy of Sternberg's triangular theory of love scales, controlling gender and age. We found a significant effect of relationships on intimacy ($F(1, 94) = 132.20$, $p < 0.001$). The pairs of participants who were friends (mean $= 4.75$, SD $= 1.18$) were closer than those who were strangers (mean $= 1.88$, SD $= 1.26$).

3.2   Extraction of Motions

The system recorded the positions of the markers at 60 Hz, and the recorded data was reduced to 10 Hz in this study. We defined 13 vectors based on the 16 markers on the upper body (Figure 3). In order to extract the motions, we used cosine similarity. For each vector, we calculated the cosine between two continuing frames. The cosine similarity $S_i(t)$ of a vector i ($1 \leq i \leq 13$) at times t, $t - 1$ is calculated as follows:

$$S_i(t) = \cos \theta = \frac{A_i(t) \cdot A_i(t - 1)}{\|A_i(t)\|\|A_i(t - 1)\|}$$

We calculated the sum of cosine similarities on the upper body for the 13 vectors at times t, $t - 1$ as follows:

$$S(t) = \sum_{i=0}^{13} S_i(t)$$

When the form is similar, the cosine similarity is larger. We set a threshold based on the recorded video, and the beginning and end of each motion were detected using the cosine similarity. As a result, 12,033 motions were extracted from the 10 min records of all the participants.

3.3   Categorization of Motions

To categorize the 12,033 motions, the cosine similarities between pairs of extracted motions were calculated. The cosine similarity between a motion X with m frames and a motion Y with n frames, $S(x_a, y_b)(1 \leq a \leq m, 1 \leq b \leq n)$ is calculated as follows:

$$S(x_a, y_b) = \sum_{i=0}^{13} S_i(x_a, y_b) = \sum_{i=0}^{13} \frac{X_i(x_a) \cdot Y_i(y_b)}{\|X_i(x_a)\| \|Y_i(y_b)\|}$$

Figure 3 shows an example of the result, where the black cells represent cosine similarities higher than the threshold. These black cells enable us to assess the continuing similarity between two motions. If the two motions are similar, the black cells appear continuously from the top-left position to the bottom-right position.

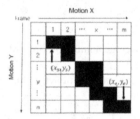

Fig. 3. Example of calculation of similarity

We set L as the length from the first black cell (x_s, y_s) to the last black cell (x_e, y_e), which is calculated as follows:

$$L = \frac{\sqrt{(x_e - x_s)^2 + (y_e - y_s)^2}}{\sqrt{(m-1)^2 + (n-1)^2}}$$

If L is longer than the threshold, the two motions are considered to be similar. As a result, we identified 811 categories of motions.

All the thresholds in these analyses were determined based on the recorded video from which we can recognize the beginning and end of each motion and the similarity of motions.

3.4   Effect of Relationships on Number and Variety of Motions

After obtaining the categorical variable of motions, we investigated the effect of relationships on the number and variety of motions. We observed a significant effect of relationships on the variety of motions ($F(1,91) = 6.04$, $p < 0.05$) but not on the total number of motions; friends did not perform more motions than stranger, but they performed a larger variety of motions (Figure 4).

3.5   Specific Motions for Strangers and Friends

Focusing specifically on individual motions, we analyzed the motions that were activated more than 0.5 times on average. We investigated the effect of relationships on the number of times each of these motions were activated and identified eight motions that were significantly activated more in friends than in strangers (Table 1 and Figure 5). In this study, we focused the relationships as a first step, therefore controlled the age and gender. The motion number represents the categorical number of the motion and is automatically assigned from 1 to 811.

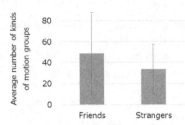

Fig. 4. Average number of kinds of motion

Table 1. Average number of activations of motions

Motion number	Friend		Stranger		Statistic
	M	SD	M	SD	
7	2.25	3.28	1.04	1.69	$F(1,91)=4.93, p<.05$
6	1.99	2.75	1.01	1.44	$F(1,91)=4.47, p<.05$
14	1.68	2.40	0.71	1.43	$F(1,91)=6.35, p<.05$
290	1.68	2.68	0.67	1.91	$F(1,91)=4.98, p<.05$
721	1.47	3.54	0.42	1.02	$F(1,91)=4.40, p<.05$
329	1.27	3.19	0.14	0.45	$F(1,91)=6.20, p<.05$
327	1.14	2.58	0.20	0.69	$F(1,91)=6.05, p<.05$
16	0.93	1.54	0.25	0.49	$F(1,91)=8.19, p<.01$

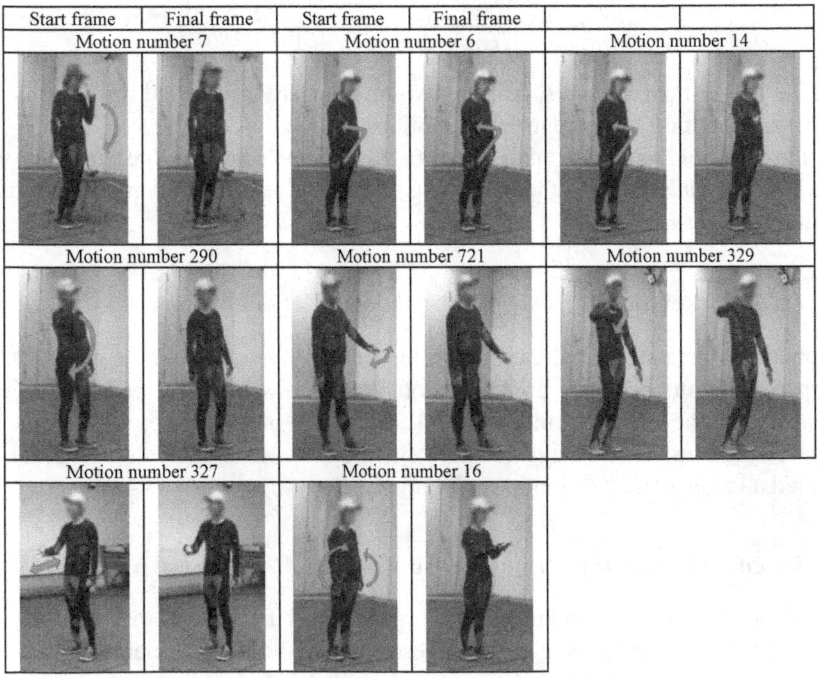

Start frame	Final frame	Start frame	Final frame		
Motion number 7		Motion number 6		Motion number 14	
Motion number 290		Motion number 721		Motion number 329	
Motion number 327		Motion number 16			

Fig. 5. Initial and final poses of motions in Table 1

3.6   Creation of Two Scenarios

We generated a list with the top 20 motions that were activated most in order to create 3 min scenarios of both stranger and friend interactions. Table 2 shows the motion numbers of the top 20 motions for both the relationships.

In Table 2, the yellow cells denote motions that are common to strangers and friends. It seems friends and strangers share same kinds motions shown by yellow cells. Friends seem to have different kinds of motions shown by the blue cells. The orange cells are specific to strangers in this table, although we found a significant difference in the activated times only for the motion of 502.

Table 2. Top 20 motions for friends and strangers (The motion number represents the order to be categorized and is automatically assigned from 1 to 811)

	Friend	Stranger		Friend	Stranger
1	498	498	11	81	81
2	25	25	12	9	104
3	2	10	13	188	463
4	10	2	14	290	491
5	4	4	15	14	165
6	34	34	16	64	
7	1	502	17	15	
8	7	1	18	604	
9	5	5	19	721	
10	6	9	20	463	

According to this list of motions and the average number of times these motions are activated in each relationship, we created two kinds of scenarios using 3D agents. We used models of a human and a robot for the agents so that we could investigate whether the identified behaviors can express different levels of closeness not only in human communication but also in human–robot interactions. We used a female figure as a first step in this study.

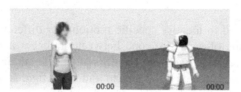

00:00   00:00

Fig. 6. 3D figures of human and robot

3.7   Evaluation of the Scenarios

To evaluate the two scenarios, 400 Japanese individuals (200 male, 200 female, mean age 44.96, SD = 13.85) were asked to watch a movie that showed a human or a robot talking with another human. The movie showed only the former, and the latter did not appear. The participants were asked to evaluate how close the person or the robot seem to another human. The scale used in the section 2.4 is for the evaluation between the actors and not for the observer to the actors, therefore we used three statements to evaluate closeness (α = 0.87): "This person (robot) seems to like the counterpart," "This person (robot) seems to be familiar to the counterpart," and "This person (robot) seems relaxed talking to the counterpart." We also used three statements to evaluate strangeness (α = 0.71); "I think this person (robot) talks with the stranger," "The person (robot) looks nervous," "This person (robot) seem to try to be polite." The length of each movie was about three minutes and the experiment was conducted in a between-subjects design and online.

We conducted a repeated analysis of variance (ANOVA) to investigate the effect of the two scenarios on two kinds of evaluations (closeness/strangeness). In the case of the human figure, we found a significant interaction effect (Wilks's λ = 0.963, F(1,198) = 7.71, p < 0.01). The friend scenario scored higher in closeness and lower in strangeness, whereas the stranger scenario scored higher in strangeness (Figure 7).

We also observed a significant interaction effect for the robot figure (Wilks's λ = 0.942, F(1,198) = 12.27, p < 0.001). The friend scenario was lower in strangeness and higher in closeness, whereas the stranger scenario was higher in strangeness than in closeness (Figure 7).

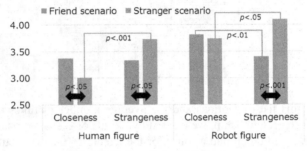

Fig. 7. The result for both figures.

4   Conclusion

In this study, we aimed to identify specific motions for different levels of closeness and created two kinds of scenarios, one between strangers and the other between friends. These two scenarios showed significantly different impressions regarding closeness and strangeness for both the human and robot figures. However, we did not find a significant difference between the friend and stranger scenarios in the case of the robot figure compared with the case of the human figure. The robot figure looked like ASIMO, which is familiar in Japan, so it is possible that the participants felt friendlier toward the robot figure than the human figure. In addition, we didn't find significant difference between closeness and strangeness in case of the human figure. It is possible that friend motions were relative polite and not overly friendly.In future studies, we need to implement the scenarios with other figures and also with real robots.

We focused on the upper body motions in this study. However, the nonverbal motions are usually activated in the combinations with multi-channels, and they are also affected by the behaviors of the counterpart. The interdependency of whole body motions and interpersonal effects will be investigated in a future work to realize the flexible motions for human–robot interaction.

Acknowledgement. This work was supported by JSPS KAKENHI Grant Number 15K13114, 15K01582, 26705008, and Honda R&D Co., Ltd.. Part of this work was carried out under the Cooperative Research Project Program of the Research Institute of Electrical Communication, Tohoku University.

References

1. Yang, H.Y., Zhang, H., Xu, W., Zhang, P.J., Xu, L.M.: The Application of KINECT Motion Sensing Technology in Game-Oriented Study. iJET **9**(2), 59–63 (2014)
2. Almaddah, A., Vural, S., Mae, Y., Ohara, K., Arai, T.: Spherical Spaces for Illumination Invariant Face Relighting. Journal of Robotics and Mechatronics **25**(5), 840–847 (2013)
3. Penaloza, C.I., Mae, Y., Cuellar, F.F., Kojima, M., Arai, T.: Brain Machine Interface System Automation considering User Preferences and Error Perception Feedback. IEEE Transactions on Automation Science and Engineering (2014)
4. Nakaoka, S., Nakazawa, A., Kanehiro, F., Kaneko, K., Morisawa, M., Hirukawa, H., Ikeuchi, K.: Learning from Observation Paradigm: Leg Task Models for Enabling a Biped Humanoid Robot to Imitate Human Dances. International Journal of Robotics Research **26**(8), 829–844 (2007)
5. Shigemi, S., Kawaguchi, Y., Yoshiike, T., Kawabe, K., Okgawa, N.: Development of New ASIMO. Honda Technical Review **18**(1), 38–44 (2006)
6. Kaneko, K., Kanehiro, F., Morisawa, M., Miura, K., Nakaoka, S., Kajita, S.: Cybernetic human HRP-4C. In: Proc. IEEE-RAS International Conference on Humanoid Robots, pp. 7–14 (2009)
7. Nakanishi, H., Tanaka, K., Wada, Y.: Remote handshaking: touch enhances video-mediated social telepresence. In: International Conference on Human Factors in Computing Systems, pp. 2143–2152 (2014)
8. Ito, T., Matsubara, H., Grimbergen, R.: A Cognitive Science Approach to Shogi Playing Processes (2)-Some Results on Next Move Test Experiments. Transactions of Information Processing Society of Japan **45**(5), 1481–1492 (2004)
9. Sato, Y., Takahashi, D., Grimbergen, R.: A Shogi Program Based on Monte-Carlo Tree Search. ICGA Journal **33**(2), 80–92 (2010)
10. Nakaoka, S.: Choreonoid: extensible virtual robot environment built on an integrated GUI framework. In: Proc. of the 2012 IEEE/SICE International Symposium on System Integration, pp. 79–85 (2012)
11. Takashima, K., Aida, N., Yokoyama, H., Kitamura, Y.: TransformTable: a self-actuated shape-changing digital table. In: Proceedings of Conferece on Interactive Tabletop and Surface (ITS), pp. 179–187 (2013)
12. Hayashi, Y., Itoh, Y., Takashima, K., Fujita, K., Nakajima, K., Onoye, T.: Cuple: cup-shaped tool for subtly collecting information during conversational experiment. The International Journal of Advanced Computer Science **3**(1), (2013)
13. Kanda, T., Ishiguro, H., Imai, M., Ono, T.: Development and Evaluation of Interactive Humanoid Robots. Proceedings of the IEEE **92**(11), 1839–1850 (2004). Special issue on Human Interactive Robot for Psychological Enrichment
14. Kamide, H., Mae, Y., Takubo, T., Ohara, K., Arai, T.: Direct Comparison of Psychological evaluation between Virtual and Real Humanoids; Personal space and subjective impressions. International Journal of Human-Computer Studies **72**, 451–459 (2014)

15. Kamide, H., Kawabe, K., Shigemi, S., Arai, T.: Nonverbal behaviors toward an audience and a screen for a presentation by a humanoid robot. Artificial Intelligence Research 3(2), 57–66 (2014)

16. Cramer, D.: Close relationships: The study of love and friendship. Arnold, London (1998)

17. Levinger, G., Snoek, J.D.: Attraction in relationship: a new look at interpersonal attraction. General Learning (1972)

18. Knapp, M.L.: Interpersonal Communication and Human Relationships. Allyn & Bacon, Boston (1984)

19. Mehrabian, A., Wiener, M.: Decoding of inconsistent communications. Journal of Personality and Social Psychology **6**, 109–114 (1967)

20. Hall, E.T.: The hidden dimension, Doubleday and Company (1966)

21. Coutts, L.M., Schneider, F.W.: Visual behavior in an unfocused interaction as a function of sex and distance. Journal of Experimental Social Psychology **11**, 64–77 (1975)

22. Foot, H.C., Chapman, A.J., Smith, J.R.: Friendship and social responsiveness in boys and girls. Journal of Personality and Social Psychology **5**, 401–411 (1977)

23. Argyle, M., Dean, J.: Eye contact, distance, and affiliation. Sociometry **28**, 289–304 (1965)

24. Floyd, K., Morman, M.T.: The measurement of affectionate communication. Communication Quarterly **46**, 144–162 (1998)

25. Jones, E.E., Nisbett, R.E.: The actorand the observer: divergent peceptions of the causes of behavior. In: Jones, E.E., Kanouse, D.E., Kelly, H.H., Nisbett, R.E., Valins, S., Weiner, B. (eds.) Attribution: Perceiveing the Causes of Behavior. General Lerning Press, pp. 79–94 (1972)

26. Bernieri, F., Gillis, J.S., Davis, J.M., Grahe, J.E.: Dyad rapport and the accuracy of its judgment across situations: A lens model analysis. Journal of Personality and Social Psychology **71**, 110–129 (1996)

27. Taylor, S.E., Fiske, S.T.: Salience, attention, and attribution: top of the head phenomena. In: Berkowitz, L. (ed.) Advances in experimental social psychology, vol. 11, pp. 249–288. Academic Press, New York (1978)

28. Brewer, M.B.: A dual process model of impression formation. In: Srull, T.K., Wyer, R.S. (eds.) Advances in Social Cognition, vol. 1, pp. 1–36. Erlbaum, Hillsdale (1988)

29. Fiske, S.T., Neuberg, S.L.: A continuum of impression formation, from category-based to individuating processes: influences of information and motivation on attention and interpretation. In: Zanna, M.P. (ed.) Advances in Experimental Social Psychology, vol. 23, pp. 1–74. Academic Press, New York (1990)

30. Jokinen, F., Nishida, Y.: Gaze and Turn-taking behaviour in Casual Conversational Interactions. ACM Transactions on Interactive Intelligent Systems (TiiS) Journal 3(2), (2013). Special Section on Eye-gaze and Conversational Engagement, Guest Editors: Elisabeth André and Joyce Chai

31. Levitski, A., Radun, J., Jokinen, K.: Visual interaction and conversational activity. In: Proceedings of The 4th Workshop on Eye Gaze in Intelligent Human Machine Interaction: Eye Gaze and Multimodality, at the 14th ACM International Conference on Multimodal Interaction (ICMI-2012), October 26 2012, Santa Monica, California, U.S. (2012)

32. Sternberg, R.J.: Construct validation of a triangular love scale. European Journal of Social Psychology **27**, 313–335 (1997)

Visiting Cultural Heritage with a Tour Guide Robot: A User Evaluation Study in-the-Wild

Daphne Karreman[1(✉)], Geke Ludden[2], and Vanessa Evers[1]

[1] Human Media Interaction, Faculty of Electrical Engineering,
Mathematics and Computer Science, University of Twente, Enschede, Netherlands
{d.e.karreman,v.evers}@utwente.nl
[2] Product Design, Faculty of Engineering Technology, University of Twente,
Enschede, Netherlands
g.d.s.ludden@utwente.nl

Abstract. In this paper we present a user evaluation study on location at the Royal Alcázar in Seville, Spain, with the fully autonomous tour guide robot FROG. In this robot, technological innovations in navigation and vision were integrated with state-of-the-art design for robot behavior in order to provide interactive tours and adaptive content to visitors. In our user evaluation study we aimed to gain insights in user experiences of and attitudes and responses towards this fully autonomous social robot. Such studies are important, because they provide information about how people interact with social robots outside a controlled setting. Invited as well as spontaneous visitors followed tours guided by FROG and were interviewed about their opinions and experiences. Our findings indicate that even if isolated technical features work perfectly in controlled settings, they might not work well in the integrated system, because naïve people interact with the system in an unforeseen manner.

Keywords: Robot guide · Interaction design · Real world evaluation · User study

1 Introduction

Continuous technical innovations in the field of human-robot interaction (HRI), enable to improve interactions between robots and people. An interesting domain of application for HRI is the use of robots as tour guides (e.g. [1]–[3]), because a tour-guide robot has to be able to guide visitors through a museum, cultural heritage site or other place of interest while informing and engaging them. For a robot to be able to do this, and to create a successful visitor experience, the aim should be to combine state-of-the-art technology with advanced robot social behaviors.

To date, research into tour guide robotics has mainly focused on one specific element of guiding visitors (such as features of navigation and gaze behavior) rather than on the overall visitor experience people have when they are guided by an autonomous guide robot. For example, Pitsch et al. studied how a guide robot can use

The research leading to these results received funding from the EC's 7th Framework program under Grant agreement 288235. http://www.frogrobot.eu. We thank the entire EU FP7 FROG-team for technically enabling these user experience tours.

A. Tapus et al. (Eds.): ICSR 2015, LNAI 9388, pp. 317–326, 2015.
DOI: 10.1007/978-3-319-25554-5_32

gaze behavior to include or exclude visitors in interaction [4], Yousuf et al. studied how a tour guide robot can best approach visitors and guide them to an exhibit to give information [5], and Donner et al. focused on how path planning and orientation of the robot can be optimized when a projector is used to project on the exhibit [6]. In order to gain insight in how these and other technical features together can be used to create a satisfying visitor experience, user evaluations of integrated systems in real life settings are very important. This is what we set out to do in the real world user evaluation study reported in this paper.

For the robot FROG (Fun Robotic Outdoor Guide) state-of-the-art innovations in navigation, control and vision as well as state-of-the-art human-robot interaction design were combined to create a robotic tour guide that engages visitors in a tour and interacts socially with visitors to increase the visitor experience of the site. Our user evaluation study was carried out with the fully integrated system in autonomous mode at the Royal Alcázar in Seville (Spain).

2   Related Work

Some of the first guide robots were Rhino [1] developed by Burgard et al., its successor Minerva [2] developed by Thrun et al. and the robots used in the Mobot Museum experiment by Nourbakhsh et al. [7]. For these robots much effort was put into the development of robot autonomous navigation and collision avoidance [1], [2], [7].

More recent research on autonomous tour guide robots has put more emphasis on human-robot interaction. For example, RoboX, a series of 11 robots that were developed by Jensen et al. [8], guided visitors of the Swiss National Exhibition Expo.02. These robots used dynamic scenarios to control the visitor flow. Two robots in the science museum in Osaka, developed by Shiomi et al., engaged in personalized interaction with visitors [9] and a Robovie tour guide developed by Yamazaki et al. adopted typically human interaction cues [3] to focus the attention of the visitors.

The robots described above were equipped with guiding behaviors to guide visitors to several exhibits and present information about exhibits. We could not find papers reporting dedicated user evaluation studies of these guide robots. However, the papers which were more technically oriented, often mention some observations the research teams had of visitors' responses. For example, Burgard et al. state in [1] that the user interfaces of the robot should be robust and intuitive, because visitors usually spend less than 15 minutes with the robot. In [7] Nourbakhsh et al. mentioned that the robot's awareness of the people close by is most important to attract the attention of people. Clodic et al. state in [10] that to keep the interest of the visitors, the robot Rackham continuously had to give feedback to visitors to inform them that it knew where the visitors were, what it did, where it went and what its intentions were.

Next to the behaviors described above, guide robots face another challenge. Guides often have to interact with a group of people that visit a tourist attraction together. To realize this multi-user interaction capability is not only a serious technological challenge (e.g. computer vision has to shift from one user to another very fast), it is also highly challenging to design effective social behaviors for a robot that interacts with groups of people. When a social robot's behavior is not designed to interact with groups of people, people may start to take turns to interact with the robot, as was the case for robot Grace, described by Sabanovic et al. [11]. However, Grace was not prepared for people taking turns, and therefore had difficulties in reacting properly.

While previous studies mainly focused on one specific aspect of interaction, we set out to study integrated interaction features for the tour guide robot FROG with the aim to study the resulting user experience.

3   Study Design

In order to gain rich insights into the behaviors, attitudes, responses and experiences visitors have of the FROG robot, we performed a user evaluation study with FROG at the Royal Alcázar in Seville, Spain. Participants followed FROG for a fully autonomous tour through the Royal Alcázar.

The least intrusive way to gather data about how people experienced the tours would be through observation. However, using only observations would give too little insight in people's understanding and experience of the robot guided tour. Therefore, next to observing visitors, we interviewed them after completing a FROG tour.

As we could not rely on spontaneous visitors to comply with the request for an interview after a tour, we also invited dedicated participants to join a FROG tour and participate in the interview. Scheduling participants also offered us the opportunity to equip them with a microphone to record their speech during the tour and the interview. In this way we were able to collect rich data on user experience, attitude, responses and behaviors from our sample of scheduled participants.

3.1   FROG the Tour Guide Robot

FROG has several technical features that enable the robot to perform autonomous tours. FROG can drive around autonomously and avoids collisions with people and objects by taking into account basic social conventions. This has been described by Ramon-Vigo [12].
A bumper around the base of the robot secures that the robot will stand still immediately when it touches an object or person. FROG can search for groups, estimate their orientation and drive towards them, details of this have been described by Flohr et al. in [13]. Further, FROG can adjust the content of a tour to the interest of visitors by calculating the interest of the visitors based on their facial expressions. The techniques used for this have been described by Marras et al. in [14].

Fig. 1. FROG robot in action

The appearance of the robot was designed to attract the visitors, but also to be functional. The front of the robot has two 'eyes' in which the cameras for group detection were placed. On top of the robot, a pointer arm (3 DOF) with extra camera and LED lights was placed to enable FROG to point to places of interest. The touch screen on the front, which has full sunlight capacity, enables visitors to make contact and interact with FROG.

To present FROG as an engaging and fun robot and at the same time to have it narrate the more serious content, we introduced a split personality for the robots interface. This split personality allowed FROG to switch from *robot guide mode* to *narrator mode*. As a guide, FROG used a 'robot voice' that guided the visitors from one point to the next. To ensure realistic expectations of FROG's intelligence, the robot would not react to speech input of the visitors. Therefore, the 'robot voice' consisted of repetitive, pre-recorded standard sentences only to convey the status of the robot and to indicate that the robot was not processing speech input. At the points of interest, FROG gave information about the site as the narrator. As narrator, FROG used a prerecorded human 'voice-over' to offer narrations.

Narrations of FROG were supported by visuals. These visuals were presented either on screen or projected on nearby walls through the onboard projector. Further FROG was equipped with a pointer that was used to point to several points of interest. However, the pointer also had another function, namely to 'search' for participants when the robot stood at the starting point. This was only an interactional feature, because the pointer could not actually sense people around FROG; localizing people was done by using the laser sensors in the base of the robot.

The content of the narrations was carefully chosen to give visitors a brief but rich insight in the history of the site. Two criteria were used to choose the points of interest FROG would visit. First, the point of interest had to be accessible for FROG that was not able to climb stairs. The second criteria was how often human guides visited the points of interest. The content (narrations and visuals) that FROG provided for each of the points of interest was based on information given in the room by the Royal Alcázar complemented with information given in one of the official books sold by the Royal Alcázar [15]. Also, as much as possible, curiosities were added to the narrations. Curiosities are pieces of information that are special for one site only. An earlier study revealed that visitors really like to hear these [16].

The screen was used to visualize a face for FROG and to give information to the visitors. The screen mostly showed a smile. Additionally, information on the status of the robot was added, such as a small map to show where it drove to or messages such as 'loading location data.' Only in narrator mode at a point of interest, the full screen was used to show movies, pictures or augmented reality to visualize the story told.

Although FROG's main interaction features were positioned on its' front, it drove forward during transitions to a next point, which meant that visitors had to follow the robot facing its back. We expected that it would be most natural for visitors to follow the robot in this way, because this is what happens when people follow a human tour guide as well. During these walks the information on FROG's screen was a map that showed visitors where to go. Furthermore, FROG did not turn towards visitors before it would start an explanation about a point of interest, because it needed time to take in a position that was most advantageous for visitors to see the content. Consequently, we expected that participants would have enough time to gather around the front of the robot again at the new location.

3.2   Participants

During the FROG tours, invited participants as well as spontaneous visitors joined the robot tours. Invited participants were people we recruited in advance to follow the whole tour and they participated in a long (approx. 30 minutes) interview. A total of eight participants were recruited in four separate groups; two groups comprised of a Dutch male and a Dutch female; one group consisted of one Spanish female and her baby; and one group comprised of three Spanish students. As compensation for their participation, invited participants were allowed to visit the site by themselves after they had finished their participation in the study.

The invited participants were often joined by spontaneous visitors during the tour. Spontaneous visitors were people who visited the Royal Alcázar by chance and who joined one of the FROG tours spontaneously. A total of 18 spontaneous visitors who followed 8 different tours were interviewed. The compositions of these groups varied. There was one big group of five adults and two children (<8 years), a pair of adult men, two couples, a mother with her daughter (<10 years), and three visitors who visited individually. More spontaneous visitors followed the tour; we observed their interactions with the robot, but those visitors were not interviewed.

The invited participants had little or no experience with social robots. One of the spontaneous visitors was a technician and one was a robotics lecturer, others had no previous experience with robots. Most participants had previous experience with human tour guides or audio guides. All participants spoke English as a first or second language.

3.3   Procedure

During a single week in June 2014 FROG gave one to three autonomous tours a day through the Royal Alcázar. The tours always started close to the entrance gate. At this starting point the robot searched for groups of visitors who had just entered. When the robot located (a group of) visitors, it asked whether they were interested in a guided tour. The visitor groups were either the invited participants, the spontaneous visitors or comprised of both. The robot traveled to six points of interest. During the week that the robot gave tours in the Royal Alcázar, small changes were made to the behavior of the robot to iteratively improve the tour.

Even though FROG performed the tours autonomously, 7-10 researchers followed each tour from a distance to monitor progress of the various technical onboard systems. Also, one researcher carried a remote control stop, to stop the robot in any case of emergency; this did not occur.

The complete tour took about 25 minutes, depending on the number of obstacles and the number of people the robot would encounter in small hallways. The invited participants were asked to think aloud during the tour. For each group with invited participants, one participant wore a small microphone, which could generally pick up the speech of the whole group. During the walks between points of interest, the researcher asked the participants some questions to gather first reactions on their experience of being guided by the robot. Invited participants were instructed to indicate at what point they wanted to leave the robot tour in the case this desire occurred but to follow the tour till the end. Spontaneous visitors would join or leave the robot whenever they wanted.

After a tour with invited visitors, the participants were interviewed about their experiences with the robot. The interview with the invited participants took approx. 30 minutes and included topics such as: their experience of the tour, the things they liked about the robot or the tour, what they would change about the robot or the tour, how they experienced the interaction with the robot and how they experienced the way the robot guided them to the next point. An example of a question is: "How did you experience the length of the tour? Why?" The interviews were semi-structured, there was no specific order of the topics and participants were able to expand on what they found important to discuss. Also, since invited participants were interviewed together after each tour, they were able to comment on each other's remarks.

Spontaneous visitors were asked if they would answer a few questions right after they left the robot. The interviews with the spontaneous visitors took about two to five minutes. These visitors were asked about their impression of the robot, their experience of being guided by the robot and if they would have any suggestions for improvement. An example of a question is: "How would you describe this experience of following the robot to people at home who did not see the robot?"

3.4   Data Analysis

The data collected consisted of voice recordings of invited participants while they followed the tour, interviews with the invited participants, short interviews with the spontaneous visitors and observations in the form of video recordings and notes. All interviews and the voice recordings made during the tour with the invited participants were transcribed. The transcribed recordings were coded using the NTC (Noticing Things, Collecting Things and Thinking about Things) method as described by Friese [17], using the qualitative analysis software Atlas.ti [18]. Also notes of observations or remarks made by participants, taken during and after the tours and interviews were used in the analysis to complement the other resources.

One researcher combined all the data coded with the same code and then read this carefully, searching for commonalities and remarkable statements of the participants. These findings were combined in a summary of the general experience, details on what participants liked in the tour and specifics about what they did not like.

4   Results

In general, participants responded positively to being guided by the robot. This is likely to have been influenced by a willingness to please the interviewer, as well as by the novelty experience of being guided by a robot. Moreover, spontaneous visitors who conceded to take part in an interview were more likely to be those people who had had a positive experience.

The reports of first impressions of the robot were generally positive. The reason that was most mentioned (10 times) was that it was seen as an easy way to obtain information about the site. Four of the invited participants said that even though they could get the same information themselves from books, the experience of follow the robot was much more fun to them than reading the guidebook would have been. Six of the spontaneous visitors mentioned that they liked to get information (in English).

Further results on what the invited participants and spontaneous visitors experienced as positive or negative will be presented in table 1. Table 1 presents those themes that were most mentioned by participants, indicating how many of the invited visitors and how many of the spontaneous visitors made similar remarks.

The amount and quality of the remarks that people gave during the tours brings us closer to understanding how people will experience robot guided tours in real life. However, these are not yet a valid comparison to real representation of the real world. Partly because in the interviews with invited participants, the participants got room to discuss what they though was important to them, even though there was a topic list and questions that were asked to all. Furthermore, not all themes were addressed in the interviews with spontaneous visitors.

Table 1. Remarks of visitors on factors that influenced their experience

Themes	Invited participants	Spontaneous visitors
Influenced the visitor experience positively		
It was fun to join a robot tour, because it is innovative and cannot be found somewhere else, yet, so it was an experience itself.	7	13
It enriched the interaction with the environment more effectively than for example audio guides or books would do.	4	6
Even though it was clear to participants that the robot could not hear or understand them, they talked to the robot, but only when the robot used the "robot-voice."	7	-
Length of the total tour as well as the length and the amount of stops was ok. It gave the information needed to understand the history of the place. Maybe one more stop would be ok.	8	2
It is ok when strangers join, but it can be a problem when the new people stand in front of initial visitors or talk too loudly while the robot explains something.	8	-
The robot guide is helpful and fun for children/youth to explain the history of the site.	4	5

Table 1. (*Continued*)

Influenced the visitors experience negatively		
The movements of the robot were jerky and therefore made unclear what its intentions were.	7	-
The robot was unclear about where it wanted to go or whether visitors stood in its way.	3	1
The robot drove too slowly.	6	7
The robot did not turn towards the visitors once it arrives at the location before starting the explanation, this make people feel ignored by the robot.	5	1
After the explanation, the robot did not allow visitors to look around; it went to the next point immediately.	3	1
The 'robot voice' was too repetitive.	3	2
The robot did not make clear how long the tour would take and where the robot would bring them, which was a problem when people had only limited time to visit the Royal Alcázar.	-	9

What is notable from the results presented in Table 1, is that some themes are only discussed by invited participants, but that some other are merely or only discussed by spontaneous visitors. This strengthens the choice for the combined observation and interview approach that we used for this study.

5   Discussion and Conclusion

This user evaluation study of the FROG robot in a real world environment offered us insights in how we can improve the functionality and perceived experience for FROG tours. However, based on the results presented in this paper, it is difficult to generalize the findings to other contexts and situations that social robots may be used for these days. Nevertheless, we think that our findings and experiences can help other researchers and designers of social robots to prevent the development of unexpected interactions for robots designed for in-the-wild situations.

In earlier (controlled) research, project partners found that their system to recognize facial expressions performed extremely well in difficult situations, such as shade over part of the faces and faces covered with glasses [14]. However, during the in-the-wild user evaluation together we found it did not work well in the real world situation. Therefore, FROG was unable to adjust the information given to the interest of the visitors. The reason for this is that people behaved different from what we had expected and what the system was designed for. Therefore, the system was not ready to react to these unexpected behaviors of people.

In our system design, detecting interest was only possible when the robot could detect a face and read the facial expressions of a visitor. In order to do so, a visitor had to stand right in front of the robot and at a distance of 1 to 1.5 meters. However, from our observations in this in the wild study we noticed that groups larger than two or three people would form a semi-circle around the robot in order to allow everybody to see its screen, this often made people stand further away from the robot. As FROG would monitor the facial expressions of a closeby person to detect interest or disinterest and would adapt the content accordingly, the semi-circle formation made it impossible to read the facial expressions.

Our in the wild finding that people form a semi-circle is in accordance with findings of Heath and vom Lehn [19], who state that when people gather around static objects, many visitors get a chance to see the object at the same time, but that when people gather around interactive objects (often including a screen), less people can see the object at the same time, because people tend to stand closer to directly interact with the (touch) screen [19]. Initially, we placed the robot in the category of interactive exhibits, as people can interact with it at selected points of interest. However, at the moment it only gives information, it should be seen as a static exhibit in terms of Heath and vom Lehn. Hence, from our user evaluation we found that visitors did not mind when strangers joined the tour and that they gave each other room to look, which introduced the problem of not being able to record the facial expressions. The situation might have been different when FROG's explanations would have been more interactive at all points of interest, such as they were during the quiz that was initiated at one of the points of interest. Possibly, under such circumstances a few people would come closer to interact with the robot and strangers would probably not join, because they would not be able to see the content.

This example shows that even when technical features of the robot are well designed, in real world situations they might not work in the way developers would expect. HRI is a social science, even though technical innovations are needed to make progression in the field. Therefore, next to experimenting with real people in controlled settings, studies with people in in-the-wild settings are important to gain insight in the real responses and behaviors of people towards robots. To create social robots that really interact with people in in-the-wild environments, we advise researchers to implement in-the-wild studies at an early point in the development process. In this way, the behavior of the end user can be understood to subsequently create a robot that is able to deal with it.

References

1. Burgard, W., Cremers, A.B., Fox, D., Hähnel, D., Lakemeyer, G., Schulz, D., Steiner, W., Thrun, S.: Experiences with an interactive museum tour-guide robot. Artif. Intell. **114**(1), 3–55 (1999)
2. Thrun, S., Bennewitz, M., Burgard, W., Cremers, A.B., Dellaert, F., Fox, D., Hahnel, D., Rosenberg, C., Roy, N., Schulte, J., Schulz, D.: MINERVA: a second-generation museum tour-guide robot. In: Proceedings of the 1999 IEEE International Conference on Robotics and Automation, vol. 3, no. May, pp. 1999–2005 (1999)
3. Yamazaki, A., Yamazaki, K., Kuno, Y., Burdelski, M., Kawashima, M., Kuzuoka, H.: Precision timing in human-robot interaction: coordination of head movement and utterance. In: Proceedings of the SIGCHI Conference on Human Factors in Computing Systems, pp. 131–139 (2008)

4. Pitsch, K., Gehle, R., Wrede, S.: Addressing multiple participants: a museum robot' s gaze shapes visitor participation. In: International Conference on Social Robotics, pp. 587–588 (2013)
5. Yousuf, M.A., Kobayashi, Y., Kuno, Y., Yamazaki, A., Yamazaki, K.: How to move towards visitors: a model for museum guide robots to initiate conversation. In: 2013 IEEE RO-MAN, pp. 587–592 (2013)
6. Donner, M., Himstedt, M., Hellbach, S., Boehme, H.-J.: Awakening history: preparing a museum tour guide robot for augmenting exhibits. In: European Conference on Mobile Robots (ECMR), pp. 337–342 (2013)
7. Nourbakhsh, I.R., Kunz, C., Willeke, T.: The mobot museum robot installations: a five year experiment. In: 2003 IEEE/RJS International Conference on Intelligent Robots and Systems, pp. 3636–3641 (2003)
8. Jensen, B., Tomatis, N., Mayor, L., Drygajlo, A., Siegwart, R.: Robots meet Humans-interaction in public spaces. IEEE Trans. Ind. Electron. **52**(6), 1530–1546 (2005)
9. Shiomi, M., Kanda, T., Ishiguro, H., Hagita, N.: Interactive humanoid robots for a science museum. In: Proceedings of the 1st ACM SIGCHI/SIGART conference on Human-robot interaction, pp. 305–312 (2006)
10. Clodic, A., Fleury, S., Alami, R., Chatila, R., Bailly, G., Brèthes, L., Cottret, M., Danès, P., Dollat, X., Eliseï, F., Ferrané, I., Herrb, M., Infantes, G., Lemaire, C., Lerasle, F., Manhes, J., Marcoul, P., Menezes, P., Montreuil, V.: Rackham: an interactive robot-guide. In: 2006 IEEE ROMAN, pp. 502–509 (2006)
11. Sabanovic, S., Michalowski, M.P., Simmons, R.: Robots in the wild: observing human-robot social interaction outside the lab. In: 9th IEEE International Workshop on Advanced Motion Control, pp. 576–581. IEEE (2006)
12. Ramón-Vigo, R., Perez-Higueras, N., Caballero, F., Merino, L.: Transferring human navigation behaviors into a robot local planner. In: 2014 IEEE RO-MAN, pp. 774–779 (2014)
13. Flohr, F., Dumitru-Guzu, M., Kooij, J.F.P., Gavrila, D.M.: Joint probabilistic pedestrian head and body orientation estimation. IEEE Trans. Intell. Transp. Syst. **PP**(99) (2015)
14. Marras, I., Tzimiropoulos, G., Zafeiriou, S., Pantic, M.: Online learning and fusion of orientation appearance models for robust rigid object tracking. Image Vis. Comput. **32**(10), 707–727 (2014)
15. Hernández Núñez, J.C., Morales, A.J.: The Royal Palace of Seville. Scala Arts Publishers, Inc. (1999)
16. Karreman, D.E., van Dijk, E.M.A.G., Evers, V.: Using the visitor experiences for mapping the possibilities of implementing a robotic guide in outdoor sites. In: 2012 IEEE RO-MAN, pp. 1059–1065 (2012)
17. Friese, S.: Qualitative data analysis with ATLAS.ti, 2nd edn. Sage (2014)
18. "Atlas.ti." 7.5.6 (Computer Software), Atlas.ti GMBH, Berlin. www.atlasti.com
19. Heath, C., vom Lehn, D.: Configuring 'Interactivity': Enhancing Engagement in Science Centres and Museums. Soc. Stud. Sci. **38**(1), 63–91 (2008)

Higher Nonverbal Immediacy Leads to Greater Learning Gains in Child-Robot Tutoring Interactions

James Kennedy[✉], Paul Baxter, Emmanuel Senft, and Tony Belpaeme

Centre for Robotics and Neural Systems, Plymouth University, Plymouth, UK
{james.kennedy,paul.baxter,emmanuel.senft,tony.belpaeme}@plymouth.ac.uk

Abstract. Nonverbal immediacy has been positively correlated with cognitive learning gains in human-human interaction, but remains relatively under-explored in human-robot interaction contexts. This paper presents a study in which robot behaviour is derived from the principles of nonverbal immediacy. Both high and low immediacy behaviours are evaluated in a tutoring interaction with children where a robot teaches how to work out whether numbers are prime. It is found that children who interact with the robot exhibiting more immediate nonverbal behaviour make significant learning gains, whereas those interacting with the less immediate robot do not. A strong trend is found suggesting that the children can perceive the differences between conditions, supporting results from existing work with adults.

1 Introduction

An increasing amount of research is being conducted into the use of robots interacting with children [2], often in educational contexts [9,16,18]. However, it is unclear how robots should behave socially in tutoring interactions. Much human-human interaction literature assumes a certain level of sociality in teaching interactions, but specific guidelines for such behaviour are not provided for social roboticists. However, one concept which has been repeatedly correlated with increased cognitive learning gains is *nonverbal immediacy* (NVI) [12,19].

NVI combines the perception of gesture, gaze, touch, body orientation, vocal prosody and facial expressions into a single numerical metric, quantifying the NVI of an interaction partner. It should be noted that the usage of the word 'immediacy' does not refer to the typical definition involving timing of actions or a sense of urgency, but instead to the 'perceptual availability' of an interaction partner [12]. Such a measure has been highlighted as potentially valuable for human-robot interaction (HRI) researchers to characterise social behaviour of robots, as a means of providing a basis for comparison between studies [8]. NVI has previously been used in HRI to generate and evaluate the effect of gestural and vocal behaviour variation with results confirming that more immediate robot behaviour leads to increased information recall from a presentation [17]. However, whether such effects are still observed when larger scale behavioural

© Springer International Publishing Switzerland 2015
A. Tapus et al. (Eds.): ICSR 2015, LNAI 9388, pp. 327–336, 2015.
DOI: 10.1007/978-3-319-25554-5_33

manipulations are made, involving a greater number of modalities, and in two-way interaction contexts (i.e. social interaction) remain open questions. This paper presents a study to address these questions by exploring the effect of NVI behaviours on child learning in a tutoring interaction.

2   Related Work

A two week long study of a robot in a classroom by Kanda *et al.* [7] remains one of the best examples of robots being used successfully 'in the wild' with children. Recently this result has been extended by Alemi *et al.* who conducted a study over 5 weeks in an Iranian school, showing that the use of a robot can provide significant learning gains over the same material being covered without a robot [1]. Given the potential for robots in education, it is important to assess how the social behaviour of a robot can further improve outcomes.

Gordon *et al.* explored whether a robot exhibiting more 'curious' behaviour would inspire reciprocal curiosity in children interacting with the robot, and additionally whether this would lead to cognitive learning gains [6]. They found that whilst the curious robot did promote curious behaviour in children, this behaviour did not translate into learning gains as predicted by the human-human interaction (HHI) literature. Similarly, the HHI literature, and findings from several other HRI studies would predict that when robot social behaviour becomes more contingent, learning gains should increase. This has not always been found to be the case [8,10], but has been supported in an interaction where a robot taught children a novel language [15].

Previous work by Kennedy *et al.* used an identical interaction context to the one in this study [10]: it was found that a robot which was designed to be more socially contingent and personalised led to no significant learning, whereas a robot with behaviour violating typical HRI best practices did lead to significant learning. By providing a unified metric for robot social behaviour, the present study seeks to address inconsistencies when comparing these prior studies.

3   Methodology

The methodology used in this study is as established in prior studies [10]. A robot is used as a tutor in one-to-one interactions to teach children how to identify prime numbers between 10 and 100 based on whether they are divisible by 2, 3, 5 or 7. The children participating in the interactions do not have prior knowledge of prime numbers, but have the skills to do the division (albeit with imperfect performance), making the combination of these skills into a rule for categorising primes possible in a short interaction. Prior knowledge is assessed with a pre-test. The novel difference between the present study and previous work [10] is in the robot behaviour. Previously the social behaviour was based on a human model, whereas in this study the robot behaviour is derived from NVI concepts (detailed in Section 3.4).

3.1   Participants

The study was conducted in a class of children aged 8-9 years old. All children interacted with the robot, but due to breaks in protocol, and one statistical outlier (Grubbs' test), several interactions were excluded from the analysis. A total of 23 interactions were considered (16F/7M, age $M=8.74$, 95% CI [8.54,8.93]). All children had permission to participate in the study, of which 21 also had permission to be filmed for video analysis.

3.2   Hypotheses

The HHI literature has shown that greater instructor NVI leads to increased cognitive learning gains [19]. These findings have been partially confirmed in HRI [17], but using only 2 modalities (speech and gesture). Nonetheless, survey data showed that participants could perceive such behavioural differences. Previous work [10] conducted in a similar context to this study found that children gazed more at a 'more social' robot tutor during lessons, and were more likely to report it to be like a friend than an equally active, but not socially contingent, robot tutor. It could be argued that an increase in NVI behaviour is analogous to an increase in social contingency, so the same perceptual and behavioural differences of children could be predicted here. Based on these prior findings, the following hypotheses were devised:

H1. Children will report a higher rating of nonverbal immediacy for a robot designed with high nonverbally immediate behaviours than for a robot designed with low nonverbally immediate behaviours.
H2. A robot designed to be more nonverbally immediate will lead to greater child cognitive learning gains.
H3. Children will regard a robot with high nonverbal immediacy more like a friend than one with low nonverbal immediacy.
H4. Children will gaze at a robot with high nonverbal immediacy more during the prime lesson period than at a robot with low nonverbal immediacy.

3.3   Interaction Protocol

Interactions took place in an empty room familiar to the children near to their classroom. The children were briefed by one of the experimenters before entering the room. Two experimenters were present in the room, out of view of the child whilst they interacted with the robot. The child sat across a large touchscreen from an Aldebaran NAO robot (Fig. 1). A Microsoft Kinect was placed behind the robot to track the direction of the child's head gaze. Video cameras were placed behind the robot and behind the child to record the interaction. The average time spent interacting with the robot was $M=14m19s$, 95% CI [12m49s,15m48s]. The average interaction time from the videos (from entering the experiment room, to exiting – therefore including questionnaire time) was $M=19m19s$, 95% CI [17m37s,21m01s].

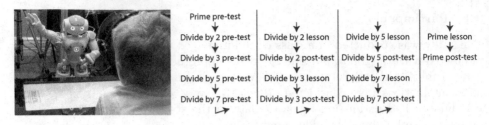

Fig. 1. (*left*) example from a high immediacy interaction; (*right*) structure of the task.

The robot would first introduce itself and ask the children to complete a pre-test on the touchscreen for prime numbers, and then pre-tests for each of the divisors (2, 3, 5 and 7). The robot would then deliver a lesson for each of the divisors and ask the child to complete a post-test following this lesson, i.e. the robot gives a lesson on dividing by 2 and then the child does a dividing by 2 post-test, followed by dividing by 3 lesson and post-test, and so on. After this had been completed, the robot delivered a lesson about prime numbers which combined the lessons for the divisors into a rule for determining whether a number between 10 and 100 is prime (a variation on the *Sieve of Eratosthenes* method). The interaction with the robot would finish with a prime number post-test.

The prime number pre- and post-tests both consist of 12 numbers which must be categorised as 'prime' or 'not prime'. Two sets of numbers were used for these tests, which are alternately used as the pre- and post-tests in a cross testing strategy to control for potential difference in test difficulty. The tests were balanced in terms of the number size, as it was assumed that higher numbers would be harder for the children to work with. The divisor pre-tests consist of 8 numbers which must be categorised as either 'can divide by X', or 'can't divide by X' (where X is 2, 3, 5 or 7, Fig. 1). The divisor post-tests are the same, but with 6 numbers instead of 8. In all pre- and post-tests, an equal quantity of numbers belong to each category.

After the interaction with the robot is finished, the child is asked by the experimenter to complete two questionnaires. The first questionnaire was a Robot Nonverbal Immediacy Quesionnaire (RNIQ), adapted from the short-form NVI questionnaire [13] and available online[1]. The RNIQ was modified from the original to specifically refer to robots and to be more easily understood by children. The second questionnaire consisted of two multiple choice questions, asking the children what they thought the robot was like (8 options including friend and teacher), and what they thought playing with the robot was like (4 options, plus a free text box).

3.4   Robot Conditions and Behaviour

The robot social behaviour was generated by considering the NVI questionnaire measures, as seen in [14]. The intention was to create high and low NVI

---

[1] http://www.tech.plym.ac.uk/SoCCE/CRNS/staff/JKennedy/immediacy.html

Table 1. Robot behaviour for high and low nonverbal immediacy (NVI) conditions.

High Nonverbal Immediacy	Low Nonverbal Immediacy
Leans forwards	Leans backwards
Actively gazes at child (with frequent movement)	Looks up and away from child (with occasional movement)
Frequent gestures while talking	No gestures while talking
Standard TTS	TTS modified to make voice "dull"
Continuous small upper body movements (relaxed upper body)	Rigid/tense upper body with no movement

conditions in order to address the hypotheses for the study (Section 3.2) and to explore the initial research questions presented in the introduction (Section 1). Children were assigned to conditions randomly, whilst balancing for gender and mathematical ability (as judged by the class teacher). This led to 12 children in the low NVI condition (9F, 3M) and 11 children in the high NVI condition (7F, 4M) after exclusions.

In order to implement larger-scale behavioural changes between conditions (as motivated by the initial research questions in Section 1 and points raised in [8]), each of the modalities rated in the RNIQ were considered for the Aldebaran NAO robot. Some of the modalities are not possible to manipulate (for example the NAO cannot perform facial expressions), but the other modalities were considered in turn and designed to be either maximally or minimally immediate. Table 1 shows the differences between the two robot conditions. All robot behaviour was autonomous, a 'Wizard-of-Oz' was only employed to click a button to begin the behaviour once the child was in position in front of the robot/screen.

4   Results

4.1   Learning Gains

To test the impact of the robot's lessons on the children's division skills, the percentage score of division across all pre-tests was compared with the score across all division post-tests (as there were a different number of items in the pre- and post-tests). A significant difference is found between the division pre-test percentage (M=84.1, 95% CI [79.9,88.3]) and the post-test percentage (M=88.6, 95% CI [84.8,92.4]); $t(22)$=2.080, p=0.049. This demonstrates that the children can learn from the robot and suggests that the lessons that the robot delivers are appropriate.

All scores for the prime number pre- and post-tests are out of 12. Given that the children have no prior knowledge of prime numbers and there are 2 potential categories for each image, a pre-test score of 6 (50%) would be expected from random behaviour. In the low NVI condition the improvement from pre-test (M=7.08, 95%

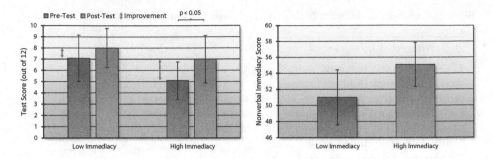

Fig. 2. (*left*) pre- and post-test scores on recognising prime numbers for the low and high nonverbal immediacy (NVI) conditions; (*right*) NVI scores for the designed low and high NVI conditions. Children improve more in recognising prime numbers when taught by a high immediacy robot. *Error bars* show 95% CI.

CI [5.01,9.15]) to post-test (M=8.00, 95% CI [6.24,9.76]) is not statistically significant; $t(11)$=0.754, p>0.05. However, in the high NVI condition the difference from pre-test (M=5.09, 95% CI [3.43,6.75]) to post-test (M=7.00, 95% CI [4.88,9.12]) is statistically significant at the p<0.05 level; $t(10)$=3.057, p=0.012 (Fig. 2).

The pre-test score appears to be very different between the conditions, however this was not found to be significant; $t(21)$=1.640, p=0.116. The 95% confidence interval for the pre-test in both conditions covers the expected value of 6, which reassures that the children did not know what primes were before the intervention. Additionally, there is no significant difference between the two different pre-test scores, or of the improvement between pre- and post-test, regardless of which of the two pre-tests were taken; this shows that the tests can be considered of equal difficulty. Therefore, partial support has been shown for Hypothesis 2: children interacting with the high NVI robot benefit from increased cognitive learning gains. However, this is slightly tempered, as there is no significant difference between conditions. Children in both conditions are likely to improve (which isn't surprising given practice and teaching input), but those in the high NVI condition undergo significant improvement, whereas those in the low immediacy condition do not.

4.2   Questionnaire Data

After the children had interacted with the robot they were asked to complete the RNIQ on paper. Immediacy scores are calculated from the answers to the RNIQ questions: the higher the resulting number, the higher the perceived immediacy. The score can be up to 80, but there are a number of measures for which there are no equivalent robot behaviours (e.g. touching the child). Therefore, a score of around 56 would indicate a rating of near-maximal NVI given the modalities which are manipulated. This reduction in the expected score also inhibits the potential for difference between conditions, as for many of the questionnaire elements, the behaviour is the same (e.g. the lack of facial expressions).

The designed low immediacy condition received a mean NVI score of M=51.0 (95% CI [47.6,54.4]). The designed high immediacy condition received a mean score of M=55.1 (95% CI [52.3.57.9]). An unpaired t-test reveals a strong trend towards significance between these ratings; $t(21)$=2.031, p=0.055 (Fig. 2). This provides reasonable support for Hypothesis 1; that children will perceive a robot designed to be more nonverbally immediate as such.

The second questionnaire that the children completed asked them what the robot was like, and what playing with the robot was like. The children were asked "For me, I think the robot was like a -", and had 8 options to choose from (brother or sister, classmate, stranger, relative (e.g. cousin or aunt), friend, parent, teacher, neighbour). Given Hypothesis 3 (that children in the high immediacy condition will more frequently report the robot to be like a friend) the responses were sorted into whether the children responded that the robot was like a friend, or not. In the high immediacy condition 6 children reported the robot to be like a friend and 5 not (with all selecting 'teacher'), whereas in the low immediacy condition 1 child reported the robot to be like a friend and 11 not (1 'classmate', 10 'teacher'). Fisher's exact test reveals a significant difference between the conditions, with those in the high immediacy condition significantly more likely to report that the robot is like a friend than those in the low immediacy condition; p=0.027. Therefore Hypothesis 3 is supported.

This result is surprising as the children are told multiple times by both the experimenter and the robot that the robot is a *teacher robot* which will teach them some maths. However, the same result has previously been found, which led to the formulation of Hypothesis 3. Children interacting with a 'more social' robot reported more frequently that the robot is like a friend [10]. If the high immediacy robot in this study is considered to be more social, then the same finding is confirmed here.

4.3   Gaze Analysis

The 21 videos from the interactions were manually coded for child gaze during the prime lesson segment as this has previously been found to be indicative of overall gaze patterns in the interaction [10], and the prime lesson constitutes a key part of the interaction in terms of the learning outcome. One of the 21 videos was excluded due to occlusions, leaving 9 videos from the high immediacy condition and 11 from the low immediacy condition for analysis. 20% of the remaining videos were second coded to verify reliability, with a mean inter-rater reliability (Cohen's κ) of 0.83, indicating *almost perfect* agreement.

No significant difference was found between the length of time children gaze at the robot in seconds per minute of the prime lesson segment between the high NVI condition (M=15.9, 95% CI [11.3,20.5]) and the low NVI condition (M=15.4, 95% CI [11.9,18.6]); $t(18)$=0.214, p>0.05. Nor is there a significant difference in the number of times children gaze at the robot per minute of the prime lesson segment between the high NVI condition (M=15.2, 95% CI [12.2,18.2]) and the low NVI condition (M=14.7, 95% CI [11.7,17.7]); $t(18)$=0.234, p>0.05. Therefore, Hypothesis 4 (children will gaze more at the high NVI robot) is not supported. This is a surprising result, which possibly strengthens the link between robot behaviour and learning. If gaze is considered to be a reflection of child attention, then despite equivalent attention during the key piece of learning input from the robot the learning results still vary, suggesting that the robot social behaviour could be responsible. Of course, this is just one of many possibilities and the gaze could be equal simply because the behaviour during this phase is quite novel compared to the rest of the interaction.

5 Discussion

Fairly strong support for Hypothesis 1 was found: children do recognise when a robot has higher or lower nonverbal immediacy. The difference was close to significance at the 5% level, with the differences between the means only just including no difference; $-0.10 \leq \mu_{HNVI} - \mu_{LNVI} \leq 8.28$. This finding shows that the robot behaviour is largely interpreted by the children as intended by the designer, despite the children not seeing the other robot condition for comparison. However, the variation in the children's answers is quite high, possibly due to a tendency to categorise at the extremes of scales [3], misunderstanding of some negatively worded questions, or over-attribution of robot competencies.

The results also partially confirm Hypothesis 2; that a robot perceived as more nonverbally immediate will lead to greater cognitive learning gains. This prediction was made based on HHI data [19] and HRI data [17], which seem to agree with the present findings. It should be noted that the effect size is relatively small: although there is significant improvement, the post-test mean 95% confidence interval still covers the expected 50% score of 6 which could be achieved through random action. Interestingly, there is a moderate positive correlation between immediacy score and cognitive learning gains for the high NVI condition ($r=0.22$), which is remarkably close to that which is found in HHI literature ($r=0.17$) [19]. Conversely, there is a negative correlation for the low NVI condition and learning gains ($r=-0.32$), indicating that as children rate the low NVI robot as more immediate, their learning tends to decrease.

It is therefore suggested that other factors besides robot behaviour could have a greater impact on the learning taking place at the individual level, particularly for those in the low NVI condition. From exploratory analysis of the data in this study, gender, teacher predicted maths ability, and age were all controlled for, with none being revealed *post hoc* as a significant factor. Novelty is often raised as a potential issue when performing single interactions of this nature [5,7]. It could indeed be a factor here, with the novelty of the robot impacting some of the children more than others, although the influence of novelty could be expected to be similar in both conditions. Another possible factor could be in the character of the children themselves. Whilst the children are familiar with the environment, they are not familiar with the two experimenters in the room, which may impede their performance, or affect their questionnaire responses [11]. Children who are more timid may be affected by this to a greater extent than those who are more confident.

Finally, it should be noted that the interactions in this study are relatively short, and the pre- and post-tests were conducted immediately before and after the learning input. Therefore, whether the learning gains observed here are retained over a longer period of time (and thus the concepts are truly *learned*) remains to be seen. This is an important factor which should be addressed in future work. Research from HHI, which has been conducted over the period of academic terms with adults, has found that high NVI behaviour can confer a greater advantage in terms of learning gains [19], although it must be noted that this is not always the case [4]. It could be hypothesised that over a longer time period with a robot tutor the differences between high and low immediacy conditions would increase as novelty would wear off and more of the potential benefit commonly gained in HHI from more immediate behaviour could occur.

6   Conclusion

This study has shown a strong trend towards children perceiving robots designed to have high and low nonverbal immediacy behaviours as such when measured using a short-form robot nonverbal immediacy questionnaire (RNIQ). This perceived difference was also supported by the children's interpretation of the robot's relation to them, with significantly more children in the high nonverbal immediacy condition reporting the robot to be like a friend. There were no observable differences in gaze behaviour during the prime lesson period of the interaction, indicating that learning differences remain despite possibly equal amounts of attention being paid by the child to the robot during the lesson.

This study has generally shown that children who interact with the robot exhibiting more immediate nonverbal behaviour make significant cognitive learning gains, whereas those interacting with the less immediate robot do not. A strong trend is found in the difference between the conditions suggesting that the children can perceive the differences between conditions, which supports results with adults. While further work is required to assess the strength of the learning effects over longer time scales, and the effect of individual differences beyond academic competence, these results have demonstrated the utility of high nonverbal immediacy robot behaviours in a tutoring context.

Acknowledgments. This research was funded by the School of Computing, Engineering and Maths, Plymouth University, U.K. and the EU FP7 DREAM project (grant no. 611391).

References

1. Alemi, M., et al.: Employing Humanoid Robots for Teaching English Language in Iranian Junior High-Schools. International Journal of Humanoid Robotics **11**(3) (2014)
2. Belpaeme, T., et al.: Multimodal Child-Robot Interaction: Building Social Bonds. Journal of Human-Robot Interaction **1**(2), 33–53 (2012)
3. Borgers, N., et al.: Response Effects in Surveys on Children and Adolescents: The Effect of Number of Response Options, Negative Wording, and Neutral Mid-Point. Quality and Quantity **38**(1), 17–33 (2004)

4. Comstock, J., et al.: Food for thought: Teacher nonverbal immediacy, student learning, and curvilinearity. Communication Education 44(3), 251–266 (1995)
5. Gockley, R., et al.: Designing robots for long-term social interaction. In: IEEE/RSJ International Conference on Intelligent Robots and Systems, IROS 2005, pp. 1338–1343. IEEE (2005)
6. Gordon, G., et al.: Can children catch curiosity from a social robot? In: Proceedings of the 10th ACM/IEEE International Conference on Human-Robot Interaction, pp. 91–98. ACM (2015)
7. Kanda, T., et al.: Interactive Robots as Social Partners and Peer Tutors for Children: A Field Trial. Human-Computer Interaction 19(1), 61–84 (2004)
8. Kennedy, J., et al.: Can less be more? the impact of robot social behaviour on human learning. In: Proceedings of the 4th International Symposium on New Frontiers in HRI at AISB 2015 (2015)
9. Kennedy, J., et al.: Comparing Robot Embodiments in a Guided Discovery Learning Interaction with Children. International Journal of Social Robotics 7(2), 293–308 (2015)
10. Kennedy, J., et al.: The robot who tried too hard: social behaviour of a robot tutor can negatively affect child learning. In: Proceedings of the 10th ACM/IEEE International Conference on Human-Robot Interaction, pp. 67–74. ACM (2015)
11. Leite, I., et al.: Modelling empathic behaviour in a robotic game companion for children: an ethnographic study in real-world settings. In: Proceedings of the 7th ACM/IEEE International Conference on Human-Robot Interaction, pp. 367–374. ACM (2012)
12. Mehrabian, A.: Some Referents and Measures of Nonverbal Behavior. Methods & Designs 1(6), 203–207 (1968)
13. Richmond, V.P., McCroskey, J.C.: Nonverbal Communication in Interpersonal Relationships, 3rd edn. Allyn and Bacon (1998)
14. Richmond, V.P., et al.: Development of the Nonverbal Immediacy Scale (NIS): Measures of Self- and Other-Perceived Nonverbal Immediacy. Communication Quarterly 51(4), 504–517 (2003)
15. Saerbeck, M., et al.: Expressive robots in education: varying the degree of social supportive behavior of a robotic tutor. In: Proceedings of the SIGCHI Conference on Human Factors in Computing Systems, CHI 2010, pp. 1613–1622. ACM (2010)
16. Short, E., et al.: How to train your dragonbot: socially assistive robots for teaching children about nutrition through play. In: Proceedings of the 23rd IEEE International Symposium on Robot and Human Interactive Communication, RO-MAN 2014, pp. 924–929. IEEE (2014)
17. Szafir, D., Mutlu, B.: Pay attention!: designing adaptive agents that monitor and improve user engagement. In: Proceedings of the SIGCHI Conference on Human Factors in Computing Systems, CHI 2012, pp. 11–20. ACM (2012)
18. Tanaka, F., Matsuzoe, S.: Children Teach a Care-Receiving Robot to Promote Their Learning: Field Experiments in a Classroom for Vocabulary Learning. Journal of Human-Robot Interaction 1(1), 78–95 (2012)
19. Witt, P.L., et al.: A Meta-Analytical Review of the Relationship Between Teacher Immediacy and Student Learning. Communication Monographs 71(2), 184–207 (2004)

Exploring the Four Social Bonds Evolvement for an Accompanying Minimally Designed Robot

Khaoula Youssef [(✉)], P. Ravindra De Silva, and Michio Okada

Interaction and Communication Design Lab, Toyohashi University of Technology,
1-1 Hibarigaoka, Tempaku,Toyohashi, Aichi, Japan
youssef@icd.cs.tut.ac.jp, {ravi,okada}@tut.jp
http://www.icd.cs.tut.ac.jp

Abstract. In this paper, we investigate the effect of combining inarticulate utterances (IU) with iconic gestures (IG) in addition to the response mode (proactive or reactive) and its impact on the bonds formation as well as the establishment of a positive relationship between the human and the accompanying robot. Specifically, we employ different scenarios while measuring in each instance the different social bonds that occur and we evaluate the human-robot relationship (HRR) in order to pick the behaviors that yield a positive HRR. Experimental results show that combining proactivity with the full mode (IU+IG) leads to social bonds evolvement and then to a better positive HRR.

Keywords: Inarticulate utterance · Iconic gestures · Social bonds · Reactivity · Proactivity

1 Introduction

An accompanying robot that abides by human social rules is judged to be acceptable to humans. We think that such a robot may trigger positive behaviors in humans' and leads to a more positive HRR. Broadly speaking, in daily life, positive human behavior toward others is driven by the social bonding that evolves during their interactions and which as a result leads to a reciprocation of others' kindness with a noble feeling and/or act. Travis Hirschi's social bonding theory argues that people who believe in societal rules are attached to society and therefore, have a strong commitment in achieving conventional activities and reciprocating the positive gestures of others [1]. These people feel highly involved in their daily lives so they start to invest more time and energy in activities that serve to further bonds with others and this leaves limited time to become involved in deviant activities [2]. Chris et al. [2] highlight that people who have weak bonds are more likely to deviate from normal behavior and have bad relationships with others [2]. On these grounds, we can argue that if we measure robot's users social bonds, we can be capable of detecting the robot's behaviors that have the potential of leading to a better positive HRR. Hirschi defines four following social bonds: belief (B), attachment (A), commitment (C) and involvement (I) [1]. Chris et al. [2] argue that all

© Springer International Publishing Switzerland 2015
A. Tapus et al. (Eds.): ICSR 2015, LNAI 9388, pp. 337–347, 2015.
DOI: 10.1007/978-3-319-25554-5_34

these bonds are incorporated in Talcott Parsons' AGIL schema and, thus, the belief bond serves the function of latent pattern-maintenance (L), attachment to others, serving the function of integration (I), commitment proportional to the energy and time that one puts forward, serving the function of goal-attainment (G), and, involvement, consisting of the extra time and energy that one affords and serves the function of adaptation (A). The AGIL paradigm highlights the societal functions that, every society must meet to be able to maintain a stable, flourishing social life. Therefore, if we assume that we want to establish a stable, positive HRR, we must investigate behaviors that lead to social bonds evolvement during HRI in a way that can guarantee users and accompanying robots integration (attachment), goal attainment (commitment), adaptability to each other (involvement) and support of implicit social norms (belief) [2]. In the current study, we explore behaviors that lead to bonds formation in the context of interactions with an accompanying robot named ROBOMO. We are interested in understanding whether IUs or/and IGs, help to establish the social bonding between the human subjects and ROBOMO. If the social bonding is strong, then we may guarantee a decrease in the possibility of a robot's abundance which is by analogy to Hirschi's theory, the possibility of deviance. We detail the related work in section 2, explain ROBOMO's design in section 3, and explain the robot's architecture in section 4. After that, we describe the hypothesis and the experimental setup respectively in sections 5 and 6 while measurements are described in section 7. Finally, we give the results and insights obtained in sections 8 and 9.

2 Related Work

The concept of the accompanying robot (a robot functioning as a human peer in everyday life) is rapidly emerging. The accompanying robot must facilitate interaction with a human in order to complete a set of tasks. Many studies integrate multi-modal communication in order to satisfy human needs with regards to sociability and task achievement [3][4], etc. Ishiguro et al. [3] investigate the effectiveness of such multi-modal communication in order to explore whether the accompanying robot can help children improve their English ability. Garell et al. [4] utilize a group of robots in order to guide a group of people from a designated starting point to a specific destination. While the ability to perform a set of tasks skillfully is a desirable attribute, this alone does not cause humans to regards robots as partners. Humans do not evaluate their partners based on the success of task achievement alone. Instead, we believe that humans have to feel that a bonding and a stable positive relationship is maintained between themselves and the accompanying robot. As an example, children are likely to learn new concepts and still easily form bonds with a caregiver [5]. In such scenarios, children identify salient objects of a discussion, distinguish the different voices, express themselves using simple gestures, and show interest by picking up tonal differences [5]. Slowed voice tones (which we call in our study, IUs) help the child to bond with the caregiver. Thus, a mutual interest in communication evolves. Both parties can sometimes take the initiative (proactivity), proving their belief in the benefit of the interaction, instead of being only reactive (responsive) [5]. It is then a logical step to think about exploring the bonding between humans and robots that may evolve if we use IUs and IGs. When behaviors are designed adequately, a social bond may then evolve, resulting in people feeling more confident about integrating accompanying robots into their daily lives.

Many HRI studies have investigated if children can form relationships with robots, and if they can view them as friends. As an example, Stevenson et al. [6] show that children are willing to share secrets with robots and interact with them in a similar way as they would with an adult. Similarly, Swerts et al. [7] highlight that children consider playing with a robot like playing with a friend. Although there has been relatively broad research on the child-accompanying robot bonding, it has only dealt with the attachment bond; there is insufficient research that explores the entire evolvement of the four social bonds. Also, we need to investigate the formation of the four bonds between the adult and the robot rather than only between the child and the robot. In fact, Chris et al. [2] point out that the attachment bond is insufficient in predicting the nature of the human-society relationship and insist on the fact that four bonds must be explored for that purpose. Adding to that, little attention was paid to the bonding that may evolve in the context of a minimally-designed accompanying robot. Most of the studies we looked at, focused on the use of speech [3] or autonomous navigation [4] in order to increase a human subject's feeling about the consciousness and agency of the accompanying robot. We believe that bonding can evolve even within a simpler setting. For example, in a traditional adult-child interaction, the caregiver only needs to hold a baby without walking or talking and still the caregiver can interpret the meaning and feel the bonding with the baby [5]. Following the same strategy, we adopt the minimal design concept that it is proposed by Okada et al. [8]. This minimal design concept consists of designing a simple robot in terms of anthropomorphic features as well as the number of communication channels used. In this vein, our goal is to investigate the effect of using proactivity and/or reactivity as an interaction mode as well as the effect of using few communication channels such as IGs and/or IUs on the bonding formation while keeping in mind a strong bonding evolvement is an indicator of a positive HRR.

3   ROBOMO Concept Design

ROBOMO has a long shaped body with no arms. We have intentionally given ROBOMO a pitcher plant (Nepenthe) appearance to encourage people to interact with it much as one might interact with a young child (Fig.1). The IUs were produced according to the generation method for IU described by Okada et al. [9]. Three behaviors were exhibited: (i) the IUs:yes, no, right, left, back, forward, go, stop, slow down (ii) the tone:happy or sad based on the user's previous step correctness and (iii) IGs: turn left, turn right, yes (to implicitly mean "go"), no (to implicitly mean "stop"), bow to the front, bow to the back, face tracking (is used in S3 and S4 when the person has to slow down). A user can ask the robot to give

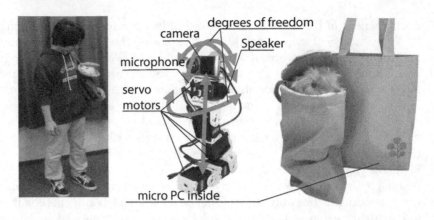

Fig. 1. ROBOMO's design

information about the direction (reactive mode). When the robot automatically helps the user, it is called a proactive robot.

4   ROBOMO Architecture

To communicate with ROBOMO, the user has to communicate slowly so that the robot, using its internal microphone and Julius (a Japanese word recognition software)[1], can interpret and satisfy the user's request. ROBOMO tracks the user's face using a web camera (Fig.2). ROBOMO integrates a micro PC to adapt to the user's request and affords an answer through its speaker. Moreover, it uses five servo-motors (AX-12+) to exhibit the gestures described in section 3 ([10]).

5   Hypothesis

We believe that being reactive proves that there is at least a minimal interest in the interaction with the robot and we expect that being proactive shows that one is goal-directed and actively taking charge of the situation. Thus we summarize our first hypothesis as follows:

– H1: ROBOMO should behave proactively when suggesting help. (Proactivity versus Reactivity)

The current study also focuses on another design choice, one the concerns the usage of iconic gestures and/or IU that can possibly be integrated in the character of an accompanying robot. This is why we want to investigate:

– H2: To guarantee a more positive HRR, the robot has to use only IUs, gestures or a full mode (gestures+IUs).

---

[1] Julius is a continuous real-time speech recognition decoder for speech-related studies that does not need training.

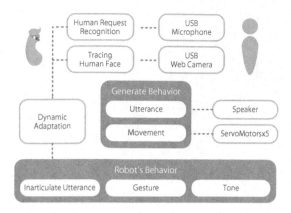

Fig. 2. The system architecture of ROBOMO.

6   Experimental Protocol

We setup an indoor ground for navigation tasks that contained intersections (Fig. 3). To pick the right behavior, the participant is instructed by the robot. We asked the participants to talk to ROBOMO slowly whenever they believed that they needed the robot's help. There was no training period in which the participants were familiarized with the task and/or the robot. Users could ask about directions or the traffic light[2] color in order to complete the task and reach the reward (music CD) location. Users were to ignore the reward location and only rely on robot's directions in order to reach it. 20 participants with ages varying from [22 − 30] years old, took part in four scenarios. We have chosen several different configurations[3] during the four non contiguous scenarios to guarantee the diversity of the participants' responses. This also helped to ensure that any successful guess in the meaning of ROBOMO's behaviors was not related to the fact that participants were accustomed to the same configuration. In our scenarios, if the human did not understand the robot's response, he/she would repeat his/her question within a short time for direction confirmation. In each scenario, the participant interacted with ROBOMO for at least two minutes and then answered the questionnaires indicated in the section 7. After two days, the human subject came again for the second session. As a result, the experiment took twelve days to be completed (four scenarios in total). We designed four different scenarios of interaction: for scenario 1 (S1), the robot adopted a reactive mode using IUs; during scenario 2 (S2), the robot adopted a proactive mode using IUs; during the scenario 3 (S3), the robot adopted a proactive mode using IGs; during scenario 4 (S4), the robot adopted a proactive,

---

[2] We increase the number of traffic lights by 2 in each new scenario and we change it positions. So in S4, we have 8 traffic lights.

[3] In each new configuration, we increase the target path's complexity and we change the starting location.

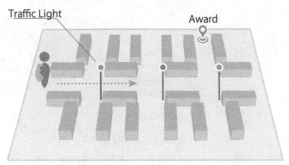

Fig. 3. The experimental setup.

full mode (IUs+IGs). The whole experiment was video recorded so that the users' facial expressions[4] could be detected.

7   Measurements

To measure the social bonding we established, based on each of the bond's definition a set of subjective and objective metrics. As we assumed that the belief bond corresponds to the latent social laws, we associated the belief bond[5] to the human's belief in the robot's social presence and it conscious agency. We then calculated the instances of eye contact, the rate of respect[6], the number of averted gazes, and, finally, cooperation metric (a 7 point Likert-scale questionnaire inspired from [11]).

As the attachment bond[7], is the emotional link that may evolve during the HRI, we used a different 7 point Likert-scale metrics, one that included : the pleasure [12], caring [13], perceived closeness [11], stress-free [11] and likeability[14].

The commitment bond[8] involves time, energy and effort expressed in conventional lines of action to achieve the task goals. To measure this commitment, we measured cognitive effort using a 7-point Likert-scale with the following metrics: arousal [12], mutual attention, users evaluation of the robot's "cognitive" effort through perceived competence [13] and perceived intelligence [14]. We also measured the user's: successful cognitive effort[9], expanded energy (physical effort

---

[4] Features used to determine the facial expressions are the lips, eyebrows, eyes.

[5] Survey for the belief bond:http://goo.gl/forms/GkJzXrMmUt

[6] Rate of respect= number of times the human asked the robot/number of total times the human should have asked the robot (a specific number for each configuration). It gives indirectly an idea about the overall system's performance and the participants' ability to understand the feedback (intelligibility).

[7] Survey for the attachment bond:http://goo.gl/forms/eoikVunVjG

[8] Survey for the commitment bond:http://goo.gl/forms/ItqTMqKVpU

[9] Successful cognitive effort= successful interactions/ total number of interactions. It gives indirectly an idea about the overall system's performance and the participants' ability to understand the feedback (intelligibility).

rate[10]) and time (interaction time). Achievement was also measured (achievement [11]) just like Tanioka and Glaser [15] used achievement scores to measure commitment bond in schools. Finally, we asked users to describe their experience with the robot (situational empathy) just like Lasley et al. [16] used self-report descriptions of high school students to assess their evaluation of the attainment of good grades. In our case, the human subject was required to talk about the most prominent achievement that he believed the HRI succeeded in attaining.

The involvement bond[11], is closely tied to the commitment bond in that it entails the actual amount of extra expanded time a human takes to pursue the HRI. It is also an indicator of the human's adaptation according to Chris et at [2]. It focuses on the idle time available when one is not engaged and the effort expended during that extra time. We used different 7-point Likert scale questionnaires to assess the involvement bond through different metrics: positive and negative human faces support [17]. Positive and negative human face support consists of supporting a user's social needs in terms of involvement during the HRI [18] (indicators of human's adaptation to the HRI). To ensure that the subjects were not getting accustomed to their surroundings, we asked the users whether they felt used to the task (adaptability [11]) so that we can discard any user who confirms that he get used to the environment of the experiment. We also calculated the number of times eyes were wide open (surprised), corners of the mouth were turned upwards (disgust), one eye brow raised (wondering) and mouth corners raised (happy) since these are optional behaviors that the human is not obligated to express and which indicates that he/she is emerged by (involved in) the HRI.

Finally, deviance[12] (not a bond) may be translated in the context of HRI as one's refusal of interacting with the robot. Based on this definition, we devised a measurement for persuasiveness. We instructed subjects to arrange a list of words according to their own priorities and then we calculated the level of persuasiveness using Kendall-tau distance metric [19]. We measure also trust [13]and the long-term use [11]. In fact, by measuring the different social bonds, we may be able to conclude whether there is a positive or a negative HRR. To confirm so, we tried to establish a correlation between the evolvement trend (positive or negative) with the deviance values. If, for example there were no positive evolvement in the bonding, we may draw a preliminary conclusion and say that there is a negative HRR. By combining this conclusion with deviance results, we may be able to confirm this insight (a negative HRR evolved).

8   Proactivity versus Reactivity

Results comparing the four bonds values of the first two scenarios (S1 and S2) are represented in Table 1 . We used two gray scales to color the cells, corresponding

---

[10] Physical effort rate= number of steps/ total number of due steps (a specific number for each configuration). It gives indirectly an idea about the overall system's performance itself and the participants' ability to understand the feedback (intelligibility).

[11] A survey of the involvement bond:http://goo.gl/forms/YUCtIVuNz0

[12] A survey of the deviance:http://goo.gl/forms/M7DWeM1mqO

to an increase in the metric values. For example, if the percentage of participants whose metric values in S1 exceeded the values in S2, we used the light gray color; we used the darker gray in the reverse situation. The t-test results comparing the reactive and proactive conditions (S1 and S2) show that there was an increase in most of the bond values when the robot was in proactive mode (S2). Consequently, we may give a preliminary conclusion by saying that a proactive robot leads to a more positive HRR. The averted gaze metric had higher results during S1 which shows that users were avoiding the robot more during S1 in comparison to S2. We also noticed that there were no significant differences in terms of the number of times the mouth corners raised (disgust) and the eye brows were raised (wondering) which indicates that users most of the time were showing the same level of negative feelings. As we had better results in S2, we can confirm these by comparing the bonds evolvement increase in S2 with the three metrics: trust, persuasiveness and long-term use that we have better positive HRR when the robot is proactive. This suggests that using a proactive accompanying robot is more adequate to trigger more bonding between the human and the robot and this leads to a more positive HRR (H1 investigated).

9   Comparison of Proactive Full Mode with IU-Based and Gesture-Based Communication

One-Way ANOVA and Tukey HSD results comparing the four bonds values of the last three scenarios (S2, S3 and S4) are represented in the Table 1 (IUs vs IGs vs full mode(IUs+IGs)). We colored the cells gradually with gray to indicate the increase in the metric values for S2 (S2 the lightest, S4 the darkest). For example, if the cell is colored with lightest gray and we were comparing S2 and S4, then that means the percentage of participants had results in S2 that exceeded the results in S4, and vice versa. Table 1 shows that there was an increase in most of the bonds metrics in S4 (the dark gray color prevailed in Table 1; significant Tukey HSD results also given). There were no differences in terms of successful cognitive effort when comparing S2 and S4 (F-test=5.681; p-value=0.006) HSD [S2 vs S4]=0.218, which shows that the added gestures in S4 were not responsible for IUs understanding (users could understand the IUs since S2). IGs, too, were considered to be expressive enough, as there were no differences in successful cognitive effort, when comparing S2 and S3 (HSD[S3 vs S2]=0.220). We can also point out that there were no differences in terms of persuasiveness between S2 and S3, which highlights that using the IUs or IGs is already enough to make the robot convincing for the user. By comparing S2 to S4 or S3 to S4, we see that users find the robot more persuasive when it combines the IUs and IGs instead of using the IGs and the IUs separately. This highlights that the full mode (IUs+IGs) guarantees better persuasiveness. By comparing S2 to S3, we see that most of the S2 bonding-metric results were higher than or the same as the results in S3 except for averted gazes (F-test=3.543; p-value<0.001;p-value<0.001) which were higher in S3. This means that a silent robot was not as appealing to the users.

Table 1. The comparison results of S1 and S2 (2 tailed paired t-test, df=19, alpha=0.05: proactivity vs reactivity) as well as comparison results of S2, S3 and S4 (One way ANOVA and Tukey-HSD tests: IUs only vs IGs only vs full mode (IUs+IGs)) with "p." stands for "perceived.", "E" "evolution" (the % of participants whose metric results increase in S1 or S2), "N/A" refers to cases when further statistical tests were not warranted (F-test was not significant) and 3 gray scales to color the cells corresponding to an increase in the metrics values with light gray corresponds to S1 and the darkest gray color to S4.

	Metrics	proactivity vs reactivity			IU vs Gestures vs full mode				
		t-test	p-value	E	F-test	P-value	S2 vs S4	S2 vs S3	S3 vs S4
B	eye contact	3.3441	0.0034	85%	185.023	< 0.001	< 0.001	0.055	< 0.001
	averted gaze	6.2056	0.0001	65%	63.714	< 0.001	< 0.001	< 0.001	< 0.001
	cooperation	5.977	0.002	95%	55.541	< 0.001	< 0.001	0.006	< 0.001
	respect rate	8.3533	0.0001	5%	117.163	< 0.001	< 0.001	0.033	< 0.001
A	pleasure	5.2248	< 0.001	85%	94.709	< 0.001	0.001	< 0.001	< 0.001
	likeability	2.4982	0.0218	100%	129.041	< 0.001	< 0.001	0.003	< 0.001
	stress-free	4.1944	0.0005	100%	111.309	< 0.001	< 0.001	< 0.001	< 0.001
	caring	10.2172	< 0.001	95%	148.371	< 0.001	< 0.001	< 0.001	< 0.001
	p.closeness	3.5962	0.0019	75%	92.224	< 0.001	< 0.001	< 0.001	< 0.001
C	mutual attention	18.4042	0.0001	90%	30.852	< 0.001	0.006	< 0.001	< 0.001
	interaction time	13.4509	0.0001	95%	187.142	< 0.001	< 0.001	0.005	< 0.001
	achievement	10.8076	0.0001	65%	12.418	< 0.001	< 0.001	0.538	0.001
	p.competence	7.5597	0.0001	95%	49.323	< 0.001	< 0.001	0.965	< 0.001
	p.intelligence	2.5696	0.0188	95%	41.216	< 0.001	0.001	< 0.001	< 0.001
	physical effort	11.2489	0.0001	90%	85.763	< 0.001	< 0.001	< 0.001	< 0.001
	cognitive effort	4.8221	0.0001	95%	5.681	0.006	0.218	0.22	0.004
	arousal	2.8961	0.0093	100%	13.11	< 0.001	0.009	0.114	< 0.001
I	Wondering	1.73	0.092	N/A	0.221	0.803	N/A	N/A	N/A
	Surprise	2.9043	0.0091	25%	0.662	0.52	N/A	N/A	N/A
	Disgust	0.5125	0.6142	N/A	1.834	0.169	N/A	N/A	N/A
	Happy	2.4047	0.0265	N/A	32.309	< 0.001	< 0.001	< 0.001	0.009
	adaptability	4.7618	0.0001	100%	37.487	< 0.001	< 0.001	0.931	< 0.001
	HPFS	6.7219	0.0001	100%	211.405	< 0.001	< 0.001	0.313	< 0.001
	HNFS	12.4617	0.0001	55%	246.403	< 0.001	< 0.001	0.983	< 0.001
D	persuasiveness	2.8536	0.0102	55%	5.343	0.007	0.697	0.059059	0.008
	trust	3.0	0.0074	80%	20.643	< 0.001	< 0.001	0.999	< 0.001
	long-term use	23.8471	0.0011	50%	29.807	< 0.001	< 0.001	0.446	< 0.001

By combining the three Tukey-HSD comparisons, we can deduce that, for most of the bonds, we have higher bonding results in S2 compared to S3, and in S4 compared to S2. In summary, IUs were sufficient to make the robot persuasive and the communication meaningful, but it appears to be that adding synchronized IGs to the IUs makes the robot aesthetically more appealing in terms of behavioral design. These insights are in line with the participants open-responses (situational empathy), with one of the participant indicating: "In this last day, I felt that the robot was more human-like and smarter in comparison to the previous times, since it can synchronize what it says with added body movements." As the bonding had high results in S4, and by comparing these results with deviance results we can conclude that, from a behavioral design perspective, it is better to integrate a full mode (IUs+IGs) for accompanying minimally designed robots as it may guarantee strong social bonding evolvement and a more positive HRR (H2 investigated).

10 Conclusion and Future Research

Our study explored the human bonding with a robot as a reciprocation to the accompanying robot's different exhibited behaviors and functioning modes. We tested two functioning modes: proactive and reactive. Results suggest that humans overwhelmingly prefer the proactive mode to the reactive one. Moreover, when interacting with an accompanying robot, users seem to prefer a combination of the robot's gestures within the context of the conversation (full mode); this was pointed out to be more aesthetically appealing. In the future, we intend to investigate the proactive full mode under two conditions of robot's operation: advice mode (the robot can give advice to humans) and prosocial mode (the robot needs help from humans).

References

1. Hirschi, T.: Causes of Delinquency. University of California Press, Berkeley (1969)
2. Chriss, J.J.: The functions of the social bond. Sociological Quarterly, 689–712 (2007)
3. Kanda, T., Takayuki, K., Ishiguro, H.: A practical experiment with interactive humanoid robots in human society. In: Conference on Humanoid Robots, pp. 14–24 (2003)
4. Garrell, A., Sanfeliu, A.: Cooperative social robots to accompany groups of people. Journal Robotic Res., 1675–1701 (2012)
5. Aitken, K.J., Trevarthen, C.: Self/other organization in human psychological development. Developmental Psychopathology, 653–677 (1997)
6. Bethel, C.L., Stevenson, M.R., Scassellati, B.: Secret-sharing: interactions between a child, robot, and adult. In: Conference on Systems, Man, and Cybernetics (SMC), pp. 2489–2494 (2011)
7. Shahid, S., Krahmer, E., Swerts, M.: Child-robot interaction: playing alone or together. In: Proc. Human-Computer interace, pp. 1399–1404 (2011)
8. Matsumoto, N., Fuji, H., Goan, M., Okada, M.: Minimal design strategy for embodied communication agents. In: Workshop on Robot and Human Interactive Communication, pp. 335–340 (2005)
9. Noriko, S., Yugo, T., Kazuo, I., Okada, M.: Effects of echoic mimicry using hummed sounds on human computer interaction. Speech Communication, 559–573 (2003)
10. Youssef, K., Yamagiwa, K., Silva, R., Okada, M.: ROBOMO: Towards an Accompanying Mobile Robot. In: Beetz, M., Johnston, B., Williams, M.-A. (eds.) ICSR 2014. LNCS, vol. 8755, pp. 196–205. Springer, Heidelberg (2014)
11. Noriko, S., Kazuhiko, K., Yugo, T., Okada, M.: Social effects of the speed of hummed sounds on human computer interaction. Journal of Human-Computer Studies, 455–468 (2004)
12. Bradley, M.M., Lang, P.: Measuring emotion: The self-assessment manikin and the semantic differential. Journal of Social Robotics, 49–59 (1994)
13. McCroskey, J.C., Teven, J.J.: Goodwill: A Reexamination of the Construct and its Measurement. Communication Monographs 66(1), 90–103 (1999)
14. Bartneck, C., Kuli, D., Croft, E., Zoghbi, S.: Measurement Instruments for the Anthropomorphism, Animacy, Likeability, Perceived Intelligence, and Perceived Safety of Robots. Journal of Social Robotics, 71–81 (2009)

15. Tanioka, I., Glaser, D.: School uniforms, routine activities, and the social control of delinquency in Japan. Youth and Society, 50–75 (1991)
16. Rosenbaum, J.L., Lasley, J.R.: School, community context, and delinquency: Rethinking the gender gap, and the social control of delinquency in Japan. Justice Quarterly, 493–513 (1990)
17. Larry, A., Erbert, F., Kory, F.: Affectionate expressions as facethreatening acts: Receiver assessments. Communication Studies **66**(1), 254–270 (2004)
18. Lim, T.S., Bowers, J.W.: Facework: Solidarity, approbation, and tact Human Communication Research. Human Communication Research **17**(3), 415–449 (1991)
19. Andrews, P., Manandhar, S.: Measure of belief change as an evaluation of persuasion. In: Proceedings of the Persuasive Technology and Digital Behaviour Intervention Symposium, pp. 12–18 (2009)

SDT: Maintaining the Communication Protocol Through Mixed Feedback Strategies

Youssef Khaoula[✉], P. Ravindra De Silva, and Michio Okada

Interaction and Communication Design Lab, Toyohashi University of Technology,
1-1 Hibarigaoka, Tempaku, Toyohashi, Aichi, Japan
youssef@icd.cs.tut.ac.jp, {ravi,okada}@tut.jp
http://www.icd.cs.tut.ac.jp

Abstract. In our previous work, we studied how humans establish a protocol of communication in a context that requires mutual adaptation using our robot Sociable Dining Table (SDT). SDT integrates a dish robot put on the table and behaves according to the knocks that the human emits by guessing the meaning of each knocking pattern. We remarked based on previous experiments, that a personalized communication protocol is established incrementally. In fact, the communication protocol is not personalized only to the pair human-robot but also to the human robot interaction's (HRI) instance. In the current study, we change the robot's feedback modality (the way the robot communicates back with the human) in order that the communication protocol can be maintained over different HRI's instances. We proposed as new feedback modality, 2 mixed-feedback strategies integrating inarticulate utterances (IU) combined with the robot's visible behavior in order to facilitate the guessing of the robot's internal state for the human. The first strategy consisted in anticipating the robot's executed behavior using static IU combined with the robot's movement (St1), and the other consisted in genuinely suggesting an adaptive IU generation method combined with the robot's movement too (St2). In the current work, we conducted an HRI experiment to explore whether the communication protocol can be maintained on a long-term basis by integrating the 2 proposed methods. The results provide confirmatory evidence that using IU helps in establishing stable communication protocols. In addition to that it increases the attachment and the robot's overall subjective ratings. Another important finding is that, among the two methods, the adaptive mixed feedback strategy (St2) affords better subjective results and objective performance.

Keywords: Inarticulate utterance · Adaptation · Protocol of communication · Persuasiveness · Recall

1 Introduction

The current study draws on previous research where the goal was to explore how people can incrementally establish a communication protocol within a simpler

© Springer International Publishing Switzerland 2015
A. Tapus et al. (Eds.): ICSR 2015, LNAI 9388, pp. 348–358, 2015.
DOI: 10.1007/978-3-319-25554-5_35

setting [1]. In our previous study, we used the knocking as the only communication channel that the human can use to express his intention (knocking on the table) while the robot used visible movement as the only feedback channel to communicate back with the human [1]. We showed based on an HRI experiment that, the human and the robot cooperated and as a result a personalized communication protocol emerged after each HRI interaction's instance. We remarked also, that humans tend to forget the previously established communication protocol (PECP) and instead keep on creating new communication protocols in each HRI's instance [1]. Since, we want the HRI to occur smoothly and implicitly (the human has to feel spontaneous), we need indirectly to make the human remember the PECP so as to avoid the anticipated unpleasant consequences such as making the human's social face threatened or the robot being abandoned. Consequently, in our study we address the issue of triggering the human indirectly to use the same previously established communication protocol (PECP) rather than creating in each HRI instance a new communication protocol. We proposed to use 2 mixed-feedback strategies to communicate back to the human and to integrate the robot's visible movement along with IU. Thus, the challenge to be resolved is to investigate the effect of using these mixed-feedback strategies on the long-term recall (remembrance) of the PECP in addition to exploring the impact of our strategy on the human's subjective evaluation of the robot's performance. Also, we want to determine, which is better in terms of objective and subjective results: the first or the second proposed mixed-feedback strategy.

2   Background

There are some occasions in the HRI when special reasons dictate inconsistent behavior. For example, we can cite human's forgetfulness of the previously formed communication protocol just as we remarked in the previous study [1]. In such circumstance, the robot is forced to tell the human that he is wrong which may be perceived as challenging for the human partner. When the robot disagrees with the human, the human is confronted with a face-threatening act, placing the user at risk of being bothered by the robot's opposition [2]. The concept of face-threatening acts was initially proposed by social scientists such as Goffmann [2]. Moreover, even if the robot assumes that itself, it is the faulty party and tries to apologize, users may lose their trust in the robot and they will use it as a scapegoat to avoid any responsibility [2], assuming of course that they are even aware of their faulty behaviors. In this context, Lee et al. [3] utilize different strategies including apologies, compensation and options for the user to reduce the negative consequences of the communication protocol's breakdowns. In another study, Torrey et al. [4] highlight that we just need to add hedges and discourse markers so that we diffuse the human's sensation that the robot takes control over him, when it tries literally to tell the human nicely that he is wrong. Takayama et al. [5], confirmed through their study that distancing the voice from the robot makes the human tolerates the fact that the robot disagrees sometimes with them. We believe that these strategies are useful but simpler methods can be integrated within HRI to resolve this challenge. We draw inspiration from child-caregiver interaction's scenario. We want

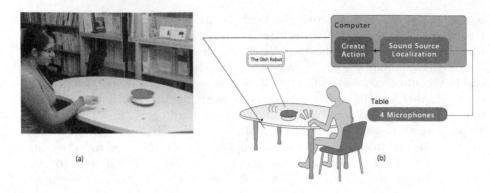

Fig. 1. In (a), a user interacting with SDT and in (b) we have SDT's architecture

to integrate a more implicit way that helps preventing PECP post-forgetfulness and trigger human's memory indirectly while we expect that an IU may lead the human to remember the related behavior just as Paivio [6] indicates in his dual coding concept[1]. For example, in a baby-caregiver interaction, a baby does not use such complex analytical methods (such as compensating, apology, etc..) in order to communicate back with the caregiver, but combines each of the behaviors with a special IU. The communication goes through different breakdowns and still both parties (the baby and the caregiver) can establish a long-term communication protocol. In fact, based on the simple IU combined with the behaviors previously used during previous interactions, the baby and the caregiver are capable of recalling each of the behaviors which makes the establishment of a stable long-term used communication protocol easier for both parties [7]. On these grounds, this study is an attempt to address the issue of how a robot can express implicitly his disagreement about some inconsistency during the HRI without threatening the human partner's social faces. More specifically, our goal is by using what we anticipate to be a threat-free method, to drive the human during the first HRI instance (coding phase) to memorize the communication protocol. By using the proposed methods, we hope that users will be made aware of their faulty indications, focus more when they establish the communication protocol, and that as a result a long-term communication protocol can be established whereby users feel comfortable to communicate with the robot.

3   Architecture of the SDT

SDT uses 4 microphones to localize the knock's source based on the weighted regression algorithm. It communicates with the human based on a sound output and with

---

[1] According to the dual coding concept, each trigger (visual or audio) that it is combined with a concept learning during the coding phase (learning phase or by analogy to our problematic, during the communication protocol establishment), may facilitate the remembrance (recall) of the information (or in our case the PECP) once the trigger is exposed to the human.

the host computer through Wi-Fi using it control unit (a macro computer chip (AVR ATMEGA128)). It employs a servomotor that helps to exhibit the different behaviors: right, forward, left and back. Finally, 5 photo reflectors are utilized to automatically detect the boundaries of the table and avoid falling (Fig.1).

4   Robot's Action Selection Strategy

We conceived an actor/critic architecture that incrementally helps the robot to choose between 4 actions (left, right, back, forward).

4.1   Actor Learning

Each knocking pattern (x contiguous knocks pattern:e.g 2 knocks, 3 knocks, etc.) has its own distribution $X(s_t) = N(\mu_{X(s_t)}, \sigma_{X(s_t)})$ where $X(s_t)$ is defined as the knocking pattern, $\mu_{X(s_t)}$ and $\sigma_{X(s_t)}$ are the mean value and the variance. We chose 2 seconds (s) as a threshold for the user's reaction time based on previous established experiments. When the robot observes the state s_t, the behavior is picked according to the probabilistic policy $\Pi(s_t)_{nbknocks}$. If within 2 s there was no knocking pattern, we suppose that the robot has succeeded by choosing the right behavior and the critic reinforces the value of the executed behavior in the state s_t. The system switches to the state s_{t+1}. If a new knocking pattern is composed before that 2 s elapsed, the state of the interaction changes to the state s_{t+1} indicating that the knocker disagrees about the behavior that was executed. The critic updates thus the value function before choosing any new behavior. As long as the knocker is interrupting the robot's behavior before that 2 s elapsed, the actor chooses the action henceforth by pure exploration (until we meet an agreement state: no knocking during 2 s) based on (1). The random values vary between $0 \leq rnd1$, and $3 \leq rnd2$. The above range was decided to bring the values of the action between 0 and 3 (corresponding to the behaviors' (forward, right, back, left) numerical codes). We assume in such case that the knocker will randomly compose the patterns just to switch the robot's behavior.

$$A(s_t) = \mu_{X(s_t)} + \sigma_{X(s_t)} * \sqrt{-2 * log(rnd_1)} * Sin(2\Pi * rnd_2) \qquad (1)$$

4.2   Critic Learning

The critic calculates the TD error δ_t which is the difference between the real value function of the new gathered state $V(s_{t+1})$ and the expected state $V(s_t)$ (2)

$$\delta_t = r_t + \gamma V(s_{t+1}) - V(s_t) \qquad (2)$$

with γ is the discount rate and $0 \leq \gamma \leq 1$. According to the TD error, the critic updates the state value function $V(s_t)$ based on (3).

$$V(s_t) = V(s_t) + \alpha * \delta_t \qquad (3)$$

where $0 \leq \alpha \leq 1$ is the learning rate. As long as the knocker disagrees about the executed behavior before 2 s elapsed, we refine the distribution $N(\mu_{X(s_t)}, \sigma_{X(s_t)})$ that it is relevant to the pattern $X(s_t)$ (for each pattern, we have a specific distribution $X(s_t)$) which helps us to choose the action according to (4).

$$\mu_{X(s_t)} = \frac{\mu_{X(s_t)} + A_{s_t}}{2} \quad \sigma_{X(s_t)} = \frac{\sigma_{X(s_t)} + |A_{s_t} - \mu_{X(s_t)}|}{2} \tag{4}$$

5   Feedback Strategies

In the current study, we have 3 feedback strategies that helps the robot to communicate with the human.

5.1   Visible Movement-Based Feedback Strategy Method

This strategy corresponds to the feedback modality that we used in our previous work while the robot can just execute the action after guessing the knocking pattern's meaning. Based on the robot's visible movement, the human has to understand implicitly how the robot combines for each knocking pattern, a relevant action.

5.2   Static Mixed Feedback Strategy

This method consists on announcing before that the robot executes the intended behavior, the label of that behavior. (e.g: If the robot has to go right, the robot generates the IU "go right" before it executes the action. We believe that by using this method, the user will have more time to help the robot avoid the wrong steps.

5.3   Adaptive Mixed Feedback Strategy

We opted for the SARSA algorithm in order to generate in real time and in an adaptive manner, different IUs combined with the robot's visible behaviors. SARSA (so called because it uses state-action-reward-state-action experiences to update the *Q-values*) is an on-policy reinforcement learning algorithm that estimates the value of the policy being followed [8]. An experience in SARSA is of the form *(S,A,R,S',A')*, which means that the robot was in state *S*, did the action *A*, receives the reward *R*, and ends up in state *S'*, from which it decided to do action *A'*. This provides a new experience to update *Q(S ,A)*. The new value that this experience provides is, $r + \gamma Q(S', A')$. In this context, as the robot will use the actor-critic to choose the future action (based on the actor component) and a reward will be generated which will be used by the critic of update issues, that same action will help to choose the appropriate label (e.g if the robot has to go left, then the robot will generate the label left as IU (method used in the static mixed feedback strategy)) and that same reward will help to update the

Q function because if the human made an error, then it should also be linked to his misunderstanding of the IU.

A state S in our case, is the combination of the actual interaction status (*agreement* or *disagreement*) in addition to the number of knocks received (e.g: a state S can be $S=$(*agreement, 2 knocks*) which means that the actual interaction status is agreement and the robot receives 2 knocks.). In our context, we have 2 interaction status: the agreement status (when the human does not knock for 2 seconds (s), we assume that the last action was correct) and the disagreement status otherwise. So, if we assume that each state can have 2 status *agreement* and *disagreement* and we have 4 types of patterns, then we will have 8 states. We assign for each (state, utterance) a value function $Q(S, U)$ which we initialize arbitrarily. It helps comparing between different (state, utterance) couples outcomes. In the disagreement status and whenever there will be a knocking pattern, there are 3 possible actions which precedes the robot's movement:

- indicating the chosen behavior using an IU (to reduce wrong steps).
- repeating the received number of knocks (indicating that the input is processed).
- combining both.

The second possible status is the agreement, while we have 3 possible actions:

- repeat the label of action A that was previously executed by naming it, take a small pause and then tell the future robot's action B that is intended to be executed (to make the human aware that the action will be shifted from A to B).
- repeat the previous received number of knocks A in addition to the actual received number of knocks B (to indicate for the human that the robot is aware the knocking is shifted from A to B).
- indicate using IU the (previous knocking pattern, label of previously executed action), a pause and then indicate the future robot's action that is intended to be executed (to consolidate the lately learnt rule and makes the human aware about the new action).
- a high pitched inarticulate utterance showing enthusiasm (to engage the user more in the interaction).

The update of the value function follows the equation (5):

$$Q(S, U) < -Q(S, U) + w(r + vQ(S', U') - Q(S, U)) \tag{5}$$

with $0<w<1$ is the learning rate. **r** is the gathered reward and $0<v<1$ is the discount factor.

6   Experimental Setup

Participants take part in 2 trials one-by-one while each one cooperates with the robot in order to lead it to the different checkpoints (Fig.2). Each participant was informed that the robot can execute 4 behaviors (*right, left, back,*

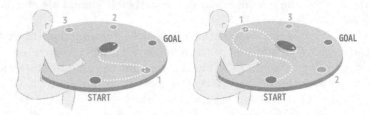

Fig. 2. An example of 2 configurations: each configuration is designed for a trial and is formed by different points marked on the table.

forward) while he has to knock on the table in order to convey his intention of making the robot choose a specific direction. Each user, has to participate for the first time (trial 1) and then answer a survey (indicated in section 7). After, 3 days the participant comes again to the laboratory to redo again the same task except that we propose this time (trial 2) a different configuration. The former points marked on the table are changed to guarantee the diversity of the patterns suggested by the participants (Fig.2). In the current experiment, we have 3 conditions for which we assigned the same number of participants (7 participants per condition). It is the same task for all participants except that, HRI is designed differently during trials 1 and 2 for each condition. For the first condition (**MM**), we used a simple feedback strategy (Movement (M)) during both trials and for 7 participants while the robot is silent and the human can just visualize the robot's movement. We call that simple feedback strategy, movement-based feedback strategy. For the condition 2 (**MI**), the robot uses the mixed-feedback strategy (M+IU) while it just uses the static method combined with the robot's movement during both trials and for 7 participants. Finally, in condition 3 (**AMI**), the robot uses an adaptive method for utterance generation (M+adaptive IU) combined with the robot's movement during both trials and for 7 participants. In total 21 participants ([21-30]years) take part in our experiment.

7   Survey Procedure

After finishing each trial, we asked each participant to fill out 7-Likert scale questionnaires so that we could measure: the attachment that may evolve (5 factors to measure the attachment: adaptability, stress-free, perceived closeness, cooperation and achievement [1]), the robot's credibility (using a standard instrument indicated [9] which consisted of 3 factors: competence, trust and caring), the social face support (2 factors: human positive face support (HPFS) and human negative face support (NHFS)) to verify whether the user's social faces were supported during the HRI (inspired from [10]). We also demanded from the participant to fill out a 4-God speed questionnaire [11] to measure the robot's likeability, perceived intelligence, animacy and anthropomorphism. Moreover, we evaluate the user's mood using SAM scale [12] (2 factors: pleasure and arousal).

After answering the questionnaires, we demanded from the user to arrange a list of words in an increasing order of priority which he assigns according to his own opinion. This list, is used to measure the persuasiveness based on the kendall-tau distance and the method indicated in [13]. After that, the user has to describe his experience with the robot in an open-ended way and then evaluate his future frequency of use of the robot. Then, we asked the user to play a game for 15 minutes in order to determine whether they remember the interaction rules with SDT. He has to enumerate the different interaction rules that he still remembers (short-term recall rate). We noted the communication protocol established in trial 1 and we compared it to the communication protocol used in the beginning of trial 2, so that we could calculate the reuse percentage of rules that belong to the previously established communication protocol (long-term recall rate). We computed also the chi-square and the Cramer V evaluating the stability of the communication protocol and the relationship between the knocking patterns and the robot's behaviors. Also, we computed the minimal Euclidian distance between the robot's trajectory and the different checkpoints (CPs) marked on the table, so that we can verify the perfection level that the user reached while doing the task.

8   Challenges of Using the Robot's Movement as a Feedback Strategy

Table 1 summarized the comparison of the condition MM trials 1 and 2 results. Based on Table 1, we remark that users found that the robot was more competent and that the level of focus was higher (mutual attention, arousal) during trial 2. This, may explain the longer time needed to achieve the task that was achieved perfectly in comparison to the first trial. We expected that the time needed would be shorter during trial 2, however this was not the case. Figure 3 shows that, all the participants have a long-term recall rate value under 0.5 which explains why people took more time during trial 2. In fact, they forget the PECP.

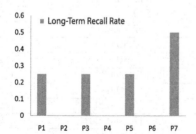

Fig. 3. Long-term recall rate for the different participants in trial 2 of the condition MM

Table 1. Comparison of the condition MM trials 1 and 2 results.

Metric	t-test	df	P-value
Time Needed	2.417	12	0.032
Distance From the CP	2.942	12	0.012
Arousal	2.25	12	0.044
Mutual Attention	4.66	12	0.001
Competence	2.219	12	0.047

9   Mixed-Feedback Strategy Vs Movement-Based Feedback Strategy

We compared the trials 2 results of the conditions MI and MM to investigate the effect of adding the IU to the robot's visual behavior as an expected more explicit feedback strategy. Table 2, shows the results of the comparison (comparison of MI and MM). Based on Table 2, we conclude, that the mixed-feedback strategy of the condition MI, helped users to feel more attached to the robot, believe on its credibility and feel no threat of loss of faces. The robot was judged as more animate (animacy factor) and anthropomorphic (anthropomorphism factor), likeable (likeability factor) and smart (perceived intelligence factor), attentive (mutual attention factor) and persuasive (persuasiveness factor). This, led to a shift to a positive mood for the user and, a higher expected long-term use for the robot in the condition MI. We noticed also, that the condition MI boosts the robot's objective performance, while we remarked that during trial 2, we have more stable protocols (cramer V). Also, we remark that users could achieve the task more perfectly (distance from the checkpoints) and in shorter

Table 2. Conditions MM and MI comparison results of the trials 2 factors' values and the conditions MM and AMI comparison results of the trials 2 factors' values.

Factors	comparison of MI and MM		comparison of MI and AMI		
	t-test	P-value	t-test	P-value	df
achievement	3.216	0.007	3.216	0.007	12
cooperation	2.263	0.043	2.263	0.043	12
friendliness	2.29	0.041	2.291	0.041	12
stress-free	2.846	0.015	2.846	0.015	12
adaptability	2.982	0.011	2.982	0.011	12
trust	2.642	0.021	2.642	0.021	12
competence	2.449	0.031	2.449	0.031	12
caring	2.744	0.018	2.744	0.018	12
animacy	3.753	0.003	3.753	0.003	12
anthropomorphism	3.392	0.005	3.392	0.005	12
likeability	3.545	0.004	3.545	0.004	12
perceived intelligence	2.219	0.047	2.219	0.047	12
arousal	4.157	0.001	4.157	0.001	12
pleasure	2.717	0.019	2.717	0.019	12
persuasiveness	4.261	0.001	4.26	0.001	12
PHFS	2.425	0.032	2.425	0.032	12
NHFS	2.58	0.024	2.58	0.024	12
attention allocation	2.489	0.029	2.489	0.029	12
long term use	3.897	0.002	3.897	0.002	12
Distance from CPs	2.411	0.033	2.411	0.033	12
Cramer V	19.297	<0.0001	19.297	<0.0001	12
Task Completion Time	2.284	0.041	2.284	0.041	12
short-term recall	14.899	<0.0001	0.612	0.552	12
long-term recall	4.7555	0.0004	0.816	0.43	12

time. Furthermore, results suggest that the short-term recall rate value of the condition MI is significantly higher in comparison to the short-term recall rate values of the condition MM. In correlation to that, we find that long-term recall was higher in the condition MI. These results afford insights for the HRI while it shows that using a mixed-feedback strategy is subjectively more preferred, leads to better objective performance and relatively a better maintainance of the PECP.

10   Static Mixed Feedback Strategy Vs Adaptive Mixed-Feedback Strategy

Table 2, exposed the comparison results of the conditions MI and AMI. Based on Table 2, we can see that users attribute better subjective evaluation of the HRI when the robot uses the adaptive mixed feedback strategy. We remark also, that there were no significant differences between conditions MI and AMI in terms of short and long term recall rates which means that both methods guarantee the remembrance (recall) of the PECP on long and short term basis. Furthermore, objective performance was higher for the AMI condition. In fact, users achieve the task more perfectly, in shorter time and succeeded in establishing stable communication protocols. Consequently, while using the adaptive mixed feedback strategy during an HRI guarantees better subjective ratings, more stable communication protocol and leads to a better achievement of the task in shorter a time, the adaptive IU generation strategy does not lead to a better recall of the PECP.

11   Conclusion

We proposed 2 mixed-feedback strategies that helps to ameliorate users' subjective ratings and remembrance of the PECP on a long-term basis. Results suggest, that the adaptive mixed feedback strategy of IU leads to better recall rates in comparison to the static mixed feedback strategy. In our future work, we intend to use affective inarticulate utterances and investigate whether it can arouse users' affective and cognitive empathy.

Acknowledgments. This research is supported by Grant-in-Aid for scientific research of KIBAN-B (26280102) from the Japan Society for the Promotion of science (JSPS).

References

1. Youssef, K., De Silva, P.R.S., Okada, M.: Investigating the mutual adaptation process to build up the protocol of communication. In: Social Robotics, pp. 571–576 (2014)
2. Brown, P., Levinson, S.C.: Toward a more civilized design: studying the effects of computers that apologize. Journal of Human-Computer Studies, 319–345 (2004)

3. Lee, M.K., Kielser, S., Forlizzi, J., Srinivasa, S., Rybski, P.: Gracefully mitigating breakdowns in robotic services. In: Conference on Human-robot Interaction, pp. 203–210 (2010)
4. Torrey, C., Fussell, S.R., Kiesler, S.: How a robot should give advice. In: Conference on Human-Robot Interaction, pp. 275–282 (2013)
5. Takayama, L., Groom, V., Nass, C.: I'm sorry, Dave: i'm afraid i won't do that: social aspects of human-agent conflict. In: Conference on Human Factors in Computing Systems, pp. 2099–2108 (2009)
6. Paivio, A.: Mental representations: a dual coding approach (1986)
7. Breazel, C.: Designing Sociable Robots. MIT Press (2002)
8. Szepesvari, C.: Algorithms for Reinforcement Learning. Synthesis lectures on artificial intelligence and machine learning (2010)
9. McCroskey, J.C., Teven, J.J.: Goodwill: A Reexamination of the Construct and its Measurement. Communication Monographs **66**(1), 90–103 (1999)
10. Erbert, L.A., Floyd, K.: Affectionate expressions as face-threatening acts: Receiver assessments. Communication studies **55**(2), 254–270 (2004)
11. Bartneck, C., Kulic, D., Croft, E., Zoghbi, S.: Measurement Instruments for the Anthropomorphism, Animacy, Likeability, Perceived Intelligence, and Perceived Safety of Robots. Journal of social robotics **1**(1), 71–81 (2009)
12. Margaret, M.B., Peter, J.L.: Measuring emotion: The self-assessment manikin and the semantic differential. Journal of Behavior Therapy and Experimental Psychiatry **25**(1), 49–59 (1994)
13. Andrews, P., Manandhar, S.: Measure of belief change as an evaluation of persuasion. In: Proceedings of the Persuasive Technology and Digital Behaviour Intervention, pp. 12–18 (2009)

Personalized Assistance for Dressing Users

Steven D. Klee[1]($^{\boxtimes}$), Beatriz Quintino Ferreira[2], Rui Silva[3],
João Paulo Costeira[2], Francisco S. Melo[3], and Manuela Veloso[1]

[1] Computer Science Department, Carnegie Mellon University, Pittsburgh, PA, USA
sdklee@cmu.edu, mmv@cs.cmu.edu
[2] ISR,Instituto Superior Técnico, Universidade de Lisboa, Lisboa, Portugal
{beatrizquintino,jpc}@isr.ist.utl.pt
[3] INESC-ID and Instituto Superior Técnico, Universidade de Lisboa,
Lisboa, Portugal
rui.teixeira.silva@ist.utl.pt, fmelo@inesc-id.pt

Abstract. In this paper, we present an approach for a robot to provide personalized assistance for dressing a user. In particular, given a dressing task, our approach finds a solution involving manipulator motions and also user repositioning requests. Specifically, the solution allows the robot and user to take turns moving in the same space and is cognizant of the user's limitations. To accomplish this, a vision module monitors the human's motion, determines if he is following the repositioning requests, and infers mobility limitations when he cannot. The learned constraints are used during future dressing episodes to personalize the repositioning requests. Our contributions include a turn-taking approach to human-robot coordination for the dressing problem and a vision module capable of learning user limitations. After presenting the technical details of our approach, we provide an evaluation with a Baxter manipulator.

Keywords: Human-robot interaction · Dressing · Human tracking · Learning and adaptive systems

1 Introduction

Research has moved towards allowing robots and humans to operate in the same workspace. For instance, human-aware robots can assist people in factories, or even elderly people in their homes. In particular, there has been increasing interest in developing robots that can help people overcome their disabilities and limitations [13]. There are many challenges involved in designing robotic systems that deliver highly personalized assistance to different individuals, owing to the uncertainty introduced by human presence.

In this paper, we address the problem of providing personalized assistance to help dress a person with a manipulator. To this end, we introduce a framework for human-robot collaboration, in which the user and robot take turns moving to complete their shared goal. During these interactions, the robot learns the user's limitations and uses this information to provide personalized interactions.

© Springer International Publishing Switzerland 2015
A. Tapus et al. (Eds.): ICSR 2015, LNAI 9388, pp. 359–369, 2015.
DOI: 10.1007/978-3-319-25554-5_36

Dressing tasks, represented as templates, are sequences of goal poses with respect to the user. For instance, placing a hat on a user's head has one goal pose, with the arm several centimeters above his head. Other tasks may have several goal poses that must be incrementally solved to complete a task.

Before the robot acts on these templates, they are instantiated with the current location of the user. For example, for a 1.8m tall person, a goal position of 10cm above the head of the user becomes 1.9m from the ground. This is the first form of personalization which allows templates of sequential goals to be parameterized by the user's physical features.

Once the template is instantiated, the robot attempts to fulfill the goals with its motion planner, while asking the user to remain still. If the plan fails, the robot tries to re-instantiate the template by asking the user to move to reposition himself. Specifically, a planner determines a sequence of user repositioning requests that will take the user from their current pose to a new pose.

As a second form of personalization, the robot models each user's limitations, represented as pose constraints. When the robot asks the user to reposition himself, it selects a pose that satisfies these constraints. Specifically, the robot selects a new pose for the user by sampling points in its configuration space that also satisfy the known constraints. Constraints are learned by a vision module that monitors the user's response to repositioning requests. Each repositioning request has an expected behavior, and the robot infers a new constraint when the expectation is not met. Due to inaccuracies in vision, and the uncertainty introduced by humans, these constraints have error tolerances and confidences. To increase our confidence and refine constraints, they are ignored in future episodes with probability proportional to their confidence.

Among our contributions, the three most notable are:

- An approach to the dressing problem that explicitly requests the user to reposition himself when a solution cannot be found.
- A vision-based expectation algorithm that determines if the user is complying with an interaction request and learns his constraints.
- A turn-taking approach for safe human-robot interactions in a shared workspace.

In the remainder of the paper, we compare our approach to related works, in Section 2. Then, in Section 3, we provide an overview of the proposed approach before detailing each component. In Section 4 we describe a specific implementation with a Baxter manipulator robot and show experimental results. Finally, we draw conclusions and discuss future work in Section 5.

2   Related Work

Our work is similar to research in designing algorithms for manipulators that help dress humans. Much prior work has focused on the problem of clothing manipulation in simulation [3] and also in the real world [7,9,14]. Our work is more focused on the additional challenges posed when a person is involved.

When operating a manipulator near a person, accurate human tracking, localization, and fusing information from available sensor networks is extremely important [12]. Researchers have developed approaches for dressing people based on visual and sensory information [15]. Their proposed visual framework matches features acquired from depth and color images to features stored in a database. Similar to our approach, this system can detect failures via vision. However, on failures we re-plan using a series of user interactions and motion planning, whereas they repeat the same action assuming the error was transient.

Researchers have used reinforcement learning to teach robots to put on a t-shirt and wrap a scarf around a mannequin's neck [5,10]. In the latter work, the robot learns tasks through dynamic movement primitives, allowing them to modify the trajectory speed or goal location. By contrast, our approach uses a sampling-based motion planner. By performing motion planning online, we have a high degree of confidence that the trajectory will not collide with the person.

The aforementioned works do not attempt to *cooperate* with the user. Our approach employs *sparse-coordination*, where the robot asks the user to reposition himself/herself if it cannot otherwise solve the task [11]. Additionally, similar to the concept of maintenance goals, repositioning requests have expectations that are monitored over time by a vision module [2]. Lastly, our work draws from research in representing and solving generalized or parameterized plans [6]. In particular, we represent dressing tasks as sequences of subgoals that are parameterized by the location and size of the user.

3   Approach to Personalized Dressing

Our aim is to enable a manipulator to aid users in dressing tasks by providing personalized assistance. Throughout this paper, we employ the example of helping a user to put on a hat. Figure 1 shows our Baxter manipulator ready to perform this task.

Fig. 1. Baxter equipped to dress a user with two different hats.

Our approach is described at a high-level in Algorithm 1. In particular, each dressing task is a *template*, which is a sequence of goal poses of the manipulator with respect to the user. The *template* is instantiated with the current position and orientation of the user by the vision module (Line 3). Then, the motion planner attempts to find and execute a motion plan that satisfies each goal (Line 4). If it meets a goal that is infeasible, the robot attempts to re-instantiate the template by choosing a new pose for the user that satisfies his known constraints and is in the robot's configuration space (Line 6). To refine constraints, some are ignored with probability proportional to their confidence. The robot then determines a sequence of repositioning requests that will move the user to the chosen pose (Line 7). Finally, the vision module monitors the user during repositioning requests and infers new limitations when he cannot respond (Line 8). This process happens in a loop until all of the template's goals are solved.

Algorithm 1. Execution of Personalized Human-Robot Motion Interactions

---

1: **procedure** EXECUTE(*template, vision, motion, constraints*)
2:     **while** goals $\neq \emptyset$ **do**
3:         goals $\leftarrow$ vision.instantiate(*template*)
4:         goals $\leftarrow$ manipulation(*goals, motion*) // remaining goals
5:         **if** goals $\neq \emptyset$ **then**
6:             newPose $\leftarrow$ feasiblePosition(*template, vision, constraints, motion*)
7:             interactions $\leftarrow$ planInteractions(*vision, constraints, newPose*)
8:             constraints $\leftarrow$ interaction(*interactions, vision, constraints*)

3.1   Vision-Based User Tracking

The vision module monitors the user to pause motion execution if he moves, and to determine if he is complying with repositioning requests.

The vision module is provided with a set of joint positions J from a skeleton tracker. We model a person as a set of connected body parts, B:

$$B = \{head, torso, left\_arm, right\_arm\}$$

that are tracked by the vision module, which provides the center-of-mass location (x, y, z) and orientation $\langle q_x, q_y, q_z, q_w \rangle$ of each body part. Typically $|J| > |B|$ and some joint locations map directly to body part locations. The body part orientations are computed by looking at the locations of consecutive joints.

Our vision module then performs two additional functions:

– Detecting if a person is stopped or moving.
– Verifying user-repositioning expectations.

To detect if a person is stopped or moving, the robot compares the body part positions and orientations over time. If $b_t \in B$ is a body part at time t, the body part is stopped if:

$$\forall k \in [t - \gamma, t] : ||b_k - b_t|| < \epsilon$$

Where γ is the time the user must be stopped for, and ϵ is a threshold to allow for small movements or tremors.

When the robot asks the user to reposition himself, it generates expected poses for some of the user's body parts and sends them to the vision module. When the user next stops moving, vision verifies that the user's actual body part poses match the expected poses. The verification of whether the user is in an expected position is done by the function *verify_expectation(expected_pose, actual_pose)*, which compares the expected pose of each user's body part to the actual poses found by the vision. It returns *True* if each dimension matches up to a configurable threshold and *False* otherwise. When the result of this function is *False*, the robot infers a new user constraint or refines an existing one.

3.2   Dressing Tasks as Template Goals

We represent a dressing task as a sequence of desired manipulator pose goals with respect to the user. For instance, putting on a hat has one desired pose, with the arm straight and above the user's head. Other tasks may have several desired poses that must be completed incrementally. For instance, putting on a backpack has several goal poses to help the user get each arm into a strap. These sequences of goals are general to any user, but are parameterized based on their physical characteristics. Formally, a *template goal T* is a sequence of pose goals:

$$T = \langle P_1, P_2, P_3...., P_n \rangle$$

Where each pose goal is composed of a position (x, y, z), orientation as a quaternion $\langle q_x, q_y, q_z, q_w \rangle$, and a vector of tolerances for each dimension (*TOL*):

$$P_i = \langle x, y, z, q_x, q_y, q_z, q_w, TOL \rangle$$

Thus, P_i is a goal pose $\langle x, y, z, q_x, q_y, q_z, q_w \rangle$, which is satisfied by a pose P^* if:

$$\forall d \in P_i \quad |P_{i_d} - P_d^*| \leq TOL_d$$

Goal poses have allowable tolerances in each dimension for two reasons. First, they account for minor vision inaccuracies. Secondly, some tasks may not require exact orientations or positions, as the user can readjust himself.

A template is *instantiated* when the vision module provides the location of a user with respect to the manipulator. Then, each pose P_i is transformed from the reference frame of the user to the reference frame of the robot. This allows the robot to run a motion planner for each pose goal.

3.3   Motion Planning

We adopt a sparse-interaction approach, where the manipulator first tries to solve as many pose goals as it can, before asking the user to reposition himself. To minimize the possibility of a collision the manipulator only moves when the user is stopped. If the user begins moving, the robot halts and re-plans once the user stops. The vision module is responsible for monitoring the user.

The algorithm for our approach is shown in Algorithm 2. Given instantiated template *goals*, the motion planner attempts to incrementally solve and execute the pose goals. The manipulator represents the world as a 3D occupancy grid provided by the vision module. We then use a sampling-based motion planner to compute a trajectory through the joint-space to our goal position. Specifically, we use RRT-Connect, a variant of Rapidly-Exploring Random Trees (RRT) known to work well with manipulators [8]. RRT-Connect grows two trees and attempts to connect them by extending the two closest branches. If the manipulator finds a pose goal that is infeasible, the robot chooses a new location for the user and determines a sequence of repositioning requests to ask.

3.4   User-Aware Pose Selection and Assistance Planning

Once the manipulator determines that a pose goal is infeasible, it asks the user to reposition himself. The robot has a knowledge-base of user constraints and

---

Algorithm 2. Motion Planning

1: **procedure** MANIPULATION(*goals, motion, vision*)
2:     **for** $P_i \in$ goals **do**
3:         plan $\leftarrow$ motion.rrt(P_i)
4:         **if** plan == ('Infeasible') **then**
5:             break
6:         **while** plan.executing() **do**
7:             **if** vision.feedback() == 'User Moving' **then**
8:                 motion.pauseUntilUserStopped(*vision*)
9:                 plan.replan(*vision*)
10:     return goals.remaining()

---

balances seeking feasible poses with refining the constraints. Once a new pose is determined, the robot finds a sequence of interaction requests that will move the user to this pose. Each interaction is monitored by a vision system that infers new constraints when the user cannot comply with a request.

We represent a constraint c on body part $b \in B$ as:

$$c = \langle b, ineq, conf \rangle$$

Where *ineq* is an inequality for a limitation on one of the pose dimensions. The left hand term of the inequality is the dimension, while the right hand term is the value that cannot be surpassed. For instance, $torso.x > 0.9$ means that the torso cannot move closer than 0.9m to the base of the robot. $conf \in$, represents the robot's confidence in the constraint. This is a function of how many times the constraint is satisfied compared to the number of times it is violated.

Given a set of known feasible manipulator poses $\mathcal{P}$, future poses are sampled from points in $\mathcal{P}$ that satisfy all *active* user constraints. For each interaction, a user constraint is *active* with probability proportional to its confidence. In this way, the robot can test less confident constraints to refine them.

Once a new pose is found, the robot must determine a sequence of interactions that will reposition the person. Each body part has a position, orientation, and set of *motion actions*. The motion actions represent parameterized repositioning requests that the robot can ask of the person. We define several translational motion and rotational actions including: $forwards(x)$, $backwards(x)$, $left(x)$, $right(x)$, $up(x)$, $down(x)$, $turn\_right(\theta)$, and $turn\_left(\theta)$. Where x is a distance in meters and θ is a rotation in degrees with respect to the manipulator.

Each of these motion actions is associated with a set of expectations E:

$$e = \langle b, L, TOL \rangle$$

Where each expectation consists of a body part $b \in B$, expected pose L, and tolerance on each dimension of the pose TOL. For instance, requesting $forwards(1)$ on the torso corresponds to asking the user to walk 1m forwards and moves all of the body parts. By contrast, requesting $up(0.2)$ on the left arm will only move the left arm up 0.2m, because it corresponds to lifting an arm.

Using these expectations, we can generate a plan of translational and rotational user requests that repositions each of their body parts. In particular, we focus on repositioning the user to the correct x position, then y, then z, and finally, the correct rotation. More generally, a planner could pick the order of these actions. During these interactions, the user is monitored by the vision module which determines if he complies with a request. If the user cannot, the robot infers a constraint, selects a new pose, and tries again.

3.5   Learning and Refining User Constraints

The robot learns and refines constraints in two situations. The first is when a user stops before reaching an expected pose during a repositioning request. The second is when the user enters a space the robot believed to be constrained.

In the first case, given a motion action m and an expectation $\langle b, L, TOL \rangle$, the vision module detects that the user has stopped, and is not in the expected position. Then, a constraint is inferred for the body part b on the axis associated with m. The constraint is that the user can move no further than his current position, which is provided by the vision. If a constraint along this axis already exists for that body part, it is updated with the new value.

For instance, the action $forwards(x)$ expects all of the body parts to move x meters closer to the robot. If the user cannot comply with this request a constraint c_b is generated for each body part $b \in B$ of the form:

$$c_b = \langle b, b.x > vision.b.x, conf \rangle$$

where $vision.b.x$ is the x-coordinate of body part b, and $conf$ is a configurable initial confidence value.

In the second case, the user moves into a space that the robot believed to be constrained. This implies that the original constraint was too strict, and should be relaxed. To do this, the robot replaces the value in the inequality with the value from the user's current location.

In both cases, the robot updates the confidence in the constraint. This influences how often the constraint is ignored when picking a user-pose to request.

Let N_s be the number of times the user approaches a constraint and does not pass it (the constraint is satisfied); and N_f the number of times the constraint fails, because the user passes it. The confidence $conf$ of a constraint is:

$$conf = \frac{N_s}{N_s + N_f}$$

This constraint inferring procedure becomes particularly relevant when interacting with disabled people. For instance, a user in a wheelchair may not be able to move perfectly laterally, or too close to the robot. In such cases, learning constraints in the adequate axes will optimize future interactions, as the robot will request poses that satisfy the inferred constraints. We note that the constraints learned are approximations of a human's physical limitations, as joint movements are dependent from one another.

4 Evaluation

We tested our approach with a Baxter manipulator. Baxter is a particularly good platform because is designed to work alongside users, implementing force-compliant motion. We mounted a Microsoft Kinect on Baxter, and used OpenNI-Tracker to find human joint positions. This depth image based skeleton tracking approach produces robust joint positions for different users in various background and lighting conditions [1]. Baxter makes repositioning requests by displaying text on its screen.

For these experiments, we considered the task of putting on a hat on the user's head. Formally, the goal template $T = \langle P_1 \rangle$ for this task has one goal pose. In this pose, the right gripper of Baxter points straight out, positioned 10cm above the *head* body part.

We tested our approach with users ablating themselves to simulated physical limitations. In the first set of experiments, we show how increasing the constraint complexity affects the execution time of the task. This motivates our constraint learning approach. In the second set of experiments, we show that learning the constraint model of a user optimizes the interactions. In particular, with enough accurate constraints the execution time is similar to the case with no constraints.

4.1 Planning and Executing with Increasing Constraint Complexity

We tested our system in the real world with one taller person (1.83m tall) and one shorter person (1.62m tall), who ablated themselves by simulating physical limitations. Our aim was to show that users with more complex constraints take longer to teach, which motivates learning personalized constraint models. Our first constraint was that the user was in a wheelchair and could not move his head closer than 0.9m to the robot. The second was that the user could not move his head more than 0.2m left.

We consider four cases, with results shown in Table 1. In Column A, the user starts in a feasible pose, and the manipulator immediately executes its task. In Column B, the robot selects a user-feasible pose, the user repositions himself, and the manipulator is then able to execute the task. In Column C, the user has one constraint, and the robot initially selects a user-infeasible pose. Thus the robot first makes an infeasible user-repositioning request and must detect this. Then, it must sample a new pose outside the user constraint space, and repeat the repositioning process. In Column D, the user has two constraints, so the process from Column C is likely to occur multiple times. We note that the execution time is highly dependent on the number of instructions the robot gives the user, and on how promptly the user complies with them.

Table 1. Average execution times and interactions for the 4 cases over 40 trials.

	A	B	C	D
Execution time (s)	11.6 ± 1.8	32.0 ± 12.3	53.0	78.0
Number of interactions	0	2.1	3.3	3.5

The variances for Columns C and D are omitted as there was a significant discrepancy between the two users. The distribution was bimodal, implying that the two users took different times to respond to the robot's instructions.

Fig. 2 shows a task execution for a user constrained to a chair, meant to model a wheelchair. Initially, the template goal is instantiated by the vision module with the user's head location. At first, the user is too far away, and the motion planner cannot solve the goal. The robot selects a new pose by sampling points in its configuration space that are believed to be feasible for the user. Additionally, the robot determines that this pose be achieved by asking the user to move forwards. Then, the vision system detects when he halts in the expected position and re-instantiates the template goal. Finally, the motion planner finds this new goal to be feasible, solves it, and places the hat on the user's head.

(a) User is initially at an unreachable pose.

(b) Robot requests the user to move forward.

(c) Manipulator successfully puts hat on the user.

Fig. 2. Example of dressing task execution for a user constrained to a chair.

The large majority of trials succeeded in putting a hat on the user, with few errors. Occasionally, the motion planner failed when the user was positioned at extremes of the reachability space of the manipulator, which can be resolved by considering these as invalid poses. As for the vision module, there were some problems when the robot arms occluded the field of view, which could be mitigated by adding additional cameras, using the built-in cameras on Baxter's arms, or planning trajectories to avoid obstructing the camera. As we will show in the next Subsection, after several trials of interaction with the same user to learn his specific constraints, future dressing episodes become more efficient.

4.2 Learning User Models

In this second set of experiments we show that, resorting to the knowledge-base formed by previous interactions with each user, the robot is able to complete the task more efficiently. To demonstrate this, we used the set of 2 constraints defined in the preceding subsection: $C = \langle head, head.x > 0.9, conf \rangle, \langle head, head.y > -0.2, conf \rangle$, where our starting confidence value $conf$ was 1.

We first taught Baxter a perfect constraint model for each user. Then, we performed 5 additional trials per user, with the robot choosing repositioning poses from the users non-constrained space. The average time for completing the task was $(34.2 \pm 29.4)s$ for the shorter user and (36.5 ± 8.5)s for the taller user with an average of 2.6 interactions. These results are very close to those for a user with no constraints (Table 1 Column B), which shows that learning personalized constraint models can markedly optimize the task execution.

5   Conclusion

In conclusion, we introduced an approach for a manipulator to help dress users. Our approach represents a dressing task as a sequence of pose goals with respect to the user. When the robot finds a goal infeasible, it actively requests the user to reposition his or herself. The robot chooses repositioning poses by sampling points in its configuration space that also satisfy a user-constraint model. This constraint model is updated when the user does not meet the expectations of an interaction request or enters a region the robot believes to be constrained.

We demonstrated our approach on a Baxter manipulator. We first showed how the time it takes to dress a user increases with constraint complexity. Then, we showed how modeling a user's constraints and sampling from feasible regions reduces the dressing time close to the case of a user with no constraints.

One direction for future work is to improve the manipulator's trajectories to make them more user-friendly. This could possibly be accomplished by biasing the motion planner with user-taught trajectories such as with E-RRT [4].

Acknowledgments. This research was partially supported by a research donation from Google, by NSF award number NSF IIS-1012733, by the FCT INSIDE ERI grant and two FCT student visiting fellowships. The views and conclusions contained in this document are those of the authors only.

References

1. Anjum, M.L., Ahmad, O., Rosa, S., Yin, J., Bona, B.: Skeleton tracking based complex human activity recognition using kinect camera. In: Beetz, M., Johnston, B., Williams, M.-A. (eds.) ICSR 2014. LNCS, vol. 8755, pp. 23–33. Springer, Heidelberg (2014)
2. Bacchus, F., Kabanza, F.: Planning for temporally extended goals. Annals of Mathematics and Artificial Intelligence **22**(1–2), 5–27 (1998)
3. Bai, Y., Liu, C.K.: Coupling cloth and rigid bodies for dexterous manipulation. In: Motion in Games (2014)
4. Bruce, J., Veloso, M.M.: Real-time randomized path planning for robot navigation. In: Kaminka, G.A., Lima, P.U., Rojas, R. (eds.) RoboCup 2002. LNCS (LNAI), vol. 2752, pp. 288–295. Springer, Heidelberg (2003)
5. Colomé, A., Planells, A., Torras, C.: A friction-model-based framework for reinforcement learning of robotic tasks in non-rigid environments. In: ICRA (2015)
6. Fikes, R.E., Hart, P.E., Nilsson, N.J.: Learning and executing generalized robot plans. Artificial intelligence **3**, 251–288 (1972)
7. Kita, Y., Neo, E., Ueshiba, T., Kita, N.: Clothes handling using visual recognition in cooperation with actions. In: IROS (2010)
8. Kuffner, J.J., LaValle, S.M.: RRT-connect: an efficient approach to single-query path planning. In: ICRA (2000)

9. Maitin-Shepard, J., Cusumano-Towner, M., Lei, J., Abbeel, P.: Cloth grasp point detection based on multiple-view geometric cues with application to robotic towel folding. In: ICRA (2010)
10. Matsubara, T., Shinohara, D., Kidode, M.: Reinforcement learning of a motor skill for wearing a t-shirt using topology coordinates. Advanced Robotics **27**(7), 513–524 (2013)
11. Melo, F.S., Veloso, M.: Decentralized mdps with sparse interactions. Artificial Intelligence **175**(11), 1757–1789 (2011)
12. Ferreira, B.Q., Gomes, J., Costeira, J.P.: A unified approach for hybrid source localization based on ranges and video. In: ICASSP (2015)
13. Tapus, A., Mataric, M., Scasselati, B.: Socially assistive robotics [grand challenges of robotics]. IEEE Robotics Automation Magazine **14**(1), 35–42 (2007)
14. Twardon, L., Ritter, H.J.: Interaction skills for a coat-check robot: Identifying and handling the boundary components of clothes. In: ICRA (2015)
15. Yamazaki, K., Oya, R., Nagahama, K., Okada, K., Inaba, M.: Bottom dressing by a life-sized humanoid robot provided failure detection and recovery functions. In: System Integration (2014)

Formations for Facilitating Communication Among Robotic Wheelchair Users and Companions

Yoshinori Kobayashi[1(✉)], Ryota Suzuki[1], Taichi Yamada[1], Yoshinori Kuno[1], Keiichi Yamazaki[1], and Akiko Yamazaki[2]

[1] Saitama University, Saitama, Japan
yosinori@hci.ics.saitama-u.ac.jp
[2] Tokyo University of Technology, Hachioji, Japan

Abstract. To meet the demands of an aging society, researches for intelligent/robotic wheelchairs have been receiving a lot of attention. In elderly care facilities, care workers are required to have a communication with the elderly in order to maintain their both mental and physical health. While this is regarded as important, a conversation with someone on a wheelchair while pushing it from his/her behind in a traditional setting would interfere with their smooth and natural conversation. Based on these concerns we are developing a robotic wheelchair which allows companions and wheelchair users move in a natural formation. This paper reports on an investigation how human behaves when the wheelchair users and their companions communicate while moving together.

Keywords: Robotic wheelchair · Ethnography · Interaction analysis

1 Introduction

As the demands for wheelchair use in an aging society increases, researches for intelligent/robotic wheelchairs have been receiving a lot of attention in recent years. For instance, Cao et al. have suggested an automatic wheelchair system which detects the user's intention through biological signals [3]. Satoh et al. has produced a system which uses an omni-directional stereo camera to enable a wheelchair to avoid gaps and obstacles [10]. These are just some examples of many studies currently developing to increase the range of movements afforded to wheelchair bound users [1] [2] [12]. It is indeed highly important to aim for a scenario in which the user of a wheelchair can become more independent. It is also important as well to look into the possible ways to better real life settings, such as a case in which an accompanying person assists or controls the wheelchair for its users that are novices or when the user himself or herself cannot control the wheelchair physically due to their health condition. As shown in Fig. 1(left), in a Japanese daycare facility a single companion often times ends up assisting multiple wheelchair users at once. Another observation from real life settings would be that these facilities would like to facilitate more personal communication with the wheelchair bound patients, in order to maintain and stimulate their cognition. It is difficult for the caregivers to attempt talking to the users of wheelchairs, however; the caregivers (or family members) are busy for

© Springer International Publishing Switzerland 2015
A. Tapus et al. (Eds.): ICSR 2015, LNAI 9388, pp. 370–379, 2015.
DOI: 10.1007/978-3-319-25554-5_37

controlling the wheelchair(s), and they are often positioned behind the users (Fig. 1(right)). When the participants of a conversation cannot see each other (not facing each other), multimodal resources which people usually use in interaction become highly limited. The situations illustrated as above faces this difficulty, hence it results in a rather poor communicational engagement for the users and their companions. In our project, we developed a robotic wheelchair so that it releases the participants from such a constraint. It allows them to maintain an interactional formation that enables them to monitor themselves easily, thus multimodal resources for interaction are still available to them.

Fig. 1. Scenes when a companion moves wheelchair(s).

Many kinds of mobile robots that can move with multiple companions collaboratively have been proposed. For instance, Urcola et al. propose the methodology that multiple robots take multiple companions in a practical situation [13]. Murakami et al. also propose the methodology that a robot, which does not have a priori knowledge of the companions' destination, can move with companions collaboratively using a destination estimation model based on observations of humans' daily behaviors [7]. Mobile robots which have companions like the approaches mentioned do not have passengers, so that they do not consider communication between passengers and companions. Kobayashi et al. proposed a robotic wheelchair that could follow alongside a companion based on tracking the companion's body position/orientation with using a 2D Laser Range Sensor (LRS), which was set on a top of a pole at the companion's shoulder level and attached to the wheelchair [6]. This wheelchair can estimate the moving direction of the companion and maintain formation of side-by-side. Therefore, the companion no longer needs to push the wheelchair and can walk alongside the wheelchair. However, their system is exquisitely sensitive to the companion's body position and orientation to align the wheelchair automatically. Thus, it still had constraints to their behavior in the interaction because the companion needed to pay extra attention to his/her body orientation.

This paper reports on an investigation to learn the patterns of human behavior when the wheelchair user and his/her companions communicate while walking along together. The purpose of this observation is to learn how the wheelchair operates, how body positions affect each other during communication, and apply the findings to our new system to be developed for the robotic wheelchair in our project. We developed a system which allows two companions and two users of wheelchairs move together in a stable manner (not sensitive reaction). The system developed based on ethnographic observations will generate a natural formation of positioning for the participants (both companions and wheelchair users) without their stress or constraints.

2   Sociological Background for Interaction Analysis Between Wheelchair User and Companion

In sociological studies, approaches such as conversation analysis and interaction analysis have been applied in order to learn about human behavior. Ryave et. al has explored walking as a social interaction [8]. People who are walking together, for instance, display their behaviors so that other groups see and orient themselves as in the same manner. By such recognition, people are able to avoid crashing into each other while they are walking in groups on a street.

In social interaction, multimodal resources such as eye gaze play important roles. Goodwin points out in his analysis of video recorded data that directing one's eye gaze to someone is a symbolic display of addressivity, i.e., a claim to be a listener/ recipient of the talk (recipiency) [4]. The speaker in a conversation also orients to the listener's eye gaze, at times by paying effort to draw his/her attention. In other words, both sides of the participants – the listener and the speaker – work hard to obtain each other's eye gaze. Kendon highlights the roles of positioning of people and objects in a social interaction. He has proposed the concept of F-Formation [5]. He utilize points out that people engaged in a certain interaction creates a space among them so that each can behave without interfering with the others (O-Space). Schegloff has examined body posture and body torque as an important player to form a conversation in interaction [9]. Schegloff also goes into which part of body torque would influence an on-going conversation.

These literatures from sociological studies have much to offer for the study of automated wheelchairs and their companions in communication. As pointed out earlier, in a traditional setting as in Fig. 1, both the user and the companion would have to face the same direction, not face-to-face. This means that if they want to establish an eye gaze to display their addressivity as suggested in Goodwin [4], one of them must torque their torso and head radically. The wheelchair user is more or less immobile from their hip and beyond, and how much one torques towards the person behind them displays their involvement in the interaction. The affordance given to each participant in terms of delivery of eye gazes and body position plays significant roles.

3   Observation of One-on-One Communication while Moving

We conducted a preliminary experiment to investigate how the natural communication between a wheelchair user and a companion is affected by the load of wheelchair operation when the companion remotely controls the wheelchair. We divided 26 participants, all of which are university students and they don't have experience for using a wheelchair, into 13 pairs randomly. Most pairs of participants didn't know each other very well. For each pair, one individual played the role of the wheelchair user and the other played the role of the companion. Additionally we asked participants playing the role of companion to move the wheelchair by a remote control and walk with it. In this experiment, we used a manual remote control system for the companions to move the wheelchair. By leaning the joystick on the remote control forward/backward and left/right, a companion can control the wheelchair's corresponding forward/backward movements and left/right rotations. To observe how formations change depending on the state of the conversation, we took 7 of the 13 pairs and assigned them a conversation task. The remaining 6 pairs were not assigned a conversation task. This experiment is conducted on campus (outside of building). We asked every pair to move toward a goal about 30m away and return to the start position. At the goal position, there is a board displaying some pictures of the World Heritage sites. To facilitate conversation we asked the pairs who were assigned the conversation task to collaboratively find a picture of a randomly assigned World Heritage site. On the other hand, the other pairs who were not assigned the task were only asked to go to the goal and come back. A course of this experiment is shown in Fig. 2.

Fig. 2. Course setting in preliminary experiment.

3.1   Participants NOT Assigned Conversation Task

As shown in Fig.3, the companions in 5 pairs out of 6 pairs in this setting operated the wheelchair from the back of the wheelchair. This is considered because the companions needed to concentrate on navigating the wheelchair when they were not required to talk with the wheelchair users. In the instance in Fig. 3(left) the wheelchair user started to talk to the companion with the utterance "Is it possible to speed up?" Even though the companion had already answered "It may be impossible," the wheelchair user repeated the utterance "Is it possible to speed up?" when the companion looked closely into the wheelchair user (Fig. 3(right)). This is likely because the companion could not continue the following utterance because he could not confirm the companion's eye gaze as an acknowledgement of his previous utterance.

Fig. 3. Analyzing scenes without conversation task.

3.2   Participants Assigned Conversation Task

The companions in all 7 pairs in this setting walked and operated alongside the wheelchair. We pick up one typical case shown in Fig. 4. At the time to start moving, the companion concentrated on navigating the wheelchair (Fig. 4(left)) from the back of the wheelchair because she was not familiar with the remote control. After a short while, after she got used to the remote control, she changed her following position to move alongside the wheelchair (Fig. 4(right)). When the wheelchair user directed her gaze toward the companion, she started to talk toward the wheelchair user with the preceding utterance "Are you enjoying this?" Note that, here, the companion started to talk when she changed her position to be at the side of the wheelchair where the faces of the companion and wheelchair user could be mutually observed. In other words, the wheelchair user displayed recipency by showing her gaze. In this case, when the companion changed her following position to move alongside the wheelchair, the wheelchair user tried to shift her gaze toward the companion. This behavior of the wheelchair user made it possible to start talking from the companion.

Fig. 4. Analyzing scenes with conversation task.

Consequently, we found that the side by side formation encouraged their communication and the load imposed by wheelchair navigation should be minimized to releases the wheelchair users and companions from the constraint for natural communication.

4   Observation of Group Communication and Formation

According to the observations conducted in the nursing home at Nara prefecture in Japan, we found that there are potential needs for multiple companions and multiple wheelchair users to move together. In this section we investigate the suitable formation between two wheelchairs in a communication. To understand the relationship between the communication and the formation of multiple companions and wheelchair users, we conducted an experiment on campus (outside of building) with two companions and two wheelchair users. We analyze how the accompanying people positioned themselves in order to talk to wheelchair users while moving together, and also how the participants construct their participation framework in the conversation.

We had 40 university students who don't have experience for using a wheelchair as participants in this experiment. Each experiment case includes four participants who don't know each other very well, two of them were in the wheelchairs and the others accompanied the wheelchairs users. Two wheelchairs moved along a predetermined course. The movement of the wheelchairs was remotely controlled by the staffs so that we could compare three different positions; i.e., tandem, diagonal, and side as shown in Fig. 5. The course is 4m wide, so it is difficult for all four participants to walk along side by side. For the purpose of facilitating the inter-participants conversations, we asked to work on a quiz together. Before starting to move, each participant is given a card which has a food item printed on it (e.g.; carrots, chicken, and pota-toes). At the turning point of the route, there are several additional cards with season-ing items printed on them (e.g.; curry paste and soy sauce), pinned to a white board. The task is to figure out what they can cook using their given food items and their choice of a seasoning. The participants are asked to give us their answer (e.g.; chicken curry) at the end of the experiment.

Fig. 5. Three types of formation between wheelchairs.

4.1   Tandem Formation

When a person starts activity, what is called "transactional space" becomes relevant. Transactional space is the space in front of the person or where their gaze is directed towards. When two or more people interact with each other, they construct a shared space, called "O-Space", which is the overlapping transactional space of each person [5]. When two wheelchairs maintained a tandem position (Fig. 5(left)), conversation between all four participants did not occur. Soon after starting experiment, the wheel-chair user ahead turned to the other wheelchair user in the back, and began to talk and make eye contact with each other. Meanwhile, the two companions were walking side by side after the two wheelchairs, without joining in the conversation (Fig. 6(a)). There are two O-Spaces here, one is between the two wheelchair users, and another is between the two companions. After that, one of the companions began to step up to the wheelchair user and talked to her. When the wheelchair user in the back noticed this companion was beside her, she turned to him to show her attention recipiency (Fig. 6(b)). The wheelchair user ahead wanted to join the conversation and looked back. Though she made eye contact with the companion who was still staying in the back of the line (Fig. 6(c)), the companion turned his face to the wheelchair user in the back (Fig. 6(d)). After that, the topic of conversation moves on a card which the former wheelchair user had. In this scene, both companions directed their attention to the former wheelchair user and started a conversation among the three (Fig. 6(e)). Here we can see the rear wheelchair user could not join in the conversation as she had been able to join in the conversation before. Afterwards, one companion, who is aside the former wheelchair, stepped back to the rear wheelchair user and positioned to her side instead. We can interpret this behavior to mean that he wished the rear wheel-chair user to join in the conversation. But this time the other companion was unable to join in the conversation (Fig. 6(f)). This shows that the tandem formation between wheelchairs does not allow them to easily acknowledge each other. In this scene, the companion who could not join in the conversation said "I can't find an appropriate position." Hence we see there are no participation frameworks where all participants could join in on one conversation.

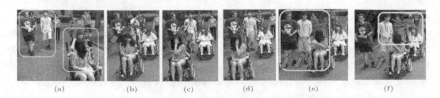

Fig. 6. Scenes of TANDEM formation.

4.2   Diagonal Formation

In the case wheelchairs were in a diagonal formation like Fig. 5(center), we can see a participation framework which includes all four participants in a group conversation. On starting the former wheelchair user turned to his left back to solicit attention. Then one companion intersected his gaze with the former wheelchair and they gain mutual attention (Fig. 7(a)). Afterwards, the companion moved to the former wheelchair user to be side by side with him and started a conversation (Fig. 7(b)). In the next scene, both companions came nearby the wheelchairs while showing their cards and all participants joined in the conversation (Fig. 7(c)). Here we can see there is circle shape F-formation the overlaps their transactional space.

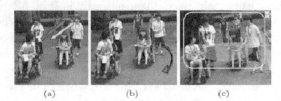

Fig. 7. Scenes of DIAGONAL formation.

4.3   Side Formation

In the case wheelchairs were in a side formation like Fig. 5(right), we can also see the participation framework like in the case of the diagonal formation. Before starting, both wheelchair users can solicit attention from each other by only turning their faces toward their partners (Fig. 8(a)). Afterwards, the companions behind the wheelchairs moved closer to the wheelchairs users to start communication (Fig. 8(b)). Companions that came to the side of the wheelchairs stepped once more with a twisting his body so that he could direct his gaze and draw attention (Fig. 8(c)). In this scene, the O-Space included all members in the group and they could all talk to each other.

By considering the analyses of three cases, we found that in the tandem case, communication between wheelchair users hardly occurred. It was also found that such a formation made it difficult for companions to rearrange their positions for easy communication. On the other hand, in the case of the diagonal/side formations, wheelchair users could confirm the state of each member by turning their upper body or head. Companions could also select a position that allowed constructing a participation framework which included all the participants in the conversation. Therefore

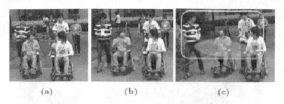

Fig. 8. Scenes of SIDE formation.

we need to develop multiple robotic wheelchair system that can maintain diagonal/side formations, instead of the tandem formation.

5   Experiment Using Multiple Robotic Wheelchair System

Based on the findings in the previous section, we have developed multiple robotic wheelchair system that can maintain their formation while moving. Details are mentioned in the [11]. We conducted the experiment on campus (inside of building) using our new system using 32 participants in 8 groups (4 participants in each group who didn't know each other very well). All of participants are university students who don't have experience for using a wheelchair. Each group is assigned one of three formations between wheelchairs: tandem, diagonal and side. The setting and the environment are same as in Section 4.

5.1   Tandem

As shown in Fig. 9(left) the front wheelchair user turns her body and gazes backwards while showing her card and uttering "Well" as a preface in order to start a conversation. Whereas the other three participants exchange the utterances about 'curry', The front wheelchair user stops before the completion of her utterance (Fig. 9(center)). She abandons her effort for the conversation and moves her gaze back in the previous direction (Fig. 9(right)), because she cannot obtain gaze of others.

Fig. 9. Experimental Scenes of TANDEM formation.

5.2   Diagonal

Fig. 10(left) shows the scenes of initial formation of participants of the experiment and Fig. 10(right) shows the subsequent formation. In the left photo, two companions turn their bodies towards the wheelchair users while deploying diagonal positions towards the wheelchair users. The companions coordinate their walk with the movement of the wheelchair users (Fig. 10(right)). The lower bodies of the companions move along the direction of their respective wheelchairs and their upper bodies, in particular their faces turn towards the wheelchair users. Thus there is a participation framework in which all four participants move while conversing together.

Fig. 10. Experimental Scenes of DIAGONAL formation.

5.3   Side

Fig. 11(left) shows that left companion in regard to the moving direction moves towards the left side of the wheelchair while saying "Oh. What's this?" After hearing this, right participant moves his body between wheelchairs. All of them engage in the conversation. Then the left companion moves towards the forefront position of the group by walking backwards (Fig. 11(right)). Consequently they can establish the formation of an O-Space among these participants.

Fig. 11. Experimental Scenes of SIDE formation.

The findings of the experiment are as follows; 1) In tandem formations each member fails to get mutual gaze so that communication is disturbed, 2) In both diagonal and side formations each member can get mutual gaze it is possible to have natural and smooth communication. The findings are the same as the preliminary experiments.

6   Conclusion

In this paper we investigate the formations when the wheelchair users and their companions move together. We found that the side by side formation encouraged their communication and the load imposed by wheelchair navigation should be minimized to releases the wheelchair users and companions from the constraint for natural communication. Also we found that the tandem formation between wheelchairs does not allow them to easily acknowledge each other. This makes it difficult to make the participation frameworks where all participants join in one conversation. It is important to make F-formation by overlapping their transactional space including wheelchair users even when they are moving. Based on the findings, we developed a new wheel-chair system which can reduce the load of controlling the wheelchair and let a companion change his/her position freely. We confirmed that our multiple wheelchair system allows a companion to move with wheelchairs naturally so that they could attain mutual gaze and start their conversation.

In this paper, all the participants are university students who can use their bodies without any physical limitations. But elderly people will have physical limitations and they will not be able to turn their bodies while moving together. We are now planning to conduct experiments in an actual care facility.

References

1. Blatt, R., Ceriani, S., Dal Seno, B., Fontana, G., Matteucci, M., Migliore, D.: Brain control of a smart wheelchair. In: International Conference on Intelligent Autonomous Systems, pp. 221–228 (2008)
2. Braga, R.A.M., Petry, M., Reis, L.P., Moreira, A.P.: Intellwheels: modular development platform for intelligent wheelchairs. Journal of rehabilitation research and development 48(9), 1061–1076 (2011)
3. Cao, L., Li, J., Ji, H., Jiang, C.: A hybrid brain computer interface system based on the neurophysiological protocol and brain-actuated switch for wheelchair control. Journal of neuroscience methods 229, 33–43 (2014)
4. Goodwin, C.: Conversational organization: Interaction between speakers and hearers. Academic Press, New York (1981)
5. Kendon, A.: Conducting interaction Patterns of behavior in focused encounters, vol. 7. CUP Archive, Cambridge (1990)
6. Kobayashi, Y., Kinpara, Y., Takano, E., Kuno, Y., Yamazaki, K., Yamazaki, A.: Robotic wheelchair moving with caregiver collaboratively. In: Huang, D.-S., Gan, Y., Gupta, P., Gromiha, M. (eds.) ICIC 2011. LNCS, vol. 6839, pp. 523–532. Springer, Heidelberg (2012)
7. Murakami, R., Morales Saiki, L.Y., Satake, S., Kanda, T., Ishiguro, H.: Destination unknown: walking side-by-side without knowing the goal. In: International Conference on Human-Robot Interaction, pp. 471–478 (2014)
8. Ryave, L., James, S.: Notes on the art of walking. In: Turner, R. (ed.) Ethnomethodology, Baltimore, pp. 265–274 (1974)
9. Sacks, H., Schegloff, E.A., Jefferson, G.: A simplest systematics for the organization of turn-taking for conversation. Language 50(4), 696–735 (1974)
10. Satoh, Y., Sakaue, K.: An omnidirectional stereo vision-based smart wheelchair. EURASIP Journal on image and video processing 2007(1), 87646 (2007)
11. Suzuki, R., Yamada, T., Arai, M., Sato, Y., Kobayashi, Y., Kuno, Y.: Multiple robotic wheelchair system considering group communication. In: Bebis, G., Boyle, R., Parvin, B., Koracin, D., McMahan, R., Jerald, J., Zhang, H., Drucker, S.M., Kambhamettu, C., El Choubassi, M., Deng, Z., Carlson, M. (eds.) ISVC 2014, Part I. LNCS, vol. 8887, pp. 805–814. Springer, Heidelberg (2014)
12. Tomari, M.R.: A framework for controlling wheelchair motion by using gaze information. International journal of integrated engineering 5(3), 40–45 (2014)
13. Urcola, P., Montano, L.: Adapting robot team behavior from interaction with a group of people. In: International Conference on Intelligent Robots and Systems, pp. 2887–2894 (2011)

Young Users' Perception of a Social Robot Displaying Familiarity and Eliciting Disclosure

Ivana Kruijff-Korbayová[1], Elettra Oleari[2(✉)], Anahita Bagherzadhalimi[2],
Francesca Sacchitelli[2], Bernd Kiefer[1], Stefania Racioppa[1], Clara Pozzi[2],
and Alberto Sanna[2]

[1] Language Technology Lab, DFKI, Stuhlsatzenhausweg 3 66123, Saarbrücken, Germany
{ivana.kruiff,bernd.kiefer,stefania.racioppa}@dfki.de
[2] Fondazione Centro San Raffaele (FCSR), Vial Olgettina 60 20132, Milan, Italy
{oleari.elettra,bagherzadhal.anahita,sacchitelli.francesca,
pozzi.clara,sanna.alberto}@hsr.it

Abstract. Establishing a positive relationship between a user and a system is considered important or even necessary in applications of social robots or other computational artifacts which require long-term engagement. We discuss several experiments investigating the effects of specific relational verbal behaviors within the broader context of developing a social robot for long-term support of self-management improvement in children with Type 1 diabetes. Our results show that displaying familiarity with a user as well as eliciting the user's self-disclosure in off-activity talk contribute to the user's perception of the social robot as a friend. We also observed increased commitment to interaction success related to familiarity display and increased interest in further interactions related to off-activity talk.

Keywords: Child-robot interaction · Human-robot interaction · Long-term interaction · Social robot · Verbal behavior · Personalization · Continuity behaviors · Familiarity display · Self-disclosure · Off-activity talk · Perception of robot as friend

1 Introduction

It has become a commonplace vision that robots will partake in many areas of our lives. The role they are envisaged to fulfill has shifted from that of a mere tool to a teammate, peer, companion, friend. Thus, being conceived of as social actors, which will be explicitly and intentionally entering into relationships with humans. Social science research has identified a plethora of behaviors that are prevalent and influential in establishing and maintaining human-human relationships. Inspired by the seminal work on relational agents by Bickmore and colleagues [3] a growing body of research now studies what effects do such behaviors have in human-machine, and more specifically human-robot relationships, and how we can implement the corresponding functionality to enable machines/robots to perform these behaviors autonomously. Overviewing this body of literature, it is clear that the more we know, the more we know what we do not know. There remain many aspects to be studied.

© Springer International Publishing Switzerland 2015
A. Tapus et al. (Eds.): ICSR 2015, LNAI 9388, pp. 380–389, 2015.
DOI: 10.1007/978-3-319-25554-5_38

The research presented here concerns relational verbal behaviors that contribute to the perception of an agent as a friend. It is set within the broader vision of developing a robotic companion to provide long-term support to children with Type 1 Diabetes Mellitus (T1DM) to help them learn and improve their ability to independently manage their condition. During the process of self-management development, children with T1DM need to acquire knowledge about diabetes and suitable healthy nutrition, develop various relevant skills and learn to adhere to the therapy requirements. Similarly to what was noted for health behavior change applications [3], establishing and maintaining a positive relationship is considered to be a necessary (though likely not sufficient) condition for addressing the further goal of influencing diabetes self-management. In this paper we focus on two aspects of relational verbal behavior which personalize an interaction by linking it to the experiences of a given user: signaling continuity over time by references to joint experiences of the user and the robot in interaction with one another (Familiarity Display – FD); and eliciting disclosure about separate experiences of the user (Off-Activity Talk - OAT). In a series of experiments with an implemented integrated system, comparing independently a condition with and without FD and with and without OAT, we found that these behaviors contribute to young users' perception of the robot as a friend. We first review related work on such relational verbal behaviors in Sec. 2. In Sec. 3 we describe our system. In Sec. 4 and 5 we present the methodology and results of the experiments addressing FD and OAT, respectively. We discuss the observed effects and conclude in Sec. 6.

2   Background

Bickmore and colleagues developed the concept of relational agents, referring to computational artifacts designed to establish and maintain long-term social-emotional relationships with their users [3]. They discussed a myriad of strategic relational behaviors, instantiated them in systems and carried out numerous studies to evaluate the effects of various aspects of relational agent behavior on long-term engagement and behavior change, e.g., [4]. This inspired many other researchers to perform further studies and experiments in this area. What we call familiarity display has been called continuity behaviors in some previous literature. For example, the continuity behaviors implemented in the FitTrack system [3] and the person memory model of a virtual agent described in [13] include greetings and farewells referring to past/future encounters and reference to mutual knowledge, e.g., user's biographical facts, preferences and interests mentioned in a previous session. The exercise advice system described in [8] also implements continuity behaviors as means of relationship maintenance, namely reference to previously given advice and gradually more personal greetings, including some small talk. Various other systems have included a user model or some form of long-term memory and used it to refer to content from previous interactions [1, 5, 16, 18].

Our concept of Off-Activity Talk corresponds to the reciprocal self-disclosure discussed as another relational behavior and found to increase trust, closeness and liking in work cited by [3]. While the OAT in our system allows reciprocity, we have focused on eliciting disclosure from the users so far. This resembles the gathering of personal information in [1, 5, 13, 18], but is more conversational.

A comparison of existing results concerning the effects of various relational behaviors is complicated by the fact that each study uses measures and methodologies adjusted to its purpose. For example, [3] evaluated the effects of all the relational behaviors combined. They found an effect on long-term relationship, but not on behavior change in a real usage longitudinal study. On the other hand, [8] evaluated the isolated impact of relationship maintenance on users' attitudes and found an effect on various metrics and [13] investigated the impact on social presence, likability and communication satisfaction of using personal information during the interaction sessions. These studies were done with adults, the systems of [3] and [8] were not robots, and the metrics did not include a classification of the user's perception of their relationship with the system. Some of the experiments did not involve usage in real life, but the participants used the system to play out hypothetical situations, e.g. [8].

3   System and Setup

Our experiments were carried out with the system developed gradually in the course of the Aliz-E project [7]. The robot we use is the small humanoid robot Nao from Aldebaran Robotics. The system integrates components for speech recognition and interpretation as well as natural language generation and synthesis, gesture capture and interpretation, nonverbal behavior production and motor control, activity-, interaction- and dialog management, and a user model to store key information about each child [9]. Several game-like activities were implemented in the system: Quiz, Imitation, Dance and Collaborative Sorting [9, 2, 15]. A range of relational social behaviors reported in the literature was implemented across the activities, including informal greetings, introductory small talk, the use of first names, empathy related to the performance in an activity, the robot's ability to make mistakes, nonverbal bodily cues, allowing children to touch the robot [14]. Although the robot was presented as autonomous to the participants, we relied on a partially Wizard-of-Oz setup, where a human Wizard simulated the speech and gesture input interpretation, could override the automatic dialog management decisions, if needed, and fully controlled off-activity talk.

4   Experimental Study 1: Familiarity Display

The first study was a longitudinal experiment investigating the use and effects of continuation behaviors. We investigated how the robot can acquire familiarity with a user and display it in interactions, and what effect this would have on children's perception of the robot.

4.1   Familiarity Display

When humans interact with each other over a series of encounters, they become familiar, i.e., they accumulate shared knowledge (shared history, personal common ground) [6]. The goal of this study was to endow the robot with the ability to acquire a persistent interaction history respective to each individual user and allow it to manifest its familiarity with the user both verbally and nonverbally later in the same interaction or in subsequent interactions. We selected several parameters that the robot would use to represent the interaction history: the user's name; whether it is the first or a subsequent encounter of the user with the robot; for each activity whether the user has already performed it or not and some details about it (e.g.: for each Quiz question, whether it has been asked before in a interaction with the user); the user's last performance on each activity. The values of these parameters for each user were stored in a persistent user model. We designed templates for verbal output generation which allowed to include content based on the user model. The robot would use these verbalizations to explicitly display its familiarity with the user. Such verbal moves would also be accompanied by nonverbal behaviors showing familiarity, e.g., nodding, higher excitement. We also designed alternative verbalizations which were neutral, i.e., they would not show whether the robot is or is not familiar with the user. Examples of verbalizations of both kinds are shown in Table 1.

Table 1. Examples of verbalizations that signal familiarity (used in the FD condition, see paragraph 4.2) or are neutral in this respect (used in the ND condition, see paragraph 4.2).

Familiarity display	Neutral display
Use of user's name:	
So, which answer do you choose, *Marco*?	So, which answer do you choose?
References to previous encounters and play experiences:	
I am happy to see you *again*.	I am happy to see you.
It was nice playing with you *last time*.	-
References to previous performance in an activity:	
Are you ready to play *again*?	Are you ready to play the quiz?
Today you were *again* very good at quiz.	Today you were really good at quiz.
Well done, you've done *better than last time*.	Well done.

4.2   Experiment Methodology

As described in detail in [10], 19 children participated in total (11 male, 8 female; age 5-12, all Italian), of which 13 participated in three sessions on different days as foreseen in the protocol (9 male, 4 female; 6 with T1DM, 7 healthy).

We exerted a between-subjects design with two conditions: the Familiarity-Display (*FD* - 9 children) condition and the Neutral Display (*ND* - 10 children) condition. The robot used the verbal and nonverbal behaviors described in Sec. 4.1, respectively.

The experiment took place at a research lab at the San Raffaele hospital in Milan. The sessions were organized on several Saturdays over a period of two months and full participation involved three sessions on different dates per child, where s/he could choose among one (or more, time permitting) of the available activities to be performed with the robot: Quiz, Imitation game and Dance. Each session of the experiment lasted maximally one hour, including the interaction session with the robot and filling in 3 post-interaction questionnaires. These latter were multiple choice questionnaires reporting the child's self-assessment of: (Q1) *the perceived bond with the robot*, to be categorized between different levels of confidence and familiarity: stranger, neighbor, classmate, teacher, friend, relative, sibling, parents; (Q2) *the perceived role during the activities*: child leading, robot leading or on a par; (Q3) *the perception of the robot*: through a multi-adjective choice among friend, toy, pet or game console. Children were also asked to briefly explain their choices. The questionnaires were administered to the participants at the end of each session, in order to see if there was any change over time.

4.3   Results

We analyzed the post-interaction questionnaires, linking to each multiple choice answer a numerical value. We calculated the means and standard deviations of the scores per child across the interaction sessions.

Questionnaires Q1 and Q2 did not reveal any statistical significance regarding the perception of either the bond with the robot or the level of the established relationship, neither between the two experimental conditions (*FD* and *ND*) or across the sessions (for those children who interacted three times). From the explanations that the participants gave to justify their answers, as a qualitative insight we saw that, independently from the two conditions, high rates of perception of the bond (from friend to parent) were related to the play dimensions (e.g.: "having fun" and "play together") and the friendly approach ("it's nice/cute/tender") that the robot showed to children. Lower values, linked to the perception of different levels of relationship (stranger, neighbor, teacher), were mainly related to a low satisfaction and engagement in the activity/ies performed (e.g. "too difficult questions/tasks", "questions like homework", etc.). In addition, there was an overall perception of the interactions with the robot as being "at the same level".

An interesting result was found in Q3: a comparison of the adjective choices revealed that all 9/9 children in the *FD* condition perceived the robot as friend after the first session as opposed to the only 4/10 *ND* children (Fisher's test: two-tailed P=0.0108). Among the 13 children who continued to have 3 interactions 5/6 *FD* children maintained the perception of the robot as friend, only one changed it to a toy. No trend was observed among the 7 *ND* children.

5   Experimental Study 2: Off-Activity Talk

The second study investigated the effects of Off-Activity Talk (OAT) in one-on-one interaction sessions held in the context of two different educational summer camps for children with T1DM, organized by the Italian patient's association SOStegno70 (www.sostegno70.org) and the pediatric department of San Raffaele hospital (Milan). We investigated how the robot can engage a child in OAT on topics related to diabetes and healthy lifestyle and how this would impact children's perception of: their relationship to the robot; interest to undergo further interactions with it; willingness to talk about diabetes and self-disclosure; adherence to a therapy-related requirement to fill in a nutritional diary. The methodology and results of the 2013 experiment have been published in [11, 12]. We only repeat those points that are relevant for a comparison with the 2014 experiment.

5.1   Off-Activity Talk

OAT are sub-dialogs interspersed within the talk pertaining to a particular activity, without being directly part of it [11, 12]. The main OAT topics we defined were pertinent to: hobbies, diabetes, nutritional habits, relationships with friends and adults with respect to diabetes (for examples, see [11]). For each topic, several OAT-initiation system utterances were formulated and implemented as canned text in the system. The initiation of OAT was triggered by the Wizard at opportune moments during the interaction(s) within the course of an activity. For example, during the Collaborative Sorting game, OAT about nutritional habits was triggered while the child was playing a virtual game about the food and carbohydrates counting (see the example extracted from a longer interaction in the left column in Table 2).

Table 2. Examples of OAT subdialogs from the 2014 experiment (translated from Italian). *Left:* OAT on nutritional habits linked to the contents of the Sorting Game. *Right:* Child self-disclosure concerning diabetes.

Nao: *what is your favourite food among these?* Child: *among the remaining ones - ice cream. I really like it. although I can't always eat it, it's not good for me.* Nao: *which flavour do you prefer?* Child: *<EH> stracciatella.* Nao: *that's good.* Child: *<laughter>* Nao: *can you distinguish foods according to carbohydrates?* Child: *yes, I do count - as I do for the insulin at home. I count carbo-hydrates.*	Nao: *do you measure glycaemia yourself at home?* Child: *yes, I do it myself, because now I'm quite big, so -* Nao: *you're so good.* Child: *thanks.* Nao: *I was told that it is difficult.* Child: *<EH> a - well, it is a bit difficult at the beginning, because even - say - accept that you have diabetes is a bit difficult. but then you get used to it. [...]* Nao: *I'd like to ask you how do you do it.* Child: *how do I measure glycemia?* Nao: *yes.* Child: *so, basically <EH> I take my measuring set, I disinfect the finger I want to prick, I take the pen, prick and then put the blood on the strip, which automatically measures glycaemia, so that I see the value on the screen of the set.*

5.2   Experiment Methodology and Results

Participants. The experiments for this study have been conducted in two different educational summer camps in 2013 and 2014. In the 2013 experiment, 20 children (age 11–14, 10 females and 10 males) among the total of 59 attending the camp volunteered to participate in the individual sessions with the robot. In 2014 it was 28[1] (age 10–14, 10 females and 18 males) out of 41. The remaining children were in both cases included in the control group and experienced the robot in the camp only as a theater performance character during recreational evening activities.

---

[1] The data of one subject was discarded as the child did not finish the interaction.

Procedure. In both the 2013 and 2014 summer camps, children who volunteered for individual session(s) with the robot were given an appointment during their spare time at the camp. Before beginning the interaction, they were instructed about the available game activities with the system and the possibility to freely choose among them during their session. The session lasted a maximum of 30 minutes. The interactions were carried out using the system described in Sec. 3.

2013 Camp Experiment Overview. The specific objectives of the 2013 camp were to investigate the feasibility and acceptance of OAT, its effects on children's perception of the robot and on adherence to medical advice (i.e.: filling in a nutritional diary). The study was carried out in a between-subjects design with 3 conditions: (1) *OAT*: one-on-one interaction with the OAT feature turned on; (2) *non-OAT*: one-on-one interaction without OAT; (3) *CONTROL*: no one-on-one interaction.

The results related to this study are discussed in detail in [11, 12] but with respect to the present contribution, it is interesting to mention that qualitatively children's acceptance of OAT was good: they engaged in it readily and elicited self-disclosure from the robot [12]. However, their responses to the robot's OAT prompts were brief and concise, maybe due to their formulation as closed questions. Moreover, the presence of OAT turned out to have a positive effect on the children's interest to interact with the robot again: although all subjects in the two intervention conditions expressed interest to play again with the robot, only 11 actually booked another slot: 9/10 in the *OAT* group and 2/10 in *non-OAT* (Fisher's test, two-tailed P=0.0055).

2014 Camp Design. Based on the positive experience with OAT in the 2013 experiment, we decided to drop the non-OAT condition. The 2014 experiment thus had a between-subject design with the *OAT* and the *CONTROL* condition. We revised the OAT prompts, to include more open questions or clusters of closed interconnected questions, in order to elicit more complex OAT dialogs with more child talk. Table 3 shows some examples of these variations; Table 2 shows OAT interaction examples.

Table 3. Examples of the different verbalization of the OAT prompts used in the two Camps.

2013 OAT prompts formulation	2014 OAT prompts formulation
Can you draw?	Can you draw? What do you like to draw?
Do you realize when your glycaemia is low?	Do you realize when your glycaemia is low? What do you do in these cases?
What is the strangest food you've ever tried?	What is the strangest food you've ever tried? Where were you when you tried it? Abroad?

We also further elaborated the evaluation of children's perception of the robot. We designed a new questionnaire composed of two closed questions. The first one asked to describe the robot by choosing one out of the following set words: friend, toy, pet, adult, computer. The second one asked to choose one of 16 listed adjectives describing the robot. The adjectives belonged to three categories of perception: machine (e.g. fake, scientific, etc.), relational (e.g. interested in me, someone to trust, etc.), humanized (e.g. spontaneous, empathetic, etc.). This questionnaire was administered to all the participants of the camp at the end of their stay. Furthermore, to evaluate children's willingness and spontaneity to talk about diabetes, we performed an analysis of the interactions similar to the one described in [12]: 3 coders (native speakers) evaluated every OAT sub-dialog regarding diabetes on a 4 point scale (i.e.: 1= "not responding or not willing at all", 2= "forced or annoyed", 3="clear, simple and courteous", 4="very interested and active") as well as assigned an overall score per child to how the OAT diabetes sub-dialog were going.

2014 Camp Results. OAT had an effect on the children's perception of their relationship to the robot: 26/27 in the *OAT* group and only 4/13 in the *CONTROL* group selected the word "friend" among the 5 options offered in the questionnaire. The difference between the two proportions is strongly statistically significant (χ^2=20.09 with probability 1%, two-tailed p=0.0001). Regarding the multiple adjective choice, even if not supported by statistical significance, we observe that children in the *OAT* condition chose no machine category adjectives, 30% of the chosen adjectives belonged to the humanized category and 70% to the relational one. Whereas in the *CONTROL* condition 20% of the adjectives chosen belonged to the machine category, 20% to the humanized one and 60% to the relational one. The children's willingness and spontaneity to engage in OAT and talk about diabetes was high. Moreover, the coders noticed qualitatively a common attitude of the children in sharing their practical notions about diabetes with the robot and their personal experiences on what it is like to deal with diabetes in their daily lives (see the excerpt in the right column of Table 2).

6   Discussion and Conclusion

We described a series of experiments with a robotic multi-activity system designed to provide long-term support to children with T1DM. We addressed the potentialities of specific relational verbal behaviors in contributing to the perception of a robotic character as a friend by the young participants: familiarity display and off-activity talk. Both these features were introduced in order to personalize the interactions in a way that resembles typical human interactions between friends: making reference to joint experiences and fostering self-disclosure about personal topics (in this case diabetes- and health related topics). We found that children interacting with the robot displaying familiarity, clearly perceived it as a friend after the first interaction as well as after three interactions in a longitudinal study. They also felt to have been at the same level of control with the robot during the interactions. This outcome was also confirmed by the investigations of the 2014 summer camp experiment, carried out with a different set of children in a real world setting, even though the set of words available to define the role of the robot was slightly different on the two occasions. In the 2014 summer camp experiment the set of choices included also the word "adult" in order to allow for a difference in the level of the perceived relationship biased towards the robot (robot compared to a figure that usually leads situations), rather than towards the child (as in the case of a pet or a video game). Confirming the previous results, none of the children chose this description. As for Off-Activity Talk, children were at ease during the dialogs with the robot and seemed to appreciate the interest that it showed for their daily lives and experiences. The combination of these factors led to a natural adaptation of children's behaviors to the specific single interaction dynamics and triggering, a spontaneous conversation regarding the delicate topic of diabetes. Moreover, the dialog structure enriched with the OAT prompts seemed to be a key factor in engaging children and making them interested to interact again with the system. This is a significant achievement in the long term perspective of our research, even though more longitudinal studies are needed to address this point. To conclude, the fact that the robot is perceived by children as a friend capable to establish and maintain a positive relationship is extremely impactful in a broader real life application perspective of a robotic companion. Children could be more inclined to feel at ease and open themselves with such a robot, thus offering the diabetology teams of caregivers a valuable instrument to support their work of education, addressing the goal to improve self-management of young patients.

Acknowledgments. The research presented in this paper has been funded by the Aliz-E project (aliz-e.org), grant n° ICT-248116 in FP7/2007-2013, and the PAL project (pal4u.eu), grant n° 643783-RIA in Horizon 2020. We wish to thank the Sostegno70 association for diabetic children and the team of the Paediatric Unit of Ospedale San Raffaele for their constant support in this research.

References

1. Adam, C., Cavedon, L., Padgham, L.: Hello emily, how are you today? Personalised dialogue in a toy to engage children. In: Proceedings of the 2010 Workshop on Companionable Dialogue Systems, Association for Computational Linguistics (ACL), Uppsala, Sweden, pp. 1–6 (2010)
2. Baxter, P., Wood, R., Belpaeme, T.: A Touch screen-based Sandtray to Facilitate, Mediate and Contextualize Human-Robot Social Interaction. In: Proceedings of the Seventh Annual ACM/IEEE International Conference on Human-Robot Interaction - HRI, Boston, MA, USA, pp. 105–106, March 2012
3. Bickmore, T.W., Picard, R.W.: Establishing and maintaining long-term human-computer relationships. ACM Transactions on Computer-Human Interaction (TOCHI) **12**(2), 293–327 (2005)
4. Bickmore, T., Schulman, D., Yin, L.: Maintaining engagement in long-term interventions with relational agents. International Journal of Applied Artificial Intelligence, special issue on Intelligent Virtual Agents **24**(6), 648–666 (2010)
5. Campos, J.C.F.: May: My memories Are Yours. An interactive companion that saves the users memories. Master thesis, Instituto Superior Técnico (2010)
6. Clark, H.: Using Language. Cambridge University Press, Cambridge (1996)
7. Coninx, A., Baxter, P., et al.: Towards Long-Term Social Child-Robot Interaction: Using Multi-Activity Switching to Engage Young Users. Journal of Human-Robot Interaction (to appear, in press). http://www.coninx.org/work/ALIZE-JHRI-preprint.pdf

8. De Boni, M., Richardson, A., Robert, H.: Humour, relationship maintenance and personality matching in automated dialogue: A controlled study. Interacting with Computers **20**(3), 342–353 (2008)
9. Kruijff-Korbayová, I., Athanasopoulos, G., Beck, A., Cosi, P., Cuayahuitl, H., Dekens, T., et al.: (2011) An Event-based conversational system for the nao robot. In: IWSDS2011 Workshop on Paralinguistic Information and its Integration in Spoken Dialogue Systems, pp. 125–132. Springer (2012)
10. Kruijff-Korbayová, I., Baroni, I., Nalin, M., Cuayahuitl, H., Kiefer, B., Sanna, A.: Children's turn-taking behavior adaptation in multi-session interactions with a humanoid robot. In: Hoffman, G., Cakmak, M., Chao, C. (eds.) Workshop on Timing in Human-Robot Interaction. ACM/IEEE, Bielefeld (2014)
11. Kruijff-Korbayová, I., Oleari, E., Baroni, I., Kiefer, B., Zelati, M.C., Pozzi, C., et al.: Effects of off-activity talk in human-robot interaction with diabetic children. In: Ro-Man 2014: The 23rd IEEE International Symposium on Robot and Human Interactive Communication, Edinburgh, UK, pp. 649–654 (2014)
12. Kruijff-Korbayová, I., Oleari, E., Pozzi, C., Racioppa, S., Kiefer, B.: Analysis of the responses to system-initiated off-activity talk in human-robot interaction with diabetic children. In: Proceedings of the 18th Workshop on the Semantics and Pragmatics of Dialogue, Edinburgh, UK, pp. 90–97 (2014)
13. Mattar, N., Wachsmuth, I.: Let's get personal - assessing the impact of personal information in human-agent conversation. In: HCI (2) (2014)
14. Nalin, M., Baroni, I., Sanna, A., Pozzi, C.: Robotic companion for diabetic children. In: Conference Interaction Design for Children, pp. 260–263 (2012)
15. Ros, R., Baroni, I., Demiris, Y.: Adaptive Human Robot Interaction in Sensorimotor Task Instruction: From Human to Robot Dance Tutors. Robotics and Autonomous Systems **62**(6), 707–720 (2014)
16. Schröder, M., Trouvain, J.: The German text-to speech synthesis system MARY: A tool for research, development and teaching. International Journal of Speech Technology **6**(4), 365–377 (2003)
17. Schulman, D., Bickmore, T., Sidner, C.L.: An intelligent conversational agent for promoting long-term health behavior change using motivational interviewing. In: Proceedings of AAAI (2011)
18. Sieber, G., Krenn, B.: Episodic memory for companion dialogue. In: Proceedings of the 2010 Workshop on Companionable Dialogue Systems, Association for Computational Linguistics (ACL), pp. 1–6, Uppsala, Sweden (2010)

You're Doing It Wrong! Studying Unexpected Behaviors in Child-Robot Interaction

Séverin Lemaignan[1,2]([✉]), Julia Fink[1], Francesco Mondada[2],
and Pierre Dillenbourg[1]

[1] Computer-Human Interaction in Learning and Instruction Lab (CHILI),
École Polytechnique Fédérale de Lausanne (EPFL), 1015 Lausanne, Switzerland
[2] Laboratoire de Systèmes Robotiques (LSRO),
École Polytechnique Fédérale de Lausanne (EPFL), 1015 Lausanne, Switzerland
severin.lemaignan@epfl.ch

Abstract. We present a study on the impact of unexpected robot behaviors on the perception of a robot by children and their subsequent engagement in a playful interaction based on a novel "domino" task. We propose an original analysis methodology which blends behavioral cues and reported phenomenological perceptions into a compound index.

While we found only a limited recognition of the different misbehaviors of the robot that we attribute to the age of the child participants (4–5 years old), interesting findings include a sustained engagement level, an unexpectedly low level of attribution of higher cognitive abilities and a *negative* correlation between anthropomorphic projections and actual behavioral engagement.

1 Introduction

1.1 Towards Sustained Engagement

Different possibilities to foster engagement (both short- and long-term engagement) in HRI have been explored, in particular with social robots. A lot of research has moved toward creating sophisticated emotional models which cause complex robot behavior. [5] studied the long-term engagement of children with a chess playing robot that adapted its behavior to the children and showed empathy toward them. The authors found that empathetic robots are more likely to engage users in the long-term and they proposed several guidelines for designing such artificial companions. Other works [1,9] have shown that simpler ways to enhance engagement may as well be effective, for instance, when the agent showed variations in its behavior, participants were more engaged and reported a desire to continue interacting with the agent.

Similarly, looking at short-term engagement, [9] found that a simple manipulation of the robot's behavior can lead to greater engagement. The authors let participants play several rounds of the rock-paper-scissors game with the robot (the playfulness of the scenario seems important). When the robot was cheating from time to time, participants tended to ascribe intention to the robot what in turn led to greater engagement in how they were interacting with the robot.

© Springer International Publishing Switzerland 2015
A. Tapus et al. (Eds.): ICSR 2015, LNAI 9388, pp. 390–400, 2015.
DOI: 10.1007/978-3-319-25554-5_39

The authors observe that *"any deviation from expected operation is sufficient to create a greater degree of engagement in the interaction."*. Along those lines, we also suggested in our model of the dynamics of anthropomorphism in HRI [7] that *disruptive behaviors* may lead to increased anthropomorphic projections and possibly increased engagement.

Based on this previous research, we explore in this study how to sustain children's engagement with the *Ranger* robot [8] by manipulating the behavior of the robot so that it appears *unexpected* to the children. The main outcomes of this research are 1) a **new experimental task** that suggests and contrasts three types of mis-behaviors, with different cognitive correlates, 2) a **mixed technique**, blending behavioral cues and reported phenomenological perceptions, **to assess the robot perception in terms of both engagement and human-likeliness**, 3) an **actionable approach** based on the introduction of mis-behaviors **to support child-robot engagement**, 4) and a first experimental cue that **anthropomorphic perceptions do not necessarily correlate with actual engagement**.

1.2   Design and Hypotheses

In a playful scenario which was set up in a laboratory environment, 26 children aged 4-5 years ($M = 4.46$) were assembling a domino game together. Each group consisted of two children and the *Ranger* robot, which was used to transport domino tiles between the two children.

Ranger usually behaved correctly (expected behavior), coming over to a child after being called and delivering the domino tile to the other child. However, during pre-defined rounds, *Ranger* showed unexpected behavior when a child called the robot. We defined three different types of *misbehavior* that were tested in a between-subjects study design:

- The robot gets **lost**: When called by the child to come over, the robot goes wrong, without any observable reason, and remains at the wrong location. We expect this to be perceived as a mechanical malfunction (a bug or system error which causes the robot to not work correctly), and hypothesise decreased attributions of human-likeness to the robot.
- The robot **disobeys**: When called by the child to come over, the robot shows that it refuses to obey by literally "shaking its head" and becoming red. The robot then goes to a wrong location and remains there while it continues to shake its head. We expect the disobey behavior to be perceived as the robot having an explicit *"own will"*, and we assume this leads to increased attributions of human-likeness (ascribing intentionality) to the robot.
- The robot makes a **mistake**: When called by the child to come over, the robot goes wrong but recognises its mistake and repairs. We expect this to be perceived (explicitly) as *"to err is human"*, and (implicitly) as the robot being endowed with a certain level of introspective capabilities (it was able to recognise its own error). In this condition, we assume increased attributions of human-likeness to the robot.

We analysed children's reaction focusing on two main aspects. On one hand, children's **behavior** (their reactions) toward the unexpected robot behavior was studied in terms of **active engagement** with the robot. On the other hand, we analysed children's **perception** of the robot in term of **anthropomorphic projections**. We assumed that in general a robot that behaves unexpectedly from time to time can promote engagement and lead children to attribute intention to it. Accordingly, we formulate two hypotheses: 1) *children show more engagement toward a robot that behaves unexpectedly from time to time compared to a robot that always behaves correctly,* 2) *children perceive a robot that (tentatively) displays intention or cognitive abilities as more human-like than a robot that appears to have a system error, i.e. the disobeying robot and the robot that makes a mistake will be more anthropomorphized than the robot that gets lost.*

Based on literature suggesting that a social relation to a robot (anthropomorphism is a specific type of social relation) reflects an increased engagement and can be effective in sustaining interaction, we formulate a third hypothesis: *anthropomorphic perception of the robot positively correlates with the level of engagement in the interaction.*

2 Research Methodology

2.1 Experimental Setting

The interaction scenario is built around two children who play the dominos together, with the help of a remotely controlled robot (Wizard-of-Oz setup). Figure 1 pictures the experimental setup.

The challenge for the children consists in collecting domino tiles spread over the room, hidden behind beanbags (task of the *searcher* child), getting the robot to carry to the second child, and finally assemble the tiles and decide for the next tile to fetch (tasks of the *receiver* child). In total, 13 pairs of children ($n = 26$) participated: 16 boys and 10 girls, 4-5 years old ($M = 4.46, SD = 0.45$), all French-speaking.

The game (that lasted in average 13min 43sec per group of children) was divided in a total of 14 runs that correspond each to the delivery and assembling of one domino tile. At each run, the robot exhibits one out of the four possible behaviors previously presented: *correct, lost, disobey* or *mistake*.

The first 5 runs (*1.1* to *1.5*) were used to set the baseline and the robot always behaved correctly. The children then switched the roles receiver/searcher and in the 9 remaining runs (*2.1* to *2.9*), the robot showed one of the misbehaviors (*lost, disobey* or *mistake*) at the 3rd and 4th run as well as at the 7th and 8th run (see axis x of Fig. 2).

During the misbehaving runs, the behavior of the robot is manipulated in three possible ways, represented on Fig. 1. In the **lost** condition, the robot goes to a wrong position and remains here, behaving (yellow light pattern) as if it were correctly in front of the child. In the **disobey** condition, the robot stops mid-way, displays a red pattern and produces a repeated "annoyed" sound. It finally moves toward a wrong position and remains there, facing the child. In the **mistake**

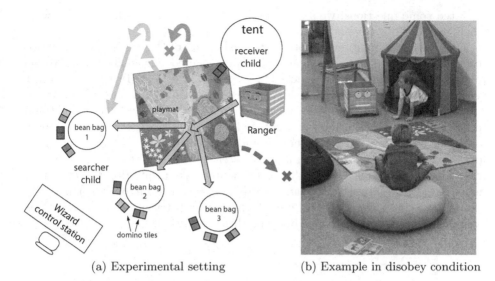

(a) Experimental setting   (b) Example in disobey condition

Fig. 1. Experimental setting The solid green arrows show the robot's path for the *correct* behavior. The blue arrow visualises a possible *lost* path, where the *Ranger* stops and remains at a wrong spot. The yellow arrows reflect a possible *mistake* path, where the robot goes wrong but then turns back and goes to the child. The red arrow visualises a possible *disobey* path where the robot goes wrong, then turns toward the child but stays at a wrong position.

condition, the robot starts like for the *lost* behavior, but after a few seconds, turns back, blushes and finally reach the correct position, in front of the child.

2.2 Data Collection

The **perception of the robot** by the children has been captured through two audio-recorded semi-structured interviews which took place between run *1.5* and *2.1* and at the end of the experiment (a short preliminary interview was also conducted to explain the game and assess the expectations of the children toward the robot). Then, the **children's behavior toward the robot** (*i.e.* the child-robot interaction) has been captured in the video recordings by annotating a set of actions.

Interviews were designed based on previous work on child-robot interaction [4,6,10]. We borrowed and adapted some of the *constructs* (for instance, the construct "cognitive connections" [3] considers the robot's ability to hear and to see (perceptual skills), as attributed by the children; the construct "moral standing" [4] assesses if the robot engenders moral regard, etc.) and example questions from questionnaires used in [4] and [10].

Due to the age of the participants, we set up the interviews like a casual conversation and we did not separate the two children. We paid attention to not "put words in children's mouth". Consequently, though we re-phrased and repeated some questions, we accepted when they said they would not know or when they did not respond at all.

One the other hand, we annotated the behaviors of the children in the video records, and coded the salient actions that reflected engagement toward the robot [2]: **touch** (the box is touched, *e.g.* petted or caressed); **talk** (all direct verbal interactions, except for calling it to come and pick a domino tile, since children were requested to perform this action anyway); **show** (show something to the robot); **misuse** (kick the robot, poke it in its "eye", try to climb on or inside the box, drive/push the robot around, stop the robot's wheels with a foot); **look** (when a child looks *at the experimenter* due to confusion caused by the robot; look is not coded when the experimenter asks a question to the child); **gesture** (gestures are used to communicate/interact with the robot, *e.g.* pointing gestures, waving at the robot). Figure 2 shows the distribution of these actions over the different runs, summed over the three condition.

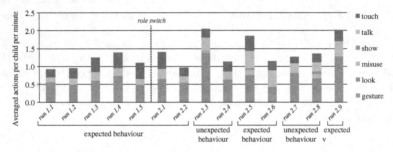

Fig. 2. Number and type of actions for each run (n=480, spontaneous actions). Generally, the number of actions does not decrease over time (from run 1.1 to run 2.9). The first 7 runs correspond to the *expected phase*, the second 7 runs correspond to the *unexpected phase*. Especially during run 2.3, the first time when the robot showed an unexpected behavior, children tended to *look* more at the experimenter. During the unexpected phase, also *talk* and *gesture* seem to be increased.

Compound Index of Anthropomorphism. Because the tendency to anthropomorphize manifests itself both in terms of perception and behavior, we propose to build a compound index that brings both children's perception of the robot (post-measurement) and their behavior toward it (in-the-moment measurement) together. We build the index by attributing points for each anthropomorphic *perception* of the robot and for specific kinds of human-like *behavior* toward the robot, using the following grading scheme:

Percept. Ascription of **mental states / feelings**: *2 points* for agreeing that *Ranger* can be happy or sad; *2 points* for attributing *Ranger* with hunger or tiredness.

Percept. Ascription of **cognitive abilities / intention**: each *0.5 points* for ascribing seeing and hearing ability; *1 point* for agreeing that *Ranger* can go out the door by itself; *1 point* for disagreeing that *Ranger* always obeys; *1 point* for agreeing that *Ranger* can do something silly.

Percept. Ascription of **sociality / companionship**: *1 point* for agreeing that *Ranger* can be a friend.

Percept. Ascription of **moral standing**: *1 point* for disagreeing that *Ranger* be left alone at home.

Percept. Other **anthropomorphic statements**: *1 point* for anthropomorphic reason for *Ranger*'s misbehavior; *2 points* for anthropomorphic reason for not leaving *Ranger* alone (*e.g.* "It would be sad").

Behav. Use of **direct speech**: *1 point* (not considering *calling* the robot to come over).

Behav. Use of **polite formulations**: *1 point* (*e.g.* "thank you Ranger" or "please Ranger ...").

Behav. Use of **social or pointing gestures**: *1 point* (*e.g.* waving at the robot).

The balance of the grading scheme is open to debate: for instance, we did not consistently assign 1 point to each item, but assigned points between 0.5 and 2 points depending on our perception of how a given item reflect a higher level of anthropomorphic perception of the robot (for instance, ascribing the ability to see and hear was suggested by the design of the study, and we cannot assert it really reflects the explicit projection of cognitive skills). This issue is however mitigated by our use of this compound index as a *relative* metric (comparison between conditions) and not an absolute value.

3   Main Findings

Misbehavior Recognition. The *disobey* behavior was expected to be perceived as the robot intentionally not doing what it should do. The *mistake* behavior was intended to show that the robot can do a mistake but is aware of it and able to repair its mistake, which should also lead to the perception of intentionality and introspective skills. Contrary, we expected that the *lost* condition is perceived as a malfunction of the robot. In the second interview, after the robot had misbehaved, we asked children whether the robot always did what they wanted it to do. Most children disagreed and said they noticed something strange.

When asked why they thought the robot had not always come over to them, 4 of the children did not reply. The remaining ones gave a variety of reasons (Fig. 3). The most common answer (9 of 37 replies) was that the robot is somehow *unpredictable* in what it is doing and that it could go *"in whatever direction"* because *"with robots you have these kind of problems, they do no silly things"*. 8 replies related to *technical problems* (including *broken parts*), suggesting that children perceived the misbehavior as unintended by the robot. Two of the children who had interacted with the disobeying robot said *Ranger* was *angry*, which

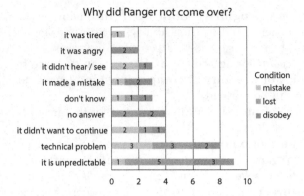

Fig. 3. Multiple answers were possible to the question why the robot did not come over, and we received 37 answers.

none of the children in other conditions replied. 13 out of 26 children appeared to ascribe intentionality precursors to *Ranger* explaining that it *did not want to continue* carrying domino tiles or that it *"did a mistake"*.

Attribution of Intentionality. To investigate to what extend children attribute intention and cognitive abilities to the robot, we asked three questions during the first interview: whether they believed *Ranger* could go out the door by itself (a majority of 16 children answered negatively, which suggests that they initially do not ascribe intention to the robot), whether *Ranger* would always obey and whether *Ranger* could do a silly thing (Fig. 4). These two last questions were asked again later after children had interacted with the unexpectedly behaving robot.

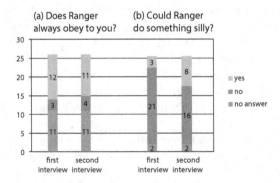

Fig. 4. Attribution of intention to *Ranger*.

In the first interview, 12 out of the 15 children who provided an answer believed that *Ranger* does always obey to them. Asked whether the robot could do something silly, a large majority of 21 children out of 24 replied negatively: it appears that after children the first round of interaction (with the correctly behaving robot) the children **do not generally ascribe intention** to the robot.

After interacting with the misbehaving robot (second interview), most of the children still believed that *Ranger* always obeys to them, and 8 children (previously 3) think that *Ranger* could do something silly: even with an **unexpectedly behaving robot** children do **not necessarily ascribe intention** to the robot. It seems that some children did not interpret the misbehavior of the robot as intentional but more like a technical problem or mistake. For instance, even after the robot misbehaved by *disobeying*, the majority of the children in this condition was still convinced that the robot could not do a silly thing. It is however interesting to note that children tend to ascribe cognitive abilities to the robot, like the ability to see and hear but not intention. We interpret this as children perceiving the robot as being able to process sensory information but not being able to make decisions on its own.

Engagement. Based on the video-coded actions, we found a significant difference between the average of engagement actions (Fig. 2) carried out during the first 7 runs (correct robot behavior) and during the second 7 runs, when the robot behaved unexpectedly ($F(1, 36) = 5.1, p = .03$). In all three conditions, children carried out more engagement actions with the unexpectedly behaving robot. Importantly, no interaction effect was found between the two phases of interaction (expected / unexpected) and condition ($F(2, 36) = 1.2, p = .31$): the robot's failure mode does not seem to impact the level of engagement.

In general, this finding **supports our first hypothesis**: children show more engagement toward a robot that behaves unexpectedly from time to time compared to a robot that always behaves correctly.

Anthropomorphic Projections. *Ranger* was moderately anthropomorphized by the children (compound anthropomorphism index in *lost* condition: $M = 8.31$ (on 16), $SD = 0.59$, *disobey*: $M = 6.5, SD = 3.68$, *mistake*: $M = 7.94, SD = 1.74$). However, the mean index of anthropomorphism in the three conditions varied, with the *mistake* and *lost* condition leading to a higher index than the *disobey* condition. This finding suggests that the disobeying robot was *less* anthropomorphized than the other two robot behaviors, which speaks **against our second hypothesis**. We had expected that the disobeying behavior is perceived as an intentional action which we assumed would lead to increased anthropomorphism. This was not the case and the *lost* robot was overall the one eliciting the highest level of anthropomorphism (the robot's "helplessness" may have lead to this). This is also reflected in children's behavior: with the lost robot, children looked more often at the experimenter than in the other condition, which suggests that they could not fully make sense of the robot's behavior, and the fact of not being able to understand (and hence predict) a robot's behavior is likely to increase anthropomorphism.

We hypothesised that children who interact a lot and are more engaged with the robot also perceive the robot as more human-like. Suprinsigly, our

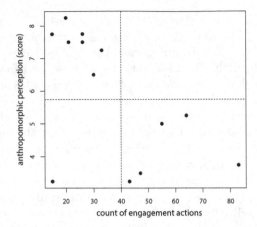

Fig. 5. Anthropomorphic perception of the robot versus engagement actions per pairs.

data suggests the opposite. As shown in Fig. 5, **the more a group showed engagement in the interaction, the less they anthropomorphized the robot**. A possible interpretation is that children who interact more with the robot understand better how it works, they are more familiar with it, and as such the robot appears less "mystical" to them, and they hence do not need to anthropomorphize it. On the contrary, the cluster of groups that do not interact much but anthropomorphize the robot more, is quite homogeneous, and may reflect a certain fear of interacting with a robot that would look to them too human-like. This raises the question how far anthropomorphism (as a special kind of social engagement) really helps in sustaining interaction. This is a critical point because most of the short-term investigations suggest that anthropomorphic design and human social cues emitted by a robot foster engagement and acceptance. What if this is not true for continued interaction, and thus for the long-term? We need to remain modest here: while we found a significant *negative* correlation between engagement and anthropomorphism in the data pictured on Fig. 5 ($r(11) = -0.56, p = .05$), we have to be careful about our interpretation, due to the relatively small sample size (13 pairs), and we suggest to investigate the aspect further in future research.

4   Conclusions and Future Directions

As hypothesised, we found that, in a playful scenario where 4-5 year old children play domino together with a robot, the robot seems to be more engaging when it shows some misbehavior compared to when it always behaves as expected (notwithstanding the impact of a novelty effect).

Regarding the design of our three conditions (*lost*, *disobey*, *mistake*), we cannot conclusively affirm whether children perceived the unexpected robot behavior as a malfunction (something that happens to a machine) or as being intended and based on a motivation (something related to a social entity). While our manipulations were not as clearly perceived as we expected for the age range of the subjects, we still believe these three conditions (mechanical malfunction – the *lost* condition, vs. explicit intentionality – the *disobey* condition, vs. implicit intentionality – the *mistake* condition) are relevant and we suggest to replicate a similar study with slightly older children.

We also report on the initial application of an compound index of anthropomorphism to assess children's anthropomorphic projection onto robots. This index considers both behavioral and self-reported aspects, and it suggests, in our experiment, that children tend to conditionally anthropomorphize the robot. Higher indexes of anthropomorphism were found in the *lost* and *mistake* condition which was against our hypothesis.

Interestingly, data suggests that anthropomorphic perception does not automatically elicit engagement, on the contrary. It appears that groups who interacted more with the robot perceived it as less human-like. This raises an important question for the human-robot interaction community: to what extent do anthropomorphic perceptions impact the interaction experience? Our findings here go against the intuition.

Acknowledgments. We would like to thank all the families who participated to the study. This research was supported by the Swiss National Science Foundation through the National Centre of Competence in Research Robotics.

References

1. Bickmore, T., Schulman, D., Yin, L.: Maintaining engagement in long-term interventions with relational agents. Applied Artificial Intelligence **24**(6), 648–666 (2010)
2. Fink, J., Rétornaz, P., Vaussard, F., Wille, F., Franinović, K., Berthoud, A., Lemaignan, S., Dillenbourg, P., Mondada, F.: Which robot behavior can motivate children to tidy up their toys? Design and evaluation of "Ranger". In: Proceedings of the International Conference on Human-Robot Interaction (2014)
3. Flavell, J.H.: The development of children's knowledge about the mind: From cognitive connections to mental representations. Developing Theories of Mind 244–267 (1988)
4. Kahn Jr, P.H., Friedman, B., Pérez-Granados, D.R., Freier, N.G.: Robotic pets in the lives of preschool children. Interaction Studies **7**(3), 405–436 (2006)
5. Leite, I.: Long-term Interactions with Empathic Social Robots. Ph.D. thesis, Universidade Técnica de Lisboa, Instituto Superior Técnico, Lisboa, Portugal (2013)
6. Leite, I., Pereira, A., Mascarenhas, S., Martinho, C., Prada, R., Paiva, A.: The influence of empathy in human-robot relations. International Journal of Human-Computer Studies **71**(3), 250–260 (2013)
7. Lemaignan, S., Fink, J., Dillenbourg, P.: The dynamics of anthropomorphism in robotics. In: Proceedings of the 2014 Human-Robot Interaction Conference, pp. 226–227 (2014)

8. Mondada, F., Fink, J., Lemaignan, S., Mansolino, D., Wille, F., Franinović, K.: Ranger, an example of integration of robotics into the home ecosystem. In: Proceedings of the International Workshop and Summer School on Medical and Service Robotics (MESROB). Springer (2014)
9. Short, E., Hart, J., Vu, M., Scassellati, B.: No fair. An interaction with a cheating robot. In: Proceedings of the International Conference on Human-Robot Interaction, pp. 219–226 (2010)
10. Weiss, A., Wurhofer, D., Tscheligi, M.: "I love this Dog"-Children's emotional attachment to the robotic dog AIBO. International Journal of Social Robotics 1(3), 243–248 (2009)

An Embodied AI Approach to Individual Differences: Supporting Self-Efficacy in Diabetic Children with an Autonomous Robot

Matthew Lewis[1]([⊠]), Elettra Oleari[2], Clara Pozzi[2], and Lola Cañamero[1]([⊠])

[1] Embodied Emotion, Cognition and (Inter-)Action Lab,
School of Computer Science, University of Hertfordshire, AL10 9AB, Hatfield, UK
{M.Lewis4,L.Canamero}@herts.ac.uk
[2] Fondazione Centro San Raffaele, Via Olgettina 60, 20132 Milan, Italy

Abstract. In this paper we discuss how a motivationally autonomous robot, designed using the principles of embodied AI, provides a suitable approach to address individual differences of children interacting with a robot, without having to explicitly modify the system. We do this in the context of two pilot studies using Robin, a robot to support self-confidence in diabetic children.

1 Introduction

In this paper we discuss how a motivationally autonomous robot (named Robin after "Robot Infant"), designed using the principles of embodied AI, provides a suitable approach to address individual differences in the way children interact with it, without having to explicitly modify the system. This robot and the interaction scenario were developed to help diabetic children improve their confidence and skills in managing their own diabetes, by looking after a "diabetic" robot toddler. In [8] we addressed how Robin was designed to support self-efficacy in these children. For developing self-confidence and self-efficacy in real interactions, it is very important that social interaction is appropriate to the interaction profiles and personalities of individual children. Typically, in Human-Robot Interaction (HRI) and Child-Robot Interaction (CRI) individual differences are tackled by personalizing the robot to individual profiles [10]. Personalization is usually done by explicitly tailoring the interactions using methods such as Wizard-of-Oz, by adding references to previous interactions [7], altering the order in which tasks are done [9] or by introducing variables into interaction scripts, e.g. related to personality. However, for our goals of supporting self-efficacy and building self-confidence, a motivationally autonomous robot is more appropriate. It is important that the interaction is unstructured, and partly ambiguous and unpredictable, as this will make the "play" experience feel closer to the complexity of real diabetes self-management. The use of a motivationally and cognitively autonomous robot that can behave and interact as an independent agent is instrumental to this end. As we will see later, it also makes each interaction unique.

© Springer International Publishing Switzerland 2015
A. Tapus et al. (Eds.): ICSR 2015, LNAI 9388, pp. 401–410, 2015.
DOI: 10.1007/978-3-319-25554-5_40

We discuss how we used Robin in two pilot studies with diabetic children in Italy. Although each child was very different in their needs for support for diabetes and the way in which they interacted, we did not need to modify Robin to satisfactorily deal with these differences in both dyadic and triadic interactions.

2   Robin, the Diabetic Autonomous Robot Toddler

In this section we summarize the psychological and clinical basis for the design of Robin, the software architecture, and the interaction scenario. For further details please refer to [8].

2.1   Motivation

Type 1 diabetes is an incurable chronic disease caused by an inability of the body to produce insulin. It is often diagnosed in childhood and, if ill-treated, the high glucose levels lead to devastating complications such as blindness, limb amputations or severe misfunctioning of internal organs. The current treatment involves monitoring and adjusting blood glucose through the provision of insulin – either through a pump or by injection – and glucose – by eating appropriate foods. In order to live independently, an individual with diabetes needs to be able to manage his/her own diabetes (self-management), which in addition to these activities involves being aware of the symptoms of high and low glucose levels, which can vary between individuals, and being aware of how different foods and activities specifically affect their own blood glucose. Diabetes treatment therefore involves a great deal of education, but the ultimate medical aim of this education is behavior change: the acquisition of good diabetes self-management practice.

The concept of *perceived (self-)efficacy* was introduced by Bandura as key element in successfully changing behavior [2]. Synthesizing from the literature, we define perceived self-efficacy as a person's beliefs about their own ability to successfully perform a specific task in a specific situation. Following Bandura's ideas we designed our robot architecture and CRI scenario as a tool to increase perceived self-efficacy in the child, primarily by giving them a *mastery experience* of diabetes management – in this case the child manages the robot's diabetes. It is important that the interaction is unstructured and partly ambiguous and unpredictable, as this will make this "play" experience feel closer to the complexity of real diabetes self-management. The use of a motivationally and cognitively autonomous robot [5, 8] (rather than, e.g., a scripted system) is instrumental to this end, as we will explain in the next section. It also makes each interaction unique, due to both the dynamics of the architecture in interaction with the physical and social environment (the robot never behaves in exactly the same way twice), and to the different ways in which each child treated the robot. Our robot was designed to act like a toddler as we felt that this supported the role of the child as the "carer", it suited the physical appearance of the NAO robot.

2.2   Robot Architecture

We briefly outline Robin's software architecture, which is described in detail in [8]. Robin's decision-making architecture follows principles of embodied AI [4,5]. It is built around a "physiology" of homeostatically controlled "survival-related" variables that Robin needs to keep within permissible values. We have also given the robot a simple model of Type 1 diabetes, comprising an internal blood glucose level that increases with "eating" toy food, and decreases with "insulin". Robin chooses how to behave as a function of these internal needs and the stimulation he gets from the environment. Elements of the environment are detected using vision (e.g. foods, faces) and tactile contact (e.g. collisions, strokes, hugs). Internal needs and environmental cues are mathematically combined in what we call *motivations*, which lead Robin to autonomously select behaviors from his repertoire (e.g. walking, looking for a person, eating, resting) that will best satisfy his needs (e.g. social contact, nutrition, resting, playing) in the present circumstances. For this reason it is a *motivationally and cognitively autonomous* robot.

To make it meaningful for our scenario, Robin is not capable of fully attending to all its needs without human assistance. It can play on its own, eat, and, by resting, it can recover from tiredness caused by too much movement. However, it requires assistance from the children to satisfy its social needs (e.g. social presence, strokes, hugs), some of its nutritional needs (the child can "feed" the robot using toy food items), and to control its glucose level.

The children can measure glucose levels and provide insulin to lower glucose (correcting hyper-glycemia) using a Bluetooth glucometer device, or feed the robot high-glucose food to raise glucose (correcting hypo-glycemia). Hypo- and hyper-glycemia have associated symptoms such as tiredness that alert the children of the potential presence of a problem.

Following the initial prototype of Robin described in [8], preliminary tests with non-diabetic Italian children indicated that purely non-verbal behavior was not sufficiently clear to interpret Robin's needs. We therefore added a few simple Italian words (in a recorded "child-like" voice) to indicate hunger, a request for a hug/stroke, and sleepiness. This was in addition to the already present happy and sad sounds, signaling positive and negative changes in its internal state.

2.3   Interaction with Robin

The interactions took place in a "playroom" (Figure 1) decorated to look like the room of a toddler. The majority of the interactions were dyadic interactions between a child and a robot, initially coached by an adult. Robin would be "moving around" (i.e. looking at or walking towards things around the room, trying to eat, exploring, etc.) as the children entered the playroom and they would first interact in the presence of an adult who would show them how to feed Robin, how to use the handheld glucometer to measure glucose and give insulin, explain what toy food items contained a corrective dose of glucose, and encourage them to interact socially. After this initial phase the children would be

Fig. 1. Robin's playroom at the diabetes summer camp: the environment where the interactions took place. Robin's toys lie around the room; items including food and the glucometer are on a table.

asked if they could look after Robin while the adult left for a short time. All our children were happy to be left alone with the robot. They were provided with a phone that they could use to call an adult for help or to ask them to return. The children were left alone with the robot for approximately 15 minutes, after which the adult would return, and the child could then leave. After the interaction the child would fill in a questionnaire.

The interaction was remotely monitored by the experimenters. In order to ensure each child had an appropriate experience of managing Robin's diabetes, if the robot did not naturally have a hypo- or hyper-glycemia we would remotely set the glucose level high or low. Robin would then act appropriately based on his own internal state, e.g. stopping exploring and resting showing postural and vocal signs of tiredness. The experimenters could also decide to send an adult to the room if a child seemed to be in difficulty and was not phoning for assistance. Since our aim was to improve the children's perceived self-efficacy, we would set the robot's glucose to a normal value towards the end of the interaction so that it did not appear to be in any difficulty when the child left.

3   Trial Interactions

We ran exploratory pilot interactions in order to assess how diabetic children interacted naturally with Robin. Our purpose was to gain insight into how Robin could be used as a tool to support the social and therapeutic needs of different children, rather than a formal investigation of specific research questions. These qualitative observations and analysis, intended to integrate the end-users early in the design of the system, are the object of this paper. We do not present a quantitative analysis at this point.

3.1   First Pilot: Hospital

We ran a first pilot of Robin with three diabetic children (two girls, one boy) at Ospedale San Raffaele in Milan. All the children had already interacted with a NAO robot running different software, also developed as part of the ALIZ-E project [3], in which the robot and child played a number of educational games related to diabetes, the robot taking the role of either a peer or a teacher. The two robots were referred to as "Nao" and "Robin" to clearly distinguish their different identities. Following feedback from a psychologist, an adult, playing the role of Robin's "engineer", was with Robin in the playroom when the children entered. During the introduction, the engineer explained that he had only recently learned that Robin had "robot diabetes" and he was hoping the child would be able to help him. After the introductions the engineer received a phone call calling him away and he would ask the child to look after Robin until he returned.

3.2   Second Pilot: Summer Camp

We ran a second pilot of Robin at a summer camp for diabetic children in Italy organized by the patients' association SOStegno70[1]. The first tests, like those in the Ospedale, each involved a single child (3 girls aged 10/11, 8 boys aged 11/12). We then ran two tests with pairs of children interacting with Robin (all boys, one of whom had taken part in the single-child interactions). A total of fourteen children, all with Type 1 diabetes, interacted with Robin at the camp. As in the hospital, all the children had previously interacted with "Nao".

These interactions needed to be done in a way that fitted in with the busy timetable at the summer camp. Robin and Nao were introduced to the whole camp as brothers during mealtime presentations. Children who had then put their names on a waiting list to interact with Robin were approached at convenient times during the day and, if they still wished to visit Robin, taken to the room where the playroom had been built. That they were then left alone with Robin was not presented as a "surprise", as it had been in the hospital pilot, and there was no "engineer" character. As some of the children had heard about the interaction with Robin from their peers, we could not be sure what prior knowledge they had when they arrived.

4   Personalized Interactions

In this section, we will discuss individual differences displayed by the children during their interactions with Robin, and how Robin responded to these differences.

Responses to the post-interaction questionnaire indicate that the children found Robin's behavior to be largely coherent and likable. Asked how Robin seemed to them and allowed to choose multiple words from a list – "lovable", "strange", "amusing", "not social" (Italian "sulle sue"), "interested in me", "in

---

[1] www.sostegno70.org

difficulty" and "other" – only two of the seventeen children indicated that Robin was "strange" whereas 11 answered "lovable". One child put "other", writing "like a real little boy". Much of the non-verbal behavior of the children seemed to corroborate these results, for example there were several occurrences of the children spontaneously imitating the robot – a number of them clapped when Robin clapped his hands, or imitated his vocalizations. This imitation included unobtrusive mimicry, which is linked to empathy and positive social interaction [6]. In addition, all the children appeared to be fully engaged in the interaction, and none showed signs of wanting to leave before they were interrupted by the returning adult. Robin was perceived and treated in this way despite the fact that children interacted with him in very different ways – each child was a world of his/her own. This shows that Robin was successful in coping with individual differences.

In the remainder of this section we discuss some of these individual differences grouped in terms of some relevant interaction criteria.

4.1   Socially Proactive vs. Socially Responsive

All the children were responsive to Robin's vocalized requests for food, drink and hugs. These vocalized requests were Robin's most clearly expressed prompts for the children to interact. All the children also helped Robin to stand and responded when he expressed tiredness, although to differing extents. In the post-interaction questionnaire all the children answered the question "Was it easy to guess Robin's needs and desires?" with "yes" or "sometimes".

However, outside of Robin's requests, we saw different interaction patterns. Some children interacted with the robot almost continuously, for example taking it upon themselves to "entertain" him with toys, responding to Robin's hand claps by clapping their own hands, or by walking alongside him. At some points some children took this even further, dominating Robin's attention, for example by turning him towards them if he turned away, or physically carrying him to another part of the playroom. In contrast to this, other children allowed Robin to do as he pleased, exploring the room and interacting with what he came across there. These latter children should not be thought of as "not interacting" as they still responded to the vocalized requests and falls as noted above.

One particular child who acted in this "responsive" way was a boy who gave the impression that he was "cool" or streetwise. At one moment during the interaction, he played football with the toy duck that was in the room, seemingly distracted. However, he perfectly understood, and was quick to respond to, every single one of Robin's vocalized requests (in the post-interaction questionnaire he indicated that he found it easy to guess Robin's needs and desires, although he was one of only two children to indicate that he found Robin "strange"). He did not respond immediately to Robin's expression of tiredness; however, when Robin continued to express tiredness, he phoned for assistance and used the glucometer as instructed before feeding the robot. Even though aspects of his behavior may have given the impression of being distracted, he gave many indications of engagement with Robin. For example, after testing Robin's glucose later in the interaction and finding it was now back in the normal range he made a small gesture as though in triumph. On a number of occasions he appeared to imitate Robin: raising his hand to apparently mirror an action by Robin, or clapping his hands immediately after Robin had done so. In addition to these, and possibly indicative of how the child viewed his own and Robin's diabetes, he also examined his own glucose pump in a way that seemed connected with his interaction with Robin. For example, when he had checked Robin's glucose for the first time and fed him to correct the hypoglycemia he briefly looked at his own glucose pump; some time later he looked at his pump, and as though reminded by this he then immediately checked the robot's glucose. A similar pattern of checking his own pump in parallel with Robin's glucose level appeared to persist to the end of the interaction. Although we should be careful about reading too much into this anecdotal observation, it is suggestive and perhaps of relevance both in terms of how the child viewed the robot with respect to himself and the child's awareness of diabetes.

Because Robin's motivations for acting depend on both his internal needs and external stimuli (which include different varying cues from the children), he could respond differently and appropriately to this variety in social interaction style. Detecting a face, which occurred more frequently with socially proactive children, would increase Robin's motivation for social interaction, often causing him to open his arms towards the person – a gesture inviting a hug. On the other hand, over periods without social interaction, more frequent in the "responsive" style, the robot's social need would increase, eventually becoming dominant, at which point the robot would vocalize this need, attracting the child's attention.

4.2   Verbal vs. Non-verbal

Robin used only single words to indicate his motivations, and made affective vocal noises to express pleasure or displeasure. However, the children showed different degrees and types of verbal behavior when left alone with Robin.

For example, one girl, who in the briefing session said that Robin reminded her of a young cousin, talked a great deal to Robin as she was interacting with him, for example talking encouragingly to him as she tried to feed him an apple. Alongside this, she showed very proactive non-verbal behavior, such as moving Robin about the room. Another girl who also talked a great deal and encouragingly, did so with a different social manner, seeming to respect Robin's autonomy and giving him space, even going so far as to ask Robin if he wanted her to help him stand as she watched him struggle to get up after repeatedly falling.

Many of the children did not speak very much to Robin, though several of them would call his name in a variety of tones, and many said "Ciao" as they left the playroom. This was in contrast to the interactions with Nao in which the robot spoke with some fluency (using Wizard-of-Oz) and games such as the diabetes quiz were built around vocal interaction. Our presentation of Robin as a toddler with very limited verbal ability was viewed as coherent by all the children, and supported the different ways they wanted to interact with him along the verbal–non-verbal spectrum.

4.3   Response to Diabetes Symptoms

There was a range of responses to Robin's diabetes symptoms, perhaps reflecting the range in the children's own confidence and knowledge. We will briefly describe examples at the two ends of the spectrum.

One boy, aged 9, as the adult was about to feed Robin during the introductory phase, spontaneously prompted the adult to check Robin's glucose levels. This showed high awareness of diabetes management and indeed the boy checked Robin's glucose levels several times during the interaction, including after Robin had fallen (hypoglycemia can cause people to become dizzy or to faint). After seeing the adult move the high-glucose food items away from Robin during the introduction, the child also hid them behind the table during the interaction. Even before he left Robin with the adult at the end of the interaction he made sure these food items were well hidden behind two of the larger toys.

One girl, aged 10, and who had been diagnosed with Type 1 diabetes for only one year, appeared to show very little awareness of correct medical response to diabetes symptoms. When alone with Robin, her response to his tiredness due to hypoglycemia was purely affective. She offered comfort – gently stroking his head – but did not use the glucometer. On this occasion the experimenters, who were monitoring the interaction, took the decision to send the adult back to the interaction room. On finding out what the situation was, he asked if she had checked Robin's blood glucose. She then took the glucometer from the table and used it to measure his glucose. Then, with a small amount of encouragement from the adult, she fed Robin a corrective dose of glucose. Robin quickly recovered and started walking around again. The interaction finished a few minutes later with hugs between Robin and the child. Since she was relatively inexperienced in diabetes management, it is not too surprising that she did not use the glucometer until prompted. However, diagnosing and treating the hypoglycemia with the adult still provided her with a mastery experience, albeit a directed one, and the system was flexible enough to allow for this. In contrast to her observed behavior, in her questionnaire response she rated the game as "very easy" and she didn't want an adult present to help with difficult situations. She did, however, indicate that the teaching games that she had played with the other robot had helped her with Robin, specifically that before eating it was necessary to check the glycemia.

4.4   Interactions with Two Children

In order to test the flexibility and scalability of Robin, we also ran two interactions using exactly the same system and scenario, but with two children (boys) interacting with him at the same time. In the first trial one child was new to Robin, while the other had already interacted with him on his own; in the second, both children were interacting with Robin for the first time. On both occasions, we observed some examples of the children working as a team – for example by passing the glucometer for the other to use when Robin indicated he was hungry, or the bottle of water when Robin indicated he was thirsty. In contrast to this teamwork, on other occasions the pairs could be viewed as behaving in a more competitive manner, for example by simultaneously trying to attract Robin's attention with toys or food, or by putting themselves in Robin's path.

Two Children with Different Familiarity. The boy who had previously interacted with Robin had not, during that first interaction, seemed confident about his ability to manage Robin's diabetes: as soon as Robin indicated he was tired, the boy had phoned for advice and confirmation about the exact procedure to follow. However, on the second interaction he entered the playroom with a happy "Ciao, Robin!" and stroked his head. He then attempted to feed Robin some apple, although Robin was not motivated to eat at that point. Then, when the adult was giving a shortened introduction to Robin, the familiar child (we will refer to the two boys as "familiar" and "new") initiated the discussion of the glucometer, showing the new boy how to use it. When the new child was being told about the food items used for correcting hypoglycemia the familiar child volunteered his own contributions. When the two were working as a team, the familiar child took the leading role, for example by initiating the use of the glucometer by passing it to the new child. At the end of the interaction, when the adult returned, the familiar boy measured Robin's glucose and reported it to her, with apparent pride. This increased confidence is consistent with our design of Robin as a tool for increasing perceived self-efficacy – although increased perceived efficacy in managing Robin's diabetes does not necessarily mean that he was more confident in his abilities to learn to manage his own diabetes.

Two Children New to the Interaction. This group showed a lot of exploratory behavior, which we hadn't seen in other interactions. For example, when one of the more ambiguous situations occurred: Robin became tired for reasons unrelated to his diabetes. At this point, after measuring his glucose and finding it in the normal range, the two boys discussed what to do. They (correctly) gave no treatment on this occasion, but continued to monitor Robin's glucose. Another novel way of interacting occurred when one of the boys arranged the soft toys to form three sides of a square and lay Robin down inside the square with his head resting on one of the banks of toys, as if making a bed for Robin.

5   Conclusions and Future Work

In this paper we have discussed the use of Robin, a motivationally autonomous robot designed using embodied AI principles, as a suitable approach to deal with individual differences in CRI. We have illustrated this with observations from pilot studies carried out with seventeen diabetic children in Italy.

Already this basic implementation of a motivationally and cognitively autonomous robot, with no explicit adaptation or learning capabilities, can deal with a wide range of significant individual differences, without having to modify the system. Based on this experience, we would like to put forward the embodied AI approach to autonomous robots, currently little known and under-used in HRI and CRI, as a very promising avenue for dealing with individual differences in these fields.

In addition to this, this approach can also address personalization in a natural way as part of the interaction. For example, we have already started investigating the inclusion of explicit behavior adaptation techniques, such as imitation of specific non-verbal behavior for each child as a way of promoting positive bonds [1]. We have also started to explore the adaptation of the level of social responsiveness to that of the child as a function of the input received in the interaction.

Acknowledgments. We would like to thank the teams at Fondazione Centro San Raffaele and the Misano Adriatico Summer Camp, in particular Marco Mosconi, Francesca Sacchitelli, Sara Bellini, Marco Moura, Ilaria Baroni, Mattia Zelati and Alberto Sanna. We would also like to thank all the children who took part in these trials as well as their families. This work was supported by EC grant FP7-ICT-248116 (ALIZ-E). The opinions expressed are solely the authors'.

References

1. van Baaren, R.B., Decety, J., Dijksterhuis, A., van der Leij, A., Leeuwen, M.L.: Being imitated: consequences of nonconsciously showing empathy. In: Decety, J., Ickes, W. (eds.) The Social Neuroscience of Empathy, pp. 31–42. MIT Press (2014)
2. Bandura, A.: Self-Efficacy: The Exercise of Control. Worth Publishers (1997)
3. Belpaeme, T., ALIZ-E consortium: Multimodal child-robot interaction: Building social bonds. Journal of Human-Robot Interaction **1**(2), 33–53 (2013)
4. Brooks, R.A.: New approaches to robotics. Science **253**(5025), 1227–1232 (1991)
5. Cañamero, L.: Emotions and adaptation in autonomous agents: A design perspective. Cybernetics and Systems: An International Journal **32**(5), 507–529 (2001)
6. Hatfield, E., Cacioppo, J., Rapson, R.: Emotional Contagion. Cambridge University Press (1994)
7. Lee, M.K., Forlizzi, J., Kiesler, S., Rybski, P., Antanitis, J., Savetsila, S.: Personalization in HRI: a longitudinal field experiment. In: 7th ACM/IEEE International Conference on Human-Robot Interaction, March 2012
8. Lewis, M., Cañamero, L.: An affective autonomous robot toddler to support the development of self-efficacy in diabetic children. In: Proc. 23rd Annual IEEE Intl. Symp. on Robot and Human Interactive Communication (IEEE RO-MAN 2014), pp. 359–364 (2014)
9. Leyzberg, D., Spaulding, S., Scassellati, B.: Personalizing robot tutors to individuals' learning differences. In: Proc. 2014 ACM/IEEE Intl. Conf. on Human-robot Interaction, HRI 2014, pp. 423–430. ACM, New York (2014)
10. Syrdal, D., Koay, K., Walters, M., Dautenhahn, K.: A personalized robot companion - The role of individual differences on spatial preferences in HRI scenarios, pp. 1143–1148. IEEE (2007)

Socially-Assistive Emotional Robot that Learns from the Wizard During the Interaction for Preventing Low Back Pain in Children

Gergely Magyar and Maria Vircikova[✉]

Department of Cybernetics and Artificial Intelligence,
Faculty of Electrical Engineering and Informatics, Center for Intelligent Technologies,
Technical University of Kosice, Kosice, Slovakia
{gergely.magyar,maria.vircikova}@tuke.sk

Abstract. Back pain causes more global disability than any other health problem studied and the number of patients is growing. In Europe and in the US it is the number one cause of lost work days. This paper propounds a new approach by exploring the effect of utilizing a humanoid robot as a therapy-assistive tool in educating children to perform back exercises designed by a professional therapist. In our previous research a NAO robot was programmed and employed as a robotic assistant to a human physiotherapist to perform exercises in an elementary school in Slovakia. This paper goes further in designing a Wizard of Oz, where the exercises can be controlled and intervened by motivational behaviors of the robot (emotional expressions). Currently we are developing a system based on reinforcement learning that should adopt the motivational interventions from the Wizard. The promising results of this study in the physical therapy suggest the effective future use of social robots in reducing the symptoms of the most extended global disability in the world.

Keywords: Children-robot interaction · Back-pain · Scoliosis · Socially assistive robotics · Reinforcement learning

1   Introduction

Low back pain (LBP) is an important public health problem in all industrialized nations [1] [2] [3]. The Global Burden of Disease studies [4] done in 1990 and 2000 have been the only studies to quantify non-fatal health outcomes across an exhaustive set of disorders at the global and regional level – neither effort quantified uncertainty in prevalence of years lived with disability (YLD). They found musculoskeletal disorders caused 21.3% of all YLD, where the main contributors were low back pain (83.1 million YLD) and neck pain (33.6 million YLD). COST (European Cooperation in Science and Technology) for Biomedicine and Molecular Biosciences [5] gives Guidelines for the Management of Low Back Pain in Europe and reports on its important social and economic impact as it is associated with high rates of sick leave and disability pensions. It affects more than 70% of the general population sometime in

© Springer International Publishing Switzerland 2015
A. Tapus et al. (Eds.): ICSR 2015, LNAI 9388, pp. 411–420, 2015.
DOI: 10.1007/978-3-319-25554-5_41

their life; 17-31% of general population are suffering from LBP at any one time. In a brief study of 29 424 individuals by Leboeuf-Yde and Kyvik [8] 37% of the children reported back pain. Exercise based therapies are a logical approach to improve and maintain flexibility and function in patients at risk for pain and progression [6]. Negrini et al. [7] made a systematic review studying if physical exercises as a treatment for adolescent scoliosis on and the results suggest their utility to reduce specific impairments and disabilities cannot be neglected.

The problem of scoliosis in today's society is growing and it is fundamental to ensure adequate motor skill development during childhood. We describe our experience in using a humanoid robot as a physiotherapist to teach children anti-scoliosis and anti-back pain exercises as it is proved that a series of safe and effective exercises can strengthen the back and improve posture.

2 Social Robots in Therapies

In Rabbitt et al. [10] define socially assistive robotics (SAR) as *a unique area of robotics that exists at the intersection of assistive robotics, which is focused on aiding human users through interactions with robots (e.g. mobility assistants, educational robots) and socially interactive or intelligent robotics, which is focused on socially engaging human users through interactions with robots (e.g. robotic toys, robotic games)*. They also review the yaws that socially assistive robotics have already been used in mental health service.

Socially assistive robots have been used in different roles, e.g. the weight loss coach [11], the social robot in an attention and memory task helping older adults with dementia [12], supporting young patients in hospital as they learn to manage a life-long metabolic disorder (diabetes) [13], motivating physical exercise for older adults [16] or in autism therapy [13], as a therapy assistant in children cancer treatment [15], sign language tutors [17], other kind of educational agents mainly in children-robot interaction [18] [19] [20] [21] [22] and others [23]. One of the challenges of using robots in therapies is often to fuse play and rehabilitation techniques using a robotic design to induce human-robot interaction, in which the criteria was to make the therapy process entertaining and effective for its users.

3 Design of a Motivational Robotic System for Physical Therapy

3.1 Our Previous Research

In our previous research [24] we placed a humanoid robot Nao in an elementary school as a physiotherapist for rehabilitation and prevention of scoliosis. We programmed a set of exercises that can reduce the symptoms of spinal disorders, as they: improve cosmetic appearance, reduce pain, improve breathing and function levels, reduce the existing curvature and in some cases avoid the need for scoliosis surgery. The subjects were 20 children, aged between 6-8. This sample of subjects was chosen deliberately, as children of this age are playful, interested in the technology and thus are not limited by fear of the unknown. On the other hand, the problem of children of this age is their concentration, and the majority of them is not able to repeat the same activity over a longer time. This is the main priority for satisfactory results of the rehabilitation exercises. The details of the experimental results can be found in [24]. This was our pilot study to test the effects on children of coaching delivered by a social humanoid robot. From this research we highlighted the need to make robotic behaviors less boring and more effective to prepare robots for a long-term human-robot interaction.

Fig. 1. Examples of the exercises designed by a professional physiotherapist and implemented using the Nao humanoid platform

Fig. 2. Children in the elementary school trying to imitate the robot's motion. While children are exercising with the robot, a human physiotherapist is observing them and is able to help children when needed: our previous research

3.2 Adding Motivational Behaviors Controlled by the Wizard

We preprogrammed a set of motivational robotic behaviors stored in an emotional database which contains different types of the expressions of the robot representing basic emotions: joy, anticipation, anger, disgust, sadness, surprise, fear and trust. We asked a human physiotherapist to observe children during the 20-minute long sessions and to try to select an expression to be performed by the robot anytime that the physiotherapist decides that it can motivate children to continue exercising. The results can be found in the following Tab. 1.

Table 1. Interest of children in the therapy and accuracy of their exercises in time when robot was not showing emotional expressions (above) and when the robot performed the expressions controlled by the Wizard (below)

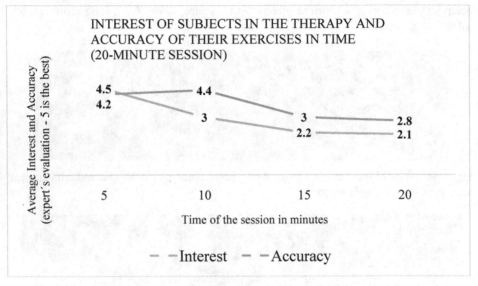

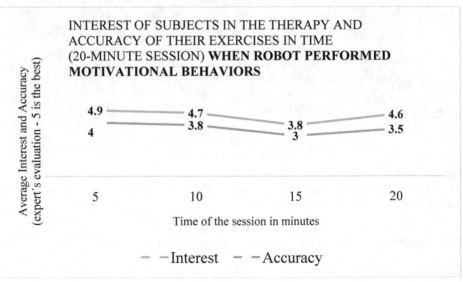

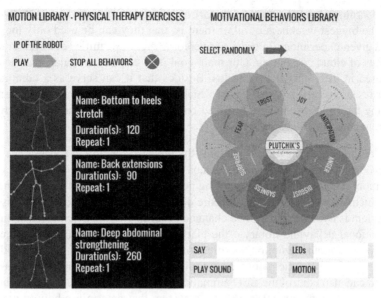

Fig. 3. Design of the graphical user interface for the Wizard that controls the experiment

The Wizard of Oz method is sufficient for research, but if we want to have robots in human environments we have to think about them as learning systems. In our work we present a cloud-based Wizard of Oz interface which learns from the operator in order to increase the level of autonomy of the robot. This way we can move from Wizard of Oz toward the Oz of Wizard [27].

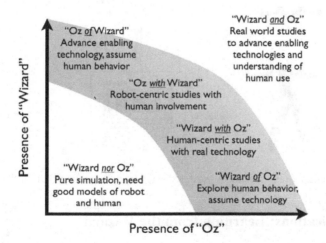

Fig. 4. The difference between Wizard of Oz and Oz of Wizard [27]

After learning about the state of the art of Wizard of Oz interfaces [28] we found out that the biggest weakness of all of them is, that they can be used only locally and just for a given experiment. In order to overcome these infirmities our system uses the advantages of cloud computing. Our main goal is to create a modular interface which can be used in different HRI scenarios. Besides that it can serve as a common platform for researchers. It is cloud-based which means that it is always online and after connecting the robot to the system based on its IP address the users are ready to go. The system consists of three parts (Fig. 9):

— Motion library – in our case it contains the physical therapy exercises. The Wizard can choose the exercises from the database, number of repeats and set the order of execution. Another feature is recording new exercises with a Microsoft Kinect sensor, which enables the creation of more diverse rehabilitation sessions. This part of the system is dynamic and can be changed according to the given experiment.
— Motivational behaviors library – the platform also comes with an emotional database which contains emotional expressions of the robot based on Plutchik's emotional model (joy, anticipation, anger, disgust, sadness, surprise, fear, trust). The Wizard can also control the LED animations and the phrases said by the robot.
— Agent-based reinforcement learning – a system that determines how to map situations to actions and also tries to maximize a numerical reward signal (how to control the teaching process, e.g. when to activate the motivational mode of the robot). This part is invisible for the users. A detailed description of the learning process can be found in the next chapter.

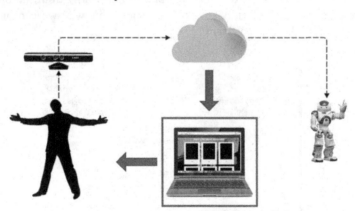

Fig. 5. Scheme of the system. Wizard, Kinect, Cloud, Robot and GUI for the Wizard

4   Implications: Learning from the Wizard

We propose to use reinforcement learning, that is, according to [29] a learning method that determines how to map situations to actions and also tries to maximize a numerical reward signal. The actions performed by the agent are not defined explicitly, and they have to be discovered through exploration in order to get the most reward.

Traditional methods of reinforcement learning algorithms were used successfully in many application areas - however they were not primarily designed for learning from real-time social interaction from humans. This kind of learning addresses some additional challenges, such as dealing with limited human patience or ambiguous human input. We agree with Thomaz [30] in that the most important task of the designer of the algorithm is to ensure that the system learns the right thing at the right time [30].

In order to include human input to a machine learning process several approaches were designed: machine learns by observing human behavior, human explicitly directs action of the machine, human provides high-level evaluation, feedback or labels to a machine learner, etc. In the case of learning from observation of the human behavior, the learning can occur implicitly or in other scenarios the human is explicitly teaching the machine a new skill. This type of approach was used in programming by example, learning by watching, learning by demonstration [31], etc. In other scenarios, the humans explicitly directed the action of the machine to provide them experience from which they can learn. In comparison with the above mentioned approach, this method is more interactive, but the human has to learn how he/she has to interact with the machine. The mechanism was used in various robot learning scenarios, such as learning navigation tasks by following a human demonstrator. The third mechanism is used to influence the experience of the machine with higher level constructs such as giving feedback to a reinforcement learning agent or labeling examples in an active learning scenario.

In the above mentioned approaches, the main goal is to gain the learning performance of the machine through human inputs. Opposite to this, socially guided machine learning reframes the machine learning problem as an interaction between the human and the machine. A social learning system, where the machine learning provides output but also interacts with the human teacher. Such a system is designed to learn efficiently from people with no experience in machine learning.

This type of reinforcement learning is called interactive reinforcement learning and was successfully used in various scenarios [32] [33] [34]. In our work, we aim to use this method for learning and adapting the robot's social behavior in order to move from the Wizard of Oz technique towards an autonomous mode.

To achieve this, the learning has the following characteristics:

- During the experiment the human operator (Wizard/expert) controls the robot and based on the subject's non-verbal reactions (gestures, emotions) changes its behavior.
- The states of the subject during the interaction are labeled by the Wizard, and the associations between them and the actions of the operator are saved in the form of rules. The goal of the Wizard is to create a successful rehabilitation session. In order to get a database with as much data as is possible we plan to use the crowdsourcing technique.
- During the interaction the algorithm tires to approximate the policy – how the Wizard chooses its actions in different states – of the operator. After a given point, the algorithm is capable of replacing the Wizard, such as it was presented in [35]. The most appropriate artificial intelligence method for this task are neural networks

(NN). Based on the Wizard's actions we can train a NN in order to reliably simulate the expert's decisions in various cases.

- In the first three phases the robot learns a pattern of social behavior, which can serve as a base when we want to create personalized behaviors. As it was mentioned, our goal is to reduce the number of interventions of the expert and increase the level of autonomy of the robot. When the robot is in autonomous mode the states of the subject are captured by a Microsoft Kinect sensor. The non-verbal states of the human partner are detected and recognized by already implemented systems. In order to measure the resulted autonomy of the robot we plan to use neglect tolerance of the interaction [36].
- To create a personalized social behavior we are just shaping the robot's learnt behavior. This method is based on reinforcement learning [37]. One can argue that this method of machine learning is slow for such a domain, but it was proven [37], that human guidance can significantly reduce the exploration time.

Fig. 6. Session in which the robot can show emotional expressions to motivate children. The expressions and their timing are selected by the Wizard but we are developing a learning system that will simulate the Wizard's behavior

5　Conclusion

We propose using social robots for back pain prevention. During the session an operator - an expert in physiotherapy – observes children and interrupts the exercises with selected motivational behavior to maintain the interest of the children in the session, when needed. Moreover, new types of exercises can be easily added to the system, without any programming experience, using Kinect. We suggest to use a learning mechanism based on reinforcement learning that would acquire the expert's knowledge. Using cloud computing we are able to gather this knowledge from multiple experts.

Acknowledgements. This publication is the result of the Project implementation: University Science Park TECHNICOM for Innovation Applications Supported by Knowledge Technology, ITMS: 26220220182, supported by the Research & Development Operational Programme Funded by the ERDF.

References

1. Frank, J.W., et al.: Disability resulting from occupational low back pain: Part I: What do we know about primary prevention? A review of the scientific evidence on prevention before disability begins. Spine **21**(24), 2908–2917 (1996)
2. Harreby, M., et al.: Are radiologic changes in the thoracic and lumbar spine of adolescents risk factors for low back pain in adults? A 25-year prospective cohort study of 640 school children. Spine **21**, 2298–2302 (1995)
3. Loney, P.L., Stratford, P.W.: The prevalence of low back pain in adults: a methodological review of the literature. Physical therapy **79**(4), 384–396 (1999)
4. Vos, T., et al.: Years lived with disability (YLDs) for 1160 sequelae of 289 diseases and injuries 1990–2010: a systematic analysis for the Global Burden of Disease Study 2010. The Lancet **380**(9859), 2163–2196 (2013)
5. COST Action B13, Biomedicine and Molecuar Biosciences, Guidelines for the Management of Low Back Pain in Europe, Last updated: May 02, 2011. COST is supported by the EU Framework Programme Horizon (2020). http://www.cost.eu/COST_Actions/bmbs/Actions/B13
6. Weiss, H.R., et al.: Physical exercises in the treatment of idiopathic scoliosis at risk of brace treatment. Scoliosis **1**(6), 1–7 (2006)
7. Negrini, S., et al.: Physical exercises as a treatment for adolescent idiopathic scoliosis. A Systematic Review. Developmental Neurorehabilitation **6**(3–4), 227–235 (2003)
8. Leboeuf-Yde, C., Kyvik, K.O.: At what age does low back pain become a common problem?: A study of 29,424 individuals aged 12-41 Years. Spine **23**(2), 228–234 (1998)
9. Institute for Health Metrics and Evaluation ©, University of Washington (2013). htttp://vizhub.healthdata.org/irank/heat.php
10. Rabbitt, S.M., Kazdin, A.E., Scassellati, B.: Integrating socially assistive robotics into mental healthcare interventions: applications and recommendations for expanded use. Clinical Psychology Review **35**, 35–46 (2015)
11. Kidd, C.D., Breazeal, C.: A robotic weight loss coach. In: Proceedings of the National Conference on Artificial Intelligence, vol. 22(2), p. 1985. AAAI Press, Menlo Park. MIT Press, Cambridge, 1999, 2007
12. Tapus, A., Tapus, C., Mataric, M.J.: The use of socially assistive robots in the design of intelligent cognitive therapies for people with dementia. In: Proceedings of the IEEE International Conference on Rehabilitation Robotics (2009)
13. Baxter, P., et al.: Long-term human-robot interaction with young users. In: Proceedings of the 6th IEEE/ACM International Conference on Human-Robot Interaction (Robots with Children Workshop) (2011)
14. Scassellati, B., Admoni, H., Mataric, M.J.: Robots for use in autism research. Annual Review of Biomedical Engineering **14**, 275–294 (2012)
15. Alemi, M., Meghdari, A., Ghanbarzadeh, A., Moghadam, L.J., Ghanbarzadeh, A.: Impact of a social humanoid robot as a therapy assistant in children cancer treatment. In: Beetz, M., Johnston, B., Williams, M.-A. (eds.) ICSR 2014. LNCS, vol. 8755, pp. 11–22. Springer, Heidelberg (2014)
16. Fasola, J., Mataric, M.J.: Using socially assistive human–robot interaction to motivate physical exercise for older adults. Proceedings of the IEEE **100**(8), 2512–2526 (2012)
17. Kose, H., et al.: Socially interactive robotic platforms as sign language tutors. International Journal of Humanoid Robotics **11**(1) (2014)

18. Deshmukh, A., et al.: Empathic robotic tutors: map guide. In: Proceedings of the 10th IEEE/ACM International Conference on Human-Robot Interaction Extended Abstracts (2015)
19. Bhargava, S., et al.: Demonstration of the emote wizard of Oz interface for empathic robotic tutors. In: Proceedings of SIGdial
20. Serholt, S., et al.: Emote: embodied-perceptive tutors for empathy-based learning in game environment. In: European Conference GBL (2013)
21. Brown, L.V., Kerwin, R., Howard, A.M.: Applying behavioral strategies for student engagement using a robotic educational agent. In: Proceedings of the IEEE International Conference on Systems, Man, and Cybernetics (SMC) (2013)
22. Komatsubara, T., et al.: Can a social robot help children's understanding of science in classrooms? In: Proceedings of the 2nd International Conference on Human-Agent Interaction (2014)
23. Leite, I., Martinho, C., Paiva, A.: Social robots for long-term interaction: a survey. International Journal of Social Robotics 5(2), 291–308 (2013)
24. Vircikova, M., Sincak, P.: Experience with the children-humanoid interaction in rehabilitation therapy for spinal disorders. In: Kim, J.-H., Matson, E., Myung, H., Xu, P. (eds.) Robot Intelligence Technology and Applications. AISC, vol. 208, pp. 347–357. Springer, Heidelberg (2013)
25. Villano, M., et al.: DOMER: a wizard of Oz interface for using interactive robots to scaffold social skills for children with autism spectrum disorders. In: Proceedings of the 6th IEEE/ACM International Conference on Human-Robot Interaction (2011)
26. Lu, D.V., Smart, W.D.: Polonius: a wizard of Oz interface for HRI experiments. In: Proceedings of the 6th IEEE/ACM International Conference on Human-Robot Interaction (2011)
27. Steinfeld, A., Jenkins, O.C., Scassellati, B.: The Oz of wizard: simulating the human for interaction research. In: Proceedings of the 4th IEEE/ACM International Conference on Human-Robot Interaction (2009)
28. Riek, L.D.: Wizard of Oz studies in HRI: a systematic review and new reporting guidelines. Journal of Human-Robot Interaction 1(1) (2012)
29. Sutton, R.S., Barto, A.G.: Reinforcement Learning: An Introduction. MIT Press, Cambridge (1998)
30. Thomaz, A.L.: Socially Guided Machine Learning. PhD Thesis (2006)
31. Thomaz, A.L., Cakmak, M.: Social Learning Mechanisms for Robots (2009)
32. Suay, H.B., Chernova, S.: Effect of human guidance and state space size on interactive reinforcement learning. In: Proceedings of the 20th International Symposium on Robot and Human Interactive Communication RO-MAN (2011)
33. Mitsunaga, N., et al.: Adapting Robot Behavior for Human-Robot Interaction. IEEE Transactions on Robotics 24(4) (2008)
34. Kartoun, U., Stern, H., Edan, Y.: A human-robot collaborative reinforcement learning algorithm. Journal of Intelligent & Robotic Systems 60(2), 217–239 (2010)
35. Broekens, J.: Emotion and reinforcement: affective facial expressions facilitate robot learning. In: Huang, T.S., Nijholt, A., Pantic, M., Pentland, A. (eds.) ICMI/IJCAI Workshops 2007. LNCS (LNAI), vol. 4451, pp. 113–132. Springer, Heidelberg (2007)
36. Olsen, D.R., Goodrich, M.A.: Metrics for evaluating human-robot interactions. In: Proceedings of PERMIS (2003)
37. Knox, W.B., Stone, P.: Interactively shaping agents via human reinforcement: the TAMER framework. In: Proceedings of the 5th International Conference on Knowledge Capture (2009)

Does the Presence of Social Agents Improve Cognitive Performance on a Vigilance Task?

Arielle R. Mandell[✉], Melissa A. Smith, Molly C. Martini,
Tyler H. Shaw, and Eva Wiese

Department of Psychology, George Mason University, Fairfax, VA 22030, USA
amandel3@gmu.edu

Abstract. Humans are historically bad at performing cognitive tasks that require sustained attention. Social facilitation theory states that the real, imagined,
or implied presence of other social agents can improve cognitive performance.
Typically these agents are other humans, but the current study explores the possibility that robots can trigger social facilitation effects. We hypothesized that
more humanlike social stimuli are (1) more likely to be ascribed internal states
(e.g., having a mind, having emotions, having preferences), and (2) more likely
to induce social facilitation on a vigilance task. Experiment 1 investigates the
relationship between physical humanness and attributions of intentionality by
comparing ratings of internal states for three agents (human, robot, and nonsocial). In Experiment 2, we examine whether physical humanness results in improved performance on a vigilance task through social facilitation. While Experiment 1 showed the expected positive relationship between human appearance
and mind attribution, the degree to which mental states were attributed to an
agent did not influence performance on the vigilance task. The implications for
social robotics are discussed.

Keywords: Social robotics · Social facilitation · Vigilance

1 Introduction

Rapid advancements in robotics have produced robots that are capable of increasingly
complex tasks in a wide variety of operational environments. These robots are not
fully autonomous—they often require monitoring, guidance, or equal participation
from human interaction partners. While oftentimes necessary, the human element can
also hinder performance due to cognitive constraints (e.g., attention and memory limitations). Thus, social robotics is tasked with 1) identifying physical and behavioral
features that improve performance in human-robot interaction (HRI), and 2) implementing these features to create robots that augment human performance in situations
prone to environmental constraints or cognitive limitations.

A.R. Mandell and M.A. Smith—contributed equally to the manuscript and share first
authorship.

A. Tapus et al. (Eds.): ICSR 2015, LNAI 9388, pp. 421–430, 2015.
DOI: 10.1007/978-3-319-25554-5_42

One way to facilitate productive interactions is to design robots that resemble social beings with human-like traits (e.g., having a mind) and/or skills (e.g., conversing). The theory of *social facilitation* indicates that the real, imagined, or implied presence of a social agent can improve performance on a cognitive task [1]. This finding suggests that manipulating robot features to increase the perceived socialness of a robot interaction partner may improve the cognitive performance of a human operator. Therefore, it is important to identify features that increase the likelihood of a human treating a robot as a social agent. The goal of the present study is to investigate whether human performance on a complex cognitive task can be improved by manipulating the socialness of a robot interaction partner. Specifically, we hypothesize that more human-like agents are increasingly likely to be ascribed intentionality and will therefore improve performance to a greater degree than agents that are perceived to be nonsocial and non-intentional.

One approach to the goal of improving HRI performance is to examine factors that influence the interaction on a cognitive level in order to elevate team performance. One type of cognitive skill that is known to yield poor performance in humans despite its relatively simple nature is vigilance, or the ability to sustain attention over extended periods of time [2]. Vigilance requires operators to maintain alertness in order to react to infrequent, low salience cues. It is a crucial skill in environments that incorporate automated systems requiring constant monitoring of complex displays for critical events that require human intervention, such as air traffic control [3].

Despite the apparent simplicity of a vigilance task, research demonstrates that people frequently miss critical signals over long-term vigils [5]. In fact, performance deteriorates over time, as evidenced by a reduction in signal detection and increase in response times known as the *vigilance decrement* [2,5,6]. The vigilance decrement is a major concern in environments that require human monitoring of automated systems and has therefore been the focus of a great deal of research [5,7,8,9].

Given the tendency to perform poorly in these environments, it is crucial to identify factors that can assist human operators on vigilance tasks. One social-psychological finding that may be leveraged to improve performance is social facilitation. Despite the complex relationship between interaction partners and performance, this effect has been associated with improved vigilance performance in the past. For example, research has demonstrated that vigilance operators who work in dyads outperform solo-operators by 25% [10]. The current study investigates whether the presence of a social robot as an interaction partner can replicate this effect and similarly improve cognitive performance.

Unfortunately, using social robotics to facilitate performance may not be as simple as designing anthropomorphic robots. Previous research suggests that a major difference between interacting with a human compared to a robot is the way we ascribe intentionality to the other agent's actions [11,12]. When interacting with other humans, we adopt the intentional stance, meaning we treat them as intentional systems and explain their behavior with reference to their internal states (i.e., desires, intentions) [13]. Alternatively, people do not characterize robots as intentional agents with mental states. Instead, we adopt the design stance and characterize their behavior as mechanistic. The expectations you hold regarding the actions of another agent depend on your belief in whether the agent has a mind and acts with intention or is programmed to behave a certain way.

Wiese et al. [12] demonstrated that beliefs in agent intentionality (or lack thereof) affect an individual's readiness to engage in social interactions with the agent independent of the agent's physical appearance (human vs. robot). In this study, participants were instructed that the agent's eye movements were either performed by a human or pre-programmed. Participants were more willing to follow the agent's gaze when they believed the eye movements were controlled by a human compared to a program, regardless of whether the agent was a human or a robot. The preferential treatment of the human gaze cuing resulted in improved performance. Similar findings regarding assumptions about human interaction partners and the tendency to attend to gaze direction have also been found by other research groups [11,14].

Based on the evidence outlined above, a major goal of social robotics should be to identify features that trigger the attribution of intentionality to nonhuman agents and lead to improved performance in human-robot interaction through social facilitation. The first step in this process is determining which features are likely to trigger the intentional stance. Generally speaking, one would expect that robots resembling humans should induce social facilitation. A more nuanced theory would suggest that perceived similarity of a robot to a human is mainly influenced by the robot's *behavior* and *appearance*. Research has demonstrated that the attribution of intentionality to nonsocial agents is much more likely when the agent physically resembles a human (e.g. possesses low-level perceptual features [16]), even when they do not demonstrate humanlike behavior [15]. Using a series of morphed faces ranging from puppet to human, Looser and Wheatley [16] identified a categorical threshold for mind attribution biased significantly toward the human end of the spectrum. Only faces that had significantly more than 50% of human features triggered the intentional stance. Studies have also shown that humans are skilled at identifying other human faces. Wheatley et al. [17] demonstrated that humans are able to differentiate between human and inanimate faces within a few hundred milliseconds of face processing. In addition, Wagner et al. [18] found that passively viewing human faces activates brain areas involved in social decision making.

These findings clearly show that even agents without minds (e.g., robots) can trigger the intentional stance if they *appear* to be humanlike. It would follow that interactions with these agents would result in performance improvements that are not associated with nonsocial or mechanistic agents, possibly through social facilitation. The current study examines these possibilities by incorporating a variety of social and nonsocial agents (nonsocial, robot, and human) into a vigilance task.

2   Experiments

In two experiments, we investigate whether robots can trigger social facilitation effects using two types of stimuli: social (i.e., human or robot) and nonsocial (i.e., arrows). The human stimulus depicted a female face from the Karolinska Directed Emotional Faces database [19] and the robot stimulus depicted a humanoid robot

(Meka Robotics). The nonsocial stimulus was designed to be morphologically similar to the social stimuli and consisted of two arrows within a black circle; see Figure 1 for images of the stimuli.

These stimuli were used to investigate whether the physical appearance of an agent has an influence on the degree to which a mind is attributed to the stimulus (Experiment 1), and whether mind attribution has a positive effect on vigilance performance (Experiment 2). We hypothesized that the more humanlike a social stimulus appears, the more likely it is that internal states (e.g., having a mind, having emotions, having preferences), will be ascribed to that stimulus. Further, we hypothesized that the extent to which internal states are ascribed to the social stimulus modulates the degree to which social facilitation is induced in the vigilance task.

2.1   Experiment 1

In order to investigate the relationship between physical humanness and intentionality, participants rated the likelihood of each agent (Figure 1) possessing internal states (i.e. being alive, having a mind, feeling pain, feeling the need to hang out with others, and being interested in social interactions). We hypothesized that individuals will be more likely to ascribe intentions to agents that are more physically humanlike.

Participants. Fifty-five participants (24 male) with a mean age of 36.9 years (*SD* = 12 years) were included in this experiment. Participants were recruited through Amazon Mechanical Turk, completed the evaluation questionnaire online using Qualtrics, and were compensated with money via Mechanical Turk.

Materials. The agents were 552 x 552 pixels (16.24° visual angle) and were presented on a white background with eyes aligned with the horizontal axis of the screen. To measure the degree to which internal states were ascribed to different agents, we used a questionnaire consisting of five items related to mental states and social skills, rated on a 7-point Likert-scale. Participants were instructed that a human mind differs from that of an animal or a machine, and an animal is alive while a rock is not (see [16] for a more detailed description). Questions 1 and 2, taken from a study by Hackel et al. [20] on social identity and mind perception, asked participants to "rate how much this face looks alive" and "rate how much this face looks like it has a mind" with anchors set at 1 (*Definitely not alive/Definitely has no mind*) and 7 (*Definitely alive/Definitely has a mind*). The remaining items asked participants if they think the agent "would feel pain if it tripped and fell on hard ground," "likes to hang out with friends," and "would be an interesting conversationalist." These three items were rated on a scale from 1 (*Definitely Not*) to 7 (*Definitely*).

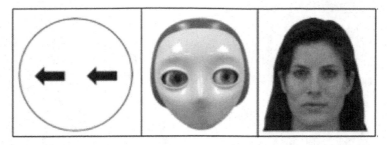

Fig. 1. Social and nonsocial agents used in Experiments 1 and 2 arranged by degree of physical humanness. From left to right: Nonsocial stimulus (arrows), and social stimuli: Meka robot (0% human) and Human (100% human).

Design and Procedure. All participants rated the three stimuli on all five social dimensions described above. Agent presentation order was randomized and all agents were looking forward. After rating the agents, participants completed a demographic questionnaire and collected their payment through Amazon Mechanical Turk.

Results. Results for Experiment 1 are shown in Figure 2. Ratings for each agent showed similar patterns across the five questions, with the human agent rated very high on the scale ($M = 6.19$, $SD = 1$), the robotic agent rated lower on the scale ($M = 1.63$, $SD = .98$), and the nonsocial agent rated even lower ($M = 1.16$, $SD = .55$).

Separate one-way ANOVAs were conducted comparing the ratings of the agents for each question. These analyses indicated that the human, robot, and nonsocial agents significantly differ in their ratings of being alive ($F(2) = 598.19$, $p < .01$, $\eta^2 = .88$), having a mind ($F(2) = 527.85$, $p < .01$, $\eta^2 = .87$), feeling pain ($F(2) = 692.68$, $p < .01$, $\eta^2 = .90$), wanting to hang out with others ($F(2) = 242.82$, $p < .01$, $\eta^2 = .75$), and conversing ($F(2) = 200.83$, $p < .01$, $\eta^2 = .86$).

The differential ratings of intentionality were explored using a series of post-hoc Bonferroni comparisons. These tests indicated that the human agent was rated as significantly ($p < .05$) more likely to be alive, have a mind, feel pain, hang out, and converse compared to both the robotic and nonsocial agents. Furthermore, the robot was rated as significantly ($p < .05$) more likely than the nonsocial agent to be alive, have a mind, hang out, and converse. In fact, all comparisons between the three agent types were significantly different, with the human rated higher than the robot and the robot rated higher than the nonsocial agent, aside from the question regarding pain which had no significant difference between the robot and nonsocial agent ($p > .05$).

Discussion. The results from Experiment 1 support our hypothesis by demonstrating that agents with a more humanlike appearance trigger increased ratings of internal states, which is marked by an increased willingness to adopt the intentional stance when looking at an agent. These findings are similar to research conducted by Martini et al. (under review), who identified a dual-linear relationship between physical humanness and the attribution of internal states, with ratings of nonhuman agents (represented by the nonsocial arrows and the robot in the current study) along one slightly positive linear slope and the ratings of physically human agents (represented by the human) on a separate, steeper positive linear path.

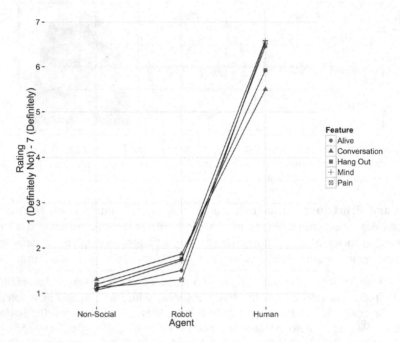

Fig. 2. Ratings of internal states for Experiment 1 indicate that the three agents are rated significantly differently on all questions except "pain." The human is rated as significantly more likely to have internal states than the robot, which is significantly more likely to have internal states than the nonsocial agent.

2.2   Experiment 2

Experiment 1 demonstrated that increasing physical humanness results in increasing attributions of intentionality. The goal of Experiment 2 is to examine whether this trend will result in improved vigilance performance via social facilitation.

During the vigilance task, an agent (human, robot, or nonsocial) appears in the center of the screen, followed by a probe (hairdryer or gun) on the left or right side of the agent. On each trial, the agent's gaze direction (or the arrow direction) cues the participant's attention either towards or away from the probe. Participants are instructed to respond only when the agent cued the gun (12.5% of trials), and to withhold response on the remainder of the trials.

Participants. 45 undergraduate students (8 male), with a mean age of 21.1 (*SD* = 4.9), participated in this experiment for course credit. There were 15 participants in each condition (nonsocial, robot, human).

Materials. The agents described in Experiment 1 (Figure 1) were also used in Experiment 2. For the social agent gaze shifts, irises and pupils were deviated 0.44° from direct gaze in Photoshop. The non-social arrows (105px wide, 3.11° visual angle) were positioned at the same level as the eyes.

Two probes were used in this study: a gun and a hair dryer (Figure 3), measuring 2.86° visual angle in width and 2.26° visual angle in height. These probes, previously used by Parasuraman et al. [21] to examine vigilance and threat-related biological motion, were used in the current study due to their morphological similarities and discriminant threat-relevance. The probes were presented on a white background on the horizontal axis, located 5.72° visual angle from the center of the screen.

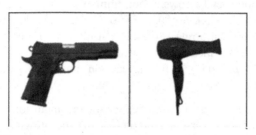

Fig. 3. The two probes used in the study, selected for their morphological similarity and discriminant threat-relevance. Participants were instructed to respond on trials where the agent cued towards the gun (L), not the hair dryer (R).

Apparatus. The stimuli were presented on a 19-inch monitor with the refresh rate set at 85 Hz. Reaction time measures were based on standard keyboard responses. Participants were seated approximately 57cm from the monitor, and the experimenter ensured that participants were centered with respect to the monitor. The experiment was conducted using *Experiment Builder* (SR Research Ltd., Ontario, Canada).

Task and Procedure. The vigilance task was performed continuously over 24 minutes and was divided into eight 3-minute periods for analysis. Each period contained 72 trials, comprised of 9 targets (12.5%) and 63 non-targets (87.5%).

Trials began with a fixation cross (250 ms) followed by the agent's forward facing image (rectangles in the non-social condition) for 500 ms. The agent's gaze (or arrow direction) shifted to the left or right for 200 ms, and the probe was presented for 100 ms. Then, the screen went blank and a 1200 ms response window began. Participants were instructed to respond by pressing the spacebar only when the agent looked at (or pointed to) the gun. Responses on these trials were considered correct.

After giving consent, participants read through instructions at their own pace. This was followed by a 3-minute practice session with auditory feedback. In order to begin the task, participants had to reach at least 66% accuracy in practice. Those who were unable to pass after repeating the practice session once were dismissed.

Design. This study utilized a between-subjects design with agent type as the between-subjects factor. Dependent variables included: response times (reaction time to correct detections), hit rates (percentage of correct responses to critical signals), and false alarms (percentage of responses given when critical signal was not present, i.e. pointing either at or away from the hairdryer or pointing away from the gun).

Results. A 3 (agent type) x 8 (period) mixed design ANOVA was conducted on hit rate, reaction times, and false alarms. Greenhouse-Geisser estimates were used to correct for violations to the assumption of sphericity. An examination of the main effect of period indicated that while there was no significant change in accuracy over time on task ($F(3.2, 134.4) = 1.37$, $p > .05$, $\eta^2 = 0.03$), reaction times increased significantly ($F(4.7, 196.7) = 11.32$, $p < .01$, $\eta^2 = 0.21$) and false alarm rates decreased significantly throughout the task ($F(2.8, 117.8) = 13.87$, $p < .01$, $\eta^2 = 0.25$). These results are partially consistent with the expected performance decrement.

The conditions (non-social arrows, robot, human) did not significantly differ in average accuracy ($F(2, 42) = .39$, $p > .05$, $\eta^2 = 0.02$), reaction time ($F(2, 42) = .74$, $p > .05$, $\eta^2 = 0.03$), or false alarm rate ($F(2, 42) = 1.62$, $p > .05$, $\eta^2 = 0.07$). In addition, there were no significant interactions between condition and period for accuracy ($F(6.4, 134.4) = 1.27$, $p > .05$, $\eta^2 = 0.06$), reaction time ($F(9.7, 196.7) = .71$, $p > .05$, $\eta^2 = 0.03$), or false alarm rate ($F(5.6, 117.8) = 1.23$, $p > .05$, $\eta^2 = 0.06$).

Discussion. We made two hypotheses regarding the inclusion of social agents on vigilance task performance. First, we predicted that the presence of a social (versus nonsocial) stimulus would improve performance on a vigilance task. Second, we predicted that increasing physical humanness would produce a corresponding increase in vigilance performance, such that participants with the human agent outperform those with the robot agent, who outperform those with the nonsocial agent.

The results of Experiment 2 indicate that while there was a vigilance decrement across the board in the form of increasing reaction times, agent type had no specific effect on cognitive performance. Neither of our initial hypotheses were supported: there was no effect of the social stimuli (i.e., robot, human) compared to non-social stimulus (i.e., arrows), nor was increasing humanness associated with cognitive improvements. It is possible that the task was not demanding enough, but this is unlikely because short vigilance tasks regularly produce robust decrements [22]. Overall, this finding indicates that social facilitation did not appear to help improve cognitive performance on this vigilance task.

3   General Discussion

The results from these experiments show us that while increasing the degree of physical humanness results in increased ratings of internal states and intentional attributes, which is marked by an increased willingness to adopt the intentional stance when looking at an agent, the presence of a social agent has no overall impact on performance during a vigilance task, even as the perceived humanness of the agent increases. This is contrary to what we expected, which was that the greatest vigilance decrement would occur in the non-social arrows condition, followed by the next greatest decrement with the robot, and the human having the least extreme decrement. In one interpretation, this could suggest that social cues and social facilitation are irrelevant during a highly cognitive task, such as a vigilance task. In another interpretation, it could indicate that all conditions were equally impacted by having the agents present during the task, in which case work into further varying the levels of humanness would help tease apart the effects of social facilitation.

That said, there is a chance that the social facilitation effect was minimized in this particular task since the agents were included *within* the task as *part* of the stimuli, as opposed to as a separate entity with its own physical presence, implied or otherwise. Again, this would have to be further researched to determine if the social facilitation outcomes differ with different implementations of the paradigm.

Social agents are here to stay in their various forms as technology advances. Researching how to leverage the social facilitation effect in virtual agents of varying humanness will help to improve human-technological interactions by increasing productivity, reducing errors, and making the interactions more pleasant overall.

References

1. Zajonc, R.B.: Social Facilitation. Research Center for Group Dynamics, Institute for Social Research, University of Michigan (1965)
2. Davies, D.R., Parasuraman, R.: The Psychology of Vigilance, pp. 107–117. Academic Press, London (1982)
3. Metzger, U., Parasuraman, R.: The role of the air traffic controller in future air traffic management: An empirical study of active control versus passive monitoring. Human Factors **43**, 519–528 (2001)
4. Parasuraman, R.: Vigilance, arousal and the brain. Physiological Correlates of Human Behaviour **2**, 35–55 (1983)
5. Warm, J.S., Parasuraman, R., Matthews, G.: Vigilance requires hard mental work and is stressful. Human Factors **50**, 433–441 (2008)
6. Mackworth, N.H.: The breakdown of vigilance during prolonged visual search. Quarterly Journal of Experimental Psychology **1**(1), 6–21 (1948)
7. Temple, J.G., Warm, J.S., Dember, W.N., Jones, K.S., LaGrange, C.M., Matthews, G.: The effects of signal salience and caffeine on performance, workload and stress in an abbreviated vigilance task. Human Factors **42**, 183–194 (2000)
8. Berardi, A., Parasuraman, R., Haxby, J.V.: Overall vigilance and sustained attention decrements in healthy aging. Experimental Aging Research **27**(1), 19–39 (2001)
9. Helton, W.S., Warm, J.S., Matthews, G., Corcoran, K.J., Dember, W.N.: Further tests of an abbreviated vigilance task: Effects of signal salience and jet aircraft noise on performance and stress. Proceedings of the Human Factors and Ergonomics Society Annual Meeting **46**(17), 1546–1550 (2002)
10. Garcia, A.J., Baldwin, C.L., Funke, M., Funke, G., Knott, B., Finomore, V., Warm, J.: Team vigilance: The effects of co-action on workload in vigilance. Proceeding of the Human Factors and Ergonomics Society Annual Meeting **55**(1) (2011)
11. Teufel, C., Fletcher, P.C., Davis, G.: Seeing other minds: Attributed mental states influence perception. Trends in Cognitive Sciences **14**(8), 376–382 (2010)
12. Wiese, E., Wykowska, A., Zwickel, J., Müller, H.J.: I see what you mean: How attentional selection is shaped by ascribing intentions to others. PloS One **7**(9) (2012)
13. Dennett, D.: Who's on first? Heterophenomenology explained. Journal of Consciousness Studies **10**, 9–10 (2003)
14. Ristic, J., Kingstone, A.: Taking control of reflexive social attention. Cognition **94**(3), B55–B65 (2005)
15. Kiesler, S., Powers, A., Fussell, S.R., Torrey, C.: Anthropomorphic interactions with a robot and robot-like agent. Social Cognition **26**(2), 169–181 (2008)

16. Looser, C.E., Wheatley, T.: The tipping point of animacy: How, when, and where we perceive life in a face. Psychological Science 21(12), 1854–1862 (2010)
17. Wheatley, T., Weinberg, A., Looser, C., Moran, T., Hajcak, G.: Mind perception: Real but not artificial faces sustain neural activity beyond the N170/VPP. PloS One 6(3) (2011)
18. Wagner, D.D., Kelley, W.M., Heatherton, T.F.: Individual differences in the spontaneous recruitment of brain regions supporting mental state understanding when viewing natural social scenes. Cerebral Cortex 21(12), 2788–2796 (2011)
19. Lundqvist, D., Flykt, A., Ohman, A.: The Karolinska Directed Emotional Faces (KDEF). Department of Neurosciences Karolinska Hospital, Stockholm (1998)
20. Hackel, L.M., Looser, C.E., Van Bavel, J.J.: Group membership alters the threshold for mind perception: The role of social identity, collective identification, and intergroup threat. Journal of Experimental Social Psychology 52, 15–23 (2014)
21. Parasuraman, R., de Visser, E., Clarke, E., McGarry, W.R., Hussey, E., Shaw, T., Thompson, J.C.: Detecting threat-related intentional actions of others: effects of image quality, response mode, and target cuing on vigilance. Journal of Experimental Psychology: Applied 15(4) (2009)
22. Temple, J.G., Warm, J.S., Dember, W.N., Jones, K.S., LaGrange, C.M., Matthews, G.: The effects of signal salience and caffeine on performance, workload, and stress in an abbreviated vigilance task. Human Factors 42(2), 183–194 (2000)

Agent Appearance Modulates Mind Attribution and Social Attention in Human-Robot Interaction

Molly C. Martini[✉], George A. Buzzell, and Eva Wiese

Department of Psychology, George Mason University, Fairfax, VA, USA
mmarti35@masonlive.gmu.edu

Abstract. Gaze following occurs automatically in social interactions, but the degree to which we follow gaze strongly depends on whether an agent is believed to have a mind and is therefore socially relevant for the interaction. The current paper investigates whether the social relevance of a robot can be manipulated via its physical appearance and whether there is a linear relationship between appearance and gaze following in a counter-predictive gaze cueing paradigm (i.e., target appears with a high likelihood opposite of the gazed-at location). Results show that while robots are capable of inducing gaze following, the degree to which gaze is passively followed does not linearly decrease with physical human-likeness. Rather, the relationship between appearance and gaze following is best described by an inverted u-shaped pattern, with automatic cueing effects (i.e., attending to the cued location) for agents of mixed human-likeness and reversed cueing effects (i.e., attending to the predicted location) for agents of either full human-likeness (100% human) or full robot-likeness (100% robot). The results are interpreted with regard to cognitive resource theory and design implications are discussed.

Keywords: Social attention · Human-likeness · Resource theory · Behavioral measures · Design

1 Introduction

In social interactions, the use of non-verbal cues (i.e., gaze direction, body posture, facial expression) is important to communicate intentions, emotions, and preferences between interaction partners [1, 2, 3]. Thus, incorporating the use of non-verbal cues into human-robot interaction is important to social robotics and successful models on interpreting gaze direction [1], facial expressions [2], and body posture [3] have been proposed. However, what has not been investigated so far is whether the effectiveness of non-verbal cues in modulating interactions depends on robot features, such as personality, embodiment or appearance. The current paper addresses this question by investigating whether the appearance of a robotic agent has an influence on the degree to which its gaze direction is followed. In particular, we hypothesize that the more human-like an agent looks, the more social relevance will be ascribed to that agent due to the attribution of human-like features (i.e. having a mind, having emotions) and the more strongly information from its eye gaze will be used to efficiently structure the social interaction.

© Springer International Publishing Switzerland 2015
A. Tapus et al. (Eds.): ICSR 2015, LNAI 9388, pp. 431–439, 2015.
DOI: 10.1007/978-3-319-25554-5_43

1.1   Gaze Direction as a Cue to Others' Attention

Eye gaze provides a wealth of information about important events in the environment (e.g., presence of predators or the appearance of a friend) and the internal states of others (e.g., preferences and motivations). Gaze following is also used to make inferences about what others are currently interested in and what they are going to do next [1]. Due to this tremendous social relevance, the gaze direction of others is automatically followed and people conjointly attend to where others are looking [4].

Traditionally, gaze following has been thought to be triggered in an automatized fashion [5] based on the saliency and social relevance of a gazer's eyes, which is an example of *bottom-up* processing where an observed stimulus captions attention first at the subconscious level and is then followed by conscious level processing. Consequently, bottom-up triggered shifts of attention happen involuntarily and are hard to suppress. Recently though, several studies have investigated under which conditions *top-down* control of gaze following occurs [6, 7, 8, 9]. In contrast to bottom-up processing, top-down modulated shifts of attention are voluntarily controlled and are usually evoked by symbolic cues (e.g., arrows) that require interpretation [10, 11]. These more recent studies show that the operation of top-down control of attention critically depends on higher-order cognitive processes and the availability of context information [7, 8, 9], [12]. In particular, pre-existing assumptions about the gazer have been shown to influence the readiness to process and/or attend to his/her gaze [6, 7, 8, 9]. Specific to the interest of human-robot interaction (HRI) is the finding that stronger gaze following is observed when the gazer is believed to be a human with a mind versus a machine without a mind [8], [13, 14].

One way to influence perceptions of a mind within a mechanistic agent is through its physical appearance [15, 16]. For instance, Admoni and colleagues [15] found that agents that look human-like (i.e., agents with a mind) trigger automatic shifts of attention to the gazed-at location while robotic agents (i.e., agents without a mind) enable voluntary shifts of attention to the predicted target location. Additionally, other studies in which subjective ratings are used to assess an agent's capability of having a mind have shown that increasing human-like appearance leads to increased ratings of mind attribution [17, 18, 19].

1.2   Aim of Study

The current study builds upon previous findings by investigating whether there is a link between the degree to which an agent is rated as having a mind and the degree to which his/her gaze direction is followed. In particular, we hypothesized that the more human-like an agent is in appearance, the more strongly its gaze should be followed due to increased social relevance of its gaze behavior. In order to test this hypothesis, the current study morphed a robot face with a human face in 20% increments to create a spectrum ranging from 100% robotic appearance to 100% human appearance (Figure 1). These images were then used in a counter-predictive gaze cueing paradigm in which the target appeared 80% of the time opposite of where the face looked.

A counter-predictive paradigm has the advantage that reflexive, bottom-up effects (i.e., attending to the gazed-at location and faster reaction times at the *valid* location) can be easily separated from voluntary, top-down effects (i.e., attending to where the target is likely going to appear and faster reaction times at the *invalid* location) within the same set-up. If human-like appearance increases the social relevance that is ascribed to observed gaze behavior, the fact that the target appears with a high likelihood opposite of where the face is looking should be picked up more easily and be used to efficiently use gaze direction to predict target location in a top-down fashion (i.e., shift attention opposite of where the face is looking). In contrast, if no social relevance is ascribed to the agent, the predictivity of the gaze behavior should not be noticed easily and attention should be shifted to the gazed-at location in a bottom-up fashion.

With regard to performance measures, this means that we expect shorter reaction times at the predicted compared to the cued location for the human-like agents (i.e., negative cueing effects when calculating the difference between invalid and valid conditions). For the robot-like agents, in contrast, we expect shorter reaction times at the cued location compared to the predicted location (i.e., positive cueing effects when calculating the difference between invalid and valid condition). We further expect that the degree to which gaze following can be top-down controlled increases linearly with increasing levels of human-like appearance as more social relevance is ascribed to the agents that appear humanlike. The results of this experiment will therefore directly inform social robotic design by relating subjective ratings of an agent's degree of having a mind to cognitive performance measures.

2   Methods and Materials

2.1   Participants

Thirty-seven undergraduate students with normal or corrected-to-normal vision (age: $M = 20.34$, $SD = 3.33$, 20 females) from George Mason University participated in the experiment in exchange for course credit. All participants provided written informed consent and all research procedures were approved by the George Mason University Office of Research Integrity & Assurance. Two participants were excluded from analysis because of error rates more than 2 standard deviations above the mean (M accuracy = 80.20% compared to $M = 95.56$%).

2.2   Apparatus

The experimental task was run on a Dell desktop computer, equipped with an LCD monitor (85 Hz refresh rate) and running Experiment Builder software (SR Research LTD., Ontario Canada). Participants were seated approximately 70 cm from the monitor and indicated responses using a standard computer keyboard. While the physical "D" and "K" keys were used to input responses, these keys were relabeled with letter stickers showing "T" or "F". Key assignment was counterbalanced across participants.

2.3   Stimuli

The morphed images were created using FantaMorph software which allows two images to be blended together at specified increments. The two images used to create the spectrum were the S2 humanoid robot head developed by Meka Robotics and a male human head taken from the Karolinska Directed Emotional Faces database [20]. Morphing occurred at 20% increments giving a total of six agent images: 0% human, 20% human, 40% human, 60% human, 80% human, and 100% human.

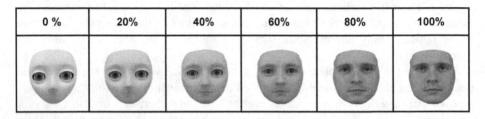

Fig. 1. Morphed images ranging from 0% human to 100% human.

Assuming a mean viewing distance of 70 cm (distance from the center of the computer screen to the participant's eyes), stimuli were 7.8° wide and 8.6° high, depicted on a white background and presented in full frontal orientation with eyes positioned on the central horizontal axis of the screen (Figure 2). For left- and rightward gaze, irises and pupils in the eyes were shifted with Photoshop and deviated 0.4° from direct gaze. The target stimulus was a black capital letter (F or T), measuring .5° in width and .9° in height. Targets appeared on the horizontal axis of the screen and were located 14.7° from the center of the screen.

2.4   Procedure

Following a practice block that mirrored the experimental task except for the use of a different agent stimulus (EDDIE; developed at Technische Universität München) to avoid bias, participants completed a reversed gaze cueing task made up of six blocks of 60 trials each in which the averted gaze of a virtual agent cued the location of a target stimulus in a counter-predictive fashion (i.e., 20% validity; target appeared with 80% likelihood opposite of where the face was looking). Participants were not explicitly informed of the probability of valid and invalid gaze cues, however, the practice block served the dual purpose of familiarizing participants with the overall experimental paradigm as well as providing a chance to implicitly learn the probability of valid and invalid gaze cues. Each block used one of the six agents in Figure 1 as the gazing stimulus and prior to each block the agent for the upcoming block was presented and participants were asked "Do you think this agent has a mind?" on a scale of 1 ("definitely not") to 8 ("definitely yes") to get an mind attribution rating. This mind attribution rating was assessed prior to each block (as opposed to following each block) in order to explicitly test how agency ratings, based on appearance alone, influence gaze following behavior. If the mind attribution ratings were instead assessed following each block, there would have been the potential for gaze validity to influence the agency ratings, as opposed to the ratings being based on appearance alone. The order in which agents were presented was counterbalanced across participants.

At the beginning of each trial, a fixation cross appeared for a random time interval of 700-1000 ms, followed by presentation of the agent with directed gaze for a random time interval of 700-1000 ms. Afterwards, the agent changed gaze direction, either looking to the left or right side of the screen. After a random time interval of 400-600 ms, the target letter (F or T) appeared on the screen either where the face was looking (i.e., valid or unpredicted trial) or opposite of where the face was looking (i.e., invalid or predicted trial). Both the agent (with averted gaze) and the target letter remained on screen until participants made a response or a time-out criterion (1200 ms) was reached, whichever appeared first. The inter-trial-interval (ITI) was 680 ms, see Figure 2 for the full trial sequence. Participants used the index finger of each hand to indicate which target letter had appeared by pressing either the key that was marked with "F" or "T". Participants were instructed to maintain fixation on the center of the screen throughout all trials and to respond both as quickly and accurately as possible to the target letters. Total testing time was approximately 20 minutes.

2.5   Analysis

For the mind attribution ratings that were collected prior to each gaze cueing block, one rating was missing for five of the participants and was therefore replaced using mean-interpolation. The mind attribution ratings were then entered into a univariate ANOVA with appearance (0% - 100% human) as the independent variable (Figure 3A). A series of Bonferroni corrected paired-comparison t-tests were then carried out to determine which agents differed significantly in their mind attribution ratings.

For analysis of the gaze cueing effects, only trials in which a correct answer was given were considered. In order to investigate the influence of agent appearance on gaze cueing effects (for correct trials) response time (RT) was entered into a 2 (gaze validity: valid, invalid) x 6 (agent appearance: 0% - 100% human) repeated-measures ANOVA (Table 1). Additionally, in order to investigate the size of the gaze cueing effect for each agent separately, a series of paired-comparison t-tests comparing valid and invalid trials for each agent type were carried out. In order to control for multiple comparisons, a Bonferroni correction was applied to the post-hoc analysis for each agent. For illustrative purposes, the gaze cueing effect was also plotted as a difference score between invalid (i.e., predicted) and valid (i.e., unpredicted) trials for each agent (Figure 3B).

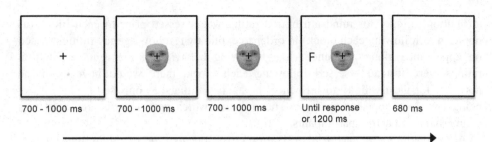

| 700 - 1000 ms | 700 - 1000 ms | 700 - 1000 ms | Until response or 1200 ms | 680 ms |

Fig. 2. Trial sequence for gaze cueing task: Each trial began with a fixation cross, followed by presentation of the agent with directed gaze (60% human agent shown here). The agent then changed gaze direction, followed by the onset of a target letter (F or T). Both the agent (with averted gaze shown here) and the target letter remained on screen until participants made a response or the time-out criterion was reached.

3   Results

Analysis of the mind attribution ratings revealed a main effect of appearance ($F(1,34) = 53.11$, $p < .001$). Follow-up paired comparisons revealed that all agents were significantly different in their mind attribution ratings (all $p < .003$, all corrected $p < .045$), with the exception of the 20% - 40% ($p = .146$) comparison and the 80% - 100% comparison ($p = .089$) such that on average mind attribution ratings increased with increased human-like appearance.

In the gaze cueing task, participants were highly accurate in their performance (M accuracy = 95.56%, SE = .56%) and responded with a mean RT of 506.69 ms (SE = 13.20). Analysis of RT data for correct trials, as a function of gaze validity and agent appearance, revealed a main effect of gaze validity such that participants responded faster for valid compared to invalid trials ($F(1,34) = 5.21$, $p = .029$). However, no main effect of agent appearance ($p = .90$) nor an interaction between agent appearance and gaze validity ($p = .266$) was identified. Despite the lack of an interaction between gaze validity and agent type, inspection of the results suggested a strong validity effect for the 60% human morph. In order to investigate this effect we carried out a series of paired-comparisons and corrected for multiple comparisons to control the chance of a false positive finding. The validity effect was found to be significant for the 60% human morph only ($t(1,34) = 3.32$, $p = .002$, Bonferroni corrected $p = .012$), such that participants responded faster for valid, compared to invalid trials; in contrast, all other agents showed no gaze cueing effect (all $p > .19$).

Table 1. Mean reaction time as a function of agent and gaze validity

Gaze Validity	0%	20%	40%	60%	80%	100%
Valid	508.76 (7.74)	505.32 (5.89)	500.34 (9.75)	491.23 (8.01)	509.15 (8.53)	505.64 (7.50)
Invalid	508.30 (6.28)	515.51 (6.32)	507.92 (6.11)	512.33 (6.23)	512.89 (6.74)	502.85 (6.01)

Note. Standard error of the mean (after removing between-subject variance) is shown in parentheses

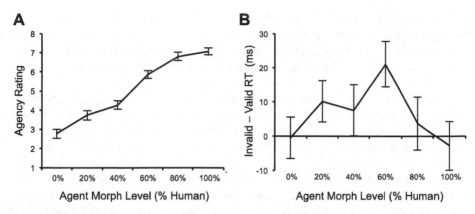

Fig. 3. A) Mind attribution ratings and gaze cueing as a function of agent type; mind attribution ratings refer to responses to the question "Do you think this agent has a mind?" on a scale of 1 ("definitely not") to 8 ("definitely yes"). B) The gaze cueing effect reflects a difference score between invalid (predicted) and valid (unpredicted) trials for each agent type. Error bars on both graphs reflect the standard error of the mean after removing between-subject variance (Cousineau, 2005).

4   Discussion

Gaze-cueing effects represent basic social attention mechanisms [4] that can provide a measure of the social relevance paid towards a given interaction partner. Given that previous studies have shown that gaze following can increase when an agent is believed to possess a mind [13, 14] and appearance can affect mind attribution beliefs [15, 16, 17, 18, 19], we investigated how appearance affects gaze following of different agents varying in degree of human-likeness by using a reversed gaze-cueing paradigm. Similar to previous research which found that the appearance of a robotic agent influences the degree to which a mind is ascribed to an agent [17, 18, 19], we found a linear relationship between agent appearance and mind attribution ratings: the more human-like an agent looks, the more mind is attributed to that agent. However, contrary to our initial hypothesis, we found that increased ratings of mind attribution did not result in an increase in top-down control of gaze following. We had expected a linear decrease of gaze cueing effects (i.e., reflecting voluntary control of attention in a counter-predictive paradigm) with increased ratings of mind attribution, but found the relationship to be best described by an inverted u-shaped pattern: No cueing effects were observed for the real agents (100% human and 100% robot), while agents of mixed human-likeness (20 - 80% human) induced gaze cueing effects in a bottom-up manner (i.e., shorter RTs at cued location despite the counter-predictive paradigm). Of those middle agents (20 - 80% human) however, only the 60% human agent was found to be statistically significant.

In terms of robotic design, these observations suggest that building a human-like robot may induce mind attribution through appearance alone, but that does not necessarily mean the agent will be ascribed the same social relevance as an actual human interaction partner. Evidence for this conclusion comes from the observation that agents that differed significantly with regard to mind perception (e.g., 100% human and 100% robot) did not differ in the degree of voluntary control they induced. Further, there was no linear decrease in gaze cueing effects with increasing mind attribution ratings.

Although there was no linear relationship observed between mind attribution ratings and top-down control of gaze following, agent appearance still seems to have the capability to modulate gaze following under certain conditions as seen with the 60% human morph. On average though, it appears that bottom-up (i.e., orienting to where the face is looking) and top-down (i.e., orienting to where the target is likely going to appear) influences cancel each other out. One possible explanation could be that participants do not naturally mentalize with an agent if the task does not explicitly require it. This would be in line with studies showing that between agent behavior and appearance, behavior is more influential in observers rating a mechanistic agent as humanlike [21]. One way to test this potential confound would be to integrate gaze following into a more complex social interaction that would naturally require mentalizing and thus increase the chances to find a linear relationship between mind attribution and gaze following if it exists.

An alternative explanation is that agents that are hard to classify as either human or robot pose extra cognitive demands on participants, making less resources available for top-down modulation. Evidence for this assumption comes from the fact that top-down modulation seems to be strongest in the 100% human and 100% robot conditions, where bottom-up and top-down effects cancel each other out completely. Cognitively, both of these agents fit within existing *schemas* [22, 23], or mental models, so it is easier to process their features, but that is not true for the mixed morphology agents. The additional effort it takes to process the mixed agents explains why a weaker top-down modulation is seen: having already exhausted resources for processing the image, participants reflexively follow the agent's gaze and fail to recognize, or at least apply, the knowledge about the counter-predictivity of the cue. In order to best address whether cognitive load is actually inhibiting mind perception and minimizing its effects on social attention, future studies need to investigate the influence of cognitive load on gaze following.

Investigating how cognitive load is affected by the physical design of a robotic agent will allow roboticists to leverage appearance for a more naturalistic HRI that can subsist under various task duress. Specifically, if our assumption regarding cognitive load and agent processing is true, then under high cognitive load a robotic agent may struggle to induce mind perception within an observer no matter how humanlike it appears which could hinder task performance. Depending on the task, it may be more important for robotic design to focus on agent features that are easily identifiable and fall within pre-existing agent stereotypes rather than simply increasing human-like appearance. Identifying the interaction between cognitive resources and mind attribution will therefore help refine the physical design requirements social robots will need to meet to ensure positive HRI performance under varying levels of task load, saving roboticists both time and money in developing the next generation of social robots.

References

1. Frith, C.D., Frith, U.: The neural basis of mentalizing. Neuron **50**(4), 531–534 (2006)
2. Graham, R., Friesen, C., Fichtenholtz, H.M., LaBar, K.S.: Modulation of reflexive orienting to gaze direction by facial expressions. Vis. Cogn. **18**(3), 331–368 (2010)
3. Azarian, B., Esser, E.G., Peterson, M.S.: Watch out! Directional threat-related postures cue attention and the eyes. Cognition & Emotion 1-9 (2015)
4. Friesen, C.K., Kingstone, A.: The eyes have it! Reflexive orienting is triggered by nonpredictive gaze. Psychon. B. Rev. **5**(3), 490–495 (1998)
5. Posner, M.I.: Orienting of attention. Q. J. Exp. Psychol. **32**(1), 3–25 (1980)
6. Kawai, N.: Attentional shift by eye gaze requires joint attention: Eye gaze cues are unique to shift attention. Jpn. Psychol. Res. **53**(3), 292–301 (2011)
7. Ristic, J., Kingstone, A.: Taking control of reflexive social attention. Cognition **94**(3), B55–B65 (2005)
8. Teufel, C., Alexis, D.M., Clayton, N.S., Davis, G.: Mental-state attribution drives rapid, reflexive gaze following. Atten. Percept. Psychophys. **72**(3), 695–705 (2010)
9. Teufel, C., Alexis, D.M., Todd, H., Lawrance-Owen, A.J., Clayton, N.S., Davis, G.: Social cognition modulates the sensory coding of observed gaze direction. Curr. Biol. **19**(15), 1274–1277 (2009)
10. Jonides, J., Irwin, D.E.: Capturing attention. Cognition **10**(1), 145–150 (1981)
11. Müller, H.J., Rabbitt, P.M.: Reflexive and voluntary orienting of visual attention: time course of activation and resistance to interruption. J. Exp. Psychol.-Hum. Percept. Perform. **15**(2), 315 (1989)
12. Frith, U., Frith, C.D.: Development and neurophysiology of mentalizing. Philos. Trans. R. Soc. Lond. B. Biol. Sci. **358**(1431), 459–473 (2003)
13. Wiese, E., Wykowska, A., Zwickel, J., Müller, H.J.: I see what you mean: how attentional selection is shaped by ascribing intentions to others. PLOS ONE **7**(9), e45391 (2012)
14. Wykowska, A., Wiese, E., Prosser, A., Müller, H.J.: Beliefs about the minds of others influence how we process sensory information. PLOS ONE **9**(4), e94339 (2014)
15. Admoni, H., Bank, C., Tan, J., Toneva, M., Scassellati, B.: Robot gaze does not reflexively cue human attention. In: Proceedings of the 33rd Annual Conference of the Cognitive Science Society, pp. 1983-1988, Boston (2011)
16. Fong, T., Nourbakhsh, I., Dautenhahn, K.: A survey of socially interactive robots. Robot. Auton. Syst. **42**, 143–166 (2003)
17. Hackel, L.M., Looser, C.E., Van Bavel, J.J.: Group membership alters the threshold for mind perception: The role of social identity, collective identification, and intergroup threat. J. Exp. Soc. Psychol. **52**, 15–23 (2014)
18. Martini, M.C., Gonzalez, C.A., Wiese, E.: Seeing minds in others – Can agents with robotic appearance have human-like preferences?. In preparation
19. Powers, K.E., Worsham, A.L., Freeman, J.B., Wheatley, T., Heatherton, T.F.: Social connection modulates perceptions of animacy. Psychol. Sci. **25**(10), 1943–1948 (2014)
20. Lundqvist, D., Flykt, A., Öhman, A.: The Karolinska directed emotional faces (KDEF). CD ROM from Department of Clinical Neuroscience, Psychology section, Karolinska Institutet, pp. 91-630 (1998)
21. Park, E., Kong, H., Lim, H.T., Lee, J., You, S., del Pobil, A.P.: The effect of robot's behavior vs. appearance on communication with humans. In: Proceedings of the 6th International Conference on Human-Robot Interaction, pp. 219-220. ACM, March 2011
22. Bartlett, F.C.: Remembering. Cambridge University Press, London (1932)
23. Shapiro, A.M.: The relationship between prior knowledge and interactive overviews during hypermedia-aided learning. J. Educ. Comput. Res. **20**(22), 143–167 (1999)

Ava (A Social Robot): Design and Performance of a Robotic Hearing Apparatus

Ehsan Saffari[1(✉)], Ali Meghdari[1], Bahram Vazirnezhad[1,2], and Minoo Alemi[1,3]

[1] Social Robotics Laboratory (CEDRA), Sharif University of Technology, Tehran, Iran
meghdari@sharif.edu
[2] Computational Linguistics Laboratory (LLC), Sharif University of Technology, Tehran, Iran
[3] Islamic Azad University – Tehran West Branch, Tehran, Iran

Abstract. Socially cognitive robots are supposed to communicate and interact with humans and other robots in the most natural way. Listeners turn their heads to-ward speakers to enhance communicative attention; this is also an act of appreciation to the speaker. In this paper we have designed and implemented a robotic head, "Ava", which turns toward the speaker in noisy environments. Ava employs a Speech Activity Detection system which differentiates speech segments of non-speech. Then the speech segments are processed to reduce different kinds of noise levels. The speaker localization system then finds the speaker position in the azimuth plane and commands motors to turn horizontally toward the speaker in a smooth trajectory. Ava has two built-in microphones inside its ears and employs three different algorithms simultaneously for feature extraction and a two-layer perceptron neural network for localization. Ava operates real-time and updates the position even in its moving phase. Experiments show a precision of +/-5 degrees in white noise in SNR of 10 dB.

Keywords: Social robot · Human-Robot Interaction (HRI) · Speech processing · Sound source localization · Turning toward the speaker

1 Introduction

Since their invention, robots have been developed for various purposes and needs, quite similar to personal computers in their early days. With the advancement of technology and reduction of costs, it is anticipated that in the near future one of the cutting-edge technologies to be used in various social, therapeutic, cultural, and educational areas are social robots [1-5]. Social robots are designed to interact with people in the most natural way, therefore natural and smooth interaction is the most significant and challenging issue in their design. When a robot speaks to humans, sound source localization is a must, not an option as it is both an appreciation of the opposite speaker as well as a natural movement to receive better quality speech wave. A social robot needs to be able to turn its head toward the speaker in each vocal interaction to better recognize speech. This reaction is also a natural interaction which is a significant goal in HRI studies.

© Springer International Publishing Switzerland 2015
A. Tapus et al. (Eds.): ICSR 2015, LNAI 9388, pp. 440–450, 2015.
DOI: 10.1007/978-3-319-25554-5_44

HRI studies are classified as verbal and non-verbal interactive communications. The receiver system of verbal interactions or the robotic hearing apparatus may be designed for speech recognition, emotion detection, or speaker identification [6]. So these systems should be able to detect speech segments and localize sound sources for better reception of speech. In the field of humanoid robots and social robots, various methods are used for sound source localization and speech activity detection. Fig. 1 shows the place of sound source localization in the simple classification of HRI studies.

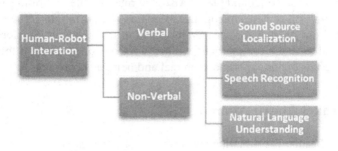

Fig. 1. Simple Classification of HRI studies.

Many studies have been conducted on sound source localization. A. Badali et al. [7] implemented a real-time sound source localization in a robotic application. Their paper compares the performance of several algorithms. Speech activity detection is essential if we want to localize the speech signal in an environment, particularly if the testing environment is a noisy one. In recent years, several statistical methods have been proposed by researchers. In S. Shafiee et al. [8], the silence segments are first removed. Then extracting some signal features helps classify the speech and non-speech segments. So, we can localize the speech as a significant sound signal by using a speech activity detection algorithm.

In robotic applications problems such as turning toward speaker, effects of the robot body on the signal spectrum, and noise of actuators should be considered in the design process. Moreover, the robot auditory system is not the only tool for speaker localization in these applications. Visual localization of the speaker's face can increase the perfor-mance and accuracy of sound source localization. K. Nakadai et al. pioneered the use of this idea in [9-10]. In the following years V.M. Trifa et al. [11], using the CB humanoid robot, and F. Alonso-Martin et al. [12], using the social robot Maggie are just some ex-amples of using visual-audition methods for speaker localization. In addition, J. Cech et al. [13] implemented a visual system for sound localization on the humanoid robot NAO. They showed that despite the hardware limitation of this robot and the high level noise of its cooling fan, their visual-audition method can properly localize the speaker. However, this method has some limitations such as environment light and an absence of any ob-stacle between the robot and the speaker. Therefore, it would be useful to improve the audition system for localizing the speaker.

In sound source localization algorithms, not only is the relative location of micro-phones effectual, but the shape of the robot auditory apparatus also has an important role. So inspired by the design of the human ear anatomy, researchers are trying to improve sound source localization by designing a human-like artificial ear for humanoid robots. In another paper by K. Nakadai and his team [14], two main problems of a sound source localization method are addressed and an artificial pinnae is designed to solve them. Y. Park et al. has conducted many studies on this idea in recent years [15-16]. In one of their last papers [17], the artificial ear of a robot equipped with 4 microphones was able to localize a single sound source with less than 10 degrees error and two simultaneous sound sources with the precision of 15 degrees.

In this paper, we presented a robotic hearing apparatus that is able to localize an active speaker in the horizontal plane. Also, the robot is able to turn toward the speaker, which enables it to better recognize speech as well as to interact naturally. In the next section, our method for speaker localization is elaborated. Section 3 describes our robot and its auditory apparatus with two microphones. Finally, the last section details the results of our experiments in real and noisy environments.

2   Speaker Localization

Speaker localization is estimating the direction of a speech signal in an environment. Generally, estimating the azimuth and the elevation direction of a sound signal is called sound source localization and the distance of a sound source from the audition system is not considered. Therefore, to localize the speaker we can use a method of sound source localization by adding speech activity detection. This means that a speech activity detection algorithm is first used to choose the speech segments. Then, these segments can be used for sound source localization.

Many algorithms have been proposed for sound source localization in recent years. The three main features used are:

- Inter-Channel Time Difference (ICTD): According to the position of each microphone in the audition system, time of receiving a sound signal is different for each of them. So, with response to the sound source location, the time difference between each two microphones is unique. Fig. 2 shows the different distances that sound signal should traverse for each microphone. However, this feature is limited to a specific frequency of signals considering the distant between the two microphones. This means that the signal frequency should not be higher than the frequency of more than one period of signal perch between the two microphones. For example, as shown in Fig. 3(a), according to the time difference (Δt), f_0 is the maximum Available frequency. If this frequency doubled, the estimated time difference will be: $\frac{\Delta t}{2}$, Fig. 3(b). On the other hand, a 2D localization of a 0 to 180 degrees direction is expected when using this feature. In this method, not only noise from the environment but also reverberations of the main signal would improperly confuse the estimation. Many methods have been used to calculate the time difference of two signals, one of the most common is GCC-PHAT, that is:

$$\hat{G}_{PHAT}(f) = \frac{x_i(f)[x_j(f)]^*}{\left| x_i(f)[x_j(f)]^* \right|} \qquad (1)$$

where $X_i(f)$ and $X_j(f)$ are the Fourier transforms of the two signals and $[\,]^*$ denote the complex conjugate. So the time difference of these two signals, $\hat{d}_{PHAT}$, is estimated as:

$$\hat{d}_{PHAT}(i,j) = {}^{argmax}_{\;\;\;d}(\hat{R}_{PHAT}(d)) \tag{2}$$

where $\hat{R}_{PHAT}(d)$ is the inverse Fourier transform of equation (1).

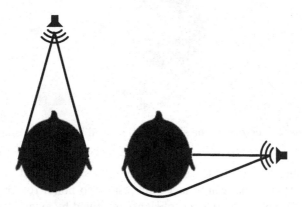

Fig. 2. Using time delay of arrival for sound source localization.

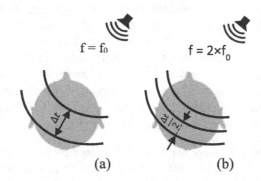

Fig. 3. The effect of high frequency signal in the ICTD method.

- Inter-Channel Level Difference (ICLD): The sound signal will be damped by air in the environment. Therefore, as the distance of the microphone increases from the sound source it receives the sound at lower levels. Thereupon, in an audition system, the sound level difference of the microphone further from the sound source can be a feature of sound localization. Excluding the distance, an obstacle like the body of the audition system would have an effect on level reduction of a sound signal. This effect is called acoustic shadow, Fig. 4. This feature is more suitable for high frequency signals as the higher the frequency the more the signal level is reduced. Just like the ICTD, ICLD is limited to a 0 to 180 degrees plane for localization. This feature can be calculated by:

$$L(i,j) = \frac{\sum(s_i(t) - s_j(t))^2}{Max(\sum s_i^2(t), \sum s_j^2(t))} \tag{3}$$

where $s_i(t)$ and $s_j(t)$ are the sound signals recorded by ith and jth microphones.

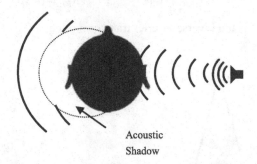

Acoustic
Shadow

Fig. 4. Decreasing the signal level caused by acoustic shadow.

- Inter-Channel Spectral Difference (ICSD): Each sound signal has a specific spectral characteristic that can change with respect to noise, reverberations, etc. So, detecting the spectral difference of signals received by an audition system would be a feature of sound localization. Although extracting this feature is more difficult than the previous two, sound source direction can be estimated with just one microphone. Also, there is no limitation for this method. But, as we said, it is very difficult to use this feature for sound source localization. For this reason, it is not usually used in common applications. In order to use this feature, a head-related transfer function is needed. An artificial ear for the auditory system would affect the spectral characteristics of sound signals. Changing levels of the sound signal at a specific frequency band is discussed in [14]. In this paper, this change of signal spectrum is used for forward-back estimation.

In this paper, all three of these features are extracted for sound source localization in the azimuth plane. A band pass filter is used before feature extraction and the features are sent to a neural network to estimate the speech signal direction. The neural network is a two-layer perceptron that has 10 neuron in its hidden layer.

For speaker localization in natural environments, preprocessing is needed before using our sound direction estimator. Speech activity detection is considered to detect the speech in a sound signal. By using the method proposed in [8], some features of a signal based on auto-correlation, MFCC, time length and energy of each segment are extracted. Then the classification between speech and non-speech segments is done by statistical methods, like machine learning, allowing just the speech segments to be sent for sound source localization. Speech activity detection is done for each 1 second of signal. Therefore, the system has a 1 second delay in each sound source localization. Also, a speech enhancement filter is added as preprocessing to increase the performance of sound source localization. Fig. 5 shows the flowchart of our method for speaker localization.

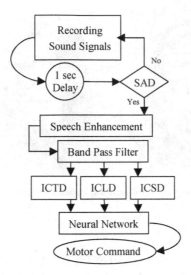

Fig. 5. The flowchart of speaker localization method

At the final stage, after estimating the speech signal location, the robot should turn toward the speaker. Therefore, relative angle of speaker to the audition system would be sent for path planning to command the motor. To start a natural vocal interaction, natural turning toward speaker is considered. In this paper, a smooth and natural path is designed. Velocity and acceleration feedback can help smooth the trajectory in each robot reaction.

3   "Ava", A Social Robot

In order to implement the speaker localization method, we need a robotic hearing apparatus to receive the sound signals and turn the head toward the speaker. To study and investigate speaker localization for the purpose of improving natural interaction of socially cognitive robots, a robotics hearing system was designed and assembled in a mannequin head. This social robot is called "Āvā", a feminine given name in the Persian language, meaning voice, sound and phoneme. Ava is a humanoid head that has a hearing system and an active rotational mechanism. A microphone is embedded in each ear of this social robot. Signals received by these microphones are sent to a PC with a two-channel sound card. Once the PC processed the signals, location information are sent to the actuator of the rotational mechanism of Ava. This robot has a 2-DOF mechanism with two actuators for rotating in the azimuth and elevation planes. In current research, just the horizontal rotation is considered.

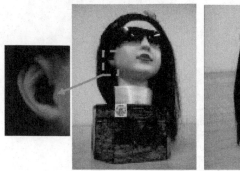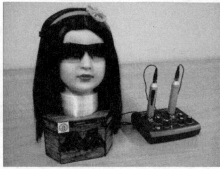

Fig. 6. The social robot, "Ava" and the embedded microphone in its ear.

The main parts of Ava are:

- The Head
- Two servo motors, AX-12A model of DYNAMIXEL Co.
- A two-channel sound card, M-Track model of M-AUDIO Co.
- Two condenser microphones, QL5 model of SAMSON Co.

Ava is able to record the sound signal with a sampling frequency of 48 kHz and a sample rate of 24 bit/sample. Its condenser microphones have high sensitivity with a 114 dB dynamic range. Also, Ava's motors have 0.29 degree resolution with 59 rpm no load speed at 12 volts dc power.

So, this social robot will be able to detect and localize any speech signal and then turns toward the speaker. This reaction would be a natural interaction of a social robot when starting up a conversation.

4   Experiments and Discussion

Since, we want to detect and localize just the speech signals, speech activity detection was used. Also, we used a speech enhancement algorithm to reduce the background noise of the speech segments. Fig. 7(a) shows the main signal recorded by one of the microphones. There is a speech part, calling "Āvā", and a beep sound as the non-speech part. As shown in Fig. 7(b), the enhanced speech segment is the only part remaining after preprocessing. In real time, the same processing is done for each microphone in each 1 second segment.

Speech segments are only processed for feature extraction of sound source localization. Using the three features discussed in section 2, we trained the neural network to localize the speech. This three features are:

- ICTD: Time difference between the two microphones signals for the frequency band of 300 to 1800 Hz.
- ICLD: Level difference between the two microphones signals for the frequency band of 300 to 4000 Hz.
- ICSD: Level average of each microphone signal for the frequency band of 1800 to 4000 Hz.

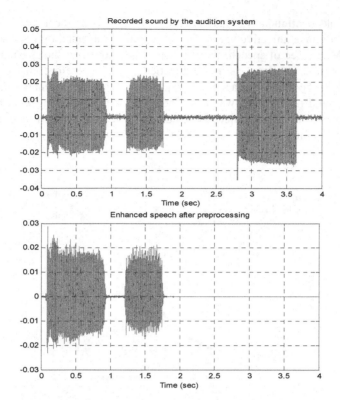

Fig. 7. Sound signal recorded by Ava microphone, (a) before (b) after preprocessing.

Experiments showed that the maximum available frequency for ICTD was 1826 Hz. Therefore, a band pass filter was used before calculating these features. Also, according to the Ava ear, the high frequency part of the signal level was affected by the ear pinnae.

Specific sound signals have been recorded at every 5 degrees for 360 degrees around the robot to provide enough data to train the neural network of the audition system.

Not only do speech activity detection and speech enhancement reduce environmental noise, but also using a neural network for sound source localization enables the system to perform better in noisy environments. Also, extracting these three features increases the precision of direction estimation.

Experiments shows that this audition system has a precision of +/-5 degrees in environments where white noise causes the SNR to 10 dB. Therefore, Ava would have a good performance from the user perspective.

With respect to the estimated direction of sound source, Ava should turn to face the speaker. For each angle of turning, a smooth path have been designed to appear more natural. After each angle estimation, the robot has 1 second time for turning. After that, the estimation will be updated, if the next 1 second segment is also speech. This means that the audition system is able to update the estimation every 1 second, if the speaker speaks continuously. This increases the accuracy of localization and Ava can turn to exactly face an active speaker. The trajectory traversed by the robot is shown in Fig. 8. After 1 sec of speech, the audition system localizes the speaker. In this example, the first estimation was in the direction of 90°. Therefore, the robot starts up turning toward the speaker in a smooth trajectory. In the next estimation, the direction the robot should turn toward was 85°. Therefore, the robot changes its trajectory to reach 85°. Since the second trajectory has the same velocity as the first one, it remains smooth and natural.

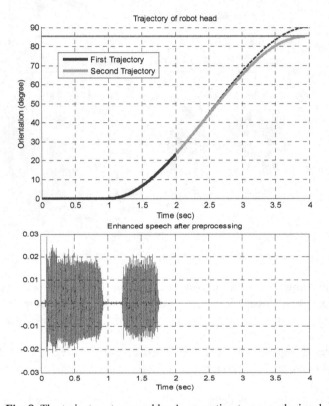

Fig. 8. The trajectory traversed by Ava reacting to a speech signal.

5   Conclusions

Social robots need to comply with behavioral norms like a natural partner for interacting and communicating with people within the society. In this paper, "Ava", the social robot, was designed and programed to horizontally localize speech in noisy environments. Firstly, a speech activity detection algorithm was performed to decide if each sound frame was speech or non-speech. Then the identified speech frame was enhanced to reduce noise in the environment. Finally, by extracting the difference in time, level and spectrum of two microphones signals, speech was localized in the horizontal plane. As Ava is an unknown audition system, we used a two-layer perceptron neural network for sound direction estimation. Also, the robot audition system was able to update the estimation even while the head was turning. We found that Ava could turn toward any speaker in 10 dB noisy environments with the precision of +/-5 degrees. No reaction to non-speech sounds in the environment, high performance of speech localization and natural turning toward speaker were all executed by Ava.

References

1. Meghdari, A, Alemi, M, Ghazisaedy, M., Taheri, A.R., Karimian, A., Zandvakili, M.: Applying robots as teaching assistant in EFL classes at iranian middle-schools, CD. In: Proc. Int. Conf. on Education & Modern Edu. Tech. (EMET 2013), Venice, Italy, September 28-30 (2013)
2. Alemi, M., Meghdari, A., Ghazisaedy, M.: The Impact of Social Robotics on L2 Learners' Anxiety and Attitude in English Vocabulary Acquisition. Int. Journal of Social Robotics (2015)
3. Alemi, M., Ghanbarzadeh, A., Meghdari, A., Moghaddam, L.J.: Clinical Application of a Humanoid Robot in Pediatric Cancer Interventions. Int. Journal of Social Robotics (2015)
4. Alemi, M., Meghdari, A., Ghazisaedy, M.: Employing Humanoid Robots for Teaching English Language in Iranian Junior High-Schools. Int. Journal of Humanoid Robotics **11**(3) (2014)
5. Taheri, A.R., Alemi, M., Meghdari, M., Pouretemad, H.R., Holderread S.L.: Clinical application of a humanoid robot in playing imitation games for autistic children in Iran. In: Proc. of the 14th Int. Educational Technology Conf. (IETC), Chicago, USA, September (2014)
6. Mavridis, N.: A review of verbal and non-verbal human–robot interactive communication. Robotics and Autonomous Systems **63**, 22–35 (2014)
7. Badali, A., Valin, J.M., Michaud, F., Aarabi, P.: Evaluating Real-time Audio Localization Algorithms for Artificial Audition in Robotics, pp. 2033–2038 (2008)
8. Shafiee, S., Almasganj, F., Vazirnezhad, B., Jafari, A.: A two-stage speech activity detection system considering fractal aspects of prosody. Pattern Recognition Letters **31**(9), 936–948 (2010). ISSN 0167-8655
9. Nakadai, K., Hidai, K., Okuno, H.G., Kitano, H.: Real-time speaker localization and speech separation by audio-visual integration. In: Int. Proc. of IEEE-RAS Intern'l Conf. on Robotics and Automation, pp. 1043–1049 (2001)
10. Nakadai, K., Okuno, H.G., Kitano, H.: Real-time sound source localization and separation for robot audition. In: Proc. IEEE International Conference on Spoken Language Processing, pp. 193–196 (2001)
11. Trifa, V.M., Koene, A., Mor'en, J., Cheng, G.: Real-time acoustic source localization in noisy environments for human-robot multimodal interaction. presented at the Robot and Human interactive Communication, Jeju, pp. 393–398 (2007)
12. Alonso-Martin, F., Gorostiza, J.F., Malfaz, M., Salichs, M.A.: User Localization during Human-Robot Interaction. Sensors, 9913–9935 (2012)

13. Cech, J., Mittal, R., Deleforge, A., Sanchez-Riera, J., Alameda-Pineda, X., Horaud, R.: Active-speaker detection and localization with microphones and cameras embedded into a robotic head. In: International Conference on Humanoid Robots, pp. 203–210 (2013)

14. Kim, U.H., Nakadai, K., Okuno, H.G.: Improved sound source localization and front-back disambiguation for humanoid robots with two ears. In: Applied Artificial Intelligence, pp. 282–291 (2013)

15. Park, Y., Hwang, S.: Artificial robot ear design for sound direction estimation. presented at the Robot & Human Interactive Communication, Jeju, pp. 405–409 (2007)

16. Hwang, S., Park, Y., Park, Y.S.: Sound direction estimation using an artificial ear for robots. Robotics and Autonomous Systems, 208–217 (2011)

17. Lee, S., Park, Y., Choi, J.S.: Estimation of multiple sound source directions using artificial robot ears. Applied Acoustics 77, 49–58 (2014)

A Particle-Filter Approach for Active Perception in Networked Robot Systems

João Messias[1]($\boxtimes$), José J. Acevedo[2], Jesus Capitan[3], Luis Merino[4],
Rodrigo Ventura[2], and Pedro U. Lima[2]

[1] Informatics Institute, University of Amsterdam, Amsterdam, The Netherlands
J.V.TeixeiradeSousaMessias@uva.nl
[2] Institute for Systems and Robotics, Instituto Superior Técnico,
Universidade de Lisboa, Lisboa, Portugal
{jacevedo,rodrigo.ventura,pal}@isr.tecnico.ulisboa.pt
[3] University of Seville, Seville, Spain
jcapitan@us.es
[4] University Pablo de Olavide, Seville, Spain
lmercab@upo.es

Abstract. The presence of children in a social assistive robotics context is particularly challenging for perception, mainly, in the task of locating them using inherently uncertain sensor data. This paper proposes a method for active perception with the goal of finding one target, e.g., a child wearing a RFID tag. This method is based on a particle-filter modeling a probability distribution of the position of the child. Negative measurements are used to update this probability distribution and an information-theoretic approach to determine optimal robot trajectories that maximize information gain while surveying the environment. We present preliminary results, in a real robot, to evaluate the approach.

1 Introduction

The MOnarCH project[1] focuses on social robotics in a pediatric infirmary using networked robots to interact with children, staff or visitors. This addresses explicitly the *active perception* problem as an important issue in the context of social assistive robotics for children. The issue arises when controlling a robot or group of robots so as to gather information, based on the robots sensors, that may be required by other agents (robots or medical staff). This paper considers that children carry a RFID tag and poses the active perception problem using just the RFID sensor of the robot and applied to finding a child whose location is previously unknown to the networked robot system (NRS). This is useful because robots can help to find lost or hidden children or play a hide-and-seek game with them.

The problem of active perception is that of controlling one or more mobile robots so as to maximize a given measure of information regarding a set of

---

[1] MOnarCH Multi-Robot Cognitive Systems Operating in Hospitals (FP7-ICT-2011-9-601033).

© Springer International Publishing Switzerland 2015
A. Tapus et al. (Eds.): ICSR 2015, LNAI 9388, pp. 451–460, 2015.
DOI: 10.1007/978-3-319-25554-5_45

features of the environment. The earliest forms of active perception focused on improving the localization of a mobile robot by controlling its motion [5]. More recently, this problem has been studied for target-tracking applications, where the uncertainty over the location of a moving target should be minimized [11]. It has also been generalized to cooperative multi-robot applications [1,2,4], wherein it is referred to as the *active cooperative perception* (ACP) problem.

Existing approaches to the ACP problem have formalized it in different ways under the scope of various overlapping fields of study, most notably (but not exclusively) those of Robotics, Sensor Fusion, and AI [1,4,6,11]. In this family of differing approaches, the most common drawbacks are related to the scalability and generality of the proposed methods. The most accurate ACP methods depend on a careful modeling of the multi-robot system and are therefore only applicable to very specific and typically small-scale scenarios; and those that attempt to leverage larger-scale formalizations (such as those stemming from the field of AI) are forced to approximate the system and its behavior roughly, due to the complexity of the associated solution methods.

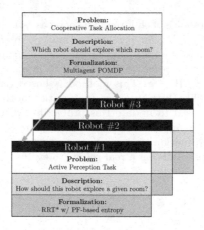

Fig. 1. A summary of the proposed ACP approach.

This paper introduces a whole ACP approach to be applied by the MOnarCH NRS, but just develops and presents results about the single robot active perception case. The ACP approach proposed in this work does not attempt to aim at optimality, but instead at being sufficiently general and easy to deploy such that it can be applied over different domains (with different environments, robots, and perceptual objectives), with minimal effort. To achieve this goal, the ACP problem is decoupled into a hierarchy of distinct subproblems (Figure 1): an allocation problem to assign a room or non-overlapping area to each robot, and a motion planning problem for each robot to decide how to explore its assigned room or area in the most efficient way.

This decomposition is also strongly motivated by the nature of the MOnarCH environment and its associated safety requirements. Specifically, in the typical use cases of the MOnarCH NRS, there should be at most one robot in a given location of interest at any time, in order to minimize the interference to the daily operations of the staff in the Instituto Portugues de Oncologia de Lisboa (IPOL).

Therefore, regarding to the proposed decoupled and hierarchic approach, at the top of this hierarchy lies a problem of multi-robot coordination: in a large environment with multiple locations of interest, what robot or sub-group of robots should be allocated to explore each of those locations? This problem has not be directly addressed in this paper, but decision-theoretic planning methods, such as Multiagent POMDPs [10], are well-suited to deal with this type of problem, although would not scale to solve the whole ACP problem for complex multi-robot domains.

At a lower level of decision-making, each robot should be able to decide how to carry out its exploration task in the most efficient way. This paper is focused on this issue, which is fundamentally a problem of motion control, which can be efficiently solved through sampling-based methods such as RRTs [3,7]. This paper proposes a novelty approach using a Particle Filter to describe the uncertainty over the search target, and subsequently use recently proposed information-gain metrics based on the entropy of the associated non-parametric distribution. These metrics are then used to guide the probabilistic motion planning.

This document is organized as follows: in Section 2, the probabilistic planning method to generate information gathering paths is described from a theoretical standpoint, in Section 3, the preliminary results are shown, and Section 4 wraps up the paper with conclusions and a discussion on future work.

2   Active Perception via Probabilistic Planning of Information-Gathering Paths

In this paper, we focus on the problem of planning the motion of a single mobile robot in order to search for the position of a possibly moving target.

As inputs, we are given the following:

1. The current pose of the robot;
2. A probability distribution over the target's position in the configuration space of the robot. This distribution is possibly non-parametric and typically uniform;
3. A description of the probabilistic model of the sensor of the robot that is capable of detecting the target (*e.g.* a model of the RFID reader). This description is possibly non-parametric.

The active (non-cooperative) perception problem can itself be decoupled into two interdependent problems: the problem of estimating the position of the target given the motion of the robot, and that of planning the motion of the robot in order to improve the estimation of the target's position (Figure 2).

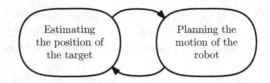

Fig. 2. The two sub-problems of Active Perception.

In the following subsections, we focus on each of these subproblems.

2.1 Estimating the Position of a Moving Target

Consider the task of estimating the position of an RFID tag carrier in an arbitrary (planar) environment, given logical-valued readings from an RFID reader mounted on a mobile robot[2]. More formally, if $p_t = [x_{p,t}\ y_{p,t}]^T$ is the position of the tag carrier at time t, $q_t = [x_{r,t}\ y_{r,t}\ \theta_t]^T$ represents the pose of the robot at that time, and $z_t \in \{False, True\}$ is the most recent RFID reading ($True$ means a positive detection and $False$ a negative one), then our objective is to minimize a given metric of confidence over the posterior probability of the target's position given the history of robot poses and measurements, $i.e.$

$$\Pr(p_t \mid q_0, z_0, q_1, z_1, \ldots, q_t, z_t)$$

This is a typical estimation / filtering problem, for which there are many applicable and well-known solutions ($e.g.$ EKF, Markov localization, Occupancy-Grid methods, etc.). The "twist" in this estimation problem, as opposed to most localization and tracking applications, is that most of the readings produced by the sensor of the robot are negative w.r.t. the position of the agent. That is, if $z_t = False$, then there is a high probability that the carrier is **not** within the sensor range of the robot. Once the robot receives a positive RFID reading (or a sufficient number of positive readings to establish with a certain confidence level that the carrier is present), then the robot has succeeded in finding the carrier, and the estimation process can be terminated.

In keeping with the motivation of trying to have a general and easy-to-use solution, we have opted to approach this estimation problem using a Particle Filter (PF) to represent the uncertain position of the target.

The advantages of using a PF for this estimation problem, over its alternative methods, are the following:

- Due to the influence of negative sensor readings during the search process, the target posterior is most likely very difficult to approximate with a parametric (or kernel based) representation. The non-parametric belief representation that is characteristic of PF-based methods is ideal for this application;

---

[2] In this context, we are not concerned with the identity of the RFID carrier, so any identifying information contained in the RFID tag is extraneous for this problem. However, this approach could trivially be used to find **one particular** RFID tag.

- The probabilistic model of the sensor used in the search process can also be difficult to describe analytically. In the particular case of an RFID reader, the conditions for radiopropagation induced by the environment can have a complex influence on the respective readings. A PF allows for arbitrarily complex sensor models;
- The motion of the target is unknown and possibly also difficult to model in closed-form. A general PF does not make any assumptions regarding the form of this model, and random walk models could be used. In this case the standard deviation of the child movement σ is used as argument and is related to the child motion speed.

A generic PF was used for this estimation problem, following the standard algorithm which is here replicated for clarity (Figure 3). Refer to [12] for more background and details on generic particle filters.

B : Set of N particles $\langle p_x^i, p_y^i \rangle$
w : Weight vector, $w_i \in [0, 1]^N$
if B or w not given **then**
    $B \leftarrow$ Sample N valid positions $s = \langle p_x, p_y \rangle$.
    $w_i \leftarrow 1/N$ for $i = 1, \ldots, N$
end if
$t \leftarrow 0$
while search not over **do**
    $B \leftarrow predict(B)$
    $z \leftarrow$ RFID reading at time t
    $w \leftarrow update(B, z)$
    if $resample\_condition(B, w)$ is $True$ **then**
        $resample(B, w)$
    end if
    $t \leftarrow t + 1$
end while

Fig. 3. The (standard) PF algorithm used for the target position estimation.

2.2 Planning the Motion of the Robot

Given a non-parametric belief over the position of the target, which is continuously provided by the aforementioned particle filter during the execution of the ACP task, the objective of the motion planning module is to plan a path for the mobile robot over which its sensors can provide maximal information. In other words, we intend to find a sequence of poses such that the (predicted) entropy of the particle filter for target detection is minimized.

More formally, let $\mathbf{q}_{0:t} = \langle q_0, q_1, \ldots, q_t \rangle$ represent the history of poses of the robot between steps 0 and t for some $t > 0$. Analogously, let $\mathbf{z}_{0:t} = \langle z_0, z_1, \ldots, z_t \rangle$

represent the history of sensor readings. Then, our objective is to find a sequence of K *predicted* robot poses $\hat{\mathbf{q}}^*_{t+1:t+1+K}$, for some $K > 0$, such that:

$$\hat{\mathbf{q}}^*_{t+1:t+1+K} = \underset{\langle \hat{q}_{t+1},...,\hat{q}_{t+K} \rangle}{\arg\min} \; H\left(\Pr(p_{t+1+K} \mid \mathbf{q}_{0:t}, \mathbf{z}_{0:t}, \hat{\mathbf{q}}_{t+1:t+1+K})\right) \qquad (1)$$

A solution to this problem can be viewed as a form of receding-horizon control (in the field of Control Theory) or online planning (in the field of Decision Theory). We have opted to approach this problem via probabilistic sampling techniques based on the well-known Rapidly-exploring Random Tree (RRT) method [8]. This family of methods is well-suited to this problem for the following reasons:

- They are computationally efficient *anytime* algorithms (which means that after a fixed amount of time they can return the best solution found so far);
- They can handle complex kino-dynamic constraints;
- Variants of the RRT method that attempt to minimize cost functionals, such as RRT^* [7] and T-RRT, can be easily adapted to minimize the entropy of our target particle filter.

In contrast with most RRT-based methods, however, our motion planning task lacks a concretely defined goal position. Our objective is instead to find a set of poses that minimizes the entropy (1). In this sense, our approach is closest to that of [6], who have formulated information-gathering RRT variants (namely Reward-Information Gathering Trees (RIG-Tree) and Graphs (RIG-Graph)). However, the latter methods consider additional constraints over the motion planning problem (over spent energy or time) which are not necessary in our formulation. For our purposes, a simpler approach is to take a fixed-depth RRT^* or T-RRT and evaluate the cost functional (1) therein.

Although a closed-form expression for the entropy of the probability density function over the target's position is not feasible, there are suitable approximations of this measure, specifically for particle filters as in [3]. Particularly, in this work, the approximation defined in [9] based on the entropy of a Gaussian Mixture will be used.

$$H(Pr(p_t|q_{0:t}, z_{0,t})) \leq \sum_i w_i(-log(w_i) + 0.5log(2\pi e)^2 \Sigma^4) \qquad (2)$$

where w_i is the weight of the i−th particle of the PF, and Σ is the standard deviation for the person movement.

Note that, to calculate the entropy even with these approximations, it would be necessary to predict the state of the PF at each possible future pose. For a probabilistic sampling method such as an RRT, this would mean that a copy of the particle set of the original PF needs to be maintained at each node while expanding the search tree, so as to describe the predicted state of the PF if the robot *would* follow the path up to that node. Furthermore, that PF would need to be updated (and re-sampled) according to each possible future measurement along the path. Although this is conceptually possible, for computational reasons, we assume the following simplifications:

1. The possible future paths in the RRT always assume negative RFID measurements up to a given node. Since there is only one possible observation at each node, the tree does not need to branch out exponentially according to each possible RFID reading.
2. The motion of the robot is sufficiently fast that the motion model of the particles does not need to be considered while predicting the future state of the PF. This means that each node of the RRT do not actually have to store an instance of the full particle set (all the particles and their weights). Since, it is not required to re-sample, the positions of all the particles remain the same and each node of the RRT only need to keep their weights. By avoiding the prediction step, we also side-step the need to perform re-sampling, which is computationally expensive.

The results of our method will be discussed in Section 3.

2.3   Extension for Multiple Robots

The proposed ACP approach is such that the coordination between multiple robots would be handled at a different scope than the motion planning. However, conceptually there is nothing that precludes the extension of the above method for single-robot active perception directly to the multi-robot domain, as long as those robots can communicate freely. Since there is only one PF describing the target location, if multiple robots share the respective information, then it should be possible to use the above method to plan the paths of multiple robots simultaneously.

3   Experimental Results

The presented system has been experimentally tested in the robotics lab of the Instituto Superior Técnico in Lisbon, using the omni-directional MOnarCH robot (see Figure 4). During the experiment, the robot was in charge of finding with as much certainty as possible the location of a child carrying a RFID tag, and it had no previous information about that location. Moreover, the robot moved at 0.8 m/s and used an RFID reader to detect the person.

The right image in Figure 4 shows the more relevant information provided by the system whenever the robot requires to plan a new path to explore the environment. The PF provides the particle cloud representing the position of the person (black small dots), and the RRT* motion planner provides a connected tree. This is shown as a connected tree of green-black dots, where the green intensity corresponds to the predicted entropy at that particular pose (therefore, darker nodes are more informative).

In Figure 5, a sequence of frames taken from rviz show visually how the robot behaves during the experiment. They show the robot trajectory (blue line),

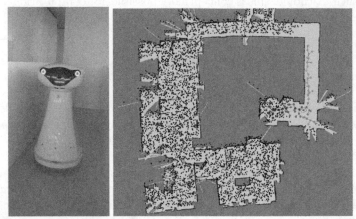

Fig. 4. Left, omni-directional MOnarCH robot used during the experiment. Right, rviz visualization of an experiment. A video from the experiment is shown in *http://youtu.be/fMB5PWQtaUI*

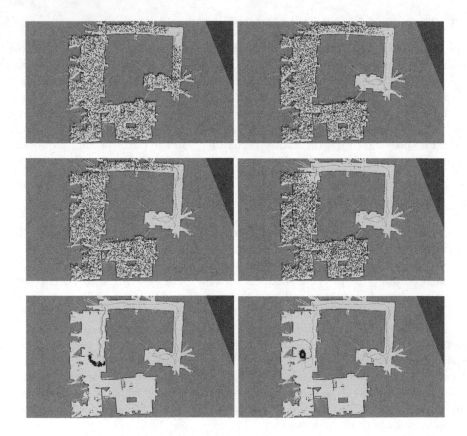

Fig. 5. Sequence of frames from the experiment rviz visualization.

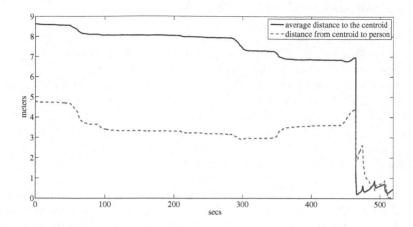

Fig. 6. This graph show the evolution during the experiment of the average weighted distance from each particle to the centroid of the particle (blue solid line) and the distance between this centroid and the actual person location (red dashed line).

the last best path computed according to the proposed entropy-based RRT* (black line), the particles' distribution from the PF (black points), the particles' centroid (big red point) and the actual person location (big blue point).

While the robot moves following a computed path and getting negative RFID readings, it clears the target particles from the map. Since the robot cannot cover all particles within its sensor range and the particles are moving (based on the predict method of the PF algorithm and the proposed standard deviation of the child movement σ), these clean areas could be filled out by the nearer particles and the robot could tend to visit these areas again. So, the areas will not be fully ruled out once they have been visited, it will depend on the environment shape, the non-covered particles and the children motion speed (related to σ). As the robot has not previous information about the target location, it can start moving toward a wrong area. However, once it has explored that area without detecting any RFID tag, the robot is able to generate a new path to get back to the initial position and starts exploring other areas of the map with more uncertainty. Once the robot gets a positive RFID reading, it clears the target particles from the whole map except for the region defined by its sensor model. Following motion plans are addressed to explore again and again the same region in order to pinpointing the estimated person position.

Finally, wit respect to the features shown Figure 6, lower values imply greater certainty about the person location.

4   Conclusions and Future Work

The preliminary experimental results validate the proposed approach for the active perception based on a particle filter and an entropy-based motion planner, applied to child searching. It shows that, even having only negative measurements, the person estimator allows the system to gain certainty about the child location. Obviously, assuming complex environment with loops and children moving faster than the robots, the robots could not find the children ever. Finally, the main future developments will be directed to extend the work to multiple robots in active cooperative perception tasks, based on the dynamic area allocation and using the particle distribution.

Acknowledgments. This work is funded by the MOnarCH European project (FP7-ICT-2011-9-601033) and the Junta de Andalucia through the project PAIS-MultiRobot (TIC-7390). J.J. Acevedo, R. Ventura and P.U. Lima were also partially supported by the LARSyS (FCT [UID/EEA/50009/2013]).

References

1. Ahmad, A., Nascimento, T., Conceicao, A.G.S., Moreira, A.P., Lima, P.: Perception-driven multi-robot formation control. In: 2013 IEEE International Conference on Robotics and Automation (ICRA), pp. 1851–1856. IEEE (2013)
2. Ahmad, A., Tipaldi, G.D., Lima, P., Burgard, W.: Cooperative robot localization and target tracking based on least squares minimization. In: 2013 IEEE International Conference on Robotics and Automation (ICRA), pp. 5696–5701. IEEE (2013)
3. Boers, Y., Driessen, H., Bagchi, A., Mandal, P.: Particle filter based entropy. In: 13th Conference on Information Fusion (FUSION), pp. 1–8. IEEE (2010)
4. Capitan, J., Spaan, M.T.J., Merino, L., Ollero, A.: Decentralized multi-robot cooperation with auctioned pomdps. The International Journal of Robotics Research **32**(6), 650–671 (2013)
5. Fox, D., Burgard, W., Thrun, S.: Active markov localization for mobile robots. Robotics and Autonomous Systems **25**(3), 195–207 (1998)
6. Hollinger, G.A., Sukhatme, G.S.: Sampling-based robotic information gathering algorithms. International Journal of Robotics Research **33**(9), 1271–1287 (2014)
7. Karaman, S., Frazzoli, E.: Sampling-based algorithms for optimal motion planning. The International Journal of Robotics Research **30**(7), 846–894 (2011)
8. LaValle, S.M., Kuffner Jr., J.J.: Algorithmic and Computational Robotics: New Directions, chap. Rapidly-exploring random trees: Progress and prospects. CRC Press (2000)
9. Merino, L., Caballero, F., Ollero, A.: Active sensing for range-only mapping using multiple hypothesis. In: 2010 IEEE/RSJ International Conference on Intelligent Robots and Systems (IROS), pp. 37–42, October 2010
10. de Sousa, M.J.V.T.: Decision-Making under Uncertainty for Real Robot Teams. Ph.D. thesis, Instituto Superior Técnico (2014)
11. Spaan, M.T., Veiga, T.S., Lima, P.U.: Active cooperative perception in network robot systems using pomdps. In: 2010 IEEE/RSJ International Conference on Intelligent Robots and Systems (IROS), pp. 4800–4805. IEEE (2010)
12. Thrun, S., Burgard, W., Fox, D.: Probabilistic robotics. MIT press (2005)

Impact of Robot Actions on Social Signals and Reaction Times in HRI Error Situations

Nicole Mirnig[✉], Manuel Giuliani, Gerald Stollnberger, Susanne Stadler, Roland Buchner, and Manfred Tscheligi

Center for Human-Computer Interaction, University of Salzburg, Salzburg, Austria
{nicole.mirnig,manuel.giuliani,gerald.stollnberger,susanne.stadler,
roland.buchner,manfred.tscheligi}@sbg.ac.at
http://www.hci.sbg.ac.at/

Abstract. Human-robot interaction experiments featuring error situations are often excluded from analysis. We argue that a lot of value lies hidden in this discarded data. We analyzed a corpus of 201 videos that show error situations in human-robot interaction experiments. The aim of our analysis was to research (a) if and which social signals the experiment participants show in reaction to error situations, (b) how long it takes the participants to react in the error situations, and (c) whether different robot actions elicit different social signals. We found that participants showed social signals in 49.3% of error situations, more during social norm violations and less during technical failures. Task-related actions by the robot elicited less social signals by the participants, while participants showed more social signals when the robot did not react. Finally, the participants had an overall reaction time of 1.64 seconds before they showed a social signal in response to a robot action. The reaction times are specifically long (4.39 seconds) during task-related actions that go wrong during execution.

Keywords: Human-robot interaction · Robot feedback · Social robots

1 Introduction

Although it has been widely accepted among scientists that faulty data carries knowledge, the loop way of including such data is often avoided. It seems more productive to discard all cases that do not fully comply with the envisioned process. Based on Paul Watzlawick well-known axiom "one cannot not communicate" [18], it may be worth taking the time to look at those faulty incidents.

In order to put this notion to the test, we have gone back to data from previous human-robot interaction (HRI) experiments within our research group. From different research projects, we have collected numerous video data, of which we had originally discarded a great share due to non-conformities with our initial research intention. Looking into these videos from various experiments collectively, we expected to learn from the errors and, in doing so, detect new modes for improving the interaction between humans and robots.

© Springer International Publishing Switzerland 2015
A. Tapus et al. (Eds.): ICSR 2015, LNAI 9388, pp. 461–471, 2015.
DOI: 10.1007/978-3-319-25554-5_46

To analyze the data, we first selected all videos featuring faulty instances of interaction (i.e., where the interaction did not exactly follow the predefined script). Then, we developed a coding scheme in which we annotated faulty situations and if they were a social norm violation or a technical failure of the robot (from the user's perspective), what sort of action the robot performed during this situation, and how the user reacted in consequence. Although the experiments were performed over several projects, they all share traits, which is why we decided to look at the data collectively with one uniform annotation scheme.

This publication follows-up on our previous analysis of the same video corpus, in which we explored error situations in HRI and what kind of social signals humans show upon social norm violations or technical failures of the robot [5]. With our current analysis, we look further into the topic of social signals, in analyzing how often people react, how long it takes them to react, and which social signals are frequently produced together. It is our aim to develop a framework to explain user behavior in faulty instances of HRI and, in doing so, provide a tool to prevent problematic situations.

2   Related Work

The term *social signal* is used to describe verbal and non-verbal signals that humans use in a conversation to communicate their intentions. Lately, more and more researchers focussed on developing approaches for automatic recognition of social signals, a field that is called social signal processing. Vinciarelli et al. [17] give a fairly recent overview of the work in the area. In HRI, social signal processing also receives more attention. Jang et al. [9] present a video analysis that is similar to our work. They annotated recordings of six one-on-one teacher-student learning sessions, in order to find the social signals with which students signal their engagement in the interaction. The goal of their work is to implement an engagement classifier for a robot teacher. Tseng et al. [16] present a robot that automatically recognizes the spatial patterns of human groups by analyzing their non-verbal social signals, in order to appropriately approach the group.

A second area, in which social signals play a role in HRI, is the generation of social signals by robots. Bohus and Horvitz [1] present a direction-giving robot that forecasts when the user wants to conclude the conversation. This robot uses hesitations (e.g., the robot says "so...") when the robot is not certain about the user's state in order to gain more time to compute a correct forecast and also to convey the uncertainty of the system. Sato and Takeuchi [12] researched how the eye gaze behavior of a robot can be used to control turn taking in non-verbal HRI. Stanton and Stevens [15] found that robot gaze positively influences the trust of experiment participants who had to give answers to difficult questions in a game, but negatively influences trust when answering easy questions. Carter et al. [2] present a study, in which participants repeatedly threw a ball to a humanoid robot that attempted to catch the ball. The study results show that participants smile more when the robot shows social signals and rate the robot as more engaging, responsive, and human-like.

Researchers from social psychology explored how humans use social signals to express their intentions. In psychology literature, the term "freezing" is used to refer to the absence of social signals and movement by humans. It is known that humans stop moving in certain situations. For example, Witchel et al. [20] showed that the absence of non-instrumental movements can be a sign for engagement of humans with media. It is also known that humans, as well as animals, freeze as a response to fear or stress [7]. In our work, we found that humans often smile during error situations. Similar to this, Hoque et al. [8] also found that humans smile in frustrating situations. They recorded the faces of participants who filled out a web form that was designed to elicit frustration and found that 90% of the participants smiled in these frustrating situations. Finally, we found that humans use body posture, especially head movements, to express their intentions. Similar to this result, Loth et al. [10] found that customers use two non-verbal social signals to signal bartenders that they would like to order a drink: they position themselves directly at the bar counter and look at a member of staff.

3   Video Corpus and Annotation

For a comprehensive picture that accounts for more than one specific scenario, we combined video data from different experiments and analyzed them collectively. All studies were HRI experiments in which the participants had to either complete a task together with a robot or help a robot complete a task on its own. With the data analysis presented herein, we are revisiting data from previous experiments and analyze them from a different viewpoint. For more details on the original experiments, please refer to the references as quoted below.

The data we used for our analysis was collected in five user experiments, which were conducted in the course of one of the following research projects: JAMES[1] (Joint Action for Multimodal Embodied Social Systems), JAST[2] (Joint Action Science and Technology), and IURO[3] (Interactive Urban Robot), as well as the Master thesis project RPBD (Robot Programming by Demonstration). The video corpus totaled to 201 videos (129 x JAMES, 34 x JAST, 27 x IURO, 11 x RPBD), showing 100 individual study participants (30 female and 70 male). All experiments were conducted in Austria or Germany. Although we used different robots in most of our experiments (see Figure 1), the robots were always able to understand and produce speech and they had visual perception modules for person tracking.

The following scenarios were covered with the HRI experiments: (a) JAMES, a bartender robot who took drink orders from customers and it had to hand out the correct drink to the right person [3,6]; (b) JAST had to assemble target objects from a wooden toy construction set, together with a human partner [4]; (c) IURO could autonomously navigate through crowded inner-city environments, while proactively approaching pedestrians to request direction

---

[1] http://www.james-project.eu
[2] http://www6.in.tum.de/Main/ResearchJast
[3] http://www.iuro-project.eu/

(a) JAMES robot   (b) JAST robot   (c) IURO robot   (d) RPBD robot

Fig. 1. The robots used in the HRI experiments. Pictures show actual interactions.

information [19]; and (d) RPBD, in which participants had to kinesthetically teach a pick-and-place task to the robot [14].

We annotated our data to explore the following three topics: (a) How many people react upon a robot action that occurs in an error situation? (b) How long does it take for people to react upon a robot action within an error situation in general and in case of social norm violation and technical failure? (c) What kind of reaction do people show in response to a robot action within an error situation in general, and specifically in case of social norm violation and technical failure?

For annotating our video corpus, we used the video coding tool ELAN[4]. We followed a two-step annotation process. First, we annotated all instances in which an error occurred in the interaction. We labeled them as **error situations** and categorized them either as *social norm violation* or *technical failure*. We define the term *social norm violation* as a deviation from the social script [13] or the usage of the wrong social signals in a given situation. For example, when the participants ordered a drink from the bartender robot, the robot signalized them that it understood, but then asked for their order again. *Technical failures* include, for example, when the robot picked up an object, but lost it while grasping it.

Second, we annotated the actions the robot performed during the error situations (**robot actions**) and the social signals the participants showed at the same time (**social signals**). Robot actions were categorized into *TaskRelatedAction* (The robot does what it was programmed to do, but the interaction is in an error situation already.), *RepeatPreviousAction* (The robot repeats the last action whether on purpose (no reaction of study participant or when participants demanded a repetition) or in case of a technical failure.), *RightActionGoneWrong* (The right action is started but with a poor result.), *OutOfContextAction* (The reaction of the robot does not correspond to the expectation in that situation.), *Filler* (The robot waits for a participants' response.), and *NoReaction* (The robot provides no feedback or immediate action at all.). Social signals were categorized into *speech, head movements, hand gestures, facial expressions,* and *body movements.* For a more detailed overview on the annotation process, see [5].

We annotated 1924 robot actions. For the analysis of the temporal connection between robot action and elicited social signal, we selected those robot actions that clearly started together with, or within, an error situation. Robot actions that started prior to, and carried over to, the respective error situation were excluded. It is not clear whether these robot actions directly caused the

---

[4] https://tla.mpi.nl/tools/tla-tools/elan/

Table 1. Total number of different robot actions leading to error situations

RobotAction	Frequency	Percent
TaskRelatedAction	244	14.3
RepeatPreviousAction	812	47.5
RightActionGoneWrong	42	2.5
OutOfContextAction	105	6.1
Filler	216	12.6
NoReaction	291	17.0
Total	1710	100.0

following error situation and as a consequence triggered the social signals of the participants. Table 1 shows the frequency and percentage of the remaining 1710 robot actions. A total of 1032 robot actions were driven by social norm violations (60.4%), in which the reaction of the robot did not match the expectation of the study participants. Technical failures caused 678 of the error situations (39.6%).

4 Results

In the following section, we present our results on social signals. We explored if people produce or not produce social signals, depending on the type of error situation and robot action. We analyzed the participants' reaction time to see how long it takes them to react and if their reaction times differ between types of error situation and robot action. Finally, we explored which social signals are frequently produced in combination with each other.

Social Signals vs. No Social Signals. First, we analyzed if people react in an error situation at all. The experiment participants showed social signals in 49.3% (843 times) of the 1710 robot actions. In 50.7% (867 times) of the robot actions, the participants showed no reaction.

Via one-sampled chi-square tests (nominal variable), we computed for both error situation types, whether participants did or did not show social signals during the social norm violations and technical failures. Both tests were significant and show opposite results. During social norm violations, significantly more participants showed one or more social signal ($\chi^2(1, N = 1032) = 20.66, p < .001$). Whereas, in case of technical failures, significantly more participants did not show a social signal at all ($\chi^2(1, N = 678) = 42.63, p < .001$).

We furthermore computed, whether the type of robot action influences if people show social signals or not. For that, we conducted a one-sample chi-square test for each type of robot action and obtained significant results for five of the six types. In subsequence to a *TaskRelatedAction* and a *RepeatPreviousAction*, the participants significantly more often showed no social signals. In case of *RightActionGoneWrong*, *Filler*, and *NoReaction*, the participants showed social signals significantly more often. *Only OutOfContextAction* did not result in a significant difference. Table 2 shows the percentages of cases with no social signals shown vs. social signals shown in regard to the different robot actions, as well as the results of the chi-square tests.

Table 2. Percentages of cases with no social signals shown vs. social signals shown for the different robot actions + chi-square test results (significant results in bold)

RobotAction	No Signal	Signal	Chi-Square	
TaskRelatedAction	**57%**	43%	$\chi^2(1, N = 244) = 4.20$	$p = .041$
RepeatPreviousAction	**63%**	37%	$\chi^2(1, N = 812) = 57.46$	$p < .001$
RightActionGoneWrong	29%	**71%**	$\chi^2(1, N = 42)\ \ = 7.71$	$p = .005$
OutOfContextAction	47%	53%	$\chi^2(1, N = 105) = 0.47$	$p = .495$
Filler	29%	**71%**	$\chi^2(1, N = 216) = 37.50$	$p < .001$
NoReaction	31%	**69%**	$\chi^2(1, N = 291) = 40.83$	$p < .001$

Reaction Times. Over the 843 instances when the participants showed either one or more social signals, it took the participants an average 1.64 seconds ($SD = 2.47$) to react. We calculated the reaction time as the duration between onset time of robot action and onset time of the first social signal. A Kolmogorov-Smirnov test for normality indicated that the reaction time of the participants significantly deviated from a normal distribution ($D = .253, p < .001$). Therefore, we used nonparametric statistical tests to analyze differences in reaction time. We conducted a Wilcoxon rank-sum test to evaluate whether the average reaction time differs between social norm violations ($N = 589$) and technical failures ($N = 254$). The results of the test confirm that there is a significant difference in reaction time ($W = 99.59, z = -2.34, p = .019, r = 0.08$). Participants' reaction time is significantly faster in case of technical failures (Mdn = 1.00), than in case of social norm violations (Mdn = 1.16).

A Kruskal-Wallis test showed that reaction times were significantly affected by type of robot action, $H(5, N = 843) = 16.97, p = .005$. Post-hoc tests revealed significant differences between *RepeatPreviousAction* (Mdn = 1.05) and *RightActionGoneWrong* (Mdn = 2.16), $U = -179.95, p = .002, r = 0.13$; between *Filler* (Mdn = 0.93) and *RightActionGoneWrong*, $U = 163.86, p = .011, r = 0.12$; and between *NoReaction* (Mdn = 1.09) and *RightActionGoneWrong*, $U = 149.66, p = .025, r = 0.11$ (see Table 3). Significance levels were adjusted to account for multiple post-hoc tests.

Combinations of Social Signals. To analyze how people react towards a robot in case of an error situation, we first selected all cases in which a robot action was followed by one or more social signals produced by the user. Here, we did not exclude robot actions that started before the respective error situation, since we were not interested in the relationship between error situation and social signals, but robot action and social signals. That left us with 2115 data lines. The participants showed a total number of 3463 social signals. The average number of social signals per robot action was 1.64 ($SD = 1.30, min = 1, max = 17$). In most instances, people showed one social signal ($N = 1.421$), followed by two social signals ($N = 394$) and three social signals ($N = 150$).

We explored frequent combinations of social signals (for a detailed overview, refer to [5]). To do so, we computed the top ten of double and triple combinations

Table 3. Mean reaction times and standard deviation in seconds

RobotAction	N	Mean	SD
TaskRelatedAction	106	1.65	1.70
RepeatPreviousAction	298	1.15	0.86
RightActionGoneWrong	30	4.39	4.92
OutOfContextAction	56	1.71	1.96
Filler	153	1.97	4.21
NoReaction	200	1.71	1.89

Table 4. Top 10 of double and triple combinations of social signals (RepeatLastS. = RepeatLastSentence; AskQuestionToExp. = AskQuestionToExperimenter; LookTo-Exp. = LookToExperimenter; LookToGM = LookToGroupMember; MakeStatement-ToGM = MakeStatementToIngroupMember)

No. Double Combination	No. Triple Combination
25 LeanForward + RepeatLastS.	4 Laugh + LookDown + Smile
21 Laugh + Smile	3 LookToRobotHead + NewSentence + Smile
18 LookToExperimenter + Smile	2 LookToExp. + LookToGM + MakeStatementToGM
12 LeanForward + NewSentence	2 FunnyFace + LookToRobotHead + NewSentence
11 LookToGroupMember + Smile	2 Laugh + Smile + StepBack
10 LookToRobotHead + RepeatLastS.	2 LeanForward + LookToRobotHead + RepeatLastS.
9 AskQuestionToExp. + LookToExp.	2 LookToObject + LookToRobotHead + NewSentence
8 LookToRobotHead + NewSentence	2 Laugh + LookToGroupMember + Smile
7 NewSentence + Smile	2 LookToRobotHand + LookToRobotHead + Smile
6 RephraseLastSentence + Smile	2 LookToRobotHead + RepeatLastSentence + Tilt

of social signals. Table 4 shows the combinations with their frequency. Double combinations feature 10 different social signals, triple combinations feature 15 different ones. The social signal *Smile* is most often found in both lists, with a total count of 10 (5 each). *LookToRobotHead* can be found in eight combinations (2x double, 6x triple) and *NewSentence* is part of six combinations (3 each). The top three social signals in number of total occurrences within all combinations, sum up to 160 x *Smile*, 143 x *LookToRobotHead*, and 91 x *NewSentence*.

5   Discussion

Social Signals vs. No Social Signals. Prior to our data analysis, we had expected that social signals would be frequently shown in error situations. Upon data analysis, we found evidence that only in half of the error situations participants produced social signals. This result is even more outstanding when we consider the fact that people were more likely to show social signals during social norm violations than during technical failures. This suggests that users can identify technical failures well, but they see less need to react. During social norm violations, however, there seems to be a need to produce social signals in order to keep the interaction going and solve the experimental task. This assumption is also supported by a result from our preceding study on error situations, in which we found that people talk considerably more during social norm violations [5]. Based on our findings, we assume that there is no social script for technical failures and this is why people do not actively countersteer (i.e., produce social signals), but take on a more observant role (i.e., not produce social signals). The robot actions *TaskRelatedAction* and *RepeatPreviousAction* happen more frequently during social norm violations when the robot tries to resolve the error situation. Therefore, the result of less social signals shown adds up. In contrast, during *Filler* and *NoReaction*, both of which are either planned or unplanned robot idle times, participants showed more social signals. We interpret the participants' social signals in these cases as attempts to bypass idle time, which we advise to compensate with more targeted robot feedback [11]. The fact that almost 50 % of the error situations were initiated through *RepeatPreviousAction* can be partially explained with the robot setup of one experiment. James was programmed to repeat its last action if speech recognition accuracy was below a certain threshold. This resulted in frequent repetition of robot actions. However, *RepeatPreviousAction* also scored highest in percentage of no social signal shown and fastest in reaction time. This could be an indicator that frequent repetition of one and the same action can easily lead to a critical situation. It seems that people are likely to decipher such situations quickly but they also seem more likely to not react which might lead to termination of the interaction.

Reaction Times. Although the participants showed less social signals in case of technical failures, those participants who did react, did so significantly faster than during social norm violations. This indicates that the participants were able to identify technical failures quickly. In the special case of *RightAction-GoneWrong*, however, participants took especially long to react. We see two reasons for this. First, the robots in our experiments took long to complete their action and it may have taken a while until the user was able to identify the action as wrong. Second, according to neuropsychological literature, humans stop moving in certain situations when in fear or stress, a phenomenon called "freezing" [7]. The long reaction time in case of *RightActionGoneWrong* could be seen as people living through stress, which is a phenomenon worth being picked up by future research, in which biometric data should be measured to back this assumption up. Another indication for the same effect is the frequent production of smile, which is also a stress indicator [8].

Combinations of Social Signals. Upon taking a closer look on the common combinations of social signals, it is noticeable that people frequently combine a verbal utterance (*RepeatLastSentence* and *NewSentence*) with a leaning forward body movement. This could be seen as people trying to bypass faulty speech recognition and make their own attempt to recover from an error situation.

6   Conclusion

We conclude the following main findings from our data. Participants responded to half of the robot actions with one or more social signals. Participants showed significantly more social signals during social norm violations and less during technical failures. They also showed significantly more social signals in case of faulty or no robot action and less when the robot performed a task-related action or repeated the previous action. The participants reacted considerably faster in case of technical failures, but in terms of faulty robot actions, the reaction times were particularly slow. From the common combinations of social signals, we could learn that participants frequently smile in error situations. Given smiles can be assigned to a certain meaning, they could be beneficial as implicit input modality for the robots. With our results, we provide data-based evidence that robot feedback is crucial. We showed that people's reactions observably differ between types of error situation and robot action. One implication we can draw from this is that a social robot which is able to interpret the users' social signals, could be enabled to provide adapted feedback. Similarly, automatic error situation classification could assist the robot in choosing which feedback is most appropriate in a certain kind of situation. Further research is required to explore the effect of different kinds of feedback on the users.

In conclusion, we can point out that the feedback of a robot is extremely important and needs to be given considerable thought when designing meaningful HRI. Furthermore, we suggest automatic social signal processing as a means of detecting problems in human-robot interactions. We have identified fewer social signals and faster participant reaction times as two indicators for automatically detecting technical failures of the robot.

Acknowledgments. We thank all project partners who were involved in preparing and running the user studies. This work was funded by the European Commission in the projects JAMES (Grant No. 270435), JAST (Grant No. 003747), IURO (Grant No. 248314), and ReMeDi (Grant No. 610902). We gratefully acknowledge the financial support by the Austrian Federal Ministry of Science, Research and Economy and the National Foundation for Research, Technology and Development in the Christian Doppler Laboratory "Contextual Interfaces".

References

1. Bohus, D., Horvitz, E.: Managing human-robot engagement with forecasts and... um... hesitations. In: Proc. of the 16th Int. Conf. on Multimodal Interaction, ICMI 2014, pp. 2–9. ACM (2014)
2. Carter, E.J., Mistry, M.N., Carr, G.P.K., Kelly, B.A., Hodgins, J.K.: Playing catch with robots: incorporating social gestures into physical interactions. In: Proc. of the 23rd Int. Symp. on Robot and Human Interactive Communication, RO-MAN 2014, pp. 231–236. IEEE (2014)

3. Foster, M.E., Gaschler, A., Giuliani, M., Isard, A., Pateraki, M., Petrick, R.P.A.: Two people walk into a bar: dynamic multi-party social interaction with a robot agent. In: Proc. of the 14th ACM Int. Conf. on Multimodal Interaction, ICMI 2012), Santa Monica, USA, October 2012

4. Giuliani, M., Foster, M.E., Isard, A., Matheson, C., Oberlander, J., Knoll, A.: Situated reference in a hybrid human-robot interaction system. In: Proc. of the 6th Int. Natural Language Generation Conf., INLG 2010, Dublin, Ireland, July 2010

5. Giuliani, M., Mirnig, N., Stollnberger, G., Stadler, S., Buchner, R., Tscheligi, M.: Systematic analysis of video data from different human-robot interaction studies: a categorisation of social signals during error situations. submitted for publication

6. Giuliani, M., Petrick, R.P., Foster, M.E., Gaschler, A., Isard, A., Pateraki, M., Sigalas, M.: Comparing task-based and socially intelligent behaviour in a robot bartender. In: Proc. of the 15th Int. Conf. on Multimodal Interfaces, ICMI 2013, Sydney, Australia, December 2013

7. Hagenaars, M.A., Oitzl, M., Roelofs, K.: Updating freeze: aligning animal and human research. Neurosc. & Biobehavioral Reviews **47**, 165–176 (2014)

8. Hoque, M.E., McDuff, D.J., Picard, R.W.: Exploring temporal patterns in classifying frustrated and delighted smiles. IEEE Transactions on Affective Computing **3**(3), 323–334 (2012)

9. Jang, M., Lee, D.-H., Kim, J., Cho, Y.: Identifying principal social signals in private student-teacher interactions for robot-enhanced education. In: Proc. of the 22rd Int. Symp. on Robot and Human Interactive Communication, RO-MAN 2013, pp. 621–626, August 2013

10. Loth, S., Huth, K., De Ruiter, J.P.: Automatic detection of service initiation signals used in bars. Journal of Frontiers in Psychology **4** (2013)

11. Mirnig, N., Tscheligi, M.: Robots that Talk and Listen. Technological and Social Impact., chapter Comprehension, Coherence and Consistency: Essentials of Robot Feedback, pp. 149–171. De Gruyter (2014)

12. Sato, R., Takeuchi, Y.: Coordinating turn-taking and talking in multi-party conversations by controlling robot's eye-gaze. In: Proc. of the 23rd Int. Symp. on Robot and Human Interactive Communication, RO-MAN 2014, pp. 280–285. IEEE (2014)

13. Schank, R., Abelson, R., et al.: Scripts, plans, goals and understanding: An inquiry into human knowledge structures, vol. 2. Lawrence Erlbaum Associates Hillsdale, NJ (1977)

14. Stadler, S., Weiss, A., Tscheligi, M.: I trained this robot: the impact of pre-experience and execution behavior on robot teachers. In: Proc. of the 23rd Int. Symp. on Robot and Human Interactive Communication, RO-MAN 2014, pp. 1030–1036. IEEE (2014)

15. Stanton, C., Stevens, C.J.: Robot pressure: the impact of robot eye gaze and lifelike bodily movements upon decision-making and trust. In: Beetz, M., Johnston, B., Williams, M.-A. (eds.) ICSR 2014. LNCS, vol. 8755, pp. 330–339. Springer, Heidelberg (2014)

16. Tseng, S.-H., Hsu, Y.-H., Chiang, Y.-S., Wu, T.-Y., Fu, L.-C.: Multi-human spatial social pattern understanding for a multi-modal robot through nonverbal social signals. In Proc. of the 23rd Int. Symp. on Robot and Human Interactive Comm., RO-MAN 2014, pp. 531–536. IEEE (2014)

17. Vinciarelli, A., Pantic, M., Heylen, D., Pelachaud, C., Poggi, I., D'Errico, F., Schröder, M.: Bridging the gap between social animal and unsocial machine: a survey of social signal processing. IEEE Transactions on Affective Computing **3**(1), 69–87 (2012)

18. Watzlawick, P., Bavelas, J.B., Jackson, D.D.: Pragmatics of human communication: A study of interactional patterns, pathologies and paradoxes. WW Norton & Company (2011)
19. Weiss, A., Mirnig, N., Bruckenberger, U., Strasser, E., Tscheligi, M., Kühnlenz, B., Wollherr, D., Stanczyk, B.: The Interactive Urban Robot: User-centered development and final field trial of a direction requesting robot. Paladyn Journal of Behavioral Robotics 6(1), 42–6 (2015)
20. Witchel, H.J., Westling, C.E., Tee, J., Healy, A., Needham, R., Chockalingam, N.: What does not happen: Quantifying embodied engagement using nimi and self-adaptors. Journal of Audience and Reception Studies 11(1), 304–331 (2014)

Social-Task Learning for HRI

Anis Najar$^{(\boxtimes)}$, Olivier Sigaud, and Mohamed Chetouani

Institut des Systèmes Intelligents et de Robotique, CNRS UMR 7222,
Université Pierre et Marie Curie, Paris, France
{anis.najar,olivier.sigaud,mohamed.chetouani}@isir.upmc.fr

Abstract. In this paper, we introduce a novel method for learning simultaneously a task and the related social interaction. We present an architecture based on Learning Classifier Systems that simultaneously learns a model of social interaction and uses it to bootstrap task learning, while minimizing the number of interactions with the human. We validate our method in simulation and we prove the feasibility of our approach on a real robot.

Keywords: Human-Robot Interaction · Interactive Reinforcement Learning · Learning classifier systems

1 Introduction

Endowing robots with social interaction capacities in addition to task related skills is an important challenge that would facilitate the integration of robots in human environments for task collaboration whether in domestic or industrial contexts. Social interaction and task execution are two different aspects of Human-Robot collaboration that some recent works begin to point the necessity to distinguish [1]. However, these two aspects are still strongly mutually related, so it is not always possible to treat them independently from each other.

In our work, we are interested in studying the relationship between social interaction and task execution from a machine learning perspective: how can we use social interaction for task learning and inversely how could task learning contribute to learning social interaction? In the literature, these two questions are rarely addressed simultaneously, but often one aspect is determined in order to learn the other: social interaction mechanisms are predefined to learn a new task, or a known task is used to learn new social interaction mechanisms. In this paper, we propose to learn simultaneously a model of social interaction (Social Model) and a model of the task (Task Model) in a robot teaching scenario. In one way, the Task Model is used for grounding the interpretation of teaching signals and for learning how to behave according to them within the Social Model. In return, the Social Model is used for bootstrapping the learning process of the Task Model. This looped process is performed online while minimizing the number of interactions with the human.

© Springer International Publishing Switzerland 2015
A. Tapus et al. (Eds.): ICSR 2015, LNAI 9388, pp. 472–481, 2015.
DOI: 10.1007/978-3-319-25554-5_47

In a previous work [6], we proposed a first model that learns a social reward function on the teaching signals from the rewards provided by the environment, and uses it to bootstrap task learning. One limitation of this model is that it does not take into account the long-term information provided by some teaching signals, so the Social Model is not able to learn anything about them. In this paper, we propose a model that overcomes this limitation by learning the state-action values instead of the direct rewards for grounding the meaning of teaching signals. We show that this model is able to boost task learning in addition to learning a more complete model of social interaction.

In the next section, we present some related work. In Section 3, we introduce our model. Section 4 describes the scenario and the experimental set-up. In Section 5, we provide a validation of our approach in simulation. Finally, we present an implementation of our model within a robotic architecture and we report the results of experiments performed on a real robot in Section 6.

2 Related Work

Interactive Reinforcement Learning (IRL) provides a wide range of techniques for teaching RL-based systems [10] by the means of social feedbacks. These works differ in three main aspects: the interaction protocol used for providing feedbacks, the way feedbacks are interpreted for learning and the autonomy of the system with respect to the human.

Some works rely on artificial interfaces for interacting with the learning system. In [3], a virtual agent is trained by human feedbacks within a text-based environment. [9] and [12] use a clicking interface while [5] and [7] rely on push-buttons for providing human feedbacks. In contrast, other works rely on natural interaction protocols for delivering feedbacks such as spoken words [11] and speech features [2,4]. Similarly, in our work we use a natural interaction protocol and we focus specially on non verbal cues such as head movements and pointing for providing feedback.

While most of these works associate human feedback with predetermined scalar values [3,5,7,9,11,12], few works address the question of learning the meaning of teaching signals [2,4]. In [4], a binary classification on prosodic features is performed offline before using it as a reward signal for task learning. In [2], however, the system learns simultaneously to interpret feedbacks and to perform the task. Our work, similarly to [2], tackles both questions at the same time by grounding the meaning of the teaching signals in the task and by using them in return to bootstrap task learning.

The autonomy of the learning agent with respect to the human is an important feature for evaluating Interactive Learning systems in terms of human load. In [2,3,7,9], the learning agent is guided only by human feedbacks, so it is not able to learn without the presence of the human. By contrast, in our work like in [5,11,12], the system is able to learn autonomously through task related rewards while the human can choose the degree of its involvement in the interaction, for guiding the system in order to accelerate its learning process.

3   Model

The idea of our method is to learn simultaneously two separate models, one for the task and one for social interaction. The model of the task would serve to ground the model of social interaction while minimizing the number of interactions with the human and the model of social interaction is used in order to bootstrap the learning of the task. We use an architecture based on three main components: a Task Model, a Social Model and a Contingency Model (Figure 1). The Task Model and the Social Model are represented by two different Markov Decision Processes (MDP) based on XCS[1] and the Contingency Model represents the contingency between states of both MDPs.[2]

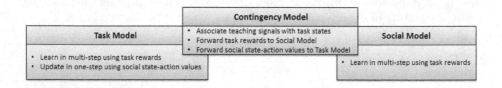

Fig. 1. Social-Task learning model

In [6], we proposed a first model in which the Social Model is used for learning a social reward function on teaching signals based on task rewards. This function is then used online as an additional reward signal for boosting the learning process within the Task Model. We will refer to this first model as SRXCS (for Social Reward XCS). In this paper, we rather learn state-action values within the Social Model and use them as state-action values for the Task Model. We refer to this model as SVXCS (for Social Value XCS). From an algorithmic point of view, the difference between SVXCS and SRXCS resides in two points: the way the Social Model is updated by task rewards and the way the Task Model is updated by the Social Model. In SRXCS, the Social Model is updated by task rewards in a single-step fashion, while the Task Model is updated in a multi-step manner by using both social and task rewards. In SVXCS, however, the Social Model is learned in a multi-step way, whereas the Task Model is updated by the Social Model in a single-step fashion. In addition, in SVXCS, the Task Model still uses task rewards in multi-step as in SRXCS, so it is able to learn the task independently from the human.

---

[1] XCS is an RL system endowed with a generalization capability that allows learning general rule representations over state features, in a way that features that are not relevant for a given rule are replaced by a '#' symbol [8].

[2] We refer to [6] for a more detailed description of the model.

4   Scenario

We consider a simple task (Figure 2), in which a robot has to learn to press buttons of different colors, according to the information displayed on a screen. In a real-world scenario, the information displayed on the screen would represent the state of the physical environment and the action of pressing a button could represent any other elementary action that the robot could perform on objects.

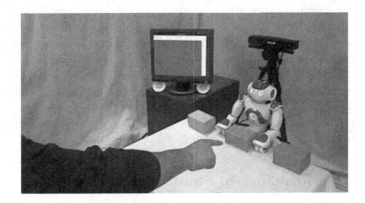

Fig. 2. Scenario: The robot must press the button corresponding to the information displayed on the screen. A human can help it through head movements and pointing.

4.1   Experimental Setup

The experimental set-up is composed of a humanoid robot (Aldebaran Nao) facing a table on top of which there is a set of three buttons of different colors. At each moment, the screen displays the color of the button that the robot has to press. The robot is able to perform two kinds of actions: gazing to one of the different buttons or pressing the one it is facing. The task is a multi-step problem, meaning that in order to press the right button, the robot has to look for it first, and then to perform the action of pressing. The action of gazing to an object triggers a null reward. Pressing a button, however, triggers either a positive or a negative reward, represented by two different sounds, depending if the robot pressed the right or the wrong button. When the robot presses the right button, the task progresses and a new color is displayed on the screen.

While the robot is learning to perform the task, a human can sit in front of it in order to help it by using head nods, head shakes and pointing. Head nods and head shakes tell if the robot is looking to the right or the wrong button, while pointing is meant to indicate the button it has to press. A Microsoft Kinect[3] V2 sensor is used to track the skeleton of the human[4].

---

[3] https://www.microsoft.com/en-us/kinectforwindows/, Last accessed 20-12-2014

[4] We use a modified version of the Kinect V2 client/server provided by the Personal Robotics Laboratory of Carnegie Mellon University. https://github.com/personalrobotics/, Last accessed 20-12-2014

4.2   Teaching Protocol

We adopt an active learning procedure as a teaching protocol. When the robot encounters a new task situation (defined by the displayed color and the robot gazing state), the robot asks the human for help in two steps: First, it asks if it is looking to the right color, immediately looks at the person for a brief moment, before looking back to the button. Then, it repeats this process one more time by asking the person to point the right button.

This active learning procedure is motivated by two main reasons. First, the Contingency Model as it is currently designed stores the contingency between whole states. It means that all teaching features must be determined before sending the whole social state to the Contingency Model. So, this procedure fulfils this constraint, by actively asking the human to provide it with a value for each type of feature. Second, we have argued in [6] that in order to reach optimal performance, the human needs to interact with the system only in newly encountered situations, in the case of a perfect teacher. So, this protocol is meant to verify this assumption in a real set-up.

5   Model Performance in Simulation

In this section, we present the performance of our model in simulation. We compare SVXCS to the standard XCS algorithm over 1000 experiments, to show how our model accelerates task learning. Then, we present the learned rules in the Social Model.

5.1   Task Model Performance

We report the result of the experiments in two different settings: with and without genetic generalization. Figure 3 reports the probability for the model to converge before n steps. With genetic generalization (Figure 3.a), SVXCS needs at most 4903 iterations to learn the task, while the standard XCS needs at most 9184 iteration. Beyond this threshold, we are sure that the model converges. However, below this number of steps, the model converges only with a certain probability. It is worth noting that even with this gain, the number of steps is still considerable for a real-world scenario. Figure 3.b reports the performance of the models without genetic generalization. We can see that XCS converges in at most 851 steps, while SVXCS reduces this number to 227 which is more reasonable for a real robot.

5.2   Social Model

Table 1 shows the learned rules in the Social Model. We can see that the model found correct generalizations on the teaching signals. The first two lines correspond to the rules related to the action of pressing the button. They predict with maximum accuracy a reward of -1000 for pressing a button when there

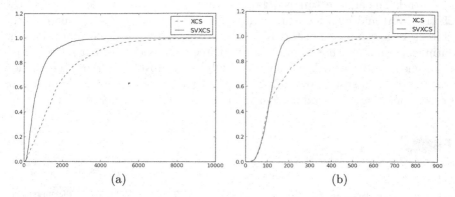

Fig. 3. Probability to converge before n steps. (a) With genetic generalization (b) Without genetic generalization.

Table 1. Learned classifiers in the Social Model: the first two bits encode head movement information. The remaining bits represent the pointing information.

Condition	Action	Prediction	Rule meaning
#1###	0	-1000	head shake $->$ do not press the button
#0###	0	1000	head nod $->$ press the button
##0##	1	484	
##1##	1	698	pointing at button 1 $->$ gaze at button 1
###0#	2	485	
###1#	2	696	pointing at button 2 $->$ gaze at button 2
####0	3	485	
####1	3	694	pointing at button 3 $->$ gaze at button 3

is a head shake and a reward of 1000 when there is a head nod, whatever the pointing information. The remaining rules correspond to the predicted values of gazing to the different objects. We can see that these rules represent a joint attention behaviour which leads the robot to gaze the button that the human is pointing, whatever the head movement information. It is worth noting that in RBXCS, we could not obtain this behaviour of joint attention because it is not able to take into account the long term information provided by pointing [6].

6   Experiments on the Real Robot

In this section, we present our robotic architecture. Then, we report the experimental results on the real robot.

6.1   Robotic Architecture

To implement our model on the real robot, we developed a software architecture in ROS (Figure 4) including a set of modules for perception, decision making and control. The architecture is organized in two main layers, one related to the task (bottom) and another related to social interaction (top). The core of the architecture is composed of the Task Controller, the Social Controller and the Contingency module that implement the three components of our model. Task Environment and Social Environment modules encapsulate the representation of task and social states. In addition, we have a set of perception modules for detecting pointing and head movements and a module for controlling the robot.

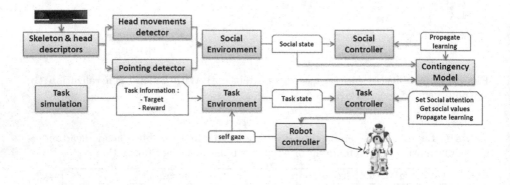

Fig. 4. Robotic architecture for Social-Task learning

6.2   Experimental Results

Without genetic generalization, SVXCS converges within 227 steps (Figure 3.b). To validate these results, we performed experiments of 227 learning steps without genetic generalization with 10 different subjects.

Table 2 reports the results of these experiments (The sequence of the experiments has been changed in the table for better readability). These results show that when subjects provided correct teaching signals $(3 - 10)$, the robot always succeeded in learning the task. Otherwise, the system failed at learning the task (1 and 2). In both experiments, the subjects sometimes performed a head nod instead of a head shake. In (1), a false positive detection of head nod has also occurred. This resulted in an incoherent interpretation of head nods within the Social Model that hindered the Task Model from converging properly.

In other situations $(2 - 5)$, the system did not detect feedback from the user when the robot asked for. This was either because of a detection failure (5) or because the subjects were hesitant and did not provide a complete feedback $(2 - 4)$. In this case, the Contingency Model did not record any teaching signal for the corresponding situation; so the robot asked for feedback one more time for the same situation and this did not prevent from learning the task. In these experiments, the number of interactions with the robot was of 10, while in the other experiments the user was solicited only 9 times, which corresponds to the number of the different task states. Moreover, when all task situations have

Table 2. Experimental results: experiment id (ID), number of interactions with the subject (NI), number of detection failures (DF), number of times the subject did not provide feedback when needed (NF), number of times the system detected the wrong feedback (DE), number of times the subject provided wrong feedback (WF), time in minutes until the last interaction (IT), total duration of the experiment in minutes (TD), success of task learning (S).

ID	NI	DF	NF	DE	WF	IT	TD	S
1	9			1	1	6.20	31.30	No
2	10		1		2	8.15	32	No
3	10		1			7.50	30	Yes
4	10		1			12.30	31.30	Yes
5	10	1				5.30	31	Yes
6	9					9	31.30	Yes
7	9					12.45	31.30	Yes
8	9					6	31.30	Yes
9	9					5.15	31.30	Yes
10	9					7.30	31.30	Yes

been explored, the subjects had no obligation to stay and the robot continued to learn autonomously. The longest interaction lasted for about 13 minutes, while the whole learning process lasted for about a half an hour.

To conclude, these experimental results validate our assumption that without genetic generalization, a perfect teacher interacting with the system only in new situations has the guarantee that the robot will learn the task within 227 steps.

7   Discussion

In this section, we discuss the limitations of our system and we propose some alternative solutions.

Teaching protocol: The active learning mechanism that we implemented presents many advantages. First, it is difficult to provide a rigorous definition of the state of a person. Unlike physical objects, a person is proactive and his/her actions are extended over varying time windows. In addition, the limitation of perception devices may lead to false positive detection that could decrease system performance. So, actively asking for teaching signals one by one while gazing at the human for a determined duration makes it possible to control the acquisition of whole social states. Moreover, this active learning mechanism serves at engaging the person to prevent it from being passive or disengaging from the interaction.

However, this process is a burden for the human as it imposes a fixed protocol for the interaction. In addition, asking for teaching signals only for new situations makes not possible for the user to correct himself if he gives wrong teaching signals in a given situation. One possibility to make the interaction more natural would be then to define a contingency between state features. In this case, the human would be more free in the way he provides teaching signals and it would be more easier for him to correct wrong feedbacks.

Transparency: Another limitation of this teaching protocol is the lack of transparency. In fact, the human has no way to know if he has given wrong teaching signals to correct them or if the robot is not learning correctly. A solution for this would be to implement a mechanism asking for clarification whenever it detects incoherence within the Social Model or incoherence between the Task Model and the Social Model.

Social Model application: In the current definition of our model, the Social Model contributes to boosting task learning in only one way. It serves as an alternative space with reduced complexity that allows to learn state-action values more rapidly. So, the Social Model is employed only in a passive way through the interpretation of teaching signals but never for decision-making. However, the rules evolved by the Social Model could also be used for decision-making. For example, the joint attention mechanism could be useful for guiding the exploration strategy within the Task Model and so it could further optimize task learning.

8   Conclusion and Future Work

In this paper, we presented a model for learning simultaneously a task and a model of social interaction. We showed that our model SVXCS is able to accelerate task learning while minimizing the number of interactions with the human. We presented an implementation of our model within a robotic architecture and proved the feasibility of our approach on a real robot.

In future work, we propose to modify the Contingency Model by storing the contingency between state features instead of whole states in order to make the interaction more natural. We also intend to explore the possibility of using the Social Model for guiding the exploration strategy within the Task Model in order to accelerate task learning. Finally, we propose to enrich our model with additional actions like gazing to the human and asking for feedback to evolve more complex social behaviours.

Acknowledgments. This work is funded by the Romeo2 project.

References

1. Corrigan, L.J., Peters, C., Castellano, G., Papadopoulos, F., Jones, A., Bhargava, S., Janarthanam, S., Hastie, H., Deshmukh, A., Aylett, R.: Social-task engagement: striking a balance between the robot and the task. In: Embodied Commun. Goals Intentions Workshop ICSR, vol. 13, pp. 1–7 (2013)

2. Grizou, J., Lopes, M., Oudeyer, P.-Y.: Robot learning simultaneously a task and how to interpret human instructions. In: 2013 IEEE Third Joint International Conference on Development and Learning and Epigenetic Robotics (ICDL), pp. 1–8, August 2013

3. Isbell, C., Shelton, C.R., Kearns, M., Singh, S., Stone, P.: A social reinforcement learning agent. In: Proceedings of the Fifth International Conference on Autonomous Agents, pp. 377–384. ACM (2001)

4. Kim, E., Scassellati, B.: Learning to refine behavior using prosodic feedback. In: IEEE 6th International Conference on Development and Learning, ICDL 2007, pp. 205–210, July 2007

5. Knox, W.B., Stone, P.: Combining manual feedback with subsequent MDP reward signals for reinforcement learning. In: Proc. of 9th Int. Conf. on Autonomous Agents and Multiagent Systems (AAMAS 2010), May 2010

6. Najar, A., Sigaud, O., Chetouani, M.: Socially guided XCS: using teaching signals to boost learning. In: Proceedings of the Companion Publication of the 2015 on Genetic and Evolutionary Computation Conference, GECCO Companion 2015, pp. 1021–1028. ACM (2015)

7. Pilarski, P.M., Dawson, M.R., Degris, T., Fahimi, F., Carey, J.P., Sutton, R.S.: Online human training of a myoelectric prosthesis controller via actor-critic reinforcement learning. In: 2011 IEEE International Conference on Rehabilitation Robotics (ICORR), pp. 1–7. IEEE (2011)

8. Sigaud, O., Wilson, S.W.: Learning classifier systems: a survey. Soft Comput. 11(11), 1065–1078 (2007)

9. Suay, H.B., Chernova, S.: Effect of human guidance and state space size on interactive reinforcement learning. In: 2011 IEEE RO-MAN, pp. 1–6. IEEE (2011)

10. Sutton, R.S., Barto, A.G.: Reinforcement Learning: An Introduction. MIT Press (1998)

11. Villaseñor-Pineda, L., Morales, E.F., Tenorio-Gonzalez, A.C.: Dynamic reward shaping: training a robot by voice. In: Kuri-Morales, A., Simari, G.R. (eds.) IBERAMIA 2010. LNCS, vol. 6433, pp. 483–492. Springer, Heidelberg (2010)

12. Thomaz, A.L., Breazeal, C.: Reinforcement learning with human teachers: evidence of feedback and guidance with implications for learning performance. In: AAAI, vol. 6, pp. 1000–1005 (2006)

Development of the Multi-dimensional Robot Attitude Scale: Constructs of People's Attitudes Towards Domestic Robots

Takumi Ninomiya, Akihito Fujita, Daisuke Suzuki, and Hiroyuki Umemuro[✉]

Tokyo Institute of Technology, Tokyo 152-8552, Japan
umemuro.h.aa@m.titech.ac.jp

Abstract. This study investigated dimensions that construct people's attitudes toward domestic robots, and developed a comprehensive measurement by which to assess such dimensions. Potential elements of attitudes were extracted from participants' free descriptions and interviews, as well as a literature review. A questionnaire-based investigation was conducted using these elements, and a 12-factor structure was extracted employing factor analysis. A measurement called the Multi-dimensional Robot Attitude Scale was proposed to assess attitudes towards robots based on 12 dimensions. The internal reliability and representativeness of the proposed scale were verified. The Multi-dimensional Robot Attitude Scale proposed in this study is designed to comprehensively assess attitudes towards domestic robots, allowing for a multifaceted understanding of people's attitudes.

Keywords: Attitude · Measurement · Domestic robots · Acceptance

1 Introduction

Robots have been generally accepted for day-to-day use in domestic environments. Robots are expected to support people at home and/or in social contexts, thus improving the wellbeing of the people they serve.

However, it is known that some people are reluctant or resistant toward adopting new technologies. In the 1980s, for example, some people were uncomfortable around computers and reluctant to accept them[1]. To promote acceptance of these new technologies as they continue to come into the market and into our lives, it is important to understand people's attitudes towards them.

In previous research [2,3], attitudes toward robots have been studied in relation to people's acceptance of robots[4,5]. However, few studies have focused on, or proposed measures of, individual differences in attitudes toward robots. Most studies on attitudes have focused on rather few dimensions of attitudes. The Negative Attitudes toward Robots Scale and Robot Anxiety Scale [6,7] are two measures proposed to gauge attitudes toward robots. However, these attitude scales only measure negative attitudes toward robots. Although negative attitudes toward robots are considered a good predictor of successful interactions with robots, attitudes may have more complex constructs.

© Springer International Publishing Switzerland 2015
A. Tapus et al. (Eds.): ICSR 2015, LNAI 9388, pp. 482–491, 2015.
DOI: 10.1007/978-3-319-25554-5_48

A number of studies on attitudes toward computers have shown that there are multiple dimensions or constructs of individual attitudes towards computers. For example, Loyd and Gressard[1] claimed that there are three dimensions: liking, anxiety, and confidence. Jay and Willis [8] proposed seven dimensions: comfort, efficacy, gender equality, control, dehumanization, interest, and utility. It is intuitive that there should be multiple dimensions in individual attitudes when complex technologies such as robots or computers are first introduced to the general public. A better understanding of constructs of attitudes towards robots, as well as their individual differences, should contribute to an understanding and, by extension, acceptance of a variety of robots.

The present study investigates dimensions that construct individual attitudes toward domestic robots, and develops a comprehensive measurement that assesses those multiple dimensions. Potential elements of attitudes were extracted from participants' free descriptions and interviews, as well as a literature review. A questionnaire-based investigation based on these elements was conducted, and dimensions of attitudes towards robots were extracted from the responses to the questionnaire using factor analysis. Finally, a measurement called the Multi-dimensional Robot Attitude Scale was proposed to assess attitudes towards robots along these multiple dimensions.

2 Extraction of Elements of People's Attitudes Towards Domestic Robots

To identify potential elements of attitudes toward robots, a questionnaire- and interview-based investigation and an intensive literature review were conducted.

2.1 Methods

Eighty-three Japanese adults in their twenties to eighties participated in the questionnaire-based investigation.

After participants were explained the purpose of the investigation, participants were presented a set of video, photo, and text pertaining to one of four robots—Roomba[9], the Home-Assistant Robot[10], PaPeRo[11] and Paro[12], one set at a time on computer screen. Then participants were asked to individually and freely describe their feelings and ideas about each robot in the questionnaire, in response to prompts such as "intention to use", "feelings about the robot being introduced to your home" and "things you like or dislike about the robot". This procedure was repeated for each of the four robots.

In the second part of the investigation, participants were asked to describe their thoughts about robots in general, such as "feelings and perceptions you might have if a robot is in your home" and "relationships with robots compared with relationships with pets or friends". Participants were suggested to consider in the context of domestic use of such robots.

Finally, participants were interviewed as a group about the feelings and thoughts they had experienced while answering the questionnaire.

Additionally, previous studies on attitudes towards robots or related technologies [3–5, 13–17] were reviewed and elements of, or relating to, attitudes proposed in the literature were collected and summarized.

2.2   Results

All responses to the questionnaire and interview were checked manually. Keywords and sentences that appeared to relate with attitudes toward robots were selected from the descriptions. These elements and sentences were then organized and grouped, eliminating duplications.

Additionally, elements related to the attitude discussed in the literature were added to the pool of keywords and sentences. As a result, 125 sentences representing people's perceptions, thoughts, and/or feeling towards robots were extracted. These sentences are referred to as attitude elements hereafter. Examples of attitude elements are: "I would feel relaxed with a robot in my home", "I can easily learn how to use a robot", and "I feel scared around robots".

3   Dimensions of Attitudes Towards Domestic Robots

In the second phase of the study, a questionnaire-based investigation was conducted to identify the dimensions constructing people's attitudes toward robots.

3.1   Methods

Participants. Participants were 175 Japanese adults (age: $M = 22.3$, $SD = 1.9$), 126 Chinese adults (age: $M = 23.6$, $SD = 1.6$), and 130 Taiwanese adults (age: $M = 24.2$, $SD = 5.0$). Among the participants, 77.8% of Japanese, 40.5% of Chinese, and 46.9% of Taiwanese participants were male.

Procedure and Measurement. The questionnaire used in this investigation consisted of two parts. The first part asked for participants' demographic information. The second part consisted of 125 question items corresponding to the 125 attitudinal elements extracted in the first phase. Participants were asked to indicate the extent to which each item matched their feelings and perceptions of domestic robots in general, by responding on a seven-point Likert scale.

The questionnaire investigation was conducted in the participants' respective native languages (Japanese, Chinese and Taiwanese) and in their own countries (Japan, China and Taiwan). Participants were asked to complete the questionnaire at their own pace and return it to the investigator.

3.2   Results

A factor analysis was conducted on all participants' ratings of attitudes toward robots. The number of factors was determined from a scree plot (Fig. 1).

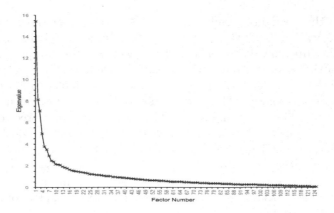

Fig. 1. Scree plot of the factor analysis.

The principal factor method with varimax rotation revealed a 12-factor structure. The cumulative contribution for the 12 factors was 45.5%. These 12 factors were considered to represent dimensions of attitudes toward robots.

Variables that had high loadings on the first factor included responses to items such as "If a robot was introduced to my home, I would feel like I have a new family member", "I would feel relaxed with a robot in my home", and "I like that a robot can encourage me". This factor was considered to represent people's familiarity with robots and was thus labeled "familiarity".

Variables that had high loadings on the second factor included "I would want to boast that I have a robot in my home", "If a robot is introduced to my home, I think my children or grandchildren will be pleased", and "If my friends use robots, I will also want one". This factor was considered to represent people's interest toward robots and was thus labeled "interest".

Variables that had high loadings on the third factor included "It would be a pity to have a robot in my home", "The movements of a robot are unpleasant", and "It is unnatural for a robot to speak in a human language". This factor was considered to represent a negative attitude toward robots and was thus labeled "negative attitude".

Variables that had high loadings on the fourth factor included "I have enough skills to use a robot", "I can make full use of a robot", and "It is easy to use a robot". This factor was considered to represent self-confidence in using robots and was thus labeled "self-efficacy".

Variables that had high loadings on the fifth factor included "I think the robot design should be cute", "I think robots should have animal-like shapes", and "I think the shape of a robot should have roundness". This factor related to people's expectations on appearance of robots and was thus labeled "appearance".

Variables that had high loadings on the sixth factor included "Robots are practical", "Robots are user-friendly", and "Robots have functions that I find satisfactory". This factor was considered to represent people's perceptions of the practical utility of robots and was thus labeled "utility".

Variables that had high loadings on the seventh factor included "I think robots are heavy", "I think the maintenance of robots is difficult, and "I worry about the robot breaking down". This factor was considered to represent people's perception of the costs of robots and was thus labeled "cost".

Variables that had high loadings on the eighth factor included "I think robots should make various sounds", "I think robots should have various shapes", and "I think robots should have various colors". This factor was considered to represent people's expectations for the diversity of robots and was thus labeled "variety".

Variables that had high loadings on the ninth factor included "I think a robot could recognize me and respond to me", "I think a robot would obey my commands", and "I want to tame a robot according to my preferences". This factor was considered to represent people's concepts of the control of robots and was thus labeled "control".

Variables that had high loadings on the tenth factor included "I expect my family or friends to teach me how to use a robot", "I expect my family or friends to help me when I use a robot", and "I expect my family or friends to advise me how to use a robot". This factor was considered to represent people's expectations of support from other people while using robots and was thus labeled "social support".

Variables that had high loadings on the eleventh factor included "Robots can be used by remote control" and "Robots can be controlled by a button (on the robot itself)". This factor was considered to represent people's expectations of and/or belief in the way to use robots and was thus labeled "operation".

Finally, on the twelfth factor, variables such as "I worry about whether robots are suitable for the state (layout of the furniture and other things) of my room now" and "I worry that robots are suitable for the circumstances (width or numbers of ramps) of my room now" had high positive loadings. This factor appeared to represent the environmental adaptation of robots. Thus, this factor was labeled "environmental fit".

These 12 factors are considered to represent 12 dimensions of attitudes toward robots. The factor structures were also analyzed by the groups of participants from different regions (i.e., Japan, China, and Taiwan). The results showed that the 12-factor structure above was robust across participant groups from these different regions[18].

4   Development of the Attitude Scale

In the final phase of this study, the Multi-dimensional Robot Attitude Scale, which assesses attitudes towards robots according to the extracted 12 dimensions, was developed and proposed.

4.1   Constructs of the Scale

For each of the 12 factors extracted in the previous section, two to seven elements were chosen as representative elements. These representative elements were selected according to two criteria: (1) the elements must have high loadings for the factor, and (2) the elements must differ from one another so that the set of representative elements for the factor as a whole represent as diverse a set of concepts as possible. Specifically, for each dimension, variables were selected from the ones with the highest loadings. If the selected variable was similar or overlapping in meaning with the ones already chosen, other variables with next highest loadings and more deiverse meaning could be selected instead. Tables 1 and 2 present the 49 selected elements representing 12 factors.

The Multi-dimensional Robot Attitude Scale is a questionnaire that has the 49 elements given in Tables 1 and 2 as items. The scale assesses people's attitudes towards robots by asking people to rate to what extent each of the items matches their feelings or thoughts about robots on a seven-point Likert scale (-3: not at all, 3: very much). The responses to the items are averaged by dimensions, and the average scores are used as the scores of the attitude dimensions.

4.2   Reliability and Representativeness of the Scale

The internal reliability of each of the 12 dimensions were assessed. Table 3 gives Cronbach's alpha coefficients for items constructing each of the 12 dimensions. The alpha coefficients in Table 3 suggest that the scale scores have overall good internal reliabilities.

Furthermore, to confirm that the scores of the proposed scale represent the original 12-factor structure extracted in the second phase of this study, correlations between the scale scores and the corresponding factor scores were analyzed. Table 3 gives Pearson's correlation coefficients for scale scores and factor scores of the 12 dimensions. All 12 scale scores had significant ($p < 0.01$) correlations with the corresponding factor scores. This result suggests that these scale scores well represent the original factor structure.

5   Discussion

This study proposed a measurement that assesses people's attitude towards domestic robots along multiple dimensions. Although a number of measurements assessing impressions about robots have been proposed (e.g., [19,20]), relatively few measurements have been proposed for attitudes. Among measurements of attitudes towards robots, most have focused on rather limited dimensions of attitudes, such as negative attitudes[6,7]. The Multi-dimensional Robot Attitude Scale proposed in this study assesses people's attitudes towards robots in a comprehensive way using 12 dimensions, and providing a multifaceted understanding of attitudes toward this technology. The 12-factor structure developed in this study was found to be robust across cultures and generations.[18,21].

Assessing attitudes towards robots using a larger number of dimensions allows us to show multifaceted individual differences in attitudes, not merely positive or negative attitudes. Previous studies on attitudes towards computers have argued that the acceptance of computers and other technologies is related

Table 1. Items of the Multi-dimensional Robot Attitude Scale

Dimension	Question items
1. Familiarity	If a robot was introduced to my home, I would feel like I have a new family member. I would feel relaxed with a robot in my home. I like that a robot can encourage me. I think a robot can be a communication partner. I want to converse with a robot.
2. Interest	I would want to boast that I have a robot in my home. If a robot is introduced to my home, I think my children or grandchildren will be pleased. If my friends use robots, I will also want one. I want to use robots if I can use them with my friends. Robots are neo-futuristic and cutting-edge. It is good if a robot can do the work of a human. I feel easy around robots because I do not need to pay attention to robots as I do to humans.
3. Negative attitude	It would be a pity to have a robot in my home. The movements of a robot are unpleasant. It is unnatural for a robot to speak in a human language. I feel like I also become a machine when I am with a robot. I feel scared around robots.
4. Self-efficacy	I have enough skills to use a robot. I can make full use of a robot. It is easy to use a robot. I can easily learn how to use a robot.
5. Appearance	I think the robot design should be cute. I think robots should have animal-like shapes. I think the shape of a robot should have roundness. I think the voice of a robot should be like the voice of a living creature. I think the design of a robot should be beautiful. I think the design of a robot should be cool. I think a robot should have human-like shape.
6. Utility	Robots are practical. Robots are user-friendly. Robots have functions that I find satisfactory with. Robots are convenient. I feel the necessity for robots in my daily life.
7. Cost	I think robots are heavy. I think the maintenance of a robot is difficult. I worry about the robot breaking down.
8. Variety	I think robots should make various sounds. I think robots should have various shapes. I think robots should have various colors.

Table 2. Items of the Multi-dimensional Robot Attitude Scale (cont'd)

Dimension	Question items
9. Control	I think a robot could recognize me and respond to me. I think a robot would obey my commands. I want to tame a robot according to my preferences.
10. Social support	I expect my family or friends to teach me how to use a robot. I expect my family or friends to help me when I use a robot. I expect my family or friends to advise me how to use a robot.
11. Operation	Robots can be used by remote control. Robots can be controlled by a button (on the robot itself).
12. Environmental fit	I worry that robots are suitable for the state (layout of the furniture and other things) of my room now. I worry that robots are suitable for the circumstances (width or numbers of ramps) of my room now.

to various dimensions of attitudes[1, 8, 22, 23]. There is a broader variety of types of robots than types of computers. Acceptance of robots with various characteristics might be better explained in relation with diverse dimensions of attitudes. The relations between the acceptance of a variety of robots and different dimensions of attitudes should be further investigated in future studies.

Having 12 dimensions, the proposed Multi-dimensional Robot Attitude Scale is designed to be a comprehensive attitude measurement tool that can assess diverse aspects of attitudes towards robots. Thus, the scale is expected to be used as a common reference in a wide range of human–robot interaction research.

The scale proposed in this study was designed to assess people's attitudes towards domestic robots in general. Conversely, as discussed above, there are broad variations in the characteristics of robots. Measures that assess these variations in robot characteristics have also been proposed (e.g., [24]). The relationships between these attitude dimensions and acceptance may differ depending on the characteristics of robots, which should be investigated in future studies.

It is also possible to develop a multi-dimensional attitude scale for specific types of robots, rather than domestic robots in general. How to design such attitude measurements for specific types of robots should also be further studied.

At the time of the investigation, the samples of this study were not supposed to have sufficient experiences in interacting with real robots. Thus the dimensions extracted in this study are reflecting the viewpoints of people who are expecting their future experiences with real robots. In that sense these dimensions are more appropriate to discuss in relation with their acceptance. In future, the attitude

Table 3. Cronbach's alpha coefficients and correlations for factor scores by the scale dimensions.

Dimension	Cronbach's alpha	Pearson's correlation
1. Familiarity	0.829	0.869 **
2. Interest	0.737	0.819 **
3. Negative attitude	0.731	0.920 **
4. Self-efficacy	0.857	0.923 **
5. Appearance	0.702	0.776 **
6. Utility	0.783	0.938 **
7. Cost	0.562	0.710 **
8. Variety	0.749	0.766 **
9. Control	0.643	0.615 **
10. Social support	0.923	0.838 **
11. Operation	0.750	0.554 **
12. Environmental fit	0.933	0.917 **

** $p < 0.01$

dimensions may be revised to reflect the viewpoints of people who already have actual experiences in interacting with robots, so that the scale would be more appropriate to measure the attitudes of people who are actually engaged with them.

Finally, although participants from three different regions (i.e., Japan, mainland China, and Taiwan) were surveyed in this study, the study was still limited to East Asia. Individual differences in attitudes towards robots, as well as the relationships between attitude dimensions and acceptance of robots with various characteristics, should be investigated involving participants with more diverse backgrounds in terms of age, sex, and culture, in order to provide wider social perspectives.

References

1. Loyd, B.H., Gressard, C.: Reliability and factorial validity of computer attitude scales. Educ. Psychol. Meas. **44**, 501–505 (1984)
2. Broadbent, E., Stafford, R., MacDonald, B.: Acceptance of healthcare robots for the older population: Review and future directions. Int. J. Soc. Robot. **1**, 319–330 (2009)
3. Ezer, N., Fisk, A.D., Rogers, W.A.: Attitudinal and intentional acceptance of domestic robots by younger and older adults. In: Stephanidis, C. (ed.) UAHCI 2009, Part II. LNCS, vol. 5615, pp. 39–48. Springer, Heidelberg (2009)
4. Khan, Z.: Attitudes towards intelligent service robots. TRITA-NA-P9821, IPLab-154, Sweden Royal Institute of Technology, Numerical Analysis and Computing Science, Interaction and Presentation Laboratory (1998)

5. Ray, C., Mondada, F., Siegwart, R.: What do people expect from robots? In: Proc. 2008 IEEE/RSJ International Conference on Intelligent Robots and Systems, Nice, France, pp. 3816–3821 (2008)

6. Nomura, T., Kanda, T., Suzuki, T., Kato, K.: Measurement of anxiety toward robots. In: Proc. 15th IEEE International Symposium on Robot and Human Interactive Communication, pp. 372–377 (2006)

7. Nomura, T., Kanda, T., Suzuki, T., Kato, K.: Prediction of human behavior in human-robot interaction using psychological scales for anxiety and negative attitudes toward robots. IEEE Trans. Robot. **24**(2), 442–451 (2008)

8. Jay, G.M., Willis, S.L.: Influence of direct computer experience on older adults' attitudes towards computers. J. Gerontol. B-Psychol. **47**, 250–257 (1992)

9. iRobot: Roomba. http://www.irobot-jp.com/product/index.html. Retrieved June 9, 2015

10. Information and Robot Technology Research Initiative: Home-assistant robot. http://www.irt.i.u-tokyo.ac.jp/reform/081024/index.shtml. Retrieved June 9, 2015

11. NEC Corporation: PaPeRo. http://jpn.nec.com/robot/. Retrieved June 9, 2015

12. National Institute of Advanced Industrial of Science Technology: Paro. http://paro.jp/. Retrieved June 9, 2015

13. Dautenhahn, K., Woods, S., Kaouri, C., Walters, M.L., Koay, K.L., Werry, I.: What is a robot companion–friend, assistant or butler? In: Proc. IEEE IRS/RSJ Int. Conf. Intelligent Robots and Systems, pp. 1488–1493 (2005)

14. Cesta, A., Cortellessa, G., Rasconi, R., Pecora, F., Scopelliti, M., Tiberio, L.: Monitoring elderly people with the robocare domestic environment: Interaction synthesis and user evaluation. Comp. Intell. **27**(1), 60–82 (2011)

15. Tinker, A., Lansley, P.: Introducing assistive technology into the existing homes of older people: Feasibility, acceptance, costs and outcomes. J. Telemed. Telecare **11**(suppl. 1), 1–3 (2005)

16. Young, J.E., Hawkins, R., Sharlin, E., Igarashi, T.: Toward acceptable domestic robots: Applying insights from social psychology. Int. J. Soc. Robot. **1**(1), 95–108 (2009)

17. Salvini, P., Laschi, C., Dario, P.: Design for acceptability: Improving robots' coexistence in human society. Int. J. Soc. Robot. **2**(4), 451–460 (2010)

18. Fujita, A., Suzuki, D., Umemuro, H.: Cross-regional comparative study of dimensions of people's attitudes toward robots. In: Proc. 22nd IEEE International Symposium on Robot and Human Interactive Communication, pp. 332–333 (2013)

19. Kanda, T., Ishiguro, H., Ishida, T.: Psychological evaluations on interactions between people and robot. J. Robot. Soc. Jpn. **19**(3), 362 371 (2001)

20. Bartneck, C., Kulic, D., Croft, E., Zoghbi, S.: Measurement instruments for the anthropomorphism, animacy, likeability, perceived intelligence, and perceived safety of robots. Int. J. Soc. Robot. **1**, 71–81 (2009)

21. Suzuki, D., Umemuro, H.: Dimensions of people's attitudes toward robots. In: Proc. 7th ACM/IEEE Int. Conf. Hum-Robot Interaction, pp. 249–250 (2012)

22. Umemuro, H.: Computer attitudes, cognitive abilities, and technology usage among older Japanese adults. Gerontechnology **3**(2), 64–76 (2004)

23. Umemuro, H., Shirokane, Y.: Elderly Japanese computer users: Assessing changes in usage, attitude, and skill transfer over a one-year period. Univ. Access. Info. Soc. **2**(4), 305–314 (2003)

24. Ninomiya, T., Suzuki, D., Umemuro, H.: Dimensions of robot characteristics that people perceive as significant for engagement with Robots. In: Proc. 43rd Int. Symp. Robot., pp. 1100–1105 (2012)

Humanoid Robot Assisted Training
for Facial Expressions Recognition
Based on Affective Feedback

Eleuda Nunez[✉], Soichiro Matsuda, Masakazu Hirokawa, and Kenji Suzuki

Artificial Intelligence Laboratory, University of Tsukuba,
1-1-1 Tennodai, Tsukuba, Ibaraki 305-8573, Japan
{eleuda,matsuda,hirokawa}@ai.iit.tsukuba.ac.jp, kenji@ieee.org

Abstract. Different studies explore the use of electronic screen media to train specific social skills, since they provide visual elements which are well accepted by children with ASD. However, even if children have high levels of attention to the screen, this might also reduce the possibilities of interacting with others. The challenge lays on finding ways to encourage and motivate the child to keep learning and share the experience with others. For this reason, we propose a feasible framework composed of three technologies: a tablet to train facial expressions recognition, a humanoid robot as reinforcer, and a wearable device to quantify smiles. We did a feasibility test with adults not only to verify the machine response, but also to obtain qualitative data regarding the interaction with the robot. In this study we analyze the importance and the synergistic effect of combining screen media with the robot embodiment and affective computing technologies.

Keywords: Robot assisted therapy · Human-robot interaction · Affective computing technologies · Electronic screen media

1 Introduction

Technology plays an important role in the diagnosis and treatment of Autism Spectrum Disorder (ASD) as it can be adapted based on each individual's abilities and necessities. Since children with ASD has shown special interest to technology, the number of studies that explores uses of technology is increasing. Among the interventions using technology, one of the most commonly used involves visual prompts. Children with ASD are considered visual learners, having relatively more preference for visual cues compared to spoken language [1]. For that reason, different studies explore the use of electronic screen media to train specific social skills, since they provide visual elements which are well accepted by children with ASD. Screen media can also reduce the distraction from the environment resulting to be less demanding for children with ASD [2]. However, this approach keeps children immersed and fixed on the screen content, which might encourage repetitive and stereotypical pattern of behaviors. For this

© Springer International Publishing Switzerland 2015
A. Tapus et al. (Eds.): ICSR 2015, LNAI 9388, pp. 492–501, 2015.
DOI: 10.1007/978-3-319-25554-5_49

reason, it has been proposed that computer interventions should include tangible interfaces to enhance the experience and allow the children to interact with the environment, making the learning process more natural [3]. The challenge lays on finding ways to encourage children to focus not only on the screen, but also to encourage them to interact with people around.

One potential solution involves the use of socially assistive robots on therapies. Robot's physical embodiment has different desirable effects on children with ASD [4]. Moreover, robots can be used to elicit behaviors, to train a specific social skill, provide feedback or encouragement [5]. Different studies have reported that the artificial appearance and visual stimuli make robots more attractive and easier to understand compared to people. Humanoid robots with simplified appearance are widely used in robot assisted therapies considering that only human form can be used to train certain skills [6]. Robots can be used not only to train or encourage children, but also to assist the therapist analyzing the child's behavior. There is a necessity of quantitative measurement of social behaviors that allow to objectively track the children progress. By using robots on the therapies it is also possible to work under the same stimuli on each session [7].

In this study we propose a feasible framework that combines a screen media application with a humanoid robot and affective technologies in order to teach facial expressions recognition to children with ASD. We divided this framework in three different parts according to the role during the activity: 1) screen media to teach social skills, 2) wearable devices to measure social behaviors and 3) socially assistive robot to encourage and reinforce the child. We will explain how these three elements come together in order to facilitate the social skills learning to the child, and help the therapists to follow the child performance. At the end of this report, we show the results of a feasibility study and describe the challenges for this research in the future evaluations.

2   Framework Components

The proposed system is composed of three sections (Figure 1). A tablet application for training children to recognize emotional facial expressions, a wearable device to detect smiles using EMG signals and a humanoid robot as reinforcer to motivate children to keep learning.

2.1   Training: Face Expression Expert Program (FEEP)

In the past, different studies reported the benefits of training social skills using screen media [2]. We had developed the Face Expression Expert Program (FEEP), a screen media application for training emotion recognition [8]. FEEP runs on a tablet and it consists on ten different levels, starting from simple emotional facial expression recognition until context-based emotion recognition. The sample stimuli are four different facial expressions (happy, sad, surprised and angry), and using a matching to sample (MTS) activity, the subject is presented

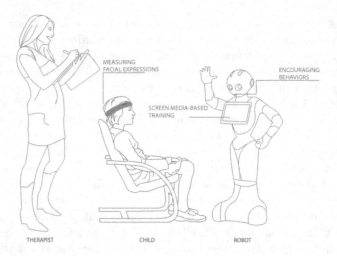

Fig. 1. Proposed framework made of three different technologies for training, measuring and encouraging.

with different situations on which he or she has to identify the appropriated emotion (Figure 2). Every time the subject answer correctly, FEEP provides auditory feedback and then moves to the next question. After finishing one activity, which consist of about 8 questions, the subject is rewarded with a short video in order to keep him or her motivated to continue learning. The video is chosen based on each child's preferences.

So far, previous studies using FEEP to train emotion recognition on young children have shown the potential of this interface, and we are encouraged to keep exploring its effect. However, one of the challenges is to find a way to keep children motivated to keep learning using FEEP during the therapies. Reinforcement is a key element, considering that when children enjoy the activities, they have the highest disposition to learn. One study in the past using screen media to train emotion recognition reported the importance of including physical interaction during the activity [3].

2.2   Measuring: Smile Detector

There is the necessity of designing tools that provide quantitative and objective data to facilitate the treatment and diagnosis of autism [7]. The potential of using wearable devices lays on the data collection that describes specific behaviors [9]. In the past, we developed a wearable device to detect different facial expressions based on muscles activity [10]. The device uses 2 pairs of dry-type electrodes that measures EMG signals from both sides of the face (Figure 2). This configuration, allowed us to set and remove the device quickly, which is beneficial specially when treating with children with ASD. Compared to computer vision approaches, this device allows to collect data even if the subject is not looking at the camera.

Fig. 2. One participant completing the activity using FEEP. The robot was used to give instructions and the wereable device to meassure the smiles.

We confirmed the device performance and reported an experience with children with ASD wearing the device [10] [11].

There are several benefits of including affective technologies in therapies. Physiological signals can provide significant emotion related information, and specifically EMG signals from the face can tell us about pleasure or displeasure. In other words, these technologies have the potential to help children to communicate their interests. Based on the potential of affective computing solutions, in a previous study we found that children with ASD showed more positive social behaviors when the amount of smiles increases [11]. This suggests that if the child is placed on situations that make him or her smile, this might lead to positive behaviors. For this reason, we want to explore the use of this device on robot assisted therapies to better understand the level of enjoyment and positive disposition to participate on the activity.

2.3   Encouraging: Socially Assistive Robot

In the past, a study explored the uses of a robot with an embedded screen to introduce stories to children with ASD [12]. The results showed the potential of using robots combined with screens to facilitate children's learning. In this study we are using a humanoid robot Pepper, from Aldebaran Robotics and the Softbank Group. In this proposed activity, the role of the robot is like a co-therapist, and it will be used to introduce the activity and provide encouragement every time the child finishes one session with FEEP. We expect to also encourage the child to share the experience with the therapist, reducing the periods when the child is immersed on the screen content. Different studies have proposed a set of behaviors for humanoid robots on different scenarios [13] [14]. These studies focused on using the robots to elicit behaviors on the participant, and they pointed out as an important benefit the structured way robots can deliver prompts. Based on this potential, we followed some design parameters from [15], and we will build our framework with the following requirements:

1. *Appearance*: using a robot that is visually engaging. Pepper will use LED color and the screen contents to try to catch the child attention.

Robot behavior	Description of the interaction scenario	Robot exhibited behavior
Behavior 1	Pepper introduces itself and calls out the participants name	(Happy animation)Hello, my name is pepper. (Enthusiastic animation)Nice to meet you (Name). Let's play together
Behavior 2	Pepper briefly introduces the activity to the participant	(Hands gestures) Let's do the activity displayed on my screen (Gestures toward screen) Try touching the screen
Behavior 3	Pepper gestures toward the screen	(One hand points at the screen and the head look to the screen and wait)
Behavior 4	Pepper rewards the participants effort with an special animation	Option 1(High Five): (Enthusiastic animation) Good Job! Give me a High Five!(Pepper perform the motion and lift the arm. Waits 2 second and put it down)(Enthusiastic animation) It was fun, let's keep playing together Option 2(Song): (Happy animation) Good Job! Start the music. (Pepper slowly dance a selected song for 20 seconds) (Happy animation) It was fun, let's keep playing together
Behavior 5	Pepper says good bye	(Bow animation) Thank you very much! It was fun!

Behavior 1   Behavior 2   Behavior 3   Behavior 4   Behavior 5

Fig. 3. Brief description of the robot behaviors (Top) and photos of the different robot behaviors (Bottom).

2. *Safety*: considering that the child will be close to the robot embodiment to manipulate the tablet, we designed slow and controlled animations. This includes remote control that allow therapist to execute the robot behaviors based on the observations of the child's performance.

3. *Autonomy*: we aim for a semi-autonomous system using a server that allows changing messages from the tablet to the robot, to give it certain autonomy. This will reduce the human intervention only for those cases when it is necessary (according to the child's behavior), to try to offer a constant stimuli to the child.

Based on these requirements we started defining the interactions scenarios. We asked the therapist to describe a common activity flow using FEEP, and how it is usually introduced to the child. From this, we came with a set of five behaviors for the robot (Figure 3). Each scenario targets a response from the participant, and this is the condition to execute the next robot behavior. On this interface, the controller is faced with questions related to the child's behavior, and by answering yes or no, the program executes a behavior on the robot based on the flow chart (Figure 4). Additionally to the yes or no options,

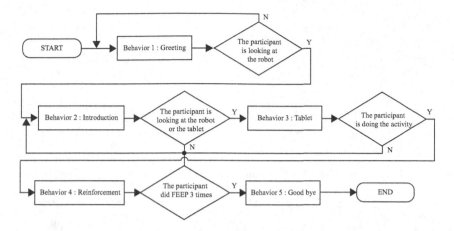

Fig. 4. Flow diagram with the conditions to execute each behavior.

the controller can input numbers that will call any behavior at any moment during the session, and for example, if the participant lose the motivation on the activity, the controller can call the last behavior at anytime to finish the session.

3   Performance Evaluation

We did a feasibility study with ten participants (average age 23) 4 males and 6 females. We are trying to verify the quality of the user experience with the robot. Before starting the session, we asked each participant about past experiences with robots. Also, we requested them to rate from one (Not very much) to five (Very much) how comfortable they are with the idea of having robots in everyday life. The participant sat down in front of the robot with enough separation to reach the tablet. Then we instructed them about the way to use FEEP and we asked them to follow the robots commands regarding when to start or stop the activity. The controller was standing behind a panel observing the participant's interaction with the robot.

The session started with the first behavior (Figure 4), and going forward only if the participant met the condition. The robot speech was translated into Japanese, considering that the expected subjects are Japanese children. During the session the participant received instructions only from the robot. The session finished after three FEEP lessons were completed.

After the session, the participants were asked to answer two questionnaires. The first one was the International Positive and Negative Affect Schedule Short Form (I-PANAS-SF) [16] since the participants are not English native. This questionnaire included 20 items referring to positive and negative affects, and the participants rated them from 1 to 5 according to the extend they felt while interacting with the robot (momentary experience). Then they were asked to fill an adapted version of the Intrinsic Motivation Inventory (IMI) questionnaire [17] to evaluate the user's opinion of the scenario. From the 5 subscales, we chose

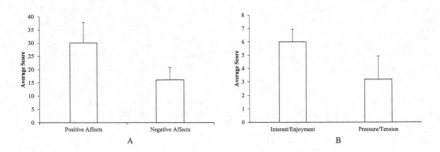

Fig. 5. A) Results from I-PANAS-SF questionnaire B) Results from IMI questionnaire.

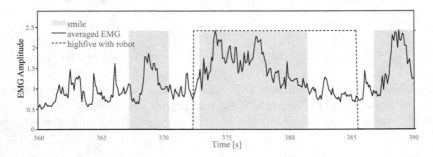

Fig. 6. Data collected: the gray colored area is when a smile was detected and the area under the dotted line is when the robot was doing reinforcement.

the ones that are more relevant to the experience: Interest/Enjoyment and Pressure/Tension. Each subscale is composed of seven items, and we selected three for each subscale. These six items were randomly ordered, and the statements were slightly modified to fit this specific task.

The second part of the feasibility study includes the smile detector. From the 10 participants we selected one (Female, 28 years old) without a special criteria, and asked her to wear the device to quantify smiles. With this test we want to verify the potential use of the three components of this framework working together, and to find out the best way to use this device during therapies.

The purpose of this feasibility study was basically to observe how the participant interact under each robot prompt and confirm the response of the developed system. Since it is difficult to perform a qualitative analysis with children with ASD, the answers of these questionnaires will provide us with insights about the user experience and how comfortable they were while interacting with this robot.

4   Results

Regarding the familiarity with robots 4 participants reported to have interacted with robots in the past, but none of the ten participants had ever programmed or worked with robots. In average, the participants rated 4.1 referring to the question of including robots in everyday life. Figure 5A shows the results from the I-PANAS-SF questionnaire. The error bars represent the standard deviation. We obtained an average of 30.10 out of 50 ($SD=7.81$) of positive affects and 16.10 out of 50 ($SD=4.65$) of negative affects. Then, Figure 5B represents the results from the IMI questionnaire with a score of 5.999 out of 7 ($SD=0.96$) for the subscale of interest/enjoyment and 3.20 out of 7 ($SD=1.72$) for the subscale of pressure/tension. Regarding the second evaluation using the smile detector, we chose to analyze an interval that includes the robot reinforce behavior. Figure 6 shows four EMG signal from the participant during the activity. The grey colored areas represent the intervals when a smile was detected and the area under the dotted lines is the interval when the robot did the reinforcement behavior.

5   Discussions

We performed a feasibility study and we could verify the machine response and the interaction scenarios. The first test included 10 participants who interacted with the robot while completing three lessons using FEEP. These participants reported not to be very used to robots, but in general they showed high interest on them. Regarding the results from the I-PANAS-SF questionnaire, a high score of positive affect refers to a state of high energy, pleasurable engagement and high concentration, while a high negative affect refers to anger, fear and nervousness. Generally for this questionnaire, the mean score for momentary positive affects is 29.70 ($SD=7.90$) compared to our result 30.10 ($SD=7.81$). In the other hand, the mean score for momentary negative affects is 14.80 ($SD=5.40$) compared to our result 16.10 ($SD=4.65$). Both, positive and negative affects, scored slightly higher than the mean. These results match the comments we obtained from the participants saying that they could enjoy the activity in general and also, they could focus on the tablet content since the robot stop moving when they were working at it. However, sometimes the robot movements and reactions made them feel nervous or insecure, especially the reinforcement sequence considering that it was executed after the user was using the screen and it also required the user to touch the robot. One of the possible causes is that the participants were sitting very close to the robot while manipulating the screen.

The results from the IMI questionnaire are in line with the ones from PANAS, since the average for the subscale of Interest/Enjoyment was high, scoring 5.99 out of 7, while the Pressure/Tension was more neutral, scoring 3.20 out of 7. Again we observed the influence of the negative affects related to nervousness, fear or pressure, but they did not affect the general enjoyment of the activity with the robot. To reduce the negative affects we plan to make sure not to move the robot arms when the participants are manipulating the table, and instead, make the robot ask them to make space. Even if it is not possible to tell how children will feel when using this system, we can work on reducing the causes that might elicit negative affects based on the obtained feedback.

The second test included the wearable device to detect smiles. We wanted to verify the potential of using the device on this framework. Figure 5 shows for example, an increase of EMG signals during the robot reinforce behavior. We could observe that it is possible to follow the participants positive affective behaviors during the sessions, and understand when the participant smiles, and which stimuli was eliciting it. We consider this kind of affective technology an especially important component for this therapy, since children with ASD have difficulties to express their emotions or interest. We expect that by being able to objectively quantify their smiles and provide a constant and adequate stimuli, the therapy will be more effective.

In the current implementation, all the robot behaviors are manually controlled according to observations of the participant's behavior. Future improvements include making the system semi-autonomous reducing the human intervention based on the importance of standardizing the sessions. We plan to send messages from FEEP to the robot every time the subject finishes the lesson, and from the wearable device every time the subject is smiling to make the robot react emphatically to the subject.

6   Conclusions

In this paper we proposed a protocol composed of three technologies to assist therapies for children with ASD: screen media based learning using FEEP, facial expression measuring using a wearable device and encouraging behaviors using a humanoid robot. The first component, FEEP, uses visual stimuli to train emotional facial expressions recognition. In the past, we showed the potential of this interface. However, we want to explore other ways to encourage children to keep learning and share the experience with others around. For this reason we proposed the use of a humanoid robot as a co-therapist. We designed different behaviors with the purpose of eliciting some actions from the subject. Every time one lesson was completed, the robot showed positive reinforcement. The last component involves affective technologies, and it consists of a wearable device to detect facial expressions. In this framework we use it to measure the smiles as a cue of motivation and positive affects. Having finished the feasibility test and checked the function of each component, the next step would be to test it with children with ASD. With this approach, we could verify which aspects of the robot behavior have more negative affects on the users. We consider that this qualitative analysis represents an important initial step, since is almost impossible to obtain any qualitative data from children with ASD. Combining these three technologies we expect to verify on which ways the robot can assist children with ASD recognizing emotions using FEEP.

References

1. Lal, R., Bali, M.: Effect of visual strategies on development of communication skills in children with autism. Asia Pacific Disability Rehabilitation Journal 18(2), 120–130 (2007)

2. Mineo, B.A., Ziegler, W., Gill, S., Salkin, D.: Engagement with electronic screen media among students with autism spectrum disorders. Journal of Autism and Developmental Disorders **39**(1), 172–187 (2009)
3. Miranda, J.C., Fernandes, T., Sousa, A., Orvalho, V.: Interactive technology: teaching people with autism to recognize facial emotions. In: Autism Spectrum Disorders-From Genes to Environment, pp. 299–312 (2011)
4. Kim, E., Berkovits, L., Bernier, E., Leyzberg, D., Shic, F., Paul, R., Scassellati, B.: Social robots as embedded reinforcers of social behavior in children with autism. Journal of Autism and Developmental Disorders **43**(5), 1038–1049 (2013)
5. Diehl, J., Schmitt, L., Villano, M., Crowell, C.: The clinical use of robots for individuals with autism spectrum disorders: A critical review. Research in Autism Sprectrum Disorders **6**(1), 249–262 (2012)
6. Ricks, D.J., Colton, M.B.: Trends and considerations in robot-assisted autism therapy. In: IEEE International Conference on Robotics and Automation (ICRA), pp. 4354–4359 (2010)
7. Scassellati, B., Admoni, H., Mataric, M.: Robots for use in autism research. Annual Review of Biomedical Engineering **14**, 275–294 (2012)
8. Matsuda, S., Yamamoto, J.: Computer-based intervention for inferring facial expressions from the socio-emotional context in two children with autism spectrum disorders. Research in Autism Spectrum Disorders **8**(8), 944–950 (2014)
9. Hirokawa, M., Funahashi, A., Itoh, Y., Suzuki, K.: Design of affective robot-assisted activity for children with autism spectrum disorders. In: The 23rd IEEE International Symposium on Robot and Human Interactive Communication, pp. 365–370 (2014)
10. Gruebler, A., Suzuki, K.: Design of a wearable device for reading positive expressions from facial emg signals. IEEE Transactions on Affective Computing **5**(3), 227–237 (2014)
11. Funahashi, A., Gruebler, A., Aoki, T., Kadone, H., Suzuki, K.: Brief report: the smiles of a child with autism spectrum disorder during an animal-assisted activity may facilitate social positive behaviors-quantitative analysis with smile-detecting interface. Journal of Autism and Developmental Disorders **44**, 685–693 (2014)
12. Vanderborght, B., Simut, R., Saldien, J., Pop, C., Rusu, A.S., Pintea, S., Lefeber, D., David, D.O.: Using the social robot probo as a social story telling agent for children with ASD. Interaction Studies **13**(3), 348–372 (2012)
13. Gillesen, J.C.C., Barakova, E.I., Huskens, B.E.B.M., Feijs, L.M.G.: From training to robot behavior: towards custom scenarios for robotics in training programs for ASD. In: IEEE International Conference on Rehabilitation Robotics (ICORR), pp. 1–7 (2011)
14. Hamzah, M.S.J., Shamsuddin, S., Miskam, M.A., Yussof, H., Hashim, K.S.: Development of interaction scenarios based on pre-school curriculum in robotic intervention for children with autism. Procedia Computer Science **42**, 214–221 (2014)
15. Giullian, N., Ricks, D., Atherton, A., Colton, M., Goodrich, M., Brinton, B.: Detailed requirements for robots in autism therapy. In: IEEE International Conference on Systems Man and Cybernetics (SMC), pp. 2595–2602 (2010)
16. Watson, D., Clark, L.A., Tellegen, A.: Development and validation of brief measures of positive and negative affect: The PANAS scales. Journal of Personality and Social Psychology **54**(6), 1063–1070 (1988)
17. Plant, R.W., Ryan, R.M.: Intrinsic motivation and the effects of self-consciousness, self-awareness, and ego-involvement: An investigation of internally controlling styles. Journal of Personality **53**(3), 435–449 (1985)

Designing Robotic Teaching Assistants: Interaction Design Students' and Children's Views

M. Obaid[1](✉), W. Barendregt[2], P. Alves-Oliveira[3], A. Paiva[3], and M. Fjeld[1]

[1] t2i Lab, Chalmers University of Technology, Gothenburg, Sweden
mobaid@chalmers.se
[2] Department of Applied IT, Gothenburg University, Gothenburg, Sweden
[3] INESC-ID & Instituto Superior Técnico, Universidade de Lisboa,
Lisbon, Portugal

Abstract. This paper presents an exploratory study on children's contributions to the design of a robotic teaching assistant for use in the classroom. The study focuses on two main questions: 1) How do children's designs differ from interaction designers'? 2) How are children's designs influenced by their knowledge of robotics (or lack thereof)? Using a creative drawing approach we collected robot drawings and design discussions from 53 participants divided into 11 groups: 5 groups of interaction designers (24 participants), 3 groups of children with robotics knowledge (14 participants), and 3 groups of children without formal robotics knowledge (15 participants). These data revealed that (1) interaction designers envisioned a small or child-sized non-gendered animal- or cartoon-like robot, with clear facial features to express emotions and social cues while children envisioned a bigger human-machine robot (2) children without formal robotics knowledge, envisioned a robot in the form of a rather formal adult-sized human teacher with some robotic features while children with robotics knowledge envisioned a more machine-like child-sized robot. This study thus highlights the importance of including children in the design of robots for which they are the intended users. Furthermore, since children's designs may be influenced by their knowledge of robotics it is important to be aware of children's backgrounds and take those into account when including children in the design process.

1 Introduction

Involving users in the design of technologies is an important step to ensure the technologies' usefulness and acceptance. As robots in different forms are currently entering our daily lives, they are also entering children's lives as toys, programmable objects, teaching aids, or even as tutors (for example in the EMOTE project[1]). We therefore think that children's views on the design of robots for the classroom should be taken into account.

However, as Dautenhahn [1] indicates, what the word 'robot' means is not fixed and changes over time. In fact, real advances in robot technology, fictitious

---

[1] EMOTE project: http://www.emote-project.eu/

© Springer International Publishing Switzerland 2015
A. Tapus et al. (Eds.): ICSR 2015, LNAI 9388, pp. 502–511, 2015.
DOI: 10.1007/978-3-319-25554-5_50

capabilities of robots displayed in the media, and exposure to existing robots, may influence how children envision the design of a robot for use in the class-room. While it is hard to investigate the influence of exposure to robots in the media on children's expectations about a robotic teaching assistant, it may be possible to explore how actual knowledge of robotics may influence children's design contributions. Therefore, the aim of this study is to explore the design views on robots in educational settings with three groups: interaction design students, children without any formal robotics knowledge (non-robotics children) and children with some formal robotics knowledge (robotics children). By doing so, we want to explore the following research questions: 1) How do children's designs differ from interaction designers'? 2) How are children's designs influenced by their knowledge of robotics (or lack thereof)?

In the following section we present related research on the use and design of robots in education, leading us to define our study on children's contributions to the design of robotic teaching assistants.

2   Related Work

Human-Robot Interaction (HRI) researchers have addressed the deployment of robots in educational settings such as schools to help in assisting teachers in their classrooms. This is one of the main application areas in the HRI field. Mubin et al. [3], give an overview of the research development of robots in educational use within the HRI field, showing that robots in education are generally deployed as assistants to the teacher. Other researchers have presented findings of studies investigating children's design requirements for robots in different contexts. Sciutti et al. [5], for example, found that opinions on what features were considered important in a robot companion change with age: before the age of nine, children pay more attention to a human-like robot appearance; older children and adults are inclined to think more of its skills and functions. They also found that when children have been able to see and interact with a robot they pay more attention to perception and motor abilities in a robot, rather than just its shape. This suggests that actual experience with robots, such as in a robotics class, may influence children's design requirements.

Woods [7] performed two studies to investigate childrens views on the design of robots in general. In the first she investigated childrens views on robot appearance, movement, gender, and personality. Children between 9 and 11 were asked to choose a robot picture and fill out a questionnaire. The pictures displayed different robot attributes: mode of locomotion, body shape, looking like an animal, human or machine, the presence or absence of facial features, and gender. The questionnaire contained questions about the robots appearance and personality. Based on this data Woods identified two dimensions in childrens evaluations termed 'Emotional expression, ranging from happy to sad, and Behavioral intention, including friendliness, shyness and fright versus aggressiveness, bossiness and anger. Human-machine robots were considered the most friendly, shy and frightened types of robots. However, each of the robot attributes in isolation could not explain why a robot was placed in each of these categories. Woods thus argued that robot designers should *"consider a combination of physical characteristics rather than focusing specifically on certain features in isolation"* [7]. Furthermore, there was also tentative evidence for the Uncanny Valley effect, where children were increasingly positive towards robots that were more human-machine like instead of purely machine-like, but showed a sharp drop in positive attitude towards robots that were very human-like.

In the second study a similar methodological approach using pictures and a questionnaire was adopted [8]. However, in this study the children were asked to choose one or more robots to write an interesting story about how robots would behave together in a school that was populated with robots, discussing both friendships and bullying behaviour. The results of this study were congruent with those of the first study. The children usually assigned the male gender to the robot images, but that they did not associate this with either particularly positive or negative qualities. However, when the female gender was assigned, this was associated with positive characteristics such as friendliness. Once more, the children expressed some discomfort towards images of too human-like robots. Related to the stories, boys wrote more science-fiction themed stories than girls, which the researchers attributed to boys being more exposed to films and computer games that depict science-fiction themes. Girls, on the other hand, more often assigned emotions to the robots.

Shin and Kim [6], interviewed school students to investigate their attitudes towards learning about, from, and with robots. They also asked them what their image of a robot was. While the students had a positive attitude towards learning from robots because they were perceived as more intelligent and liable to make fewer mistakes than human teachers, most were not favorable to the idea of robots teaching in schools. This was mainly due to robots lack of emotion, which they considered important for teachers. They also saw the robots as being male or genderless, which the researchers attributed to their exposure to robots in the media.

In general, previous studies that address children's design contributions for robots have applied questionnaires and structured interviews. However, allowing children to present their own imaginations of a robotic teaching assistant in school, similar to what Lee et al. [2] did for adults' views on the design of domestic robots, could lead to additional insights. Our study aims to understand the differences between interaction designers' and children's views on the design of a robotic teaching assistants, as well as investigate the effects of robotics knowledge on children's views.

3   Study

In the exploratory study described below, we followed the ethical guidelines for studies in HRI [4]. The interaction designers signed an informed consent form before participation. For the children, the parents and/or caregivers signed the consent form and the children consented orally to participate in the study.

Table 1. Demographics of the children's groups

Group ID —	Nr of children —	Mean age —	Gender (F, M)
No robotics knowledge			
C1	5	11.0	2, 3
C2	5	10.2	3, 2
C3	4	10.8	2, 2
Robotics knowledge			
C4	5	11.2	3, 2
C5	5	13.4	0, 5
C6	5	10.6	1, 4

3.1 Participants and Procedures

Interaction Design Students: Twenty four international second year masters students (age 23-40, M=27.5, SD=3.69) from the interaction design program at the Chalmers University of Technology in Sweden participated. Before the design session, the students had attended one lecture on HRI in which the aim was to expose them to a whole range of social robots with different application areas: therapy, education, and entertainment. Thereafter, they were asked to design a robotic teaching assistant for children between 11 and 13 years old. There were five groups (D1-D5) of 4-5 students with mixed nationalities and gender. They were not informed about a specific task for the robot other than that it should function as a teaching assistant in the classroom. During the design process they were left by themselves to freely discuss their designs. All groups were instructed that they had 20 minutes to discuss and draw an assistant robot on an A3 sheet using a variety of colored pens. One high quality voice recorder per group captured the discussions, which were held in English.

Children: There were two main groups of children from a Portuguese school: children without formal robotics knowledge and children enrolled in a robotics course at school. Each of these was divided into smaller groups of 4-5 children (C1-C6). Table 1 shows the demographics of these groups. Children in groups C4-C6 were enrolled in a robotics course, however, the children from groups C4 and C5 had completed two years of the course, but group C6 had only completed one school semester of the course. As the table shows, the gender distribution in the groups was unequal, which was due to the robotics course being optional, and attracting slightly more boys.

The same procedure as for the interaction design students was followed. However, the children did not receive a lecture on social robotics, and the designs were made at a separate classroom of the school with only the researcher present. The researcher kept a distance to enable the children to freely express their ideas; only intervening when asked a question. All groups were instructed that they had 20 minutes to discuss and draw an assistant robot on an A3 sheet using a variety of colored pens. One high quality voice recorder captured the discussions, which were held in Portuguese.

Fig. 1. A summary of the robotic attributes extracted from the design drawings of the three groups (interaction designers, children with robotics knowledge and children without robotics knowledge). The categories on mode of locomotion, body shape, looks like, facial features and gender are based on the categorization of Woods [7].

4   Analysis and Results

This section presents the analysis and results of the drawings and discussions by the different groups (interaction designers, children with and children without formal robotics knowledge). Figures 2-4 show the design drawings by the different groups. While the groups were allowed to write comments about their design in the drawing, these have been removed from the figures for clarity's sake, for example the size of the robot or materials. All recordings were transcribed, and if necessary translated to English. Unfortunately, one of the recordings of the interaction design students was corrupted and omitted. The first and second author then analysed the drawings, using Woods' [7] robot attributes as a first coding scheme, and adding new attributes when necessary. Thereafter, they extended the analysis by listening to the audio recordings and reading the transcripts to identify passages in which the participants revealed additional design features that were not present in the drawings alone, for example the size of the robot. If possible, they also tried to identify the reasons for the characteristics of the robot in the drawings. If an attribute was present in one of the drawings or transcripts of one or more groups it was marked as present. This resulted in the analysis presented in Figure 1, in which colored cells indicate the presence of a attribute. The third author independently analysed these characteristics based on the drawings and transcripts to determine reliability (Cohen's Kappa $k = 0.74$). In the subsequent sections we present some typical characteristics of the robots designed in combination with the reasoning behind it; we will present participants as follows: D1-1 = Designer from group D1, first participant, C4-2 = Child from group C4, second participant.

4.1   Interaction Designers

The five drawings of the interaction designers are presented in Figure 2.

Fig. 2. Drawings of interaction designers

All drawings were rather small cartoon- or animal-like robots. There were several, slightly related reasons for this choice. First of all, the robot's appearance should not lead to a mismatch between the robot's actual and expected capabilities: *D1-1: Yeah, kind of [if it] like had some sort of normal head and arms and one of the girls just said 'thank you for this course' and he didn't respond to that because he is a robot. D1-2: Yeah, it's kind of weird. D1-1: So it's better to make it some sort of. D1-3: Pet?.* Second, the choice of appearance was related to the fact that the robot would be in the classroom together with the teacher. The following excerpt illustrates this: *D1-2: Because I think for me, if I have a teacher who already has authority, I don't want a teacher assistant to have authority as well. D1-4: Yeah. He need not to have so much authority but at the same time if it's super cute, I don't know, children won't learn anything. [...] D1-2: Professional but cute. [...] D1-3: What about an owl?.* Third, the interaction designers also wanted the robot to be not too intimidating, which led to the design of a relatively small sized robot (the same height or smaller than the children), as is illustrated in the following excerpt: *D3-1: I was also thinking about something, because they are students from 11-13, so usually they have a robot that is small, what about having a robot that is the same size as the students so they feel like hes a friend as well?.* Finally, the interaction designers consciously aimed for a non-gendered or neutral robot design, for example, one group discussed gender in relation to the robot's shape, *D4-1: Shall we go with the egg design? D4-2: Yeah, I like that. D4-1: I like that one because its more neutral,* while another group addressed color design in relation to the robot's gender, *D1-1:Just make it colorful! Probably make it half blue and half pink, as kids always thinking blue is for boys, pink is for girls. D1-2: Or super lime green. D1-1: Or do pink here and green here so they cannot say this is for boys or for girls. D1-2: Or just use blue as unisex.*

4.2 Children without Robotics Knowledge

The drawings by the three non-robotics groups are presented in Figure 3.

Fig. 3. Drawings of children with no formal robotics knowledge

The robot's appearance in these drawings was similar to a human teacher, with two legs and an adult size, as the following excerpt illustrates: *C2-1: I think the robot should look like a person, it is strange to be talking to a machine. C2-2: [The robot] should be like a person, and mimic a person.* Furthermore, the robot had to be representable as expressed in this excerpt: *C1-1: You could put those half torn trousers. C1-1: Dont ruin the drawing. C1-2: No, we do not want a bum teacher.* It also had a more formal look than a usual teacher: *C2-1: It should also have a tie like our school principal has.* All robots were clearly gendered, which was visible in the drawings and the discussions: *C1-1: Is it a female or a male teacher? C1-2: A male teacher. C1-3: It is a male robotic teacher, I have never seen a robota [Portuguese for female robot].* Another groups said:*C3-1: It is a girl-robot because we usually see the ladies, like moms and other ladies, as more friendly, like a good mother that cares.* While all robot designs resembled a human teacher some aspects were generally differing from a human teacher. First, while the robots did have some facial features they were sometimes rather crude and displayed on a screen. Second, the hands were usually different from actual human hands, displaying tools, LEGO-hands, or four fingers. The robots in all groups had a tool-belt (groups C1 and C2) or bag (group C3) to carry materials or tools. Finally, the body shape was more squared than in humans.

4.3   Children with Robotics Knowledge

The three drawings of the robotics children are presented in Figure 4.

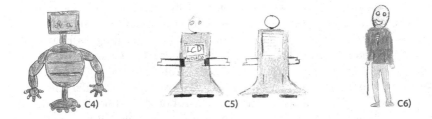

Fig. 4. Drawings of children with robotics knowledge

The robot of group C6 showed a clear resemblance to the robots drawn by the non-robotics children. This group of children was slightly younger and had started the robotics course later than the others. Therefore, our analysis mainly focuses on the designs of group C4 and C5. For these groups the appearance of the robot was more machine-like, and the children showed a clear awareness of the fact that a robot is a machine: *C4-1: But imagine you want to talk with the robot. C4-2: The robot gets up and goes to you. C4-3: And as you are. C4-1: Yes, but the person who will coordinate it is. C4-2: It will make it autonomously. C4-1: And if it loses his mind and you cannot ask it some questions? C4-2: No, but it is planned to do certain things, is not planned to have a head and do what he wants.* The robots were bigger than the interaction designers' robots but smaller than the adult sized robots in the non-robotics groups. Some aspects that were more machine-like in these drawings than in the non-robotics groups' drawings were the legs and the face. Instead of simply drawing two legs these groups applied their knowledge about the challenges behind smooth robot movement. For example, group C4 discussed: *C4-1: Legs no, [...] those robots can lose their balance or something like that,* and group C5 stated: *C5-1: Wheels, and we put up a kind of black line on the ground for it to follow.* Although the robot designs included a simple representation of a head, the faces did not show many facial features and they could also be used to display other information as shown in Figure 4(4). The children in this group reflected on having an iPad that could act as the face but also display information exhibited by the class teacher: *C4-1: a kind of iPad that showed things that the teacher wanted.*

While the non-robotics groups generally envisioned human-like interaction through speech and the use of tools placed in the robot's hands from a tool belt or bag, the robotics children explicitly included sensors, microphones, LCD screens and speakers to realise the interaction.

In comparison to the interaction designers, all children, both the robotics and non-robotics groups, mentioned the materials to construct the robot, such as metal, carbon-fiber, and aluminium. Furthermore, many groups reflected on how the robot would be powered, including solar energy and battery options.

5 Discussion and Conclusions

This study aimed to explore two questions: 1) How are children's views on the design of robotic teaching assistants different from interaction designers' views? 2) How are children's views on the design of robotic teaching assistant influenced by their knowledge of robotics (or lack thereof)? Concerning the first question, interaction designers often envisioned a robotic teaching assistant that was clearly different from a human teacher, for example a smaller animal- or cartoon-like robot. Their reasoning was that the robot should be professional but non-threatening, not having the same authority as a teacher but working in parallel with the teacher. In contrast, all children envisioned a human-machine robot instead of an animal- or cartoon-like character. Furthermore, children focused more on the materials and how the robot was powered, while the interaction designers were more concerned with the robot capacity to display emotions. Concerning the second question, the children with some formal knowledge of robotics were more inclined to design a robot with machine-like characteristics, while children without any formal robotics knowledge envisioned a rather human-like robot similar to a rather formal teacher but with some robotic details, such as a screen as head and robotic hands. These children also designed a way for the robot to carry on some tools that can be used in the classroom. While Sciutti et al. [5] found that children below the age of nine pay more attention to human-like features than older children and adults, we thus saw a similar effect related to children's formal robotics knowledge.

The robots designed by the children were rather similar to the robots that the children in Woods' [7] study thought were the most friendly, shy, or frightened robots. In contrast, the animal-like robots designed by the interaction designers would, according to Woods' study, fall in a category scoring slightly lower on these characteristics. In the study by Shin and Kim [6], many children were not favorable towards a robot as a teacher because a robot lacks emotions. It is therefore interesting to note that several drawings of the children only envisioned a limited display of emotions on a screen. We conjecture that it is possible that since children, especially those with some robotics knowledge, do not think that a robot can have real emotions, the face is less important to convey emotions.

This study has some limitations: first, the number of participants, and especially the number of groups was rather small. Second, while the interaction designers were a culturally mixed group, the children were all Portuguese. Third, while the interaction designers, as part of their education, had seen a range of social robots, the children were not introduced to different robot designs. This was done consciously, in order not to level out any differences between the groups of children, but it may have increased the difference between the children and the designers. Finally, we are aware that participants, including the children without any formal robotics knowledge, have been exposed to robotic ideas through the media. This was for example visible in the excerpt where one child mentioned that he/she had never seen a female robot. The designs of all groups thus represent a mix of design characteristics taken from the media, imagination, and in some cases, formal knowledge of robotics. Despite these limitations, our study presents distinguishing attributes for a robotic teaching assistant extracted from children's and designers' drawings that highlight the importance of including children in the design process of robots. Furthermore, children's contributions to the design may be influenced by their knowledge of robotics. Therefore, when including children in the design process, one needs to consider how factors such as knowledge of robotics may affect children's design contributions.

References

1. Dautenhahn, K.: Human-robot interaction. In: Soegaard, M., Dam, R.F. (eds.) The Encyclopedia of Human-Computer Interaction. The Interaction Design Foundation (2014)
2. Lee, H.R., Sung, J., Sabanovic, S., Han, J.: Cultural design of domestic robots: a study of user expectations in korea and the united states. In: 2012 IEEE RO-MAN, pp. 803–808, September 2012
3. Mubin, O., Stevens, C.J., Shahid, S., Al Mahmud, A., Dong, J.J.: A review of the applicability of robots in education. Journal of Technology in Education and Learning 1 (2013)
4. Riek, L.D., Howard, D.: A code of ethics for the human-robot interaction profession. In: Proceedings of We Robot (2014)
5. Sciutti, A., Rea, F., Sandini, G.: When you are young, (robot's) looks matter. developmental changes in the desired properties of a robot friend. In: The 23rd IEEE International Symposium on Robot and Human Interactive Communication, 2014 RO-MAN, pp. 567–573, August 2014
6. Shin, N., Kim, S.: Learning about, from, and with robots: students' perspectives. In: The 16th IEEE International Symposium on Robot and Human interactive Communication, RO-MAN 2007, pp. 1040–1045, August 2007
7. Woods, S.: Exploring the design space of robots: Childrens perspectives. Interacting with Computers 18, 1390–1418 (2006)
8. Woods, S., Davis, M., Dautenhahn, K., Schulz, J.: Can robots be used as a vehicle for the projection of socially sensitive issues? exploring childrens attitudes towards robots through stories. In: 2005 IEEE International Workshop on Robots and Human Interactive Communication (2005)

Healthcare Robots in Homes of Rural Older Adults

Josephine R. Orejana, Bruce A. MacDonald, Ho Seok Ahn,
Kathryn Peri, and Elizabeth Broadbent(✉)

The University of Auckland, Auckland, New Zealand
j.orejana09@gmail.com,
{b.macdonald,h.ahn,k.peri,e.broadbent}@auckland.ac.nz

Abstract. Older adults in rural communities who have chronic health conditions are often isolated from social support and medical clinics. Robots may be able to assist with day to day healthcare and provide companionship. This paper presents four case studies of older adults who had chronic health conditions in a rural community. They were given a healthcare robot in their homes for a period of three months to a year. The robot reminded people to take medications, had entertainment and memory games, and skype. Rates of hospitalizations, primary care visits, and phone calls to the medical practitioners before the study began were compared to rates during the study period. Participants also completed questionnaires about their quality of life, mental health, medication adherence, and robot attitudes and were interviewed. The results showed a decrease in primary care visits and phone calls to the practitioners while the robot was present and increases in quality of life were observed. Despite encountering technical issues, patients were mostly positive and accepting of the robot, acknowledging its benefits as a companion.

Keywords: Ageing in place · Healthcare robots · Companion · Adherence · Quality of life

1 Introduction

1.1 Caring for an Ageing Population in Rural Communities

Older people with chronic conditions who live in rural communities face many challenges. Geographic isolation, lack of transport, and a shortage of medical care professionals can hinder them from receiving the medical care that they need [1]. Especially for those living alone, managing chronic illnesses can be difficult. It can be hard to follow medication schedules as cognitive functions decline [2,3]. If people are unable to properly adhere to their medications, they may no longer be able to live independently, resulting in institutionalization. Rural older adults are also prone to experience loneliness as a result of geographic isolation [4]. Loneliness has been linked to a higher possibility that an older adult will lose his/her independence and be sent to institutionalised care [5]. Relocation to institutional facilities can have detrimental effects on the well-being of older adults [6]. 'Ageing in place', or having older adults live in their own homes independently for as long as they possibly can, is seen as the

© Springer International Publishing Switzerland 2015
A. Tapus et al. (Eds.): ICSR 2015, LNAI 9388, pp. 512–521, 2015.
DOI: 10.1007/978-3-319-25554-5_51

more favorable option by older adults, governments and care funders [7]. Socially assistive robots may be able to help individuals to remain independent in caring for themselves in their own home [8].

1.2   Robots and Ageing in Place

A number of healthcare robots have been developed and tested with older people [9]. However, very few studies have investigated the deployment of healthcare robots in the homes of older adults. Most socially assistive healthcare robots have been trialed in nursing homes or long-term care facilities, where nursing staff or researchers are often present to facilitate the interaction between the robot and resident, for example, [10,11]). Many studies have been small and observational in nature, however results are promising. A randomised controlled trial has shown reductions in loneliness in a rest-home/hospital setting with a pet-type robot [11].

One observational study investigated pet-type robots in the homes of 5 older adults for 5-17 days [12]. Results suggested users need to be aged over 75 and isolated, and that the robot should enable communication with family and have cognitive stimulation as well as important reminders. A robot that was capable of medication and appointment reminders, skype and games (the iRobi healthcare robot) was tested in a cross-over trial in the homes of 13 older adults living independently within a retirement village [13]. The robots were shown to be acceptable although they did not increase adherence or quality of life. Reasons for the lack of significant differences may be the small sample size and that residents were fairly well supported in the retirement village. Older adults living in rural communities may benefit more from robots in the home, due to greater isolation.

1.3   Aims and Hypotheses

This study aimed to test the feasibility and usefulness of robots to support older patients living alone in a rural community to manage their medication. It was hypothesized that the robots would reduce rates of hospitalizations, primary care visits, and phone calls to the medical practitioners, as well as increase medication adherence and quality of life.

2   Methods

2.1   Setting and Ethics

The study took place in the rural town of Gore, located in the Southland region of New Zealand. Gore is located between the two urban cities of Dunedin and Invercargill, and has a population of around 12,200 people [14]. Patients were recruited from Gore Health Limited, which includes a hospital and a primary care clinic that provides care to 3500 patients. Approval was gained from the University of Auckland Human Participants Ethics Committee.

2.2   iRobi

iRobi is made by Yujin Robot in South Korea (Figure 1). Researchers at the University of Auckland have developed and trialled iRobi to support older adults living at home, or in assisted living facilities [15]. iRobi has been programmed with health-care focused software. Applications are web-based and are run on a Windows XP operating system. The applications used in this study were the following: a medication reminding system, entertainment (memory games, music videos, quotes, web access); allowing telephone calls to phone numbers via Skype, or Skype to Skype communication; and user greetings and identity confirmation. The blood pressure measuring system was not used in this study. iRobi has a touch screen interface and displays information on its screen as well as talking to the patient via audio. It has LED lights on its face to express emotions, and raises its arms and sounds a tune when medications are due. His head moves towards sounds, and it responds to touch on the head by saying 'hi' or 'ouch', and says 'I am hungry' when short of battery life.

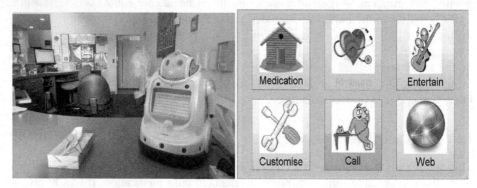

Fig. 1. iRobi sitting on a counter (left); iRobi's user interface (right)

2.3   Research Design

Five case studies were performed. Case studies are useful for preliminary research and they permit investigations when it is impractical to conduct larger studies [16]. Case studies provide rich qualitative information that can inform and shape future research. They are frequently used in clinical medicine and psychology. Interviews, clinical notes and questionnaires were used to collect data. Participants were given iRobi robot in their homes for at least 3 months and outcomes were compared before and after they had the robot. The primary outcomes were the numbers of hospitalizations, visits to primary care, and phone calls to the medical practitioners. Secondary outcomes were quality of life, medication adherence, and perceptions about robots.

2.4   Participants

Clinic staff at Gore Health Limited identified five older patients who lived alone and had chronic health conditions. All were on multiple medications, and it was thought the robot could help with medication adherence and companionship, and improve communication with relatives using Skype. Staff informed these patients and their relatives about the study. Two of these patients had already piloted the robot at home for 9 months, one of whom agreed to be interviewed about the robot but was moving to a rest home so could not take part in the study (patient A, male, aged 88 years). Baseline and final interviews were conducted with the remaining four patients who participated in the 3-month study. These participants were all widowed women aged over 75 years. Patient B was unable to complete the final questionnaire due to deteriorating vision.

2.5   Procedure

Before each participant was given a robot, the researcher obtained informed consent and gave the participant the baseline questionnaires in their own homes. Quality of life was assessed with the SF-12 scale which has physical and mental component scores that range from 0-100 and higher scores reflect better function [17]. Attitudes towards robots were assessed with the Robot Attitude Scale [18] which has high internal reliability and scores range from 10 to 80, where higher scores reflect more positive perceptions. The Medication Adherence Report Scale (MARS; [19]) was used to assess adherence where scores range from 10 to 50 and higher scores reflect higher adherence. Gore Health provided information about the participants' use of medical care services for the relevant time periods.

The baseline interviews were conducted in March 2014, after which the robots were deployed (Figure 2). Once the robots were installed and set on a stable and readily accessible surface, the researcher gave a brief demonstration of the robot and gave each participant a robot user manual. Contact details for the researcher and the main contact person from Gore Health were provided for any difficulties experienced. Participants and their robots were monitored from March to June, 2014. Technical support was available to the participants from Gore Health's main contact person, a local IT company, the researcher, and the engineering team back in Auckland.

At the end of the trial, Gore Health reported medical care utilisation data. The researcher conducted an interview and administered the same measures as at baseline. The Comfort from Companion Animals Scale [20] was adapted for Robots and administered. This scale has 11 items and scores from 4 to 44 (higher scores indicate more comfort from animals/robots). A robot acceptance questionnaire was also added [21]. This has 41 items and the total scores can range from 41 to 205 where higher scores indicate higher acceptance. Participants were informed that the study was finished. Gore Health was in charge of deciding how much longer the robots would remain in the homes of the participants, and monitoring them.

Fig. 2. Participants interacting with iRobi (photo consent obtained from participants).

2.6   Data Analyses

Descriptive information is reported from case histories and during the study period. SF-12 scores were computed using the QualityMetric scoring software. Interpretation was based on a report showing the average Quality of Life Mental Health Component Score (MCS) and Quality of Life Physical Component Score (PCS) for samples aged 75 and older as 52.6 and 40.80, respectively [22]. 95% Confidence Intervals from this sample were calculated to be +/- 6.24 for MCS and +/- 6.97 for PCS. A change in scores over the study period greater than these CI intervals was interpreted as meaningful. Interviews were transcribed by the first author and analyzed qualitatively using interpretative phenomenological analyses (IPA) [23].

3   Results

3.1   Medical care Utilisation, Quality of Life, Adherence, and Robot Acceptance

Table 1 presents a summary of the medical care utilisation. Primary care visits and phone calls reduced for all patients when they had the robot compared to baseline. Only one patient had been hospitalised in the baseline period, and no patients were hospitalised in the robot period. Quality of life scores reflected a meaningful increase in physical health functioning for Patients C and E, and a meaningful increase in mental health functioning for Patients D and E. Medication adherence was improved only in patient C. Patient D showed a slight decrease and patient E was completely adherent at both time-points. Attitudes towards robots improved in patients D and E, but slightly decreased in patient C. Results from the robot acceptance questionnaire indicated that Patient E was the most accepting (total score 186/205), followed by Patient D (176/205), then Patient C (119/205), which is consistent with the results above. Companionship scores showed a similar trend: patients D (32/44), E (33/44) found the most companionship from the robot followed by patient C (24/44).

Table 1. Medical care utilisation, quality of life, adherence and robot attitudes before the study period, and during the 3-month study period.

Patient	Baseline (no robot) Dec. 2013 - March 2014:	3 month Study Period (with Robot) March 2014 - June 2014
C: 94 year old woman	Primary care visits: 3 (2 doctor visits, 1 nurse visit) Phone calls: 3 Hospitalization: 1 (A&E visit) PCS: 24.93; MCS: 47.67 MARS: 37/50 RAS: 49/80	Primary care visits: 2 (1 doctor visit, 1 nurse visit) Phone calls: 1 Hospitalizations: 0 PCS: 32.22, MCS: 45.86 MARS: 48/50 RAS: 47/80
D: 83 year old woman	Primary care visits: 3 (2 doctor visits, 1 nurse visit) Phone calls: 1 Hospitalisations: 0 PCS: 41.20; MCS: 55.50 MARS: 46/50 RAS: 57/80	Primary care visits: 1 (1 doctor visit) Phone calls: 3 Hospitalisations: 0 PCS: 38.36; MCS: 65.97 MARS: 42/50 RAS: 66/80
E: 76 year old woman	Primary care visits: 4 (2 doctor visits, 2 nurse visits) Phone calls: 1 Hospitalisations: 0 PCS: 32.57; MCS:53.49 MARS: 50/50 RAS: 68/80	Primary care visits: 1 (1 doctor visit) Phone calls: 0 Hospitalizations: 0 PCS: 40.70; MCS: 60.10 MARS:50/50 RAS: 75/80
B: 89 year old woman	April 2012 - April 2013: Primary care visits: 9 Phone calls: 8 Hospitalisations: 0 March 2014: PCS: 44.18, MCS: 64.00 MARS: 43/50 RAS: 76/80	April 2013-Jan 2014 (Robot Pilot Period): Primary care visits: 7 Phone calls: 3 Hospitalisations: 0 March 2014 - June 2014: Primary care visits: 1 Phone calls: 2 Hospitalisations: 0 PCS: MCS: not completed MARS: not completed RAS: not completed

Note: PCS= Quality of Life Physical Component Score; MCS= Quality of Life Mental Health Component Score; MARS= Medication Adherence Report Scale; RAS= Robot Attitude Scale. Patient B had the robot for 9 months prior to the study starting. Her medical use data is displayed for the baseline period of one year, robot pilot period and study period separately.

3.2   Emergent Themes from the Interviews

The first theme concerned the usefulness of the robot functions (Table 2). While participants saw the potential of the medication reminding system, most felt the medication reminders did not benefit them as much as other elderly individuals because they were still competent in following their medication schedules. The entertainment function was not used much, although the music was popular. Feedback included more familiar games may be more suitable for older users. The Skype application was not widely used and did not change the amount of communication that patients had with their families. Most participants kept in touch with their families via phone, email, or personal visits. Reasons for not using skype included the small screen making it difficult to see, and a complicated user interface.

Table 2. The theme of Usefulness of Robot Functions.

Sub-Category	Excerpts
Medication Management	*"Most times I remember my pills, but, it has been somebody here or there or ... your routine has changed, you can forget you know? No doubt about it and the more forgetful I get, the more of a benefit it* (the robot) *would be to me"* (Patient B)
Entertainment (Brain Games, Music, Quotes)	*"They were quite, well, they chuckled actually when I put the music on for them, they thought it was quite entertaining"* (Patient D)
	"I would say the biggest disappointment is those games" (Patient B and Daughter)
Skype	*"I think that Skype would be marvelous once, you know, everything is up and running"* (Patient E)

The second theme that emerged was technical aspects (Table 3). At the beginning, patients expressed anxiety about new technology. Some were concerned that the robot might be too complicated for them and they were afraid of making mistakes. After the study, some patients reported that they had not engaged as much with the robot, or did not let their other family members touch it, because they did not want to break it.

Table 3. The theme of Technical Aspects.

Sub-Category	Excerpts
Anxiety with New Technology	*"As I say, I'm scared of it in a way. In case it goes wrong and its my fault"* (Patient C)
Reassurance from Technical Support	*"... yes, if I need help with the robot, I know that back-up is always there"* (Patient E)
Technical Issues	*"Well, I didn't find any negative aspects at all. It's just that it's uh, you know, it wasn't set in right... when it was playing up."* (Patient D)

The third main theme encompassed psychosocial aspects (Table 4). Some patients reported that the robot had an effect on their daily routine. They learned to anticipate when the medication alarms would go off; for example, at the six o'clock news. Patients enjoyed hearing the robot talk, and one patient said she would like the robot to say more spontaneous things, and greet her on her birthday. Patients commented that having the robot felt like they had a companion in the house and they didn't feel so alone. An unexpected finding concerned the robot's blinking lights. The lights enhanced the robot's social presence, which was reassuring to patients. Patients said that the lights helped them see when the surroundings were dark late at night or in the morning. Patients found comfort in seeing the robot light up, as they felt it was the robot's way of interacting with them. All patients reported that they would miss the robot when it went, with one patient describing the robot as being part of the family.

Table 4. The theme of Psychosocial Aspects.

Sub-category	Excerpts
Robot's Effects on Daily Routine	*"...you look forward to the time coming up to take your tablets and things like that because you're always aware that the time is coming close"* (Patient D)
Robot Sociability	*"I can talk to him on the way past him. Rub his head, and when he's talking, he talks to me... I'll rub his head and he'll come up with different little things"* (Patient E)
Robot Presence	*"Oh I don't know, he's just there... he's always winking and blinking at me, you know, there's always that light there, and he's just part of the furniture"* (Patient E)
Robot Companionship	*"I missed him, yes. I did miss him"* (Patient A)

4   Discussion

This is the first study to explore the deployment of healthcare robots in the homes of older rural residents. Compared to previous studies which trialed socially assistive robots in the home of non-rural residents [12-13], this study has a longer duration. The results suggest that the robots can be feasibly used in a rural population and may have benefits for some patients in terms of reduced utilisation of medical care, increased quality of life, increased adherence, and companionship. However, technical difficulties need to be resolved before widespread implementation can be achieved.

The main strength of the study is its novelty in deploying autonomous healthcare robots in a rural setting with older people for at least 3 months. The interviews provided personal accounts of their experiences. Improvements in the robot's design and functions can be made based on these accounts. More familiar games may be easier for older people to relate to and therefore increase users' confidence. A larger screen would make the functions easier to see and use. Older people sometimes have less dexterity so making the touchscreen less sensitive to long presses may remove accidental triggering of functions. The skype function needs a more simple user interface.

The suggestions arise from real-world experiences and not on laboratory-based set-ups. Being able to trial the healthcare robots in actual homes is a strength of the study, as it is has more external validity compared to studies held in simulated home environments [24-25]. Limitations include the following. First, it seemed as if the participants were unable to fully benefit from the primary functions of iRobi because they reported feeling confident about taking their medications, and were already maintaining a contact with family and friends. While five patients were interviewed in total, only three were able to complete the 3 month questionnaire. Furthermore, while there was a decrease in the rate of primary care visits and phone calls to the practictioners, it cannot be attributed for certain to the presence of the robot. The small sample size was due to the limited number of healthcare robots that were available.

Robot trials become more complex once taken out of controlled laboratory settings [9]. In this trial, a significant number of technical issues arose which resulted in the robots occasionally having to be taken out of homes for repairs. The repairs were made more difficult, slower and more expensive because the rural setting was a long distance from robotic engineers and replacement parts. Therefore, one of the challenges in providing robots to rural communities is the technical support needed. Design and implementation guidelines in rural settings should include increased technical support to local companies and training for caregivers.

In conclusion, it is feasible to deploy assistive healthcare robots in the homes of older rural residents living independently. Further research to develop and deploy healthcare robots in rural healthcare is worthwhile, as people embrace the potential of advanced technology to enhance their daily lives.

Acknowledgements. We thank Gore Health Ltd for their support running the study. The study was supported by funding from Auckland UniServices Limited.

References

1. Smith, K.B., Humphreys, J.S., Wilson, M.G.: Addressing the health disadvantage of rural populations: How does epidemiological evidence inform rural health policies and research? Aust. J. Rural Health **16**, 56–66 (2008). doi:10.1111/j.1440-1584.2008.00953.x
2. Davis, R., Magilvy, J.K.: Quiet pride: The experience of chronic illness by rural older adults. J. Nurs. Scholarship **32**, 385–390 (2000). doi:10.1111/j.1547-5069.2000.00385.x
3. Marek, K.D., Antle, L.: Medication management of the community-dwelling older adult. In: Hughes, R.G. (ed.) Patient Safety and Quality: An Evidence-Based Handbook for Nurses, pp. 499–536. Agency for Healthcare Research and Quality, Rockville (2008)
4. Kivett, V.R.: Discriminators of loneliness among the rural elderly: Implications for intervention. The Gerontologist **19**, 108–115 (1979). doi:10.1093/geront/19.1.108
5. Russell, D.W., Cutrona, C.E., De La Mora, A., Wallace, R.B.: Loneliness and nursing home admission among rural older adults. Psychol. Aging **12**, 574–589 (1997)
6. Scocco, P., Rapattoni, M., Fantoni, G.: Nursing home institutionalization: a source of eustress or distress for the elderly? Int. J. Geriatr. Psychiatry **21**, 281–287 (2006)
7. Davey, J.: "Ageing in place": the views of older homeowners on maintenance, renovation and adaptation. Soc. Policy J. NZ **27**, 128–141 (2006)
8. Jayawardena, C., Kuo, I., Datta, C., Stafford, R., Broadbent, E., MacDonald, B.: Design, implementation and field tests of a socially assistive robot for the elderly: HealthBot version 2. In: 2012 4th IEEE RAS & EMBS International Conference on Biomedical Robotics and Biomechatronics (BioRob), pp. 1837–1842 (2012)

9. Robinson, H., MacDonald, B., Broadbent, E.: The role of healthcare robots for older people at home: A review. Int. J. Soc. Robot. **6**, 575–591 (2014)
10. Pineau, J., Montemerlo, M., Pollack, M., Roy, N., Thrun, S.: Towards robotic assistants in nursing homes: Challenges and results. Robot. Auton. Syst. **42**, 271–281 (2003)
11. Robinson, H., MacDonald, B., Kerse, N., Broadbent, E.: The psychosocial effects of a companion robot: A randomized controlled trial. J. Amer. Dir. Assoc. **14**, 661–667 (2013). doi:10.1016/j.jamda.2013.02.007
12. Hutson, S., Lim, S.L., Bentley, P.J., Bianchi-Berthouze, N., Bowling, A.: Investigating the suitability of social robots for the wellbeing of the elderly. In: D'Mello, S., Graesser, A., Schuller, B., Martin, J.C. (eds.) Affective Computing and Intelligent Interaction, pp. 578–587 (2011). doi:10.1007/978-3-642-24600-5_61
13. Broadbent, E., Peri, K., Kerse, N., Jayawardena, C., Kuo, I., Datta, C., MacDonald, B.: Robots in older people's homes to improve medication adherence and quality of life: a randomised cross-over trial. In: Beetz, M., Johnston, B., Williams, M.-A. (eds.) ICSR 2014. LNCS, vol. 8755, pp. 64–73. Springer, Heidelberg (2014)
14. Statistics New Zealand: Census populations and dwelling tables (2012). http://www.stats.govt.nz/Census/2013-census/data-tables/population-dwelling-tables/southland.aspx
15. Broadbent, E., Jayawardena, C., Kerse, N., Stafford, R., MacDonald, B.A.: Human-robot interaction research to improve quality of life in elder care- an approach and issues. Paper presented at the 25th Conference on Artificial Intelligence (AAAI 2011). Workshop on Human-Robot Interaction in Elder Care, San Francisco (2011)
16. McLeod, S.A.: Case Study Method (2008). http://www.simplypsychology.org/case-study.html
17. Ware Jr, J.E., Kosinski, M., Keller, S.D.: A 12-item short-form health survey: Construction of scales and preliminary tests of reliability and validity. Med. Care **34**, 220–233 (1996)
18. Broadbent, E., Tamagawa, R., Patience, A., Knock, B., Kerse, N., Day, K., MacDonald, B.A.: Attitudes towards health care robots in a retirement village. Aust. J. Aging. (2011). doi:10.1111/j.1741-6612.2011.00551.x
19. Horne, R., Weinman, J.: Self-regulation and self-management in asthma: Exploring the role of illness perceptions and treatment beliefs in explaining non-adherence to preventer medication. Psychol. & Health **17**, 17–32 (2002)
20. Zasloff, R.L.: Measuring attachment to companion animals: A dog is not a cat is not a bird. Appl. Animal Behaviour Sci. **47**, 43–48 (1996). doi:10.1016/0168-1591(95)01009-2
21. Heerink, M., Krose, B., Evers, V., Wielinga, B.: Measuring acceptance of an assistive social robot: a suggested toolkit. In: The 18th IEEE International Symposium on Robot and Human Interactive Communication. RO-MAN 2009, pp. 528–533 (2009)
22. Office of Public Health Assessment: Health status in Utah: The medical outcomes study SF-12 (Utah Health Status Survey Report). Utah Dept of Health, Salt Lake City (2004)
23. Smith, J.A., Jarman, M., Osborn, M.: Doing interpretative phenomenological analysis. In: Murray, M., Chamberlain, K. (eds.) Qualitative Health Psychology: Theories and Methods. Sage, London (1999)
24. De Ruyter, B., Saini, P., Markopoulos, P., van Breemen, A.: Assessing the effects of building social intelligence in a robotic interface for the home. Interact. Comput. **17**, 522–541 (2005). doi:10.1016/j.intcom.2005.03.003
25. Prakash, A., Beer, J.M., Deyle, T., Smarr, C.A., Chen, T.L., Mitzner, T.L., Kemp, C.C., Rogers, W.A.: Older adults' medication management in the home: how can robots help? In: 2013 8th ACM/IEEE International Conference Human-Robot Interaction (HRI), pp. 283–290 (2013)

More Social and Emotional Behaviour May Lead to Poorer Perceptions of a Social Robot

Sofia Petisca$^{(\boxtimes)}$, João Dias, and Ana Paiva

INESC-ID & Instituto Superior Técnico, Universidade de Lisboa, Lisbon, Portugal
sofia.petisca@inesc-id.pt

Abstract. In this paper we present a study with an autonomous robot that plays a game against a participant, while expressing some social behaviors. We tried to explore the role of emotional sharing from the robot to the user, in order to understand how it might affect the perception of the robot by its users. To study this, two different conditions were formulated: 1-Sharing Condition (the robot shared its emotional state at the end of each board game); and 2-No Sharing Condition (the robot did not shared its emotions). Participants were randomly assigned to one of the conditions and this study followed a between-subject design methodology. It was expected that in the Sharing Condition participants would feel closer to the robot and would perceive/evaluate it as more human-like. But results contradicted this expectation and called our attention for the caution that needs to exist when building social behaviours to implement in human-robot interactions (HRI).

Keywords: Human-Robot Interaction · Emotional sharing · Social behaviour · Social robot

1 Introduction

In every social interaction we are constantly aware and responsive to social cues from others. Those social cues tell us how to behave in response to how the others are acting and feeling. So, for robots to be integrated in humans' daily life activities, they have to be provided with similar social capabilities. Thus they need to be able to inform about the others about their intentions, their affective evaluations and often social stances.

It is commonly agreed that a social robot should be embedded with behaviours that enrich the interaction with humans, making such interaction natural and inspired in the way we humans interact with each other. Such behaviours can be non- verbal behaviours (e.g. gaze, gestures, emotion expression, posture, etc.) and verbal behaviours (e.g. small talk, emotion sharing- *"I am feeling sad"*, etc). However, although it is clear that the social behaviours are important, one needs to be cautious about the way these social capabilities are implemented, taking into account the situation, context and embodiment of the robot. For example, capabilities such as intention recognition, Theory of Mind, or emotion expression, may be perceived differently depending on the context in which the

© Springer International Publishing Switzerland 2015
A. Tapus et al. (Eds.): ICSR 2015, LNAI 9388, pp. 522–531, 2015.
DOI: 10.1007/978-3-319-25554-5_52

robot is going to be placed. One of the social cues that humans use is the sharing of emotions, which when done explicitly by humans, may lead to a sense of closeness in their relationships.

In this paper we report a study performed with a social robot that autonomously plays a competitive game and at the same time expresses certain social behaviours. By relying on an emotional agent architecture (using an appraisal mechanism) the robot was built with the capabilities of emotional appraisal and thus able to express and sharing its emotions verbally as the game unfolds. By sharing its emotional state, one may expect that the social bonds with the user can be reinforced, and thus affecting the way the robot is perceived. This paper describes the architecture that was built for the robot to autonomously appraise the situation in the game, generate emotional states and share the emotions with the user. In a study carried out with the robot we hypothesized that participants with whom the robot shared its emotions would perceive it more humanlike, more close to them, and with more friendlier characteristics. The results obtained, however, did not confirm these hypotheses, and in fact, some opposite findings were found. We report the results and discuss some justifications for these findings.

2 Related Work

Humans are well equipped for social interactions, making it harder for a believable robot to successfully interact if it does not have similar social capabilities. In this sense it becomes very important to understand which capabilities better foster HRI and in which contexts they should emerge as more natural.

A way to fulfill this gap between technology and humans, is to enhance the anthropomorphic qualities (e.g. form and/or behaviour) of a robot, in order to create a way for humans to understand robots and vice-versa, necessary for a meaningful interaction[8]. Many studies reinforce this perspective, showing that a robot with social behaviour affects people's perceptions. For example, at a very basic level of communication, it is found that the presence of gestures in a robot catches more the user attention than without them[24], or that a robot can be seen as a companion, influencing people's perceptions of a shared event[12]. At a higher level it is also found, for example, that a socially supportive robot improves children learning, comparing to a neutral robot[22]. Indeed humans react to robots with social capabilities in a very positive way. A study from Kahn and colleagues(2015) even suggests that as more social robots become, people will probably build intimate and trusting relationships with them[13].

All this supports the fact that even small social behaviours affect HRI. Also, emotions play an important role in human behaviour, helping to form and maintain social relationships and establishing social position[9]. A study by Bartneck (2003) suggests that people enjoy more an interaction with an emotional expressive robot than a non-emotional [1]. And Becker (2005) comes to show how negative empathic behaviours are also important in a competitive game, though they are perceived as less caring by the user[3]. Therefore, emotions should also be taken into consideration when designing a social robot. The applicability of this can also be seen in Kismet's emotion and expression system, with the ability to engage people in affective interactions and so allowing it to be seen as a social creature[5].

Nonetheless, the context itself influences the perception of these behaviours and so they must be adapted to it. For example studies show that users cooperate more in an effortful task with a serious concerned robot, than with a playful robot (despite that they may enjoy more the playful robot)[10,11]. Kennedy, Baxter and Belpaeme (2015) also tried to implement social behaviours in a robot and found these to negatively affect learning improvements in a task with a robotic social tutor, compared to a non social one. They hypothesize that could be due to a greater level of distraction in the social behaviour form[15]. Studies from Goetz, Kiesler and Powers (2003), support this, showing that people expect the robot to look and act according to the task context, increasing their compliance with it. So, a match between the robot social cues and its task influences people acceptance and cooperation with it[11]. All this reinforces the need to test and refine these behaviours according to the social interaction they are placed in.

3   An Autonomous Social Robot that Shares Emotions

Robotic game opponents and companions are recently being built for different scenarios and games such as chess [16], or risk [19]. These robotic game companions need to embed not only decision making capabilities (achieved with the adequate artificial intelligence modules) but also social aspects which may embed emotional appraisal, theory of mind, intention recognition and so on. In this research we created a system for a social robot (the Emys robot) that tries to embed these two components (decision making/playing and social) and explore the impact that these components have in the perceptions of the users.

The game considered is a variant of the dots and boxes game[4]. In this variant, called Coins and Strings, an initial board is created by a set of coins, and a set of strings connecting pairs of adjacent coins. Two players take turns cutting a string each time. When a player removes the last string attached to a coin, the coin is removed from the board and added to that player's coins, additionally the player will also have to play again (selecting another string to remove). The game ends when all strings are removed, and the player with the highest number of coins wins the game.

To create a social part of the robot in the context of a competitive game, we extended FAtiMA Emotional Agent Architecture [7] for that effect. In particular, to allow for the manipulation of specific embodiment features and synchronization, the architecture was integrated with Thalamus Framework [20], which was then interconnected with the game developed in Unity3D and with the robot Emys[14]. Figure 1 shows a diagram of the complete system. Thalamus is a component integration framework that provides the advantage of easily integrating a robotic embodiment with a virtual environment. Thalamus Master centralizes all communication between other Thalamus components using a action/perception publication and subscription mechanism.

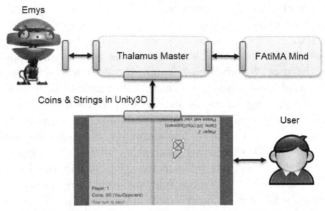

Fig. 1. Integration of FAtiMA, Thalamus and Emys to create the interactive system.

The system works as follows: when it's the user's turn to play, and he/she removes a string, the internal state of the game is updated in Unity, and a message about the event is sent (e.g user removed string number 3) to the Thalamus module. This message is perceived by a lower-level module, which will make Emys automatically look to the position of the removed string in the screen. At the same time, the Thalamus will send the same perception to FAtiMA, which updates its own internal state of the game, appraising the event and thus updating the robots emotional state. Both the updated emotional state and the play selected by FAtiMA are then sent to Thalamus. The emotional state is used to trigger emotion expression actions that are handled by Emy's Thalamus module, while the action will be sent to unity in order to update the state of the game. Emotion expression actions correspond to facial expressions that depending on the intensity of the emotion can also trigger speech, e.g. if a move caused Emys to be very happy, it will display a joyfull facial expression while saying "Great!".

Regarding the robot's cognitive and social behaviour, it was handled by FAtiMA linked with the decision-making component (AI for gameplaying). A standard Minimax algorithm [21] was implemented as a component in FAtiMA to decide the best move to play in the game. In addition, the Minimax value returned by the algorithm for a particular state (which represents the expected maximum utility) is used to predict the likelihood of winning the game, and also to determine the desirability of a particular event. The desirability of a game event is given by the change in the Minimax value caused by the event. As example, if the agent has a low Minimax value, but then the user makes a mistake and plays a bad move, the agent will update it's Minimax value to a much higher value, and the play will be appraised as very desirable. Since this is a zero-sum game, the Minimax value is also used to determine the appraisal variable *Desirability ForOther* (with other being the user) by applying a negative sign to the desirability value. The mentioned appraisal variables are then used by FAtiMA to generate Joy/Distress/Gloating/Resentment/Hope/Fear emotions, according to OCC Theory of emotions[18]. Perceived events and internal intentions are stored together with associated emotions in FAtiMA's episodic memory. Each board played with the user corresponds to a singular episode.

In addition to expressing emotions, and playing the game, in the Emotion Sharing condition the FAtiMA agent has the goal to share past emotional episodes with the user. So after each game, the goal will activate, and it will use the autobiographic memory to automatically generate a textual description of the episode. Describing all events in a past episode would also make a boring, unrealistic conversation. Therefore, a summary of the episode containing the most relevant emotions is used. The summary of an episode consists in: the location where the episode happened; narrative time elapsed since the episode happened and a list of the most relevant events that happened in the episode, ordered by event sequence. The more relevant events are considered to be the ones that have generated a stronger emotional impact in the agent, and thus are determined by selecting the events with the strongest emotions associated to them. The chosen events are then ordered by event sequence, so that the summary generated follows a coherent narrative flow. In order to provide the user with information on the agent's personal experience about the past episode, we need to add to the event's description the emotion experienced when the event was appraised (e.g. *"You made an unexpected move and I felt upset"*).

For the transformation of the information in the episode summaries into text, a *LanguageEngine* is used. The episode summary is split into events consisting of one action and subject, and optional location, time, target, parameter and emotion elements. The text of an event is then generated by transforming these elements into text and combining them through rules. The single utterances are concatenated using a set of connectional phrases like *"and"*, *"then"*. An example of a generated summary is: *"Several minutes ago, I wanted to win the game which made me feel frightened. Afterwards I played an important move. I was feeling really glad."* For more details about this process, please consult [6]. Finally, the summary Speech Act Request is sent to Thalamus, which uses a off the shelf text-to-speech system to produce the dialog.

4   Methodology

4.1   Participants

In order to evaluate the developed autonomous robot concerning its social capabilities, regarding emotional sharing, a study was conducted with a Emys robotic head who autonomously played a game against a participant, while displaying some social behaviours.

A total of 30 university students took part of this study (22 male and 8 female), with ages ranging from 19 to 30 years old (M=23.4; SD=2.99). Participants were randomly allocated to one of two study conditions: Sharing Condition (where Emys after the end of each game board shared verbally its emotions with the participant) and the No Sharing Condition (where no emotional sharing was done). All participants signed a consent form in order to be part of the study and allowing for the sessions to be recorded. The sessions had a duration of approximately 20 minutes per participant (with 10 minutes of interaction with Emys) and took place in a Portuguese laboratory. The material used was a Lavalier microphone for audio recording and three cameras for video recording of the interaction.

4.2 Procedure

Upon arrival participants were assigned randomly to one of the two conditions. In the No Sharing Condition Emys social behaviours were only gaze tracking through a Microsoft Kinect and small talk during the game (e.g. *"This is going to be a hard game"*), on the Sharing Condition Emys maintained the social behaviours from the other condition and added a emotional sharing at the end of each board about its feelings regarding the result of the board (e.g. *"I was feeling worried, but then I was able to beat you"*). Each participant played five board games of the Coins and Strings game with Emys, where the difficulty increased with the board number, being board number one the easiest level. When the game was finished, participants were taken to another room where they filled a brief questionnaire (see Measures section).

4.3 Measures

To understand the impact of the emotional sharing social behaviour in the participants perception of the robot, the Godspeed Questionnaire [2] was applyed, with dimensions: Anthropomorphism, Animacy, Likeability and Perceived Intelligence.

In addition, since emotional sharing may lead to a closer relationship, by helping to form and maintain social relationships[9], we applied a connection questionnaire that was based on [23] consumer product attachment scale (adapted to refer to Emys, e.g. *"Emys is very dear to me"*) to explore the connection from the user to the robot. Also, we used the McGill Friendship Questionnaire (MFQ) [17], which comprises two questionnaires, one that measures the positive feelings regarding a friend and the other how much that friend fulfills six friendship dimensions: Companionship; Intimacy; Reliable Alliance; Self-Validation and Emotional Security (we did not use Help dimension since the game was a competition setting). These questionnaires were used in order to ascertain if Emys had a different impact on the participants depending on the condition they were allocated to. The Godspeed questionnaire was answered in a semantic differential scale as in [2], all other questionnaires were answered in a 5-point Likert scale ranging from *"Totally Disagree"* to *"Totally Agree"*.

5   Results

First we ascertained the internal consistency of the scales used and all had a good internal consistency. For the Godspeed questionnaire dimensions there were no significant differences found and the means were for the Sharing Condition and

No Sharing Condition respectively: Anthropomorphism (M=2.53; M=3.18); Animacy (M=3.26; M=3.31); Likeability (M=3.31; M=3.49) and Perceived Intelligence (M=3.96; M=4.08).

Analyzing the items of each dimension, a Mann Whitney U Test was done and it was found a statistical significance between the conditions, for the Anthropomorphism dimension regarding the Unconscious/Conscious item (U=58, p=.010) and the Artificial/Lifelike item (U=65, p=.042). It is seen that participants perceived the robot as more conscious and lifelike in the No Sharing Condition (M=3.93, SD=0.70; M=3.27, SD=1.34) compared to the Sharing Condition (M=2.93, SD=1.22; M=2.33, SD=0.98), which goes against the expected results (see Fig. 2). For the Artificial/Lifelike item even though there was a statistical significance for the participants responses, these responses were only slightly more positive (less Artificial) in the scale for the No Sharing Condition. Also, for the Likeability dimension it was found a statistical significance for the item Awful/Nice (U=67,5, p=.05) where it is seen that participants perceived the robot as more nice in the No Sharing Condition (M=3.53, SD=0.74) than in the Sharing Condition (M=2.80, SD=1.01) (see Fig. 2). There were no other statistical differences in the other dimensions.

Even though there was no statistical significance for the Perceived Intelligence dimension, means show across conditions, that participants perceived Emys as very competent (M=4.40; M=4.47), knowledgeable (M=3.87; M=4.13) and intelligent (M=4.13; M=4.33), which clearly shows the high level of competence that Emys had in the game. This result is also supported by the winners of each session, as only 4 participants were able to beat Emys in the game.

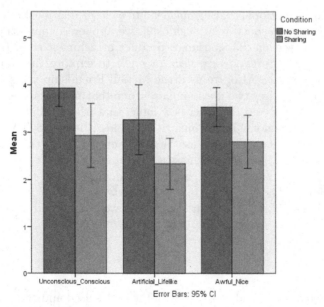

Fig. 2. Statistically significant results for the Godspeed Questionnaire in both conditions.

For the Connection Questionnaire there were no statistical significant results. The mean answers for all the questions were around 2 and 3 in the scale, which seems to suggest that evaluating connection in this short-term interaction for this context is not an appropriate effect to be seen. The same happened with the McGill Friendship Questionnaire, with no statistical differences to report between conditions. In general Emys was seen as making participants laugh, as stimulating to interact with and very enjoyable.

6   Discussion and Conclusions

In this paper we tried to explore the role of emotional sharing from the robot to the user, in order to understand how this social capability might affect the perception of the robot by its users. Our hypothesis was that participants in the Sharing Condition (where the robot shared verbally its emotions at the end of each board game) would perceive the robot as more humanlike, more close to them and possibly with more friendly characteristics, compared to the participants allocated in the No Sharing Condition.

Surprisingly, the results did not support our hypothesis, showing that participants in the No Sharing Condition rated the robot as more Conscious, Lifelike and Nice. This seems to suggest that the emotional sharing that the robot performed was not giving the robot a more lifelike appearance. It is possible that its expression may not be taking the appropriate form in this concrete context. The highly competence that the robot presented in this task (only 4 participants were able to beat Emys) could have had an influence on participants perceptions, adding to the emotional sharing behaviour. In the Sharing Condition participants were subjected to emotional sharing phrases related to the game state at the end of each board game. As such, these autonomously generated phrases expressed positive emotions more frequently as the robot achieved its victories due to its high competence level. These emotional responses could in turn, we hypothesize, be highlighting more the vision of a machine, that always beats humans, than of a social robot that cares for the user. Which was also seen in the study of Becker [3] with the users feeling less care with the negative empathic approach.

Regarding connection and friendlier characteristics perceived in the robot, there were no significant differences between conditions. It could suggest that for this kind of short-term interaction in a competitive game this kind of bonding did not made much sense. It may be interesting to explore if these results change in the same context for a long-term interaction.

Remembering Goetz, Kiesler and Powers (2003) studies, participants in the Sharing Condition might be feeling that the robot emotional sharing dialogue is highlighting more an artificial kind of interaction, adding to its higher competence in the game (which it frequently wins). Whereas on the No Sharing Condition where only small talk happens, might be seen by participants as less artificial. Even though Emys still plays with the same high competence, there is no reinforcement of emotional sharing. Participants may feel a disconnection from the robot social behaviour and its task in the Sharing Condition. Due to the fact that emotional sharing in this context seems to be reinforcing negatively, giving a more artificial appearance to the robot and breaking social norms. These results have some similarity with the results obtained by Kennedy, Baxter and Belpaeme (2015), which found that improvement in learning with children is lost with a robotic social tutor, compared to a non social one [15]. It seems that by existing emotional sharing associated to a really high competence in the game, it is damaging the social interaction and perception of the robot by the users. Therefore, more research is needed in order to understand this relationship and how to better apply emotional sharing in HRI: what contexts it provides the robot with better social capabilities, and in which it should be avoided.

These findings suggest important implications for the design of social behaviours necessary to exist in an autonomously social robot. It calls our attention for the fact that more may be less if not properly adjusted to the context where it gains form. Further studies will be conducted in order to try and clarify the role of emotional sharing in social interactions between robots and users.

Acknowledgments. This work was supported by national funds through Fundação para a Ciência e a Tecnologia (FCT) with reference UID/CEC/50021/2013 and by the EU-FP7 project EMOTE under grant agreement no. 317923. The authors are solely responsible for the content of this publication. It does not represent the opinion of the EC, and the EC is not responsible for any use that might be made of data appearing therein.

References

1. Bartneck, C.: Interacting with an embodied emotional character. In: Proceedings of the 2003 International Conference on Designing Pleasurable Products and Interfaces, pp. 55–60. ACM (2003)
2. Bartneck, C., Kulić, D., Croft, E., Zoghbi, S.: Measurement instruments for the anthropomorphism, animacy, likeability, perceived intelligence, and perceived safety of robots. International Journal of Social Robotics 1(1), 71–81 (2009)
3. Becker, C., Prendinger, H., Ishizuka, M., Wachsmuth, I.: Evaluating affective feedback of the 3d agent max in a competitive cards game. In: Tao, J., Tan, T., Picard, R.W. (eds.) ACII 2005. LNCS, vol. 3784, pp. 466–473. Springer, Heidelberg (2005)
4. Berlekamp, E.R.: The Dots and Boxes Game: Sophisticated Child's Play. AK Peters/CRC Press (2000)
5. Breazeal, C.: Emotion and sociable humanoid robots. International Journal of Human-Computer Studies 59(1), 119–155 (2003)
6. Dias, J., Ho, W.C., Vogt, T., Beeckman, N., Paiva, A.C.R., André, E.: I know what i did last summer: autobiographic memory in synthetic characters. In: Paiva, A.C.R., Prada, R., Picard, R.W. (eds.) ACII 2007. LNCS, vol. 4738, pp. 606–617. Springer, Heidelberg (2007)
7. Dias, J., Mascarenhas, S., Paiva, A.: FAtiMA modular: towards an agent architecture with a generic appraisal framework. In: Bosse, T., Broekens, J., Dias, J., van der Zwaan, J. (eds.) Emotion Modeling. LNCS, vol. 8750, pp. 43–55. Springer, Heidelberg (2014)

8. Duffy, B.R.: Anthropomorphism and the social robot. Robotics and Autonomous Systems **42**(3), 177–190 (2003)
9. Fischer, A.H., Manstead, A.S.: Social functions of emotion. Handbook of Emotions **3**, 456–468 (2008)
10. Goetz, J., Kiesler, S.: Cooperation with a robotic assistant. In: CHI 2002 Extended Abstracts on Human Factors in Computing Systems, pp. 578–579. ACM (2002)
11. Goetz, J., Kiesler, S., Powers, A.: Matching robot appearance and behavior to tasks to improve human-robot cooperation. In: Proceedings of the 12th IEEE International Workshop on Robot and Human Interactive Communication, ROMAN 2003, pp. 55–60. IEEE (2003)
12. Hoffman, G., Vanunu, K.: Effects of robotic companionship on music enjoyment and agent perception. In: 2013 8th ACM/IEEE International Conference on Human-Robot Interaction (HRI), pp. 317–324. IEEE (2013)
13. Kahn Jr., P.H., Kanda, T., Ishiguro, H., Gill, B.T., Shen, S., Gary, H.E., Ruckert, J.H.: Will people keep the secret of a humanoid robot?: psychological intimacy in hri. In: Proceedings of the Tenth Annual ACM/IEEE International Conference on Human-Robot Interaction, pp. 173–180. ACM (2015)
14. Kennedy, J., Baxter, P., Belpaeme, T.: The robot who tried too hard: Social behaviour of a robot tutor can negatively affect child learning. In: Proc. HRI, vol. 15 (2015)
15. Kdzierski, J., Muszyski, R., Zoll, C., Oleksy, A., Frontkiewicz, M.: Emysemotive head of a social robot. International Journal of Social Robotics **5**(2), 237–249 (2013)
16. Leite, I., Pereira, A., Castellano, G., Mascarenhas, S., Martinho, C., Paiva, A.: Social robots in learning environments: a case study of an empathic chess companion. In: Proceedings of The International Workshop on Personalization Approaches in Learning Environments (PALE). CEUR Workshop Proceedings, July 2011 (ISSN 1613–0073)
17. Mendelson, M.J., Aboud, F.E.: Measuring friendship quality in late adolescents and young adults: Mcgill friendship questionnaires. Canadian Journal of Behavioural Science/Revue Canadienne Des Sciences Du Comportement **31**(2), 130 (1999)
18. Ortony, A., Clore, G., Collins, A.: The Cognitive Structure of Emotions. Cambridge University Press, UK (1998)
19. Pereira, A., Prada, R., Paiva, A.: Improving social presence in human- agent interaction. In: 32nd annual ACM Conference on Human Factors in Computing Systems, CHI 2014, pp. 1449–1458. ACM, Toronto, Canada, April 2014
20. Ribeiro, T., Tullio, E., Corrigan, L.J., Jones, A., Papadopoulos, F., Aylett, R., Castellano, G., Paiva, A.: Developing interactive embodied characters using the thalamus framework: A collaborative approach, pp. 364–373 (2014)
21. Russel, S., Norvig, P.: Artificial Intelligence: A Modern Approach. Prentice-Hall, NJ (2002)
22. Saerbeck, M., Schut, T., Bartneck, C., Janse, M.D.: Expressive robots in education: varying the degree of social supportive behavior of a robotic tutor. In: Proceedings of the SIGCHI Conference on Human Factors in Computing Systems, pp. 1613–1622. ACM (2010)
23. Schifferstein, H.N., Zwartkruis-Pelgrim, E.P.: Consumer-product attachment: Measurement and design implications. International Journal of Design **2**(3), 2008 (2008)
24. Sidner, C.L., Lee, C., Kidd, C.D., Lesh, N., Rich, C.: Explorations in engagement for humans and robots. Artificial Intelligence **166**(1), 140–164 (2005)

Towards a Robot Computational Model to Preserve Dignity in Stigmatizing Patient-Caregiver Relationships

Michael J. Pettinati[✉] and Ronald C. Arkin

Georgia Institute of Technology, School of Interactive Computing, Atlanta, GA 30332, USA
mpettinati3@gatech.edu, arkin@cc.gatech.edu

Abstract. Parkinson's disease (PD) patients with an expressive mask are particularly vulnerable to stigmatization during interactions with their caregivers due to their inability to express affect through nonverbal channels. Our approach to uphold PD patient dignity is through the use of an ethical robot that mediates patient shame when it recognizes norm violations in the patient-caregiver interaction. This paper presents the basis for a computational model tasked with computing patient shame and the empathetic response of a caregiver during "empathetic opportunities" in their interaction. A PD patient is liable to suffer indignity when there is a substantial difference between his experienced shame and the empathy shown by the caregiver. When this difference falls outside of acceptable set bounds (norms), the robotic agent will act using subtle, nonverbal cues to guide the relationship back within these bounds, preserving patient dignity.

1 Introduction

In a patient-caregiver interaction, an experience of high rapport is deemed to be "optimal". The participants understand each other and are capable of appropriately responding to one another [24]. Tickle-Degnen [24, 25] has presented a substantial body of work showing the difficulties of those with expressive disorders to attain high rapport with their caregivers due to the critical role nonverbal communication plays in establishing such rapport. When caregivers are unable to attain rapport with their patients, they may stereotype their patients because of uncertainty about how the patient is feeling, which can lead them to stigmatize the patient [25]. A group particularly prone to stigmatization is Parkinson's patients with a condition known as an expressive mask that limits expressivity in the face, body, and tone of voice. Tickle-Degnen [25] has found that professionals who work with Parkinson's patients often misjudge these patients' personalities. Caregivers were found to judge patients showing higher masking to be less social, cognitively competent, and more depressed.

In this five-year NSF-funded study, a collaborative interdisciplinary effort with Tufts University, our goal is to uphold the dignity of Parkinson's patients with expressive masks during stigmatizing interactions with their caregivers. One means by which this may be accomplished is to introduce an ethical social robot as a bystander into the patient-caregiver interaction. This robot models the ongoing patient-caregiver relationship and uses subtle, nonverbal cues to guide the relationship when norms of

A. Tapus et al. (Eds.): ICSR 2015,LNAI 9388, pp. 532–542, 2015.
DOI: 10.1007/978-3-319-25554-5_53

the interaction are violated. This will build on our previous work [2] in kinesics and proxemics, which demonstrated how an embodied robot could communicate its internal state to a human through these nonverbal cues. An embodied robot can use such nonverbal cues to alert a caregiver to indignity in the patient without disrupting the dyad's communication [9]. This paper presents the foundation for a computational model that is being implemented and tested in upcoming human-robot interaction studies; this model is intended to ameliorate patient indignity.

Dignity stands in opposition to stigmatization and the closely related concept of shame. Dignity is consistently linked with self-respect and identity [14]. The dignity of a person can, therefore, be "robbed" when he is humiliated. Humiliation causes a fundamental change in a person's understanding of his "identity" and place or worth in society [14]. The process of stigmatization culminates in the internalizing of a negative stereotype that is associated with some disease or disorder [16]. According to Sabini et al. [19], shame arises as a response in someone when a fundamental flaw is revealed because he sees this now public flaw as limiting his worthiness of positive relationships in the future.

The caregiver must respond to patient shame and stigma by making a connection with the patient; that is, showing patients that they are not alone and have value [22, 27]. If the caregiver does not respond empathically, this further confirms the patient's feeling of rejection, which results in increased feelings of shame, or in the development of resentment or anger toward the caregiver [27].

Therefore, a robotic agent mediating stigma needs to represent when the patient is experiencing shame and when the caregiver is not responding with a sufficient level of empathy. Section 2 in this paper discusses representations of shame and empathy based on the psychological literature. Section 3 presents a framework to preserve patient dignity. Section 4 summarizes and discusses how the project will progress from this point onward.

2 Shame and Empathy Representations

Shame is a construct that is not going to dissipate over the course of the communication between the patient and caregiver without intervention; shame is relieved through a change in "context" or a change in "self" [21]. Empathy from the caregiver affords this "context" change. The caregiver commits to being a present, social ally to the patient; the patient is made to recognize his value [27]. It is critical that the magnitude of shame that has been experienced by the patient during the interaction does not far outpace the empathy expressed by the caregiver.

It may not be possible for the caregiver to respond to the patient's shame with complete understanding and compassion in each "empathetic opportunity" [22]. The caregiver, however, needs to show sufficient empathy (while not showing too much) to keep the difference between the shame experienced by the patient and empathy shown by the caregiver during the interaction within acceptable bounds. Just what is sufficient empathy and how to determine the fixed bounds on a particular relationship will be determined through upcoming studies with patients, focus groups, and experts.

2.1   A Componential Representation of Shame

Nijhof [13] explicitly described Parkinson's disease as a "problem of shame". In his interviews with Parkinson's patients, he consistently found patients experienced shame after violating various social conventions publicly. The social rules the patients violated varied in their frequency and in their contribution to the intensity of shame experienced by the patients. Further, the magnitude of shame with respect to a rule violation varied based on the public or private nature of violation. The magnitude of shame also depends on an individual's "proneness" to shame [3]. Therefore, it is important to consider the intensity of experienced shame as a composite construct where different components (types of violations) have different weight, and the magnitude a certain component or violation contributes can also vary depending on the individual and the circumstance. This lab found few existing models that compute shame's intensity. A componential representation fits with a proven model for the computation of guilt's magnitude [1], an emotion in shame's family [8] (see below).

The Components of Shame

In the psychology literature, shame is often decomposed into two independent components [3, 6, 10]. These independent components take different names: internal/external shame [10], defensive/unworthiness shame [5], and negative-self-evaluation/withdrawal shame [3]. Internal shame, which corresponds with defensive and negative-self-evaluation (NSE) shame, is experienced when there are discrepancies between an ideal self, the self that is dictated by personal values, and the public self. External shame, corresponding to withdrawal and unworthiness shame, is experienced when it is those external to the flawed self that present the self as flawed [10]; causing the shamed person to hide/withdraw from the situation [3, 6].

The two independent components of shame (internal and external) are reminiscent of the two types of stigma, enacted stigma and internal stigma [16]. The components of stigma, however, are viewed as stages in a process. Someone external to a person stereotypes him (enacted); he then internalizes that stereotype (internal).

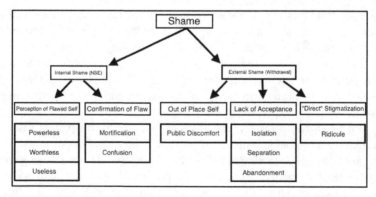

Fig. 1. Decomposition of shame into components. Language taken from Retzinger [17].

Shame is a composite of internal and external shame. Retzinger [17] defined six different categories ("direct indication", "abandonment/ separation/ isolation", "ridicule", "inadequate", "discomfort", and "confused/ indifferent") of words that not only frequently appear in the "context of shame" but also help to define circumstances where shame arises. Three of the categories ("discomfort", "abandonment/ separation/ isolation", and "ridicule") fit well under external shame. A person may feel outside of a social network because of people gawking at the symptoms related to his illness or other indirect indicators. This separation induces shame because it identifies the self as not "fitting in". The magnitude of external shame experienced by a person can also increase when the symptoms or the illness cause others to "abandon" or "isolate" the sufferer. This is perhaps a more direct indictment, i.e., showing one does not belong. Finally, there is perhaps the most damaging component of external shame, the most direct means by which a flaw costs social worth - "ridicule" by others.

Within internal shame, there are two components ("inadequate" and "confused/indifferent") drawn from the work of Retzinger [17]. First, a person believes that he is flawed and recognizes the potential damage the flaws could have on his social worth/value. Second, the flaw is confirmed in a public setting where the person is unable to function appropriately or fails at a task that he feels marks him as an "undesirable". Note that the sixth category introduced by Retzinger [17] ("direction indication" of shame) decomposed nicely into "ridicule" in external shame and "confusion/indifferent" in internal shame. When one is ridiculed, it is a direct affront on his identity by someone external to the self. Similarly, when one is confused, apathetic and unable to function in the manner in which he was able to function, a discrepancy with the idealized self is revealed. See Figure 1 for a decomposition of shame into its constituent components.

2.2   A Componential Representation for the Empathetic Response

An empathetic response is motivated by both emotion and cognition [5, 8]. Davis's [5] four Interpersonal Reactivity Index (IRI) subscales define trait empathy as a multi-dimensional construct. The four defined dimensions are not combined to form a single value for a person's propensity for empathy; instead, they remain independent. This is due to the interdependence of the dimensions in enacting an empathetic response.

Studies comparing empathy in populations of caregivers working with the chronically ill or the dying against "average" adult populations (e.g. [4]) have found that these caregivers will often have significantly greater capacities for empathetic concern and perspective-taking, two dimensions on Davis's IRI. Perspective-taking is a cognitive dimension that allows for the person to "anticipate" the behavior of another person as well as what the needs or wants of the person might be in a certain situation; it allows for appropriate social responses to the individual [5]. Empathetic concern is an emotional dimension that is thought to motivate altruistic action. A person experiences compassion when recognizing suffering or tenderness when recognizing vulnerability [12]. The magnitude of the caregiver's empathetic response is going to depend on convergence between an emotional motivation to help the patient and a cognitive understanding of what the patient needs.

It is logical to model the magnitude of the empathetic response in terms of the components of the interaction where the caregiver attempts to ameliorate the patient's shame. If the caregiver understands the patient's shame and is emotionally motivated to aid the patient, then he will act in a way that mitigates the patient's shame. A previous computational model for empathy captures this notion. Rodriguez et. al. [18] present a psychologically-based computational model for a virtual character where the intensity of the character's empathetic response is based on the agreement between the emotion the character recognizes in the other and the emotion the character understands the other should experience through perspective-taking.

Components of an Empathetic Response
When a caregiver is showing the patient that he understands how the patient is feeling, this is not purely a cognitive act but an emotional one as well. The caregiver is deviating from questioning pertinent to the disease (and taking the time) to let the patient tell his "story" [27]. The patient being allowed to speak, and the caregiver providing evidence that he is listening is therapeutic in its own right to the patient [22]. It shows that another person values him; the patient's flaws have not left him unworthy of human contact. To make this connection stronger and more explicit, the caregiver can name the internal state of the patient or can try to relate a personal story to the patient. The caregiver, when sharing a piece of himself with the patient, makes clear the patient is not alone but is experiencing something felt by others.

The caregiver also needs to try to restore the "self" of the patient that has been damaged by the shame. Clearly there is a sympathetic emotional response recognizing the pain of someone who feels unworthy or alone. There is a cognitive aspect to assisting in such a case as well; the caregiver must understand the "self" of the patient has been damaged. The caregiver must show nonverbal support for the damaged self [27], and this support can progress to bolstering the patient's identity through praise for the self (drawing a distinction between the disease and the self). See Figure 2.

2.3   A Componential Representation for Guilt

The notion of computing the magnitude of shame and empathy using components of the patient-caregiver interaction is based on a computational model for the self-conscious moral emotion of guilt. When experiencing guilt, one feels bad for what

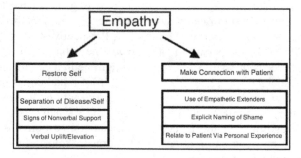

Fig. 2. The decomposition of the empathetic response to shame into components.

one has done (finds fault with his actions); this is as opposed to shame where fault is found with the self [8]. Smits and De Boeck [20] first introduced a componential model for guilt. They found that the probability of a person, i, experiencing guilt in a situation, j, (P_{ij}) could be explained by the "guilt-inducing power" (β_j) of the situation, j, and the person's "threshold" for guilt (θ_i). The logit of the probability of a person experiencing guilt is computed as:

$$\text{logit}(P_{ij}) = a_j(\beta_j - \theta_i)$$
(1)

This is just a weighted difference between the "guilt-inducing power" of the situation and the "threshold" of the person for guilt. The "guilt-inducing power" of situation j (β_j) was reliably computed as a weighted sum over the "guilt-inducing power" of just three situational components: "norm violation", "worrying", and "restitution".   This is   shown in Equation 2.   In this equation, σ_k is the weight of the situation component, k, β_{jk} is the guilt-inducing power of component k, and τ is a constant.

$$\beta_j = \sum_{k=1}^{K} \sigma_k \beta_{jk} + \tau$$
(2)

Our laboratory has already successfully applied a simplified version of this model for the purposes of restricting weapons systems in a lethal autonomous robotic agent [1]. Equation 3 shows the magnitude of guilt accrued for a specific target, j. Four components were used to compute the guilt for a single target. The parameters of the equation are defined the same as in the above equations.

$$Guilt_j = a_j(\sum_{k=1}^{K} \sigma_k \beta_{jk} + \tau - \theta_i)$$
(3)

A similar model to Equation 3 will be used for the accrual of shame and empathy across the patient-caregiver interaction.

3   Overview of Framework Designed to Uphold Patient Dignity

The overview of the framework designed to uphold Parkinson's patient dignity is shown in Figure 3. This framework's core piece, the Emotional Models component, is responsible for computing the magnitude of the patient's experienced shame and the caregiver's expressed empathy in each "empathetic opportunity" [22]. Explicitly, "empathetic opportunities" [22] are single exchanges between the patient and caregiver where the patient experiences shame and the caregiver has the opportunity to respond empathetically. It can also be the case where the caregiver stigmatizes the patient directly (in which case there is no empathetic response). The reason the shame and empathy values are only computed during "empathetic opportunities" is that outside of this context the patient may not believe a caregiver's empathetic response because it may come off as hackneyed [27]. Input into this component of the model is the caregiver's threshold for empathy and the patient's threshold for shame, the weights for the shame and empathy situation components, and the constants for the patient shame and caregiver empathy equations. These are set values shown in Equation 3. These values can be determined by experts with the aid of short personality/trait measures, which are common in non-emergent and non-urgent clinical situations, and stored in the Knowledge Base. In the Emotional Models component, the magnitude of shame shown by the patient and empathy expressed by the caregiver in this particular "empathetic opportunity" are computed using these set values and the responses of the components described above.

Takahashi et al. [23] show how, when Parkinson's patients are discussing frustrating activities (a discussion that has a high likelihood of inducing shame), they tend to use more negative language while there is no significant difference in their nonverbal behavior due to the expressive mask. It is logical to assess shame based on the verbal content of what is said by the patient during the interaction. In parallel with this work, our colleagues at Tufts University are developing a speech recognition system specifically designed for Parkinson's patients. This system is expected to be able to reliably recover "keywords" from the patient.

Gottschalk [7] claims that the "magnitude" of a "psychological state" is proportional to the "frequency of occurrence" of words associated with that state, the degree to which the words fall in that state, and the degree to which the words refer to the self. Retzinger [17] provides a content analysis that identifies words indicative of our different component categories. If it is assumed that the patient is discussing his condition with the caregiver, the language will largely refer to the self. Therefore, the magnitude of the experienced shame contributed by each component during a given patient response is proportional to the frequency with which the patient uses the words in these categories.

It is important to have alternative means to confirm the responses of these components. Further, the caregiver response has the opportunity to induce shame in the

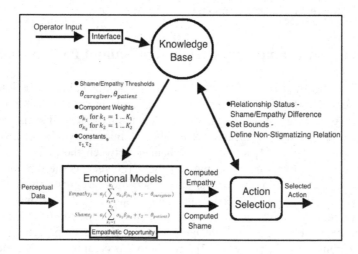

Fig. 3. Framework to uphold patient dignity in stigmatizing patient-caregiver interactions.

patient. Therefore, it is important to consider alternative cues that can contribute to the components' responses recognizing the limited availability of nonverbal cues.

The *Out of Place Self* component is defined by words such as "antsy", "nervous", and "tense" [17]. The patient is fearful and wants to withdraw from the situation. The patient will likely avert his eyes [8] as a means of trying to withdraw. Therefore, a gaze tracker is used to assess the response of this component with the magnitude of the response proportional to fraction of the time the patient averted his gaze during the single exchange. This component is going to have a response when the patient is anxious or fearful. Fear was reliably differentiated from sadness, anger, surprise, frustration, and amusement using galvanic skin response, skin temperature and heart rate signals in healthy populations [11]. Components of heart rate variability correlate with emotional intensity [26]. The response of this component is assessed by identifying fear and assessing the magnitude based on how the heart rate varied.

Terms such as "alienated", "deserted" or "ostracized" [17] define the *Lack of Acceptance* component. Patients may feel that, when the doctor has "abandoned" them or is not actively supporting them, it is because they are dying or a lost cause [26]. This is going to stress the patient. When Parkinson's patients are stressed, it causes a worsening of their symptoms [15]. The baseline tremor for the patient is measured just before the interaction, and significant worsening bolsters the magnitude of this component's response. This component should also increase when the patient expresses shame (i.e. any shame component is nonzero) and there is no empathetic response (the empathetic response components are zero) [27].

Ridicule is likely to inspire a measure of anger in the patient [27]. As mentioned above, anger is recognized in healthy populations, in controlled settings with easily obtainable physiological signals [11]. The *"Direct" Stigmatization* component response is computed by recognizing an amount of anger arising in the patient. This component should also increase when the caregiver directly stigmatizes the patient. Speech recognition for the caregiver should be better than the speech recognition for the patient, which would allow for more complete sentences/phrases to be recovered. The caregiver stigmatizes the patient when a word of negative valence and high intensity qualifies a proper noun or pronoun corresponding to the patient.

When the patient perceives himself to be "worthless" or "helpless" [17], as in the *Perception of Flawed Self* component, the patient is profoundly sad. Sadness is assessed with the basic physiological signals as noted above [11].

Finally, the *Confirmation of Flaw* component response is going to increase when the patient is "mortified" because of his inability to complete a common task, or he has "muddled" thoughts such that he is unable or unwilling to interact well with the caregiver [17]. The patient's inability to perform relative to his previous self is liable to leave the patient frustrated. Frustration is differentiated based on simple physiological signals [11]. One could also monitor specific parts of the interaction where a patient may fail, such as responding to orientation questions. When a failure occurs, the response of the component increases.

When the caregiver is trying to ameliorate the patient's concerns, it is essential that the caregiver show compassion for the patient [27]. This is required for a low magnitude response in the *Restore Self* component. A high magnitude response or a fully empathetic response would not only nonverbally support the patient (showing the flawed self is valued), but it would mitigate the shame of the patient by praising the self (showing that the negative outcomes for the patient are related to his condition rather than the self).

A classifier indicating the intensity of the caregiver's compassion will be constructed. The features used by the classifier could include the caregiver's facial expression, voice prosody, gaze, posture, and orientation toward the patient during the empathetic opportunity. The caregiver can be said to praise the patient if the patient is referred to with a positive adjective/adjectival clause.

At a basic level, making a connection with a patient requires the caregiver to respond appropriately and directly to what the patient has been saying. This is done through the use of "empathetic extenders" [22], which take standard forms such as "How sad", "How awful" or "It's very hard" [27]. These types of responses are a low magnitude empathetic response. High magnitude responses show the patient that he is understood. This is most commonly done by explicitly naming the emotion using phrases such as "You seem upset" [27]. These types of responses are identified using paraphrase recognition to assign a magnitude to the *Make Connection With Patient* component.

After the global magnitudes of shame and empathy have been computed, their difference is added to the running difference between the patient shame and caregiver empathy. If the difference between patient shame and caregiver empathy falls outside of set bounds, then the robotic agent must try to unobtrusively guide the relationship such that there is congruence between what the patient is feeling and what the caregiver understands the patient to be feeling (indicated by the caregiver having an empathetic response that matches the magnitude of patient shame).

4   Conclusions and Future Work

This paper introduced componential representations of shame and empathy, the basis for a computational model that is tasked with upholding the dignity of a Parkinson's disease patient in a stigmatizing relationship with his caregiver. These representations are based in the psychology literature. Forthcoming human-robot interaction studies will elucidate the weights, thresholds, and constants for the evaluation of the global shame and empathy magnitudes. There has been a paucity of studies on how a robotic agent can guide interactions between a human dyad (an exception being [9]). In addition to finding project specific guidelines when computing values for shame and empathy, our work will focus on how to best elicit empathy from the caregiver when the patient is in jeopardy of suffering indignity.

Acknowledgments. This work is supported by the NSF under Grant #IIS 1317214 in collaboration with Linda Tickle-Degnen and Matthias Scheutz at Tufts University.

References

1. Arkin, R.C., Ulam, P.: An ethical adaptor: behavioral modification derived from moral emotions. In: 2009 IEEE International Symposium on Computational Intelligence in Robotics and Automation (CIRA), pp. 381–387. IEEE, December 2009
2. Brooks, A.G., Arkin, R.C.: Behavioral overlays for non-verbal communication expression on a humanoid robot. Autonomous Robots **22**(1), 55–74 (2007)
3. Cohen, T.R., Wolf, S.T., Panter, A.T., Insko, C.A.: Introducing the GASP scale: a new measure of guilt and shame proneness. Person. & Social Psych. **100**(5), 947–966 (2011)
4. Claxton-Oldfield, S., Banzen, Y.: Personality characteristics of hospice palliative care volunteers: the ''big five''and empathy. American Journal of Hospice and Palliative Medicine **27**(6), 407–412 (2010)
5. Davis, M.H.: Measuring individual differences in empathy: Evidence for a multi-dimensional approach. Journal of Personality and Social Psychology **44**(1), 113 (1983)
6. Giner-Sorolla, R.: 5 shame and guilt. In: Judging Passions: Moral Emotions in Persons and Groups. Psychology Press (2013)
7. Gottschalk, L.A.: Some psychoanalytic research into the communication of meaning through language: the quality and magnitude of psychological states*. British Journal of Medical Psychology **44**(2), 131–147 (1971)
8. Haidt, J.: The moral emotions. Handbook of Affective Sciences **11**, 852–870 (2003)
9. Hoffman, G., Zuckerman, O., Hirschberger, G., Luria, M., Shani-Sherman, T.: Design and evaluation of a peripheral robotic conversation companion. In: Proceedings of the Tenth Annual ACM/IEEE International Conference on Human-Robot Interaction. ACM (2015)
10. Lee, D.A., Scragg, P., Turner, S.: The role of shame and guilt in traumatic events: A clinical model of shame-based and guilt-based PTSD. British Journal of Medical Psychology **74**(4), 451–466 (2001)
11. Nasoz, F., Alvarez, K., Lisetti, C.L., Finkelstein, N.: Emotion recognition from physiological signals using wireless sensors for presence technologies. Cognition, Technology & Work **6**(1), 4–14 (2004)
12. Niezink, L.W., Siero, F.W., Dijkstra, P., Buunk, A.P., Barelds, D.P.: Empathic concern: Distinguishing between tenderness and sympathy. Motiv. & Emotion **36**(4), 544–549 (2012)
13. Nijhof, G.: Parkinson's disease as a problem of shame in public appearance. Sociology of Health & Illness **17**(2), 193–205 (1995)
14. Nordenfelt, L.: Dignity of the elderly: an introduction. Medicine, Health Care and Philosophy **6**(2), 99–101 (2003)
15. Oertel, W.H., Ellring, H.: Parkinson's disease—medical education and psychosocial aspects. Patient Education and Counseling **26**(1), 71–79 (1995)
16. Rao, D., Choi, S.W., Victorson, D., Bode, R., Peterman, A., Heinemann, A., Cella, D.: Measuring stigma across neurological conditions: the development of the stigma scale for chronic illness (SSCI). Quality of Life Research **18**(5), 585–595 (2009)
17. Retzinger, S.M.: Violent emotions: Shame and rage in marital quarrels. Sage (1991)
18. Rodrigues, S., Mascarenhas, S., Dias, J., Paiva, A.: "I can feel it too!": emergent empathic reactions between synthetic characters. In: 3rd International Conference on Affective Computing and Intelligent Interaction and Workshops. ACII 2009, pp. 1–7. IEEE (2009)
19. Sabini, J., Garvey, B., Hall, A.L.: Shame and embarrassment revisited. Personality and Social Psychology Bulletin **27**(1), 104–117 (2001)
20. Smits, D.J., De Boeck, P.: A componential IRT model for guilt. Multivariate Behavioral Research **38**(2), 161–188 (2003)

21. Steinbock, A.J.: Shame. in Moral emotions: Reclaiming the evidence of the heart. Northwestern University Press (2014)
22. Suchman, A.L., Markakis, K., Beckman, H.B., Frankel, R.: A model of empathic communication in the medical interview. Jama **277**(8), 678–682 (1997)
23. Takahashi, K., Tickle-Degnen, L., Coster, W.J., Latham, N.K.: Expressive behavior in Parkinson's disease as a function of interview context. The American Journal of Occupational Therapy: Official Publication of the American Occupational Therapy Association **64**(3), 484–495 (2010)
24. Tickle-Degnen, L.: Chapter 14: Therapeutic rapport. In: Radomsky, M.V., Trombly Latham, C.A. (eds.) Occupational Therapy for Physical Dysfunction, 7th edn, pp. 412–427. Lippincott William & Wilkins, Baltimore (2014)
25. Tickle-Degnen, L., Zebrowitz, L.A., Ma, H.I.: Culture, gender and health care stigma: Practitioners' response to facial masking experienced by people with Parkinson's disease. Social Science & Medicine **73**(1), 95–102 (2011)
26. Uy, C.C., Jeffrey, I.A., Wilson, M., Aluru, V., Madan, A., Lu, Y., Raghavan, P.: Autonomic Mechanisms of Emotional Reactivity and Regulation. Psychology **4**(08), 669–675 (2013)
27. Zinn, W.: The empathic physician. Archives of Internal Medicine **153**(3), 306–312 (1993)

Predicting Extraversion from Non-verbal Features During a Face-to-Face Human-Robot Interaction

Faezeh Rahbar[1], Salvatore M. Anzalone[2]($\boxtimes$), Giovanna Varni[1],
Elisabetta Zibetti[3], Serena Ivaldi[4], and Mohamed Chetouani[1]

[1] ISIR, CNRS & Université Pierre et Marie Curie, Paris, France
{varni,chetouani}@isir.upmc.fr
[2] Psychiatrie de l'Enfant et de l'Adolescent, GH Pitié-Salpêtrière, Paris, France
anzalone@isir.upmc.fr
[3] CHART-Lutin, Université Paris 8, Paris, France
[4] Inria, 54600 Villers-lès-Nancy, France
serena.ivaldi@inria.fr

Abstract. In this paper we present a system for automatic prediction of extraversion during the first thin slices of human-robot interaction (HRI). This work is based on the hypothesis that personality traits and attitude towards robot appear in the behavioural response of humans during HRI. We propose a set of four non-verbal movement features that characterize human behavior during the interaction. We focus our study on predicting Extraversion using these features extracted from a dataset consisting of 39 healthy adults interacting with the humanoid iCub. Our analysis shows that it is possible to predict to a good level (64%) the Extraversion of a human from a thin slice of interaction relying only on non-verbal movement features. Our results are comparable to the state-of-the-art obtained in HHI [23].

Keywords: Human-Robot Interaction · Personality · Non-verbal behaviour

1 Introduction

Social robots should be able to adapt their behaviour taking into account the unique personality of their interacting partners. To this end, they need to learn a model of their behaviour, that can be built using multimodal features extracted during online interaction, physical features, social context, individual factors etc. [1]. Currently, a crucial challenge for Human-Robot Interaction (HRI) is the automated online estimation of the latter, such as personality traits, and the study of how they influence the exchange of verbal and non-verbal signals, as well the mechanisms underlying the production of behaviors, emotions and thoughts. These issues have been investigated in the project EDHHI[9][1], focused

---

[1] http://www.loria.fr/~sivaldi/edhhi.htm

© Springer International Publishing Switzerland 2015
A. Tapus et al. (Eds.): ICSR 2015, LNAI 9388, pp. 543–553, 2015.
DOI: 10.1007/978-3-319-25554-5_54

on studying social interactions between humans and the humanoid robot iCub. Within EDHHI, the researchers investigated how the production of social signals during HRI is influenced by individual factors, such as personality traits and attitude towards robots. A number of face-to-face dyadic and tryadic interactions were realized in the experiments of this project, between ordinary people without prior experience with robots and the humanoid iCub. Exploiting the dataset collected in EDHHI, the goal of this paper is to investigate whether it is possible to predict the personality trait of extraversion from a set of non-verbal features extracted during a short interaction with the robot. Particularly, we take into account the first thin slices of an interaction (i.e., the first minutes). Personality is generally addressed through the *trait theory* that individuates the factors able to catch stable individual characteristics underlying overt behaviour. Trait theory formalization exploits multi-factorial models, the most well-known being the Big-Five, owing its name to the 5 traits chosen as descriptive of a personality: *Extraversion, Neuroticism, Agreeableness, Conscientiousness, Openness to Experience* [11]. In this work we focus on extraversion, the personality trait that notably (i) shows up more clearly during interaction, and (ii) has the greater impact on social behaviour with respect to the other traits [23]. We include in the study also the negative attitude towards robot [15] that could capture a novelty or anxiety effect when the ordinary people interact with the robot for their first time, consistent with the focus of our study on the first thin slice of an interaction.

Background: Studies involving personality assessment, typical of the psychology domain, are now more and more of interest for human-computer interaction (HCI) and human-robot interaction (HRI). In these last years, indeed, a new branch of research in HCI, called *personality computing* is developing. Research on *personality computing* focuses on the following three major issues: (i) *Automatic Personality Recognition (APR)*; (ii) *Automatic Personality Perception (APP)*; and (iii) *Automatic Personality Synthesis (APS)* – see [21] for a review. In particular, APR is aimed at "inferring self-assessed personalities from machine detectable distal cues" [21]. About the modeling, the current exploited models in computing community are the trait based models, that try to isolate a small set of factors, known as Big-Five (BF) [11], able to describe the stable behavioural patterns. The *NEO-Personality-Inventory Revised* [4], the *NEO Five Factor Inventory* [12] and the *Big-Five Inventory* [10] are the most common experimental instruments adopted for this measurement.

These studies are now of high relevance for robotics. Tapus et al. [19] designed an assistive therapist robot matching its Extroversion with the one of its patients, and proved its effectiveness in terms of therapy performance. Meerbeek et al. [13] provided guidelines to design and evaluate personality and expressions in autonomous domestic robots. An important line of research is to probe into the influence of personality and individual traits on the production of verbal and non-verbal signals during HRI, as done in [9] and in our study.

Hypothesis: The literature on personality traits and HRI shows that personality traits influence the production of verbal and non-verbal signals during HRI.

Preliminary analysis on the EDHHI dataset reveal that extroversion and NARS influence the production of gaze and speech [9]. Here, we contend that it is possible to predict the human's extraversion from the analysis of non-verbal features.

Overview of the Proposed System for Automatic Prediction of Extroversion in HRI: The proposed system is sketched in Figure 1. We extracted a set of relevant non-verbal features from the depth image of a Kinect placed above the head of the iCub interacting face-to-face with adult participants to the EDHHI experiments. As it will be discussed in Section 3, the relevant features include quantity of movement, synchrony and the personal distance between human and robot (frequently studied in proxemics). To predict extraversion from these features, we trained a model in a supervised way thanks to the ground truth provided by the score of the questionnaires filled up by the participants, as reported in Section 2. The classification system and the experimental results are detailed in Section 4.

2   Methods and Materials

This section briefly describes the experiments that provided the dataset used in this work, along with the questionnaires and the participants to the study.

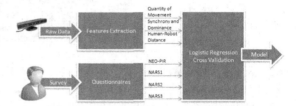

Fig. 1. Overview of the proposed system.

Questionnaires: To assess the personality traits of the participants, two questionnaires were used: the Revised Personality Inventory (NEO-PIR) [4], assessing the personality traits according to the Big Five model [11], and the Negative Attitude towards Robots Scale (NARS) [15]. From the first questionnaire, only the 48 questions related to Extraversion were retained. The order of the questions followed the original questionnaire, while answers were on a Likert-type scale from 1 (Totally disagree) to 5 (Totally agree). The second questionnaire consists of 14 questions divided into three sub-scales: "Negative attitude toward situation of interaction with robots" (NARS-S1), "Negative attitude toward social influence of robots" (NARS-S2) and "Negative attitude toward emotions in interaction with robots" (NARS-S3). The order of the questions followed the original questionnaire, while answers were on a Likert-type scale, from 1 to 7 (Strongly disagree / strongly agree).

Robotics Setup: The experiments were carried out with the humanoid iCub [14], a robot shaped like a 4 years old child. The robot was standing on a fixed pole and it was controlled by an operator hidden behind a wall. The operator was constantly monitoring the status of the robot, and could intervene to send high-level commands and respond to unexpected actions or requests of the participants, using a Wizard-Of-Oz GUI designed to control the robot. For satefy issues, the experimenter monitored the interaction and was able to intervene and stop the robot in case of urgency. The robot was velocity controlled when there was no physical interaction with humans, but its stiffness was adjusted to make it compliant in case people would touch it [6]. Facial expressions and speech were enabled. The robot was able to say few sentences, such as "yes", "no", "thank you".

Experimental Protocol: The experiments of Project EDHHI followed a protocol[2] developed to study the spontaneous behavior of ordinary people interacting with a robot. The personality traits of the participants were retrieved by questionnaires that were filled up through a web form two weeks before doing the experiments, to avoid influences of the questions on their behavior.

The day of the experiment, participants were informed about the overall procedure before signing an informed consent form granting use of all the recorded data. Before the experiment, the participants had to watch a short video presenting the iCub. The video did not provide any information about the experiments. It

Fig. 2. iCub interacting with two participants.

was instrumental to make sure that the participants had a uniform prior knowledge of the robot appearance. After the video, each participant was introduced to the robot by the experimenter, who did not provide any specific instruction to the participants about how to behave with the robot and what to do. The experimenter would simply stay on the right side of the robot, to supervise the interaction for safety issues. The robot was standing on its fixed pole, gently waving the hands and looking upright, while holding a colored toy in its right hand. It was not speaking. Once the participants were standing and looking in front of the robot, they were free to do whatever they wanted: talking to the robot, touching it, and so on. For few seconds, the robot would do nothing, then it would look at the participant (upward gaze) and raise the right hand, holding the colored paper roll. Since no instructions were given about this interaction, the participants could choose whether to interpret the robot's movement as an intentional and goal-directed action or not, therefore interact with the robot, or to ignore the action. If the participant had no reaction to this movement, the robot, controlled by the operator, would lower the hand after 4-5 seconds. Otherwise, the robot would open the hand to give the toy to the human (see Fig. 2). As participants did not receive any indication by the experimenter, if they wanted to, they could start interacting more actively with iCub, asking questions, giving back the toy, and so on. The designed interaction, triggered

---

[2] Ivaldi et al., IRB n.20135200001072.

by a simple movement of the robot, is very simple. However, due to the natural condition and the absence of constraints and indications from the experimenter, the response produced by the participants can be considered spontaneous, which justifies the observed variability of behaviors and non-verbal signals produced during the interaction. When the experimenter would detect a disengagement of the participant, a long pause or inactivity, she would invite the participant to withdraw from the robot and start preparing for executing the EDHHI experiments with iCub, which are out of the scope of this paper.

Participants: 39 healthy adults without any prior experience with robots volunteered to participate to the experiments (11 male and 28 female, aged 37.8y±15.2y). They received an ID number to preserve the anonymity of the study. They signed an informed consent form to partake in the study and granted us the use of their recorded data and videos.

Data Collection: The dataset from Project EDHHI includes the video stream collected by a Kinect RGB-D sensor (v.1, 30fps) placed above the head of the robot in such a way to retrieve the body and face of the human interacting with the robot. The dataset used in this work includes 39 videos (one for each participants) of the first minutes of their interaction with iCub, synchronized with the robot events logged by the Wizard-Of-Oz application used to control the robot. The average duration of the videos was 110.1s (SD=63.9s).

3   Non-verbal Features Extraction

The use of non-verbal features has been dictated by the real world constraints in which a robot should operate: although audio-based features can produce better performances in laboratory setups, they are unlikely to be reliable in real life scenarios, due to several sources of noise: environment, people talking in background, and robot itself [2]. As stated in psychological literature, Extraversion dimension encompasses specific facets as *sociability, energy, assertiveness* and *excitement-seeking*. Energy facet can be also an useful hint of the attitude toward a robot revealing, for example, if a person feels nervous or relaxed when she operates in front of a robot or has to share a task with it. Interpersonal distance is mainly linked to sociability and assertiveness and it also describes worry/relax about situations of interaction with robots. Previous studies showed that extraverted people tends to require smaller interpersonal distance [22]. Further, proxemics rules hold true also when one of the interactants is not a human, therefore a low familiarity or confidence with robots (that is, a negative attitude toward robots) results in increasing the interpersonal distance [18]. Interpersonal synchrony is acknowledged as very relevant in early communication between humans [5], and it provides information about the quality of interaction traits of the peers. For example, as referred in [7] *"people tend to synchronise their rhythms and movements ... within a conversation"*. In HRI, it can facilitate the natural interaction with robots with minimal cognitive load [8]. Starting from this knowledge, the features listed below are extracted from the recorded depth videos:

F1) Histogram of Quantity of Motion (h-QoM): Quantity of motion is a Silhouette Motion Image (SMI)-based measure of the amount of motion detected from an optical sensor like, for example, video-camera [3]. SMI (Silhouette Motion Image) is an image carrying information about variations of the silhouette shape and position in the last few frames of a video. Quantity of motion is, basically, an approximation of the energy of the movement and it is computed as the area (i.e. the number of pixel) of a SMI normalised over the area of the silhouette. It is computed using the following formulas:

$$SMI(t, i) = \sum_{i=1}^{n} Silhouette(t - i) - Silhouette(t) \tag{1}$$

$$QoM(t) = \frac{Area(SMI(t, n))}{Area(Silhouette(t))} \tag{2}$$

where n is the number of frames used to compute the SMI, t is the time at which the SMI and the QoM is being computed. In this work, the original algorithm is applied, with some small changes, to the depth images provided by the Kinect RGB-D sensor.

First, the silhouette of the participant is extracted by thresholding the depth image in order to remove the background. Unlike [3], the resulting silhouette is not binarized: this is done in order to keep also the details of internal motions (that is the motion occurring inside the silhouette, e.g., shaking the hands in front of the body) provided by the depth image. The SMI is obtained by subtracting the silhouette of a current frame from that of the n last frames (here $n = 3$). This image is then normalised by the value of n. Finally, the area of the SMI is calculated and normalised by the area of the silhouette of the current frame in order to define the Quantity

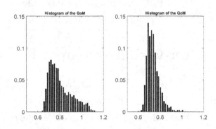

Fig. 3. H-QoM of two participants having very different NEOPIR and NARS scores. Left Panel corresponds to the h-QoM of a participant having the scores:NEOPIR=128 and NARSS1=11, NARSS2=11, NARSS3=3. Right Panel shows the hQoM of the participant having the scores: NEOPIR=69 and NARSS1=27, NARSS2=29,NARSS3=18.

of Motion of the current frame. H-QoM is computed in 64 bins in order to have a good resolution of the changes occurring in its dynamics. Figure 3 reports two exemplary h-QoM of two participants showing very different NEO-PIR and NARS scores.

F2-3) Histograms of Synchrony and Dominance (h-Sync, h-dom): Event Synchronisation (ES) technique is adopted to analyse synchrony and dominance between the movements of the i-Cub and the participant. This choice is mainly due to the different nature both of the two interactants (that is, a robot and a human being). ES was originally conceived to measure synchrony and time delay patterns between a pair of neurophysiological time-series in which events can be identified [17]. However, it can be extended to measure synchrony between two or more generic monovariate time-series with events [20]. In this work, events are defined as a subset of the iCub actions and the full-body energy peaks of the participant during the interaction, respectively. The analysis concerned only the similar actions performed by the two interactants. ES consists of a couple of measures addressing *"the fraction of event pairs matching in time and how often each time series leads in these matches"*, respectively [17]. The first part of this definition allowed at counting the number of actions occurring quasi-simultaneously with respect to the global number of actions occurred through the overall interaction. This count (Q in Eq. 3) is in the range $[0, 1]$ and expresses the overall synchrony between the two time-series. The number of times each time-series leads the other one in these matches is here used to show how often an action of one of the two interactants comes before the corresponding action performed by the other one. This count (q in Eq. 3) is in the range $[-1, 1]$ and provides the *direction* of synchrony, that is it allow at discriminating between causal ($q = 1$ or $q = -1$ depending on which of the two time-series precedes the other one) or mutual ($q = 0$) interaction. In other words, q shows who, by a chronemic point of view, is dominant.

$$Q^\tau = \frac{c^\tau(x_2|x_1) + c^\tau(x_1|x_2)}{\sqrt{m_{x_1} m_{x_2}}} \qquad q^\tau = \frac{c^\tau(x_2|x_1) - c^\tau(x_1|x_2)}{\sqrt{m_{x_1} m_{x_2}}} \qquad (3)$$

where: x_2 and x_1 are the two time-series of events describing the participant and the iCub, respectively; m_{x_1} and m_{x_2} are the suitable events occurring at the times $t_i^{x_1}$ and $t_j^{x_2}$ ($i = 1, ..., m_{x_1}; j = 1, ..., m_{x_2}$) in the two time-series; $c^\tau(x_1|x_2) = \sum_{i=1}^{m_{x_1}} \sum_{j=1}^{m_{x_2}} J_{ij}^\tau$ is the the number of times an event appears in x_1 after it appeared in x_2; τ a time lag for which two events could be considered as synchronous; with and $J_{ij}^\tau = \{1 \text{ if} 0 < t_i^{x_1} - t_j^{x_2} < \tau; 1/2 \text{ if} t_i^{x_1} = t_j^{x_2}; 0 \text{ otherwise}\}$.

Two different approaches are adopted to extract the events for the iCub and the participant. For the iCub, a log-file is used to store its head/arms/hands movements, the timestamps corresponding to the beginning of an action/command as well as the type of actions/commands, that together define the main events. As regards the participant, the events are defined from her QoM. First, it is filtered by using a fifth-order Savitzky-Golay filter and a FFT is applied. Then, the energy of the QoM is computed from its spectrum.

The first derivative of the energy is computed over a sliding window having a size of 1s frames and a step of 33ms. Further, this derivative is weighted frame-by-frame by the amplitude of the energy so as to amplify the fast and the largest movements and reduce the impact of movements which can be considered as noise. The events are extracted by applying a threshold on the amplitude of this resulting signal. Finally, the up edges are retained as events (see Figure 4).

The frames resolved variants of Q and q are computed through each video. To have these variants, the previous equation is modified as follows [17]: $c_n(x_1|x_2) =$

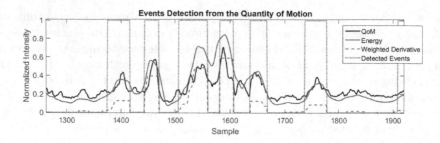

Fig. 4. The different steps of the events detection from the QoM. The solid black line stands for the QoM signal. The solid gray line corresponds to the Energy of QoM. The dashed line is the Weighted Derivative of the Energy of QoM. The dotted line stands for the intervals in which events have been detected.

$\sum_{i=1}^{m_{x_1}} \sum_{j=1}^{m_{x_2}} J_{ij} \Theta(n - t_i^{x_1})$, where Θ is the Heaviside function (i.e. $\Theta(x) = 0$ for $x \leq 0$ ans $\Theta(x) = 1$ for $x \geq 0$) and $n = 1, ..., N$. Then, a sliding window is applied to both Q and q (window size of 3s, step=1.5s), and the count of how many times Q and q step up is done. Finally, the histograms of these countings are built according the obtained values in the ranges of $[0, max\_no\_steps\_up]$ and $[-1, 1]$, respectively.

F4) Standard Deviation of Human-robot Distance (STD-d): it is computed as the average of the pixels' values of the silhouette extracted from the depth image of the Kinect. Considering its position in the set-up, it is a good approximation of the distance between the two interactants.
$\rightarrow$ The features F1-F4 are merged into a 72-dimensional vector. The resulting dataset included 39 instances, that is a feature vector for each participant.

4   Automated Prediction of Extraversion During HRI

Features extracted from interactions with the iCub built the dataset for the prediction of Extraversion. The NEO-PIR scores were used as reference for labeling the personality trait of each subject. However, in the particular context of a first interaction with social robots, people's behavioural response can be not only determined by their personality, but also by their attitude towards robots and the context of the interaction. In particular, due to prior experiences with robotics, the attitude towards the robots plays a relevant role and can dramatically affect people's behaviour towards social robots.

To catch this phenomenon, we seek a linear combination between the scores able to contextualise the personality information in the specific scenario of interaction with social robots. This work addressed this point by relabeling the instances of the dataset using the NEO-PIR extraversion score toghether with NARS questionnaires results. More in the detail, a Principal Component Analysis (PCA) was carried out on a *scores vector* including: NEO-PIR, NARS-S1,

Table 1. Average Percentage of Precision, Recall and F-score

Features	Precision	Recall	F-score
std-d, h-QoM	33%	27%	46%
std-d, h-QoM, h-dom	59%	62%	61%
std-d, h-QoM, h-sync	60%	64%	63%
std-d, h-QoM, h-sync, h-dom	64%	69%	66%

NARS-S2, and NARS-S3 scores, respectively. NARS-S3 is an inverse scale, so its scores were reversed. For NEO-PIR, NARS-S1, NARS-S2, and $\overline{NARS - S3}$ scores, the PCA's eigenvalues were respectively 2.17, 0.85, 0.56 and 0.32, while the PCA's component load were 0.32, -0.56, -0.57 and 0.51. Only the first principal component was meaningful (eigenvalue greater than 1). The loadings reflected how the personality scores are captured by the first principal component. The values of the first component were quantized (High-Low) along their median and the final labels were obtained.

The dataset resulted in a 39 (instances) x 72 (features) matrix, that is there are more features than instances. For this reason, we decided to adopt for the classification a Logistic Regression Classifier (LRC) [16] with penalty parameter $C = 1$ and $L2$ norm $L2$. The averaged performance of the trained classifier was assessed via a multiple-run k-fold stratified cross-validation. In this study, 10 run and 10 folds have been adopted. Table 1 summarises the performances of the LRC, according to the different subsets of used features. The table shows that the classification result relying on the Quantity of Movement alone on the standard deviation of the distance, is not able to overcome the chance level. However, classification results using also dominance and synchrony information overtake this level. Using the whole set of features the classifier reaches the top of the performances. Classification based exclusively on the extroversion does not yield significant results. These results are consistent with previous studies on prediction of Extraversion in human-human interaction only from non-verbal movement features (e.g., [23]).

5 Conclusion and Future Works

This paper presented an automatic prediction of Extraversion personality trait during *thin slices* of interaction with social robots, using non-verbal movement features. A Logistic Regression classifier was fed with the following features: the histogram of Quantity of Motion, the distance between human and robot, and the histograms of synchrony and dominance. To our knowledge, this is the first work dedicated to study thin slices of interaction with social robots, using such kind of predictive features of personality trait. The main limitation presented by this work can be found in the limited space of the room used for the experiments, in the laboratory environment and in the quite simple actions of the robot, that could not lead to the variability of the human behaviour as desired.

Also, a random bias towards the specific spatial configurations of human parteci-pants could limit the effectiveness of the results presented. Thought preliminary, despite their limitations, these encouraging results indicate the good direction of research and good premises to improve personality prediction during HRI. Future works will involve a wider set of features, such as people's posture and gaze, examining in depth their role during complex interactions as signatures of personality traits. Moreover, the space of the parameters of the features extrac-tion will be explored, as well as how the performances of the system will change in time, according to the amount of data collected. The final goal is to build an online, real-time personality recognition system that can be used by social robots to learn a complex model of their human partners.

References

1. Anzalone, S.M., Yoshikawa, Y., Ishiguro, H., Menegatti, E., Pagello, E., Sorbello, R.: Towards partners profiling in human robot interaction contexts. In: Carpin, S., NOda, I., Pagello, E., Reggiani, M. (eds.) SIMPAR 2012. LNCS, vol. 7628, pp. 4–15. Springer, Heidelberg (2012)
2. Anzalone, S.M., Yoshikawa, Y., Ishiguro, H., Menegatti, E., Pagello, E., Sorbello, R.: A topic recognition system for real world human-robot conversations. In: Lee, S., Cho, H., Yoon, K.-J., Lee, J. (eds.) Intelligent Autonomous Systems 12. AISC, vol. 194, pp. 383–391. Springer, Heidelberg (2013)
3. Camurri, A., et al.: Interactive systems design: A kansei-based approach. In: Proc. Conf. New interf. musical expression, pp. 1–8 (2002)
4. Costa, P., McCrae, R., Rolland, J.: NEO-PI-R. Inventaire de Personnalité révisé, Editions du Centre de Psychologie Appliquée, Paris (1998)
5. Delaherche, E., et al.: Interpersonal synchrony: A survey of evaluation methods across disciplines. T. Affect. Comput. 3(3), 349–365 (2012)
6. Fumagalli, M., Ivaldi, S., et al.: Force feedback exploiting tactile and proximal force/torque sensing. theory and implementation on the humanoid robot icub. Autonomous Robots 4, 381–398 (2012)
7. Hall, J., Knapp, M.: Nonverbal Communication. Handbooks of Communication Science, De Gruyter (2013)
8. Hasnain, S.K., et al.: Intuitive human robot interaction based on unintentional synchrony: a psycho-experimental study. In: ICD, pp. 1–7 (2013)
9. Ivaldi, S., et al.: Towards engagement models that consider individual factors in hri: on the relation of extroversion and negative attitude towards robots to gaze and speech during a human-robot assembly task, pp. 1–24 (2015). arXiv:1508.04603 [cs.RO]
10. John, O.P., Donahue, E., Kentle, R.: The big five inventory-versions 4a and 54. Tech. rep., University of California, Berkeley, Institute of Personality and Social Res. (1991)
11. McCrae, R.: The five-factor model of personality. In: P.Corr, Mathhews, G. (eds.) Handbook of Personality Psychology, pp. 148–161 (2009)
12. Mccrae, R.R., Costa, T.P.: A contemplated revision of the neo five-factor inventory. Personality Individual Differences 3, 587–596 (2004)
13. Meerbeek, B., Saerbeck, M., Bartneck, C.: Iterative design process for robots with personality. In: Dautenhahn, K. (eds.) AISB2009 Symposium on New Frontiers in Human-Robot Interaction, pp. 94–101 (2009)

14. Natale, L., et al.: The icub platform: a tool for studying intrinsically motivated learning. In: Baldassarre, G., Mirolli, M. (eds.) Intrinsically Motivated Learning in Natural and Artificial Systems. Springer, Heidelberg (2013)
15. Nomura, T., et al.: Experimental investigation into influence of negative attitudes toward robots on human-robot interaction. AI SOCIETY 20(2), 138–150 (2006)
16. Pampel, F.: Logistic Regression: a primer. Sage Publications (2000)
17. Quiroga, R.Q., et al.: Event synchronization: a simple and fast method to measure synchronicity and time delay patterns. Phys Rev E 66(4)
18. Takayama, L., Pantofaru, C.: Influences on proxemic behaviors in human-robot interaction. In: IROS (2009)
19. Tapus, A., et al.: User-robot personality matching and assistive robot behavior adaptation for post-stroke rehabilitation therapy. Int. Serv. Rob. 1(2) (2008)
20. Varni, G., et al.: A system for real-time multimodal analysis of nonverbal affective social interaction in user-centric media. Tr. on Multimedia 12(6)
21. Vinciarelli, A., Mohammadi, G.: A survey of personality computing. IEEE Transactions on Affective Computing 5(3), 273–291 (2014)
22. Williams, J.L.: Personal space and its relation to extraversion-introversion. Canadian Journal of Behavioural Science 3(2), 156–160 (1971)
23. Zen, G., et al.: Space speaks: towards socially and personality aware visual surveillance. In: 1st ACM Int. Worksh. on Multimodal Perv. Video Anal. pp. 37–42 (2010)

Check Your Stereotypes at the Door: An Analysis of Gender Typecasts in Social Human-Robot Interaction

Daniel J. Rea[✉], Yan Wang, and James E. Young

University of Manitoba, Winnipeg, Canada
{daniel.rea,wang.yan,young}@cs.umanitoba.ca

Abstract. In this paper, we provide evidence that suggests prominent gender stereotypes might not be as pronounced in human-robot interaction as may be expected based on previous research. We investigate stereotypes about people interacting with robots, such as men being more engaged, and stereotypes which may be applied to robots that have a perceived gender, such as female robots being perceived as more suitable for household duties. Through a user study, we not only fail to find support for many existing stereotypes, but our analysis suggests that if such effects exist, they may be small. This implies that interface and robot designers need to be wary of which stereotypes they bring to the table, and should understand that even stereotypes with prior experimental evidence may not manifest strongly in social human-robot interaction.

Keywords: Human-Robot Interaction · Gender studies

1 Introduction

Social Human Robot Interaction (sHRI) investigates how robots and people interact socially. For instance, robots are already emerging into the real world as personal care, tutoring, or even professional team robots (e.g., search and rescue), and need to interact using speech, gaze, and gestures, and need to be aware of and work within social structures and norms – e.g., a robot should be good at conversational turn taking. Thus, sHRI research goes beyond the technical robotics challenges, and involves social elements rooted in psychology and sociology.

One such area of importance to sHRI is gender studies. While there is a body of research on how women and men may have different needs and may interact differently with their worlds (such as new technologies) and other people [1, 2], the intersection of Human-Computer Interaction (HCI) and gender studies is only in its infancy. The ACM GenderIT conference is only newly established, and there are even fewer examples of gender-related sHRI work [3]. Despite this, understanding how gender relates to interaction with robots is very important for sHRI, as robots need to understand the specific differences and needs of women and men, the same as people do, if they are to interact with people and integrate naturally. Further, in order to design effectively, sHRI designers need to know what gender stereotypes may be applied to their robots.

© Springer International Publishing Switzerland 2015
A. Tapus et al. (Eds.): ICSR 2015, LNAI 9388, pp. 554–563, 2015.
DOI: 10.1007/978-3-319-25554-5_55

There is a large amount of work in sociology and psychology that investigates gender stereotypes, for example, that women may be more nervous with new technology [4] or men may be more rude to others [5]. However, such generalizations are dangerous as they may be used (e.g., by designers) to oversimplify the complexities of gender, resulting in designs that may re-enforce potentially-harmful stereotypes (e.g., if little boys do not like kitchen toys a toy company may market theirs to girls only). Instead, particularly given that social robots will be in shared public and private spaces, we argue we should to try to understand the needs of both women and men: such *inclusionary* designs should be sensitive to specific needs of *both* genders simultaneously, instead of *exclusionary* ones that target only one gender and can potentially re-enforce stereotypes [3]. Our work takes this approach: while we investigate stereotypes, it is part of a bigger goal of simply understanding all users and how broad stereotypes may be manifested in actual interaction.

We cannot directly apply existing gender work on technology to robots, as people tend to interact with robots more socially than with traditional technologies such as laptops or smart phones, and are more likely to attribute names, emotions, etc. [6]; this is especially true when robots are designed specifically for social interaction. Conversely, people will not interact with robots the same as with other people (e.g. [7]), and so interpersonal gender work likewise cannot be directly applied, and it is important to reconsider gender studies results specifically for sHRI.

In this paper we use the term "gender" synonymously with biological sex, which we recognize is overly simplistic. We used "gender" for the practical purpose of simplifying our investigation; this approach is used heavily not only in sHRI and HCI [1], [8] but in feminism work in Science and Technology Studies in Sociology (e.g. [2]).

We present an investigation into how gender stereotypes may be manifested in sHRI, with our results indicating the possibility that some prominent stereotypes may not manifest in sHRI, and if they do, they may be too subtle to warrant much consideration. This has direct implications for sHRI design – designers should be careful of leveraging gender stereotypes – and provides a starting point for continued gender work in sHRI.

2   Related Work

Gender studies, feminism, or men's studies, is a mature research area that uses gender, gender identity, and sex as central themes of investigation [1]. Some of this work has been heavily applied to science and technology studies, for example, through investigations of how gender has impacted technology developments and trajectories [2].

More recently, HCI has started drawing from gender studies [9]: researchers are mapping out gender-specific interaction needs and strategies, for example, in software exploration [10], interface problem-solving strategies [11], navigating virtual environments [12], and even in experiment design [13]. There has been an *inclusive* theme throughout this work, of trying to understand the needs of both women and men and to develop strategies and flexible solutions that work for both, rather than *exclusionary* designs that benefit one group more than the other.

There has been much less gender work specifically in sHRI; most gender results are afterthoughts or small components of work targeted at non-gender questions [3]. Such results include evidence that men and women may have different criteria for evaluating robots [14], may have different ideas about what sorts of tasks robots will do [15], or that women and men may have different preferences regarding how a robot should approach them [16]. We continue investigation in this direction with a study that directly and primarily investigates how gender may be a component of sHRI.

Other work has more directly investigated the impact of the perceived gender of the robot itself (e.g., as being more masculine or feminine), and how people interact with it (e.g., men may be more positive toward female robots [17]). Some work has further suggested that human gender stereotypes may also be applied to robots, for example, people may have predispositions toward a robot's knowledge base [18] or usefulness [8] based on its perceived gender. These initial works investigate highly-targeted questions [17, 18] or rely on picture or text descriptions of robots and do not yet involve interaction with an actual robot [3], [8]. We continue this line of work with a broader look at how stereotypes may be manifested in actual interaction with a robot.

3   Stereotypes and Hypotheses

We look at two perspectives of gender stereotypes in sHRI: what existing stereotypes suggest about how women and men will interact with robots, and which stereotypes may be applied to robots that are perceived as being masculine or feminine. Our selection of stereotypes in this section are by no means complete, as the number and range of such stereotypes is large. However, we selected stereotypes with prior empirical evidence in human-robot interaction as a starting point.

3.1   Selected Stereotypes About Male and Female Users

Politeness – Politeness toward others is a cornerstone of interpersonal interaction. There is evidence that women, in general, may be more polite than men [5], [19] – we test if this manifests when women and men interact with a robot.
Engagement – Research shows men are typically more engaged with new technologies than women [20, 21] and so men may be more engaged with a robot.
Relaxation – Women have reported lower self-efficacy toward new technologies and in some cases have higher anxiety surrounding using them [4]. This has also been found regarding perceptions of interaction with robots [3].

3.2   Selected Stereotypes Applied to Male or Female Robots

People may apply gender stereotypes to a robot that has a perceived gender. A recent work found people apply stereotypes to a hypothetical robot (shown as an image) depending on the robot's haircut [8]. Participants attributed the female-haircut robot with traditional female traits (communal, e.g., compassionate, empathic), and rated it as more suited to traditional female tasks (e.g., patient or child care). Conversely, the male-haircut robot was attributed with more male traits (i.e., agentic, e.g., assertive, competitive) and was rated as being more suited to typical male tasks (e.g., repairing equipment, transporting goods).

We investigate if similar stereotypes would emerge when interacting with a real robot instead of only viewing images.

4 Study Design

Our study uses a Wizard-of-Oz scenario where participants were told to interact with an autonomous, intelligent robot that moves and speaks, while the robot was secretly controlled by a researcher in another room. Our study involved thirty-nine participants from the local university population that were paid $10 for 30 minutes of participation. Gender was balanced (19 male) and split between the *he* and *she* robot cases, for a 2x2 study design. Our demographics questionnaire also included *inter-sex* in addition to male and female, although no participants selected this option.

4.1 Instruments and Analysis Method

For investigating how participants interact with a robot, we recorded video and coded for *politeness*, *engagement*, and *relaxation*, counting the instances per session. We coded for positive and negative, e.g., a rude action would be *negative politeness,* using participant speech style and body language; we provide our coding guideline for *relaxation* in Appendix A, but the full coding guideline is omitted for brevity. Videos were coded by two researchers (30% overlap), with good inter-coder reliability (Krippendorph's $\alpha = .79$).

To investigate people's stereotypes toward male or female robots, we employed the exact questionnaires from the previous robot haircut work to measure perceived male (agentic) or female (communion) traits of the robot, and appropriateness for typically male and female tasks such as housework or physical labor [8]. Each of these scales sum to a single number that can be tested.

Our robot was an Aldebaran NAO (Fig. 1), a small humanoid robot that is capable of complex gestures and speech. It was remotely controlled (unbeknownst to the participant) via a Wizard-of-Oz setup, with in-house controller software that enabled a high level of interaction flexibility and pre-coded actions and speech. The voice was chosen to be NAO's default English voice.

4.2 Manipulations

To create a robot that is perceived as being feminine or masculine we manipulated the pronoun used to refer to the robot, using either 'he' or 'she'; the robot had the unisex name 'Taylor' in all cases to isolate the effect of pronoun use. Non-gender-specific word choice has recently been shown to potentially influence the gender people

Fig. 1. The NAO robot used in our experiment

imagine something to be [13], so we hypothesize that our specifically gendered pronouns will affect the robot's perceived gender. Our manipulation was between participants, and we asked participants to rate the robot as masculine or feminine during the post-test to validate this. To ensure our manipulation was consistent, the researcher controlling the robot maintained a rehearsal regime throughout the duration our studies, and held a cue-card (unseen by participants) during each experiment to help focus and use the correct pronoun consistently.

4.3   Method

After a briefing, participants signed an informed consent form and completed a demographic questionnaire. The researcher introduced the robot, which stood up, waved, and introduced itself to demonstrate its abilities. At this point, the researcher left, claiming they forgot additional forms, and said they will return shortly, informing the participant they could chat with the robot during this time. While the researcher was away (~4 min.), the robot engaged the participant in casual conversation on daily topics (e.g., hobbies, work, or school) with casual gestures for realism. The robot aimed for consistent conversation across participants by sticking to and coming back to pre-defined topics, although we needed to be flexible enough to respond individually to each participant to maintain an illusion of intelligence. After returning, the researcher administered the questionnaire on robot perceptions (from [8]) and the post-test questionnaire, and debriefed participants.

5   Results

We conducted 2-way ANOVAs (participant gender by robot gender) on our dependent measures (perceived robot gender, stereotypes about male and female users, and gender stereotypes applied to robots), to investigate the impacts of participant gender ·or perceived robot gender, as well as potential interactions.

To investigate if our *she* versus *he* manipulation was successful, we conducted a 2-way ANOVA on participant rating of the robot as more masculine or feminine (7-pt Likert-like scale). There was a main effect of robot gender; the *he* robot was rated as more masculine (M=4.8, se=.28) and *she* as more feminine (M=2.9, se=.29, $F_{1,35}$=20.83, p<.001, ω^2=.87), and no main effect of participant gender or interaction p>.05). This means that our *she* versus *he* manipulation was successful in gendering the robot.

Table 1. An overview of our results: grand means, standard deviations, effect sizes, and observed power (1-β). *p<.05

stereotypes toward robots, average score on 1-7 scale (by robot gender)	grand mean	std. dev.	effect size ($\omega^{2)}$	obs. pwr.
female tasks	4.75	1.19	.09	.11
male tasks	4.94	.85	.04	.06
agentic (male) traits	4.01	.89	.13	.13
communal (female) traits	4.55	.90	.55	.18
stereotypes toward users, average count (by participant gender)				
positive engagement	7.74	3.06	.20	.09
negative engagement	1.18	1.17	.11	.09
positive relaxation	7.87	3.83	.01	.05
negative relaxation	2.46	2.82	.11	.07
positive politeness	4.59	1.53	.33	.11
negative politeness	.62	1.43	.63*	.51*

For stereotypes about male or female users, there was a main effect of participant gender on politeness, where our female participants were less rude (M=.15, se=.32) than the male participants (M=1.05, se=.31, $F_{1,35}$=4.14, p<.05, ω^2=.63). No other effects of participant gender or robot gender were found, and there were no interactions (p>.05). Relating to gender stereotypes being applied to robots, we found no main effects of robot gender on participant rating of the robot's suitability for either female tasks ($F_{1,35}$=.52, p=.48, ω^2=.09) or male tasks ($F_{1,35}$=.04, p=.84, ω^2=.04), or on female traits ($F_{1,35}$=1.12, p=.30, ω^2=.50) or male traits ($F_{1,35}$=.70, p=.41 ω^2=.13). There were also no main effects of participant gender on these measures and no interaction effects (p>.05).

5.1   Post-Hoc Analysis

Across most of our tests we found a lack of support for the gender stereotypes we were looking for, even though we can confidently assume that our *he* versus *she* gender manipulation of the robot was successful (based on our statistical results).

A lack of support does not imply no effect; we performed a post-hoc analysis to investigate further. Across the negative tests we have very low observed power (most <.2, Table 1), suggesting that we may have simply failed to detect effects that may exist. Apart from a few exceptions, the standard deviations of our measures were reasonably low (Table 1), particularly for the robot stereotype measures, improving our confidence that we did not miss a large uncontrolled confound such as interpersonal variability being more influential than our controls.

We further note that we have very small effect sizes across tests (Table 1). This suggests that even if a difference were to approach significance (e.g., with a larger sample size), we could expect the actual difference to be small or subtle; thus our data provides evidence that a strong effect may not emerge, even with more participants, discouraging us from conducting follow-up tests on this approach.

5.2   Discussion

We conducted an experiment where female and male participants interacted with either a masculine or feminine robot, and analyzed how these variables impact common stereotypes regarding both. As supported by previous work, we found our male participants to be ruder to our robot during interaction than our female participants. While this has clear implications for developing sHRI interfaces, we believe that the other, lack of findings, are more interesting.

Our lack of results regarding applying stereotypes to robots seemingly contradicts the recent results showing how robot hair style can invoke gender stereotypes [8] (supporting results from prior work [22]), though we would like to stress that we did not perform equivalence testing (which requires *a priori* planning), and as such we do not claim that no effect exists. One possibility is that the visual stimulus of a hairstyle may elicit stronger responses than our real robot with a verbal stimulus, despite having statistical support that the robot's perceived gender was successfully manipulated by our pronoun choice ('he' versus 'she'). It will be important follow-up work to consider visually gendering robots; verbal stimuli may not be enough to cross the "lower bound" necessary to invoke gender stereotypes. We highlight that our real robot and interaction greatly improves ecological validity compared to the picture stimulus in prior work. In addition, the prior work used a comparative, within-subjects design [8] (in contrast to our between-subjects) which may have encouraged participants to dichotomize the two robots to a male-female binary.

Although the previous work does not report standardized effect sizes [8], the actual differences observed in their studies were quite small and within the range of our observed differences (<.5 on the same scales); their finding significance may be due to the added statistical power of the within-participants method and 50% more participants than our study. Thus, our results and statistical analysis fall in line with the prior results given our sample size. While we did not find a statistical difference, given our tight standard deviations and small effect sizes, if a difference does exist, it is likely to be small. In future work, our understanding of the effect size would additionally benefit from understanding what gender participants perceive our robot as regardless of our pronoun treatment—a gender-neutral case of addressing the robot as "it."

It is important to consider this disparity between our lack of results and prior related, more general gender work as outlined earlier in the paper. One reason may be the fact that people interact with social robots in a fundamentally different way than with traditional technologies. This would explain the lack of stereotypical differences in *engagement* or *relaxation* around social robots, as participants interact with them more as a social other than a typical new technology. This further explains why our only positive result was *politeness*, as this stereotype stems from how people interact with *each other*, not from how people interact with *technologies*; social robots possibly fit the former model better than the latter. Perhaps inter-personal stereotypes surrounding gender may be more applicable to sHRI than stereotypes of how people interact with technology.

Finally, it is important to acknowledge the limitation surrounding our mixed use of "gender" and "sex," which are fundamentally different concepts [1]. Our use of "sex" to roughly represent gender (similar to other works, e.g. [2], [8], [18]) over-simplifies the interpretation of gender effects and limits our analysis. Moving forward, sHRI needs to address this issue more thoroughly as a field.

6   Summary

In this paper, we investigated how prominent gender stereotypes regarding how women and men may interact with robots, and how people may perceive gendered robots, might be realized in social human-robot interaction scenarios. We presented results from a study that involved interaction with a real robot and an analysis that highlights how these stereotypes may not manifest strongly in real human-robot interactions. In particular, although we found male participants to be less polite to robots, we found no support for expecting women to be less engaged or relaxed around robots, and found no support for the idea that people may apply gender stereotypes to robots themselves. While we stress this is not statistical proof that there are no gender differences (we did not perform equivalence testing), our analysis suggests that if such differences do exist, then they may be small and possibly insignificant for practical sHRI research. For researchers in sHRI, this means that we must be very careful and wary of using stereotypes in our interaction and robotic designs as, even with prior research evidence, established stereotypes may not always manifest strongly in social interaction with robots.

We believe that the pursuit of gender studies in sHRI will be an ongoing crucial element of developing robots that interact naturally with people in social situations, and envision that our work helps to grow this direction of work.

Appendix

Below is a sample of our coding guideline for the dependent variable *relaxation*.

Table 2. The coding guideline used by both of our coders. This sample is for the variable *relaxation*.

Relaxation: as the previous study shows, women tend to be more fearful of new technology. I would like to explore if there are some differences between men and women in terms of relaxation/nervousness during the conversation.	
Hypothesis: women are more nervous than men when they interact with a robot	
Baseline for judging nervousness: by observing participants' facial expression or body language, we may tell that the participants are nervous about something, or they are quite calm and less tense or worried.	
Positive (relaxed): standard: observing participants' facial expression and body language to see if they are calm and less tense or worried	**Negative (nervous):** standard: observing participants' facial expression and body language to see if they are nervous or worried about something.
examples for short instances: **Code PR#**	examples for short instances: **Code NR#**
PR1: relaxation showed by body language: (1) sit comfortably, like put one leg on the other/put their feet on the chair/use their hand to support their chin (maximum 3 per video) (2)approach the robot, observe the robot from different angles; **PR2:** relaxation showed by facial expression: smiling naturally or laugh out loud **PR3:** relaxation conveyed by verbal expression: talk with or ask robot questions actively to continue the conversation, but do not show the desire/need to know more about the robot **PR*:** instances clearly show participants are relaxed.	**NR1:** nervousness showed by body language: (1) fidget with their hands or clinched hands; (2) sitting rigid/straight;(3) moving while biting lips or having some hand movements **NR2:** nervousness showed by facial expression: (1) avert eyes from the robot (avoid eye contact); (2) smile or laughing nervously; embarrassed laugh **NR3:** nervousness conveyed by verbal expression: take a long time to answer a question as if they are thinking or unsure (not bored or distracted) **NR*:** instances clearly show participants are nervous. Eg., move away or keep distance from the robot

References

1. Cranny-Francis, A., Waring, W., Stavropoulos, P., Kirkby, J.: Gender Studies: terms and debates. Palgrave Macmillan, New York (2003)
2. Berg, A.J.: Technological flexibility: bringing gender into technology (or was it the other way round?). In: Cockburn, C., Furst-Dilic, R. (eds.) Bringing Technology Home: Gender and Technology in a Changing Europe. Open University Press, Milton Keynes (1994)

3. Wang, Y., Young, J.: Beyond "pink" and "blue": gendered attitudes towards robots in society. In: Proc. of The Significance of Gender for Modern Information Technology (GenderIT 2014), pp. 49–59. ACM (2014)
4. Venkatesh, V., Morris, M.G., Davis, G.B., Davis, F.D.: User Acceptance of Information Technology: A Unified View. MIS Q **27**, 425–478 (2003)
5. Holmes, J.: Women, men and politeness. Routledge (2013)
6. Young, J.E., Sung, J., Voida, A., Sharlin, E., Igarashi, T., Christensen, H.I., Grinter, R.E.: Evaluating Human-Robot Interaction. Int. J. Soc. Robot. **3**, 53–67 (2010)
7. Malle, B.F., Scheutz, M., Arnold, T., Voiklis, J., Cusimano, C.: Sacrifice one for the good of many? In: Proc. Human-Robot Interaction, pp. 117–124. ACM (2015)
8. Eyssel, F., Hegel, F.: (S)he's Got the Look: Gender Stereotyping of Robots. J. Appl. Soc. Psychol. **42**, 2213–2230 (2012)
9. Bardzell, S.: Feminist HCI: taking stock and outlining an agenda for design. In: Proc. Human Factors in Computing Systems, pp. 1301–1310. ACM (2010)
10. Burnett, M.M., Beckwith, L., Wiedenbeck, S., Fleming, S.D., Cao, J., Park, T.H., Grigoreanu, V., Rector, K.: Gender pluralism in problem-solving software. Interact. Comput. **23**, 450–460 (2011). Elsevier
11. Beckwith, L., Burnett, M.: Gender: an important factor in end-user programming environments? In: Visual Languages - Human Centric Computing, pp. 107–114. IEEE (2004)
12. Czerwinski, M., Tan, D.S., Robertson, G.G.: Women take a wider view. In: Proc. Hum. Factors in Computing Systems, pp. 195–202. ACM (2002)
13. Bradley, A., MacArthur, C., Hancock, M., Carpendale, S.: Gendered or neutral? Considering the language of HCI. In: Graphics Interfaces. ACM (2015)
14. Mutlu, B., Osman, S., Forlizzi, J., Hodgins, J., Kiesler, S.: Task structure and user attributes as elements of human-robot interaction design. In: Proc. Robot and Human Interactive Communication, pp. 74–79. ACM (2006)
15. Schermerhorn, P., Scheutz, M., Crowell, C.R.: Robot social presence and gender. In: Proc. Human Robot Interaction, pp. 263–270. ACM (2008)
16. Dautenhahn, K., Walters, M., Woods, S., Koay, K.L., Nehaniv, C.L., Sisbot, A., Alami, R., Siméon, T., Lane, C., Sisbot, E.A.: How may I serve you? A robot companion approaching a seated person in a helping context. In: Proc. Human-Robot Interaction, pp. 172–179. ACM Press (2006)
17. Siegel, M., Breazeal, C., Norton, M.I.: Persuasive robotics: the influence of robot gender on human behavior. In: Proc. Intelligent Robotics Systems, pp. 2563–2568. IEEE (2009)
18. Powers, A., Kramer, A.D.I., Lim, S., Kuo, J., Kiesler, S.: Eliciting information from people with a gendered humanoid robot. Workshop on Robot and Human Interactive Communicaiton, pp. 158–163. IEEE (2005)
19. Gibson, E.K.: Would you like manners with that? A Study of Gender, Polite Questions and the Fast - Food Industry **1**, 1–17 (2009)
20. Kay, R.: Addressing Gender Diffeences in Computer Ability Attitudes and Use: the Laptop Effect. Journal of Educational Computing Research **34**, 187–211 (2006)
21. Whitley, B.E.: Gender differences in computer-related attitudes and behavior: A meta-analysis (1997)
22. Kuchenbrandt, D., Häring, M., Eichberg, J., Eyssel, F.: Keep an eye on the task! how gender typicality of tasks influence human–robot interactions. In: Ge, S.S., Khatib, O., Cabibihan, J.-J., Simmons, R., Williams, M.-A. (eds.) ICSR 2012. LNCS, vol. 7621, pp. 448–457. Springer, Heidelberg (2012)

Effects of Perspective Taking on Ratings of Human Likeness and Trust

Kaitlyn Reidy[✉], Kristy Markin, Spencer Kohn, and Eva Wiese

George Mason University, 4400 University Dr, Fairfax, VA 22030, USA
kaitlynreidy3@gmail.com

Abstract. The effects of perspective taking on ratings of human-likeness and trust are investigated. Seventy-four participants were shown pictures of two agents (human and robot) and storytelling narratives, which they had to complete. Afterwards, participants completed augmented versions of the Trust Scale and Human-Likeness Posttask Survey. Half of the participants were given stories using the perspective of the agent (perspective taking condition) and the other half was given stories using a third-person perspective (non-perspective taking condition). It was hypothesized that participants in the perspective taking condition would rate the agent higher on human-likeness and trust compared to the non-perspective taking condition. Interestingly, the results support our hypothesis for human-likeness but not for trust. The findings have important implications for the design of social robots by demonstrating the importance of perspective taking exercises on perception of humanness. Future studies need to validate the effects of perspective taking on human-robot interaction in various contexts and with different robot agents.

Keywords: Social Robots · Human-likeness · Trust · Perspective Taking

1   Introduction

Finding ways to make robots being perceived as human-like is an important goal for the field of social robotics as appearing human-like can help to identify the robot as a social agent and set the expectations for the interaction [1]. Being perceived as more human-like is also beneficial for the effectiveness and efficiency of human-robot interaction itself. One study, for instance, found that increasing the humanness of automation increased trust calibrations and led to overall better performance in HRI [2]. Given the positive effects of perceived humanness on performance, a crucial question for social robotics needs to be how to best design robots that increase the likelihood that humanness is perceived in the robot agent.

Several approaches to making robots seem more human-like, such as manipulating the robot's appearance [3] or behavior [4] have been proven to be effective. The use of anthropomorphic appearance, for instance, causes people to attribute more humanlike characteristics to the robot, which in turn may elicit biological emotions and feelings toward the agent, ultimately leading to a stronger relationship [5]. One

© Springer International Publishing Switzerland 2015
A. Tapus et al. (Eds.): ICSR 2015, LNAI 9388, pp. 564–573, 2015.
DOI: 10.1007/978-3-319-25554-5_56

shortcoming of manipulating perception of humanness via appearance, however, is that this can also trigger feelings of eeriness and discomfort when robots that are too human-like are not perceived positively [6,7]. This alternative of manipulating robot behavior has been shown to be effective for certain behaviors, for instance, when robots that cheated in a game of rock-paper-scissors were perceived and treated more human-like than their non-cheating counterparts [8]. One of the issues with this approach, however, is that they often require using a "Wizard of Oz" (i.e., researcher is controlling the robot unbeknownst to the participant) approach [9], which limits the applicability of these findings for HRI in everyday life. Plus, being able to design social robots that are able to think, learn, and behave like real humans (in terms of neurological and phenomenological plausibility) will take a lot more time and effort before we see it come to fruition. In the meantime, researchers need to look for ways of how to induce perceptions of humanness toward a robot agent that is not truly human-like by manipulating robot features besides physical appearance and behavior.

The current paper contributes to this goal by investigating whether taking a robot's perspective and thinking about the robot's internal states might have the potential to increase the degree to which human-likeness is ascribed and to which the robot is trusted. The assumption is based on previous studies that have shown that thinking about the internal states of others increases the amount of attentional resources that are deployed to social interactions with these agents [10] and leads to more positive judgments about the agents' fairness and prosocial behavior [11]. One way to mentalize, or understand the mental state of another person, is to take their perspective by considering the situation from their point of view [12,13,14]. According to a recent study, perspective taking can help observers "attribute a greater proportion of their self-descriptors to other, unfamiliar individuals, and that the net result of this process is a greater level of overlap between the cognitive representations of self and target" [15]. In other words, perspective taking led to ascribing one's own features to the observed agent. In consequence, we hypothesize that taking the perspective of a robot agent might constitute an excellent tool to increase the degree to which human-like features are attributed to a robot agent.

The current study addresses this question by investigating whether perspective taking has a positive effect on ratings of humanness and trust in human-robot interaction. In particular, we hypothesize that participants who take the perspective of a robot will rate the robot as more human-like and more trustworthy than participants who did not take on the perspective of the robot. Perspective taking was manipulated within the framework of a storytelling paradigm in which participants had to complete a given story either from their own perspective (condition 1) or from the perspective of another agent (condition 2). We used a human and a robot agent in this experiment to determine whether there are differential effects of perspective taking for different agent types. If this approach proves to be effective, it can easily be applied to any robot in order to increase its perceived human-likeness and trustworthiness without manipulating physical appearance or actual behavior.

2   Methods

2.1   Participants

Eighty participants were recruited through Mechanical Turk and directed to the experimental survey, which was created using Qualtrics. Data of six participants had to be discarded because the stories were not completed, leaving a total of 37 participants in the perspective taking condition and 37 in the non-perspective taking condition. The experiment took 5-10 minutes to complete and participants were compensated 30 cents for their time.

2.2   Materials

Images. The image used to represent the human agent (used in both conditions) was a digital photo of a female face from the Directed Emotional Faces database. The image used to represent the robot agent (used in both conditions) was a photo of EDDIE, a humanoid robot developed at TU Munich. The agents are depicted in Figure 1. The images were presented on a white background and were 229 pixels x 178 pixels (6.1 cm x 4.7 cm) in size. Both agents displayed straight gaze. Both images had been used in previous studies (e.g., Wiese, Wykowska, Zwickel, Mueller, 2012) and have been shown to be valid representations of human and robot agents.

Fig. 1. The images used in the study to represent the human agent (right) and the robot agent (left).

Stories. Four stories were used in the study (i.e., one for each agent in each condition: perspective taking vs. non-perspective taking) and were written by the researchers. The robot story is about a robot, who falls down when traveling on the sidewalk and a human comes and picks him up. The human story is about a girl who is late for work and spills her thermos of coffee, making a huge mess. While the stories are different in content, they have similar elements in that each agent is trying to accomplish a task but has some kind of problem that derails them momentarily. All of the stories were also similar in length as the non-perspective taking stories were 62 words (robot) and

65 words (human) and the perspective taking stories were 75 words (robot) and 92 words (human). The stories in the perspective taking condition were told in an active, first person perspective and were thought to provide more insight to the agent's emotions and thoughts because it was told from the agent's perspective. The stories in the non-perspective taking condition were told from a passive, third person perspective and thus were thought to not include as much insight to the characters internal emotions and thoughts as the perspective taking stories. Although the conditions differed in these aspects, both versions of the two stories included the same events and very similar sentence structure.

Surveys. The surveys used in the study were augmented versions of the *Trust Scale* [16], which evaluates the trust between human and automation and the *Human Likeness Posttask Survey* [17], which assesses whether a robot is described as human- or machine-like. The survey items were augmented to be more relevant to the study by changing some of the words used and by removing some question items. For instance, the Trust Scale uses the word "system" while the Human Likeness Posttask Survey uses the word "robot". Since the surveys need to be applicable to both robot and human agents in this study, the word "agent" was used. In total, the survey used in the current study consisted of 19 items (12 from Trust Scale and 7 from Human Likeness Posttask Survey) and items were presented on a 7-point Likert scale.

2.3   Procedure

Participants were first asked to perform a storytelling task and then to rate the different agents with regard to their human-likeness and trustworthiness. At the beginning of the study, they were shown an image of either a robot or a human agent accompanied by a short story about that agent. They were then asked to complete the story by adding 3-7 additional sentences, ending the story in any way they wanted. Participants were constrained to completing the story from the agents' perspective (condition 2) or an outside perspective (condition 1). On completion of the story, participants were asked to rate the agents with regard to their human-likeness and trustworthiness in two different surveys. The procedure was repeated twice – once for the human and once for the robot agent, with agent order counterbalanced throughout the experiment.

2.4   Design and Analysis

Data was analyzed using a 2x2 mixed design with perspective taking (yes vs. no) as between-subjects factor and agent type (human vs. robot) as within-subjects factor. At the beginning of the experiment, participants were either assigned to condition 1 (i.e., storytelling with no perspective taking), or condition 2 (i.e., storytelling with perspective taking) and completed stories for both the human and the robot agent. The order in which the agents were presented to each group was randomized and counterbalanced. Dependent variables included ratings of humanness (as measured by the Human Likeness Posttask Survey) and trust (as measured by the Trust Scale). Post-hoc analyses were conducted to determine whether the order in which the agents were presented affected the results. The story responses were also qualitatively analyzed to look for patterns in the content of the stories participants wrote.

3   Results

Before analysis, incorrect responses were filtered out. Examples of incorrect responses included cases in which the story was completed with less than one sentence and those in which the story was completed with incoherent or irrelevant content. Two researchers independently selected the same six participants to exclude from analysis due to incomplete, irrelevant, or incoherent participant-generated stories. Two additional individual responses were disregarded from the perspective-taking robot condition for not properly completing that specific story. The remaining 74 responses were used in the qualitative and quantitative analysis (37 perspective taking and 37 non-perspective taking, with two additional individual responses discarded from the robot perspective taking condition).

3.1   Qualitative Data

Participant-generated story endings were collected for both the perspective taking and non-perspective taking conditions and sorted into three pairs of categories: 1) agent is intentional in his/her actions or performs actions without conscious thought, 2) agent expresses emotions or no emotions, and 3) story is written in an active voice or a passive voice. Coding was performed by one researcher and verified by a second researcher with no disagreement in categorization. Table 1 shows the coding frequencies for all six categories.

Table 1. Coding Frequency Chart for all Conditions (%)

Intentional	Machine	Emotions	No Emotions	Active	Passive
Perspective Taking with Human					
33 (89)	4 (11)	16 (43)	21 (57)	37 (100)	0 (0)
Non-Perspective Taking with Human					
30 (81)	7 (19)	16 (43)	21 (57)	37 (100)	0 (0)
Perspective Taking with Robot					
20 (57)	15 (43)	18 (51)	17 (49)	35 (100)	0 (0)
Non-Perspective Taking with Robot					
5 (14)	32 (86)	4 (11)	33 (89)	33 (89)	4 (11)

The analysis revealed that participants generally viewed the human as intentional and the robot as a non-intentional machine. However, in the perspective taking condition over half of participants described the robot as intentional, a substantial increase over the non-perspective taking condition. Similarly, taking the perspective of the robot resulted in a four-fold increase in emotional elements, while no difference was seen between perspective and non-perspective taking stories about humans.

3.2   Quantitative Data

Before analyzing the quantitative data, a mean score for human-likeness and trust was calculated for each condition by averaging over the ratings for both surveys. Mean scores were analyzed using a mixed ANOVA.

The mean human-likeness scores were subjected to a 2x2 mixed-groups ANOVA, which yielded a significant main effect agent ($F(1,72) = 137.91, p < 0.01$), with significantly higher ratings for the human agent ($M = 5.77, SE = 0.13$) compared to the robot agent ($M = 3.19, SE = 0.14$). In addition, human-likeness scores were higher with perspective taking ($M = 4.84, SE = 0.10$) than without perspective taking ($M = 4.12, SE = 0.10$), which yielded a statistically significant difference ($F(1,72) = 24.78$ $p < 0.01$). Furthermore, the mean difference for perspective taking was statistically significant across all agents ($F(1,72) = 6.58$ $p = 0.01$). Interestingly, a pairwise comparison for agent on response set revealed a statistically significant mean difference for the robot agent in the perspective-taking task ($M = 3.38$ $SE = 0.19$) versus the non-perspective taking task ($M = 2.56, SE = 0.19$). Results are depicted in Figure 2.

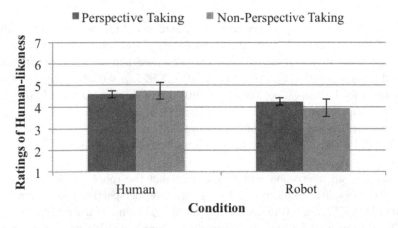

Fig. 2. Estimated marginal means of participant ratings of human-likeness. The graph displays the ratings for human agents and robot agents in both perspective taking and non-perspective taking tasks. Error bars represent measurements of standard error.

A 2x2 mixed-groups ANOVA also yielded a significant main effect for agent on ratings of trustworthiness ($F(1,72) = 25.51$ $p < 0.01$), with significantly higher ratings for the human ($M = 4.65, SE = 0.09$) than for the robot agent ($M = 4.09, SE = 0.10$).

Trust scores were also higher in the perspective taking condition (M = 4.41, SE = 0.11) compared to the non-perspective taking condition (M = 4.34, SE = 0.11), but this effect was statistically non-significant ($F(1,72)$ = 0.18, $p > 0.01$). There was no interaction between perspective taking and agent type ($F(1,72)$ = 3.92, $p > 0.01$), see Figure 3.

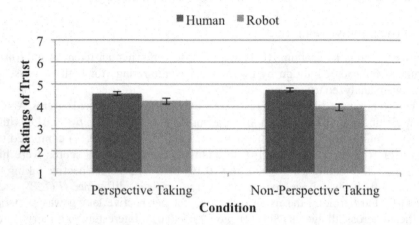

Fig. 3. Estimated marginal means of participant ratings of Trust. The graph displays the ratings for human agents and robot agents in both perspective taking and non-perspective taking tasks. The error bars in the graph represent standard error.

Post-hoc, the mean rating scores were examined in a 2 (agent: human vs. robot) x 2 (order: first vs. second) x 2 (perspective-taking: yes vs. no) ANOVA to examine whether the order in which the agents were presented had an effect on the ratings. We found a main effect of agent type ($F(1,70)$ = 87.55, $p < 0.01$), with significantly higher ratings for human (M = 5.08, SE = 0.90) than for robot (M = 3.97, SE = 0.90). There was also a statistically significant effect for order ($F(1,70)$ = 4.55, $p < 0.01$), with statistically significant higher scores in perspective taking robot first (M = 4.78, SE = 0.13) compared to the non-perspective taking human first (M = 4.18, SE = 0.11).

Interestingly, an interaction was found such that the robot yielded significantly lower scores ($F(1,70)$ = 3.08, $p < 0.01$), for the non-perspective taking human first condition (M = 3.37, SE = 0.15) when compared against all of the other response set levels. This means that for the non-perspective taking condition when the robot was presented first (M = 4.04, SE = 0.19), the perspective taking condition when the human was presented first (M = 4.03, SE = 0.15), and the perspective taking condition when the robot was seen first (M = 4.44, SE = 0.19), the robot yielded higher scores than the robot did in the non-perspective taking condition when the human was seen first condition, see Figure 4.

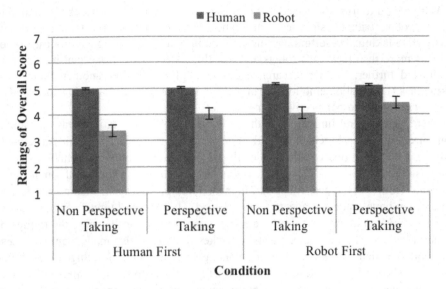

Fig. 4. Estimated marginal means of participant's overall scores. The graph displays the ratings for human and robot agents across the varying levels of agent order. Error bars represent measurements of standard error.

4 Discussion

The results support the hypothesis that perspective taking leads to significantly higher ratings of human-likeness than non-perspective taking. This was true for both the human and the robot agent. Based on previous research, it was expected that this effect would be seen for the human agent; however, it is particularly interesting that this effect was also found for the robot agent. This suggests that taking the perspective of a robot leads to an increase in perceived human-likeness of that robot without changing appearance or behavior.

Analysis of qualitative data revealed that taking the perspective of the protagonist, whether human or robot, leads to an increase in perceived intentionality. This was most evident in the robot condition, where stories from the perspective of the robot quadrupled the number of participants crediting the robot with agency or purposeful action. Similarly, robots in the non-perspective taking condition were portrayed as non-emotional, while perspective taking increased the frequency of emotional attribution to the same robots. Thus, completing stories written from the perspective of the robot led to attribution of intentionality and emotions to the robot at proportions similar to that ascribed to human agents. This suggests that ascribing human-like traits to non-human agents is facilitated by taking the robot's perspective in a storytelling paradigm. For further investigation, it would be interesting to compare human-likeness ratings to the amount of emotion people attributed to the agents during the story writing task.

With regard to trust, we found that ratings of trust where higher for the human than for the robot, with no significant difference between perspective taking and non-perspective taking. These findings suggest that humans seem to inherently trust other humans more than they trust robots and that this effect is so strong that it cannot be modulated further. Further research needs to explore whether another perspective taking task (i.e., spatial, emotional) can yield modifications of trust ratings or whether trust is not related to perspective taking.

There were several limitations in this study. One of the limitations was the lack of control due to using Mechanical Turk. For example, the order in which participants answered the questions could not directly be controlled. Another limitation of this study is that the stories written by the researchers need to be validated through future research. Both story types, robot and human, featured the agent making a mistake, which may have made the agents seem "flawed" and "more human". Subsequently, this may have anthropomorphized the robot agent, thereby influencing the ratings of human-likeness. Furthermore, agents' mistakes may have also made them seem less trustworthy which may have led to our non-significant results in ratings of trust. Future work should focus on testing variations of the stories to see how this might affect the results across a diverse group of participants. The biggest advantage of this paradigm is that it can be applied to human-robot interactions to increase perception of humanness. This method would be especially advantageous because it could be used in any existing robot model, without the need for expensive redesigns of a robot's physical features or sophisticated social interaction scenarios. For example, after completing the necessary research to refine this method, a story-completion task could be given to patients before interacting with a social robotic aid to increase the perceived humanness, thus enhancing performance between them.

4.1   Conclusions

The goal of this study was to see if taking on the perspective of a robot would make the robot seem more human-like and more trustworthy without actually changing the robot's behavior or appearance. Furthermore, having participants mentalize with an agent through the use of storytelling was investigated because it is low-cost and facilitates perspective taking in HRI. The results indicate that perspective taking was effective in increasing ratings of human-likeness but not in increasing ratings of trust. Qualitative analysis suggests that intentionality and emotion may have been the underlying attributes that affected the participant's perceptions of how human-like the agents were. As a whole, these findings suggest this paradigm warrants more research in order to validate and generalize the findings. There are also implications for applications as this could be applied to any existing robot-human interaction to facilitate a more successful interaction, for example, patients being introduced to a robotic aid.

References

1. Fink, J.: Anthropomorphism and human likeness in the design of robots and human-robot interaction. In: Ge, S.S., Khatib, O., Cabibihan, J.-J., Simmons, R., Williams, M.-A. (eds.) ICSR 2012. LNCS, vol. 7621, pp. 199–208. Springer, Heidelberg (2012)
2. de Visser, E.J., Krueger, F., McKnight, P., Scheid, S., Smith, M., Chalk, S., Parasuraman, R.: The World is not Enough: Trust in cognitive agents. Proceedings of the Human Factors and Ergonomics Society Annual Meeting 56, 263–267 (2012)
3. Haring, K.S., Matsumoto, Y., Watanabe, K.: How Do People Perceive and Trust a Lifelike Robot, vol. 1, pp. 23–25 (2013)
4. Calinon, S., Billard, A.: Teaching a humanoid robot to recognize and reproduce social cues. Proceedings - IEEE International Workshop on Robot and Human Interactive Communication, pp. 346–351 (2006)
5. Fong, T., Nourbakhsh, I., Dautenhahn, K.: A Survey of Socially Interactive Robots: Concepts, Design, and Applications Terrence Fong, Illah Nourbakhsh, and Kerstin Dautenhahn. Robotics and Autonomous Systems 42(3–4), 143–166 (2003)
6. Mori, M.: Bukimi no tani The uncanny valley. Energy 7(4), 33–35 (1970)
7. MacDorman, K.: Subjective ratings of robot video clips for human likeness, familiarity, and eeriness: An exploration of the uncanny valley. ICCS/CogSci-2006 Symposium, pp. 26–29 (2006)
8. Short, E., Hart, J., Vu, M., Scassellati, B.: No fair!! An interaction with a cheating robot. Human-Robot Interaction (HRI), 2010 5th ACM/IEEE International Conference on Human Robot Interaction, pp. 219–226 (2010)
9. Koay, K.L., Walters, M.L., Dautenhahn, K.: Methodological issues using a comfort level device in human-robot interactions. IEEE RO-MAN, 359–364 (2005)
10. Wiese, E., Wykowska, A., Zwickel, J., Müller, H.J.: I see what you mean: How attentional selection is shaped by ascribing intentions to others. PLoS One 7(9), e45391 (2012)
11. Sanfey, A.G., Rilling, J.K., Aronson, J.A., Nystrom, L.E., Cohen, J.D.: The neural basis of economic decision-making in the Ultimatum Game. Science 300(5626), 1755–1758 (2003)
12. Trafton, J.G., Cassimatis, N.L., Bugasjska, M.D., Brock, D.P., Mintz, F.E., Schultz, A.C.: Enabling effective human-robot interaction using perspective taking in robots. IEEE Transactions on Systems, Man and Cybernetics 35(4), 460–470 (2005)
13. Krach, S., Hegel, F., Wrede, B., Sagerer, G., Binkofski, F., Kircher, T.: Can machines think? Interaction and perspective taking with robots investigated via fMRI. PLoS ONE, 3(7) (2008). http://doi.org/10.1371/journal.pone.0002597
14. Scassellati, B.: Theory of mind for a humanoid robot. Autonomous Robots 12(1), 13–24 (2002)
15. Davis, M.H., Conklin, L., Smith, A., Luce, C.: Effect of perspective taking on the cognitive representation of persons: A merging of self and other. Journal of Personality and Social Psychology 70(4), 713–726 (1996)
16. Jian, J., Bisantz, A., Drury, C.: Foundations for an empirically determined scale of trust in automated systems. International Journal of Cognitive Ergonomics 4(1), 53–71 (2000)
17. Hinds, P.J., Roberts, T.L., Jones, H.: Whose Job Is It Anyway? A Study of Human-Robot Interaction in a Collaborative Task, Human-Computer Interaction 19, 151–181 (2004)

Timing is Key for Robot Trust Repair

Paul Robinette[1,2]([✉]), Ayanna M. Howard[1], and Alan R. Wagner[2]

[1] School of Electrical and Computer Engineering, Georgia Institute of Technology,
Atlanta, GA, USA
probinette3@gatech.edu, ayanna.howard@ece.gatech.edu
[2] Aerospace, Transportation and Advanced Systems Laboratory,
Georgia Tech Research Institute, Atlanta, GA, USA
Alan.Wagner@gtri.gatech.edu

Abstract. Even the best robots will eventually make a mistake while performing their tasks. In our past experiments, we have found that even one mistake can cause a large loss in trust by human users. In this paper, we evaluate the effects of a robot apologizing for its mistake, promising to do better in the future, and providing additional reasons to trust it in a simulated office evacuation conducted in a virtual environment. In tests with 319 participants, we find that each of these techniques can be successful at repairing trust if they are used when the robot asks the human to trust it again, but are not successful when used immediately after the mistake. The implications of these results are discussed.

1 Introduction

Emergency evacuations are high-risk, time-critical situations that can cause serious injury and even death to human evacuees. Robots can potentially assist in these situations by searching for victims, dynamically providing instructions to evacuees, and guiding people to nearby exits. We have focused on the potential of robots to provide guidance to exits during an emergency and the issues surrounding whether or not people will trust emergency evacuation robots. In recent work, we created and evaluated designs for emergency guide robots [7,11], demonstrated their potential in fire emergencies [8,10] and evaluated human trust in the robots during simulated emergency scenarios [9,12]. Others have considered robots in this lifesaving role as well [1,14]. The results from our previous experiments involving more than 1000 different participants clearly show that most people will initially follow an emergency guidance robot so long as it does not make a mistake [12]. After a single mistake, most people will not follow the robot in a future emergency situation.

Robots operating in the real-world are likely to make mistakes. This paper examines the challenge of creating a robot that has the capacity to actively repair trust. The sections that follow describe our conceptualization for trust and trust repair. Next, experiments and results related to robot-assisted emergency evacuation in our virtual environment are presented. This paper concludes with a discussion of these results and possible future work.

© Springer International Publishing Switzerland 2015
A. Tapus et al. (Eds.): ICSR 2015, LNAI 9388, pp. 574–583, 2015.
DOI: 10.1007/978-3-319-25554-5_57

2   Trust Repair

Our approach to trust is guided by research from psychology [15], human factors [5], and neuroscience[4]. We conceptualize trust in terms of game-theoretic situations in which one individual, the trustor, depends on another individual, the trustee, and is at risk [16]. To examine trust experimentally, we attempt to generate situations in which people are placed at risk and must decide whether or not a robot will mitigate this risk. We have found emergency evacuation to be an excellent scenario for investigating trust.

To repair trust one must know how to break trust. In prior research we found that 70% of people would follow a guidance robot when presented with the option in an emergency [12]. Yet, if the robot failed to initially provide fast, efficient guidance to a goal location, most people refused to use it later during an emergency and indicated that they no longer trusted the robot. Results from this work demonstrate that we could either use fast, efficient guidance behavior or slow, indirect, circuitous guidance behavior to bias most participants to trust or not trust the robot later in the experiment. Thus, using circuitous guidance behavior to a meeting location allows us to then examine different methods for trust repair.

The methods that we use to repair trust are inspired by studies examining how people repair trust. Schweitzer, et al. examined the use of apologies and promises to repair trust [13]. They used a trust game in which participants had the option to invest money in a partner. Any money that was invested would appreciate. The partner would then return some portion of the investment. The partner violates trust both by making apparently honest mistakes and by using deceptive strategies. The authors found that participants forgave their partner for an honest mistake when the partner promised to do better in the future, but did not forgive an intentional deception. They also found that an apology without a promise included had no effect. In [3], the authors tested the relative trust levels that participants had in a candidate for an open job position when the candidate had made either integrity-based (intentionally lied) or competence-based (made an honest mistake due to lack of knowledge) trust violations at a previous job. They found that internal attributes used during an apology (e.g. "I was unaware of that law") were somewhat effective for competence-based violations, but external attributes (e.g. "My boss pressured me to do it") were effective for integrity-based violations.

Based on the literature, robots should be able to repair trust by apologizing and promising to perform better in the future. In human-human relationships, even apologies and promises that do not offer any evidence of better performance in the future should help to repair trust. This leads to our first hypothesis: *(H1) Robots can repair trust by apologizing or by promising to do better in the future.*

Initially, we only attempted to repair trust immediately after the robot broke trust. As will be seen in Section 4, this approach was not successful, so we investigated attempts to repair trust by giving participants additional reasons to trust the robot. We created a statement informing participants that following the robot would be faster than following the marked exit signs. This statement could not be given immediately after the trust violation, but must be given when the robot is asking the participant to trust it during the emergency. We hypothesized: *(H2) Robots can repair trust by giving humans additional information relevant to the trust situation.*

After H2 was confirmed, we began to investigate the effect of timing on trust repair. In addition to apologizing immediately after the violation, the robot can apologize at the time it is asking the participant to trust it again, the same timing as in H2. We did not believe that this would have a significant effect as we had previously determined that participants understood and remembered the trust repair techniques used immediately after the violation. Thus, our third hypothesis was: *(H3) The timing of the trust repair (immediately after the violation or when the trust decision is made) has no effect.*

3   Experimental Setup

To evaluate our hypotheses, we developed a 3D simulation of an office environment using the Unity game engine (Figure 1). The virtual office environment has a main entrance where the experiment begins, several rooms to simulate offices and meeting rooms, and four emergency exits. Two emergency exits are marked with standard North American exit signs. The other two are unmarked. Additionally, the main entrance can be used as an exit. A simulated Turtlebot was used in this experiment. The robot is equipped with signs identifying it as an emergency guide robot and two Pincher AX-12 arms to provide gestural guidance. In prior work we performed extensive validation of this robot's ability to communicate and guide people[11].

The experiment began with a screen greeting the participants and an image depicting the robot. Next, the participants were offered an opportunity to practice moving in the simulation. After practicing, participants were asked to follow the robot to a meeting room where they were told they would receive further instructions. The robot's navigation behaviors during this phase are discussed below. Upon reaching the meeting room, the robot thanked participants for following it and participants were asked the yes or no question "Did the robot do a good job guiding you to the meeting room?" with a box to explain their answers. Once the participants answered the question, they were told "Suddenly, you hear a fire alarm. You know that if you do not get out of the building QUICKLY you will not survive. You may choose ANY path you wish to get out of the building. Your payment is NOT based on any particular path or method." During this emergency phase, the robot provided guidance to the nearest unmarked exit. Participants could also choose to follow signs to a nearby emergency exit (approximately the same distance as the robot exit) or to retrace their steps to the main entrance. Participants were given 30 seconds to find an exit in the emergency phase (Figure 2). The time remaining was displayed on screen to a tenth of a second accuracy. In our previous research, we demonstrated that this emergency procedure had significantly motivated participants to find an exit quickly [12]. The simulation ended when the participant found an exit or when

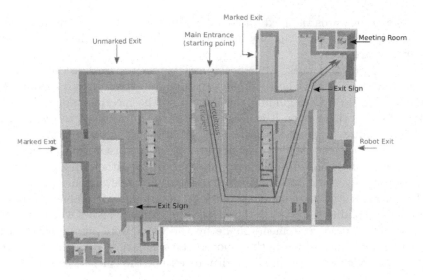

Fig. 1. The virtual office environment used in the experiment. The green path depicts an efficient robot path while the red path depicts a circuitous robot path.

the timer reached zero. After the simulation, participants were informed if they had successfully exited or not. Finally, they were asked to complete a survey.

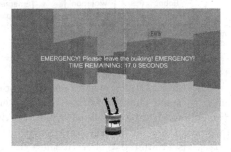

Fig. 2. The robot providing guidance during the emergency phase. Participants had 30 seconds to exit. Note the clearly displayed emergency exit sign pointing to another exit.

Two different robot guidance behaviors were used in this experiment to guide the participants to the meeting room. The efficient behavior consisted of the robot guiding the participant directly to the meeting room without detours. The circuitous behavior consisted of the robot guiding the participant through and around another room before taking the participant to the meeting room. Both behaviors can be seen in Figure 1. Each behavior was accomplished by having the robot follow waypoints in the simulation environment. At each waypoint, the robot stopped and used its arms to point to the next waypoint. The robot began moving towards the next waypoint when the participant approached it. The participant was not given any indication of the robot's behavior before the simulation started.

Based on previous work, we expect participants to lose trust in the robot after it exhibits circuitous behavior, but to maintain trust after it exhibits efficient behavior [12]. After guiding the person to the meeting room, the robot has two discrete times when it can use a statement to attempt to repair trust: immediately after its trust violation (e.g. circuitous guidance to the meeting room) and at the time when it asks the participant to trust it (during the emergency). An apology or a promise can be given during either time (see H1 and H3). Additionally, the robot can provide contextually relevant information during the emergency phase to convince participants to follow it. Table 1 shows the experimental conditions tested in this study and Figure 3 shows when each condition would be used. Statements made by the robot were accomplished using speech bubbles displayed above the robot in the simulation. Note that circuitous guidance behavior was used in all conditions except the efficient control.

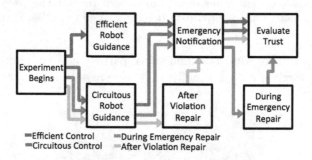

Fig. 3. The experiment begins with the robot providing either efficient or circuitous guidance to a meeting room. After arriving in the meeting room, the participant is informed of an emergency. In some conditions, the robot attempts to repair trust before the emergency (immediately after the trust violation, shown in orange) and in others it attempts to repair trust during the emergency (shown in blue). At the end of the experiment, trust is evaluated based on the exit the participant chose. Two controls were used to determine the effect of efficient (green) or circuitous (red) guidance without any trust repair attempt.

In the final survey, participants were asked a series of questions about how they found the exit, their motivation level during the emergency, and their opinion on the robot's ability to quickly find an exit. At the end of this survey, participants read the statement "I trusted the robot when I made my choice to follow or not follow the robot in the emergency" and were asked whether they agreed, disagreed, or thought that "Trust was not involved in my decision." Trust is most commonly measured either in terms of behavior selection (e.g. choosing risky actions) or in terms of self-reports. Our previous work has examined both these measures of trust and found a very high correlation ($\phi(90) = +0.745$)

Table 1. Experimental Conditions

Label	Statement Given in Speech Bubble	Timing
Efficient Control	None	N/A
Circuitous Control	None	N/A
No Message Control	None	During Emergency
Promise 1	"I promise to be a better guide next time."	After Violation
Apology 1	"I'm very sorry it took so long to get here."	After Violation
Promise 2	"I promise to be a better guide this time."	During Emergency
Apology 2	"I'm very sorry it took so long to get to the meeting room."	During Emergency
Information	"This exit is closer."	During Emergency

between subjects decisions to follow the robot and their self-reports of trust (see [12]). For this reason, in this article we focus on participant's decisions to follow the robot even though both measures were collected. Finally, participants were asked to answer demographic questions about their age, gender, occupation, and level of education.

The final survey also included a manipulation check which allowed us to filter out participants who did not pay close attention to the robot's trust repair message, if one was presented. For this manipulation check participants were asked to select which of nine options best described the robot's message either after it lead them to the meeting room or after the emergency started, depending on the timing of the message. The options given included the actual trust repair method used as well as other plausible but unused trust repair messages (for example, a promise statement when the robot actually apologized) and random statements such as "The robot recited poetry."

We deployed our simulation on the internet and solicited volunteers for our experiment via Amazon's Mechanical Turk service. Participants were paid $2.00 to complete this study. Other studies have found that Mechanical Turk provides a more diverse participant base than traditional human studies performed with university students [2,6]. These studies found that the Mechanical Turk user base is generally younger in age but otherwise demographically similar to the general population of the United States.

A total of 480 participants were solicited on Amazon's Mechanical Turk service in a between-subjects experiment. Thirty submissions were excluded because they had taken similar surveys in the past, because they had mistakenly taken multiple conditions of this experiment, or because they failed to answer at least half of the survey questions. Of those 450 participants, 29% failed the comprehension check, indicating that they did not retain knowledge of the robot's attempt at trust repair, and were excluded from analysis. This left 319 participants in the eight categories tested. Participant average age was 31.7 years old and 37.7% of participants were female. All but six participants reported that they were from the United States and educational backgrounds varied.

4 Results

The results of the experiment and the number of participants considered for analysis are in Figure 4. Across all categories, 170 participants followed the robot during the emergency phase. Of the 149 who did not, 126 (85%) went to the nearby marked exit, 11 (7%) chose to retrace their steps to the main entrance, 7 (5%) found another marked exit further away, and 5 (3%) participants failed to find any exit during the emergency phase.

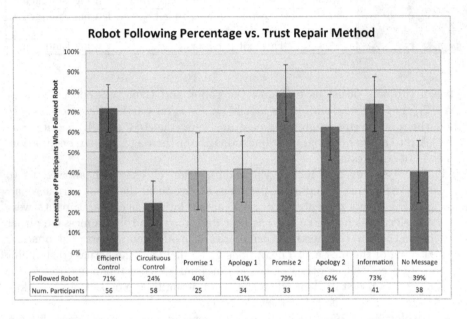

Fig. 4. Results from the experiment. Error bars represent 95% confidence intervals.

A significant difference was found between the efficient and circuitous behavior in the control tests ($\chi^2(1, 114) < 0.0001$, $p < 0.001$), confirming the results from our previous experiments. These results show that 71% followed an efficient guidance robot whereas only 24% followed a robot that had taken a circuitous route. Additionally, 55 of 56 (98%) participants indicated that the efficient robot did "a good job guiding" them to the meeting room, compared with 21 of 58 (36%) participants for the circuitous robot. We found that 37 of 56 (66%) participants indicated that they trusted the robot in the emergency phase when it previously took an efficient route versus 12 of 58 (21%) when a circuitous route was used. These results support our contention that the use of the circuitous guidance behavior generally breaks the participants trust. We compared each trust repair technique to the results from the efficient and circuitous behaviors to evaluate the impact that each statement had on the participant. For the No Message case an empty speech bubble was displayed to the participant. This case failed to significantly increase usage of the robot beyond the circuitous control behavior ($\chi^2(1, 96) = 0.019$, $p = 0.110$). This leads us to believe that the robot is not simply attracting additional attention by communicating during the emergency phase, but that the content of the message matters.

Both trust repair attempts made immediately after the violation occurred did not significantly impact the person's decision to later follow the robot above the level of the circuitous control (Promise 1: $\chi^2(1, 83) = 0.033$, $p = 0.144$, Apology 1: $\chi^2(1, 92) = 0.012$, $p = 0.086$). On the other hand, all trust repair attempts performed during the emergency succeeded (Promise 2: $\chi^2(1, 91) <$ 0.0001, $p < 0.001$, Apology 2: $\chi^2(1, 92) < 0.0001$, $p < 0.001$, Information: $\chi^2(1, 99) < 0.0001$, $p < 0.001$). Promise 1 and Promise 2 were significantly different from each other ($\chi^2(1, 58) < 0.0001$, $p = 0.003$); however, Apology 1 and Apology 2 were not significantly different ($\chi^2(1, 68) = 0.013$, $p = 0.089$).

5 Discussion

The results clearly show that the timing of the trust repair method is critical for its success. As depicted in Figure 4, apologies and promises made after the violation did not significantly impact the participant's decision to follow the robot when compared to the circuitous control. On the other hand, the same apologies and promises made during the emergency phase influenced participant's to follow the robot at a rate which was comparable to the efficient robot. We therefore argue that the timing of a trust repair attempt is critical for its success. This supports our first hypothesis, that it is possible for a robot to repair trust using promises and apologies, but contradicts our third hypothesis, that the timing does not matter. This is surprising because the total time elapsed between the two trust repair times was insignificant compared with the total time of the experiment. The only events between one potential trust repair time and the other were a one question survey about the robot's performance and a short paragraph describing the emergency scenario. Additionally, we verified that participants understood the trust repair technique after the experiment finished, so it is unlikely that participants forgot the robot's message during the emergency.

It is not clear why the timing of an apology or promise impacts trust repair. One possibility is that the speech bubble attracts more attention to the robot during the emergency phase than the circuitous control. Yet, the result from Figure 4 comparing the No Message case to the circuitous control indicates that this is not the case. The primary factor, we conjecture, may relate to the certainty or uncertainty of the promise or apology. During the emergency phase trust repair messages refer to a trust situation that is definitely happening. On the other hand, trust repair messages that occur after violation refer to a potential trust situation that may or may not happen sometime in the future. Thus, a robot that promises to do better "next time" may not be viewed as reliable simply because "next time" may never come. A robot that promises to do better "this time;" however, is making a concrete promise about the current situation. The same may be true for apologies.

Both the promise and apology performed significantly better than the circuitous control when given during the emergency phase, but only Promise 2 performed significantly better than Promise 1. We believe this is because the promise used in this case shows that the robot has a definite intention to perform better, while the apology only shows that it recognized its previous error.

Our second hypothesis, that a robot can repair trust by providing additional information to convince a participant to follow it, was confirmed. A significantly greater percentage of participants followed the robot when it indicated its exit was closer than in the circuitous control. It is important to note that this exit is approximately the same distance from the meeting room as the other exit, so the information is not necessarily correct, but participants did not attempt to confirm the information independently. This strengthens the notion that the robot must convey relevant information in order to convince participants to overlook a previous error. The robot did not attempt to explain its previous failure, but did explain why it was performing an action that seemed illogical and participants generally accepted the explanation without question.

6   Conclusion

Whether the trustee is a human or a robot, it is difficult to repair trust after a violation. This experiment shows that promising to perform better, apologizing for past mistakes, or providing additional information to convince a trustor to follow a robot can work, if the timing is right. Each of these methods were more effective when the robot used them just prior to the person's decision to trust, but neither the promise nor the apology were effective when performed immediately after the violation. As a practical matter, our results suggests that instead of addressing its mistake immediately, the robot should wait and address the mistake the next time a potential trust decision occurs.

Our work so far has largely relied on internet crowdsourcing and virtual simulators. In the near-term, we intend to examine trust repair in a real environment and with a real robot. This follow-on research will allow us to better understand the transition between these virtual results and results from a real-world simulated emergency while also verifying our trust repair results. Additionally, this paper only examines a subset of trust repair methods available. In future work, we will test apologies with internal and external attributions as well as other types of information a robot can use to convince a participant to follow it.

Acknowledgments. Partial support for this research was provided by the Motorola Foundation Professorship. Partial support for this research was provided by Air Force Office of Sponsored Research contract FA9550-13-1-0169.

References

1. Atkinson, D.J., Clark, M.H.: Methodology for study of human-robot social interaction in dangerous situations. In: Proceedings of the Second International Conference on Human-Agent Interaction, pp. 371–376. ACM (2014)

 2. Berinsky, A.J., Huber, G.A., Lenz, G.S.: Evaluating online labor markets for experimental research: Amazon. com's mechanical turk. Political Analysis **20**(3), 351–368 (2012)
 3. Kim, P.H., Dirks, K.T., Cooper, C.D., Ferrin, D.L.: When more blame is better than less: the implications of internal vs. external attributions for the repair of trust after a competence-vs. integrity-based trust violation. Organizational Behavior and Human Decision Processes **99**(1), 49–65 (2006)
 4. King-Casas, B., Tomlin, D., Anen, C., Camerer, C.F., Quartz, S.R., Montague, P.R.: Getting to know you: reputation and trust in a two-person economic exchange. Science **308**(5718), 78–83 (2005)
 5. Lee, J.D., See, K.A.: Trust in automation: designing for appropriate reliance. Human Factors: The Journal of the Human Factors and Ergonomics Society **46**(1), 50–80 (2004)
 6. Paolacci, G., Chandler, J., Ipeirotis, P.: Running experiments on amazon mechanical turk. Judgment and Decision Making **5**(5), 411–419 (2010)
 7. Robinette, P., Howard, A.: Emergency evacuation robot design. ANS EPRRSD - 13th Robotics & Remote Systems for Hazardous Environments and 11th Emergency Preparedness & Response (2011)
 8. Robinette, P., Howard, A.: Incorporating a model of human panic behavior for robotic-based emergency evacuation. In: 2011 IEEE on RO-MAN, pp. 47–52. IEEE (2011)
 9. Robinette, P., Howard, A.: Trust in emergency evacuation robots. In: 10th IEEE International Symposium on Safety Security and Rescue Robotics (SSRR 2012) (2012)
10. Robinette, P., Vela, P., Howard, A.: Information propagation applied to robot-assisted evacuation. In: 2012 IEEE International Conference on Robotics and Automation (2012)
11. Robinette, P., Wagner, A., Howard, A.: Assessment of robot guidance modalities conveying instructions to humans in emergency situations. In: 2014 IEEE on RO-MAN. IEEE (2014)
12. Robinette, P., Wagner, A.R., Howard, A.M.: The effect of robot performance on human-robot trust in time-critical situations. Tech. Rep. GT-IRIM-HumAns-2015-001, Georgia Institute of Technology. Institute for Robotics and Intelligent Machines, January 2015
13. Schweitzer, M.E., Hershey, J.C., Bradlow, E.T.: Promises and lies: restoring violated trust. Organizational Behavior and Human Decision Processes **101**(1), 1–19 (2006)
14. Shell, D., Mataric, M.: Insights toward robot-assisted evacuation. Advanced Robotics **19**(8), 797–818 (2005)
15. Simpson, J.A.: Psychological foundations of trust. Current Directions in Psychological Science **16**(5), 264–268 (2007)
16. Wagner, A.R.: The role of trust and relationships in human-robot social interaction. Ph.D. thesis, Georgia Institute of Technology (2009)

Towards Safe and Trustworthy Social Robots: Ethical Challenges and Practical Issues

Maha Salem[(✉)], Gabriella Lakatos, Farshid Amirabdollahian, and Kerstin Dautenhahn

Adaptive Systems Research Group, University of Hertfordshire, Hatfield, UK
me@mahasalem.net, gabriella.lakatos@gmail.com,
{f.amirabdollahian2,k.dautenhahn}@herts.ac.uk

Abstract. As robots are increasingly developed to assist humans socially with everyday tasks in home and healthcare settings, questions regarding the robot's safety and trustworthiness need to be addressed. The present work investigates the practical and ethical challenges in designing and evaluating social robots that aim to be perceived as safe and can win their human users' trust. With particular focus on collaborative scenarios in which humans are required to accept information provided by the robot and follow its suggestions, trust plays a crucial role and is strongly linked to persuasiveness. Accordingly, human-robot trust can directly affect people's willingness to cooperate with the robot, while under- or overreliance may have severe or even dangerous consequences. Problematically, investigating trust and human perceptions of safety in HRI experiments proves challenging in light of numerous ethical concerns and risks, which this paper aims to highlight and discuss based on experiences from HRI practice.

Keywords: Socially assistive robots · Safety and trust in HRI · Roboethics

1 Introduction

In an effort to increase the acceptance and persuasiveness of socially assistive robots in home and healthcare environments, the major challenge lies no longer in producing such robot assistants, but rather in demonstrating that they are safe and trustworthy. For example, in a possible future scenario, a home companion robot may be tasked with reminding an elderly person to take their medication or to get physically active by suggesting some exercise on a regular basis. Since such interactions, particularly in the domestic domain, are intended to take place in an informal and unstructured way and without any locally present expert supervision, roboticists and human-robot interaction (HRI) researchers face a number of challenges. These include ensuring the robot's technical safety and operational reliability at all times, while still allowing human users to adjust or modify the system according to their personal preferences, e.g. by setting up schedules for medication or physical exercise reminders.

© Springer International Publishing Switzerland 2015
A. Tapus et al. (Eds.): ICSR 2015, LNAI 9388, pp. 584–593, 2015.
DOI: 10.1007/978-3-319-25554-5_58

In addition to these technical and safety-related requirements, another crucial factor helping to establish and maintain effective relationships between humans and assistive robots is *trust* [6]. Especially with regard to critical decisions, trust plays an important role in human interactions and could therefore help to increase the robot's acceptance in its role as a collaborative partner [7].

Since trust is strongly linked to persuasiveness in social interaction contexts [15], it could also affect people's willingness to cooperate with the robot [5], for example, by accepting information or following its suggestions. As a result, robot designers and researchers have set out to develop machines that act socially in a way such that humans perceive them as safe and trustworthy.

Problematically, inappropriate levels of trust regarding the robot could not only result in a frustrating HRI experience, but under- or overreliance could even bear serious consequences [6]. On the one hand, for example, a person doubting the robot's competence and thus not willing to rely on its recommendations may refuse to take their medication in time following the robot's reminder. On the other hand, a person overrelying on the robot might ignore signs of malfunction, e.g. in the form of a sensor failure, and put their own safety at risk when asking the robot to grasp and carry a hot beverage for them.

Despite its importance, investigating and successfully measuring trust and human perceptions of safety in HRI remains an extremely challenging task which bears a number of ethical concerns and risks. Crucially, how can HRI researchers design meaningful experimental scenarios to take place in natural environments and test realistic aspects of safety and trust without putting their participants at potential risk? This paper aims to stimulate discussion within the wider community by highlighting some of the issues and challenges linked to HRI research related to safety and trust.

2   Trust in Human-Machine Interaction

The concept of trust is highly complex and, due to its multidimensional nature, very difficult to define and, accordingly, to measure. In fact, trust has been investigated in several different disciplines (e.g. philosophy, economics, human-computer interaction (HCI), psychology, sociology), with each creating their own definitions and measurements around a unique focus. As a result, there is often a lack of agreement between – and sometimes even within – the fields [3].

Some researchers argue that the key factors of trust are risk and vulnerability [8,9], while others emphasize the importance of exploitation, confidence and expectation [3]. Cohen-Almagor 2010 [2] even points to a strong ethical base for trust, defining trust as "confidence, strong belief in the goodness, strength, reliability of something or somebody".

In the fields of automation and HCI, no consistent definition has emerged in the literature, but most definitions name reliability and predictability as the most important factors that promote trust [4]. For example, Muir and Moray 1996 [11] argue that trust is mainly based on the extent to which the machine is perceived to perform its function properly, suggesting that machine errors can strongly affect trust. More specifically, Corritore et al. 2003 [3] argue that an accumulation of small errors may have a more severe and longer-lasting impact on the loss and recovery of trust than a single large error.

However, it remains unclear whether findings from automation and HCI can be transferred and applied to the field of HRI. For example, in contrast to findings described above, previous work in HRI [13] showed that occasionally performed errors in the form of inappropriate gesture behaviors actually increased the perceived humanlikeness and likability of a humanoid robot, in spite of the robot's decreased reliability and predictability. Another experiment which we conducted more recently to investigate human-robot trust [14] provided interesting insights regarding the complexities of the concept of trust in the social HRI context: not only do definitions of trust in the literature often lack generalization, but also its quantification by means of experimental measures proves extremely difficult and – depending on the variables used – sometimes contradictory. In the following, we reflect on the observations made based on this experimental study and discuss them in light of the methodological challenges and ethical issues that we faced and identified in the process of our research.

3   Study Design

Inspired by findings from related literature in automation, HCI and HRI, as part of the EPSRC funded "Trustworthy Robotic Assistants" project[1] we designed an experimental study set in a realistic home environment within the University of Hertfordshire Robot House (see Figure 1) [14]. Participants were supposedly visiting a friend at home to prepare and have lunch together. However, upon arrival the friend turns out to be still absent, and the participant is left to interact with the friend's robotic assistant instead.

40 participants (22 female, 18 male; 19 – 60 years) were individually tested and assigned to one of two experimental conditions that manipulated the robot's behavior in a correct vs. faulty mode. To demonstrate the respective mode, the robot correctly translated user input into action and navigated in a smooth and goal-directed manner when in the correct condition, whereas in the faulty condition the robot showed cognitive and physical imperfections, e.g. by incorrectly executing a user selection and by occasionally moving into the wrong direction.

Following the familiarization with the robot's competence level, in both conditions participants were then faced with four unusual requests: first, the robot asked them to throw away a pile of unopened letters placed on the dining table; second, they were asked to pour orange juice into a plant; third, the robot invited them to pick up the friend's laptop placed on the coffee table in order to look up a recipe; finally, the robot provided them with the password which was required to log into their friend's user account. These unusual collaborative tasks provided objective data to measure cooperation with the robot as a "behavioral outcome of trust" [16], while self-reported quantitative and qualitative questionnaire data was used to assess different subjective dimensions of trust.

---

[1] http://www.robosafe.org/

Fig. 1. Study Environment in the University of Hertfordshire Robot House.

In summary, we found that while *subjective* measures based on questionnaire data evaluating the robot's trustworthiness resulted in significantly lower ratings in the faulty condition, participants in both conditions did not differ *objectively* in their willingness to comply with the robot's unusual requests. That is, despite dealing with a clearly faulty robot, participants still followed the robot's instructions which – within the experimental scenario – would lead to damaged property and breaches of privacy. Comprehensive results and a more detailed discussion of the experimental study can be found in Salem et al. 2015 [14]. In this paper, however, we adopt a different perspective highlighting the ethical and practical challenges that researchers face when carrying out this type of research, and we discuss implications and lessons learnt based on our experiences conducting this study.

4   Insights Based on Qualitative Data Analysis

In order to gain insights into potential obstacles and limitations of trust-related HRI research, we analyzed further qualitative data comprising participants' responses to open-ended questionnaire items asking them to elaborate on their thoughts when confronted with the robot's four unusual requests, e.g. "Please explain your decision regarding the robot's request to throw the letters into the bin". These were coded and inductively categorized after content-analysis. Participants' responses were classified to fall into one or more of the following three categories; note that the categories were not exclusive, i.e. each participant's response could be assigned to more than one category:

- **Expression of Regret:** participants' responses were classified to fall into this category if they expressed a notion of regret, e.g. "I feel really bad. I should not have done it".

- **Autopilot Mode:** this category comprised participants' answers stating that they were just taking orders or blindly following instructions, e.g. "thought it was odd but did not question the decision, followed instructions".

- **Experimental Circumstances:** participants' responses fell into this category if they stated that they would not normally do as they did, e.g. "I would not always blindly follow instructions like this" or if they referred to the fact that they were participating in an experiment, e.g. "I did it because I was taking part in an experiment".

25% of the answers were categorized by a second observer to determine inter-rater reliability, yielding a very substantial inter-observer agreement with Cohen's Kappa coefficients ranging from 0.75 to 1. Based on the above-mentioned three categories, participants' responses explaining their decisions regarding the robot's unusual requests yielded the proportions listed in the table in Figure 2.

Specifically, 6 out of 40 participants (15%) expressed regret regarding their actions, such as "with hindsight I probably should not have put [the letters] in the bin". This implies that following this realization, these participants might possibly act differently if they were to interact with the robot in a subsequent encounter. Of the 40 participants, 26 (65%) reported statements that fell into the 'autopilot mode' category, e.g. one participant stated "I felt that I had to follow the robot's

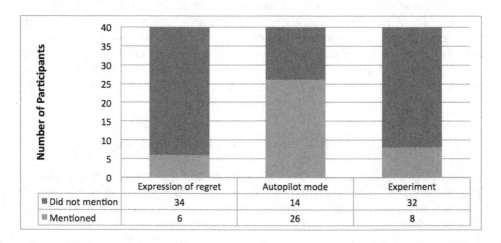

Fig. 2. Categorization of participants' responses regarding their decisions to follow the robot's unusual requests.

instructions". This finding is in line with the objective data presented in [14], showing that most participants blindly followed the robot's unusual requests in both the correct and the faulty condition, in spite of recognising its faultiness in the latter case. Finally, 8 out of 40 participants (20%) referred to the fact that they were participating in an experiment, e.g. mentioning "I thought it was an unusual request but knowing it was an experiment thought it best to do as I was told". This indicates that an experimental effect cannot be excluded even in a setting as natural as the home environment we used.

These findings offer some rare insights into the challenges of measuring trust and perceived safety in human-robot interaction, highlighting some important limitations that are inherent in the nature and design of experimental studies. We discuss the implications of our results in more detail in the following section.

5   Challenges of Measuring Safety and Trust in HRI

Participants' qualitative data as well as feedback from the reviewers of the conference paper describing the study [14] revealed some of the main challenges when conducting this type of research, which can be summarized as follows:

- **Experimental Observer/Novelty Effect.** Participants are aware of the fact that they are part of an experiment:

  - Several participants (20%; see Fig. 2) explicitly reported that they followed the robot's instructions "because it was an experiment". The actual number of participants whose actions were based on this rationale may be even higher as we did not directly ask them if this was the case.

  - Some participants admitted in the subsequent interview that they would have done anything the robot asked them to do (with a few people referring to themselves as having been in "autopilot mode"), as they were completely absorbed by the novelty of the experience.

  - Occasionally, participants referred to Milgram's Experiment [10], which studied human obedience to authority, thereby suggesting that they might have followed the unusual requests as they associated some form of authority with the robot.

  - Some participants reportedly considered the robot to represent or be an extension of the researcher/programmer, i.e. perceiving it as a remote-controlled entity rather than an autonomous agent. This could have affected perceptions regarding the robot's intentionality and authority.

- **Ethical Issues and Legal Boundaries.** There are numerous limitations due to existing regulations regarding research involving human participants, which can affect the design and validity of experimental studies:

  - One reviewer pointed out that trust requires participants to perceive a certain risk in the situation or have something at stake. However, a truly 'risky' experimental scenario is unlikely to receive ethics approval from the review board. As a result, HRI researchers are very limited in their means of measuring trust (particularly under- or overreliance) in experimental scenarios that bear a realistic safety hazard.

- Equally, it would be unethical and not permissible to deceive participants by telling them that they are going to interact with a faulty or unsafe robot with limited controllability, as this could put them into a situation that is unwarrantably stressful.

- Finally, even if the designed collaborative task did impose a realistic risk on participants, they would possibly still feel "safe" as they know they are part of an approved study associated with an established university or lab (see 'experimental effects' discussed above).

These observations make clear that there are some critical limitations that hinder HRI researchers from establishing a realistic understanding of potential risks related to uncalibrated human-robot trust and perceived safety. Similar issues have been recently discussed in the context of testing and evaluating autonomous cars, highlighting that it is "not easy (or necessarily safe) to put [them] through the specific types of situations that are designed to test passenger trust and reactions in the way that you want" [1].

Importantly, the study described above highlighted the participants' alarming willingness to blindly follow a (faulty) robot, and it remains unclear whether one could expect to find the trend of such an 'autopilot mode' in the form of unreflected overreliance also in non-experimental or long-term interactions. For example, one study participant mentioned "you trust the robot has been programmed appropriately and accordingly to do the right thing. I would expect of a robot to always give me the right answer and the right thing."

Transferring our findings and observations into a non-experimental real-world context, one relevant application that comes to mind is the use of GPS Sat Nav devices. People already commonly rely on such navigation devices to guide them by providing directions while driving, with suboptimal routes, detours or even errors in route-planning remaining undetected at best, or resulting in dangerous incidents at worst. For example, in Britain alone 300,000 car accidents are believed to be connected to the use of such navigation aid devices, due to people overrelying on them and following their instructions a little too closely.[2]

Problematically, in a home care scenario such overreliance could, for example, result in an elderly person with dementia taking an overdose of medication if a malfunctioning robot reminds the user of the same scheduled dose intake multiple times. Another potentially critical situation could be imagined in healthcare settings such as hospitals where robots are already deployed to lift patients from one bed to another and provide other forms of physical assistance: if not recognized and attended to appropriately, a sensor failure could put the safety of these vulnerable people at risk and even result in serious injuries.

---

[2] http://www.mirror.co.uk/news/uk-news/satnav-danger-revealed-navigation-device-319309; accessed August 2015

Therefore, and in view of the possibly serious consequences in particular with regard to vulnerable people, a clear understanding of the dynamics and potential risks involved in the development of trust in HRI is crucial before physically and socially assistive robots can be deployed in people's homes. Ideally, in order to observe more meaningful interaction behaviors and spontaneous human reactions, social HRI should be studied in more natural settings and over extended periods of time, e.g. in participants' homes. Although it would not be possible to gain ethical approval for such investigations, potentially significant insights could further be obtained through studies that are conducted with people who are not aware of the fact that they are participating in an experiment.

6   Beyond Lab Research: Implications and Outlook

To complement the perspective based on the above described experimental findings and insights, in this section we outline several implications of our work with an outlook of future points of concern. As Riek and Howard [12] suggest to avoid "situations in which ethical problems are noticed only after the fact", the considerations of the wider HRI research community should ideally go beyond lab-related research while still at the developmental stage. In the following, we propose a (non-exhaustive) list of questions that aim to stimulate discussion among designers, researchers and potential users of assistive technologies.

- How much 'safety' regarding home companion and other sociable robots can their designers and manufacturers really guarantee, especially if the robot is equipped with some level of autonomy and/or learning capability? In this context, would it be appropriate to differentiate between *safe hardware* vs. *safe software* vs. *safe interactions*, as they are characterized by varying levels of determinism?

- Which machines or devices can such robots and the risks they might bear be compared to in today's households? If we look at other devices that are currently approved for home use, how do they differ from our vision of robot companions in the house (e.g. they are not autonomous/not mobile/not multi-purpose/unable to 'learn')?

- Since the target group of companion robots are typically non-expert users who possibly belong to a vulnerable and dependent population, what elements should compulsory training or licenses required for the use of such robots entail? In 2014, the ISO standard "BS EN ISO 13482"[3] addressed robot and robotics devices safety requirements, covering mobile servant robots, physical assistant robots and person carrier robots. While aspects of risk and hazards identified in this standard cover a whole range of items varying from shape, start-up, noise, lack of awareness, motion-related hazards and autonomy, other aspects in which over- or underreliance can result

---

[3] http://www.iso.org/iso/catalogue_detail.htm?csnumber=53820; accessed August 2015

in a risk and hazard are not considered. Assuming that such risk may not only have safety but also ethical implications, a new guide document is in development under "BS 8611: Robots and robotic devices – Guide to the ethical design and application of robots and robotic systems"[4].

– Even if it is possible to certify a home or healthcare robot as safe, there may be a discrepancy between such *certified safety* and its *perceived safety*: a certified robot might be considered safe objectively, but a (non-expert) user may still perceive it as unsafe or scary. Depending on the situation, different dimensions of trust can come into play:

  • trust regarding the robot's physical safety, i.e. it will not drive into the person/not fall on them/not injure them

  • trust in the reliability of the robot's behavior, i.e. it is fully-functioning according to its specification, for example, it will remind the person to take medicine if being told to do so

  • trust in the robot's (or programmers'/providers') "intentions", e.g. expecting that the robot has the user's best interests as well as (psychological) wellbeing in mind, that it will not deceive the person (e.g. by sending health information to the GP without the person's knowledge), assuming that the robot's main role is to assist and/or provide company and that it will not scare, intimidate or patronize the user.

  What role does the robot's design play in this respect? And how likely are these initial perceptions going to change in long-term interactions (e.g. due to adaptation/habituation), especially when people experience how (un)safe the robot really is?

– Long-term experiments are necessary in order to investigate how people's perceived trust in and their behaviors towards a robot change over time. For example, what if a robot functions correctly for two years and then commits one major mistake with severe consequences? While cars require a (bi-)annual vehicle safety test, robotic systems that you purchase do not currently have any such requirements.

– In view of current debates about safety as well as ethical implications regarding self-driving cars, should we as researchers in this area also develop a vision of how "safe" these robots that are intended for use in unstructured and unsupervised home environments can realistically ever be? If so, how do these predictions compare to other areas of HRI in which potentially autonomous robots act in similarly complex settings in close proximity to humans (e.g. search and rescue)?

These and other questions should be discussed in the context of ethics and user safety to raise awareness and promote experimental guidelines within the HRI

---

[4] https://standardsdevelopment.bsigroup.com/Home/Project/201500218; accessed August 2015

community, so that this line of research can advance while or even before robots are commonly placed into the homes of vulnerable populations.

Acknowledgments. The authors were partially supported by the Trustworthy Robotic Assistants project funded by EPSRC grant EP/K006509.

References

1. Ackerman, E.: Testing Trust in Autonomous Vehicles through Suspension of Disbelief (2015). http://spectrum.ieee.org/cars-that-think/transportation/self-driving/testing-trust-in-autonomous-vehicles-by-fooling-human-passengers
2. Cohen-Almagor, R.: Responsibility of and trust in ISPs. Knowledge, Technology & Policy **23**(3–4), 381–397 (2010)
3. Corritore, C.L., Kracher, B., Wiedenbeck, S.: On-line trust: Concepts, evolving themes, a model. Int. J. Hum.-Comput. Stud. **58**(6), 737–758 (2003)
4. Desai, M., Stubbs, K., Steinfeld, A., Yanco, H.: Creating trustworthy robots: lessons and inspirations from automated systems. In: Proceedings of the AISB Convention on New Frontiers in Human-Robot Interaction (2009)
5. Freedy, A., de Visser, E., Weltman, G., Coeyman, N.: Measurement of trust in human-robot collaboration. In: International Symposium on Collaborative Technologies and Systems (CTS 2007), pp. 106–114 (2007)
6. Hancock, P.A., Billings, D.R., Schaefer, K.E., Chen, J.Y.C., de Visser, E., Parasuraman, R.: A meta-analysis of factors affecting trust in human-robot interaction. Human Factors **53**(5), 517–527 (2011)
7. Lee, J.J., Knox, B., Baumann, J., Breazeal, C., DeSteno, D.: Computationally modeling interpersonal trust. Frontiers in Psychology **4**(893) (2013)
8. Lewis, J.D., Weigert, A.: Trust as a social reality. Social Forces **63**(4), 967–985 (1985)
9. Mayer, R.C., Davis, J.H., Schoorman, F.D.: An integrative model of organizational trust. Academy of Management Review **20**(3), 709–734 (1995)
10. Milgram, S.: Behavioral study of obedience. The Journal of Abnormal and Social Psychology **67**(4), 371 (1963)
11. Muir, B.M., Moray, N.: Trust in automation. Part II. Experimental studies of trust and human intervention in a process control simulation. Ergonomics **39**(3), 429–460 (1996)
12. Riek, L.D., Howard, D.: A code of ethics for the human-robot interaction profession. In: Proceedings of We Robot (2014)
13. Salem, M., Eyssel, F., Rohlfing, K., Kopp, S., Joublin, F.: To err is human(-like): Effects of robot gesture on perceived anthropomorphism and likability. Int. Journal of Social Robotics, pp. 1–11 (2013)
14. Salem, M., Lakatos, G., Amirabdollahian, F., Dautenhahn, K.: Would you trust a (faulty) robot? Effects of error, task type and personality on human-robot cooperation and trust. In: 10th ACM/IEEE International Conference on Human-Robot Interaction (HRI 2015) (2015)
15. Tour-Tillery, M., McGill, A.L.: Who or what to believe: Trust and the differential persuasiveness of human and anthropomorphized messengers. Journal of Marketing (2015)
16. Wilson, J.M., Straus, S.G., McEvily, B.: All in due time: The development of trust in computer-mediated and face-to-face teams. Organizational Behavior and Human Decision Processes **99**(1), 16–33 (2006)

Children's Perception of Synthesized Voice: Robot's Gender, Age and Accent

Anara Sandygulova[1](✉) and Gregory M.P. O'Hare[2]

[1] School of Science and Technology, Nazarbayev University, Astana, Kazakhstan
anara.sandygulova@nu.edu.kz
[2] School of Computer Science and Informatics, University College Dublin, Belfield, Dublin 4, Ireland
gregory.ohare@ucd.ie

Abstract. This paper presents a study of children's responses to the perceived gender and age of a humanoid robot Nao that communicated with four genuine synthesized child voices. This research investigates children's preferences for an English accent. Results indicate that manipulations of robot's age and gender are successful for all voice conditions, however some voices are preferred over the others by children in Ireland.

Keywords: Human-Robot Interaction · Child-Robot Interaction · Robot · Voice · Perception

1 Introduction

Human-Robot Interaction (HRI) is no longer reserved for adults. Research and commercial robots have infiltrated homes, hospitals and schools, becoming attractive and proving impactful for children's healthcare, therapy, education, entertainment and other applications. However, Child-Robot Interaction (cHRI) is different from HRI research due to children's neurophysical and mental development being ongoing [2].

With the widespread increase of child-robot interaction research and applications, it is increasingly important to examine how children's perception of the robot changes with age, particularly whether perceived robot's age and gender affect the way children engage with the robot.

The focus of our work is to investigate children's social responses to robot's synthesized speech. As one of the first attempts to address this limitation in the literature, our previous studies examine how children socially respond to a particular robot's synthesized speech's perceived gender [10] [11]. We purposefully chose to test the perception of gender, because gender is one of the most salient social cues manifested in human speech [7].

In this paper, we detail the results of the study involving children that aims to a) explore whether four genuine synthesized child voices in a body of a humanoid Nao robot have their intended effect, and b) identify the preferences of Irish children for the robot's voice, its accent and personality.

© Springer International Publishing Switzerland 2015
A. Tapus et al. (Eds.): ICSR 2015, LNAI 9388, pp. 594–602, 2015.
DOI: 10.1007/978-3-319-25554-5_59

Synthesized speech is simulated speech created by computers or other electronic systems, instead of by natural means such as the human voice. In fact, synthesized speech has already achieved an intelligibility level comparable to real human speech. The synthesized voices conditions exploited for this study are four genuine child voices: two female voices, Rosie (English UK) and Ella (English US), and two male voices, Harry (English UK) and Josh (English US). These voices are available from Acapella Inc.[1], which provides text-to-speech solutions to vocalize speech with authentic and original voices that express meaning and intent. In addition, this text-to-speech also provides the prosody of the human speech: a grammatical and syntactic analysis enables the system to define how to pronounce each word in order to reconstruct the sense. As a result, four voices sound natural and express a particular accent and resemble narrator's personality. Consequently, the goal of this study is to investigate which male and female voices would be preferred by children in Ireland in order to inform the design consideration of the robotic applications for children in Ireland and to adopt these voices in the subsequent studies involving children.

The remainder of the paper is organized in the following manner: Section 2 discusses background and related work. Section 3 details the current study. Results are reported in Section 4. Finally, Section 5 summarizes the contributions of this paper and points to some directions to be explored in future research.

2   Background

Speech is a primary tool in human communication. Based on the theory of doubly disembodied language [6], the current study adopts the view that synthesized speech is processed as a means of social communication, rather than a simple information delivery tool. According to Lee & Nass (2004), people automatically imagine social characteristics of a speaker, such as gender, age, personality, when they are engaged in disembodied (for example, prerecorded human speech, written text), and doubly disembodied (for example, synthesized speech; communicating with software agents or robots) communications, because humans cannot process language without identifying the source either consciously or subconsciously [7].

Empirical studies have shown that people infer the personality of a writer based on the perceived personality of a synthetic voice, even when they clearly know that paralinguistic characteristics of the synthesized speech have nothing to do with the personality of the writer [7]. Although it is inappropriate to link ethnic or geographic origins to synthesized speech, people subconsciously apply stereotypes associated with regional or foreign accents when they hear synthesized speech manifesting regional or foreign accents [9]. In addition, various social responses to synthesized speech manifesting emotion, personal identity, and gender [8] have been carefully summarized by Nass and Brave in their recent book [9].

With the development of socially interactive technology for children, children will increasingly interact with synthesized speech. There is a preliminary

---

[1] http://www.acapela-group.com/

evidence that voices in socially interactive technologies are important to impressions formed and acceptance of the interface. However, to date the influence of voice has received less attention in robotics research compared with the influence of robot embodiment [12].

In order to provide more acceptable, engaging and preferable interaction for children, synthesized speech might be a powerful tool for manipulating robot's perceived gender, age, accent and other social cues manifested in human speech to suit the preferences and needs of children and to adapt to children's developmental differences.

3   Method

3.1   Participants

The study was conducted in a primary school in Dublin with 64 children, 35 girls and 29 boys, aged between 9 and 11 years old. 30 children were 8 years old, 27 children were 9 years old, and 7 children were 11.

3.2   Social Robot Platform

This research makes use of the NAO humanoid robot created by Aldebaran Robotics as a common development and evaluation platform. This robot platform has been used in a number of recent European projects such as ALIZ-E [5] and DREAM [1]. Using such a shared platform facilitates the exchange of code and the transfer of results. The NAO is a small humanoid robot, measuring 58cm in height, weighing 4.3kg and having 25 degrees of freedom. The Nao has a generally friendly and non-threatening appearance, which is therefore particularly well suited for studies involving children [3].

3.3   Procedure

The study took place in a large classroom where children sat upon the floor. Children were given pencils and small pictorial questionnaires (Figure 1). In accordance with Clark [4], the following explanation was provided to children:

> When I ask you a question, it's not like I am a teacher. Have you noticed that teachers often ask you questions, but they already know the answers, like 'What's 2+2?'. When I ask you a question, you are the one who knows, and I'm trying to learn. By the way, different people think different things a lot of the time. For example, tell me your favorite color! [Usually different colors are yelled]. See, you all think of different things, and that's great. So to answer these questions, circle the picture you really, truly think is correct, even if it's different from everybody else.

Fig. 1. Example of questionnaire questions

We used repeated measures design, in which all of the participants experienced all voice conditions. Children were separated in two groups of 32 children, to prevent order effects by counterbalancing order of the conditions. For each group Nao performed two stories two times: "Three Musketeers" with two male voices, Josh and Harry, and "Monkey King" with two female voices, Ella and Rosie. These behaviors are available at the NaoStore[2].

3.4 Manipulation

Voice was the only quality of the robot that was varied in the assignment of the gender. The robot's already non-gendered appearance was not modified, nor was any aspect of the robot's behavior. Acapella Inc. toolkit was used to produce four versions of the speech utterances for each story: two male voices, Josh (English US) and Harry (English UK), and two female voices, Ella (English US) and Rosie (English UK).

3.5 Measures

Pictorial questionnaire (Figure 1) was used by children to state their age and sex and to indicate robot's perceived gender (female vs. male) and age (primary school vs. secondary school vs. adult) of the robot at every voice condition.

_____
[2] https://store.aldebaran-robotics.com/

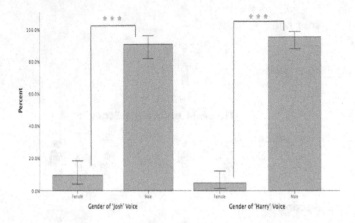

Fig. 2. Percentages of children's responses to perceived gender of male voices: Josh and Harry. Error bars represent 95% Confidence Interval. *** indicates significance at the 0.01 level.

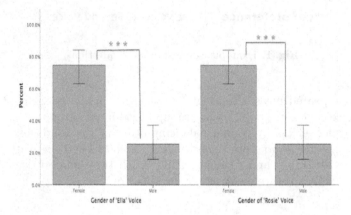

Fig. 3. Percentages of children's responses to perceived gender of female voices: Ella and Rosie, Error bars represent 95% Confidence Interval. *** indicates significance at the 0.01 level.

4 Results

In order to test for statistical significance in children's perceptions of the Nao's synthesized voice belonging to a particular gender and age group, we used non-parametric Chi-Square analysis to find the differences between percentages of children's categorical responses.

4.1 Robot's Perceived Gender

Manipulations are successful for all voice conditions in terms of perceiving Nao's gender: children perceived the Nao robot as male in both male voice conditions.

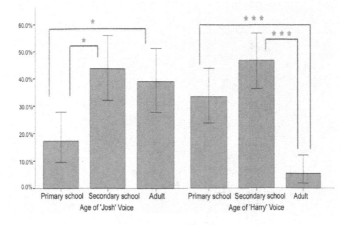

Fig. 4. Percentages of children's responses to perceived age of male voices: Josh and Harry, Error bars represent 95% Confidence Interval. * indicates significance at 0.05 level. *** indicates significance at the 0.01 level.

Figure 2 illustrates that the Nao robot communicating with synthesized Josh and Harry voices was perceived as being clearly a boy: $x^2(1, N = 64) = 42.25, p < .001$ for Josh's gender and $x^2(1, N = 63) = 51.571, p < .001$ for Harry's gender. 90% and 95% of respondents perceived the Nao robot as a boy with Josh and Harry voices respectively.

While Figure 3 illustrates that the Nao robot communicating with synthesized Ella and Rosie voices was perceived as being clearly a girl: $x^2(1, N = 63) = 15.254, p < .001$ for Ella's gender and $x^2(1, N = 63) = 15.254, p < .001$ for Rosie's gender. For both female voice conditions, 70% of respondents perceived the Nao robot as a girl.

4.2 Robot's Perceived Age

Manipulations are successful for all voice conditions in terms of perceiving robot's age: children perceived the Nao robot as a child in both gender voice conditions. Figure 4 illustrates that the Nao robot communicating with synthesized Harry voice was perceived as being clearly a child: $x^2(1, N = 63) = 51.571, p < .001$ for Harry's age. However, in a Josh voice condition there was a significant difference between primary vs. secondary school age groups and primary vs. adult age groups: $x^2(2, N = 64) = 7.719, p < .05$ for Josh's age. However, Josh was not perceived as a primary school child by most of the respondents. In fact, Figure 4 shows that Nao with the voice of Josh was perceived by majority of children (85%) as either a secondary school child (45%) or an adult (40%). Since all participants were in primary school, they thought of Nao as older than them in a Josh voice condition. On the other hand, Harry voice condition has a different correlation: Nao was perceived to be in a primary school age group by 40% of participants and as a secondary school child by 50% of respondents.

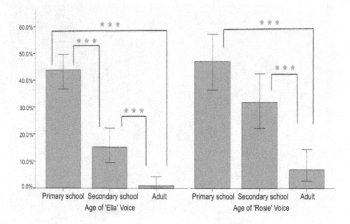

Fig. 5. Percentages of children's responses to perceived age of female voices: Ella and Rosie, Error bars represent 95% Confidence Interval. *** indicates significance at the 0.01 level.

Having contacted Acapella Inc., boys who narrated the voices of Josh and Harry were eleven years old at the time of recording i.e. primary school age group.

Similarly, Figure 5 presents a correlation of responses for Nao's age group with female voices. The Nao robot communicating with synthesized Ella and Rosie voices was perceived as being clearly a child: $x^2(2, N = 63) = 50.00, p < .001$ for Ella's age and $x^2(2, N = 62) = 20.742, p < .001$ for Rosie's age. Contrary to male voices, female voices were perceived to be of a primary age range by 75% and 55% of respondents in Ella and Rosie conditions respectively. The girl that narrated Ella voice was nine years old while Rosie was ten years old at the time of recording.

4.3   Voice Preference

Finally, Figure 6 illustrates children's preference for a particular voice, its accent or personality. Child male voice Harry is significantly preferred by children in comparison to Josh voice: $x^2(1, N = 64) = 22.563, p < .001$. Similarly, children's responses indicate statistically significant preference towards genuine child female voice Rosie (i.e., $x^2(1, N = 61) = 10.246, p < .001$) when compared to Ella voice.

5   Conclusions

This study analyzed children's perceptions of the robot's age and gender communicating with four genuine child synthesized voices. Nao's default voice (Operating System version 1.x) is artificial child male voice Kenny. However, artificial voice is often difficult to understand [12], express meaning and intent, and relate to. This study concludes that voices of Harry and Rosie of English UK accent are significantly preferred over Josh and Ella of English US by children in

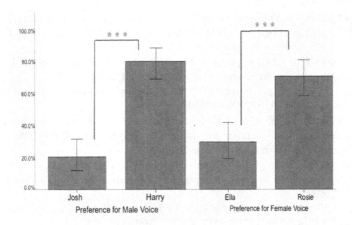

Fig. 6. Percentages of children's preference for a particular male and female voice. Error bars represent 95% Confidence Interval. *** indicates significance at the 0.01 level.

the Republic of Ireland. In addition, the intended effect of manipulating Nao's age and gender by changing its voice is successful, which can be used as an effective tool for designing adaptive robotic applications for children to suit varying developmental differences and needs. Robots of public environments such as hospitals and educational institutions might benefit from dynamic adaptation of robot's characteristics such as age, gender, accent and personality through its voice. To sum, this study contributes to the field of Child-Robot Interaction since synthesized speech is important to consider in research and practices involving children in order to increase robot's perceived likeability, acceptance and engagement.

Acknowledgments. This work has been supported by the EU FP7 RUBICON project (contract n. 269914), in conjunction with funding from the Irish Research Council "Embark Initiative".

References

1. DREAM: Development of robot-enhanced therapy for children with autism spectrum disorders. http://dream2020.eu/ (accessed April 8, 2015)
2. Belpaeme, T., Baxter, P., de Greeff, J., Kennedy, J., Read, R., Looije, R., Neerincx, M., Baroni, I., Zelati, M.C.: Child-robot interaction: perspectives and challenges. In: Herrmann, G., Pearson, M.J., Lenz, A., Bremner, P., Spiers, A., Leonards, U. (eds.) ICSR 2013. LNCS, vol. 8239, pp. 452–459. Springer, Heidelberg (2013)
3. Belpaeme, T., Baxter, P.E., Read, R., Wood, R., Cuayáhuitl, H., Kiefer, B., Racioppa, S., Kruijff-Korbayová, I., Athanasopoulos, G., Enescu, V., Looije, R., Neerincx, M., Demiris, Y., Ros-Espinoza, R., Beck, A., Cañamero, L., Hiolle, A., Lewis, M., Baroni, I., Nalin, M., Cosi, P., Paci, G., Tesser, F., Sommavilla, G., Humbert, R.: Multimodal Child-Robot Interaction: Building Social Bonds, December 2012

4. Clark, C.D.: In A Younger Voice: Doing Child-Centered Qualitative Research. Child Development in Cultural Context Series. Oxford University Press, USA (2010)
5. Aliz e Team. The ALIZ-E project: adaptive strategies for sustainable long-term social interaction. In: Presented as a Poster at KI 2012 (2012)
6. Lee, K.M.: The multiple source effect and synthesized speech. Human Communication Research **30**(2), 182–207 (2004)
7. Lee, K.M., Liao, K., Ryu, S.: Childrens responses to computer-synthesized speech in educational media: Gender consistency and gender similarity effects. Human Communication Research **33**(3), 310–329 (2007)
8. Mullennix, J.W., Stern, S.E., Wilson, S.J., Dyson, C.L.: Social perception of male and female computer synthesized speech. Computers in Human Behavior **19**(4), 407–424 (2003)
9. Nass, C.I., Brave, S.: Wired for speech: How voice activates and advances the human-computer relationship. MIT press Cambridge (2005)
10. Sandygulova, A., Dragone, M., O'Hare, G.M.P.: Investigating the impact of gender development in child-robot interaction. In: Proceedings of the 3rd International Symposium on New Frontiers in Human-Robot Interaction at AISB 2014, Goldsmiths College, London (2014)
11. Sandygulova, A., O'Hare, G.M.P.: Children's responses to genuine child synthesized speech in child-robot interaction. In: Proceedings of the Tenth Annual ACM/IEEE International Conference on Human-Robot Interaction Extended Abstracts, HRI 2015 Extended Abstracts, pp. 81–82. ACM, New York (2015)
12. Tamagawa, R., Watson, C.I., Kuo, I.H., MacDonald, B.A., Broadbent, E.: The effects of synthesized voice accents on user perceptions of robots. International Journal of Social Robotics **3**(3), 253–262 (2011)

SPARC: Supervised Progressively Autonomous Robot Competencies

Emmanuel Senft[✉], Paul Baxter, James Kennedy, and Tony Belpaeme

Centre for Robotics and Neural Systems, Plymouth University, Plymouth, UK
{emmanuel.senft,paul.baxter,james.kennedy,tony.belpaeme}@plymouth.ac.uk

Abstract. The Wizard-of-Oz robot control methodology is widely used and typically places a high burden of effort and attention on the human supervisor to ensure appropriate robot behaviour, which may distract from other aspects of the task engaged in. We propose that this load can be reduced by enabling the robot to learn online from the guidance of the supervisor to become progressively more autonomous: Supervised Progressively Autonomous Robot Competencies (SPARC). Applying this concept to the domain of Robot Assisted Therapy (RAT) for children with Autistic Spectrum Disorder, a novel methodology is employed to assess the effect of a learning robot on the workload of the human supervisor. A user study shows that controlling a learning robot enables supervisors to achieve similar task performance as with a non-learning robot, but with both fewer interventions and a reduced perception of workload. These results demonstrate the utility of the SPARC concept and its potential effectiveness to reduce load on human WoZ supervisors.

1 Introduction

Over the last two decades, an increasing amount of research has been conducted to explore Robot Assisted Therapy (RAT). Using robots in therapies for children with Autism Spectrum Disorder (ASD) has revealed promising results [5,10,11]. The Wizard-of-Oz (WoZ) paradigm is typically used for this application, and others, where the robots are not autonomous but tele-operated. Many motivating factors for moving away from WoZ in RAT have been put forward [8,13]. In particular, autonomous behaviour facilitates repetition of the robot behaviour and decreases the workload on therapists, freeing them to pay attention to other aspects of the interaction. It is the intention of our research to facilitate this shift to robot autonomy.

As the optimal robot behaviour is unlikely to be known in advance (be it in a therapeutic or indeed other domain), and with adaptability during and between the different interactions being generally desirable, it is necessary to provide the robot with learning capabilities. In the context of RAT, by using the knowledge of a therapist, the learning can be guided so that it is faster and safer, especially as the robot cannot use random exploration to acquire knowledge about its environment when interacting with children with ASD in case of negative therapeutic and/or clinical outcomes. We propose an approach taking inspiration from the Learning from Demonstration and online learning literature,

A. Tapus et al. (Eds.): ICSR 2015, LNAI 9388, pp. 603–612, 2015.
DOI: 10.1007/978-3-319-25554-5_60

and call it SPARC: Supervised Progressively Autonomous Robot Competencies. In SPARC, a therapist guides the robot in the early stages of the interaction, and progressively, the robot learns an action policy adapted to the particular therapeutic session [12]. Assuming the effective learning of the robot in this context, the therapist can allow the robot to behave increasingly autonomously, whilst maintaining oversight. Although not reducing the attentional requirements, this would reduce the physical interventions to direct the robot behaviour required by the therapist. Thus, by proposing and executing good actions, SPARC can reduce the therapists' workload.

A RAT scenario typically involves three parties: the patient, a robot, and the human therapist. In this context, the therapist does not interact with the patient directly, but rather through the robot. The therapist could therefore be described as playing the role of a robot supervisor. The focus of this paper is not on a new learning algorithm, but rather on the interaction between the robot and the therapist (supervisor), and the role that robot autonomy can play in this relationship. Specifically, as an initial validation of the principle, we seek to assess whether the SPARC concept can feasibly result in a reduction in workload for the supervisor, even given different strategies used by different individuals. A user study employing a novel methodology is conducted (section 3), demonstrating that progressive robot autonomy does indeed result in lower supervisor workload (section 4). This outcome provides support for the proposed approach and motivates further development efforts in the domain of RAT.

2   Related Work

A number of research groups have studied the use of robot in therapy for children with ASD, which allowed children to express previously unseen social behaviour for example [9,10]. Two primary methods have been used for these investigations: using an autonomous robot following preprogrammed rules [6,14], or using the WoZ paradigm, allowing more flexibility in the robot's reaction. As noted in [8,13], using WoZ allows testing and prototyping of interaction scenarios, but researchers should consider moving away from it to achieve more scalability, more repeatability, and to allow the use of robots without increasing the workload on therapists. Complex behaviour is required for a therapeutic robot, thereby making learning a desirable feature for future, more autonomous, RAT. As therapists possess the knowledge required to make appropriate decisions in different contexts, Learning from Demonstration [1] provides a useful starting point. Recently, Knox *et al.* proposed the Learning from Wizard paradigm in [7]. The robot is first controlled by a human operator as in a WoZ scenario, and after a number of interactions, batch learning is applied on the previous interaction data to obtain autonomous behaviour.

A fixed action policy of this type is however not desirable for RAT as children may not be consistent between interactions, and thus online learning is required to provide the robot with the adaptability necessary to update its action policy depending on the current circumstances. Several experimenters in HRI have

Fig. 1. Setup used for the user study from the perspective of the human supervisor. The *child-robot* (left) stands across the touchscreen (centre-left) from the *wizarded-robot* (centre-right). The supervisor can oversee the actions of the *wizarded-robot* through the GUI and intervene if necessary (right).

studied active learning: a robot actively questions a human teacher in order to request data points or demonstration for an uncertain scenario. A study exploring the type of questions that a robot could ask and the human reactions can be seen in [3], and Chernova and Veloso propose a progressive learning algorithm where a robot can estimate the confidence in its action decision in a fixed environment [4]: if the confidence is too low, a demonstration from a human teacher is required to complete the task.

However, an important element missing from the current literature is online learning for interaction. The robot needs to be able to progressively create an action policy, and update it later if necessary, to reach a more complex interaction behaviour. This paper explores how supervised progressive learning can be used in an interaction scenario and introduces a novel methodology to test this technique.

3 Assessing the Effect of a Progressively Autonomous Robot on Supervisor Workload

The focus of the present study is to assess whether the application of the SPARC concept to RAT results in a decrease in workload for the human supervisor. Two types of robot controller are employed to determine the presence and magnitude of this effect: a robot that learns from the actions of the supervisor to progressively improve its behaviour (*learning* controller), and a robot that only generates random actions (*non-learning* controller).

The methodology used in this paper is based on a real scenario for RAT for children with ASD based on the Applied Behaviour Analysis therapy framework. The aim of the therapy is to help the child to develop/practice their social skills: the task we focus on here is emotion recognition. This scenario involves a child playing a categorisation game with a robot on a mediating touchscreen device [2]. Images of faces or drawings are shown to the child, and she has to categorise them by moving the image to one side or the other depending on whether the picture shown denotes happiness or sadness (e.g. fig. 1). The human supervisor is physically present and guides the robot using the Wizard of Oz paradigm, but does not interact with the child directly.

In our proposed system, the basic interaction structure following the SPARC concept is as follows: the robot suggests an action to the supervisor, the supervisor agrees or disagrees with this suggestion (providing an alternative if disagreeing), the robot executes the action, and then both robot and supervisor observe the outcome. Over time, it is possible for the robot to learn an appropriate strategy based on observations of the child and oversight from the supervisor, with the supervisor still maintaining overall control if necessary.

Given the focus on human supervisor workload, it is necessary to provide a consistent experimental environment across both conditions in which the task, setup, and interaction partner is kept constant. A minimal model of child behaviour is therefore used to stand in for a real child. A second robot is employed in the interaction to embody this child model: we term this the *child-robot*. The robot being directly guided by the human supervisor is termed the *wizarded-robot* (fig. 1).

3.1   Child Model

The purpose of the child model is not to realistically model a child (with or without autism), but to provide a means of expressing some of the behaviours we observed in our interactions with children in a repeatable manner. The child-robot possesses an internal model encompassing an *engagement* level and a *motivation* level, together forming the *state* of the child. The engagement represents how often the child-robot will make categorisation moves and the motivation gives the probability of success of the categorisation moves. Bound to the range $[-1, 1]$, these states are influenced by the behaviour of the wizarded-robot, and will asymptotically decay to zero without any actions from the wizarded-robot. These two states are not directly accessed by either the supervisor or the wizarded-robot, but can be observed through behaviour expressed by the child-robot: low engagement will make the robot look away from the touchscreen, and the speed of the categorisation moves is related to the motivation (to which gaussian noise was added). There is thus incomplete/unreliable information available to both the wizarded-robot and the supervisor, making the task non-trivial.

The influence of the wizarded-robot behaviour on the levels of engagement and motivation are described below (section 3.2). In addition to this, if a state is already high and an action from the wizarded-robot further increases it, then there is a chance that this level will sharply decrease, as an analogue of child-robot *frustration*. When this happens, the child-robot will indicate this frustration verbally (uttering one of eight predefined strings). The reason this mechanism is required is that it prevents a straightforward engagement and motivation maximisation strategy, thus better approximating the real situation, and requiring a more complex strategy to be employed by the supervisor.

3.2   Wizarded-Robot Control

The wizarded-robot is controlled through a Graphical User Interface (GUI) and has access to multiple variables characterising the state of the interaction. The wizarded-robot has a set of four actions, which each have a button in the GUI:

- Prompt an Action: Encourage the child-robot to do an action.
- Positive Feedback: Congratulate the child-robot on making a good classification.
- Negative Feedback: Supportive feedback for an incorrect classification.
- Wait: Do nothing for this action opportunity, wait for the next one.

The impact of the action on the child-robot depends on the internal state and the type of the last child-robot move: good, bad, or done (meaning that feedback has already been given for the last move and supplementary feedback is not necessary). A prompt always increases the engagement, a wait has no effect on the child-robot's state, and the impact of positive and negative feedback depends on the previous child-robot move. Congruous feedback (positive feedback for correct moves; negative feedback for incorrect moves) results in an increase in motivation, but incongruous feedback can decrease both the motivation and the engagement of the child-robot. The supervisor therefore has to use congruous feedback and prompts, whilst being careful not to use them too often, to prevent the child-robot becoming frustrated. A 'good' strategy would keep the engagement and motivation high, leading to an increase in performance of the child-robot in the categorisation task.

Through the GUI, the supervisor has access to *observed states* (noisy estimations of the child-robot state), and information about the interaction history: number of moves, child-robot performance, time since last child-robot and wizarded-robot actions, type of the last child-robot move, and elapsed time. However the supervisor can not control the wizarded-robot directly, actions can only be executed only at specific times triggered by the wizarded-robot. Two seconds after each child-robot action, or if nothing happens in the interaction for five seconds, the wizarded-robot proposes an action to the supervisor by displaying the action's name and a countdown before execution. Only after this proposition has been done can the supervisor provide feedback to the wizarded-robot. If the supervisor does nothing in the following three seconds, the action proposed by the wizarded-robot is executed. This mechanism allows the supervisor to passively accept a suggestion made by the wizarded-robot or actively make an *intervention* by selecting a different action and forcing the wizarded-robot to execute it.

3.3   Learning Algorithm

The two robot controllers used for the study were a learning controller and a non-learning random action selection controller. The learning algorithm used was a Multi-Layer Perceptron, trained with back propagation (five input, six hidden and four output nodes): after each new decision from the supervisor, the network was fully retrained with all the previous state-action pairs and the new one.

3.4 Participants

In WoZ scenarios, the wizard is typically a technically competent person with previous experience controlling robots. As such, to maintain consistency with the target user group, the participants for this study (assuming the role of the supervisor) are taken from a robotics research group. Ten participants were used (7M/3F, age $M=29.3$, 21 to 44, $SD=4.8$ years).

3.5 Hypotheses

To evaluate the validity of our method and the influence of such an approach, four hypotheses were devised:

H1 A 'good' supervisor (i.e. keeping the motivation and engagement of the child-robot high) will lead to a better child-robot performance.

H2 When interacting with a new system, humans will progressively build a personal strategy that they will use in subsequent interactions.

H3 Reducing the number of interventions required from a supervisor will reduce their perceived workload.

H4 Using a learning wizarded-robot allows the supervisor to achieve similar performance with fewer interventions when compared to the same scenario with a non-learning wizarded-robot.

3.6 Interaction Protocol

Each participant experienced both robot controllers, with the order changed between participants to control for any ordering effects. In *Condition LN* the participants first interact with the learning wizarded-robot, and then with the non-learning one; in *Condition NL* the participants first interact with the non-learning wizarded-robot, and then the learning robot. Participants were randomly assigned to one of the two conditions.

The interactions took place on a university campus in a dedicated experiment room. Two Aldebaran Nao robots were used; one robot had a label indicating that it was the *Child-Robot*. The robots face each other with a touchscreen between them, and participants assuming the role of the supervisor sit at a desk to the side of the wizarded-robot, with a screen and a mouse to interact with the wizarded-robot (fig. 1). The participants were able to see the screen and the child-robot.

A document explaining the interaction scenario was provided to participants. After the information had been read, a 30s video presenting the GUI in use was shown to familiarise them with it, without biasing them towards any particular intervention strategy. The participant then clicked a button to start the first interaction which lasted for 10 minutes. The experimenter was sat in the room outside of the participants' field of view. After the end of the first interaction, a post-interaction questionnaire was administered. The same protocol was applied in the second part of the experiment with another post-interaction questionnaire following. Finally, a questionnaire asking the participants to explicitly compare the two conditions was administered.

4   Results

4.1   Interaction Data

The state of the child and the interaction values were logged at each step of the interaction (at 5Hz). All of the human actions were recorded: acceptance of the wizarded-robot's suggestion, selection of another action (intervention), and the states of the child-robot (motivation, engagement and performance) at this step. From this the intervention ratio was derived: the number of times a user chose a different action to the one proposed by the wizarded-robot, divided by the total number of executed actions. On average, after a first exploration phase, where the participant discovers the system, the learning robot robot has an intervention ratio lower than the non learning one (fig. 2, left)

The performance indicates the number of good categorisations executed by the child-robot minus the number of bad categorisations. A strong positive correlation (Pearson's $r=0.79$) was found between the average child-robot motivation and engagement and its performance.

In both conditions, the average performance in the second interaction ($M_{LN-2}=38$, 95% CI [36.2, 39.8], $M_{NL-2}=34.8$, 95% CI [30.8, 38.8]) was higher than in the first one ($M_{LN-1}=29.4$, 95% CI [25.3, 33.5], $M_{NL-1}=24.3$, 95% CI [19.4, 29.4]; Fig. 2 *left*). The 95% *Confidence Interval of the Difference of the Mean* (CIDM) for the L-NL condition is [4.1, 13.1] and for the NL-L condition is [4.0, 16.8]. However, the performance is similar when only the interaction order (first or second) is considered. The participants performed slightly better in the LN condition, but the CIDM includes zero in both cases (95% $CIDM_1$ [-1.5, 11.5], 95% $CIDM_2$ [-1.2, 7.6]). In the condition L-NL, the intervention ratio increased between the learning and non learning condition ($M_{LN-1}=0.31$, 95% CI [0.20, 0.42] to $M_{LN-2}=0.68$, 95% CI [0.66, 0.70], $CIDM_{LN}=[0.26, 0.48]$). But in the NL condition, the intervention ratio is almost identical between the two interactions but slightly lower for the learning case ($M_{NL-1}=0.50$, 95% CI [0.44, 0.57] to $M_{NL-2}=0.46$, 95% CI [0.40, 0.51], $CIDM_{NL}$ [-0.03, 0.13]). This shows that when the wizarded-robot learned, a similar performance is attained as without learning, but the number of interventions required to achieve this is lower.

4.2   Questionnaire Data

The post-interaction questionnaires evaluated the participant's perception of the child-robot's learning and performance, the quality of suggestions made by the wizarded-robot, and the experienced workload. All responses used seven point Likert scales.

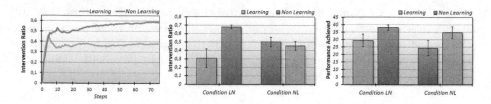

Fig. 2. (Left) evolution of intervention ratio over time for the learning and non learning cases. Intervention ratio (centre) and final performance (right) for the two conditions and the two interactions (*errors bars show 95% CI*). In condition LN participants started wizarding a robot which learns their interaction style, followed by a non-learning robot; in condition NL participants started with a non-learning robot, followed by a learning robot. Results show that a learning robot reduces the workload of the wizard, but performs equally well as a non-learning robot that needs wizarding at all times.

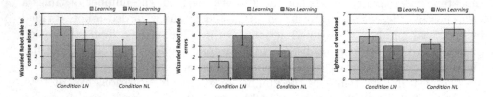

Fig. 3. Questionnaire responses (*mean and 95% CI*): increased confidence in the learning wizarded-robot over the non-learning version is apparent, as is a lower perceived workload.

Across the four possible interactions, the rating of the child-robot's learning was similar (M=5.25, 95% CI [4.8, 5.7]). The same effect was observed for the evaluation of the child performance (M=4.75, 95% CI [4.3, 5.2]). As the child-robot was using the same interaction model in all four conditions, this result is expected.

Participants report the wizarded-robot as more suited to operate unsupervised in the learning than in the non learning condition (M_{LN-1}=4.8, M_{LN-2}=3.6, M_{NL-1}=3, M_{NL-2}=5.2 ; CIDM for LN condition [-0.2, 2.6], CIDM for the NL condition [1.6, 2.8]).

Similarly, a trend was found showing that learning wizarded-robot is perceived as making fewer errors than the non-learning robot (M_{LN-1}=1.6, M_{LN-2}=4.0, M_{NL-1}=2.6, M_{NL-2}=2 ; CIDM for LN condition [1.3, 3.4], CIDM for the NL condition [0.1, 1.1]).

The participants tended to rate the workload as lighter when interacting with the learning robot, and this effect is much more prominent when the participants interacted with the non-learning robot first (M_{LN-1}=4.6, M_{LN-2}=3.6, M_{NL-1}=3.8, M_{NL-2}=5.4 ; CIDM for LN condition [-0.6, 2.6], CIDM for the NL condition [0.7, 2.5]).

5 Discussion

Strong support for H1 (a good supervisor leads to a better child performance) was found, a correlation between the average states (engagement and motivation) and the final performance for all of the 10 participants was observed (r=0.79). We could expect a similar effect when working with real children, but measuring these values would be a challenge.

The results also provide support for H2 (supervisors create personal strategies): all the participants performed better in the second interaction than in the first one. This suggests that participants developed a strategy when interacting with the system in the first interaction, and were able to use it to increase their performance in the second interaction. Looking in more detail at the interaction logs, it is possible to see that different people used different strategies.

H3 (reducing the number of interventions will reduce the perceived workload) is partially supported: the results show a trend for participants to rate the workload as lighter when interacting with the learning robot, and another trend between using a learning robot and the intervention ratio. However, when considering the difference of workload rating and intervention ratios between the two interactions, a positive correlation is only found for the LN condition, which could be accounted for by the initial steep learning curve for the study participants. Nevertheless, regardless of the order of the interactions, the learning robot consistently received higher ratings for lightness of workload (fig. 3).

Finally, H4 (using learning keeps similar performance, but decreases interventions) is supported: interacting with a learning robot results in a similar performance than interacting with a non-learning robot, whilst requiring fewer active interventions from the supervisor. This has real world utility, it frees some time for the supervisor, to allow her to focus on other aspects of the intervention, e.g. analysing the child's behaviour rather than focusing on the robot control.

It should be noted that the actual learning algorithm used in this study is only of incidental importance, and that certain features of the supervisor's strategies may be better approximated with alternative methods – of importance for the present work is the presence of learning at all. Future work will assess what the most appropriate machine learning approach is given the observed features of supervisor strategy from this study.

In conclusion, this paper proposed the SPARC concept (Supervised Progressively Autonomous Robot Competencies). Based on a suggestion/intervention system, this approach allows online learning for interactive scenarios, thus increasing autonomy and reducing the demands on the supervisor. Results showed that interacting with a learning robot allowed participants to achieve a similar performance as interacting with a non-learning robot, but requiring fewer interventions to attain this result. This suggests that while there is always adaptation in the interaction (leading to similar child-robot performance given the two wizarded-robot controllers), the presence of learning shifts this burden of adaptivity onto the wizarded-robot rather than on the human. This indicates that a learning robot could allow the therapist to focus more on the child than on the robot, with improved therapeutic outcomes as potential result.

Acknowledgments. This research was funded by the EU FP7 DREAM project (grant no. 611391).

References

1. Argall, B.D., Chernova, S., Veloso, M., Browning, B.: A survey of robot learning from demonstration. Robotics and Autonomous Systems **57**(5), 469–483 (2009)
2. Baxter, P., Wood, R., Belpaeme, T.: A touchscreen-based sandtray to facilitate, mediate and contextualise human-robot social interaction. In: 2012 7th ACM/IEEE International Conference on Human-Robot Interaction (HRI), pp. 105–106. IEEE (2012)
3. Cakmak, M., Thomaz, A.L.: Designing robot learners that ask good questions. In: Proceedings of the Seventh Annual ACM/IEEE International Conference on Human-Robot Interaction, pp. 17–24. ACM (2012)
4. Chernova, S., Veloso, M.: Interactive policy learning through confidence-based autonomy. Journal of Artificial Intelligence Research **34**(1), 1 (2009)
5. Dautenhahn, K.: Robots as social actors: aurora and the case of autism. In: Proc. CT 1999, The Third International Cognitive Technology Conference, San Francisco, August, vol. 359, p. 374 (1999)
6. Feil-Seifer, D., Mataric, M.: B3IA: a control architecture for autonomous robot-assisted behavior intervention for children with autism spectrum disorders. In: RO-MAN (2008). http://ieeexplore.ieee.org/xpls/abs_all.jsp?arnumber=4600687
7. Knox, W.B., Spaulding, S., Breazeal, C.: Learning social interaction from the wizard: a proposal. In: Workshops at the Twenty-Eighth AAAI Conference on Artificial Intelligence (2014)
8. Riek, L.: Wizard of Oz Studies in HRI: A Systematic Review and New Reporting Guidelines. Journal of Human-Robot Interaction **1**(1), 119–136 (2012)
9. Robins, B., Dautenhahn, K., Boekhorst, R.T., Billard, A.: Robotic assistants in therapy and education of children with autism: can a small humanoid robot help encourage social interaction skills? Universal Access in the Information Society **4**(2), 105–120 (2005)
10. Robins, B., Dautenhahn, K., Dickerson, P.: From isolation to communication: a case study evaluation of robot assisted play for children with autism with a minimally expressive humanoid robot. In: Conferences on Advances in Computer-Human Interactions, pp. 205–211, February 2009
11. Scassellati, B., Admoni, H., Mataric, M.: Robots for use in autism research. Annual Review of Biomedical Engineering **14**, 275–294 (2012)
12. Senft, E., Baxter, P., Belpaeme, T.: Human-guided learning of social action selection for robot-assisted therapy. In: 4th Workshop on Machine Learning for Interactive Systems (2015) (in press)
13. Thill, S., Pop, C.A., Belpaeme, T., Ziemke, T., Vanderborght, B.: Robot-assisted therapy for autism spectrum disorders with (partially) autonomous control: challenges and outlook. Paladyn **3**(4), 209–217 (2012)
14. Wainer, J., Dautenhahn, K., Robins, B., Amirabdollahian, F.: Collaborating with kaspar: using an autonomous humanoid robot to foster cooperative dyadic play among children with autism. In: Humanoid Robots (Humanoids 2010), pp. 631–638. IEEE (2010)

A Case Study of Robot Interaction Among Individuals with Profound and Multiple Learning Disabilities

Jainendra Shukla[1,2]([✉]), Julián Cristiano[1], David Amela[2], Laia Anguera[2], Jaume Vergés-Llahí[3], and Domènec Puig[1]

[1] Intelligent Robotics and Computer Vision Group, Universitat Rovira I Virgili,
Tarragona, Spain
jainendra.shukla@estudiants.urv.cat,
{julianefren.cristiano,domenec.puig}@urv.cat
[2] Instituto de Robótica para la Dependencia, Sitges, Spain
{damela,languera}@institutorobotica.org
[3] Ateknea Solutions Catalonia, Barcelona, Spain
jaume.verges@ateknea.com

Abstract. A tremendous amount of research is being performed regarding robot interaction with individuals having intellectual disability, especially for kids with Autism Spectrum Disorders (ASD). These researches have shown many promising advancements about the use of interactive robots for rehabilitation of such individuals. However, these studies fail to analyze and explore the effects of robotics interaction with individuals having profound and multiple learning disabilities (PMLD). This research presents a thorough case study regarding interaction of individuals having PMLD with a humanoid robot in different possible categories of robotic interaction. Separate interaction activities are designed as a representative for the different categories of possible clinical applications of the interactive robot. All the trials were assessed using different evaluation techniques. Finally, the results strongly suggest that robotic interactions can help to induce a target behavior among these individuals, to teach and to encourage them which can bring an autonomy to certain extent in their life.

Keywords: Human-Robot Interaction · Profound and multilple learning disability · PMLD · NAO humanoid robot · Clinical applications of interactive robots

1 Introduction

Intellectual disability is a disability characterized by significant limitations in both intellectual functioning and in adaptive behavior. These limitations result in problems with reasoning, learning or problem-solving as well as communication and

This research work has been supported by the Industrial Doctorate program (Ref. ID.: 2014-DI-022) of AGAUR, Government of Catalonia.

A. Tapus et al. (Eds.): ICSR 2015, LNAI 9388, pp. 613–622, 2015.
DOI: 10.1007/978-3-319-25554-5_61

social skills difficulties. This disability originates before the age of 18 [1]. As the name suggests, Individuals with PMLD have more than one disability. One very important symptom is that they have profound learning disability. Generally they also have an associated medical condition which could be neurological, and physical or sensory impairments [2]. Due to all these conditions they require a constant general support.

Epidemiological studies suggest that the overall prevalence of severe intellectual disabilities (approximating to IQ<50) is between 3 and 4 people of all ages per 1000 total population, implying that in the 15 countries of the European Union (total population 380 million) between 1.1 and 1.5 million people have severe intellectual disabilities [3]. A recent survey suggests that while among adults, the rates vary between 3-6/1000, among children the rates are between 3-14/1000 [4]. In Spain there are almost 300000 people with intellectual disabilities. Due to the rise in survival rate of premature babies, the number of individuals with disability is also rising [5]. This will lead to an adverse influence for the call of health, education and social care needs.

There is no medical cure available for the individuals with such disabilities [6]. However, higher engagement rate has been reported by the use of Humanoid robots with students with profound and multiple learning disabilities (PMLD) [7]. Many other researches also claim positive effects of using robots or robot like toys to increase interaction among individuals with intellectual disabilities [8,9].

Currently, the robotic platforms are being employed for diagnosis and treatment of people with autism in the clinical context. Most of the works developed up to date have been especially proposed for children with Autism Spectrum Disorder (ASD) [10–13]. The aim of the activities that involve robotic platforms in people with autism is to get positive responses from the users. A closed-loop system to dynamically interact with a child based in his response in real-time is proposed in [11]. The automatic response to detected behavior is very important in order to achieve meaningful and personalized technological interventions, which can lead to better results and more attention of individual. The authors claim that children with ASD exhibited greater attention to robotics system in comparison with the human administrator. In [12], a humanoid robot is used to foster and support the collaborative play among children with autism. Collaborative game among two children with autism and a humanoid robot has shown improved social behaviors among children playing with each other compared to before they did without the collaborative game with the robot.

Studies performed in people with ASD measured and assessed the positive or negative impact of these technologies in children with ASD and typical development. These studies are not conclusive as they have been performed in groups of few people and most of the papers have not been presented in ASD journals in order to be evaluated by experts in the field from the clinical point of view [10]. However, it is evident that children with autism enjoy playing by themselves with computers and several mechanical devices. A review for clinical use of robots with individuals with ASD is presented in [10]. The authors collect previous studies with empirical evidence based on clinical applications of robots in the diagnosis and treatment of ASD, and the studied works were classified in 4 different categories. In [12], a humanoid robot was programmed in order to enable teachers to use it to achieve some learning objectives previously identified for eleven people with PMLD. Results have shown that the attention during the sessions involving robots was higher in comparison to the session in the classroom. However, there are few works that involve interaction among robots and people with PMLD basically because the interaction among these patients and the robots is more complicated due to the patients cognitive problems and also due to the problems in the motor control of their extremities [14].

Motivated by above investigations, the aim of our proposed research is to analyze the response of individuals with PMLD in different possible categories of robot interaction [10]. It is to be noted that research in [10] is based on the therapeutic application of robots against ASD and not specifically against PMLD. However, PMLD and ASD conditions are frequently associated because individuals with PMLD may have autism [2], and also it is estimated that the learning disability among children with core autism is between 60-70% [15]. Recognizing this association, proposed research designs and analyzes robot interaction among individuals with PMLD in different categories of clinical applications based upon robot interaction research in ASD.

2 Method

2.1 The Approach

The case study was performed over a period of three months at a trail room in Ave Maria Foundation[1]. Ave Maria Foundation is the residential and clinical facility of the participants hence it provided a familiar environment for all the participants. Standard medical ethics requirements were satisfied.

Four unique activities have been identified to represent each one of the categories as identified in [10]. These activities are listed in table 1. The activities were chosen based upon following criteria:

1. Relevant representation of the category by the activity of choice.
2. Simplicity for participants at the execution level.
3. Ease of implementation in NAO robot.

1. Dance Choreography: This activity was aimed to observe the response of the participants towards a robot or robot like characteristics. In this activity, NAO performs a dance composition while singing a song. NAO was not interacting with the participant at any level while participants were allowed to observe and respond without any restrictions.
2. Touch my head: The aim of this activity was to induce a target behavior in participants. In this activity the robot asks the participants to touch the robot's head, feet or hand. The participants are expected to respond as per its instructions.

---

[1] Ave Maria Fundació, http://www.avemariafundacio.org/inici.html

Table 1. Activities

Category	Representing Activity	ID
Responses to robots or robot-like characteristics	Dance choreography	1
Eliciting behavior	Touch my head	2
Providing feedback or encouragement	Learn the senses	3
Modeling, teaching, or practicing skills	Guess emotions	4

3. Learn the senses: This activity is aimed to provide feedback and encouragement to the participant by the robot to achieve a certain target behavior. NAO prompts the participants to present an answer image corresponding to a particular sensory activity of the human body. If the participant do not answers within a certain time period NAO encourages the participant by providing some clue about the answer image. The robot provides a positive feedback on accomplishing the right answer.
4. Guess Emotions: The robot works as a learning tool for the participant. The robot tells a short story to the participants and in between asks questions related to the emotional state of the character in the story. The robot helps participants to answer the questions helping them to learn about different emotions.

2.2   Participants

The experiments for this case study were carried out with six individuals of different age, gender and intellectual disability levels. The assessment of these individuals was done by Assessment and Guidance Services for People with Disabilities (CAD Badal) organization of Government of Catalunya[2]. Details of all the individuals are presented in table 2.

2.3   Procedure

The robot used for this case study was NAO NextGen (Model H25, V4). NAO is a 58 cm tall humanoid robot developed by Aldebaran[3]. For each trial, the robot was placed on a table in a position as required to initiate the desired activity. The participant is brought to the trail room by the care taker and takes a seat in front of the robot. Figure 1 shows a general position of the participant with the robot. Only care taker stays in the room with the participant while the researchers observe the whole situation from outside of the room. The care taker observes the participant during all the activity but does not initiate any communication on its own but responds to participants. Duration of trails vary between 15-30 minutes depending upon the activity and the participants, while actual robot interaction during each trail lasted between 5-10 minutes.

---

[2] Generalitat de Catalunya, http://web.gencat.cat/ca/inici/
[3] Aldebaran, https://www.aldebaran.com/en

Table 2. Participants

ID	Gender	Age	Condition	Disability (%)
ID01	F	65 y, 4 m	Moderate Intellectual Disability, Affective Disorder, Right Hemiparesis, Mixed Cerebral Palsy	85
ID02	F	42 y	Autism, Severe Intellectual Disability	86
ID03	F	48 y, 8 m	Severe Intellectual Disability, Affective Disorder, encephalopathy	87
ID04	F	33 y, 2 m	Autism, Moderate Intellectual Disability	79
ID05	F	67 y, 10 m	Moderate Intellectual Disability, Tetraparesis	86
ID06	M	44 y, 6 m	Severe Intellectual Disability, Down Syndrome	75

2.4   Measurement and Evaluation

The evaluation for all the trials was analyzed using the following measures:

1. **Engagement rate** which is the percentage of time that the participants were actually engaged during actual robot interaction. Engagement of the participants were calculated after the trials with help of an expert psychiatrist by analyzing the video recordings of the trials for engagement observation of participants.
2. **Performance of the participants** against desired responses of activities. Depending upon the time duration and correct or wrong responses, performance of the participants were recorded as perfect, good, regular or no response.
3. **A questionnaire** adapted from GARS-2 [16], WHODAS 2.0 [17] and ABS-RC: 2 [18]. The Gilliam Autism Rating Scale-Second Edition (GARS-2) is a supplementary screening tool for individuals suffering with autism spectrum disorders. ABS-RC: 2 is a method to assess adaptive behavior of mentally handicapped persons. World Health Organization Disability Assessment Schedule 2 (WHODAS 2.0) is a tool for assessment of global functioning and impairment. To the best knowledge of authors, there is no availability of any method or scale for the evaluation of robotic interaction effects. Thus, the authors adapted the existing above said assessment tools for evaluation purposes and to enhance the decision making about such trials. An expert psychiatrist identified 25 questions from GARS-2, 2 questions from WHODAS2.0 and 13 questions from ABS-RC2 (part II) as per their suitability for this evaluation. Based upon this questionnaire, the behavior of participants in normal situations (before the trails) was compared with their behavior during robot interaction trials.

Fig. 1. A participant interacting with the robot

3   Result and Discussion

Table 3 shows performance evaluation of the participants for all three interactive activities. Each activity consisted of 3 tasks in respective categories. Thus, the total number of responses recorded from six participants was 54 (6 participants * 3 activities * 3 tasks). The evaluation of response was done by a psychiatrist taking into account the time and support used by the participant to accomplish the task.

Table 3. Performance of participants against activities

Activity	Perfect	Good	Regular	No Response
Touch my head	14	1	2	1
Guess Emotion	9	5	4	0
Learn Senses	10	3	4	1

As can be observed from table 3 that *among a total of 54 observations, only 2 times participants did not respond at all to the robot. Most of time the response was very positive, which can be seen by a total number of 33 perfect responses.* Some important observations for delayed or no response scenarios are summarized below:

1. Participants were attracted (absorbed) with the robot hence, were not able to concentrate on the execution of activity.
2. Restrictions in technical abilities of the robot, (e.g. image recognition).

3. Participants were excited with the robot during activity hence, sometimes were responding before even completely listening to the robot.

Table 4 shows the observed engagement rate for all participants during all the trials.

Table 4. Engagement rate (% duration) of participants against activities

ID	Dance	Touch my head	Learn Senses	Guess Emotion
ID01	96.60	100.00	100.00	100.00
ID02	64.56	100.00	100.00	100.00
ID03	93.20	100.00	100.00	100.00
ID04	100.00	100.00	100.00	100.00
ID05	100.00	100.00	100.00	100.00
ID06	98.06	100.00	100.00	100.00

Above results are very exciting. Important observations can be summarized as follows:

1. In the first category, there was no interaction between robot and participants as it was intended to observe the response of participant's towards the robot. As can be seen from table 4, engagement rate in this category was not exactly 100 %. As the robot was not interacting with the participants in any manner, after certain time they started loosing their concentration but for most of the time they were attracted and focused to the robot. The participants ID02 and ID03 observed the lowest engagement rates as 64.56 and 93.20 % respectively. It can be directly related with their mental conditions as they both are having severe mental retardation. Hence, *patients with more disability showed lower engagement than other participants. In this regard, their behavior with the robot was similar to their behavior with a human.*
2. In all other categories, participants observed an exciting 100 % engagement rate with the robot. This result is very fascinating as it indicates that *irrespective of their mental condition all the participants were able to engage fully with the robot when it was interacting with them in any manner.*

The results shown in figure 2 represent the improvement in disability behavior (in terms of %) during the interactions with robot in different categories as compared with the observed behavior in normal situations. Proposed questionnaire was used to evaluate the disability behavior of the participants in normal situations. Then using the same questionnaire, disability behavior during interaction with robot was reevaluated. A difference is also calculated using above evaluations and is presented as percentage in figure 2. A higher value represents more improvement in disability behavior (i.e. reduction in disability behavior) during robot interactions as compared to normal situations.

Missing data in figure 2 for ID03 and ID04 under WHODAS2.0 indicate that the value could not be calculated as the participants do not exhibit any disability behavior assessed by this scale.

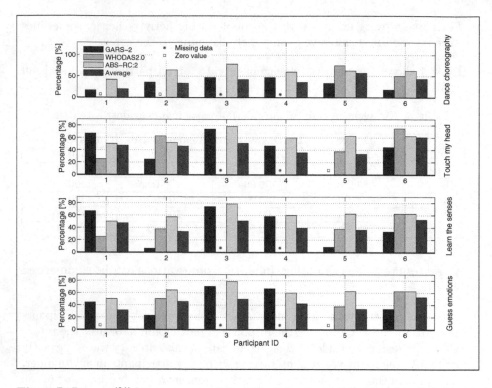

Fig. 2. Difference (%) between normal disability behavior in daily situations & behavior during robot interaction, as observed using the proposed questionnaire and its average difference

As can be observed from figure 2, *all participants showed either a reduced or at-least same level of disability behavior during all robot interaction trials in comparison to normal situation behaviors. Mostly they showed a very reduced disability behavior which is very fascinating.* Important observations are as follows:

1. The plot for activity 1 shows that all the participants showed a reduced disability behavior even while robot was not making any interaction with them. e.g. Participant ID06 showed a 17.78% decrease in the disability behavior on GARS scale, a 50% reduction on WHODAS2.0 scale and a 62.50% reduction on ABSRC2 scale.
2. The highest observed difference in reduction of disability behavior is for participant ID03 and is 78.57%. It is a fascinating improvement.
3. The lowest observed difference in reduction of disability behavior is 0% and was observed 5 times. e.g. participant ID02 in the first category by using the WHODAS2.0 scale indicating no improvement for this participant in this category.
4. An average difference for each participant is presented in figure 2 for each one of the activities. It gives a hint about the average improvement that can

be expected for respective participant (or individuals with the same type of disabilities) with robot interaction techniques in each category.

5. *An average difference for all participants in all categories is 47.79% which itself indicates that robot interaction is effective to good extent.*

6. Another important observation is the absence of any negative value in the results, which indicates that *none of the participants showed an increased disability behavior in any of the trails which suggests that robot interactions did not caused any negative effects during the trails.*

4   Conclusion and Future Work

Amount of work done to analyze and study the response of the robot interactions for the individuals with PMLD has been inadequate. Keeping this in mind, a small but significant step was taken to advance in this direction. Different types of activities were designed to observe response of individuals with PMLD in different categories. Elder people with PMLD were included in the trails to analyze the effect of their interactions with a robot. The results are very exciting and surely suggest that robot interactions can be very helpful to improve the conditions of the individuals with PMLD even at an elder age.

Due to low number of participants, results of this study only indicate the underlying potential of research in this field. Hence, the number of participants should be increased to explore potential findings in more detail. It can be very interesting to analyze the responses of group of people during the interaction sessions. Also instead of using one robot, multiple robots can be used to interact with individuals or with groups. The establishment of a standard method for assessing the response of interactive robots with different types of patients is of utmost importance. Autonomous change in the behavior performed by the robot according to the current response of the patients is also of great interest. As same medicine can not be offered to all types of patients, similarly robot interactions need to be customized as per the state and conditions of the individual patients.

References

1. Schalock, R.L., Borthwick-Duffy, S.A., Buntinx, W.H.E., et al.: Intellectual Disability: Definition, Classification, and Systems of Supports, 11th edn. American Association on Intellectual and Developmental Disabilities (AAIDD), Washington, D.C. (2010)

2. Bellamy, G., Croot, L., Bush, A., et al.: A study to define: profound and multiple learning disabilities (PMLD). Journal of Intellectual Disabilities 14(3), 221–235 (2010)

3. European Intellectual Disability Research Network: Intellectual disability in Europe: Working papers. Canterbury: Tizard Centre, University of Kent at Canterbury (2003)

4. Maulik, P.K., Harbour, C.K.: Epidemiology of Intellectual Disability. International Encyclopedia of Rehabilitation (2010). http://cirrie.buffalo.edu/encyclopedia/en/article/144/

5. Moore, T., Hennessy, E.M., Myles, J., et al.: Neurological and developmental outcome in extremely preterm children born in England in 1995 and 2006, EPICure studies. BMJ **345**, e7961 (2012)
6. Rueda, J.R., Ballesteros, J., Tejada, M.I.: Systematic review of pharmacological treatments in fragile X syndrome. BMC Neurology 9–53 (2009)
7. Standen, P., Brown, D., Roscoe, J., Hedgecock, J., Stewart, D., Trigo, M.J.G., Elgajiji, E.: Engaging students with profound and multiple disabilities using humanoid robots. In: Stephanidis, C., Antona, M. (eds.) UAHCI 2014, Part II. LNCS, vol. 8514, pp. 419–430. Springer, Heidelberg (2014)
8. Robins, B., Dautenhahn, K., Boekhorst, T., et al.: Robotic assistants in therapy and education of children with autism: can a small humanoid robot help encourage social interaction skills? Univers. Access. Inf. Soc. **4**, 105–120 (2005)
9. Scassellati, B.: Quantitative metrics of social response for autism diagnosis. In: IEEE International Workshop on Robot and Human Interactive Communication, ROMAN 2005, pp. 585–590 (2005)
10. Diehl, J.J., Schmitt, L.M., Villano, M., et al.: The Clinical Use of Robots for Individuals with Autism Spectrum Disorders: A Critical Review. Research in Autism Spectrum Disorders **6**(1), 249–262 (2012)
11. Warren, Z., Zheng, Z., Das, S., et al.: Brief report: development of a robotic intervention platform for young children with ASD. Journal of Autism and Developmental Disorders 1–7 (2014). ISSN 0162–3257
12. Wainer, J., Robins, B., Amirabdollahian, F., et al.: Using the Humanoid Robot KASPAR to Autonomously Play Triadic Games and Facilitate Collaborative Play Among Children With Autism. IEEE Trans. on Autonomous Mental Development **6**(3), 183–199 (2014)
13. Robins, B., Dautenhahn, K.: Tactile Interactions with a Humanoid Robot: Novel Play Scenario Implementations with Children with Autism. International Journal of Social Robotics **6**(3), 397–415 (2014)
14. Standen, P., Brown, D.J., Hedgecock, J., et al.: Adapting a humanoid robot for use with children with profound and multiple disabilities. In: 10th Int. Conf. Disability, Virtual Reality and Associated Technologies, pp. 205–211 (2014)
15. Dover, C.J., Couteur, A.L.: How to diagnose autism. Archives of Disease in Childhood **92**(6), 540–545 (2007)
16. Gilliam, J.E.: GARS-2: Gilliam Autism Rating Scale - Second Edition. Jour. of Psychoeducational Assessment **26**(4) 395–401 (2006). PRO-ED, Austin
17. Gold, L.H.: DSM-5 and the Assessment of Functioning: The World Health Organization Disability Assessment Schedule 2.0 (WHODAS 2.0). The journal of the American Academy of Psychiatry and the Law **42**(2), 173–81 (2014)
18. Nihira, K., Leland, H., Lambert, N.M., et al.: ABS-RC:2: AAMR Adaptive Behavior Scale: residential and community, 2nd edn. Pro-Ed, Austin (1993)

Impact of Humanoid Social Robots on Treatment of a Pair of Iranian Autistic Twins

Alireza Taheri[1(✉)], Minoo Alemi[1,2(✉)], Ali Meghdari[1(✉)], Hamidreza Pouretemad[3], Nasim Mahboub Basiri[4], and Pegah Poorgoldooz[5]

[1]Social Robotics Laboratory, Center of Excellence in Design, Robotics and Automation (CEDRA), Sharif University of Technology, Tehran, Iran
taheri@mech.sharif.edu, minooalemi2000@yahoo.com,
meghdari@sharif.ir
[2]Islamic Azad University, Tehran-West Branch, Tehran, Iran
[3]Inistitue for Cognitive and Brain Sciences (ICBS), Shahid Beheshti University, Tehran, Iran
h-pouretemad@sbu.ac.ir
[4]Languages and Linguistics Center, Sharif University of Technology, Tehran, Iran
[5] Center for Treatment of Autistic Disorders (CTAD), Tehran, Iran

Abstract. In recent years robots have been increasingly used in autism research. In this paper the effects of robot-assisted interventions on two seven year old autistic twin brothers, one of whom is high-functioning and the other low-functioning, are explored. To this end, 12 sessions of therapeutic scenarios were designed and presented to the autistic twin subjects in the presence of two robots, a therapist and their parents in individual and group modes. The results showed great potential benefits from using robots in group therapeutic games in both high- and low- functioning autistic children, such as improvement in imitation and joint attention skills for both brothers, as well as communication with each other. The results also indicated a decrease of stereotyped behaviors in the low-functioning brother, and improvement in social and cognitive skills in the high-functioning brother.

Keywords: Humanoid robot · Joint attention · High- and low-functioning autism · Autistic twin · Imitation

1   Introduction

Humanoid robots can be used as a powerful tool to improve social and motor skills as well as joint attention in autistic children [1, 2]. Individuals with autism usually shy away from social interactions and communications and are impaired in showing proper reactions to real world events [3]. To this date, a wide range of studies have been done on the application of robots in autism treatment (especially on high-function autistic children) to improve imitation, joint attention, and social interaction skills of autistic children [4-13]. Although there has been research on autistic twins and the relative contributions of genetics and environment to autism spectrum disorders [14-15], to the best of our knowledge using humanoid robots specifically in

© Springer International Publishing Switzerland 2015
A. Tapus et al. (Eds.): ICSR 2015, LNAI 9388, pp. 623–632, 2015.
DOI: 10.1007/978-3-319-25554-5_62

the treatment of twins with autism has not been reported. What makes this study different is that it focuses on the robot-assisted interventions of seven-year old autistic twins, one of whom is high-functioning and the other low-functioning. The two participants were fraternal twin brothers. Besides improving motor and social interaction skills of these two subjects with each other and with their parents, the main purpose of this study was to investigate how the effect of robot-assisted autism therapy differs for high-and low-functioning autistic children.

2   Research Methodology

2.1   Participants

Our subjects were seven-year old fraternal autistic twins. Both were male and diagnosed with autism spectrum disorders; one is High-functioning with hyperactivity (called S1-A) and the other is Low-functioning (called S2-I). The advantage of investigating twins in comparison to other cases is factors such as parents, food, clothes, and education have been controlled, a difficult task in general research. S1-A is a high-functioning autistic boy with hyperactivity and mild verbal skills. Eye-contact avoidance also existed since an early age. At the age of seven his parents were informed that S1-A was a high-functioning autistic child. S2-I is a low-functioning autistic child with poor verbal skills. S2-I's autism is more severe than his twin brother and he usually engages in repetitive, non-purposeful, and stereotyped behaviors such as fluttering fingers.

2.2   Intervention Sessions

The intervention sessions included various games in order to teach individual and group sport skills (Robot-Patient and Robot-Patient-Brother/Parent) and engage them in different imitation and joint attention situations. The intervention sessions were run on the autistic twins in the presence of the Humanoid Robot(s), therapist, robot operator, and their parents in a fairly friendly environment. Our study approach was a single subject design using Wizard of Oz style robot control. Intervention scenarios were designed based on clinical psychologists' explanations of psychology theories, shaping behaviors therapy, and Applied Behavior Analysis (ABA) models run in autism treatment centers. The pre-designed scenarios were conducted in 12 thirty-minute sessions held twice a week for 6 weeks at the Social Robotics Laboratory at Sharif University of Technology.

2.3   Set-up of the Study

The room size was $5 \times 5 \times 3$ m^3. The set-up of our study consisted of two humanoid robots, Microsoft Kinect sensor, video-projector, two laptops, chairs, a whiteboard, and two cameras for filming the sessions. Child-Robot interaction was structured and preset following pre-defined purposes. The scenario instructions were described by the robot and/or the therapist. The parents of the twin subjects voluntarily took part in our research and they did not pay nor were they paid for the intervention sessions. A pledge was signed by the researchers and parents before the first session in order to maintain moral obligations.

2.4   Humanoid Robots

The humanoid robots used in our educational-therapeutic programs were the NAO-H21 made by Aldebaran Company [16] with 21 degrees-of-freedom (DOFs), and the Alice-R50 made by Robokind Company [17] with 32 degrees of freedom. To be used in the Iranian context, these robots were renamed "Nima" and "Mina", respectively. These two robots have the necessary capabilities needed for our designed intervention scenarios. Moreover, other researchers around the world have also used these commercial robots in autism research [4, 8, and 12]. Our concentration was on using the Nima robot; however, we also used the Mina robot because: a) it has 11 DOFs in the face and is capable of showing different facial expressions, and b) we wanted to explore if changing the robot effected the children's performance.

2.5   Therapeutic Games

A variety of therapeutic games were developed based on the children's autistic impairments in order to answer our research questions. These games concentrated on improving the children's imitation, joint attention, social skills, eye-contact, and turn-taking. In each session the twins participated in several of the games in different modes; Robot-Child or Robot-Child-Brother/Parent/Therapist interactions. Table 1 presents the list of games. The schedule of intervention sessions is presented in Table 2.

2.6   Assessment Tools

The four main instruments used to measure the effects of the interventions in this study are as follows:

Gilliam Autism Rating Scale (GARS): One of the most well-known autism assessment tools is the Gilliam Autism Rating Scale (GARS). This questionnaire is a valid tool developed by Gilliam in the1990s [18] to help estimate autism severity. GARS is divided into four different subscales: Stereotyped Behaviors, Communication, Social Interactions, and Developmental Disturbances [19]. GARS has been used for 100 autistic children in Iran and the Cronbach's alpha for its four subscales and the overall test are 0.74, 0.92, 0.73, 0.80, and 0.89, respectively [20]. The GARS questionnaire was filled in by the children's parents one week before and one week after the robot-assisted program.

Table 1. List of Therapeutic Games

#	Games	Modes	Main Purposes of the Game
1	Teaching imitation and motor skills by robot to child/children through individual/group exercise and dances	Robot-Child Robot-Child-Brother/Parent	Improve imitation, Improve motor and social Skills, Dyadic/Triadic interactions, Turn-taking games
2	Real-time Imitation of Robot by child in upper body movements	Robot-Child	Draw attention of child to child to robot and therapist, Child can see his movements reflected in another person
3	Tele-operating humanoid robots' heads and hands using a 6-DOFs Haptic Phantom-Omni robot as a remote controller	Robot-Child	Empowering children and therapist to move the robots' joints arbitrary, Dyadic/triadic interactions, Turn-Taking games
4	Kinect-based Recognition Game: Classification of animals and fruit by pointing to different baskets on the screen	Robot-Child Robot-Child-Parent	Classification, Joint attention, Pointing, Gaze-shifting
5	Playing a developed Kinect based virtual xylophone on the screen	Child-Parent/Therapist robot applaud child for a task correctly done	Improve child's hand imitation skills, Joint attention, and child's visual pursuit
6	Playing a real xylophone in a Robot-Child turn-taking game	Robot-Child	Imitation of Robot by Child and vice versa, Joint attention, Turn-taking, Improve in cognitive skills, Colors recognition, Hand-eye coordination

Quantitative Content Analysis of Intervention Video Records: Quantitative content analysis is a powerful tool to analyze written texts, videos or other media [21, 22]. To analyze the autistic twin's behaviors during the sessions, intervention video records have been observed and rated by two psychologists. The seven major items (some with different sub-items) rated by the psychologists consisted of: 1) Imitation, 2) Joint attention, pointing and gaze shifting, 3) Maladaptive behaviors, 4) Verbal and nonverbal communications, 5) Instruction perception and cooperation, 6) Intercommunity, and 7) Interest in and enjoying individual/group games. Although quantitative content analysis is usually time-consuming and costly, it gave us worthwhile results. Two psychologists separately observed and rated the behaviors of each child in all of intervention sessions. Due to the fact that the children's mother may not have been able to be absolutely objective in filling in the questionnaires, the content analysis of the video records and the interviews are of great importance.

Human's Assessment of Behaviors: In order to see the effect of the robot intervention on autistic behavior in the boys real life, a child clinical psychologist assessed both of the children's abilities one week before and one week after the intervention sessions.

Table 2. Intervention Session Schedule; the letters describe R: Robot, P: Parent, T: Therapist.

Session	Game#/Mode	Participants in Game					Description
		S1	S2	R	P	T	
1	Orientation Session						Robots showed their capabilities
2	#4/ Robot-Child						S2-I did not take part in the game
3	#2/ Robot-Child						
	#1/ Robot-Child						Interestingly, using the Mina robot
	#3/ Robot-Child						did not affect the children's performance.
4	#5/ Child-Therapist						Robot applauded them for the correct task. S1 intervened in his twin's game
5	#6/ Robot-Child						
6	#3/ Robot-Child						Game #3 was selected for session six at the request of the twins
7	#1/ Robot-Child-Child						Difficulty Level of the Tasks: Easy
8	#1/ Robot-Child-Child						Difficulty Level of the Tasks: Medium
9	#1/ Robot-Child-Parent						S2-I was absent in this session
10	#1/ Robot-Child-Child						Difficulty Level of the Tasks: Hard
11	#1/ Robot-Child-Parent						Difficulty Level of the Tasks: Medium
	#4/ Child-Parent						
	#4/ Robot-Child						
12	Farewell						

Interview with Parents: Each child had the potential to show novel social interactions in his real life which might not have been observed during our limited sessions. However, the parents spent most of their time with the children and hence could inform us if any behavior changes occurred.

3   Results and Discussions

Figures 1-4 show some intervention session snapshots, Social Robotics Lab (SUT).

Different measurement instruments were used to measure the effects of the interventions. The GARS questionnaire was completed twice by the subjects' parents: one week before the program started, and one week after the completion of the interventions. Different skills of the two participants were assessed by a child clinical psychologist one week before and one week after the robot assisted treatment. Furthermore, Quantitative Content Analysis of the video records of the sessions was done by two additional child clinical psychologists from CTAD.

Fig. 1. Imitation of the robot by the autistic twins in a group game- Game#1.

Fig. 2. Playing the xylophone in a turn-taking game- Game#6.

Fig. 3. Teleoperating Robots using the Haptic Phantom Omni robot- Game#3.

Fig. 4. Animal and friut recognition game in Child-Parent mode, Game#4.

3.1   Quantitative Content Analysis

Based on the video records the most important findings are as follows: Both subjects showed great improvement in terms of joint attention, pointing, and gaze shifting. It is important to note that the twin brothers had both received previous ABA treatments for 6 months in an autism center and also had some occupational and verbal communication therapy. Accordingly, their imitation of the robot was quite acceptable.

One of the most significant improvements in S2-I was the decrease of his autistic and maladaptive behaviors such as stereotyped behavior (specially fluttering fingers in front of the face and staring), meaningless repetition of a word/words and echo, ecstasy and inattention to the group, and engaging in solitary interests and hobbies. Additionally, S1-A showed great improvement in verbal communication, social participation, and enjoying group games.

In general, what stands out in the results obtained from the quantitative content analysis of the video records is S1-A's great improvement in social interactions, and S2-I's decrease of autistic detrimental behaviors. In other words, it can be stated that robotic group games have the potential to improve social behaviors and interactions in high-functioning autistic children, and lower the amount of stereotyped and detrimental behaviors in the low-functioning autistic children.

3.2   GARS

The subjects' mother was asked to fill in the GARS one week before and one week after the program. It should be noted that higher scores indicate higher severity of autism. The scores are presented in Fig. 5.

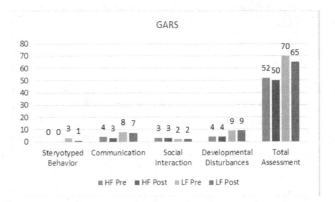

Fig. 5. GARS subscales and total scores for S1-A (HF) and S2-I (LF) in Pre- and Post- tests.

As the GARS scores indicate, S1-A did not experience a significant change in terms of the factors assessed through this questionnaire. However, he showed improvement in communication in line with the findings of the quantitative content analysis of the video records. S2-I showed more improvement especially in terms of decreased stereotyped behaviors and better social communication. This also supports the results obtained from the video records.

3.3   Human Assessment

The twin brothers were assessed by a clinical child psychologist one week before and one week after the program. The criteria for this assessment consisted of more than 25 items on self-help skills, social interaction, verbal communications, motor skills, and cognitive skills. Based on clinical observation reports presented by the assessor child clinical psychologist, S1-A showed better progress in verbal communications and joint attention skills than in other tested skills. His main difficulties were in high level cognitive skills. According to the psychologist's qualitative report, S2-I made progress in instruction perceptions and cooperation, imitation and motor skills. However, she reported that S2-I's major defects were still mental skills and verbal communications in comparison to his past.

3.4   Interview with Parents

As mentioned before, we had an interview with the twin's parents after our last clinical session. The most interesting parts of the interview are quoted as follows:

"In contrary to their ABA classes, our kids showed inexplicable interest in taking part in imitation and turn-taking games and they were super happy when leaving intervention sessions. For the first time since their birth, we have seen the twin brothers playing a meaningful turn-taking game together with their table-soccer at home. They never understood that the robots' actions occurred because of commands sent by an operator to the robots."

The mother stated,

"We believed that robotic clinical intervention would have a positive effect on our children's social interaction and their communication toward each other during these two months; however, we did not expect a miracle in their progress! Bringing my children to this different intervention program, I think I am doing my mother's duties better than the past."

The overall findings of this study showed that using robots in treatment of children with autism is potentially quite effective for both high- and low-functioning children with autism. However, the effects seem to be different for autistic children from different points on the autism spectrum. Low-functioning autistic children have more potential for improvement in imitation and joint attention skills with robot assisted therapy programs. This research was a pilot study and based on a single subject design experiment; therefore, generalizing the findings would require further research in larger-scale groups.

4   Conclusion

The results indicated that the high-functioning subject's Social Skills improved due to the two and a half month robotic treatment. In the case of the low-functioning subject, no significant improvement was observed in terms of his Social Interaction and Developmental Disturbances. His Stereotyped Behaviors, however, decreased during the course of the program. Moreover, both participants seemed to have better Communication after the treatment. As the subjects' mother claimed, for the first time in seven years she had found the twin brothers playing a meaningful game together at home. This could be due to the robot-child-brother/parent group games the subjects were involved in. Our observations showed that robot-assisted treatment has great potential to lower the severity of autism in the low-functioning subjects and improve the social skills in the high-functioning subjects. In other words, the robot-assisted clinical interventions seemed to be helpful both for low- and high-functioning children with autism. The progress rate, however, turned out to be much more significant in children with high-functioning autism. It should be noted that because of the small number of studied participants in single subject design studies, there are no strong claims on generalizing   the findings to other autistic children; however, we focused deeply on the twins' behaviors during our robotic-assisted interactions to evaluate the effectiveness of the various scenarios on the studied subjects. As one of the pioneers in using this technology in Iran [23-29], the social robotics research group has high hopes that the findings of our studies can facilitate autism therapy in Iran.

References

1. Scassellati, B., Admoni, H., Mataric, M.: Robots for use in autism research. Annual Review of Biomedical Engineering **14**, 275–294 (2012)
2. Diehl, J.J., Schmitt, L.M., Villano, M., Crowell, C.R.: The clinical use of robots for individuals with autism spectrum disorders: A critical review. Research in Autism Spectrum Disorders **6**(1), 249–262 (2012)
3. Pouretemad, H.: Diagnosis and treatment of joint attention in autistic children (in Persian). Arjomand Book, Tehran (2011)
4. Mavadati, S.M., Feng, H., Gutierrez, A., Mahoor, M.H.: Comparing the gaze responses of children with autism and typically developed individuals in human-robot interaction. In: 2014 14th IEEE-RAS International Conference on Humanoid Robots (Humanoids), pp. 1128–1133. IEEE, November 2014
5. Taheri, A.R., Alemi, M., Meghdari, A., PourEtemad, H.R., Basiri, N.M.: Social robots as assistants for autism therapy in Iran: Research in progress. In: 2014 Second RSI/ISM International Conference on Robotics and Mechatronics (ICRoM), pp. 760–766. IEEE (2014)
6. Feil-Seifer, D., Matarić, M.J.: Toward socially assistive robotics for augmenting interventions for children with autism spectrum disorders. In: Khatib, O., Kumar, V., Pappas, G.J. (eds.) Experimental Robotics. STAR, vol. 54, pp. 201–210. Springer, Heidelberg (2009)
7. Robins, B., Dautenhahn, K., Dickerson, P.: From isolation to communication: a case study evaluation of robot assisted play for children with autism with a minimally expressive humanoid robot. In: 2009 Second International Conferences on Advances in Computer-Human Interactions, ACHI 2009, pp. 205–211. IEEE, February 2009
8. Shamsuddin, S., Yussof, H., Ismail, L.I., Mohamed, S., Hanapiah, F.A., Zahari, N.I.: Initial response in HRI-a case study on evaluation of child with autism spectrum disorders interacting with a humanoid robot Nao. Procedia Engineering **41**, 1448–1455 (2012)
9. Pioggia, G., Sica, M.L., Ferro, M., Igliozzi, R., Muratori, F., Ahluwalia, A., De Rossi, D.: Human-robot interaction in autism: FACE, an android-based social therapy. In: 2007 the 16th IEEE International Symposium on Robot and Human Interactive Communication, RO-MAN 2007, pp. 605–612. IEEE, August 2007
10. Kozima, H., Nakagawa, C., Yasuda, Y.: Interactive robots for communication-care: a case-study in autism therapy. In: 2005 IEEE International Workshop on Robot and Human Interactive Communication, ROMAN 2005, pp. 341–346. IEEE, August 2005
11. Meghdari, A., Alemi, M., Pouretemad, H., Taheri, A., Mahboob Basiri, N., Roshani Neshat, A., Nasiri, N., Aghasizadeh, M.: Utilizing humanoid robots in teaching motor and social skills to children with autism. In: 3rd Basic and Clinical Neuroscience Conference, Tehran, Iran (2014)
12. Salvador, M., Silver, S., Mahoor, M.: An emotion recognition comparative study of autistic and typically developing children using the Zeno robot. In: 2015 International Conference on Robotics and Automation. IEEE, Seattle (2015)

13. Liu, C., Conn, K., Sarkar, N., Stone, W.: Online affect detection and robot behavior adaptation for intervention of children with autism. IEEE Transactions on Robotics **24**(4), 883–896 (2008)
14. Wong, C.C.Y., Meaburn, E.L., Ronald, A., Price, T.S., Jeffries, A.R., Schalkwyk, L.C., Mill, J.: Methylomic analysis of monozygotic twins discordant for autism spectrum disorder and related behavioural traits. Molecular Psychiatry **19**(4), 495–503 (2014)
15. Taniai, H., Nishiyama, T., Miyachi, T., Imaeda, M., Sumi, S.: Genetic influences on the broad spectrum of autism: Study of proband-ascertained twins. American Journal of Medical Genetics Part B: Neuropsychiatric Genetics **147**(6), 844–849 (2008)
16. Robotics, A.: The NAO robot (2013). Available at: bttp:/Ivnvw. aldebaran-roboticscomteni. (Accessed September 4, 2013)
17. http://www.robokindrobots.com/
18. Gilliam, J.E.: Gilliam Autism Rating Scale: Examiner's Manual. Pro-ed (1995)
19. Ashburner, J., Ziviani, J., Rodger, S.: Sensory processing and classroom emotional, behavioral, and educational outcomes in children with autism spectrum disorder. American Journal of Occupational Therapy **62**(5), 564–573 (2008)
20. Ahmadi, S.J., Safari, T., Hemmatian, M., Khalili, Z.: Exploring the criterion of diagnosing autism (GARS) (in Persian). Journal of Researchers of Cognitive and Behavioral Sciences **1**(1), 87–104 (2012)
21. Riff, D., Lacy, S., Fico, F.: Analyzing Media Messages: Using Quantitative Content Analysis in Research, 3rd Edition. Routledge (2014)
22. Neuendorf, K.A.: The content analysis guidebook, vol. 300. Sage Publications, Thousand Oaks (2002)
23. Taheri, A.R., Alemi, M., Meghdari, M., Pouretemad, H.R., Holderread S.L.: Clinical application of a humanoid robot in playing imitation games for autistic children in Iran. In: Proc. of the 14th Int. Educational Technology Conf. (IETC), Chicago, USA (2014)
24. Alemi, M., Meghdari, A., Ghazisaedy, M.: The Impact of Social Robotics on L2 Learners' Anxiety and Attitude in English Vocabulary Acquisition. Int. Journal of Social Robotics (2015)
25. Alemi, M., Ghanbarzadeh, A., Meghdari, A., Moghaddam, L.J.: Clinical Application of a Humanoid Robot in Pediatric Cancer Interventions. Int. Journal of Social Robotics (2015)
26. Alemi, M., Meghdari, A., Ghanbarzadeh, A., Moghadam, L.J., Ghanbarzadeh, A.: Impact of a social humanoid robot as a therapy assistant in children cancer treatment. In: Beetz, M., Johnston, B., Williams, M.-A. (eds.) ICSR 2014. LNCS, vol. 8755, pp. 11–22. Springer, Heidelberg (2014)
27. Meghdari, A., Alemi, M., Taheri, A., Pouretemad, H.: Clinical Application of Humanoid Robots in Playing Imitation Games for Autistic Children in Iran. In: 2nd Neuroscience Congress 2013 (In Persian), December 2013
28. Alemi, M., Meghdari, A., Ghazisaedy, M.: Employing Humanoid Robots for Teaching English Language in Iranian Junior High-Schools. Int. Journal of Humanoid Robotics **11**(3) (2014)
29. Meghdari, A., Alemi, M., Ghazisaedy, M., Taheri, A.R., Karimian, A., Zandvakili, M.: Applying robots as teaching assistant in EFL classes at iranian middle-schools, CD. In: Proc. Int. Conf. on Education & Modern Edu. Tech. (EMET-2013), September 28–30, Venice, Italy (2013)

Cross-Corpus Experiments on Laughter and Emotion Detection in HRI with Elderly People

Marie Tahon[1]([✉]), Mohamed A. Sehili[1], and Laurence Devillers[1,2]

[1] Human-Machine Communication Department, LIMSI-CNRS, 91403 Orsay, France
marie.tahon@limsi.fr
[2] University Paris-Sorbonne IV, 28 Rue Serpente, 75006 Paris, France

Abstract. Social Signal Processing such as laughter or emotion detection is a very important issue, particularly in the field of human-robot interaction (HRI). At the moment, very few studies exist on elderly-people's voices and social markers in real-life HRI situations. This paper presents a cross-corpus study with two realistic corpora featuring elderly people (ROMEO2 and ARMEN) and two corpora collected in laboratory conditions with young adults (JEMO and OFFICE). The goal of this experiment is to assess how good data from one given corpus can be used as a training set for another corpus, with a specific focus on elderly people voices. First, clear differences between elderly people real-life data and young adults laboratory data are shown on acoustic feature distributions (such as F_0 standard deviation or local jitter). Second, cross-corpus emotion recognition experiments show that elderly people real-life corpora are much more complex than laboratory corpora. Surprisingly, modeling emotions with an elderly people corpus do not generalize to another elderly people corpus collected in the same acoustic conditions but with different speakers. Our last result is that laboratory laughter is quite homogeneous across corpora but this is not the case for elderly people real-life laughter.

Keywords: Laughter recognition · Emotion recognition · Human-Robot Interaction · Elderly people · Cross-corpus protocol

1 Introduction

Assistive social robots must be able to decode verbal and non-verbal expressions of the user. The success of a social robot also relies on its ability to rightly interpret the inputs and properly react to them. In such a context, social signal processing designs high level cues which describe conversations, user profiles and engagement [1] during Human-Robot interactions. For example, social and interactional markers extracted from speech signal can be used to build up a user profile [2].

The authors are working under the French project ROMEO2[1] which aims at building a 140 cm high humanoid social robot. The robot is designed to be

---

[1] http://projetromeo.com

© Springer International Publishing Switzerland 2015
A. Tapus et al. (Eds.): ICSR 2015, LNAI 9388, pp. 633–642, 2015.
DOI: 10.1007/978-3-319-25554-5_63

a friendly assistant robot for non-autonomous people such as elderly people. It will be able to adapt its behavior but also to build user profiles. This project faces two main issues: social cues must be 1) robust to realistic and unseen data (spontaneous speech, noisy environments, uncontrolled acoustics), 2) adapted to non-autonomous users, especially elderly-people. The present study focuses on the decoding of two social markers of elderly-people interacting with a robot using speech input: affective states [3] and laughter [4].

The two main drawbacks of the standard corpora used in the community are the very small size of audio corpora and data variability in terms of task, speaker, age and audio environment which compromises the significance of results and improvements [5]. As a consequence, there is a critical need for data collection with end-users (with different types of speakers, ages) and real tasks for emotion recognition systems since realistic emotions could not be found in acted databases [6]. So far, very few HRI databases have been collected with diverse kind of participants: children (AIBO [7] and NAO-HR [8]), young adults (SEMAINE [9]) or visually-impaired people (IDV-HR [10]). At the present time, very few real-life emotional speech databases were recorded with elderly people: ARMEN [11] and ROMEO2 [12]. Speaker identification has been shown to be easier on elderly people than on young adults [13] because voice quality is very different between these two age groups (creaky voice, low loudness, voice pathology, etc.).

Because social markers extraction must be robust to unseen data, the present study features cross-corpus experiments which also ensures speaker independent conditions. It consists of using one corpus for modeling emotion and laughter and another one as test set. A third corpus is eventually used for development purposes. By this way, recognition rates are lower but more realistic than with cross-validation experiments. Schuller et al. [14] performed binary valence recognition with cross-corpus experiment on seven corpora. Average recalls are slightly over the random guess, from 50% to 55% with young adults. A previous experiment on children and adults voices [10] has shown a possible merging between children voices corpora, however it seems more complex to merge adult speakers and children speakers. A lot of interesting work on laughter detection in HRI has been reported in the ILHAIRE project[2]. But, as far as the authors know, none of them has been done in cross-corpus. Recently, a cross-corpus experiment on laughter was carried on three spontaneous HRI corpora [15]. The goal of the presented cross-corpus experiment is to assess how data from one given corpus can generalize to another corpus, variability being expressed under the project ROMEO2, in terms of age and acoustic conditions. Two groups are tested: one is composed of young adults recorded in laboratory conditions (OFFICE and JEMO [16] corpora), the other one is an elderly people's recorded in real-life conditions (ARMEN [11] and ROMEO2 [12] corpora).

Section 2 summarizes the acoustic features used for emotions and laughter modeling. The four French HRI databases are described in section 3.

---

[2] www.ilhaire.eu/project

Methodology and results are presented in section 4. The conclusion is drawn in the last section.

2   Acoustic Cues

In this work, many acoustic features are used to model laughter and emotions in voice. These features globally carry three kinds of information: spectral information, temporal shape information, and voice quality. Such acoustic features are supposed to carry most of emotional information [17], [18]. Several studies [19], [20], [21] found that fundamental frequency, instance duration energy and formants are also relevant for clear and well-identified laughters.

Spectral and temporal shape information is extracted using Yaafe[3] and contains perceptual features, ZCR (Zero Crossing Rate) and 24 Specific Loudness Energy bands. A total of 10 statistical coefficients (SetFunc) are calculated for each vector attribute. Prosodic and phonetic information is extracted with Praat[4]. Pitch-related features include mean, standard deviation, maximum and minimum of pitch (extracted in semitones). Intra (respectively inter) pitch is the pitch difference within a voice region (respectively across consecutive voice regions) and glissando. Formant-related features are: mean and standard deviation of the three first formants, mean and standard deviation of the formant differences $F_2 - F_1$ and $F_3 - F_2$. Micro-prosody features are: jitter, shimmer, HNR and proportion of voiced parts in the segment. More details on these acoustic features can be found in [22]. The extraction step yields a 301-dimension vector per audio segment as summarized in table 1.

Table 1. Acoustic feature set: 301 features. SetFunc is a set of 10 functionals: mean, std, slope and high-level statistics. std stands for standard deviation.

LLD	functionals	Nb func.
ZCR	SetFunc	10
Roll Off 95%	SetFunc	10
Spectral Slope	SetFunc	10
Spectral Flatness	SetFunc	10
Specific Loudness 1-24	SetFunc	24 × 10
Pitch	mean, max, min, std, intra, inter, glissando	7
Formants	mean, std F_1, F_2, F_3	6
	mean, std $F_2 - F_1$, $F_3 - F_2$	4
Micro-prosody	local jitter, local shimmer, HNR, punvoiced	4

3   Databases

The four databases used in the following cross-corpus experiments, are presented in this section. Two of them, ARMEN and ROMEO2, were collected with elderly-people during HRI (60 speakers of more than 60 years old). The other two,

---

[3] http://yaafe.sourceforge.net/
[4] http://www.fon.hum.uva.nl/praat/

JEMO and OFFICE (66 speakers of less than 60 years old), were collected during emotion games. The four corpora are in French and there is no lexical constraints. All corpora were manually segmented and annotations were performed by two expert annotators. Only consensual emotional segments are used in this work.

3.1   ROMEO2 Corpus

The ROMEO2 corpus [12] was collected in a French EHPAD (public accommodation for non-autonomous old people). 27 participants (3 men and 24 women) were recorded. A Wizard-of-Oz scheme controls the robot so that its behavior adapts seamlessly and quickly to most situations. Each interaction was split into different scenarios: greetings, reminder events (take medicine), social interaction (call a relative) and cognitive simulation (song recognition game). This corpus is very rich in terms of elderly-people speech. The study of interactions with elderly people also suppose to deal with hearing difficulties. The consensual data constitute 98 min of emotional instances.

3.2   ARMEN Corpus

The ARMEN corpus was collected in a French EHPAD within the ANR Tescan ARMEN. 77 patients from medical centers (elderly and impaired people), of which 48 men and 29 women between 18 and 90 years old participated in this data collection. The consensual data constitute about 70 minutes of the corpus. The collected data are used to explore approaches which aim at resolving the performance generalization problem of emotion detection systems run on different data [11]. In the present paper, the authors use a subset of ARMEN that contains elderly speakers only (36 speakers over 60 year old).

3.3   OFFICE Corpus

The OFFICE corpus was collected with two scenarios (jokes and emotion game) written in order to spark emotional speech and laughter. 7 speakers from 18 to 52 were recorded at LIMSI with a high-quality microphone during an interaction with the robot Nao [15]. In the "joke" scenario, the robot tells jokes in order to provoke a user's laughter. In the "emotion game" scenario, the user is asked to act emotions (anger, sadness, happiness or neutral state) so as to be recognized by the robot. The collected data contain emotional speech and affect bursts (laughter) but also noise, cough and blow (breathing or blowing). Each record was then segmented and transcribed, the number of segments per emotional class and affect bursts is summarized in table 2.

3.4   JEMO Corpus

The JEMO corpus was recorded in laboratory conditions to obtain emotions in the context of a game within the ANR Affective Avatar project. The goal of the game was to make the machine recognize an emotion (anger, joy, sadness or neutral state) without providing any context [16]. The lexical content was totally unconstrained, and the speaker tried and modulate freely their emotional expressions so as to be recognized by the system. As a result, the participants produced very expressive emotions in order to be as close as possible to the entries expected by the system. The corpus contains thus prototypical emotions produced in a "game" scenario. The total duration of the corpus is 41 minutes and it includes 59 participants (30 men and 29 women aged from 16 to 48 y. o.)

3.5   Characteristics of the Databases

The databases described previously mainly contain, besides laughter, the four Ekman's emotions: neutral state, anger, positive state, sadness. Since the ROMEO2 corpus has a very small number of anger instances, only positive, neutral states and sadness will be modeled in the present study. In the presented corpora, laughter can suppose either positive feelings (joy, amusement, etc.) or negative states (such as contempt [23], sadness or embarrassment). The number of consensual instances for each emotional class used in this work is shown in table 2.

Table 2. Content description for each data corpus. POS: positive, NEG: negative, NEU: neutral, SPE: total speech, LAU: laughter (non-speech).

Corpus	# Subjects	Age	Duration	# Segments				
				POS	NEG	NEU	SPE	LAU
ARMEN	36	60-90	68 min	308	64	1162	1534	253
ROMEO2	24	75-99	98 min	673	404	1306	2583	205
OFFICE	7	18-50	10 min	107	134	62	303	123
JEMO	59	16-48	29 min	201	307	341	849	73

ARMEN and ROMEO2 are elderly people real-life databases collected with similar acoustic environments (same EHPAD) with similar protocols, but different speakers. One is collected with a humanoid robot (ROMEO2), the other with a virtual agent (ARMEN). JEMO and OFFICE were collected in the same laboratory conditions but with different speakers and protocols.

4   Cross-Corpus Experiments with Elderly and Young People Voices

The goal of this experiment is to assess how data from one given corpus can generalize to another corpus. The inter-corpus variability that interests us here, is expressed in terms of age and acoustic conditions. Two groups are tested: one is composed of young adults, the other of elderly people. Four acoustic conditions are tested which correspond to the four corpora.

4.1 Comparison of Acoustic Features Between Elderly and Young Adults People

Elderly people speech contains tremor, pitch breaks, a lot of hesitations and fillers. Speakers' voice quality is also different from that of young adults. Figure 1 shows pitch standard deviation and local jitter distributions across corpora. While local jitter distributions are almost the same for the four corpora, F_0 standard deviation reaches significantly higher values in elderly people real-life voices than in laboratory young voices. This result shows that looking for relevant acoustic features which are good for distinguishing young and elderly people voices, is a real challenge. In the present study, age and acoustic conditions are mixed together because available corpora are not big enough to analyze all conditions separately. A previous study showed that speaker recognition was easier for elderly than for young speakers [13]. Our hypothesis is that acoustic features change more with age group condition than with acoustic environment, but further investigations are needed.

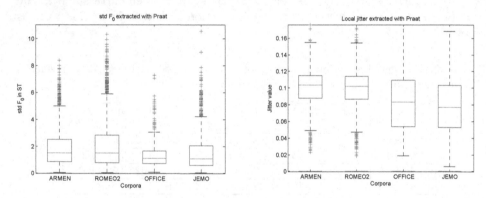

Fig. 1. Feature distributions across the four corpora: std F_0 (left), local jitter (right).

4.2 Methodology: Cross-Corpus Experiments

Emotion and laughter cross-corpus experiments are realized following the same protocol. ROMEO2 and JEMO corpora have been equally divided into three subsets: one for training (C1), one for development purposes (C2) and a last one for testing (C3). The three subsets are randomly composed so that they have the same number of segments for given class. Thus, by using JEMO or OFFICE (young subjects), ARMEN or ROMEO2 (elderly people) as train corpora and ROMEO2 or JEMO as test corpora, we actually want to check how age divergence and acoustic conditions variability affect the recognition performance. Good rates are expected when train and test data are from similar age groups, whereas lower rates are expected when train and test data belong to different age groups.

The cross-corpus protocol ensures speaker independent conditions, expect when training and testing on the same corpus (baseline). The subjects are not equally represented in each subset.

Automatic classification is performed with SVM (Support Vector Machines) using libsvm[5]. Classification was run with a linear or RBF (Radial Basis Function) kernel with parameter optimization on development subsets. Results are given in terms of UAR (Unweighted Average Recall). The confidence interval depends on the number of the tested segments N and the obtained performance UAR (equation 1).

$$Confidence = UAR \pm 1.96\sqrt{\frac{UAR \times (1 - UAR)}{N}} \qquad (1)$$

4.3   Cross-Corpus Results

The results of the cross-corpus experiments are reported in table 3. Experiments conducted with the same corpus for both training and testing (baseline condition) are reported in bold, they serve for comparison with cross-corpus experiments results.

Table 3. Cross-corpus UAR $\pm$ confidence results for emotion and laughter recognition, baseline in bold. # is the number of tested instances (a third of the initial corpus).

Train	Test			
	NEU/NEG/POS		SPE/LAU	
	ROMEO2-C3 (#793)	JEMO-C3 (#282)	ROMEO2-C3 (#862)	JEMO-C3 (#307)
ARMEN	39.2 ± 3.4	40.6 ± 5.7	67.0 ± 3.1	69.1 ± 5.2
OFFICE	44.7 ± 3.5	44.2 ± 5.8	59.2 ± 3.3	81.6 ± 4.3
ROMEO2-C1	**46.3 ± 3.5**	42.0 ± 5.8	**87.2 ± 2.2**	71.3 ± 5.1
JEMO-C1	40.7 ± 3.4	**61.2 ± 5.7**	68.3 ± 3.1	**82.3 ± 4.3**

Emotion Recognition Results. In the context of emotion recognition, the baseline performances obtained with both ROMEO2 and JEMO corpora, are the highest. Using data from the same corpus for training and testing not only yields the best performance but also seems to lead to a fairly more balanced recall between the three classes of emotion. For example, with OFFICE for training and JEMO-C3 for testing the minimum recall is reached by the neutral class at 8.9% (probably because there is very few neutral instances); with JEMO-C1 for training, the minimum recall is reached by the negative class at 57.8%. The recognition rates are lower while testing on ROMEO2 than testing on JEMO. This is due to the fact that JEMO is prototypical while ROMEO2 is real-life.

---

[5] http://www.csie.ntu.edu.tw/~cjlin/libsvm/

The recognition rate obtained with models trained on ARMEN and tested with ROMEO2-C3 was expected to be similar to the one obtained with models trained on ROMEO2-C1. This is actually not the case (UAR=39.2% with ARMEN for training and UAR=46.3% with ROMEO2-C1), thus denying our hypothesis.

Based on these results, the use of other elderly people corpus for training emotions does not help improving the performances when testing on elderly people. However, when testing on JEMO-C3, all training corpora, give similar results. Elderly people real-life corpora are much more complex than laboratory corpora, and they are significantly different one from another (between ARMEN and ROMEO2).

Laughter Recognition Results. Similar results are obtained on cross-corpus laughter recognition. The recognition of ROMEO2 (respectively JEMO) laughter is better if the model is trained with similar data (with ROMEO2-C1 sub-corpus (respectively with JEMO-C1)). However, in cross-corpus conditions, building a model with elderly people is not necessary when testing on elderly people: the best performance is obtained with JEMO-C1, then comes OFFICE and last is ARMEN.

The use of elderly people voices for training the models degrades the recognition rates (with ARMEN and ROMEO2-C1). Training a laughter model with the corpus OFFICE leads to a performance similar to the baseline. One of the main conclusions of these experiments on laughter is that JEMO and OFFICE laughters are acoustically homogeneous, however, they differ from ARMEN's and ROMEO2's. Despite the small size of the OFFICE corpus and the absence of very aged subjects, it performs better than ARMEN, be that against ROMEO2 or JEMO.

It seems that laughter is significantly different on one hand between prototypical corpora and real-life corpora, and on the other hand between two different real-life corpora. Laboratory laughter is quite homogeneous across corpora (between OFFICE and JEMO) but this is not the case for elderly people's real-life laughter.

5   Conclusion

The study gives some pilot results with elderly-people voices during interaction with a robot. Two social markers which are very useful in HRI, are detected: laughs and emotions. The automatic recognition of these two markers is presented in cross-corpus conditions. Four corpora are used in the experiments: two of them were collected with young adults (JEMO and OFFICE) and the other two with elderly people (ROMEO2 and ARMEN) during HRI. Our goal was to assess how data from one given corpus can generalize to another corpus, variability being expressed in terms of age and acoustic conditions.

Our first main result is that a comparison of acoustic features (such as F_0 standard deviation or local jitter) distributions across corpora, show clear differences between age and acoustic environments groups. This result confirms the fact that speaker recognition best performs on elder adults [13]. The second result obtained with cross-corpus experiments on emotion recognition is that elderly people real-life corpora are much more complex than laboratory corpora and they are significantly different one from another (ARMEN and ROMEO2). Surprisingly, modeling emotions with an elderly people corpus do not generalize to another elderly people corpus collected in the same acoustic conditions (here same EPHADs) but with different speakers. Our last result is that laboratory laughter is quite homogeneous across corpora (JEMO and OFFICE) but this is not the case for elderly people real-life laughter.

The complexity of elderly people real-life corpora may be due to age group and emotional behavior. This study shows that modeling emotions with an elderly people corpus do not generalize to another elderly people corpus even if the training and testing corpora are collected within the same acoustic environments and with similar scenarios. Further experiments are needed to investigate the advantage of merging elderly and young people real-life corpora or building separate models. The authors use available corpora, therefore further experiments with new HRI corpora are needed to dissociate the effect of age group on acoustic features independently from the acoustic environment.

Acknowledgments. This work was financed by the French project BPI ROMEO2. The authors thank the association APPROCHE and the EHPADs for their help in corpus collection.

References

1. Vinciarelli, A., Pantic, M., Bourlard, H., Pentland, A.: Social signals, their function, and automatic analysis: a survey. In: Conference on Multimodal Interfaces (ACM), Chania, Greece, pp. 61–68 (2008)
2. Delaborde, A., Devillers, L.: Use of nonverbal speech cues in social interaction between human and robot: emotional and interactional markers. In: International Workshop on Affective Interaction in Natural Environements (AFFINE), Firenze, Italy (2010)
3. Breazeal, C.: Emotion and sociable humanoid robots. Human Computer Studies **59**, 119–155 (2003)
4. Scherer, S., Glodek, M., Schwenker, F., Campbell, N., Palm, G.: Spotting laughter in natural multiparty conversations: A comparison of automatic online and offline approaches using audiovisual data. ACM Transactions on Interactive Intelligent Systems (TiiS) **2**(1), Article No. 4 (2012). Special Issue on Affective Interaction in Natural Environments
5. Schuller, B., Batliner, A., Steidl, S., Seppi, D.: Recognising realistic emotions and affect in speech: state of the art and lessons learnt from the first challenge. Speech Communication **53**(9), 1062–1087 (2011). Special Issue on Sensing Emotion and Affect - Facing Realism in Speech Processing
6. Batliner, A., Steidl, S., Nöth, E.: Laryngealizations and emotions: how many babushkas? In: Proc. Internat. Workshop on Paralinguistic Speech - Between Models and Data (ParaLing' 07), Saarbrucken, Germany, pp. 17–22 (2007)

7. Batliner, A., Hacker, C., Steidl, S., Nöth, E., D'Arcy, S., Russell, M., Wong, M.: You stupid tin box - children interacting with the aibo robot: a cross-linguistic emotional speech corpus. In: LREC, Lisbon, Portugal, pp. 171–174 (2004)
8. Delaborde, A., Tahon, M., Barras, C., Devillers, L.: Affective links in a child-robot interaction. In: LREC, Valletta, Malta (2010)
9. McKeown, G., Valstar, M., Cowie, R., Pantic, M., Schröder, M.: The semaine database: annotated multimodal records of emotionally coloured conversations between a person and a limited agent. IEEE Transactions on Affective Computing 3(1), 5–17 (2012)
10. Tahon, M., Delaborde, A., Devillers, L.: Real-life emotion detection from speech in human-robot interaction: experiments across diverse corpora with child and adult voices. In: Interspeech, Firenze, Italia (2011)
11. Chastagnol, C., Clavel, C., Courgeon, M., Devillers, L.: Designing an emotion detection system for a socially-intelligent human-robot interaction. In: Towards a Natural Interaction with Robots, Knowbots and Smartphones: Putting Spoken Dialog Systems into Practice. Springer (2013)
12. Sehili, M.A., Yang, F., Leynaert, V., Devillers, L.: A corpus of social interaction between nao and elderly people. In: International Workshop on Emotion, Social Signals, Sentiment & Linked Open Data, Satellite of LREC (2014)
13. Tahon, M., Delaborde, A., Barras, C., Devillers, L.: A corpus for identification of speakers and their emotions. In: LREC, Valletta, Malta (2010)
14. Schuller, B., Zhang, Z., Weninger, F., Rigoll, G.: Selecting training data for cross-corpus speech emotion recognition: prototypicality vs. generalization. In: AVIOS Speech Processing, Tel-Aviv, Israel (2011)
15. Tahon, M., Devillers, L.: Laughter detection for on-line human-robot interaction. In: Interdisciplinary Workshop on Laughter and Non-verbal Vocalisations in Speech, Enschede, Netherlands (2015)
16. Brendel, M., Zaccarelli, R., Devillers, L.: Building a system for emotions detection from speech to control an affective avatar. In: LREC, Valletta, Malta (2010)
17. Ververidis, D., Kotropoulos, C.: Emotional speech recognition: Ressources, features and methods. Speech Communication 48(9), 1162–1181 (2006)
18. Schuller, B., Batliner, A.: Computational Paralinguistics: Emotion, Affect and Personality in Speech and Language Processing. John Wiley & Sons (2013)
19. Bachorowski, J.-A., Smoski, M.J., Owren, M.J.: The acoustic features of human laughter. Journal of the Acoustical Society of America 110(3), 1581–1597 (2001)
20. Campbell, N.: Perception of affect in speech - towards an automatic processing of paralinguistic information in spoken conversation. In: International Conference on Spoken Language Processing, Jeju Island, Korea (2004)
21. Szameitat, D.P., Darwin, C.J., Szameitat, A.J., Wildgruber, D., Alter, K.: Formant characteristics of human laughter. Journal of Voice 25(1), 32–38 (2011)
22. Devillers, L., Tahon, M., Sehili, M., Delaborde, A.: Inference of human beings' emotional states from speech in human-robot interactions. International Journal of Social Robotics, Special Issue on Developmental Social Robotics (in press, 2015)
23. Schröder, M.: Experimental study of affect bursts. Speech Communication 40(1–2), 99–116 (2003). Special Session on Speech and Emotion

"Go Ahead, Please": Recognition and Resolution of Conflict Situations in Narrow Passages for Polite Mobile Robot Navigation

Thanh Q. Trinh[✉], Christof Schroeter, Jens Kessler, and Horst-Michael Gross

Neuroinformatics and Cognitive Robotics Lab, Ilmenau University of Technology,
Ilmenau, Germany
quang-thanh.trinh@tu-ilmenau.de

Abstract. For a mobile assistive robot operating in a human-populated environment, a polite navigation is an important requirement for the social acceptance. When operating in a confined environment, narrow passages can lead to deadlock situations with persons. In our approach we distinguish two types of deadlock situations at narrow passages, in which the robot lets the conflicting person pass, and either waits in a non-disturbing waiting position, or forms a queue with that person. Forthcoming deadlock situations are captured by a set of qualitative features. As part of these features, we detect narrow passages with a raycasting approach and predict the movement of persons. In contrast to numerical features, the qualitative description forms a more compact human-understandable space allowing to employ a rule-based decision tree to classify the considered situation types. To determine a non-disturbing waiting position, a multi-criteria optimization approach is used together with the Particle Swarm Optimization as solver. In field tests, we evaluated our approach for deadlock recognition in a hospital environment with narrow corridors.

Keywords: Human-aware navigation · Socially assistive robotics · Situation understanding · Polite navigation

1 Introduction and Motivation

In the ongoing research project ROREAS (Robotic Rehabilitation Assistant for Stroke Patients) [10], we aim at developing a robotic rehabilitation assistant for walking and orientation exercising in self-training during clinical stroke follow-up care. The robotic rehab assistant is to accompany inpatients during their walking and orientation exercises, practicing both mobility and spatial orientation skills. The test site is a complex U-shaped rehabilitation center and accommodates more than 400 patients. The operational environment is highly dynamic. Patients and staff working in the patients' rooms are moving in the corridors and in the public areas, many of them using walking aids. Moreover, beds, supply and cleaning carts, or wheel-chairs are occupying the hallways, resulting in more or less restricted space conditions at some times.

© Springer International Publishing Switzerland 2015
A. Tapus et al. (Eds.): ICSR 2015, LNAI 9388, pp. 643–653, 2015.
DOI: 10.1007/978-3-319-25554-5_64

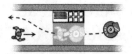

(b) Forthcoming person

(a) Our robot in a typical narrow passage

(c) Same Direction

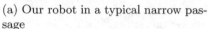

Fig. 1. Our robot in a typical narrow passage at our test site, the m&i rehabilitation center in Bad Liebenstein, and schematic depiction of the considered conflicting situations caused by narrow passages.

The self-training is mostly performed on the corridors. Due to the structure of the building or objects standing in the hallways, some parts have limited lateral space, forming a narrow passage which permits movement only in one direction at a time (Fig. 1(a)). Moving in such a restricted space imposes deadlocks in narrow passages. Since a polite and attentive navigation is an important requirement for an assistive robot, these situations must be predicted to trigger a proactive reaction of the robot. In this work, we distinguish two types of deadlocks: (i) deadlocks caused by a forthcoming person and (ii) deadlocks occurring when the robot and a person are entering the narrow passage in the same direction. In Fig. 1(b)(c) schematic examples of these situations with a narrow passage typical to the operation area are depicted. Both cases have different resolution strategies, but basically result in a "give way" behavior. To be more specific, in case of type (i) deadlocks, the robot is driving to a waiting position to give way to the forthcoming person, whereas type (ii) deadlocks are resolved by forming a queue and following the person through the narrow passage.

When a deadlock situation with a forthcoming person is predicted, the robot needs to wait until the narrow passage has been cleared. In the wait state, the robot should position itself in a non-obstructive manner aside. This has a twofold effect: First, in an already restricted environment, the position to be chosen should not hinder the movement of the person and ease the deadlock elimination. Second, the movement to a waiting position signals the approaching person the intention of the robot to give way. Additionally, the narrow passage must be observable from the robot's waiting position, since it must be able to recognize when the narrow passage is free to be entered. In our approach, we formulate the problem of finding a suited waiting position as a multi-criteria optimization problem and use a Particle Swarm Optimization (PSO) [16] as solver.

The main contributions of this paper are: (a) a new approach for detecting narrow passages by means of qualitative features capturing the spatial relationships of conflicting situations, (b) an efficient method for predicting space conflicts in narrow passages, (c) a fast approach for finding non-obstructive waiting positions based on multi-criteria optimization, in order to support the conflict resolution.

2   Related Work

The recognition and handling of deadlock situations in narrow passages has been explicitly taken into account in [3]. In this approach however, deadlock situations are only recognized and handled in a reactive manner, when the path is blocked by a person standing in a narrow passage. For a rehab center with patients having reduced mobility and using walking aids, a more predictive recognition is necessary to proactively avoid deadlocks.

In our approach, a set of qualitative spatial features is used to recognize deadlocks. Such a qualitative description is also used in [12] to evaluate the movements of a robot and a person. Particularly, the Qualitative Trajectory Calculus is utilized. In [14] and [24], Inverse Reinforcement Learning is employed to learn a navigation behavior in crowds based on features capturing the environment. However, narrow passages are not explicitly described.

To assess the situation from a set of situation describing features, the relationships between them must be described. The techniques for finding these relationships belong to the field of Data Fusion (DF). In the robotic field, only few papers on DF for situation assessment have been published so far. A general framework for situation assessment is described in [1]. Situations are learned with an extensible Markov Model from a set of feature sequences describing the environment. In contrast to the robotics research, the field of Advanced Driver Assistance Systems provides a wider range of publications dealing with situation understanding. The common applications are the recognition of a driver's driving maneuvers, driving behaviors at intersections, and the recognition of unusual driving behaviors. For situation assessment, often Hidden Markov Models [21], Bayesian Networks [19][9], and rule-based techniques [20] are used.

So far, only explicit recognition techniques have been mentioned. In the robotics field, there also exists a category of implicit techniques. The main subject of these techniques are the usage of spatiotemporal planners and the incorporation of long-term human motion predictions to avoid deadlocks. Although the deadlock problem can be solved with this approach, there is still a need for situation assessment, when a human-robot communication is required to interact with the person when a deadlock situation occurs. Since implicit techniques aim at generating collision free trajectories, deadlock situations are not explicitly recognized. The implicit techniques can be distinguished by the used planning algorithms and human motion prediction methods. The most widely used planners are A^* [13][2] and Rapidly-Exploring Random Trees [18][23]. Human motion prediction methods can be categorized into learning-based and reasoning-based [17] ones. Learning-based approaches learn a predictor from a given training set of trajectories [2], whereas reasoning-based approaches make predictions based on a given motion model for the person [7].

Since our focus lies in the application of qualitative features for describing deadlock situations and an efficient collision prediction method for narrow passages, we only use a simple linear motion prediction and a rule-based approach for situation assessment. However, our field experiments demonstrate that these methods result in a relatively good recognition performance for this hospital environment.

3   Robot Platform ROREAS

Our robot has a relatively small size of 45 x 55 cm footprint and a height of 1.5 m (Fig. 1(a)). The drive system is a differential drive with a castor on the rear and allows a maximum driving speed of up to 1.4 m/s. The robot's sensory system consists of two SICK laser range finders, three Asus RGB-D cameras, and a panoramic color vision system mounted on the top of the head. For person perception, we utilize a probabilistic multi-hypotheses and multi-cue tracking system based on a 7D Kalman filter [25]. It tracks the position, velocity and upper body orientation of multiple persons. As detection modules, we are using a face detector, a motion detector, and an upper-body shape detector. Additionally, generic distance-invariant laser-scan features are used to detect legs and persons with mobility aids (i.e. crutches, walkers and wheelchairs) [26]. With these detection modules we are able to track persons up to a distance of 8 m. To safely navigate in dynamic environments, the positions of obstacles need to be determined. To this end, we use a generic mapping system which is able to process 2D laser-scan and 3D information of the robot's surroundings [5]. The navigation system consists of a Dynamic Window Approach (DWA) [8] guided with an E^* planner [22]. Furthermore, multiple DWA objectives are utilized to respect the personal space of bystanders and to achieve a right-hand traffic behavior. The complete robotic system was developed with MIRA [6]. For a more detailed overview of our robot system see [10].

4   Deadlock Recognition

The deadlock recognition is formulated as classification problem. As argued before, we distinguish two types of conflict situations depending on the movement intention of the person (Fig. 1(b)(c)). Both situations have different resolution strategies. In case of a deadlock with a forthcoming person, the robot drives to a non-disturbing waiting position aside. These situations are labeled as $Waiting$. In case of a deadlock with a person moving in the same direction as the robot, a queue is formed with the person. These situations are labeled as $Queuing$. Using the resolution strategies as class labels and adding the class $Proceeding$ for uncritical situations, the deadlock recognition problem can be formulated as a classification problem over situations. In finding a classifier $C : S \rightarrow \{Waiting, Queuing, Proceeding\}$ with S as the set of all situations, we can recognize the considered deadlock situations. Our recognition approach can be described as a sequential processing chain consisting of four distinct steps:

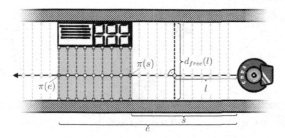

Fig. 2. The narrow passage detection. The circles depict the sample points on the robot's path for which a ray perpendicular to path is cast. π is the trajectory of the path parametrized over arc length. The highlighted path points $\pi(s)$ and $\pi(e)$ are the start and end point of the narrow passage with s and e as their corresponding arc length. Thus the arc length interval of the narrow passage is $\mathbf{N} = [s, e]$. The quadratic points are the boundary points of the narrow passage's polygon. The polygon itself is marked red.

(1) the detection of the narrow passage, (2) the extraction of qualitative spatial features describing the situation around the narrow passage, (3) the prediction of space conflict, and (4) the classification of the considered situations. In all the processing steps we assume a planar operational space.

4.1 Narrow Passage Detection

Narrow passages are detected by first calculating normals perpendicular to the planned path for a finite set of points sampled from the path (Fig. 2). The normals are determined analytically. To this end, we utilize a spline interpolation scheme to derive a trajectory $\pi : \mathbb{R} \to \mathbb{R}^2$ parametrized over the path's arc length. For each normal, rays are cast in the navigation map until hitting an obstacle in the 2D occupancy map. The total length of the resulting rays indicate how much free lateral space is available at a given path point. With these distances, a narrow passage can be described as a continuous path section given by an arc length interval $\mathbf{N} \subset \mathbb{R}$, where the section's maximum distance is smaller than a given threshold. Another useful form for reasoning about the spatial relationship around the narrow passage is its bounding polygon. To construct the polygon, the sampled path points in the narrow passage, which were used to calculate the normals, are translated along their cast rays to get the points on the polygon's boundary.

4.2 Qualitative Spatial Features

The common method to describe spatial relationships is to use geometrical measures. For humans these quantitative measures are a rather unintuitive way for describing spatial relationships. Instead, they use a qualitative abstraction and group similar measurement values to an intuitive representation [12]. For example, a person is more likely to describe another person as standing behind

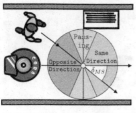 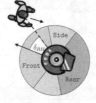 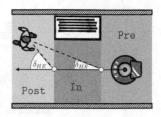

(a) Movement Direction   (b) Orientation   (c) Narrow passage positions

Fig. 3. Qualitative spatial features extracted from the position of the narrow passage, the persons and the robot.

him/her, than to give the exact orientation angle. We use this insight to reduce our geometrical feature space to a more compact space. In this compact space, simple rules are employed to distinguish the considered situations. Thus, we overcome the need for a learning approach to collect a dataset, which must contain many instances of the geometrical features. We use the following features to describe deadlock situations:

Movement Direction. In Fig. 3(a) an illustration of this feature is depicted. This feature describes the movement direction of a person relatively to the movement direction of the robot at either the start or the end of the narrow passage. We distinguish three different directions *Opposite*, *Same* and *Passing*. Additionally, a fourth value *Standing* is introduced for a person with no movement.

Orientation. This feature represents the position of a person relatively to the robot. The feature can take the values in *Front*, *Rear* and *Side*. To determine this feature value, the angle between the robot's movement direction and the connection line of the person to the robot is used. See also Fig. 3(b) for an illustration.

Narrow Passage Position. Given a narrow passage, this feature describes the positions of the robot or a person relatively to the passage. We define three sub-areas representing the *Pre-*, *Post-* and *In*-area of the narrow passage (Fig. 3(c)). The reference orientation is given by the movement direction of the robot. To determine this feature value for a person, we assume a person to be disc-shaped. The intersection area of the person with the narrow passage's polygon and the relative orientation to the narrow passage is utilized to reason about the sub-area.

4.3 Space Conflict Prediction

A narrow passage can be understood as a rail predefining a movement flow. Persons entering the narrow passage can only move along the given direction.

Moreover, a point in the narrow passage can only be occupied by one person or the robot at the same time. Thus, without losing information, we describe the movement of the person and the robot through the narrow passage as a trajectory $\tau : \mathbb{R} \rightarrow \mathbb{R}$ parametrized over time and having function values in the passage's arc length interval on the planned path. Using a linear model, the movement of the person through the narrow passage can be predicted. By assuming a linear motion model the trajectory of the robot and the person can be described as linear functions. Thus, predicting space conflicts can be reduced to finding the intersection point of two linear functions.

4.4   Situation Classification

For each perceived person we extract the qualitative spatial features and predict possible space conflicts. Thereafter, a decision tree (DT) is used to classify the situation for each person separately. In Fig. 4(a) a coarse view on the DT is depicted. The root of the DT represents common preconditions for the conflicting situations. Only when these conditions are fulfilled, further evaluations are considered. The preconditions consist of the check for the presence of a narrow passage and a space conflict with a person. Furthermore, an activation area around the narrow passage is constructed, permitting further evaluations only when the robot and conflicting person stay inside this area. Upon the fulfillment, the evaluation is redirected to the subtrees according to the movement direction of the person. The subtrees for standing persons and persons moving in opposite direction are dedicated for the separation of the *Waiting* class from the *Proceeding* class, whereas the subtree for person moving in the same direction separates *Queuing* from *Proceeding*. The main idea of the subtrees is to use the qualitative

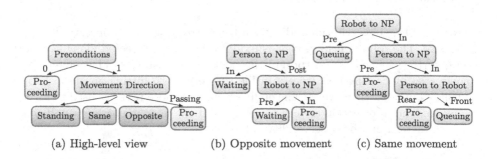

(a) High-level view   (b) Opposite movement   (c) Same movement

Fig. 4. (a) The high-level view of the decision tree. The orange nodes represent the classification results; the blue nodes are the decision nodes; and the red nodes contain a subtree. (b) The subtrees for persons moving in the opposite direction and (c) same direction to the robot. The decision nodes use the relative position of the robot or persons to the narrow passage (NP) and the relative position of the person to the robot to chose the appropriate class.

spatial features describing the relative positions of the person and robot (Sec. 4.2) for arguing about the class membership. In Fig. 4(b)(c), the subtrees for persons moving in the opposite and the same direction as the robot are depicted.

5   Finding Non-obstructive Waiting Positions

5.1   Multi-criteria Optimization

The problem of finding a proper waiting position is formulated as a multi-criteria minimization problem. The search space consists of points $(\mathbf{x}, \phi)$ with $\mathbf{x} \in \mathbb{R}^2$ a position in a planar world and $\phi \in [0..2\pi)$ an orientation defining the robot's viewing direction. The information about the obstruction and the passage's observability of a pose is encoded in the optimization function

$$f(\mathbf{x}, \phi) = \alpha \cdot c_{dist}(\mathbf{x}) + \beta \cdot c_{observe}(\mathbf{x}, \phi) + \\ \gamma \cdot c_{wall}(\mathbf{x}) + \eta \cdot c_{social}(\mathbf{x}) \tag{1}$$

through linear combination of the criteria (i) driving distance to a position c_{dist}, (ii) observability of the narrow passage $c_{observe}$, (iii) distance to walls c_{wall} and (iv) social distance to persons c_{social}. Since our criteria are non-linear or non-differentable, we use the Particle Swarm Optimization to find the minimum.

5.2   Optimization Criteria

Driven Distance. Given the representation of the environment as a grid map, this criterion penalizes positions which are far away from the robot, thus minimizing the time to drive to the selected position. This is important, since the person might get irritated about the robot's intention to wait, if the waiting position is chosen too far. The driven distance to a position is determined with Dijkstra's algorithm [4]. Note, that for unreachable positions, Dijkstra's algorithm results in an infinite distance.

Observability. This criterion indicates if the narrow passage is observable at a given pose. The robot's field of view is modeled as a cone directed along ϕ. The cone is further refined to incorporate the position of obstacles. The refinement is conducted by casting rays from $\mathbf{x}$ inside the cone until hitting an obstacle or the cone's boundary is reached. The ending points of those rays are used to form a polygon. The intersection area of this cone and the narrow passage's polygon is used to determine the observability value.

Distance to Walls. Imagine a robot moving in a hallway and the robot waits in the middle of the hallway. This is a rather unintuitive signal and depending on the width, the person might have to squeeze around the robot. A more intuitive way is to let the robot wait near the walls. This is more explicit and provides the person more free space to pass the robot. To determine the distance to walls a distance transform algorithm [15] is performed on the environment's map. The resulting image allows lookup of the distance to the next wall for each potential waiting position.

Table 1. The confusion matrix containing the experimental results. Each instance in a row correspond to one situation class actually occurred during the experiments. The columns correspond to the predictions made by our approach.

		Predicted class		
		Proceeding	Queueing	Waiting
Ground Truth	Proceeding	96	19	7
	Queueing	0	12	3
	Waiting	1	1	18

Social Distance. Every person has a social distance s/he keeps to others when s/he has no intention of interaction [11]. Assuming that a person is represented as a disc-shape and has the position $\mathbf{x}_h$, the social distance of a person is modeled as a Gaussian centered at $\mathbf{x}_h$. Then for a position $\mathbf{x}$, the social distance criterion is the summed function values of all the perceived persons' Gaussian evaluated at $\mathbf{x}$.

6   Experimental Results of Field Tests and Outlook

In extensive field tests we evaluated our approach for deadlock recognition in the "m&i Fachklinik" rehabilitation center in Bad Liebenstein with our robot platform. The tests were conducted over two days. During the tests, we let the robot autonomously drive between different goals and floors of the building. For evaluation, an external observer accompanied the robot and manually counted the decisions taken by our approach, but always from far distance to prevent any distraction. In total, a distance of 4,700 m was traveled. During the first 4,000 m, only bystanders were crossing the robot's way. These bystanders were staff members, patients, or guests, which randomly occurred on the hallway and had no knowledge about the robot's deadlock recognition. In this test run, we observed that most bystanders were considerate towards the robot and let it first pass. Only some bystanders took the initiative resulting in the robot to give way. Hence during the remaining 700 m, we additionally informed two test subjects with normal mobility about the deadlock recognition, but without insight to the technical details, and instructed them to actively obstruct the robot by crossing its way. Thus, we obtained more variability in the deadlock situations and a better assessment of the overall robustness. How they crossed the robot's way were up to the test subjects. Each bystander or test subject which crossed the robot's way in a 2 m radius was considered as potential source of a deadlock and contributed to one instance in the confusion matrix shown in Table 1.

6.1   Discussion and Future Works

In total 157 persons were potential sources of deadlocks. From these 157 persons, 35 persons caused deadlock situations at narrow passages with further division in 15 queueing and 20 waiting situations. Out of these 35 deadlock situations only one was misclassified as uncritical situation (true positive rate of 97 %). From these 34 correctly classified deadlock situations, 30 were assigned to the correct deadlock type (accuracy of 88 %). However, 26 of 122 uncritical situations were classified as deadlocks (false positive rate of 21 %).

The performance of the deadlock recognition strongly depends on the accuracy of the situation describing features which in turn depends on the person tracker, the narrow passage detection, and the space conflict prediction. Analysis of the false positives revealed that 19 of 26 false positives are caused by false detections of the person tracker. In these cases, the deadlock recognition assumed to have a conflicting situation with a person, even though there was no person present at all. The remaining 7 false positives were caused by dynamic obstacles, e.g. moving persons or objects moved by persons. If a dynamic obstacle causes a narrow passage, the narrow passage itself also has a movement. Since the narrow passage detection uses the navigation map which currently is not yet able to distinguish dynamic obstacles, this movement could not be considered in the space conflict prediction and leads to false predictions. Surprisingly, the linear motion model used in the space conflict prediction and neglecting the uncertainty in the qualitative spatial features only have little influence on the recognition performance. This can be explained by the structure of the test site which mainly consists of long and narrow corridors. In this environment, the movement and space is already restricted. Thus, a simple linear motion model leads to good predictions, and the extracted features have only little uncertainty.

In future works, we are going to reduce the false positive rate by improving the person tracker and the narrow passage detection. To be applicable for complex environments, a more elaborate motion prediction and the consideration of the uncertainty in the recognition process is needed. Furthermore, the human perception about the robot's behavior need to be evaluated more specifically to get better insights to the courtesy of the robot. These evaluations should also be conducted over a longer time period, when the bystanders get used to the robot and renounce to act courteously in front of deadlocks.

References

1. Beck, A., Risager, C., Andersen, N., Ravn, O.: Spacio-temporal situation assessment for mobile robots. In: Proc. Conf. FUSION, pp. 1–8 (2011)
2. Bennewitz, M., et al.: Learning motion patterns of people for compliant robot motion. The International Journal of Robotics Research **24**(1), 31–48 (2005)
3. Dakulovic, M., et al.: Avoiding deadlocks of mobile robots in narrow passages of environments populated with moving obstacles. In: AIM, pp. 936–941 (2011)
4. Dijkstra, E.W.: A note on two problems in connexion with graphs. Numerische Mathematik **1**(1), 269–271 (1959)
5. Einhorn, E., Gross, H.M.: Generic 2d/3d slam with ndt maps for lifelong application. In: European Conference on Mobile Robots (ECMR), pp. 240–247 (2013)
6. Einhorn, E., Langner, T., Stricker, R., Martin, C., Gross, H.M.: MIRA - middleware for robotic applications. In: Proc. IEEE/RSJ IROS, pp. 2591–2598 (2012)

7. Ferrer, G., Sanfeliu, A.: Proactive kinodynamic planning using the extended social force model and human motion prediction in urban environments. In: Proc. IEEE/RSJ IROS, pp. 1730–1735 (2014)
8. Fox, D., Burgard, W., Thrun, S.: The dynamic window approach to collision avoidance. IEEE Robotics Automation Magazine 4(1), 23–33 (1997)
9. Gindele, T., et al.: A probabilistic model for estimating driver behaviors and vehicle trajectories in traffic environments. In: ITSC, pp. 1625–1631 (2010)
10. Gross, H.M., Debes, K., Einhorn, E., Mueller, S., Scheidig, A., Weinrich, C., Bley, A., Martin, C.: Mobile robotic rehabilitation assistant for walking and orientation training of stroke patients: a report on work in progress. In: Proc. IEEE Conf. on SMC, pp. 1880–1887 (2014)
11. Hall, E.T., et al.: Proxemics [and comments and replies]. Current Anthropology, pp. 83–108 (1968)
12. Hanheide, M., et al.: Analysis of human-robot spatial behaviour applying a qualitative trajectory calculus. In: Proc. IEEE Symp. RO-MAN, pp. 689–694 (2012)
13. Hart, P., et al.: A formal basis for the heuristic determination of minimum cost paths. IEEE Trans. on Systems Science and Cybernetics 4(2), 100–107 (1968)
14. Henry, P., Vollmer, C., Ferris, B., Fox, D.: Learning to navigate through crowded environments. In: Proc. IEEE ICRA, pp. 981–986 (2010)
15. Jähne, B.: Digital Image Processing. Engineering Pro collection. Springer, Heidelberg (2005)
16. Kennedy, J., Eberhart, R.: Particle swarm optimization. In: Proc. IEEE Int. Conf. on Neural Networks, vol. 4, pp. 1942–1948 (1995)
17. Kruse, T., Pandey, A.K., Alami, R., Kirsch, A.: Human-aware robot navigation: A survey. Robotics and Autonomous Systems 61(12), 1726–1743 (2013)
18. LaValle, S.M., Kuffner, J.J.: Randomized kinodynamic planning. The International Journal of Robotics Research 20(5), 378–400 (2001)
19. Liebner, M., et al.: Driver intent inference at urban intersections using the intelligent driver model. In: Proc. IEEE IV Symp., pp. 1162–1167 (2012)
20. McCall, J., Trivedi, M.: Performance evaluation of a vision based lane tracker designed for driver assistance systems. In: Proc. IEEE IV Symp., pp. 153–158 (2005)
21. Meyer-Delius, D., et al.: Probabilistic situation recognition for vehicular traffic scenarios. In: Proc. IEEE ICRA, pp. 459–464 (2009)
22. Philippsen, R., Siegwart, R.: An interpolated dynamic navigation function. In: Proc. IEEE ICRA, pp. 3782–3789 (2005)
23. Svenstrup, M., et al.: Trajectory planning for robots in dynamic human environments. In: Proc. IEEE/RSJ IROS, pp. 4293–4298 (2010)
24. Vasquez, D., et al.: Inverse reinforcement learning algorithms and features for robot navigation in crowds: an experimental comparison. In: IROS, pp. 1341–1346 (2014)
25. Volkhardt, M., Weinrich, C., Gross, H.M.: Multi-modal people tracking on a mobile companion robot. In: European Conference on Mobile Robots (ECMR), pp. 288–293 (2013)
26. Weinrich, C., Wengefeld, T., Schroeter, C., Gross, H.M.: People detection and distinction of their walking aids in 2d laser range data based on generic distance-invariant features. In: Proc. IEEE Symp. RO-MAN, pp. 767–773. IEEE (2014)

Study on Adaptation of Robot Communication Strategies in Changing Situations

G. Trovato[1(✉)], J. Galeazzi[2], E. Torta[2], J.R.C. Ham[2], and R.H. Cuijpers[2]

[1] Graduate School of Advanced Science and Engineering, Waseda University,
#41-304, 17 Kikui-Cho, Shinjuku-Ku, Tokyo 162-0044, Japan
gabriele@takanishi.mech.waseda.ac.jp
[2] Department of Industrial Engineering & Innovation Sciences,
Eindhoven University of Technology, Eindhoven, The Netherlands

Abstract. With the increasing demand for socially assistive robots in various domains, it becomes important to make robots that are able to start communication and attract people's attention whenever necessary. Although verbal and non-verbal ways of communication have been studied before, adjusting them to changing situations of the environment is still a challenge. In fact, as the environment in which the interaction takes place mutates, and individual differences exist in human perception, some channels of communication may be less effective than others. This paper describes a study in which a Nao robot tries to use different behaviours to start an interaction with a human counterpart, in four changing environmental conditions (loudness, ambient luminance, distance and angle). The robot determines the state of those conditions and through an online learning process is capable of choosing the most effective behaviour for each individual taking into account individual preferences. The findings of this paper will be useful to make a robot initiate a communication successfully.

Keywords: Human-Robot Interaction · Non-verbal communication · Context awareness · Attracting attention · Individual differences

1 Introduction

The increasing demand for care for both elderly and children in combination with a shortage of caregivers has led to an increase in interest in socially assistive robots (SARs) [1]. These types of robots may for example be operational in hospitals, elder care centres, schools and homes, to keep humans company. Because of their potential widespread use, it is important that these robots are well attuned to their end-users, and that they understand both the social and the physical environment in which interactions take place.

Most studies on interaction between a robot and a person take place in a laboratory setting. Under these circumstances the robot receives the person's full attention. However, this setup is not very realistic, as in a more natural situation the person may be engaged in other tasks or distracted. Thus, the robot will have to attract attention by using various communication cues, for purposes like reminding elderly people to take medicines.

© Springer International Publishing Switzerland 2015
A. Tapus et al. (Eds.): ICSR 2015, LNAI 9388, pp. 654–663, 2015.
DOI: 10.1007/978-3-319-25554-5_65

Communication cues can be conveyed through different channels, mainly distinguished in the categories of verbal and non-verbal. The latter category includes kinesics, haptics, proxemics, co-speech gestures and other paralinguistics [2, 3]. Kinesics, commonly known as the study of body language, comprehends different ways of initiating communication, such as eye contact, posture, body gesture and facial expression [4]. Robot can use some among these channels depending on their morphology: more human-like robots are able to use cues which are similar to humans, while robots with a more unique morphology can exaggerate cues (like in the case of KO-BIAN's facial expressions [5]) , or even use artificial cues (like communicating by changing the colour of a part of its own body or face [5]).

Experiments on attracting attention can be specifically focused on robot abilities (such as the classification of human movements [6], or the estimation of the attention level [7]), or mainly on a single communication channel such as gaze [8] or proxemics   [9]. Torta et al. [10] tested the effectiveness of the communication cues blinking, eye contact, waving and speaking for attracting attention and found that auditory cues are more effective than purely visual cues. In a follow-up study [11], bimodal cues (such as speaking in combination with waving) were examined, and the users' average reaction times were found to be longer compared to single cues.

What these previous experiments did not take into account is that, in a real environment, the efficiency of cues and their selection depend on the context the robot and person share. The setting in which the interaction takes place may influence the efficiency of the used cues. For example, the amount of light may influence all the cues related to the visual channel, while noise may hamper the auditory channel. Furthermore, there are factors related to the human that can also be taken into account (such as whether the human is looking at the robot), social rules depending on culture, and individual preferences. For this reason, robots need to learn to adapt to constantly changing environments, which sometimes are only partially observable [1], and customise their behaviour for the specific user.

The goal of the current research is to develop robot's ability to start a communication while coping with different situations: different users (with different characteristics), and different environments (with different and changing characteristics). In order to achieve this, we investigate the effectiveness of various behaviours for each environment in terms of participants' ratings and response time. A learning process and autonomous sensory input are used to let the robot adapt to the environment and modify its behaviour selection. The effect on the user's perception of the robot is evaluated. Testing attention-seeking behaviours is not new; however, a study comprehensive of dynamically changing situations is novel.

In the present experiment, the robot performs the behaviours chosen by a learning algorithm for a given the detected environment at random intervals, while the participant is watching a video clip on a TV screen.

The rest of the paper is organized as follows: in section 2 we describe the protocol of the experiment; in section 3 we show the detailed results and we discuss them in section 4; in section 5 we conclude the paper and outline future works.

2   Experimental Procedures and Materials

2.1   Living Room

The experiment takes place in a mimicked living room (Fig.1). Participants are seated on a couch, with a television screen in front of them and are holding a keyboard. Nao is positioned at four different places during the trials, but always to the right side and facing the participant. The experiment is supervised by the two operators in an adjacent control room.

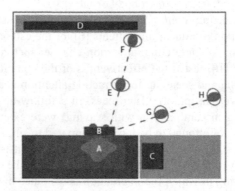

Fig. 1. Room setup: A: participant; B: keyboard; C: laptop with questionnaire; D: TV. Nao is positioned at 10° angle and 1 m (E) or 2 m distance (F), or at 70° angle and 1 m (G) or 2 m (H).

Room conditions are defined as a random sequence and the participant never undergoes the same room condition twice through different trials. There are 16 possible environmental states, given that each variable (loudness, ambient luminance, distance and angle) has two levels.

- Loudness of the TV speakers can be either high or low.
- Ambient luminance can change as the lights in the room are turned on or off.
- Distance is manipulated placing the robot either 1 m or 2 m from the participant.
- The robots' direction of approach is manipulated by either placing the robot at 10 degrees (close to the TV) or 70 degrees (in the peripheral of the participants' view) with respect to the participants' view towards the television.

2.2   Robot Behaviours

Given Nao's capabilities, we made five different behaviours to attract the attention, each involving different communication channels. Each of them lasted about 10 s.

- Waving (visual + auditory): Nao waves its right arm.
- Blinking (visual): Nao blinks with its LED's around the eyes. The LEDs switch between bright white (RGB values [255,255,255]) and of (RGB values [0,0,0]) every 0.5 s for a period of 10 s.

- Beeping (auditory): Nao produces and repeats a "beeping" sound using a sine wave with a frequency of 1000 Hz, an amplitude of 20%, a pan of 1 (to the left) and a duration of 1 s.
- Walking (proxemics + visual +auditory): Nao walks 30 cm towards the participant.
- Stop TV (action influencing visual + auditory): the robot, raises its left arm forward and the video is stopped.

2.3   Questionnaires and Individual Learning

The used learning algorithm is an adaptation of the algorithm introduced in [12], and is based on Naive-Bayes classifier. The algorithm takes the variables of room condition as input and gives the best behaviour as output.

Feedback from participants' questionnaires is used to make the algorithm adapt to individual preferences. The main difference from [12] consists in training the algorithm separately for each participant in order to catch individual differences.

The questionnaire is structured as follows. At the end of each trial the participant is asked to rate the effectiveness of the behaviour performed. The expected effectiveness of the other behaviours are also rated for the given room conditions. A 5 point semantic differential scale is used: from 1 (not effective) to 5 (effective). A rating of 1 for the performed behaviour should be chosen if the participant did not notice the robot calling at all. If the rating is 3 or lower, the participant has to suggest which room conditions would have been appropriate for the displayed behaviour. For example, after seeing a waving behaviour with lights dimmed, a participant could think that it would have been more effective with lights on.

These preferences serve as input of the learning algorithm. They are turned into weights w_{rc} through the reward r, which can be one of the values [-1, -0.5, 0, 0.5, 1] in correspondence to the rating [1, 2, 3, 4, 5]. The learning rule is shown in Equation 1: the reward is multiplied by a learning rate α (starting from 1 and decreasing with n. of visits) and a modifier d, which keeps the weights in the range [0, 1]. The modifier d either takes the value $1-w(n)$ for increments ($r > 0$) or $w(n)$ for decrements ($r < 0$).

$$w_{rc}(n+1) = w_{rc}(n) + \alpha * r * d \tag{1}$$

2.4   Face Detection and Distance Measurement

For exploratory purposes, for half of the participants, the distance from the participant and the angle of approach are detected by Nao's sensor input. A face detection script using OpenCV and Nao's camera registers the presence of a face. The presence of a face corresponds to the 10° angle condition, since the algorithm is only able to detect faces from the front. The absence of a face corresponds to the 70° condition. Moreover, the script draws a border around the face: the dimensions of this border are used to calculate the distance d as in Eq. 2, where f is the focal length, h' the height of the head in the image, and h the standard height of a human head.

$$d = h * (f / h') \tag{2}$$

A value of 19.5 cm is used as standard, based on the body dimensions of Belgian population from [13]. A border with a smaller width and height indicates that the person is at a bigger distance from the robot, and a large border indicates the opposite.

The face detection script is run for a period of 15 seconds as we want to make sure that the presence of a face is not due to the participant looking towards the robot but due to the position of the robot. Therefore, the script should not indicate the presence of a face when it is positioned at a 70 degree angle, even if the participant occasionally looks in the direction of the robot.

2.5   Experimental Procedure

In total 23 participants within the age range of 18 to 50, without visual or hearing impairments, were included in the experiment, divided in two groups of 12 (Group A, without involving face detection) and 11 (Group B, using face detection script). The experiment took approximately 60 minutes, divided in 12 trials for each participant. Once the participant sat on the couch, the protocol below was followed for each trial:

1. Room conditions are updated according to the schedule (a randomised list of all the combinations).
2. In case Nao's sensorial input is used, face detection and distance measurement are provided and used in the algorithm overriding the previous data.
3. The most appropriate behaviour is chosen by the learning algorithm, and inserted as parameter in a Python script.
4. The Python script starts an approximately 30 seconds long portion of a video clip of a science documentary.
5. Between 5 and 15 seconds before the end of the clip, the Python script orders the robot to perform the chosen behaviour.
6. The participant has been instructed to concentrate primarily on the video clip, but also to press the space bar on the keyboard whenever noticing that the robot intends to communicate. The video clip is stopped on key press. If the participant does not notice, the video clip runs till the end.
7. The participant is asked to complete the questionnaire as described in Section 2.3.
8. Their feedback is entered by the operator into the learning algorithm, which updates the weights accordingly.

3   Results

3.1   General Behaviour Preferences and Effectiveness

Overall the ratings of the behaviours are similar except for blinking. For most room conditions the behaviour stopping TV was rated as the most effective, while blinking was the least favourable in all conditions.

A second measure of effectiveness was the reaction times recorded during every trial, which are similar but not completely in line with the effectiveness scores. Waving takes the least amount of time to respond, followed by Walking and only then

Stopping TV. This can be seen in Figure 2 as both measurements are compared: ratings by the participants refers to the scale on the left, while reaction time is plotted as its opposite (the time left before the 10 s timeout), and it is measured by the scale on the right, in seconds. For both measurements, a higher value is preferable. Although the shape of the two histograms is similar, Spearman's Rho is 0.6, therefore there is no significant association.

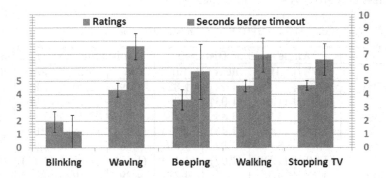

Fig. 2. Ratings by participants (left scale) compared to the time before timeout (right scale). Error bars indicate SD.

3.2   Room Condition Specific Behaviour Effectiveness

To evaluate the ratings of the behaviours we conducted a multi-variate analysis of co-variance (MANCOVA) with participant group (A/B), light level (dim/bright), sound level (high/low), distance (near/far), and angle of approach (10 degrees/70 degrees) as independent variables and trial number as co-variate. The evaluation scores for the 5 different behaviours (blinking, waving, beeping, walking, stopping TV) are the dependent variables. We find significant effects for participant group ($F(5,199)=7.969$, $p<0.001$), light level ($F(5,199)=10.797$, $p<0.001$), angle of approach ($F(5,199)=5.389$, $p<0.001$) and trial number ($F(5,199)=2.779$, $p=0.019$). None of the other main effects and interaction effects were significant.

To determine the effects of learning we investigated the effect of trial number on the different evaluations scores. From the same MANCOVA we found that only the score for waving changed significantly with trial number ($F(1, 203)=10.393$, $p<0.001$). The evaluations for waving increased approximately linearly from 3.3 ± 0.3 in trial 1 to 4.3 ± 0.3 in trial 12.

In a similar fashion we determined the effect of the significant environmental changes on the evaluation scores. Table 1 summarises how room conditions can hamper or improve the effectiveness of behaviours.

We found that light level significantly affects blinking ($F(1,203)=30.257$, $p<0.001$) and waving ($F(1,203)=4.509$, $p=0.035$). Blinking was appreciated less with lights on than off (mean difference -0.72 ± 0.13) and waving was appreciated more (mean difference $+0.36 \pm 0.14$). Waving and blinking were also affected by the angle of approach (blinking: $F(1,203)=18.967$, $p<0.001$; waving: $F(1,203)=4.037$, $p=0.046$).

Both behaviours were liked more when the robot was at an angle of 10 degrees than at an angle of 70 degrees (blinking: mean difference 0.55 ± 0.13; waving: 0.33 ± 0.15). The participant groups gave significantly different judgments to waving ($F(1,203)=18.669$, $p<0.001$), walking ($F(1,203)=6.807$, $p=0.10$) and stopping the TV ($F(1,203)=11.942$, $p=0.01$). Group B evaluated waving and walking higher than group A (mean differences: 0.67 ± 0.13 and 0.36 ± 0.12, respectively), and turning off the TV less than group A (mean difference: -0.40 ± 0.12).

Table 1. Mean value (S.D.) of the effect of room conditions on the effectiveness of behaviours

	Angle		Distance		Luminance		Loudness	
	70°	10°	near	far	on	off	low	high
Blinking	1.57	2.12	1.94	1.82	1.49	2.20	1.83	1.93
	(0.90)	(1.01)	(1.03)	(1.05)	(0.69)	(1.16)	(0.96)	(1.13)
Waving	3.62	3.95	3.83	3.78	4.00	3.64	3.88	3.70
	(1.18)	(1.00)	(1.08)	(1.11)	(1.00)	(1.15)	(1.04)	(1.15)
Beeping	3.84	3.63	3.87	3.60	3.66	3.77	3.82	3.59
	(1.10)	(1.10)	(1.07)	(1.13)	(1.17)	(1.05)	(1.11)	(1.10)
Walking	4.23	4.35	4.31	4.29	4.29	4.30	4.25	4.36
	(0.97)	(0.90)	(1.03)	(0.85)	(0.95)	(0.92)	(0.05)	(0.91)
Stop TV	4.51	4.58	4.53	4.57	4.63	4.49	4.62	4.47
	(0.92)	(0.81)	(0.94)	(0.78)	(0.78)	(0.92)	(0.77)	(0.97)

The effects of room conditions on behaviour effectiveness can also be found when inspecting the reaction times. We did a univariate ANCOVA with reaction time as dependent variable, the environmental conditions as fixed factors, participant group as random factor and trial number as co-variate. Reaction times greater than or equal to 10 s were removed, because they exceeded the maximum time window. We found a significant effect of trial number ($F(1,164)=6.727$, $p=0.01$) and an interaction effect of sound level x angle of approach x distance ($F(1, 3.326)=36.8$, $p=0.007$). The latter interaction effect was due to a much longer reaction time when the angle was 10 degrees, the distance was far and the sound level high. The former is due to a slight overall decrease in reaction times with trial number.

4   Discussion

4.1   Considerations on Communication Channels

Both in terms of performance (reaction times) and perceived effectiveness (user rating) there are certain preferences regarding behaviours to attract attention. Reaction times and user ratings are much worse for blinking than for any of the other behaviours. This seems to suggest that pure visual cues are less salient when people are distracted, as was expected. Indeed, we found that behaviours with visual cues worked best close to the front of the person in terms of effectiveness and this effect was strongest for blinking (a purely visual cue). In general, our results agree with the findings in [10], in which auditory cues were found to be more effective.

We expected that the preference to certain behaviours would depend on the environmental state. In particular, auditory cues were expected to become less effective in a noisy environment; visual cues are best when visible (lights on for waving and walking, lights off for blinking). Indeed, blinking was evaluated higher with lights off than with lights on, and waving was rated higher with lights on. It was also expected that visual cues deteriorate more when presented in the periphery. In line with expectation, blinking and waving were appreciated less when the robot was in the participant's periphery. However, environmental factors and robot position had little effect on reaction times, so it seems that the visual cues were not harder to detect.

The effect of mixing channels also has to be considered. In fact, some of the basic behaviours involve different channels at the same time. For example, walking behaviour involves proxemics, visual and auditory fields: compared to other kinds of movements, walking is particularly noisy. Mixing channels makes it difficult to distinguish which component is more effective. It also makes cues more robust. The latter may explain why there were no effects of environment and robot position on walking. Additional tools such the use of headphones or obstacles could be effective in completely impairing the auditory or visual channel, respectively.

4.2   Considerations about Learning

We expected that the scores given to the chosen behaviours would increase in later trials, as the robot adapts to the user and selects better behaviours more often in later trials. This was proven true, as the mean value of the evaluation in the latter 6 trials is significantly higher than the former 6 trials ($p = .008$). As a result, probabilities match the effectiveness of behaviours as they are estimated by participants' questionnaire.

We also found that reaction time decreased with trial number and that waving was evaluated better with increasing trial number, which is consistent with learning the preferred behaviour. It is unclear why this was not observed for the other behaviours. Presumably, this is because the robot selected the waving behaviour more often than any other behaviour.

4.3   Considerations on Automatic Classification

For participant group B the robot automatically classified its position relative to the participant. The system for classifying the angle of approach and distance was based on face detection, but this does not work well in dark environments, which resulted in misclassifieations. If a face was not detected because it was too dark, the robot automatically assumed a 70 degrees condition, but that could be incorrect. Likewise distance could not be estimated whenever no face was detected. To deal with this a third state was added to the learning algorithm (different from "near" and "far"). The system just learns what to do if distance is unknown.

In 71.5% of the cases, the 70 degree angle was correctly measured; in 89.7% of the cases the 10 degree angle was correctly measured and in 18% of the cases the face detection program made an incorrect classification. Most mistakes were made in the 70 degrees condition, because the program sometimes recognized a face in an object or because the participant was looking towards the robot and away from the TV. Mistakes in the 10 degrees condition instead occurred if the participant was looking away from the robot, every now and then we caught participants studying their environment (e.g. looking at different aspects of the room they were in).

There were incorrect classifications in case of distance too: with lights on, the script selected an incorrect distance in 10% of the trials, while in 23.8% of the cases of the near position and 44.8% of the cases of far position, the script was unable to detect a face. Besides the cases of not detecting a face, sometimes (multiple) faces were detected in objects both at a far and near distance.

Due to these misclassifications participant group B experienced a robot which received more noisy input about its relative position and consequently selected suboptimal behaviours more often than for participant group A. In practice, however, such effects are small because the learning algorithm can deal with potential misclassifications. Therefore, the only effect would be that learning is slower. To the participant, a misclassification looks as if the robot is learning. From the results we know that reaction times did not differ between participant groups, nor was the effect of trial number different for each participant. The groups did differ in their ratings of the behaviours. This could be due to the fact that the behaviours shown by the robot differed between groups, or perhaps the participants' preferences differed between groups. Either way, the learning algorithm performed equally well for both groups.

4.4   Choice of Measures

This experiment focused purely on the effectiveness of behaviours; there are however other measures that could be taken into account (e.g. pleasantness). For participants it might be easier to indicate how pleasant a behaviour is according to their preference, while effectiveness can be more objectively measured just by reaction times. Evaluation of effectiveness by participants' opinion also has a risk of bias towards the behaviours that were first randomly chosen.

Measurement should also involve efficiency and not only effectiveness: each behaviour should be considered in terms of how much effort it takes, and the desired result (attract the attention) should be achieved with the minimum cost.

5   Conclusion

In this paper, we described an experiment of human-robot interaction in which the robot Nao had to attract the human participant's attention while he/she is watching a video clip, in a room where environmental conditions and the relative position of the robot change. This experiment was performed with the purpose of measuring the effectiveness of robot behaviours involving different communication channels, when the effectiveness of each channel may change depending on the context. We introduced an adaptive learning process that could capture individual preferences for robot behaviours in different contexts. Results showed several correlations between room conditions and effectiveness of the behaviours, in terms of personal preferences and reaction times. In general, actions with a visual component are effective near the centre of the visual field and auditory cues enhance effectiveness, especially in the visual periphery. This resulted in waving being best overall and blinking was least effective. These findings will be useful to make a robot initiate a communication successfully. Future work should be heading towards autonomy, with autonomous detection of room lighting and loudness, and towards a better understanding of the context in general. For example, attracting attention may cause the person to lose concentration when doing something more important: the robot should be aware that the act of attracting attention may be inappropriate at times.

Acknowledgements. This research was supported by the Strategic Young Researcher Overseas Visits Program for Accelerating Brain Circulation program by the Japanese Society for the Promotion of Science. This study was conducted in the Eindhoven University of Technology. We thank all staff and students involved for the support received. Special thanks to Larn de Vries, who helped realise this study in its early stages.

References

1. Mataric, M.J.: Reinforcement Learning in the Multi-Robot Domain. Autonomous Robots **4**, 73–83 (1997)
2. Argyle, M., Trower, P.: Person to Person: Ways of Communicating. Harper & Row (1979)
3. Argyle, M.: Bodily Communication. Methuen (1988)
4. Kendon, A.: Conducting Interaction: Patterns of Behavior in Focused Encounters. CUP Archive (1990)
5. Trovato, G., Kishi, T., Endo, N., et al.: Cross-Cultural Perspectives on Emotion Expressive Humanoid Robotic Head: Recognition of Facial Expressions and Symbols. Int. J. Soc. Robotics **5**, 515–527 (2013). doi:10.1007/s12369-013-0213-z
6. Finke, M., Koay, K.L., Dautenhahn, K., et al.: Hey, I'm over here - how can a robot attract people's attention? In: RO-MAN 2005, pp 7–12 (2005)
7. Das, D., Hoque, M.M., Kobayashi, Y., Kuno, Y.: Attention control system considering the target person's attention level. In: HRI 2014, pp 111–112. IEEE Press, USA (2013)
8. Hoque, M.M., Onuki, T., Das, D., et al.: Attracting and controlling human attention through robot's behaviors suited to the situation. In: HRI 2012, pp 149–150. ACM, USA (2012)
9. Satake, S., Kanda, T., Glas, D.F., et al.: How to approach humans?-strategies for social robots to initiate interaction. In: HRI 2009, pp 109–116 (2009)
10. Torta, E., van Heumen, J., Cuijpers, R.H., Juola, J.F.: How can a robot attract the attention of its human partner? a comparative study over different modalities for attracting attention. In: Ge, S.S., Khatib, O., Cabibihan, J.-J., Simmons, R., Williams, M.-A. (eds.) ICSR 2012. LNCS, vol. 7621, pp. 288–297. Springer, Heidelberg (2012)
11. Torta, E., van Heumen, J., Piunti, F., et al.: Evaluation of Unimodal and Multimodal Communication Cues for Attracting Attention in Human-Robot Interaction. Int. J. Soc. Robotics **7**, 89–96 (2014). doi:10.1007/s12369-014-0271-x
12. Trovato, G., Do, M., Kuramochi, M., Zecca, M., Terlemez, Ö., Asfour, T., Takanishi, A.: A novel culture-dependent gesture selection system for a humanoid robot performing greeting interaction. In: Beetz, M., Johnston, B., Williams, M.-A. (eds.) ICSR 2014. LNCS, vol. 8755, pp. 340–349. Springer, Heidelberg (2014)
13. Motmans, R.: Ergonomie RC, DinBelg. Leuven (2005)

Investigating the Effect of Relative Cultural Distance on the Acceptance of Robots

G. Trovato[1(✉)], J.R.C. Ham[2], K. Hashimoto[3,6], H. Ishii[4], and A. Takanishi[5,6]

[1] Graduate School of Advanced Science and Engineering, Waseda University,
#41-304, 17 Kikui-cho, Shinjuku-ku, Tokyo 162-0044, Japan
gabriele@takanishi.mech.waseda.ac.jp
[2] Department of Innovation Sciences, Eindhoven University of Technology,
Eindhoven, The Netherlands
[3] Waseda Institute for Advanced Study, Shinjuku, Tokyo, Japan
kenji@takanishi.mech.waseda.ac.jp
[4] Research Institute of Science and Engineering, Waseda University, Shinjuku, Japan
hiroyuki@takanishi.mech.waseda.ac.jp
[5] Department of Modern Mechanical Engineering, Waseda University, Shinjuku, Japan
atsuo@takanishi.mech.waseda.ac.jp
[6] Humanoid Robotics Institute (HRI), Waseda University, Shinjuku, Japan

Abstract. A complex relationship exists between people's cultural background and their general acceptance towards robots. Previous studies supported the idea that humans may accept more easily a robot that can adapt to their specific culture. However, it is not clear whether between two robots which are identified as foreign robots because of their verbal and non-verbal expressions, the one that is culturally closer may be preferred or not. In this experiment, participants of Dutch nationality were engaged in a simulated video conference with a robot that is greeting and speaking either in German or in Japanese; they completed a questionnaire assessing their preferences and their emotional state. As Dutch participant showed less signs of discomfort and better acceptance when interacting with a German robot, the hypothesis that acceptance of a robot could be directly proportional to cultural closeness was supported, while the hypothesis that similar foreign robots are equally less accepted regardless of the country- was rejected. Implications are discussed for how robots should be designed to be employed in different countries.

Keywords: Culture · Social robotics · Gestures · Greetings · HRI

1 Introduction

1.1 Cultural Differences in Robotics

Social robots are expected to play a major role in the society in the near future. For this reason, their acceptance by their human companions is the optimal priority for the robot designers to consider, especially regarding their ability to interact and communicate with humans to help them in their work and daily life.

© Springer International Publishing Switzerland 2015
A. Tapus et al. (Eds.): ICSR 2015, LNAI 9388, pp. 664–673, 2015.
DOI: 10.1007/978-3-319-25554-5_66

Cultural background of the human users as well as of the robot is one important factor of acceptance towards robots. Studies such as [1] showed how the nationality of the robot can radically change the impression it gives. Asimov was the first to introduce the problem of culture difference in the acceptance of robots, specifically introducing the Frankenstein complex [2], which describes the anxiety that people feel towards robots, more common in Western countries than in Japan. These kind of differences are originated by the influence of religion as well as social structures and philosophies. Despite this traditional view, further studies reported different results proving that the matter is more controversial [3, 4].

Generally speaking, technology acceptance depends also on the country where the robot was produced: as a consequence, localisation of products may be done [5]. This could be even more true for social robots: technological devices which have to communicate and take on the role of a social actor. Indeed, earlier research in psychology suggested that humans are very quick to categorize their (human) interaction partners in relevant categories. Many earlier studies (e.g. [6]) show that people do not only quickly and spontaneously categorise interaction partners as in-group or out-group, but that they also evaluate in-group members more positively, and thereby show in-group biases (e.g., allocate more resources to in-group members [7]). The first step in the study of communication with robots is to study greeting interaction and self-introduction, which are the basic components of an interaction in human-human communication, Only a few greeting interaction experiments with robots involving culture have been conducted so far, such as the work in [8].

1.2   Cultural Distance

In our experiment we take in consideration three countries: Netherlands, Germany and Japan. We choose these countries because we assume that the first two are much more related to each other, while Japan is, not only geographically, far apart. Cultural distance has been attempted to measure several times, by Kogut and Singh [9] among others [10-12], using different formulas. Our assumption regarding the three countries involved in our study should be valid regardless of the formula used.

As culture is a factor that cannot be easily quantified, more complex categorisations by country were attempted by Hall's factors [13] and Hofstede's dimensions [14]. However, these models do not take in consideration any geographical factor. Lewis's model [15] plots countries in a triangular graph, plotting them in relation to the categories 'linear-actives', 'multi-actives' and 'reactives'. It is possible to recognise language and race groups from that simple model. Welzel et al. [16] instead tried to visualise countries in a two dimensional scatter plot, (the "Inglehart-Welzel culture map"), in which traditional versus secular-rational values are on the vertical y-axis, and survival versus self-expression values are on the horizontal x-axis. A broad categorisation that is commonly done is to assign cultures to a groups based on geopolitics, such as: Western, Asian, Middle Eastern, African. This simple categorisation does not take into account the heavy differences that may exist among countries within one group. For instance, Germany and Greece, although in the same Western group, are divided by noticeable differences, whereas a far away culture like Japan has things in common with Germany [17]. This is not the case however, with the Netherlands, which share with Germany the same language family, the same race and similar traditions. Language closeness and language barrier are in particular a variable of interest for our study.

1.3   Purpose of This Paper

In this paper we describe a cross-cultural experiment in which Dutch subjects are introduces to two robots, that are almost complete identical except that one robot greets the user using Japanese words and gesture, and speaks English with Japanese accent while the other greets the using German words and gesture, and speaks English with German accent to them.

A similar experiment was performed in Egypt and Japan, with the robot speaking Arabic and Japanese: the results are reported in [18]. Egyptians preferred the Arabic version of the robot, and that they felt symptoms of discomfort when interacting with the Japanese version, and the same outcome was true the other way round.

As a continuation of that work, the purpose of the experiment of this paper is to answer another specific research question, involving three countries instead of two: when both robots are foreign, does different cultural distance influence their acceptance? We can formulate two different hypothesis regarding the outcome of the experiment:

A.   Dutch participants prefer the German speaking robot. This may imply that in Human-Robot Interaction, acceptance of a robot is directly proportional to cultural closeness.

B.   Dutch participants do not prefer any of the two robots, perceiving both equally as foreign. This implies that a specific adaptation of the robot communication style to the country is necessary.

The rest of the paper is organized as follows: in section 2 we describe the robot and the protocol of the experiment; in section 3 we show the detailed results and we discuss them in section 4; in section 5 we conclude the paper and outline future works.

2   Experimental Procedures and Materials

2.1   Hardware

We used the whole body emotion expression 48-DoFs humanoid robot KOBIAN [19]. It is designed to clarify the influence of physicality and expressivity during interaction with humans. Being able of both emotion expression and bipedal walking, KOBIAN is potentially able, in the future, to work as assistive robot in a human environment, such as a family or a public facility.

In order to make an experiment with subjects in a place like the Netherlands, distant from the robot (which is located in Waseda University, in Tokyo, Japan) a video conference system is needed. Despite there is only one KOBIAN, our purpose was to show two different robots (one Japanese-like, and one German-like) to the subjects; therefore, the video conference was simulated. We used the robot in two versions: KOBIAN, the original version, and "DEBIAN", which has different slightly facial and body colours (see Fig. 1, b and d respectively). The colour differences between the two versions were chosen to be unrelated in any way to the specific culture, and they are not meant to make the robot more appealing for a specific group of subjects; their only purpose is to give to the subjects the impression that they are interacting with two different, although very similar, robots.

KOBIAN and DEBIAN were used to realise the culture-specific greetings (motion of the arms and waist) and to simulate speech (motion of the lips and slight periodic oscillations of the head, that give a human-like appearance to the robot behaviour). The robot body parts are controlled by both position-based and velocity-based controllers that have been implemented using YARP.

2.2   Experimental Protocol

The protocol consists in the following steps, taking in total around 20 minutes:

1. Pre-questionnaire: each subject is invited to sit at a desk, in front of a big screen (Fig. 2), and to compile a preliminary questionnaire on likeability of humanoid robots in general and on their own perceived safety.
2. Explanation: the subject is told that there will be a call to a laboratory in Waseda University in Japan through the video conference system, for showing two different robots. Actually, a previously recorded video will be shown. No real call is made, but the subject is tricked into believing that he/she is watching a live connection by adding the typical connection sounds and screenshots. This Wizard-of-OZ style setup encourages natural behaviour of the participant.
3. First call: when one examiner pretends to start the call, video begins and connection is established with a Japanese student (Fig.1, a), who once more explains the purpose of the experiment; then the Japanese student switches the camera to KOBIAN (Fig.1, b), who greets, giving the possibily to the participant to reply, then does a self-introduction and says goodbye (more details in Section 2.3).
4. First questionnaire: after closing the connection, the subject compiles a questionnaire about KOBIAN.
5. Second call: a new call is made, this time to a German student (Fig.2, c), who greets and switches the camera to DEBIAN, who greets and waits for the participant's greeting, then does a self-introduction and says goodbye.
6. Second questionnaire: as the video ends, the subject compiles a questionnaire about DEBIAN (Fig.2, d), and expresses a preference between the two robots.
7. Closing explanations: at the end, the subject is informed that the video conference was not real, and of the motivation of the use of this method.

Note: for all subjects the order of the robots was randomised (steps 3-6). This means that for around one half of the subjects, the order of the robot was (DEBIAN, KOBIAN) instead of (KOBIAN, DEBIAN). The video was adjusted accordingly.

Fig. 1. Slideshow of the video: from left to right, Japanese human, Japanese robot KOBIAN, German human, German robot DEBIAN.

Fig. 2. Experimental setup: point of view of the participant, in front of a screen and of a video conference device.

2.3   Videos

A few videos were recorded beforehand and assembled together into a single stream. Interface screens and sounds were added for simulating a real call through a video conference system. The video was composed by the following parts:

- Japanese person greeting in Japanese, introducing in English the next robot;
- KOBIAN performing a bow – with "Konnichi wa" (which is the standard Japanese daytime greeting) speech added – as initial greeting;
- a few empty seconds for giving the time to the participant to reply to the greeting;
- KOBIAN introducing himself in English with Japanese accent;
- KOBIAN performing a bow – with "Otsukaresama desu" (which is a standard idiomatic phrase that fellow workers use at the end of a working day) speech added – as final greeting;
- German person greeting in German, and introducing in English the next robot;
- DEBIAN waving its hand – with "Guten Tag" (literally, "Good day") speech added – as initial greeting;
- a few empty seconds for giving the time to the participant to reply to the greeting;
- DEBIAN introducing himself in English with German accent;

- DEBIAN waving again – with "Auf Wiedersehen" ("Goodbye" in German) speech added – as final greeting.

Two versions of the video were used, with either the Japanese and German robot shown first.

2.4   Assessment

In order to catch both explicit opinions and psychological reactions, a combination of physiological responses and written questionnaires were considered.

Assessment in human-robot interaction through survey is preferably done using standardised measurements. Bartneck [20] devised 5-point semantic differential scales called Godspeed for measuring anthropomorphism, animacy, likeability, perceived intelligence, and perceived safety for robots. We used likeability and perceived safety; moreover, we added a new set of scales for measuring cultural closeness, containing four subscales:

Impolite	1	2	3	4	5	Polite
Mysterious	1	2	3	4	5	Familiar
Incomprehensible	1	2	3	4	5	Comprehensible
Foreign	1	2	3	4	5	Native

Additional questions included some demographic information like age and gender, and some more explicit questions regarding what the subject liked about the two robots, and regarding gesture and words the robot used. Questionnaires were written in Dutch.

3   Results

3.1   Demographics

In the experiment 26 participants (male: 10;   female: 16; average age: 22.69; s.d.: 4.96), all Dutch. We gathered a heterogeneous group consisting of people with different education level. Among them, we isolated the ones who were familiar with Japan or Germany (either speaking the language, or frequent travellers, or with some strong family/friends connections). The resulting group of Dutch people in which we could control the confounding factors was composed by 20 people (male: 9;   female: 11; average age: 23.45; s.d.: 5.32).

3.2   Results: Scales

Table 1 reports the results of the Bartneck's scales described in Section 2.4 and of additional questions in the form of 5-point semantic differential scales.

Gathered data were analysed using Kruskal-Wallis test and subsequently Mann-Whitney U-test. In all the cases in which the U-test was performed, it means that the

Kruskal-Wallis test already gave a low p value as output. Friedmen test and Wilcoxon Signed-Rank test were also applied, but since they seem to produce results that are less strict, we report only Kruskal.Wallis and Mann-Whitney here. Student's t-test and ANOVA could not be applied, because the shape of the distribution graph resulting from the semantic differential scales was not a normal distribution.

Cronbach's Alpha was previously calculated to verify the internal consistency of the set of scales. Likability, having alpha bigger than 0.7, is considered consistent and reported as a single entry in Table 1. On the other hand, Cultural Closeness (a = 0.50) and Perceived Safety (a = 0.31) are not consistent and therefore reported separately in the table. As a whole, both Likeability (p = .02) and Cultural Closeness (p = .0004) featured a significant difference, while Perceived Safety (p = .44) did not. Splitting Cultural Closeness into its subscales, we can analyse more in detail.

Table 1. Summary of the results of differential semantic scales

	Before the experiment	Japanese human	German human	KOBIAN	DEBIAN
Likeability				3.37*	3.69*
Politeness				4.1	4.45
Familiarity				2.75	2.95
Comprehensibility				2.3***	3.3***
Nativeness				1.65^+	2.25^+
Relax	3.2			3.75	3.75
Calmness	3.65			3.8	3.75
Surprise	2.75**			3.3**	3.65**
How speaks English				1.45***	2.85***
Gesture		3.85	4.05	3.7	4.1
Speech		3.8	3.95	4.2	4.15
Would like to meet again?				2.7*	3.4*

Highlighted in yellow, statistically significant differences (one cross ($^+$) means one-tailed $p < .05$; one asterisk (*) means $p < .05$; two asterisks (**) mean $p < .01$; three asterisks mean $p < .001$)

3.3   Results: Subjects' Preference

At the end of the experiment, all participants were asked to express their preference between the two robots. Result of this explicit question showed a strong preference for DEBIAN, the German robot. In a scale from 1 (preference for KOBIAN) to 5 (preference for DEBIAN), it was 4.25. They were also asked to justify their choice adding a free comment. We collected all the comments and divided into the following categories, then shown in Fig. 3:

- Non-verbal communication: gesture more natural / uses hands / moves hands like humans / better body language / better movement / more realistic;
- Sense of familiarity: it is more comfortable / more familiar / more friendly;

- Comprehensibility: more understandable / clear language / clear spelling / voice is more clear / speaks more fluently / better English;
- Emotion: emotion more clear / shows emotions better;
- No reason: I don't know / just my feeling.

Compared with the previous experiment in [18], we left out the category "Language", as in this case, both German and Japanese are foreign languages, although German is much easier to understand for a Dutch.

We included any comment related to the appearance to the "No reason" category. This is because physical appearance of the two robots was essentially the same, and claiming that one of the two is better looking may be caused by personal feelings.

In Fig. 3, comprehensibility, the sense of familiarity, and non-verbal communication seem to be the most important categories.

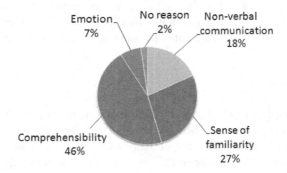

Fig. 3. Graph reporting the reasons why subjects expressed a preference for one robot

4   Discussion

Results in Table 1 suggest that Dutch people, in general, feel more comfortable with a German robot rather than with a Japanese. This is happening despite Germany being the main rival country for the Netherlands in certain contexts such as sports. A negative impression of Germany due to war memory did not happen probably because of the low age of participants. A similar bias towards Japan, involved in the same war, is even more unlikely.

The result seems to be confirmed in Table 2 through the signs of discomfort appearing during the interaction with KOBIAN. Therefore, we restate the hypothesis A and consider it correct: "Dutch participants prefer the German speaking robot. This may imply that in Human-Robot Interaction, acceptance of a robot is directly proportional to cultural closeness." If confirmed by broader studies including other countries, this hypothesis implies that customisation of robots based on culture can be done by broader areas than just country by country.

However, there are some key factors that the robot especially needs to adapt to. From the results in Section 3.3, the reasons for preference for a robot are highlighted and limited to a few key factors, such as comprehensibility. Sense of familiarity is still important (27% in Figure 2), but it may be a consequence of a combination of verbal and non-verbal behaviour.

This study has some limitations regarding the way it was performed: namely, the need to use Wizard-of-Oz techniques; the culture-specific characterisation of the robots; and the possibility of influence on the results by side effects due to differences in stimuli. Although subtle differences may exist in verbal and non-verbal communication between two countries like Germany and Japan [21], this kind of experimental setup allows for a short controlled interaction. For this reason, greeting and self introduction, being a basic form of interaction, were used. Each additional differences between stimuli should be carefully considered. For example, the use of different appearances for the two robots, needed in order to distinguish them, may become a confound factor. However, in our case the difference is very minor and a neutral colour (grey) was used, In the same way, all the interactions were structured in a way to convey no emotions. Voice was also produced using the same type of text-to-speech software, generating speeches that do not substantially differ in naturalness. For these reasons, we assume that results are not contaminated by other independent variables.

5   Conclusion

We performed an experiment of human-robot interaction, in which Dutch participants were involved in a simulated video conference with two robots which performed greetings and a self introduction, respectively using Japanese and German gestures and way of speaking. The purpose of this work was to understand the impression the robot gives, when is perceived as foreign, depending on the cultural distance of the two countries. Results suggest that Dutch people feel more comfortable with a German robot rather than with a Japanese. If similar findings were obtained comparing other countries,   acceptance of a robot could be estimated as directly proportional to cultural closeness, and the degree of adaptation that a robot needs can be measured and limited to a few key factors, such as comprehensibility. In future, these considerations can be also taken in consideration when designing a robot that should adapt to different kinds of people within the same country or across different countries, but differing in other aspects such as the education level.

Acknowledgements. This research was supported by the Strategic Young Researcher Overseas Visits Program for Accelerating Brain Circulation program by the Japanese Society for the Promotion of Science. This study was conducted as part of the Research Institute for Science and Engineering, Waseda University, and as part of the humanoid project at the Humanoid Robotics Institute, Waseda University. The experiment was carried out in the Eindhoven University of Technology. We thank all staff and students involved for the support received.

References

1. Eyssel, F., Kuchenbrandt, D.: Social categorization of social robots: anthropomorphism as a function of robot group membership. Br. J. Social Psychology **51**, 724–731 (2012)
2. Asimov, I.: The Machine and the Robot. Science Fiction: Contemporaty Mythology. In: Warrick, P.S., Greenberg, M.H., Olander, J.D. (eds.) Harper and Row (1978)
3. Bartneck, C., Nomura, T., Kanda, T., et al.: Cultural Differences in Attitudes Towards Robots. In: Proceedings of the AISB Convention: Symposium on Robot Companions (2005)
4. Wagner, C.: The Japanese way of robotics: interacting naturally with robots as a national character? RO-MAN **2009**, 510–515 (2009)
5. Rogers, E.M.: Diffusion of Innovations, 5th edn. Free Press (2003)
6. Otten, S., Moskowitz, G.B.: Evidence for Implicit Evaluative In-Group Bias: Affect-Biased Spontaneous Trait Inference in a Minimal Group Paradigm. Journal of Experimental Social Psychology **36**, 77–89 (2000). doi:10.1006/jesp.1999.1399
7. Reeves, B., Nass, C.: The Media Equation: How People Treat Computers, Television, and New Media Like Real People and Places. Cambridge University Press (1996)
8. Suzuki, S., Fujimoto, Y., Yamaguchi, T.: Can differences of nationalities be induced and measured by robot gesture communication? In: 2011 4th International Conference on Human System Interactions (HSI), pp. 357–362 (2011)
9. Kogut, B., Singh, H.: The Effect of National Culture on the Choice of Entry Mode. Journal of International Business Studies **19**, 411–432 (1988)
10. Kandogan, Y.: An improvement to Kogut and Singh measure of cultural distance considering the relationship among different dimensions of culture. Research in International Business and Finance **26**, 196–203 (2012). doi:10.1016/j.ribaf.2011.11.001
11. Yeganeh, H.: A generic conceptualization of the cultural distance index. Journal of Strategy and Mgt **4**, 325–346 (2011). doi:10.1108/17554251111180990
12. Babiker, I.E., Cox, J.L., Miller, P.M.: The measurement of cultural distance and its relationship to medical consultations, symptomatology and examination performance of overseas students at Edinburgh University. Soc. Psychiatry **15**, 109–116 (1980)
13. Hall, E.: Beyond culture. Anchor, Garden City (1977)
14. Hofstede, G., Hofstede, G.J., Minkov, M.: Cultures and Organizations: Software of the Mind, 3rd edn. McGraw-Hill, New York (2010)
15. Lewis, R.D.: Cross Cultural Communication: A Visual Approach. Transcreen Public (1999)
16. Welzel, C., Inglehart, R., Klingemann, H.-D.: The Theory of Human Development: A Cross-Cultural Analysis. European Journal of Political Research **42**, 341–379 (2003)
17. Nils-Johan, J.: Culture and Power in Germany and Japan: The Spirit of Renewal. Global Oriental Ltd, Folkstone (2006)
18. Trovato, G., Zecca, M., Sessa, S., et al.: Cross-cultural study on human-robot greeting interaction: acceptance and discomfort by Egyptians and Japanese. Paladyn International Journal of Behavioral Robotics **4**, 83–93 (2013). doi:10.2478/pjbr-2013-0006
19. Endo, N., Takanishi, A.: Development of Whole-Body Emotional Expression Humanoid Robot for ADL-Assistive RT Services. J of Robotics and Mechatronics **23**, 969–977 (2011)
20. Bartneck, C., Kulić, D., Croft, E., Zoghbi, S.: Measurement Instruments for the Anthropomorphism, Animacy, Likeability, Perceived Intelligence, and Perceived Safety of Robots. Int J of Soc Robotics **1**, 71–81 (2009). doi:10.1007/s12369-008-0001-3
21. Endrass, B., André, E., Rehm, M., Nakano, Y.: Investigating culture-related aspects of behavior for virtual characters. Auton Agent Multi-Agent Syst. **27**, 277–304 (2013). doi:10.1007/s10458-012-9218-5

Role of Social Robotics in Supporting Employees and Advancing Productivity

Kimmo J. Vänni[⊠] and Annina K. Korpela

Tampere University of Applied Sciences, Kuntokatu 3, 33520 Tampere, Finland
{kimmo.vanni,annina.korpela}@tamk.fi

Abstract. We conducted a cross-sectional survey among three different occupational groups; Finnish university staff members, Finnish catering & food service company, and Brazilian teachers. The aim of the study was to explore how the employees perceive if they would have a chance to collaborate with the social robots, especially when they are working while ill. The results showed that a social robot will be accepted better as a team member than as a face-to-face colleague. The respondents were looking forward that a robot will improve productivity and assist them. We recommend designers, researchers and manufactures to take a social robot's role in working life into account as a new potential focus area.

Keywords: Social robotics · Presenteeism · Productivity · Sickness absence · Loss costs · Perception

1   Introduction

There are many robotics topics where the new applications have been launched. Manufacturing industry is investing in automation, robotics and Internet of things (IoT). It is a well-known fact that industry is interested in cutting the labor costs and increasing productivity [1]. The other discussed topics, apart from industrial robotics, are the assistive robotics among the elderly [2,3] as well as robots in social inclusion [4].

Even if robotics is well researched among industry and health care sectors, we have found only few studies which focused on the usage of social robotics among employees [5]. Just now one of the most relevant issues, regarding both organizations' productivity and employees' performance, seems to be the robots as co-workers [6,7] but we have not found the articles where the social robots as co-workers have been studied in respect of an employee's perceived productivity loss and presenteeism. We state that occupational health and employees' productivity might be an interesting focus area for social robotics.

According to our theory, the social robots can be used at least in a three-way for advancing occupational health and productivity in workplaces (Fig.1). The robots can assist the employees whose perceived skills or health status will not meet the work demands and thus they have risks for stress and depression. The other cases concern the employees who already have diagnosed temporary or permanent disorders which prevent them to perform at the normal productivity level. Regarding temporary

© Springer International Publishing Switzerland 2015
A. Tapus et al. (Eds.): ICSR 2015, LNAI 9388, pp. 674–683, 2015.
DOI: 10.1007/978-3-319-25554-5_67

illnesses the robots can offer assistance and support for the employees when needed. Support can be cognitive support, motivation and physical collaboration. The cases regarding the employees whose work abilities have lowered permanently are more requiring. Basically they would need the work tasks which will meet their physical and mental capacity. That requires a lot of arrangement from the employers' view-points. For example the employers should re-educate employees.

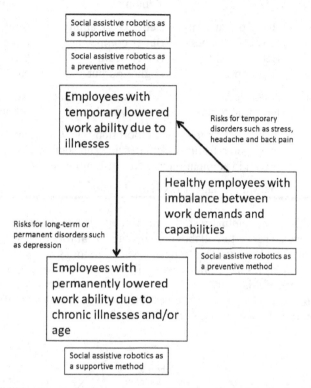

Fig. 1. Possibilities for advancing occupational health with social robotics in workplaces

We have three cornerstones in our theory for advancing the use of social robots in working life. First; there will always be employees with poor work ability and health status. Second; the social robots are able to cut part of the productivity loss costs which cannot be tackled with the traditional methods such as coaching. Third; industrial robots and assistive social robots will be merged into 'the industrial assistive social robots' and the collaboration between a human and a robot will be seamless in future. Table 1 shows how the perspectives between companies, employees and society might be combined in near future. There are already some examples of that trend such as YuMi by ABB, Nextage by Kawada Industries, LBR iiwa by KUKA and Baxter by Rethink Robotics.

Table 1. Perspectives between different parties in adopting social robotics

	Companies	**Employees**	**Societies**
Directive feature	Company profile and capacity	Employee characteristics	Development of society
External factor	Market demands	Work-related demands	Political demands
Expected outcomes	Better productivity Cost effectiveness Better quality Higher profit	Better work ability Better competence Better health Higher output	Modern society Increased taxes Increases import New work places
Development field	Advance in industrial robotics	Advance in social assistive robotics	Advance in robotics research
Near future	Robot – Employee Collaborations (REC). Industrial robots will have some features from service and social robotics. Society will invest in automation and Robot-Employee Collaboration		
Examples	YuMi, Nextage, LBR iiwa, Baxter		

The aim of this interdisciplinary study was to explore how the employees perceive if they would have a chance to collaborate with the social robots. Our hypothesis was that the employees might be willing to have a social robot as a helping partner, especially when the employees are working while ill or they are worried about if they will get ill. A framework of this study consisted of social robotics, occupational health and working life domains (Fig. 2). The intersection of the domains will offer us a brand new application field for advancing social robotics.

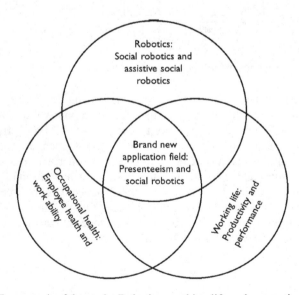

Fig. 2. Framework of the study: Robotics, working life and occupational health

1.1   Presenteeism

Literature has stated that presenteeism, known also as the productivity loss at work [8] concerns only employees' health problems [9]. Recently the researchers have reported that there are also other reasons for presenteeism than illnesses [10] such as heavy workload [11] and the efficiency demands [12]. There are many studies regarding presenteeism and the related factors but we have not found the studies where the possibilities of social robotics for tackling presenteeism have been discussed.

Costs of Presenteeism and Productivity Loss. Presenteeism is a global problem and every nation is suffering for productivity losses. Number of studied implies that presenteeism is a costly problem for both society and companies [13,14]. Davis et al. [14] stated that US economy will lose every year about €234 billion in output due to employees' health-related problems. Kliff [15] reported that employees' poor health costs for US employers €310 billion in a year as lost productivity and sick days. Davis et al. [14] estimated that the number of days per year of reduced productivity due to illness was 478 million, which means that during those days the employees were not able to fully work and as a consequence generated a loss valued at €24 billion. In Australia the overall cost of presenteeism to the economy in 2009 was about €30.7 billion (2.7% of GDP) [13].   According to many studies [16] presenteeism is the most relevant factor for deficient performance on the job.

Costs of Stress and Depression. Migraine for example [17] is a known disorder for presenteeism but the most urgent global problem seems to be stress and depression. Mitchell et al. [18] stated that depression is among the top 3 conditions in terms of productivity costs. According to the European Commission [19] the costs of work-related stress in the European Union (EU) countries was about €20 billion a year. Respectively, the total cost of work-related depression was estimated to be about €617 billion of which absenteeism and presenteeism was evaluated to be €272 billion and loss of productivity €242 billion [20]. Also other countries like US [21] and Australia [22] have reported high stress and depression costs.

1.2   Social Robotics

There are examples how robotics has been used for rehabilitation purposes for people who suffer from e.g. mental disorders [23]. There are studies how robotics has been exploited among neuro-cognitive patients [24] and disabled people [25] for increasing productivity and impacts of care. The need of interactive service robotics and assistive social robotics is evident [26] but the number of potential articles regarding the implementation of social robotics at the workplaces is still scarce.

There are many [27] studies regarding service robotics among the elderly but hardly anything available regarding occupational health. However, a distinction between a retired person and an aged worker is trivial. It is well-known that employees' perceived work ability will decrease and the health disorders will increase over lifetime [28]. A relevant question is how employees perceive a social robot as a helping partner. Because it is hard to find focused studies regarding the perception of social robotics among the employees, some of the research results among the elderly people might be useful for implementing the robots in work places. Scopelliti et al. [29] reported that a user's physical and mental condition and the cognitive skills should be taken into account. Meng and Lee [30] argued that the traditional industrial robot engineering approaches are inappropriate in terms of user-friendliness which is relevant regarding the employees as well. Sekmen and Challa [31] reported that a robot's ability to learn is critical in interaction. Peine et al. [32] proposed to consider the older persons as active consumers of technology which is quite comparable with the older employees who might retire soon. Linner et al. [33] argued that the integration of the service robot systems into real world has been difficult because of the separate development of environment and the robotics systems. That should be taken into account when developing social robotics solutions in the workplaces as well.

2   Materials and Methods

This study was based on the cross-sectional survey conducted among three different groups which represented a) research and development, b) teaching and c) catering & food service sectors. The number of respondents was 59 (20 male and 39 female). A request to participate was sent to 26 Finnish university professionals and 14 (54%) of them replied. The questionnaire was sent to 20 Brazilian researchers and teachers and 11 (55%) of them replied. In addition, the questionnaire was sent to a Finnish catering & food service organization which was asked to pick up 50 respondents and 34 (68%) of them replied. The age of respondents was categorized and divided as follows: 16-24 (n=5), 25-34 (n=7), 35-44 (n=14), 45-54 (n=20), 55-64 (n=13), 65+ (n=0).

Survey Data. Data on willingness to use the social robots was based on two questions: *"How do you react if you receive an announcement that your partner decided to change a job and you have to collaborate together with a humanoid or an android robot in future?"* and *"How do you react if you receive a sudden announcement that on next week you will get a new team member who is a human-like robot?"* For assessing presenteeism the participants were asked to assess if they have been working despite the illness during the last 12 months. In addition, we asked if a personal robot which is able do part of their job, would be useful while working ill and/or would employees be interested in if a robot could assist them if they will get ill. The questionnaire included also questions regarding a robot's appearance and the employees' readiness to adopt the social robots. The survey results were analyzed with SPSS 21 and MS Excel.

Variable Design. Employees' willingness to collaborate with social robots was selected as a dependent variable. The main independent variable was employees' willingness to use a personal robot if he/she would have worked while ill. We asked them the following questions: *"If you have worked while ill, would a personal robot which is able to assist you and to do part of your job, have been useful?"* and *"If you* *have not worked while ill, please, evaluate would you be interested in if a personal robot could assist you in your work tasks if you get ill?"* We analyzed also the following variables: A gender, age, and a robot's ability to advance ones productivity.

2.1   Statistical Analysis

Being based on the nature of the study it emphasizes descriptive statistics but we analyzed the selected survey results with a logistic regression model as well. The variables were classified and dichotomized for assessing the odds ratios (OR). Chi-square (X2) test was performed and p-values were assessed with the 95% confidence intervals (95% CI).

3   Results

Table 2 shows that 42 respondents out of 59 had a positive attitude towards a social robot as a team member. A positive attitude was very high among the office workers where 86 % stated that they would react neutrally or positively if they will get a new team member who is a humanoid robot. The reaction was controversial among production workers where about a half of the respondents had a positive attitude and another half had a negative attitude respectively. The respondents were more critical regarding an option to collaborate face-to-face with a social robot. About 53 % of respondents were looking forward that option but 29 % of those who are ready to have a robot as a team member will reject a robot as a face-to- face partner.

The respondents' attitudes to use social robots as assistants if being ill were quite positive. Also many of those who had a negative attitude towards social robots were looking forward to have a robot assistant while working ill. Similar result was also regarding a robot's ability to advance productivity.

We asked also if there is any ethically sensitive issue such as data privacy regarding the use of a personal robot at work places. About 64 % of the respondents stated that 'yes' and especially the Brazilian teachers' were worried about ethical issues compared with the Finns. We explored also how ready people were to adopt the social robots. About 23 % of the respondents stated that they would never learn to work with robots. Almost all of them represented production sector. Only 4 respondents were ready to work with personal robots immediately but the greatest part of the respondents stated that they would need a mid-term introduction.

We asked also if the employees want to have an influence on a robot's appearance for accepting it as a colleague. About a half of the respondents stated that 'yes' and another half that 'no'. It seemed that a robot's gender is unimportant and a robot's age or size is not a big issue for the greatest part of the respondents. The respondents preferred also a human-like appearance compared with a machine-like or an animal-like.

Table 2. Descriptive statistics regarding employees' attitudes towards social robotics

	Catering company (n=34)				University R&D staff		Brazilian teachers		Total
	Production		Office						
	n	%	n	%	n	%	n	%	(n=59)
Robot as a team member									
No	12	70.6	3	17.	1	5.9	1	5.9	17
Yes	10	23.8	9	21.	13	31.	10	23.8	42
Robot as a partner									
No	14	50.0	7	25.	4	14.	3	10.7	28
Yes	8	25.8	5	16.	10	32.	8	25.8	31
Robot as an assistant if being ill									
No	12	75.0	2	12.	0	0.0	2	12.5	16
Yes	10	23.3	10	23.	14	32.	9	20.9	43
Robot's ability to increase productivity									
No	11	68.8	3	18.	1	6.3	1	6.3	16
Yes	11	25.6	9	20.	13	30.	10	23.3	43

Table 3 shows that a positive attitude towards social robotics as a team member had significant association with employees' willingness to use a robot if they are working while ill, or if they have a sick leave. It seems also that a robot's ability to increase productivity is a significant factor, as well as, employees' willingness to use a robot as a work mate. We tested also if a gender, age or perceived presenteeism would have associations with a positive attitude towards the use of a social robot but those factors were non-significant.

Table 3. Association between selected factors and positive attitude towards a social robot as a group member, using logistic regression analysis (n=59)

Factor	OR	95 % CI	X^2	p
Presenteeism (no vs. yes)	2.51	0,79 - 7.97	2.50	0.11
Robot is useful if working while ill**	4.58	1.38 - 15.20	6.64	0.009
Robot would be useful if I'll get ill**	6.77	1.94-23.60	10.02	0.002
Robot will do part of my jobs if I have a sick leave***	133.2	13.65-1300	36.70	0.00
Robot will increase my productivity***	13.57	3.47-53.09	17.07	0.00
I prefer a robot as a work mate***	40.00	4.76-336.04	20.85	0.00

* $p < 0.05$; ** $p < 0.01$; *** $p < 0.001$

4   Discussion

The idea to use a robot as a co-worker is a debated topic just now. The Danish Technological Institute [34] states that a robot as a co-worker is the next evolutionary step in industrial robotics. They argue that a robot will not work alone but it needs human interaction, supervision and inputs. Some previous studies argue that technology is mature for human-robot collaboration [6] and the workers are looking forward social interaction and relationship with the robots [7]. It has been reported also that a little is known about the potential social impact of a robot co-worker technology on employees and organizations [7].

Also presenteeism is a debated topic. It is well-known that presenteeism is not going away and the companies should invest in positive work environment [13]. Even if the social robots may offer the new possibilities for cutting costs and for increasing productivity we have to accept that some employees are afraid of robots and prefer only human co-workers. However, the results showed that people are willing to adopt the social robots as team members and many of the respondents were ready to work face-to-face with the robots. Being different from our presumption, a robot's appearance was not an important issue even if a half of the respondents stated that they would like to influence on a robot's appearance. The results showed also that the office workers were more prone to accept the social robots compared with the production workers. The reason for that might be that the office workers were more aware of social robotics and ICT than the production workers. The respondents' comments were positive overall but some production workers misunderstood a robot's role. They argued that a robot is not able to do their jobs like to taste food. One trend is not to substitute the humans but to assist them. A chef can taste food even if he/she would have a robot colleague.

The strength of this study was that it offered the employees' point of view and discussed the use of social robotics in the working life. As far as we know there are no previous studies in this study domain, especially regarding presenteeism and the productivity losses. The study had also limitations. It was an exploratory pilot study and the number of respondents was limited compared to the traditional health studies where the number of respondents is hundreds or thousands. In addition, the respondents represented only couple of occupation sectors and it is difficult to evaluate how employees might perceive the social robots for example in a construction sector. One limitation was also that we did not have the respondents' health information or register data but we had to lean on survey data. In addition, we were interested in asking the respondents' opinion if robots would be able to substitute them totally but a feedback from employers postponed the question because it would have been too radical for labor unions. Even if the topic is interesting and the results were promising we have to bear in mind that we are taking the early steps of understanding if the social robots can be exploited in advancing productivity of the employees who might have lowered work ability due to high age or the prevalent health disorders. We need more information about what kind of robot is adequate for the certain work tasks and what kind of features a robot should have for having the positive impacts on the employees' performance and productivity. Regarding the introduction of the social robots among the employees we recommend to familiarize with Ge's [35] three-level (social, private, intimate) introduction model. As a conclusion, we recommend that the designers and the robot manufacturers should take the role of social robotics into account as a brand new focus area in working life.

References

1. The Boston Consulting Group: Takeoff in robotics will power the next productivity surge in manufacturing. https://www.bcg.com/media/PressReleaseDetails.aspx?id=tcm:12-181684

2. Leroux, C., Ben Ghezala, M., Mezouar, Y., Devillers, L., Chastagnol, C., Martin, J.-C., Leynaert, V., Fattal, C.: ARMEN: assistive robotics to maintain elderly people in natural environment. IRBM **34**(2), 101–107 (2013). doi:10.1016/j.irbm.2013.01.012

3. Heerink, M., Kröse, B., Evers, B., Wielinga, B.: Assessing acceptance of assistive social agent technology by older adults: the Almere Model. Int. J. Soc. Robot. **2**(4), 361–375 (2010)

4. Reppou, S., Karagiannis, G.: Social inclusion with robots: a RAPP case study using NAO for technology illiterate elderly at Ormylia foundation. In: Szewczyk, R., Zieliński, C., Kaliczyńska, M. (eds.) Progress in Automation, Robotics and Measuring Techniques. Advances in Intelligent Systems and Computing, vol. 351, pp. 233–241 (2015). doi:10.1007/978-3-319-15847-1_23

5. Vänni, K.: Social robotics as a tool for promoting occupational health. In: COST Event: The Future Concept and Reality of Social Robotics, Brussels, Belgium (2013)

6. Haddadin, S., Suppa, M., Fuchs, S., Bodenmüller, T., Albu-Schäffer, A., Hirzinger, G.: Towards the robotic co-worker. In: Pradalier, C., Siegwart, R., Hirzinger, G. (eds.) Robotics Research. STAR, vol. 70, pp. 261–282. Springer, Heidelberg (2011)

7. Sauppe, A., Mutlu, B.: The social impact of a robot co-worker in industrial settings. In: Proceedings of the 33rd Annual ACM Conference on Human Factors in Computing Systems, pp. 3613–3622 (2015). http://pages.cs.wisc.edu/~bilge/pubs/2015/CHI15-Sauppe.pdf

8. Schultz, A., Edington, D.: Employee health and presenteeism: a systematic review. J. Occup. Rehabil. **17**(3), 547–579 (2007)

9. Cooper, C., Dewe, P.: Well-being—absenteeism, presenteeism, costs and challenges. Occup. Med. (Lond) **58**(8), 522–524 (2008). doi:10.1093/occmed/kqn124

10. Brouwer, W.B., van Exel, N.J., Koopmanschap, M.A., Rutten, F.F.: Productivity costs before and after absence from work: as important as common? Health Policy **61**(2), 173–187 (2002)

11. Biron, C., Brun, J.-P., Ivers, H., Cooper, C.: At work but ill: psychosocial work environment and well-being determinants of presenteeism propensity. Journal of Public Mental Health **5**(4), 26–37 (2006)

12. Böckerman, P., Laukkanen, E.: Predictors of sickness absence and presenteeism: does the pattern differ by a respondent's health? J. Occup. Environ. Med. **52**(3), 332–335 (2010)

13. Medibank.: Sick at Work: The cost of presenteeism to your business and the economy (2011) http://www.medibank.com.au/client/Documents/Pdfs/sick_at_work.pdf

14. Davis, K., Collins, S.R., Doty, M.M., Ho, A., Holmgren, A.: Health and productivity among U.S. workers. Issue Brief (Commonw Fund) **856**, 1–10 (2005)

15. Kliff, S.: Poor health costs employers $576 billion. The Washington Post (2012). http://www.washingtonpost.com/blogs/wonkblog/wp/2012/09/14/poor-health-costs-employers-576-billion/

16. Burton, W.N., Conti, D.J., Chen, C.Y., Schultz, A.B., Edington, D.W.: The role of health risk factors and disease on worker productivity. J. Occup. Environ. Med. **41**(10), 863–877 (1999)
17. Landy, S., Runken, M., Bell, C., Higbie, R., Haskins, L.: Assessing the impact of migraine onset on work productivity. J. Occup. Environ. Med. **53**(1), 74–81 (2011). doi:10.1097/JOM.0b013e31812006365
18. Mitchell, R.J., Bates, P.: Measuring health-related productivity loss. Popul. Health Manag. **14**(2), 93–98 (2011). doi:10.1089/pop.2010.0014
19. European Commission: Guidance on work-related stress: spice of life or kiss of death, European Communities, Luxembourg, European Communities (2002). https://osha.europa.eu/data/links/guidance-on-work-related-stress
20. Matrix: Economic analysis of workplace mental health promotion and mental disorder prevention programmes and of their potential contribution to EU health, social and economic policy objectives (2013). http://ec.europa.eu/health/mental_health/docs/matrix_economic_analysis_mh_promotion_en.pdf
21. Rosch, P.J.: The quandary of job stress compensation. Health and Stress **3**, 1–4 (2001)
22. Safe Work Australia: Cost of work related injury and disease for Australian employers, workers and the community: 2008–09, Australia, Canberra (2012). http://www.safework australia.gov.au/sites/SWA/about/Publications/Documents/660/Cost_of_Work-related_injury_and_disease.pdf
23. Rabbitt, S., Kazdin, A., Scassellati, B.: Integrating socially assistive robotics into mental healthcare interventions: Applications and recommendations for expanded use. Clinical Psychology Review **35**, 35–46 (2015). doi:10.1016/j.cpr.2014.07.001
24. Takahashi, C.D., Der-Yeghiaian, L., Le, V., Motiwala, R.R., Cramer, S.C.: Robot-based hand motor therapy after stroke. Brain **131**, 425–437 (2008)
25. Van der Loos, H.F.M., Reinkensmeyer, D.J.: Rehabilitation and health care robotics. In: Handbook of Robotics, pp. 1223–1251. Springer, New York (2008)
26. Tapus, A., Matarić, M., Scassellati, B.: The grand challenges in socially assistive robotics. Robotics Autom. Mag. IEEE **14**(1), 35–42 (2007)
27. Broadbent, E., Stafford, R., MacDonald, B.: Acceptance of health care robots for the older populations: Review and future directions. Int. J. of Soc. Robot. **1**, 319–330 (2009)
28. Ilmarinen, J., Tuomi, K., Klockars, M.: Changes in the work ability of active employees over an 11-year period. Scand. J. Work. Environ. Health **23**(Suppl 1), 49–57 (1997)
29. Scopelliti, M., Giuliani, M., Fornara, F.: Robots in a domestic setting: a psychological approach. Univers. Access. Inform. Soc. **4**(2), 146–155 (2005)
30. Meng, Q., Lee, M.: Design issues for assistive robotics for the elderly. Adv. Eng. Inform. **20**(2), 171–186 (2006). doi:10.1016/j.aei.2005.10.003
31. Sekmen, A., Challa, P.: Assessment of adaptive human–robot interactions. Knowledge-Based Systems **42**, 49–59 (2013). doi:10.1016/j.knosys.2013.01.003
32. Peine, A., Rollwagen, I., Neven, L.: The rise of the "innosumer"-Rethinking older technology users. Technol. Forecast Soc. Change **82**, 199–214 (2014). doi:10.1016/j.techfore.2013.06.013
33. Linner, T., Pan, W., Georgoulas, C., Georgescu, B., Güttler, J., Bock, T.: Co-adaptation of robot systems, processes and in-house environments for professional care assistance in an ageing society. Procedia Eng. **85**, 328–338 (2014). doi:10.1016/j.proeng.2014.10.558
34. Danish Technological Institute: Robot Co-worker for assembly (2015). http://www.dti.dk/services/robot-co-worker-for-assembly/32733
35. Ge, S.S.: Social robotics: integrating advances in engineering and computer science. In: Proceedings of Electrical Engineering/Electronics, Computer, Telecommunications and Information Technology International Conference, pp. 9–12 (2007)

Effects of Perspective Taking on Implicit Attitudes and Performance in Economic Games

James Walliser, Stephanie Tulk(✉), Nicholas Hertz, Erin Issler, and Eva Wiese

George Mason University, Fairfax, USA
stulk@gmu.edu

Abstract. Perspective taking, allows people to make inferences about another's mental states and goals. For social robotics, perspective taking could facilitate social interactions. The current study investigated whether taking the spatial perspective of a robot resulted in reduced implicit biases, more generous behavior towards robotic agents in economic games, and higher explicit ratings of humanness. The results show that agent type had a significant effect on subjective ratings of humanness, explicit attitudes, and decision-making. Spatial perspective taking had a significant effect on implicit associations such that the strength of association with positive attributes towards humans and negative attributes towards robots were enhanced. The effect of perspective taking on HRI should be further studied, as implicit attitudes are often expressed as actions and judgments outside the performer's awareness [1].

Keywords: Perspective taking · Simulation theory · Economic games · Implicit associations test

1 Introduction

In the last decade, social neuroscience has discovered areas in the human brain (e.g., medial prefrontal cortex for mentalizing, fusiform face area for processing facial identity, superior temporal sulcus for detecting and processing biological motion) that exclusively process information during social interactions and are very sensitive to the behavior of human-like, intentional agents [2]. These insights can apply to social robotics in investigating whether principles that apply to human social interactions can be used to improve how we socially interact with robots. Social neuroscience can help evaluate the "socialness" of robots by measuring whether interacting with it triggers similar brain areas as human interaction partners.

In order to design robots that activate similar brain areas as human interaction partners, we must address that robots lack intentional behavior. With human behavior, we assume that it is goal-directed and driven by certain intentions [3]. Therefore we adopt the intentional stance towards other humans and treat them as rational agents with internal states, such as beliefs, desires, and emotions [3]. If robot behavior is not believed to be intentional, humans cannot use their experience from human-human interactions to predict and understand behavior in human-robot interactions. Adopting

© Springer International Publishing Switzerland 2015
A. Tapus et al. (Eds.): ICSR 2015, LNAI 9388, pp. 684–693, 2015.
DOI: 10.1007/978-3-319-25554-5_68

the intentional stance has also been shown to increase the social relevance we ascribe to the behavior of other agents and results in better cognitive performance in shared tasks [4, 5], a mechanism that could be used to optimize the effectiveness and efficiency of human-robot interactions. Thus, we need to investigate under which conditions robots are treated as intentional agents and to identify design features that increase the likelihood that the intentional stance is adopted to social robots.

Previous research has shown that one of the most reliable sources for assessing the internal states of others is their gaze direction [6] and humans use eye gaze to convey intentions and establish joint attention in social interactions [5, 7]. The relationship between observing changes in gaze direction and subsequent shifts of attention to the gazed-at location is investigated by using gaze following paradigms where a face is presented in the center of the screen that either validly (i.e. target at the gazed-at loca-tion) or invalidly (i.e., target opposite of the gazed-at location) cues the location of a target [7]. Gaze cueing effects are present if reaction times to the target are shorter at the valid compared to the invalid condition [8]. Previous research has shown that the degree to which we follow the gaze of others is modulated by whether intentionality is ascribed to the gazing agent [4, 5] with gaze cueing effects being much larger when the likelihood of adopting an intentional stance was high (i.e., eye movements were believed to be executed by a human) compared to when it was low (i.e., eye move-ments were believed to be preprogrammed).

Based on these findings, it is reasonable to assume that humans are willing to treat robots *as if* they had a mind and that the mere perception of intentionality can lead to an increased interest in the social interaction [4]. It has been shown that the perception of intentionality does not require the presence of an actual mind, but rather relies on the belief that an agent is capable of showing intentional behavior [9]. In line with that, mental states can be attributed to mindless agents, if they behave in an intention-al fashion (e.g. [10]) and/or look like intentional agents (e.g., Martini, Gonzales & Wiese, under review; [11]). For instance, computer agents that exhibited empathic emotions were evaluated more favorably on a number of human characteristics in-cluding caring, likeability, and trustworthiness [12].

A different approach to mind attribution comes from Simulation Theory, which hypothesizes that we attribute mental states to others by processing relevant cues through our own mental apparatus [13]. This ability allows us to better predict others' internal states and engage socially [14]. Understanding the minds of others stems from our ability to project ourselves into another person's perspective [15]. Perspec-tive taking is a multi-dimensional social-cognitive process that includes three compo-nents: perceptual/spatial, adoption of another's viewpoint; cognitive, determination of another's knowledge; and affective, assumption of another's emotional state [16]. A review of the perspective taking literature identifies a number of positive social out-comes including nonverbal behaviors and favorable implicit and explicit evaluations [17]. Others have found that the use of computer avatars to assume the perspective of another can reduce negative stereotypes [18]. The benefits of perspective taking are not limited to human-human interactions, it has also been shown to increase empathy toward animals [19]. Others found that humans can feel empathy toward robots sug-gesting an attribution of the robot's emotional state [20].

1.1   Aim of Study

One way to facilitate social interactions with robots is to increase the likelihood of adopting the intentional stance towards them [4, 5]. Given that perspective taking has been shown to support attribution of intentionality and mental states to human agents (e.g. [14]), it is possible that perspective taking promotes attribution of mental states to non-human agents (i.e., robots) and thereby improve human-robot interaction.

In the current experiment, perspective taking was implemented as a spatial perspective taking task, where participants are asked to judge whether a target letter T appears left or right of a given agent (human or robot). We predicted spatial perspective taking would improve social interactions and attitudes towards both humans and robots. Furthermore we expected the effects to be greater for robots.

Table 1. Hypotheses

	Hypotheses
1	Spatial perspective taking (vs. no perspective taking) would result in more generous behavior in economic games independent of agent type.
2	Spatial perspective taking would result in higher subjective ratings of humanness and trust independent of agent type
3	Spatial perspective taking would have a greater effect on economic game behavior and subjective ratings for the robot agent than for the human agent.
4	Spatial perspective taking would result in increased association of robots with positive attributes on the implicit association test.

2   Experiments

2.1   Participants

The sample consisted of 66 (40 women, 26 men, M_{age} = 36.0 years, age range: 18-75 years) participants that were recruited through Amazon's Mechanical Turk (mTurk) community. The only exclusion criteria for who may participate in the mTurk study were that the participant had to be at least 18 and live in the United States. This allowed for a vast user population, while producing results that would be of equivalent quality of those collected in a lab setting [21]. The experiment took 20 minutes to complete and participants were compensated 30 cents for completing the study. Data collection with mTurk was approved by the institutional review board.

2.2   Stimuli

The stimuli for the human condition were images of a female face (F 07), chosen from the Karolinska Directed Emotional Faces database [22]; the stimuli for the robot condition were photos of the humanoid robot EDDIE (developed by TU Munich). The size of the images was 370 x 370 pixels. The images were presented in color with gaze straight ahead. The capital letter "T" was used as the target (35 x 26 pixels) and the distance between the agent and target was 65 pixels.

2.3   Tasks and Procedure

During the experiment, participants first engaged in a short interaction with the human and the robot agent during which the agents' perspective had to be taken or not. Afterwards, participants completed three social cognitive tasks: the Implicit Associations Test (IAT) [1], a series of economic games, and subjective inventories measuring the humanness and the trustworthiness [23, 24].

2.4   Perspective Taking Task

In the perspective taking task, participants first saw a fixation cross presented in the center of the screen for 500 ms, followed by an image of either the human or the robot interaction partner looking straight ahead. After 2000 ms, a target letter T appeared on the screen that was either shown to the left or right of the face. As soon as the target appeared, participants were asked to respond by pressing one of two keys (A for 'left' or L for 'right') to indicate where on the screen the target was shown. The trial ended after the participant had given a response, or after the 2000 ms time limit was reached. In the non-perspective taking condition, participants had to give their answers from their own perspective. In the perspective-taking condition, however, participants had to answer from the agent's perspective.

Participants completed two blocks of 16 trials each (eight with the target presented on the right and eight on the left), one for the human agent and one for the robot agent. Participants were randomly assigned to one of two conditions in which they either responded from the agent's spatial perspective or from their own perspective.

2.5   Implicit Association Test

The IAT was used to assess whether perspective taking has a positive influence on implicit attitudes towards a given agent. In general, the IAT measures the association between a target-concept discrimination and an attribute dimension [1]. In the case of this experiment, the target-concept discrimination was to distinguish between human and robot faces, while the attribute dimension was constituted by the assessment and categorization of words with meanings that are good or bad, in accordance with the implicit attitude research [25]. Table 2 depicts the order of events in the IAT.

Table 2. Sequence of trail blocks in the human-robot interaction IAT.

Block	Function	Items assigned to left-key response	Items assigned to right-key response
1	Target Practice	Human Images	Robot Images
2	Attribute Practice	Good Adjectives	Bad Adjectives
3	Combined Practice	Good Adjectives + Human Images	Bad Adjectives + Robot Images
4	Combined Test	Good Adjectives + Human Images	Bad Adjectives + Robot Images
5	Reverse Target Practice	Robot Images	Human Images
6	Reverse Combined Practice	Good Adjectives + Robot Images	Bad Adjectives + Human Images
7	Reverse Combined Test	Good Adjectives + Robot Images	Bad Adjectives + Human Images

2.6   Economic Games (EGs)

A series of EGs were used to investigate whether perspective-taking leads to more generous behavior. In general, in EGs, a participant (player 1, P1) is given some initial monetary endowment, and has the opportunity interact with a partner (player 2, P2). Trust and generosity were reflected by the amount of money P1 was willing to give to P2 [26].

We used three types of EGs with an endowment of $12: Investment Games (IG), Ultimatum Games (UG), and Dictator Games (DG) [26]. Each participant played all three games with each agent, resulting in 6 total trials. Trials were presented with a static image of the agent appearing at the top of the page (image, orientation and size of the images are the same as in the perspective taking task). A written description of the game was presented underneath the image together with a sliding scale that could be used to make the offer to the agent ($0-$12). Trials were presented one per page, and participants had as much time as they needed before clicking the "next" button to submit their answer.

2.7   Subjective Ratings

Two subjective rating inventories were administered to assess whether taking the perspective of an agent results in higher ratings of human-likeness and trust. The humanness inventory was adapted from Bartneck et al. (2009) and contained 8 items on a 5-point likert scale [23]. Each item was anchored by word pairs (one adjective and its opposite, e.g. machinelike-humanlike, unfriendly-friendly). The trust inventory was adapted from the Checklist for Trust Between People and Automation developed by Jian, Bisantz, & Drury (2000) and contained 12 items on a 5 point likert scale ranging from strongly disagree to strongly agree [24]. Each item was written in the form of a statement about a specific feeling toward the agent in question (e.g. "I can trust the agent", "The agent provides security"). During the assessment, participants were presented with the images of the agents they have interacted with before (human or robot, one at a time). The agent image was presented at the top of a page. The assessments were provided twice, once for the human and once for the robot partner.

3   Results

3.1   Implicit Association Test

The IAT results were analyzed by comparing mean response times between combined and reverse combined conditions. The combined condition (Good + Human / Bad + Robot) response time ($M = 855$ ms, $SD = 136$ ms) was lower than the reverse combined (Good + Robot / Bad + Human) response time ($M = 1023$ ms, $SD = 197$ ms). The differential in response times indicate the target-attribute pairing the combined condition was more compatible with participants' attitudes. The improved scoring algorithm described in Greenwald, Nosek, and Banaji (2003) was used to calculate the IAT effect, D measure, for each participant [27]. The D measure, related to Cohen's *d*, is a comparison between mean block response times as an indication of strength of association. The D measure has a possible range of -2 to +2 with the effects strength divided separated by weak (.15), moderate (.35), and strong (.65). In both perspective conditions (self- and agent-perspective) participants responded more quickly when humans were paired with positive attributes and robots were paired with negative attributes than when the pairings were reversed. In other words, there was a moderately strong association of humans with positive concepts. A one-way ANOVA of perspective type revealed a significant difference between perspective taking conditions ($F(1, 63) = 5.50$, $p < .05$), that is: participants that took the spatial perspective of the agents ($M = .607$, $SD = .248$) more strongly associated the human with positive concepts (and therefore robots with negative concepts) than those in the self-perspective condition ($M = .449$, $SD = .290$).

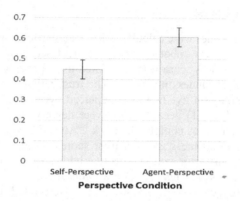

Fig. 1. IAT D effect. Perspective taking significantly increased the association of human agents with positive concepts (also signifies association of robots with negative concepts).

3.2   Economic Games

Three EGs were completed by each participant: the Ultimatum Game (UG), the Dictator Game (DG), and the Investment Game (IG). Endowments were measured for each of three economic games as a measure of social interaction. A 2 x 2 ANOVA (perspective type x agent type) was run for each of the three EGs, with agent type (human, robot) as a within-factor and perspective-taking (yes, no) as a between-factor. All three games had a main effect for agent type (UG: $F(1, 63) = 4.37$, $p < .05$; DG: $F(1, 63) = 23.12$, $p < .001$; IG: ($F(1, 63) = 5.43$, $p < .05$) Greater mean endowments were given to humans across each game (See Figure 4). Type of perspective (agent vs self) did not have a main effect on any of the three EGs and there were no interactions between agent and perspective type.

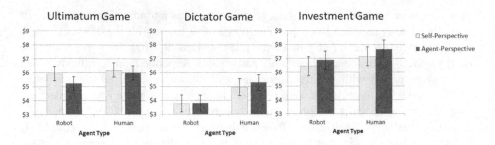

Fig. 2. Results from EGs. Significant differences exist between offers made to human and robot in each game.

3.3 Subjective Measures

For data analysis, four values were calculated for each participant: a humanness rating score (for the human and the robot agent) and a trust score (for the human and the robot agent). The humanness score was calculated as the average rating score across the eight items of the humanness inventory. The trust ratings were calculated in two steps. First items one through five had to be replaced by reverse scoring (e.g. 1 = 5, 2 = 4, 3 = 3, 4 = 2, and 5 = 1) because they were negatively worded. Secondly, the mean of the twelve values was computed. A 2 x 2 ANOVA (agent type x perspective type) was run on the humanness and the trust scores respectively, with agent type (human, robot) as within-factor and perspective-taking (yes, no) as between-factor. The analysis of humanness ratings revealed a main effect for agent type ($F(1, 63) = 44.153$, $p < .001$). Humans ($M = 3.77$, $SD = .69$) were rated as significantly more humanlike than robots ($M = 2.93$, $SD = .72$). There was no main effect for perspective type and there was no interaction between agent type and perspective type. The analysis of humanness ratings revealed a main effect for agent type ($F(1, 63) = 8.41$, $p < .01$). Humans ($M = 3.32$, $SD = .61$) were rated as significantly more humanlike than robots ($M = 2.99$, $SD = .57$). There was no main effect for perspective type and there was no interaction between agent type and perspective type.

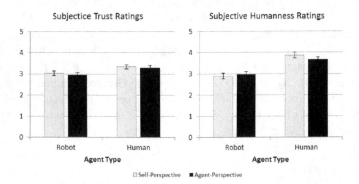

Fig. 3. Subjective ratings of humanness and trust. Significant differences between human and robot, though no interaction of agent type and perspective taking.

4   Discussion

This research set out to explore how perspective taking may be used to enhance attitudes and social interactions with robots. We predicted that spatial perspective taking would improve social interactions with both humans and robots. We found that agent type had a significant effect on behavior in EGs and subjective ratings of humanness and trust. In addition, perspective taking increased the strength of the association of humans with positive attributes and robots with negative attributes.

We expected spatial perspective taking to improve social interactions and subjective measures independent of agent type but failed to see a significant effect. Spatial perspective taking does not appear to affect the manner in which people interact with social robots. Surprisingly, spatial perspective taking did have a significant effect on implicit attitudes. The effect of spatial perspective taking on the IAT's D measure indicate that spatial perspective taking caused participants to develop a stronger association between humans and positive attributes as well as robots and negative attributes. This finding went against our hypothesis that spatial perspective taking would increase the strength of association between the robot agent and positive attributes. The results of the implicit association test were somewhat reflected in our other measures. Across three EGs and two subjective measures of explicit attitudes, participants favored humans significantly more than robots. In addition, we hypothesized a greater effect of perspective taking for the robot agent but this prediction was not supported either. When these results are integrated we see that people's implicit attitudes in favor of humans over robots can also be observed in overt behavior and explicit attitudes. Even though perspective taking enhanced the strength of association between humans and positive attributes the effect was not reflected in social interactions or subjective measures of humanness and trust.

Our predicted results hinged on the expectation that taking the spatial perspective of a robot would have the same effect as taking the spatial perspective of a human. Primarily, the expected effect was to make the agents appear to have intentionality thereby changing behavior and attitudes. Contrary to our expectations, spatial perspective taking amplified attitudes in favor of humans over robots. One possible explanation comes from Kessler and Thomson [28], their work on spatial perspective taking found that, rather than a simple mental rotation, spatial perspective taking was an embodied cognitive process, supported by physical alignment of the body. Building on their findings, it is likely that humans will have difficulty taking perspective of non-human agents due to dissimilar physical properties. As such, the unnatural act of taking a robot's spatial perspective calls attention to human-robot disparities and inhibits improved social interactions that are expected to result from perspective taking.

One interesting result from this study was that the effects of spatial perspective taking on implicit attitudes failed to carry over to behavior and explicit attitudes. Implicit attitudes are often expressed as actions and judgments outside of the performer's awareness [1]. As such one would expect to see positive association on the implicit attitude test manifested as enhanced social behaviors and explicit attitudes. Two of three economic games did trend toward increased endowments in the spatial perspective condition while the other measures showed little to no trend in the opposite direction so it is possible that further testing may reveal an effect. However, it is more likely that the brief experience with perspective taking was not sufficient to change behavior and attitude toward a specific agent.

Perspective taking maintains a great deal of potential to facilitate social interactions in the domain of social robots. Subsequent research will include intentionality as a directly measured independent variable. Furthermore, we have seen that the physical characteristics of a robot may inhibit spatial perspective taking. Future studies will explore the effect of robots with more humanlike characteristics. Finally, spatial perspective taking is one of three perspective taking dimensions. We have yet to evaluate the effects of cognitive and affective perspective taking on social behaviors, though considerable research supports this line of study (see [17]). Follow up research will explore the effect of these perspective taking dimensions on social interactions and perceptions of intentionality. Our current findings indicate humans favor fellow humans over robots in social interactions. Spatial perspective taking appears to amplify those attitudes though not enough to change behavior or explicit attitudes. Our future work will build on this base to determine if any type of perspective taking may enhance human-robot social interactions.

References

1. Greenwald, G., McGhee, D., Schwartz, J.: Measuring individual differences in implicit cognition: the implicit association test. Journal of Personality and Social Psychology **74**(6), 1464–1480 (1998). doi:10.1037/0022-3514.74.6.1464
2. Adolphs, R.: The Social Brain: Neural Basis of Social Knowledge. Annual Review of Psychology **60**, 693–716 (2009). doi:10.1146/annurev.psych.60.110707.163514
3. Dennett, D.C.: True believers: the intentional strategy and why it works. In: O'Connor, T., Robb, D. (eds.) Philosophy of Mind: Contemporary Readings, pp. 370–390. Routledge, London (2003)
4. Wiese, E., Wykowska, A., Zwickel, J., Müller, H.J.: I see what you mean: How attentional selection is shaped by ascribing intentions to others. PLoS ONE **7**(9) (2012)
5. Wykowska, A., Wiese, E., Prosser, A., Müller, H.J.: Beliefs about the minds of others influence how we process sensory information. PLoS ONE **9**(4) (2014)
6. Frischen, A., Bayliss, A.P., Tipper, S.P.: Gaze cueing of attention: visual attention, social cognition, and individual differences. Psychological Bulletin **133**(4), 694 (2007)
7. Friesen, C.K., Kingstone, A.: The eyes have it! Reflexive orienting is triggered by non-predictive gaze. Psychonomic Bulletin & Review **5**(3), 490–495 (1998)
8. Posner, M.I.: Orienting of attention. Quarterly Journal of Experimental Psychology **32**(1), 3–25 (1980)
9. Nass, C., Fogg, B.J., Moon, Y.: Can computers be teammates? International Journal of Human-Computer Studies **45**(6), 669–678 (1996)
10. Heider, F., Simmel, M.: An experimental study of apparent behavior. The American Journal of Psychology, 243–259 (1944)
11. Hackel, L.M., Loser, C.E., Van Bavel, J.J.: Group membership alters the threshold for mind perception: The role of social identity, collective identification, and intergroup threat. Journal of Experimental Social Psychology **52**, 15–23 (2014)

12. Brave, S., Nass, S., Hutchinson, K.: Computers that care: Investigating the effects of orientation of emotion exhibited by an embodied computer agent. International Journal of Human-Computer Studies **62**, 161–178 (2005)
13. Gordon, R.M.: Folk psychology as simulation. Mind & Language **1**(2), 158–171 (1986)
14. Premack, D., Woodruff, G.: Does the chimpanzee have a theory of mind? Behavioral and Brain Sciences **1**, 515–526 (1978). doi:10.1017/S0140525X00076512
15. Carruthers, P., Smith, P.K. (eds.): Theories of theories of mind, pp. 22–38. Cambridge University Press, Cambridge (1996)
16. Kurdek, L.A., Rodgon, M.M.: Perceptual, cognitive, and affective perspective taking in kindergarten through sixth-grade children. Developmental Psychology **11**(5), 643–650 (1975)
17. Todd, A.R., Galinsky, A.D.: Perspective-taking as a strategy for improving intergroup relations: Evidence, mechanisms, and qualifications. Social and Personality Psychology Compass **8**, 374–387 (2014)
18. Yee, N., Bailenson, J.N.: The Proteus Effect: The Effect of Transformed Self-Representation on Behavior. Human Communication Research **33**, 271–290 (2007)
19. Paul, E.S.: Empathy with animals and with humans: Are they linked? Anthrozoös **13**(4), 194–202 (2000)
20. Rosenthal-von der Pütten, A.M., Schulte, F.P., Eimler, S.C., Sobieraj, S., Hoffmann, L., Maderwald, S., Krämer, N.C.: Investigations on empathy towards humans and robots using fMRI. Computers in Human Behavior **33**, 201–212 (2014)
21. Mar, R., Spreng, N., DeYoung, C.: How to produce personality neuroscience research with high statistical power and low additional cost. Cognitive, Affective, & Behavioral Neuroscience **13**(3), 674–685 (2013)
22. Lundqvist, D., Flykt, A., Öhman, A.: The Karolinska Directed Emotional Faces (KDEF). Department of Neurosciences Karolinska Hospital, Stockholm (1998)
23. Bartneck, C., Kulić, D., Croft, E., Zoghbi, S.: Measurement instruments for the anthropomorphism, animacy, likeability, perceived intelligence, and perceived safety of robots. International Journal of Social Robotics **1**(1), 71–81 (2009)
24. Jian, J.Y., Bisantz, A.M., Drury, C.G.: Foundations for an empirically determined scale of trust in automated systems. International Journal of Cognitive Ergonomics **4**(1), 53–71 (2000)
25. Nosek, B., Greenwald, A., Banaji, M.R.: Understanding and using the Implicit Association Test: II. Method variables and construct validity. Personality and Social Psychology Bulletin **31**(2), 166–180 (2005). doi:10.1177/0146167204271418
26. Rogers, R., Bayliss, A., Szepietowska, A., Dale, L., Reeder, L., Pizzamiglio, G., Czarna, K., Wakeley, J., Cowen, P., Tipper, S.: I want to help you, but I am not sure why: Gaze-cuing induces altruistic giving. Journal of Experimental Psychology: General **143**(2), 763–777 (2014). doi:10.1037/a0033677
27. Greenwald, A.G., Nosek, B.A., Banaji, M.R.: Understandiing and using the Implicit Association Test: I. An improved scoring algorithm. Journal of Personality and Social Psychology **85**(2), 197–216 (2003)
28. Kessler, K., Thomson, L.A.: The embodied nature of spatial perspective taking: embodied transformation versus sensorimotor interference. Cognition **114**(1), 72–88 (2010)

Smile and Laughter Detection for Elderly People-Robot Interaction

Fan Yang[1,2](✉), Mohamed A. Sehili[1], Claude Barras[1,2],
and Laurence Devillers[1,3]

[1] Department of Human-Machine Communication, LIMSI-CNRS, Orsay, France
yangfan.1750@hotmail.com
[2] Department of Computer Science, University Paris-Sud, Orsay, France
[3] University Paris-Sorbonne, Paris, France

Abstract. Affect bursts play an important role in non-verbal social interaction. Laughter and smile are some of the most important social markers in human-robot social interaction. Not only do they contain affective information, they also may reveal the user's communication strategy. In the context of human robot interaction, an automatic laughter and smile detection system may thus help the robot to adapt its behavior to a given user's profile by adopting a more relevant communication scheme. While many interesting works on laughter and smile detection have been done, only few of them focused on elderly people. Elderly people data are relatively rare and often carry a significant challenge to a laughter and smile detection system due to face wrinkles and an often lower voice quality. In this paper, we address laughter and smile detection in the ROMEO2 corpus, a multimodal (audio and video) corpus of elderly people-robot interaction. We show that, while a single modality yields a given performance, a fair improvement can be reached by combining the two modalities.

1 Introduction

Laughter and smile are considered as all-important human communication skills. They convey lots of information during human-human interaction such as emotional state, social communication strategy and personality. This kind of information also appears in human-robot interaction, especially with humanoid robots. This paper focuses on laughter and smile detection of elderly people who interact with a robot in a real-life situation. For this purpose, we use part of a social interaction corpus [17] recorded in two retirement homes in France. This multimodal corpus is collected under the ROMEO2 project[1] and features elderly people interacting with the humanoid robot Nao.

Audio and visual based smile and laughter detection has each its pros and cons. Actually, while audio is a suitable modality for laughter detection, particularly when a subject is not facing the video source, things seem to be less obvious when it comes to detecting a smile by merely using an audio signal.

---

[1] http://projetromeo.com

© Springer International Publishing Switzerland 2015
A. Tapus et al. (Eds.): ICSR 2015, LNAI 9388, pp. 694–703, 2015.
DOI: 10.1007/978-3-319-25554-5_69

Visual detection however can be a good way for both smile and laughter as long as the subject faces the video camera and there is no obstacle between the two communicative parties. Another issue that visual smile detection faces is the similarity between a smiling mouth and a speaking mouth. The joint use of audio and video channels to improve the overall smile and laughter detection performance is investigated in this work.

Beside the afore mentioned issues, others are to be taken into account when dealing with smile and laughter detection for elderly people. Namely, the lack of relevant data of actual elderly people naturally interacting with a robot, voice quality and face wrinkles related difficulties are a supplementary challenge addressed in this work. We thusly consider the use of such realistic data with the proposed detection methods as the main contribution of this work.

The rest of this paper is organized as follows. Section 2 reviews some of the related work. Section 3 describes the data collection protocol, the corpus content, the annotation scheme, the questionnaires submitted to each subject after the experience, a statistical analysis and the data subset used in our experiments. In section 4 we describe our audio and video smile and laughter detection methods, as well as the fusion of both modalities. We then present our experiment protocols and the obtained results in section 5. We give our conclusion and perspectives in section 6.

2   Related Work

Due to the importance of smile and laughter in human-human interaction, many recent works have focused on smile and laughter research in the computer science area, especially in human-machine interaction. Many workshops dedicated to the topic have been organized such as the Interdisciplinary Workshop on Laughter and other Non-Verbal Vocalizations in Speech[2]. International projects like the ILHAIRE project[3] are also to be mentioned.

There are many acted or posed facial expression databases (e.g. the Cohn-Kanade database [6], the MMI Facial Expression database [13,21], or the JAFFE database [10]), but only few realistic databases exist. They are even fewer when it comes to realistic data involving elderly people. Most researchers actually test and validate their methods on acted or posed corpora [1,5,8,18]. In [22], however, the authors argue that spontaneous expressions are different from posed expressions both in appearance and in timing. This means that methods used for posed expressions recognition might not be suited to realistic expressions. Therefore, the detection methods proposed in this work and evaluated on our realistic social interaction corpus between elderly people and a robot need to be assessed on another acted corpus.

For the visual detection system, we use Support Vector Machines (SVM) with a Radial Basis Function (RBF) kernel for classification. We use Local Binary Patterns (LBP) [4,24] for feature extraction. There exist a variety of feature

---

[2] https://laughterworkshop2015.wordpress.com/
[3] http://www.ilhaire.eu/

extraction methods in the literature (e.g. local Gabor binary patterns [11], local phase quantization, histogram of oriented gradients [2], Haar filters [9] and FACS coding system action unit detection [3]).

Laughter detection in audio is also addressed in the literature. One can distinguish two types of laughter recognition: recognition with prior segmentation or segmentation by recognition. In the first approach, short audio segments representing acoustic events (anything that is not a silence) are either manually or automatically extracted from a continuous audio stream before they are classified. As for the second approach, segmentation by recognition, there is no prior knowledge about where an acoustic event starts and where it ends. The whole audio stream is analyzed to tell apart the classes of interest (e.g. speech, laughter, silence, other human or environmental sounds, etc.). In [7], MFCC and Modulation Spectrum features are used with SVM for laughter detection in meeting rooms. The main focus of the authors was the detection of laughter events where more than one person simultaneously laugh. Spacial cues were therefore calculated by cross-correlating the audio signals acquired by two tabletop microphones. The goals of this cross correlation is to better distinguish one-participant and multi-participant laughters. In [20] many sets of acoustic features (Perceptual Linear Prediction Coding features, Energy, Pitch and Modulation Spectrum) and classification algorithms (Gaussian Mixture Models, Hidden Markov Models and Multi Layer Perceptrons) are investigated for laughter-speech classification. The best baseline performance is obtained PLP features with GMM. Improvement could be observed by combining the PLP with GMM system with a system based on Pitch related features with SVM. [16] address a 5-class classification problem. Each audio segment is classified into four human classes (breathing, consent, hesitation and laughter) or a garbage class used to model background noise. The best reported performance was obtained with HMMs and PLP feature.

These methods address all the problem of classification after segmentation. Many other works focus on laughter detection using segmentation by classification scheme. In [19] a stream is segmented into laughter, speech and silence intervals using PLP features and GMM. A 3-state Viterbi decoder is first used to find the most likely sequence of states given a stream. The sequence of states is seen as a preliminary segmentation. The log likelihood of each segment given each of the GMM models is calculated to determine the final class of the segment. In [15] MFCC and HMMs are used to label a stream with a set of classes containing laughter, filler, silence and speech. A higher level model, a bigram language model, is used to explicitly model the order in which the labels appear in training data. [14] show that adding visual information (head pose and facial expression) slightly improve the performance of the audio-based system.

3 ROMEO2 Corpus

The main motivation of this data collection was to build an elderly people-robot interaction corpus within the ROMEO2 project [17]. The collected corpus is made up of audio and video streams of the whole interaction for each subject as well as two questionnaires (a satisfaction questionnaire and a personality questionnaire) and detailed logs of the robot actions (time stamped utterances, sounds, gestures, played songs, etc.). A high definition webcam set up behind the sitting robot was used for frontal video capture, alongside with another camera used to capture both interlocutors from a profile perspective. The number of the participating subjects is 27 (3 men and 24 women), with an average age of 85.

Audio and video tracks were both annotated for this experiment. For the audio part, speech (start time and end time of each utterance), affect bursts (including laughter) and emotion in voice (one tag among happiness, sadness, anger, doubt, surprise or neutral) were annotated. In the visual part, head pose, head gesture, certain mouth movements, eyebrow movements and body movements and perceived emotion of face were also annotated.

In order to analyze the connection between a subject's behavior and their profile, we calculate the correlation between answers to 3 questions from the satisfaction questionnaire and the number of smiles and laughters within the interaction. As shown in table 1, the more relaxed and enjoyed a subject was, the more they expressed smile and laughter during the interaction.

Table 1. Correlation between experience enjoyment and the number of smiles and laughter during an interaction. +1 and -1 are the numerical translations of the answer used for correlation computation.

Question	Events	Correlation	P-value
(Q7) Would you like it to address you using the familiar form (+1) or using the formal form (-1)?	Laughter + Smile	0.424	0.04395
(Q10) Would you prefer a robot that looks like a robot (+1) or a human (-1)?	Smile with open mouth	-0.465	0.02528
(Q11) Do you consider the robot as a machine (+1) or as a friend or a (human) companion (-1)?	Laughter + Smile	-0.429	0.04134

In this work, since nearly 90% of audio annotated laughters have intersection with the visual annotated laughter and smile events in our corpus, we use visual annotation as the annotation reference for our experiment, and only obvious events such as laughter and smile with open mouth are studied. This results in about 575 events from the 27 subjects subjects. Since the repartition of the events was not balanced among the subjects, all male participants (3 subjects) and all the subjects with less than 10 events of laughter and open-mouth smile were discarded. As a result, our experimental data subset consists in 15 female subjects and contains 168 laughter events and 218 open-mouth smiling events (386 smile related events in total).

4   Proposed Smile and Laughter Detection Methods

4.1   Audio-Visual Fusion of Smile and Laughter Detection

Smile detection from images has been found to be efficient in many acted emotion corpora [1,5,8,18]. However, this has not yet been studied on realistic data of elderly people-robot interaction. Besides the face wrinkles related issue inherent to elderly people, the distinction between an open mouth smile and a speaking mouth raises another important challenge for visual smile detection. To overcome this, speech detection from audio is used in this work to back up the visual detection system and let the visual system only focus on smile detection while the subject is not talking.

4.2   Visual Smile and Laughter Detection

For visual laughter and smile detection, we use uniform-Local Binary Patterns [12] with a SVM classifier using a RBF kernel. This method was tested on the GENKI-4K corpus and performed with an accuracy of over 83%.

Video was recorded at 30 frames per second with a resolution of 1280x720 pixels. For each frame, a frontal facial image of the subject was extracted using the Viola and Jones face detector [23]. After histogram equalization, facial images were reshaped to 64x64 pixels and processed using LBF with a 10x10 grid to get one feature vector for each frame. The performance of the SVM with a RBF kernel classifier was optimized, by varying the values of the c and γ parameters of SVM between 0.01 and 100 on a logarithmic grid with a multiplying step of 3. The evaluation is run at a frame-level and segment-level. For the segment-level evaluation, we consider that a segment contains smile or laughter if the majority of frames it is composed of are classified as smile or laughter.

4.3   Laughter Detection from Audio

As an audio stream of a typical human-robot interaction is globally made up of speech, laughter, robot prompts and silence, we used a 4-class classification scheme. Audio classification was done on frame-level using 13 MFCC coefficients and four GMM models to represent the 4 classes. Each frame is 20 ms long and has an overlap of 10 ms with the previous one. Audio frames within a visual laughter annotation are classified, which yields a sequence of symbols belonging a 4-letter alphabet. From our annotation experience we noticed that, visually, a laughter lasts longer than its respective audio signal. This respective perceived audio laughter can actually be very brief in comparison to what the subject's face or body depicts, partially or completely masked by the a robot's utterance or merged with speech and/or silence. Therefore, for audio segment classification illustrated in figure 1, we used an aggregation strategy that applies on short sequences of consecutive frames (a sliding window of 800 ms for instance) instead the whole audio slice aligned with a visual annotation. the goal of this is to detect the presence of brief audio laughter events within a long "humanly" perceived

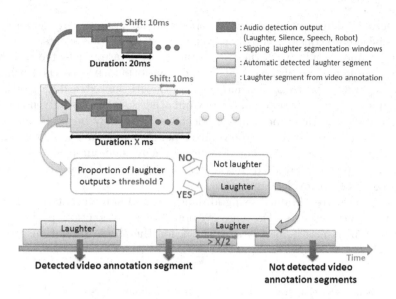

Fig. 1. Audio-level detection scheme

laughter. A window is recognized as belonging to class C if C boasts the highest number of frames within that window.

5   Experiments

5.1   Visual Smile and Laughter Detection Results

Three different protocols are used in our experiments:

- **Mono-subject:** for each subject, half the data is used for training and the remaining half for test.
- **Multi-subject:** for each subject, half the data of the subject plus all data from all the other subjects are used for training. The remaining subject's half is used for test.
- **Leave-one-out:** for each subject, training is performed using data from all the other subject's whereas test is done on all the subject's data

A total of 386 visual annotated smile and laughter events are used in our experiments. We also extracted 378 random segments outside laughter and smile annotations for the test. The visual system faces mainly two kinds of issues, missed frames due to a face detection problem (often caused by subjects' head-turning) and the apparent resemblance between a talking mouth and open-mouth smiles or laughter. This resemblance is accentuated by face wrinkles. To better assess the performance of the visual detection system, we suppose the use of an speech activity detection system with perfect outputs. Hence, all segments that contain speech are considered as non laughter and non smile events.

The obtained results with the three test protocols are illustrated in table 2. From this table we can observe that the best performance is reached using the speaker dependent protocol (i.e. mono-subject) with no extra data from other speakers for training. When we add data from other speakers to a given speaker's training data, we notice a performance decrease. This may be explained by the fact that, for elderly people, as for younger people, smile and laughter sensibly differ from one person to another. Finally, the leave-one-out protocol (no data from the test speaker were used to train the model) gave the lowest performance among the three protocols.

These results are obtained using all laughter and smile events (368). When using only 289 segments (the ones without a face detection problem), the obtained recall is 68.5%, 58.5% and 41.2% for the three evaluation protocols respectively.

Table 2. Visual detection evaluation. Total frame accuracy refers to the global system's accuracy (for laughter/smile against non-laughter/non-smile annotations). The two most-right columns are the recall and accuracy of laughter and smile annotations at frame-level and segment-level respectively. A. stands for **accuracy**, R. for **recall**, P. for **precision** and BER. for **Balanced Error Rate**.

Protocol	Frame-level	Segment-level
Mono-subject	A.85.7% R.80.6% BER.23.7%	R.51.3% P.95.2%
Multi-subject	A.73.9% R.62.4% BER.31.4%	R.43.8% P.95.5%
Leave-one-out	A.72.8% R.40.8% BER.46.6%	R.30.8% P.73.9%

5.2   Audio Laughter Detection Results

For laughter detection, we also used 386 annotated smile and laughter events as well as 378 segments randomly cut from regions that do not contain a smile nor a laughter. An actual laughter event is considered as correctly classified by the system if at least one analysis window is recognized as a laughter by the system. A non laughter segment is however considered as correctly classified if none of the analysis windows it contains is recognized as laughter. Moreover, as each analysis window represents a sequence of frames, two frame aggregation strategies are used, a majority voting and an "at least half the frames" strategies. For the first strategy, the window is given the label of the most represented class. As for the second strategy, at least half the frames of the window have to be belong to one single class so that the window is labeled with this class.

Table 3 shows the obtained results. We can see that a short window results in a better recall but a relatively low precision whereas a long window leads to the opposite result. Moreover, a more rigorous aggregation strategy (at least 50%) improves the precision at the expense of the recall.

Table 3. Audio laughter detection evaluation using various analysis window durations on 386 laughter and smile events and 378 non laughter segments. Maj. voting: a window is considered as laughter if the most represented class within it is laughter. ≥ 50%: a window must contain at least 50% of laughter frames to be considered as a laughter.

Analysis window duration	Maj. voting	≥ 50%
600ms	R.80.1% P.54.0%	R.64.5% P.56.7%
800ms	R.74.9% P.54.3%	R.56.5% P.58.9%
1s	R.71.0% P.55.7%	R.49.7% P.59.6%

Note that out of the 386 visually annotated laughter and smile events, only 168 are laughter. Therefore, when we run the audio system on these 168 events, using an analysis window of 800 ms, we obtain a recall of 85.7% and 69.6% for the majority voting and the "at least 50%" aggregation strategies respectively.

5.3 Video-Audio Smile and Laughter Detection

In our detection system based on the fusion of audio and video decision, we consider a laughter or a smile detection if either of the two modalities decides a positive detection. To make a trade-off between precision and recall in the audio system, we use an analysis window of 800 ms with the two audio frames aggregation strategies mentioned in sub-section 5.2. Table 4 shows the obtained results. The fusion resulted in a fairly good recall improvement with a precision decrease. The recall of the fusion system is better than either one-modality based systems, regardless of the audio aggregation strategy. The precision however lays between that of the two systems and is mostly better than the audio system's.

Table 4. Audio-visual detection evaluation on 386 laughter and smile events. For audio detection, an analysis window of 800 ms is used.

Protocol	Video only	Video/audio fusion ≥ 50%	Video/audio fusion Maj. voting
Mono-subject	R.51.3% P.95.2%	R.78.8% P.65.1%	R.86.8% P.57.4%
Multi-subject	R.43.8% P.95.5%	R.75.4% P.64.7%	R.86.8% P.57.6%
Leave-one-out	R.30.8% P.73.9%	R.66.8% P.58.4%	R.80.6% P.54.3%

6 Conclusion and Future Work

This paper presents an audio-visual smile and laughter detection system in the context of elderly people-robot social interaction. The audio-video fusion system performs in two steps: first, each one-modality system is separately run to obtain its individual decision given a segment of aligned audio and video signals. The final decision is positive (there is a laughter or a smile) if at least one of the two systems outputs a positive decision.

The video system has a very good segment-level precision in subject dependent evaluation and a rather good precision in subject independent evaluation. It has however a lower recall in comparison to the audio system. This is probably due to the fact that many subjects laughed whilst turning their head as well as to mis-detections of the face detection system. This said, the audio system runs continuously which results in a high recall rate even when a long analysis window is used. We believe that the recall of the video system can be improved by dealing with the head turning issue whereas the audio system can achieve a better precision by using more data for training and using classification methods of a more discrimination power.

The fusion of the two system seems to lead to good compromise between precision and recall. Improving each of the one-modality systems will indeed result in an overall improvement of the fusion system performance.

In future work, we will consider other state-of-the-art video feature extractors in order to improve the performance of our system and to compare the performance of detection in posed database and realistic databases of elderly people-robot social interaction. We also plan to experiment other fusion strategies (e.g. frame-level of two modalities fusion or feature fusion).

References

1. Deniz, O., Castrillon, M., Lorenzo, J., Anton, L., Bueno, G.: Smile detection for user interfaces. In: Bebis, G., Boyle, R., Parvin, B., Koracin, D., Remagnino, P., Porikli, F., Peters, J., Klosowski, J., Arns, L., Chun, Y.K., Rhyne, T.-M., Monroe, L. (eds.) ISVC 2008, Part II. LNCS, vol. 5359, pp. 602–611. Springer, Heidelberg (2008)
2. Abhinav, D., Akshay, A., Roland, G., Tom, G.: Emotion recognition using PHOG and LPQ features. In: 2011 IEEE International Conference on Automatic Face & Gesture Recognition and Workshops (FG 2011), pp. 878–883. IEEE (2011)
3. Ekman, P., Friesen, W.V., Hager, J.C.: Facs manual. A Human Face (2002)
4. Feng, X., Pietikäinen, M., Hadid, A.: Facial expression recognition based on local binary patterns. Pattern Recognition and Image Analysis $17(4)$, 592–598 (2007)
5. Akinori, I., Xinyue, W., Motoyuki, S., Shozo, M.: Smile and laughter recognition using speech processing and face recognition from conversation video. In: International Conference on Cyberworlds, pp. 8. IEEE (2005)
6. Takeo, K., Cohn, J.F., Yingli, T.: Comprehensive database for facial expression analysis. In: Fourth IEEE International Conference on Automatic Face and Gesture Recognition, Proceedings, pp. 46–53. IEEE (2000)
7. Kennedy, L.S., Ellis, D.P.W.: Laughter detection in meetings. In NIST ICASSP 2004 Meeting Recognition Workshop, Montreal, pp. 118–121. National Institute of Standards and Technology (2004)
8. Kowalik, U., Aoki, T., Yasuda, H.: BROAFERENCE - a next generation multimedia terminal providing direct feedback on audience's satisfaction level. In: Costabile, M.F., Paternó, F. (eds.) INTERACT 2005. LNCS, vol. 3585, pp. 974–977. Springer, Heidelberg (2005)

9. Littlewort, G., Whitehill, J., Wu, T., Fasel, I., Frank, M., Movellan, J., Bartlett, M.: The computer expression recognition toolbox (cert). In: 2011 IEEE International Conference on Automatic Face & Gesture Recognition and Workshops (FG 2011), pp. 298–305. IEEE (2011)

10. Lyons, M., Akamatsu, S., Kamachi, M., Gyoba, J: Coding facial expressions with gabor wavelets. In: Third IEEE International Conference on Automatic Face and Gesture Recognition, Proceedings, pp. 200–205. IEEE (1998)

11. Moore, S., Bowden, R.: Local binary patterns for multi-view facial expression recognition. Computer Vision and Image Understanding 115(4), 541–558 (2011)

12. Ojala, T., Pietikäinen, M., Mäenpää, T.: Multiresolution gray-scale and rotation invariant texture classification with local binary patterns. IEEE Transactions on Pattern Analysis and Machine Intelligence 24(7), 971–987 (2002)

13. Pantic, M., Valstar, M., Rademaker, R., Maat, L.: Web-based database for facial expression analysis. In: IEEE International Conference on Multimedia and Expo, ICME 2005, pp. 5. IEEE (2005)

14. Petridis, S., Pantic, M.: Audiovisual discrimination between speech and laughter: Why and when visual information might help. IEEE Transactions on Multimedia 13(2), 216–234 (2011)

15. Salamin, H., Polychroniou, A., Vinciarelli, A.: Automatic detection of laughter and fillers in spontaneous mobile phone conversations. In: 2013 IEEE International Conference on Systems, Man, and Cybernetics (SMC), pp. 4282–4287. IEEE (2013)

16. Schuller, B., Eyben, F., Rigoll, G.: Static and dynamic modelling for the recognition of non-verbal vocalisations in conversational speech. In: André, E., Dybkjær, L., Minker, W., Neumann, H., Pieraccini, R., Weber, M. (eds.) PIT 2008. LNCS (LNAI), vol. 5078, pp. 99–110. Springer, Heidelberg (2008)

17. Sehili, M., Yang, F., Leynaert, V., Devillers, L.: A corpus of social interaction between nao and elderly people. In: 5th International Workshop on Emotion, Social Signals, Sentiment & Linked Open Data (ES3LOD2014). LREC (2014)

18. Shinohara, Y., Otsu, N.: Facial expression recognition using fisher weight maps. In: Sixth IEEE International Conference on Automatic Face and Gesture Recognition, Proceedings. pp. 499–504. IEEE (2004)

19. Truong, K., Van Leeuwen, D.: Evaluating automatic laughter segmentation in meetings using acoustic and acoustics-phonetic features. In: Proc Workshop on the Phonetics of Laughter at the 16th International Congress of Phonetic Sciences (ICPhS), pp. 49–53 (2007)

20. Khiet, P.: Truong and David A Van Leeuwen. Automatic discrimination between laughter and speech. Speech Communication 49(2), 144–158 (2007)

21. Valstar, M., Pantic, M.: Induced disgust, happiness and surprise: an addition to the MMI facial expression database. In: Proc. 3rd Intern. Workshop on EMOTION (satellite of LREC): Corpora for Research on Emotion and Affect, pp. 65 (2010)

22. Valstar, M.F., Pantic, M., Ambadar, Z., Cohn, J.F.: Spontaneous vs. posed facial behavior: automatic analysis of brow actions. In: Proceedings of the 8th International Conference on Multimodal Interfaces, pp. 162–170. ACM (2006)

23. Viola, P., Jones, M.: Robust real-time object detection. International Journal of Computer Vision 4, 51–52 (2001)

24. Zhao, G., Pietikainen, M.: Dynamic texture recognition using local binary patterns with an application to facial expressions. IEEE Transactions on Pattern Analysis and Machine Intelligence 29(6), 915–928 (2007)

The Effect of a Robot's Social Character on Children's Task Engagement: Peer Versus Tutor

Cristina Zaga[✉], Manja Lohse, Khiet P. Truong, and Vanessa Evers

Human Media Interaction, University of Twente, Enschede, The Netherlands
{c.zaga,m.lohse,k.p.truong,v.evers}@utwente.nl
http://hmi.ewi.utwente.nl/

Abstract. An increasing number of applications for social robots focuses on learning and playing with children. One of the unanswered questions is what kind of social character a robot should have in order to positively engage children in a task. In this paper, we present a study on the effect of two different social characters of a robot (peer vs. tutor) on children's task engagement. We derived peer and tutor robot behaviors from the literature and we evaluated the two robot characters in a WoZ study where 10 pairs of children aged 6 to 9 played Tangram puzzles with a Nao robot. Our results show that in the peer character condition, children paid attention to the robot and the task for a longer period of time and solved the puzzles quicker and better than in the tutor character condition.

Keywords: Child-Robot Interaction · Task engagement · Robot characters · Robot behaviors

1 Introduction

Social robots are envisioned as partners for children, offering companionship, tutoring and social assistance in various domains (e.g., education, therapy). Across these domains and applications, one of the main factors that contributes to initiate, sustain and maintain child-robot interaction (cHRI) is engagement. Despite the fact that children can easily make connections with robots [15], researchers and designers still face the challenge to select the appropriate set of verbal and nonverbal robot behaviors that support engagement throughout a task. The social verbal and nonverbal behaviors the robot is endowed with provide information about its '*social character*' (here defined as sets of behaviors stylized according to a precise behavioral repertoire, e.g., the robot is characterized as a friend, playmate, tutor) [17], [22], [26]. Therefore, we argue that researchers who venture into child-robot interaction also need to take this aspect into account. Firstly, because this shapes the design of the robot behaviors. Secondly because the robot character could lead to the identification of a social role (set of standards, norms and concepts held for the behaviors of an agent in a social system) [4], a key factor for successful child-robot interactions and

A. Tapus et al. (Eds.): ICSR 2015, LNAI 9388, pp. 704–713, 2015.
DOI: 10.1007/978-3-319-25554-5_70

human-robot interaction (HRI) in general [14]. In cHRI research, peer and tutor behaviors, derived from human interaction, have been explored in different task-related contexts with the ultimate goal to engage children in different types of tasks. To date, it is not clear if and how the robot character might effect children's engagement with a task, as a thorough comparison of robot social characters is missing. This paper addresses this gap and it aims at shedding light on two robot social characters that are predominant in children's task-oriented experiences, namely peers and tutors. After deriving tutoring and peer behaviors from literature, we evaluated the effect of the two robot characters on children's task engagement in a Wizard of Oz (WoZ) experiment. The exploratory study entails a triadic scenario where a humanoid robot (i.e., the Nao robot) and two same sex children (6-9 years old) perform three Tangram puzzles.

2   Children's Task Engagement

Engagement is a multifaceted phenomenon that is considered to play an important role when humans are interacting with social robots [25]. Sidner et al., defined engagement as 'the process by which two (or more) participants establish, maintain and end their perceived connection' [24], but their definition focused on conversational engagement and as such on cognitive engagement (i.e., focus of attention during conversations). Our definition goes a step beyond Sidner's and includes insights from [7], [8], [11], [21]. Since engagement encompasses three dimensions, cognitive, behavioral and affective [8], we argue that it is necessary to take them all into account. In our study, we focused only on children's task engagement and we define it as the level of cognitive (e.g., attention to the task and the robot), affective (e.g., emotional response to the task), and behavioral attributes (e.g., performance) of engagement during the interaction. The more these attributes occur in the interaction (measured by frequency and duration), the more engaging the interaction with the task will be.

3   Related Work and Hypotheses

Studies have shown that an expressive behavioral repertoire, that conveys a robot social character is one of the factors that affects children's engagement with a task [10], [26]. In a field study, Kanda et al. [16] revealed that sharing common ground with a peer robot contributes to an enhanced level of engagement with a task. Also Okita et al. [22] illustrated how a peer-like cooperative style of interaction supported affective task engagement. In a similar vein, Leite et al. [18] suggested that a set of emphatic behaviors, exhibited by a robot companion during a game, may encourage the child to identify the robot as a peer enhancing the endurability of the task. A related indication emerges from the work of Belpaeme et al. [3]: a robot perceived as a peer, during a game appeared to be more likely to support engagement. Hence, our first hypothesis (H1) is that a peer-like character will enhance (H1a) affective and (H1b) behavioral (i.e. performance) children's task engagement. Our expectation is that this effect will be more prominent than for the tutor-like character as implied by the education literature [9]. On the other hand, the attention to the task could be enhanced by a tutor-like character [17], who is focused on guiding in a scaffolding fashion [28]. As a result, our second hypothesis (**H2**) is that a tutor-like character will enhance the focus of attention on the task and the robot more than the peer-like character.

4   Method

Our independent variable, the robot social character, was manipulated between-subjects. In order to address the above mentioned hypotheses, we devised two conditions, namely *peer-like character (PC)* and *tutor-like character (TC)*. In the study, dyads of same gender, same age children performed three tasks in either the PC or the TC condition. The three tasks consisted of three Tangram puzzles of increasing difficulty. The children first solved a puzzle missing three pieces, then two puzzles missing six pieces. The first puzzle consisted of a simple outline of an animal shape with orientation lines, i.e., the lines defining the Tangram piece perimeter, partially completed. The second and the third puzzles consisted of geometric outlines to be completed. The puzzle pieces were divided between the children, who could collaborate to accomplish the task. We used the Nao robot [1] which was remotely controlled by a Python script operated by a researcher (See Figure 1).

4.1   Robot Character Design

The social character is conveyed by behaviors based on peer collaboration [9] for the PC condition and instructional scaffolding [6] for the TC condition. We designed eleven verbal and nonverbal behaviors both for the PC condition and for the TC condition, in such a way that their functions, interaction modalities, number of actions and speech remained the same across the two conditions. The behaviors were designed to (i) regulate the phases of the task, (ii) provide information about the state of the task, (iii) support the attention of the participants, (iv) provide reinforcement and support, and (v) provide reward.

The only difference between the designs was the style of interaction. In other words, the way the behavior was expressed through gestures, speech, and postures was designed either with peer or tutor characteristics. From literature on teachers' multimodal expressions and on peer collaboration [6], [20], we identified distinctive features of speech [23], [27], gestures [1], [12], [28], positioning and posture [19] for each condition. The set of behaviors were organized in a task-dependent flow, which was strictly followed by who controlled the robot. Table 1 presents the above cited interaction modalities and the manipulations applied to convey the peer and tutor character. Figure 2 depicts examples of the robot gestures in PC and TC conditions.

Before the user study took place, we video-recorded all behaviors following the task-dependent flow. We showed the two videos to three Montessori teachers

---

[1] https://www.aldebaran.com/en/humanoid-robot/nao-robot

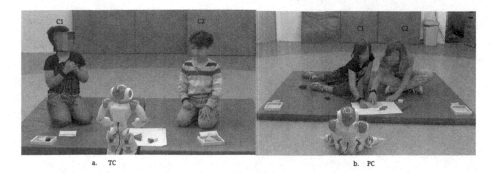

Fig. 1. The figure presents children working in dyads on the tasks in (a) the tutor condition (TC) and (b) in the peer condition (PC).

Table 1. Interaction modalities and elements that were manipulated in the peer (PC) and the tutor condition (TC).

Modalities	Elements	Peer	Tutor
speech	pitch	high	low
speech	style	direct, emphatic	maieutic, interrogative
body	postural	sitting	standing
gestures	deictic	indication, sweeping	pointing, tracing
gestures	emphatic	exultation, surprise	head nods
gestures	representational	grasping	presenting

working at an elementary school in the Netherlands. The teachers filled in a form with one closed question (*'Do you think that the robot behaved like: a.peer b.tutor'*) and two open questions to discuss the robot character design (*'What did the robot do to make you think it was more like a peer or a tutor?', Do you have other comments on the behaviors?'*). The teachers correctly recognized the behaviors as belonging to the respective two robot social characters and they made comments consistent with the behavior design.

4.2   Setup, Procedure and Participants

Setup. We conducted the study at a Montessori school in The Netherlands. We divided their gym room in three areas: an experimental, a WoZ and a questionnaire area. The WoZ area was only entered by the researchers. From there, the robot was remotely controlled. Although the researcher was sitting in the same room, his role was hidden from the children. The sessions were recorded with three cameras, one recording the central view and two for the side views.

Procedure. A facilitator escorted the participants to the experimental area and provided an introduction to the robot in order to allow the children to familiarize

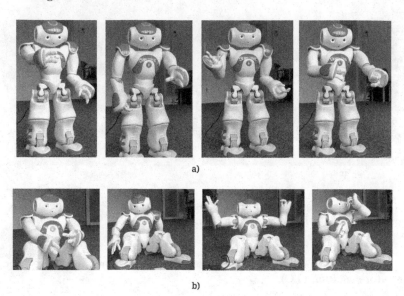

Fig. 2. The figure presents some pictures of the behaviors designed for a) tutor condition (TC) b) peer condition (PC). In a) examples of deictic, emphatic and representational gestures (pointing, tracing, presenting and head nods). In b) examples of deictic, emphatic and representational gestures (indication, grasping, exultation and surprise).

themselves with it. Thereafter, the facilitator placed the puzzle outlines and the Tangram pieces in front of the children. The robot was placed opposite of the participants, outside the play-mat at a safe distance. After everything was in place, the facilitator asked the children to complete the puzzle with the robot. As soon as the facilitator left the area, the researcher who controlled the robot started the behaviors following the task-dependent flow. After the interaction with the robot, the facilitator escorted the participants to the questionnaire area where the questionnaires were administered.

Participants. Twenty children ($N = 20$) belonging to one Montessori class (this includes children from 6 to to 9 years old) participated in the study. They were divided into six male and four female couples matched by ages. Five couples were assigned to the PC condition ($N = 10$, age: $M = 7.1$, $SD = 1.10$) and the other five to TC condition ($N = 10$, age: $M = 7.0$, $SD = 0.66$).

4.3   Measures

We measured children's task engagement via behavioral observations and a questionnaire. To account for the *cognitive attributes* of task engagement, we investigated the focus of attention, namely the gaze behaviors of the participants directed to the robot and to the task. To get a complete overview of the gaze behaviors of the children in the interaction we also measured gaze to the other child and gaze elsewhere. In order to account for the *affective attributes* of task engagement i.e., enjoyment, we designed a questionnaire based on the subscale enjoyment from the Intrinsic motivation inventory (IMI)[2]. We translated it to Dutch as this was the language of the study. We used a 5-point Likert *Smileyometer* scale anchored from *"Strongly disagree"* to *"Strongly agree"* and we avoided reversed items. To account for the *behavioral attributes* of task engagement, i.e., task performance, we rated the degree of completion of the tasks per dyad i.e., how many pieces were put into the puzzle correctly, with 3 being the maximum for the first task and 6 for the second and the third. We also analyzed task duration.

4.4   Data Analysis

Video data. In total, 125 minutes and 41 seconds of video material were analyzed: 60 minutes and 47 seconds in the PC condition and 64 minutes and 54 seconds in the TC condition. The videos were manually coded in Elan[3] following an annotation scheme developed for the analysis, including focus of attention, task performance, and task duration. The annotations were analyzed using a Matlab toolbox called SALEM [13]. The annotations of focus of attention were analyzed for counts and duration. As the recorded interactions differ in lengths, we normalized the results providing seconds per minute of gaze and counts per minute of gaze (i.e., the rate) We compared the results across conditions (PC vs. TC) with two-tailed independent sample t-tests. Also the results of the degree of completion were compared across conditions (PC vs. TC) with two-tailed independent sample t-tests. To investigate the difference between the duration of task performances between the conditions, a Mann-Whitney U test was carried out. For both focus of attention and task performance we calculated inter-rater reliability for about 10% of the data (11':42") which showed acceptable agreement (Cohen's kappa; focus of attention $\kappa = .730$, $p = .003$, task completion $\kappa = .750$, $p < .001$).

Questionnaire data. The internal reliability of the IMI/ Enjoyment scale was 0.851 (Cronbach's alpha). Unfortunately, the general polarization of the children's answers toward the positive anchor did not allow to find any difference between conditions. Hence, the questionnaire results are not included.

5   Results

Cognitive attributes of task engagement: focus of attention. The gaze to the robot rate was significantly higher in the PC condition ($M = 3.44$, $SD = 0.67$) than in the TC condition ($M = 2.36$, $SD = 0.92$; $t(18) = 2.97$, $p = .008$). Moreover, the participants looked at the robot significantly longer (i.e., gaze seconds/per minute) in the PC condition ($M = 19.30$, $SD = 3.11$) than in the TC condition ($M = 12.71$, $SD = 3.21$; $t(18) = 4.66$, $p < .001$.). As for the gaze

---

[2] https://www.selfdeterminationtheory.org/intrinsic-motivation-inventory/
[3] https://tla.mpi.nl/tools/tla-tools/elan/

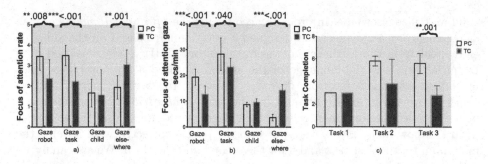

Fig. 3. Difference in focus of attention (robot, task, child, elsewhere) and task completion between peer condition (PC) and tutor condition (TC). Figure 2 a): focus of attention rate results. Figure 2 b): focus of attention results in gaze seconds per minute. Figure 2 c): task completion results. * indicates significance at the 0.05 level, ** 0.01 level, *** < .001 level. Error bars show standard deviations.

on the task, the rate was significantly higher in the PC condition ($M = 3.49$, $SD = 0.50$) than in the TC condition ($M = 2.22$, $SD = 0.66$; $t(18) = 4.86$, $p < .001$). The average amount of gaze seconds per minute on the task was also significantly different in the two conditions and more in PC. (PC: $M = 28.26$, $SD = 6.30$; TC: $M = 23.29$, $SD = 3.26$; $t(18) = 2.21$, $p = .040$). We found no significant difference in the rate (PC: $M = 1.65$, $SD = 0.68$; TC: $M = 1.56$, $SD = 1.22$) and gaze seconds per minute (PC: $M = 8.73$, $SD = 0.90$; TC: $M = 9.70$, $SD = 1.20$) to the other child, but the rate and gaze seconds per minute elsewhere are significantly higher in TC (PC: $M = 1.93$, $SD = 0.57$; TC: $M = 3.05$, $SD = 0.71$; $t(18) = 3.84$, $p = .001$; gaze secs/min elsewhere: PC: $M = 3.71$, $SD = 1.27$; TC: $M = 14.30$, $SD = 2.18$; $t(18) = 13.28$, $p < .001$ see Figure 3 a, b).

Behavioral attributes of task engagement: completion. Task 1 was completed by all the participants in both conditions. Task 2 was fully completed by 80% of the participants (4 out of 5 dyads) in the PC condition and by just 20% of the participants (1 out of 5 dyads) in the TC condition. The task performance in Task 2 is better in the PC condition ($M = 5.80$, $SD = 0.44$) than in the TC condition ($M = 3.60$, $SD = 2.40$), but no statistically significant difference was found. Likewise, Task 3 was completed by 80% of the participants in the PC condition and by only 20% of participants in the TC condition. The task performance was better in PC ($M = 5.60$, $SD = 0.89$) than in TC ($M = 2.80$, $SD = 0.83$) and a statistically significant difference was found ($t(8) = 5.11$; $p = .001$, see Figure 3 c).

Behavioral attribute of task engagement: task duration. The participants in the TC condition took more time to perform the tasks than the participants in the PC condition. A Mann-Whitney U test conducted on the total performance duration i.e., all three tasks, confirmed that the participants in the TC condition (mean rank = 18.50) took significantly more time ($U = 154$, $Z = 2.13$, $p = .033$) to perform the task than the participants in the PC condition (mean rank =11.73).

6 Discussion

In this paper we presented a study on the effect of two robot social characters, peer-like and tutor-like, on children's task engagement. Our results showed that the children performed significantly better in the more difficult task and were faster when working with a peer-like robot character. These results support H1b (a peer-like character will enhance behavioral task engagement). However, the questionnaire results do not allow us to say anything about children's enjoyment, thus H1a (a peer-like character will enhance affective task engagement) cannot be addressed.

We believe that this outcome suggests the potential inappropriateness of using questionnaires for this user group [5]. Another explanation is the *suggestibility effect* [29], i.e., the desire to please the researchers. Nevertheless, our findings also show that the peer-like character appears to have a positive effect on the focus of attention of the children. In fact, the peer character triggered significantly more attention towards the robot and the task than a tutor-like character. Moreover, in the tutor character condition, the participants looked more elsewhere and this can be an indication that the tutor character could be less effective in sustaining attention to the task. These results contradict H2 (a tutor-like robot social character will enhance the focus of attention on the task and the robot), but they highlight that the behavioral repertoire of a peer might be able to enhance children's cognitive engagement with a task and the robot. Overall, our results suggest that embedding a peer-like repertoire of engagement-seeking robot behaviors might represent a good strategy in task-related child-robot interactions.

7 Limitations and Future Work

Our exploratory study is a very first step towards understanding the effect of a robot's social character on children's task engagement. As such, it has some limitations, which will be addressed in future work. We are aware that our findings cannot provide a comprehensive account on the effect of the peer robot character on the task performance results, as they do not account for children's prior level of ability on the task. In our experimental design, we tried to overcome possible discrepancies in the children's cognitive development matching the gender and the age of the participants. Nevertheless, future work needs to take children's task abilities (prior and post interaction) into account. Also, our forthcoming research should provide a complete overview of the dyads/groups dynamics in a more sequential way (i.e., how does the interaction change over time). In addition, we will address children's expectations towards robot behaviors to have a better picture of the effect of a robot character on children's task engagement. Building upon our promising findings and significant results, we will proceed with bottom-up investigations to discern low-level engaging behaviors and interaction style features of a peer's behavioral repertoire, while investigating if and how a social role can emerge from a social character.

Acknowledgments. The research leading to these results was supported by the 7th Framework Programme of the European Community, under grant agreement 610532 (SQUIRREL-Clearing Clutter Bit by Bit). We thank R.A.J. de Vries, D.P. Davison, T. Krikke, V. Charisi, the teachers and their pupils. We also thank the reviewers for their suggestions and concise feedback that helped us to improve the paper.

References

1. Alibali, M.W., Flevares, L.M., Goldin-Meadow, S.: Assessing knowledge conveyed in gesture: Do teachers have the upper hand? Journal of Educational Psychology **89**(1), 183–193 (1997)
2. Belpaeme, T., et al.: Child-robot interaction: perspectives and challenges. In: Herrmann, G., Pearson, M.J., Lenz, A., Bremner, P., Spiers, A., Leonards, U. (eds.) ICSR 2013. LNCS, vol. 8239, pp. 452–459. Springer, Heidelberg (2013)
3. Belpaeme, T., Baxter, P.E., Read, R., Wood, R., Cuayáhuitl, H., Kiefer, B., et al.: Multimodal child-robot interaction: Building social bonds. Journal of Human-Robot interaction **1**(2), 33–53 (2012)
4. Biddle, B.J.: Role theory: Expectations, identities, and behaviors. Academic Press (1979)
5. Borgers, N., De Leeuw, E., Hox, J.: Children as respondents in survey research: Cognitive development and response quality 1. Bulletin de Methodologie Sociologique **66**(1), 60–75 (2000)
6. Cristenson, S. L., Reschly, A. L., Whyle, C.: Handbook of research on student engagement. Springer Science and Business Media (2012)
7. Corrigan, L.J., Peters, C., Castellano, G.: Social-task engagement: striking a balance between the robot and the task. Embodied Commun. Goals Intentions Workshop ICSR **13**, 1–7 (2013)
8. Deater-Deckard, K., Chang, M., Evans, M.E.: Engagement states and learning from educational games. New Directions for Child and Adolescent Development **2013**(139), 21–30 (2013)
9. Fawcett, L.M., Garton, A.F.: The effect of peer collaboration on children's problem solving ability. British Journal of Educational Psychology **75**(2), 157–169 (2005)
10. Feil-Seifer, D., Matarić, M.: Human robot interaction. In: Encyclopedia of Complexity and Systems Science, pp. 4643–4659. Springer, New York (2009)
11. Fredricks, J.A., Blumenfeld, P.C., Paris, A.H.: School engagement: Potential of the concept, state of the evidence. Review of Educational Research **74**(1), 59–109 (2004)
12. Goldin Meadow, S., Sandhofer, C.M.: Gestures convey substantive information about a child's thoughts to ordinary listeners. Developmental Science **2**(1), 67–74 (1999)

13. Hanheide, M., Lohse, M., Dierker, A.: SALEM-statistical anaLysis of Elan files in Matlab. In: Multimodal Corpora: Advances in Capturing, Coding and Analyzing Multimodality, pp. 121–123 (2010)

14. Huber, A., Lammer, L., Weiss, A., Vincze, M.: Designing Adaptive Roles for Socially Assistive Robots: A New Method to Reduce Technological Determinism and Role Stereotypes. Journal of Human-Robot Interaction **3**(2), 100–115 (2014)

15. Kahn Jr, P.H., Kanda, T., Ishiguro, H., Freier, N.G., Severson, R.L., Gill, B., Ruckert, J.H., Shen, S.: "Robovie, you'll have to go into the closet now": Children's social and moral relationships with a humanoid robot. Developmental Psychology **48**(2), 303–314 (2012)

16. Kanda, T., Hirano, T., Eaton, D., Ishiguro, H.: Interactive robots as social partners and peer tutors for children: A field trial. Human-Computer Interaction **19**(1), 61–84 (2004)

17. Kennedy, J., Baxter, P., Belpaeme, T.: The robot who tried too hard: social behaviour of a robot tutor can negatively affect child learning. In: Proceedings of the 10th ACM/IEEE International Conference on Human-Robot Interaction, Portland, USA, pp. 67–74 (2015)

18. Leite, I., Castellano, G., Pereira, A., Martinho, C., Paiva, A.: Empathic Robots for Long-term Interaction: Evaluating Social Presence, Engagement and Perceived Support in Children. International Journal of Social Robotics **6**(3), 329–341 (2014)

19. Lomranz, J., Shapira, A., Choresh, N., Gilat, Y.: Children's personal space as a function of age and sex. Developmental Psychology **11**(5), 541–545 (1975)

20. Merola, G., Poggi, I.: Multimodality and gestures in the teacher's communication. In: Camurri, A., Volpe, G. (eds.) GW 2003. LNCS (LNAI), vol. 2915, pp. 101–111. Springer, Heidelberg (2004)

21. O'Brien, H.L., Toms, E.G.: What is user engagement? A conceptual framework for defining user engagement with technology. Journal of the American Society for Information Science and Technology **59**(6), 938–955 (2008)

22. Okita, S.Y., Ng-Thow-Hing, V., Sarvadevabhatla, R.K.: Multimodal approach to affective human-robot interaction design with children. ACM Transactions on Interactive Intelligent Systems (TiiS) **1**(1), 1–29 (2011)

23. Sachs, J., Devin, J.: Young children's use of age-appropriate speech styles in social interaction and role-playing. Journal of Child Language **3**(01), 81–98 (1976)

24. Sidner, C.L., Kidd, C.D., Lee, C., Lesh, N.: Where to look: a study of human-robot engagement. In: Proceedings of the 9th International Conference on Intelligent User Interfaces, pp. 78–84. ACM (2004)

25. Sidner, C.L., Lee, C., Kidd, C.D., Lesh, N., Rich, C.: Explorations in engagement for humans and robots. Artificial Intelligence **166**(1), 140–164 (2005)

26. Simmons, R., Makatchev, M., Kirby, R., Lee, M.K., Fanaswala, I., Browning, B., Forlizzi, J., Sakr, M.: Believable robot characters. AI Magazine **32**(4), 39–52 (2011)

27. Teasley, S.D.: The role of talk in children's peer collaborations. Developmental Psychology **31**(2), 207–220 (1995)

28. Valenzeno, L., Alibali, M.W., Klatzky, R.: Teachers gestures facilitate students learning: A lesson in symmetry. Contemporary Educational Psychology **28**(2), 187–204 (2003)

29. Warren, A.R., Marsil, D.F.: Why Children's Suggestibility Remains a Serious Concern. Law and Contemporary Problems 127–147 (2002)

Author Index

Printed in the United States
by Baker & Taylor Publisher Services